Springer Proceedings in Complexity

Springer Proceedings in Complexity publishes proceedings from scholarly meetings on all topics relating to the interdisciplinary studies of complex systems science. Springer welcomes book ideas from authors. The series is indexed in Scopus.

Proposals must include the following:

- name, place and date of the scientific meeting
- a link to the committees (local organization, international advisors etc.)
- scientific description of the meeting
- list of invited/plenary speakers
- an estimate of the planned proceedings book parameters (number of pages/articles, requested number of bulk copies, submission deadline)

Submit your proposals to: Hisako.Niko@springer.com

More information about this series at https://link.springer.com/bookseries/11637

Christos H. Skiadas · Yiannis Dimotikalis
Editors

14th Chaotic Modeling and Simulation International Conference

Springer

Editors
Christos H. Skiadas
ManLab
Technical University of Crete
Chania, Crete, Greece

Yiannis Dimotikalis
Department of Management Science
and Technology
Hellenic Mediterranean University
Agios Nikolaos, Crete, Lasithi, Greece

ISSN 2213-8684 ISSN 2213-8692 (electronic)
Springer Proceedings in Complexity
ISBN 978-3-030-96963-9 ISBN 978-3-030-96964-6 (eBook)
https://doi.org/10.1007/978-3-030-96964-6

This Springer imprint is published by the registered company Springer Nature Switzerland AG
The registered company address is: Gewerbestrasse 11, 6330 Cham, Switzerland

Preface

It is our pleasure to thank the guests, participants and contributors of the 14th International Conference (CHAOS2021) on Chaotic Modeling, Simulation and Applications. We support the study of nonlinear systems and dynamics in an interdisciplinary research field, and very interesting applications were presented.

We provide a widely selected forum to exchange ideas, methods and techniques in the field of nonlinear dynamics, chaos, fractals and their applications in general science and in engineering sciences.

The principal aim of CHAOS2021 International Conference is to expand the development of the theories of the applied nonlinear field, the methods and the empirical data and computer techniques, and the best theoretical achievements of chaotic theory as well.

Chaotic modeling and simulation conferences continue to grow considerably from year to year, thus making a well-established platform to present and disseminate new scientific findings and interesting applications. We thank all the contributors for the success of this conference and especially the authors of this proceedings volume of CHAOS2021.

Special thanks to the plenary, keynote and invited presentations, the Scientific Committee, the ISAST Committee, the Conference Secretary Eleni Molfesi and the members of the Secretariat and of course Niko Hisako and all the Springer team for help and support.

Athens, Greece
Agios Nikolaos, Crete, Greece
January 2022

Christos H. Skiadas
Yiannis Dimotikalis

Contents

Contributors

Andrii V. Abramov Donetsk National University, Donetsk, Ukraine

Olga P. Abramova Donetsk National University, Donetsk, Ukraine

Valeriy S. Abramov Donetsk Institute for Physics and Engineering Named After A.A. Galkin, Donetsk, Ukraine

Agalar M.-Z. Agalarov H. I. Amirkhanov Institute of Physics of Dagestan Federal Research Center, Russian Academy of Sciences, Makhachkala, Russia

Maricel Agop Department of Physics, Gheorghe Asachi Technical University of Iasi, Iasi, Romania

Sodeif Ahadpour University of Mohaghegh Ardabili, Ardabil, Iran

Elena S. Alekseeva Nizhny Novgorod Mathematical Society, Nizhny Novgorod, Russia

A. A. Balyakin NRC Kurchatov Institute, Moscow, Russia

Sohrab Behnia Department of Physics, Urmia University of Technology, Urmia, Iran

Jean-Marie Bilbault ImViA, Université de Bourgogne, Dijon Cedex, France

Stéphane Binczak ImViA, Université de Bourgogne, Dijon Cedex, France

Bernd Binder Bernstadt, Germany

Belqassim Bouteghrine LCOMS, Université de Lorraine, Metz, France

Ilya V. Boykov Penza State University, Penza, Russia

Alla I. Boykova Penza State University, Penza, Russia

Yiannis Dimotikalis Department of Management Science and Technology, Hellenic Mediterranean University, Crete, Greece

Andrey Dmitriev Department of Business Informatics, National Research University Higher School of Economics, Moscow, Russia;
Cybersecurity Research Center, University of Bernardo O'Higgins, Santiago, Chile

Victor Dmitriev Department of Business Informatics, National Research University Higher School of Economics, Moscow, Russia

Serhii Donetskyi National Technical University of Ukraine "Igor Sikorsky Kyiv Polytechnic Institute", Kyiv, Ukraine

D. P. Dowling School of Mechanical and Materials Engineering, University College Dublin, Belfield, Dublin 4, Ireland

Samira Fathizadeh Department of Physics, Urmia University of Technology, Urmia, Iran

Tatiana F. Filippova Ural Federal University, Ekaterinburg, Russia;
Krasovskii Institute of Mathematics and Mechanics, Ural Branch of Russian Academy of Sciences, Ekaterinburg, Russia

Masumeh Garagozi Department of Physics, Urmia University of Technology, Urmia, Iran

Lori-Anne Gardi Western University, London, ON, Canada

Alina Gavrilut Department of Mathematics, Alexandru Ioan Cuza University from Iasi, Iasi, Romania

Evgenii Gradusov Department of Business Informatics, National Research University Higher School of Economics, Moscow, Russia

Katica R. (Stevanović) Hedrih Department of Mechanics, Mathematical Institute of Serbian Academy of Science and Arts, Belgrade, Serbia;
Faculty of Mechanical Engineering, University of Niš, Nis, Serbia

Nafiseh Hematpour University of Mohaghegh Ardabili, Ardabil, Iran

Aliyu Isah ImViA, Université de Bourgogne, Dijon Cedex, France;
ELE-FAENG, Kano University of Science and Technology, Kano, Nigeria

Shunji Kawamoto Osaka Prefecture University, Sakai, Osaka, Japan

Anastasiia Kazmina Department of Business Informatics, National Research University Higher School of Economics, Moscow, Russia

Tatyana S. Krasnopolskaya Institute of Hydromechanics NASU, Kyiv, Ukraine

Viktor A. Kuznetsov Air Force Academy, Voronezh, Russia

V. J. Law School of Mechanical and Materials Engineering, University College Dublin, Belfield, Dublin 4, Ireland

Jean Jacques Loiseau Laboratoire des Sciences du Numérique de Nantes LS2N, Nantes, France

Wiesław M. Macek Faculty of Mathematics and Natural Sciences, Institute of Physical Sciences, Cardinal Stefan Wyszyński University, Warsaw, Poland;
Space Research Centre, Polish Academy of Sciences, Warsaw, Poland

Oxana G. Matviychuk Krasovskii Institute of Mathematics and Mechanics, Ural Branch of Russian Academy of Sciences, Ekaterinburg, Russia

Fatemeh Nemati Department of Physics, Urmia University of Technology, Urmia, Iran

Kosmas Papadopoulos Department of Economics, Democritus University of Thrace, Komotini, Greece

Evgeniy D. Pechuk Institute of Hydromechanics NASU, Kyiv, Ukraine;
Kyiv National University of Construction and Architecture, Kyiv, Ukraine

Vasiliy D. Pechuk Kyiv National University of Construction and Architecture, Kyiv, Ukraine

Alexander A. Potapov V.A. Kotelnikov Institute of Radio Engineering and Electronics, Russian Academy of Sciences, Moscow, Russia;
JNU-IREE RAS Joint Lab. of Information Tekhnology and Fraktal Processing of Signals, JiNan University, Guangzhou, China;
Nizhny Novgorod Mathematical Society, Nizhny Novgorod, Russia

Anton N. Pototskiy Air Force Academy, Voronezh, Russia

E. Prozorova Mathematical-Mechanical Department St. Petersburg State, Peterhof, Russia

Lenka Přibylová Department of Mathematics and Statistics, Faculty of Science, Masaryk University, Brno, Czech Republic

Alexander E. Rassadin Higher School of Economics, Nizhny Novgorod, Russia;
Nizhny Novgorod Mathematical Society, Nizhny Novgorod, Russia

Volodymyr Rusyn Department of Radio Engineering and Information Security, Yuriy Fedkovych Chernivtsi National University, Chernivtsi, Ukraine

Vladimir A. Ryazantsev Penza State University, Penza, Russia

E. Rybalova Institute of Physics, Saratov State University, Saratov, Russia

Said Sadoudi Ecole Militaire Polytechnique, Algiers, Algeria

Aceng Sambas Department of Mechanical Engineering, Universitas Muhammadiyah Tasikmalaya, Tasikmalaya, Indonesia

Yuriy Sanochkin Department of Business Informatics, National Research University Higher School of Economics, Moscow, Russia

Georges Sarafopoulos Department of Economics, Democritus University of Thrace, Komotini, Greece

Narmin Sefidkar Department of Physics, Urmia University of Technology, Urmia, Iran

Jan Ševčík Department of Mathematics and Statistics, Faculty of Science, Masaryk University, Brno, Czech Republic

Alexander V. Shishulin G.A. Razuvaev Institute of Organometallic Chemistry, Russian Academy of Sciences, Nizhny Novgorod, Russia

Anna V. Shishulina R.E. Alekseev Nizhny, Novgorod State Technical University, Nizhny Novgorod, Russia;
N.I. Lobachevsky Nizhny Novgorod State University, Nizhny Novgorod, Russia

Aleksandr Shvets National Technical University of Ukraine "Igor Sikorsky Kyiv Polytechnic Institute", Kyiv, Ukraine

Christos H. Skiadas ManLab, Technical University of Crete, Chania, Crete, Greece

Siavash H. Sohrab Department of Mechanical Engineering, Robert R. McCormick School of Engineering and Applied Science, Northwestern University, Evanston, IL, USA

V. M. Somsikov Al-Farabi Kazakh National University, Almaty, Kazakhstan

G. Strelkova Institute of Physics, Saratov State University, Saratov, Russia

Camel Tanougast LCOMS, Université de Lorraine, Metz, France

Ina Taralova Laboratoire des Sciences du Numérique de Nantes LS2N, Nantes, France

A. S. Tchakoutio Nguetcho LISSAS, Faculté des Sciences, Université de Maroua, Maroua, Cameroun

Adriana Pedrosa Biscaia Tufaile Soft Matter Lab, EACH, University of São Paulo, São Paulo, Brazil

Alberto Tufaile Soft Matter Lab, EACH, University of São Paulo, São Paulo, Brazil

Sergey Varbanets Department of Computer Algebra and Discrete Mathematics, Odessa I.I. Mechnikov National University, Odessa, Ukraine

Yakov Vorobyov Department of Mathematics, Informatics and Information Activities, Izmail State University of Humanities, Odessa, Ukraine

Chunxiao Yang Laboratoire des Sciences du Numérique de Nantes LS2N, Nantes, France

A. Zakharova Institut für Theoretische Physik, Technische Universität Berlin, Berlin, Germany

V. G. Zhulego NPO ANEK, Moscow, Russia;
NRC Kurchatov Institute, Moscow, Russia

Jakub Záthurecký Department of Mathematics and Statistics, Faculty of Science, Masaryk University, Brno, Czechia

The Higgs Boson and the Higgs Field in Fractal Models of the Universe: Supermassive Black Holes, Relativistic Jets, Solar Coronal Holes, Active Microobjects

Valeriy S. Abramov

Abstract To describe the masses of black holes, their relationships with the parameters of the Higgs boson, models based on the distribution density functions of the number of quanta in the ground and excited states for relic photons, and on the basis of the density distribution functions of the radiation intensity are proposed. It is proposed to represent the central region of a supermassive black hole near the upper mass boundary as a Bose condensate from black holes. Various states for a black hole with an intermediate mass are introduced. The following estimates have been made: masses for light black holes, binary and supermassive black holes; the speeds of motion of relativistic jets (emissions of matter); widths of active regions of coronal holes on the Sun; a number of parameters of active microobjects. These estimates are consistent with experimental data.

Keywords Supermassive black holes · Bose condensate from black holes · Higgs boson · Relic photons · Relativistic jets · Coronal holes on the Sun · Active microobjects

1 Introduction

Roger Penrose, Reinhard Henzel, Andrea Ghez are the laureates of the 2020 Nobel Prize in Physics. Using the general theory of relativity, Penrose theoretically predicted the gravitational collapse of massive stars, space–time singularities, and the birth of black holes [1, 2]. Genzel and Ghez discovered and described a supermassive black hole in the center of our Milky Way galaxy [3, 4]. Earlier Thorne [5] showed that a star can collapse under the influence of its own gravity: the space around it becomes curved, the star disappears and a black hole appears. It has been experimentally established, that the merger of two black holes [6], two neutron stars [7] is accompanied by the emission of gravitational waves. In [8, 9], a description of the parameters of gravitational waves, relict photons and their relations with the parameters of the

V. S. Abramov (✉)
Donetsk Institute for Physics and Engineering Named After A.A. Galkin, Donetsk, Ukraine
e-mail: vsabramov2018@gmail.com

C. H. Skiadas and Y. Dimotikalis (eds.), *14th Chaotic Modeling and Simulation International Conference*, Springer Proceedings in Complexity,
https://doi.org/10.1007/978-3-030-96964-6_1

Higgs boson was carried out in the framework of the Dicke superradiance model. In this case, supernonradiative states of gravitational fields are possible [10, 11]. However, the mechanisms of transitions from black holes with light masses (of the order of 29 − 32 M_s [6, 7], where M_s is the mass of the Sun) to supermassive (of the order of $4 - 5 \times 10^6$ M_s [3, 4]) and relativistic (of the order 10^{11} M_s) black holes have not yet been described. The creation of such theoretical models requires taking into account stochastic processes, the mass distribution functions of black holes in the Universe, the effect of ordering operators and the presence of qubit states [12, 13] for binary black holes and neutron stars. It also becomes necessary to describe the ejections of matter (relativistic jets) from a supermassive black hole [14]. The use of experimental methods with high angular resolution [15] makes it possible to study the nature of the Higgs field by the example of the behavior of solar active regions (coronal holes). The parameters of active objects are determined by the connections with the Higgs boson and with the different nature of the Higgs field. In [16], experimental evidence was obtained for the decay of the Higgs boson into a lepton pair and a photon, which indicates to the presence of an asymmetry of matter and antimatter [16, 17]. Experimentally in [18] the processes of formation and decay of tetraquarks were investigated. The authors believe that the structure of the new tetraquark contains charmed diquark and antidiquark, which are coupled by gluon interaction. In [19], a target made of gaseous deuterium was irradiated with a proton beam and the cross section for reactions with the formation of a helium isotope was measured. The authors estimated the baryon density for the early Universe during the process of primordial nucleosynthesis. However, the contributions of nonzero rest mass antineutrinos to Higgs fields have not been described.

The aim of this work is to describe the parameters of black holes, relativistic jets, active microobjects, their connections with the Higgs boson and the Higgs field of various nature (taking into account antineutrinos with nonzero rest mass) within the framework of a number of fractal cosmology models.

2 Models for Describing Black Hole Masses

In [8, 9] the Dicke superradiance model was used to describe gravitational waves and relict photons from binary black holes and neutron stars. For the ratio of the radiation intensities (maximum I_m to initial $I(0)$) was obtained

$$I_m/I(0) = (a_0 + a_m)(a_0 - a_m + 1); \quad a_0^2 = a_m^2 + z'_\mu(z'_\mu + 2)/4;$$
$$a_m^2 = z'_{A2}; \quad N_{ra} = z'_{A2} + z'_\mu. \tag{1}$$

Here $z'_{A2} = 1034.109294$ and $z'_\mu = 7.18418108$ are the usual and cosmological redshifts; the number of relic photons $N_{ra} = 1041.293475$; intensity ratio $I_m/I(0) = 81.06580421$. Supernonradiative states (of which the radiation intensity is equal to zero) were considered within the framework of the models A_0, A_1 [8, 9]. In the model A_0, the characteristic value of the number of bosons in the equilibrium state $N_{0A} = 3.557716045 \times 10^5$ was obtained. This made it possible to determine the characteristic energy $E_{0A} = N_{0A} E_G = 4.311073329$ eV, where the rest energy of the graviton $E_G = 12.11753067$ µeV. In the model A_1, a characteristic distribution density function $n'_{zg} = 0.114317037$ is obtained, where $n'_{zg} + |n_{zg}| = 1$ for Fermi-type particles. This function allows us to determine the characteristic frequencies ν'_{zg}, ν^*_{zg}, ν_{D0}

$$\nu'_{zg} = n'_{zg}\nu_{G0}; \quad \nu^*_{zg} = \nu'_{zg}/\psi_{01};$$
$$\psi_{01} = \varepsilon_{01}/E_{H0}; \quad \nu_{G0} = N_{0A} \cdot \nu_{D0}. \tag{2}$$

Here $E_{H0} = 125.03238$ GeV and $\varepsilon_{01} = 126.9414849$ GeV are Higgs boson energies obtained without and taking into account the Higgs field; frequency $\nu_{G0} = 2.9304515$ GHz, $\psi_{01} = 1.015268884$ [8, 9]. Based on (2), we find the numerical values $\nu'_{zg} = 335.0005326$ MHz, $\nu^*_{zg} = 329.9623754$ MHz, $\nu_{D0} = 8.236889799$ kHz. Our calculated frequency ν^*_{zg} practically coincides with the frequency of 330 MHz, at which dark matter dominates from observations of radio filaments [20].

Model B$_0$. Black holes with light masses M_{bh} are described on the basis of spectra for occupation numbers $n_{Ax} = n_{A0}S'_{0x}$ and $n^*_{Ax} = n_{A0}S_{xu}$ ($x = 1, 2, 3, 4$; spectral parameters S'_{0x} and S_{xu} are determined in [8, 9]) within the framework of the anisotropic model, where the main parameter $n_{A0} = 58.04663887$ is determined based on the expressions

$$n_{A0} = (z'_{\mu\lambda})^2 - 1 = (z'_\mu + 3/2)(z'_\mu - 1/2); \quad z'_{\mu\lambda} = z'_\mu + 1/2;$$
$$1/z'_{\mu\lambda} = \sin\varphi'_{\mu\lambda}; \quad n'_{A0} = (z'_{\mu\lambda})^2; \quad n'_{A0} - n_{A0} = 1;$$
$$\varphi'_{\mu\lambda} = \varphi_a Q_{H2}; \quad (n_{h1} + n_{h2}) - 2n_{A4} = n_G. \tag{3}$$

Using the example of binary black holes in [8, 9] and expressions (3), the quanta number of the second black hole $n_{h2} = M_{h2}/M_s = n_{A0}/2 = 29.02331944$, the first black hole $n_{h1} = M_{h1}/M_s = 35.98093926$ before their merger was obtained. After the merger, a black hole is formed with a number of quanta $2n_{A4} = M_{bh}/M_s = 62.0042587$ and a number of quanta $n_G = 1/Q_{H2} = 3$ are carried away by gravitational waves. In the general case, the number of quanta n_{A0}, n_G and the cosmological redshift z'_μ determine the number of quanta of the gluon field $n_g = 2n_G/[z'_\mu(z'_\mu + 1) - n_{A0}]$. At $n_G = 3$, constant parameters z'_μ from (1), n_{A0} from (3) we obtain $n_g = 8$. If n_G, z'_μ, n_{A0} are variables, then the number of quanta of the gluon field n_g becomes a function of these three arguments, that is typical for bulk fractal structures of the Universe.

Model B$_1$. To estimate the masses of supermassive black holes, we write down the basic relations for the energies

$$E_{H0}/E_G = \nu^*_{H0}/\nu_{G0} = N_{HG}; \quad E_G/\nu_{G0} = E_{H0}/\nu^*_{H0} = 2\pi\hbar;$$
$$E_{H0}/E_{0A} = N_{0n}; \quad E_{H0}/\varepsilon_{0n} = N^*_{0n}; \quad N^*_{0n} = (1+n'_{zg})N_{0n}. \tag{4}$$

Here $\hbar$ is Planck's constant. Taking into account (4), we find the parameters $N_{HG} = 1.031830522 \times 10^{16}$, $N_{0n} = 2.900261036 \times 10^{10}$, $N^*_{0n} = 3.231810284 \times 10^{10}$, energy $\varepsilon_{0n} = 3.86880321$ eV. The parameter N_{HG} is a function of the main parameters N_{ra}, N_{0A}, N_{0n}

$$N_{HG} = N_{ra}N_{cv} = N_{0A}N_{0n} = N_{D\nu}n_{ra} = N_{ra}N_{0A}n_{ra};$$
$$N_{cv} = N_{0A}n_{ra}; \quad N_{0n} = N_{ra}n_{ra}; \quad N_{D\nu} = N_{ra}N_{0A}, \tag{5}$$

where parameters $n_{ra} = 2.785248449 \times 10^7$, $N_{D\nu} = 3.704626502 \times 10^8$, $N_{cv} = 9.909123093 \times 10^{12}$ are additional. For bulk fractal structures of the Universe, the main and additional parameters from (5) can be operators. In the general case, these operators do not commute; when describing light and supermassive black holes, the appearance of stochastic properties is possible. We introduce the distribution density functions in the ground f_{ra} and excited f'_{ra} states for relic photons

$$f'_{ra} - f_{ra} = 1; \quad f'_{ra} = \langle \hat{c}_{ra}\, \hat{c}^+_{ra} \rangle = N_{ra}/(N_{ra} - z'_\mu);$$
$$f_{ra} = \langle \hat{c}^+_{ra}\, \hat{c}_{ra} \rangle = z'_\mu/(N_{ra} - z'_\mu) \tag{6}$$

where $\hat{c}^+_{ra}$, $\hat{c}_{ra}$ are creation and annihilation operators of relic photons; $\langle \dots \rangle$ is averaging symbol. Based on (6), (1), we find the numerical values $f_{ra} = 0.006947216$, $f'_{ra} = 1.006947216$. Expressions (1)–(6) make it possible to estimate the masses M_{0B}, M_{b0}, M'_{b0} black holes by the formulas

$$M_{0B} = f'_{ra}M_{b0}; \quad M_{b0}/M_s = n_g(1+n'_{zg})n_{ra}/n_{A0};$$
$$M'_{b0} = M_{0B} - M_{b0} = f_{ra}M_{b0}. \tag{7}$$

The numerical values are equal: $M_{0B}/M_s = 4.307173111 \times 10^6$, $M_{b0}/M_s = 4.277456693 \times 10^6$, $M'_{b0}/M_s = 0.029716418 \times 10^6$. Our estimate of the mass M_{0B}/M_s practically coincides with the mass of the central body 4.31×10^6 of a supermassive black hole in the center of the Milky Way galaxy [3, 4]. The value $2M'_{b0}/M_s = 0.059432836 \times 10^6$ determines the error 0.06×10^6, associated with the error in measuring the parameters of the orbit of the S2 star, rotating around the central body [3, 4].

Model B$_2$. The fractal structure of the Universe is characterized by the distribution of masses of black holes, which are found in the center of various galaxies. So for a supermassive black hole in the core of the galaxy M87, using the Event Horizon Telescope [21, 22], a shadow image in the radio range was obtained. Using four Chandra X-ray observations [14] for the MAXI J1820 + 070 binary black hole relativistic jets were detected. To estimate the upper mass limit $M_{J0} = N_{0A}M_{b0}$, we will represent the central body of a supermassive black hole as a Bose condensate of black holes with masses M_{b0}. In this case, for the parameter $N_{0A} = \psi_{1A} N^*_{GE}$, representation is acceptable, where $\psi^2_{1A} = 1 + \Omega^*_m$, $N^*_{GE} = M_s/M_E = R_{Gs}/R_{GE}$. Here M_E is mass of the Earth; R_{Gs} and R_{GE} are Schwarzschild gravitational radii of the Sun and Earth; $N^*_{GE} = 3.32958 \times 10^5$. In this model, the density of matter near supermassive black holes $\Omega^*_m = 0.141730642$ is close to our calculated value $\Omega_m = 0.141145722$ from [23, 24] and the value of 0.141 obtained by the Planck observatory, based on the new Hubble constant H^*_0 for the attenuation of γ-rays against the intergalactic background. As a result, we find $M_{J0}/M_s = 15.21797631 \times 10^{11}$. For experimentally search of supermassive black holes near the upper mass boundary, brightness distributions, changes in stellar radiation intensity when photographing galaxies with high resolution, adaptive optical spectroscopy to compensate for fluctuations in the atmosphere, and speckle spectroscopy are used [4]. For the maximum radiation intensity I_m from (1) near the upper mass boundary, the representation is acceptable

$$\begin{aligned}
&I_m = I^*_1 + I^*_2; \quad I^*_1 = n'_{zg} I_m = \mathrm{v}^2_{1J} I_m \sin^2(\theta^*_W);\\
&I^*_2 = n_{zg} I_m = (\mathrm{u}^2_{1J} + \mathrm{v}^2_{1J} cos^2(\theta^*_W)) I_m;\\
&\mathrm{v}^2_{1J} = k^2_{1J} = 0.5(1 - I(0)/I_m); \quad \mathrm{u}^2_{1J} = (k'_{1J})^2 = 0.5(1 + I(0)/I_m);\\
&\mathrm{u}^2_{1J} + \mathrm{v}^2_{1J} = 1; \quad I^*_1/I_m = k^2_{1J} \mathrm{sn}^2(u_{1W}; k_{1J}) = n'_{zg}; \quad I^*_2/I_m = \mathrm{dn}^2(u_{1W}; k_{1J}) = n_{zg}
\end{aligned} \tag{8}$$

Here k_{1J}, k'_{1J} and u_{1W} are modulli and effective displacement for elliptic functions $\mathrm{sn}(u_{1W}; k_{1J})$, $\mathrm{cn}(u_{1W}; k_{1J})$, $\mathrm{dn}(u_{1W}; k_{1J})$; the angle θ^*_W acts as the effective Cabibo angle for supermassive black holes; parameters u_{1J}, v_{1J} depend on the initial and maximum radiation intensity and are analogous to the N. N. Bogolyubov's transformation parameters in the theory of superconductivity. Numerical values are equal: $k^2_{1J} = 0.493832171$, $(k'_{1J})^2 = 0.506167829$, $\sin^2(\theta^*_W) = 0.231489651$, $\cos^2(\theta^*_W) = 0.768510349$, intensity distribution density functions $f_{J1} = I^*_1/I^*_2 = 0.129072187$, $f'_{J1} = I_m/I^*_2 = 1.129072187$. Expressions (8) allow us to estimate the masses of black holes M'_{J1}, M_{J1} near the upper mass boundary by the formulas

$$M'_{J1} - M_{J1} = M_{J0}; \quad M'_{J1} = f'_{J1} M_{J0}; \quad M_{J1} = f_{J1} M_{J0}; \quad f'_{J1} - f_{J1} = 1 \tag{9}$$

Based on (9), we obtain a numerical value $M_{J1}/M_s = 1.964217483 \times 10^{11}$, that is close to the experimental value $1.96 \times 10^{11} M_s$ for the supermassive black hole SDSS J140821.67 + 025,733.2. For intermediate masses of black holes, the maximum radiation intensity I^*_m can change over a segment $I(0) \le I^*_m \le I_m$. These changes are described by a variable number of quanta n^*_m and an inversion parameter B_{Jm}

$$\begin{aligned}
&n^*_m = I^*_m/I_m = \mathrm{u}^2_{1J} + \mathrm{v}^2_{1J} B_{Jm}; \quad B^*_1 = B_{Jm}(n'_{zg}) = -\mathrm{u}^2_{1J}/\mathrm{v}^2_{1J} + \sin^2(\theta^*_W);\\
&-1 \le B_{Jm} \le 1; \quad B^*_2 = B_{Jm}(n_{zg}) = \cos^2(\theta^*_W); \quad n_{1J} = \mathrm{v}^2_{1J} + n'_{zg};\\
&n'_{1J} = \mathrm{u}^2_{1J} - n'_{zg}; \quad n_{1J} + n'_{zg} = 1.
\end{aligned} \tag{10}$$

From (10) it follows, that a black hole with an intermediate mass can be in different states, which are determined by a pair of parameters n_m^* and B_{Jm}. Let's introduce these states: ground $J1$ ($n_m^* = 1$, $B_{Jm} = 1$), supernonradiative $J2$ ($n_m^* = \mathrm{u}_{1J}^2$, $B_{Jm} = 0$), fully inverse state $J3(n_m^* = \mathrm{u}_{1J}^2 - \mathrm{v}_{1J}^2$, $B_{Jm} = -1$), partially inverse state $J4$ ($n_m^* = n'_{zg}$, $B_{Jm} = B_1^*$), deviated from the ground $J5$ ($n_m^* = n_{zg}$, $B_{Jm} = B_2^*$). The parameters $B_1^* = -0.793489803$, $1 + B_1^* = 0.206510197$, $B_2^*, n_{1J} = 0.391850792$, $n'_{1J} = 0.608149208$ carry information about the characteristic parameters (velocities, energies) of a relativistic jet (ejection of matter from a supermassive black hole) [14].

3 Relativistic Jets

To describe the parameters of a relativistic jet, we will use the basic model equations

$$Q_{H2}R_{AB}/2R_{AH} = n'_F + 0.5; \quad \Omega_{0\nu} = (n'_F)^2; \quad \Omega_{\tau L}E_{W0} = \Omega^*_{\tau L}E_{Z0}. \tag{11}$$

Here the parameter $Q_{H2} = 1/3$ is determined by the expression from (3) and is related to the angles $\varphi'_{\mu\lambda}$, $\varphi_a = 22.43261135°$, cosmological redshift z'_μ from our anisotropic model [8] of the expanding Universe; the number of quanta $n'_F = 0.054219932$ determines the Fermi level and the neutrino density $\Omega_{0\nu} = 0.002939801$; the lepton quantum number $\Omega_{\tau L} = 0.002402187$ is related to the quantum number $\Omega^*_{\tau L} = 0.002116741$ through the rest energies $E_{W0} = 80.35235464$ GeV and $E_{Z0} = 91.188$ GeV for $W0$ and $Z0$ bosons, respectively; Hubble radius $R_{AH} = 13.75 \times 10^9 \cdot L_{c0}$. From (11) we find the characteristic radius (horizon of matter particles) $R_{AB} = 45.72314437 \times 10^9 \cdot L_{c0}$. Note, that parameters are: L_{c0} = light year $= c_0\tau_{c0} = N_{c0}L_{ES} = 0.306597989$ pc; limiting speed of light in vacuum $c_0 = 2.99792458 \times 10^5$ km s^{-1}, $N_{c0} = 6.324043414 \times 10^4$, $\tau_{c0} = 365.2503353$ day $= 3.155762897 \times 10^7$ s, $L_{ES} = 1$au $= 1.495995288 \times 10^8$ km. Based on (11), we introduce the refractive index n_{AB} of the medium of matter particles

$$n_{AB} = Q^2_{AB}; \quad Q_{AB} = R_{AB}/R_{AH} = 2(n'_F + 0.5)/Q_{H2}. \tag{12}$$

Numerical values are $n_{AB} = 11.05775038$; $Q_{AB} = 3.325319591$. Next, we find the particle velocities υ_{AH}, $\upsilon_{0\xi}$, υ_{AW} and velocities ratios ξ_{AH}, ξ_{0J}, ξ_{AW}

$$\upsilon^2_{AH} = c_0^2/n_{AB}; \quad \upsilon_{AH}S^2_{1u} = \upsilon_{0\xi}S^2_{2u} = \upsilon_{AW}; \quad \xi_{AH} = \upsilon_{AH}\ c_0; \quad \xi_{0J} = \upsilon_{0\xi}/c_0. \tag{13}$$

Values are: $\upsilon_{AH} = 9.015447983 \times 10^4$ km s^{-1}, $\upsilon_{0\xi} = 1.803089597 \times 10^5$km s^{-1}, $\upsilon_{AW} = 196.9672387$ km s^{-1}; $\xi_{AH} = 0.300722975$, $\xi_{0J} = 0.60144595$, $\xi_{AW} = \upsilon_{AW}/c_0 = 657.0119876 \times 10^{-6}$.

From our **model B$_2$** it follows, that the density of matter near supermassive black holes $\Omega_m^* > \Omega_m$. This leads to a change in the refractive index of the medium n_{AB}, the radius R_{AB} from (12), the neutrino density $\Omega_{0\nu}$ from (11), and the particle velocities from (13). Accounting for these changes near supermassive black holes is described by new parameters

$$\overline{n}_{AB} = \overline{Q}_{AB}^2; \quad \overline{Q}_{AB} = \overline{R}_{AB}/R_{AH} = 2(\overline{n}_{0\nu} + 0.5)/Q_{H2};$$
$$\overline{\Omega}_{0\nu} = (\overline{n}_{0\nu})^2; \quad 2\overline{n}_{0\nu} = \Omega_m^* - S_{2u}; \quad \overline{\upsilon}_{AH} = c_0/\overline{Q}_{AB};$$
$$\overline{\upsilon}_{AH} S_{1u}^2 = \overline{\upsilon}_{0\xi} S_{2u}^2 = \overline{\upsilon}_{AW}. \tag{14}$$

Numerical values are equal: medium refractive index $\overline{n}_{AB} = 11.06252927$, radius $\overline{R}_{AB} = 45.73302352 \times 10^9 \cdot L_{c0}$, parameters $\overline{Q}_{AB} = 3.326038074$, $\overline{n}_{0\nu} = 0.054339679$, density of the relativistic neutrino $\overline{\Omega}_{0\nu} = 0.002952801$; velocities $\overline{\upsilon}_{AH} = 9.013500487 \times 10^4$ km s^{-1}, $\overline{\upsilon}_{0\xi} = 1.802700097 \times 10^5$ km s^{-1}, $\overline{\upsilon}_{AW} = 196.9246903$ km s^{-1}; velocities ratios $\overline{\xi}_{AH} = 1/\overline{Q}_{AB} = 0.300658013$, $\overline{\xi}_{0J} = 2\overline{\xi}_{AH} = 0.601316027$, $\overline{\xi}_{AW} = \overline{\upsilon}_{AW}/c_0 = 656.8700603 \times 10^{-6}$.

Further, taking into account (1), (2), (8), we find the energies of the jet particles E_{0J} and E_{1J}, E_{2J} in the absence and presence of the Higgs field, respectively, by the formulas

$$E_{0J}/E_{H0} = I_m/I(0) = I_1^*/I(0) + I_2^*/I(0); \quad E_{1J} = \psi_{01} E_{0J}; \quad E_{2J} = \psi_{02} E_{0J}. \tag{15}$$

Numerical values of energies are equal $E_{0J} = 10.13585044$ TeV, $E_{1J} = 10.29061357$ TeV, $E_{2J} = 9.978687329$ TeV, where $\psi_{02} = 0.984494334$.

The effective Cabibo angle θ_W^* allows us to estimate the angular width φ_W^* of the jet based on the angular parameters φ_{n0}, φ_{En}, φ_{E0} by the formulas

$$\varphi_W^* = \varphi_{n0}/\sin^2(\theta_W^*); \quad \varphi_{n0} = 2n_G\,\varphi_{En}/n_g; \quad \varphi_{En} = m_n\,\varphi_{E0}/m_p(1 + S_{1u}), \tag{16}$$

where the parameters n_G, n_g are determined in the **model B$_0$** by expressions (3);$m_n/m_p = 1.008985047$ is the ratio of the neutron mass (m_n) to the proton mass (m_p). Based on (16), we find estimates of the angular parameters $\varphi_W^* = 2.592779092''$, $\varphi_{n0} = 0.600201527''$, $\varphi_{En} = 0.800268702''$, where the parameter $\varphi_{E0} = 0.830215001''$ describes the behavior of photons near supermassive bodies in Einstein's theory [5]. The obtained estimates of the parameters ξ_{0J}, E_{0J}, φ_W^* do not contradict the experimental data [14] for a velocity ratio of 0.6, an energy of 10TeV, and an angular width 2.5928″ of jet particles. Based on the effective radii R_{AB} from (12), $\overline{R}_{AB}$ from (14), we obtain estimates of the distance R_0 from the Sun to the supermassive black hole in the center of our Milky Way galaxy and the errors δR_0 by the formulas

$$R_0 = \overline{\delta}_{AB}/n_{R0}; \quad \delta R_0 = \overline{\delta}_{AB}/N_{R0}; \quad \overline{\delta}_{AB} = (1 + \delta_Q)\delta_{AB};$$

$$\delta_{AB} = \overline{R}_{AB} - R_{AB}; \quad N_{R0} = n_g(N_{ra} + 0.5 I_m / I(0));$$
$$n_{R0} = Q_{H2}(N_{ra} + n_{A0} - n_g - \overline{\xi}_{0J}). \tag{17}$$

The numerical values of the parameters are equal: $N_{R0} = 8654.611017$, $n_{R0} = 363.5795993$, $\delta_{AB} = 9.87915 \times 10^6 \cdot L_{c0}$, $\overline{\delta}_{AB} = 9.879150543 \times 10^6 \cdot L_{c0}$. Based on (17), we find estimates of the distance $R_0 = 8.330851608$ kpc and error $\delta R_0 = 0.349978489$ kpc.

Based on the distribution density function f'_{J1} from (9), the number of quanta $\overline{n}_{0\nu}$ from (14), we find the radius r_{JB} of the central body by the formulas

$$N_{G0}\, r_{JB} = \delta'_{JB} + l_{AB}; \quad \delta'_{JB} = \overline{\delta}_{AB}\, f'_{J1}; \quad l_{AB} = \overline{\delta}_{AB} \sin(\theta_{0\nu});$$
$$N_{G0} = N_a / N_{HG}; \quad N_{G0} E_{H0} = N_a E_G; \quad \sin(\theta_{0\nu}) = \overline{n}_{0\nu}(1 - \overline{n}_{0\nu}) = \overline{n}_{0\nu} - \overline{\Omega}_{0\nu}. \tag{18}$$

Values of the parameters are equal: $N_{G0} = 5.839561703 \times 10^7$, $\theta_{0\nu} = 2.945548561°$, $\sin(\theta_{0\nu}) = 0.051386878$, $l_{AB} = 5.076587037 \times 10^5 \cdot L_{c0}$, $\delta'_{JB} = 11.15427411 \times 10^6 \cdot L_{c0}$. From (18) we obtain $r_{JB} = 0.199705618 \cdot L_{c0} = 1.262947001 \times 10^4$au.

Next, we find estimates for the semi-axes x_{0S}, y_{0S} the elliptical orbit of the star S2, rotating around the central body by the formulas

$$y_{0S} = r_{JB} / \overline{n}_{AB}(1 + \Omega^*_m); \quad x^2_{0S} / y^2_{0S} = S^2_{1u} \sin(\varphi_{0g}) / S^2_{2u};$$
$$\sin^2(\varphi_{0g}) = (n_{A0} - n_g)(E_e + E_{eh}) / E_{0g}; \quad E_{0g} = n_g E_{H0}. \tag{19}$$

Here the rest energies of the gluon $E_{0g} = 1.00025904$ TeV, electron E_e and electron hole E_{eh} are assumed to be equal $E_e = E_{eh} = 0.51099907$ MeV; $\sin(\varphi_{0g}) = 0.007150827$, the angle of polarization of the radiation $\varphi_{0g} = 0.409715696°$; semi-axes $y_{0S} = 999.9241011$ au, $x_{0S} = 119.5804463$ au. Our estimates of the parameters R_0, δR_0, r_{JB}, x_{0S}, y_{0S} agree with the experimental data [3, 4] for the distance 8.33 kpc from the Sun to the supermassive black hole in the center of the Milky Way galaxy, the error 0.35 kpc, the radius of the central body $0.2 \cdot L_{c0}$, for the semi-axes 120 au, 1000 au the elliptical orbit of the S2 star, rotating around the central body, respectively.

4 Asymmetry of Matter, Antimatter and the Higgs Field

The presence of a Higgs field of various nature (gluon, lepton, neutrino, hadronic based on the parameter $\Omega^*_{\tau L}$ from (11), gravitational, etc.) leads to changes in the rest energy of the Higgs boson E_{H0} in (18); energies of holes (antiparticles) E_{eh} in (19), $E_{\mu h}$, $E_{\tau h}$ for e, μ, τ-leptons, respectively; the appearance of asymmetry of matter and antimatter. We introduce the energy E_{0L} based on the total energy ε_{0L} of paired leptons, the number of quanta of gluons n_g

$$E_{0L} = n_g \varepsilon_{0L}; \quad \varepsilon_{0L} = (E_e + E_{eh}) + (E_\mu + E_{\mu h}) + (E_\tau + E_{\tau h}). \tag{20}$$

Here $E_\mu = E_{\mu h} = 105.658389$ MeV, $E_\tau = E_{\tau h} = 1777.00$ MeV are rest energies for μ, τ-leptons, respectively. From (20) we find the energies $\varepsilon_{0L} = 3.766338776$ GeV, $E_{0L} = 30.13071021$ GeV (close to the data from [16]).

Next, we introduce the distribution density functions of the Bose type f_{gA} (ground state), f'_{gA} (excited state) based on the number of quanta of black holes (n_{A0}), gluons (n_g). Based on E_{H0} we find the energy E_{gA}, E'_{gA}

$$\begin{aligned} &f'_{gA} - f_{gA} = 1; \quad f_{gA} = n_g/(n_{A0} - n_g); \quad f'_{gA} = n_{A0}/(n_{A0} - n_g); \\ &E_{gA} = E_{H0}\, f_{gA}/2; \quad E'_{gA} = E_{H0} f'_{gA}/2; \quad E'_{gA} - E_{gA} = E_{H0}/2. \end{aligned} \tag{21}$$

The numerical values are equal: $f_{gA} = 0.159850895$, $E_{gA} = 9.993268924$ GeV, $E'_{gA} = 72.50945893$ GeV. Taking into account the energy E_{gA} from (21), the expressions for the rest energies of leptons have the form

$$E_e = E_{gA} \sin^2(\varphi_{eg}); \quad E_\mu = E_{gA} \sin^2(\varphi_{\mu g}); \quad E_\tau = E_{gA} \sin^2(\varphi_{\tau g}). \tag{22}$$

Here angles are equal: $\varphi_{eg} = \varphi_{0g}$, $\varphi_{\mu g} = 5.901862921°$, $\varphi_{\tau g} = 24.94112323°$. To describe the interaction of μ and e-leptons, we find the energies E'_μ, E^*_μ from the expressions

$$\begin{aligned} &E'_\mu = E_{gA} \sin^2(\varphi_{\mu g} + \varphi_{eg}) = (E^2_\mu + 4\Delta^2_\mu)^{1/2}; \quad 2\Delta_\mu = n_{A0} E_{ex}; \\ &E_{ex} = E_e + E'_h; \quad E^*_\mu = E_{gA} \sin^2(\varphi_{\mu g} - \varphi_{eg}) = (E^2_\mu - 4(\Delta^*_\mu)^2)^{1/2}; \\ &2\Delta^*_\mu = n_{A0} E^*_{ex}; \quad E^*_{ex} = E_e + E^*_h; \quad E_e/E_{ex} = 0.5 + \sin(\varphi_{ex}); \\ &E'_h/E_{ex} = 0.5 - \sin(\varphi_{ex}); \quad E_e/E^*_{ex} = 0.5 + \sin(\varphi^*_{ex}). \end{aligned} \tag{23}$$

For variant I (sum of angles), the parameter values are:$\varphi_{\mu g} + \varphi_{eg} = 6.311578617°$, $E'_\mu = 120.7760733$ MeV, $E'_\mu - E_\mu = 15.11768432$ MeV, energy gap $\Delta_\mu = 29.25390878$ MeV, energy $E_{ex} = 1.007944968$ MeV, hole energy $E'_h = 0.496945898$ MeV, $\sin(\varphi_{ex}) = 0.0069712$, characteristic angle $\varphi_{ex} = 0.399423573°$. For variant II (angle difference), the parameter values are: $\varphi_{\mu g} - \varphi_{eg} = 5.492147225°$, $E^*_\mu = 91.54109182$ MeV, energy gap $\Delta^*_\mu = 26.38145028$ MeV, energy $E^*_{ex} = 0.908974259$ MeV, hole energy $E^*_h = 0.397975189$ MeV, $\sin(\varphi^*_{ex}) = 0.062171112$, characteristic angle $\varphi^*_{ex} = 3.564441086°$, $E^*_h/E^*_{ex} = 0.5 - \sin(\varphi^*_{ex})$. Note, that the values of the angle differences $(\varphi_{eg} - \varphi_{ex})/2 = 18.52582072''$, $(\varphi_{eg} - \varphi_{ex})/4 = 9.26291036''$ are characteristic of the angular widths of coronal holes on the Sun [15]. From (23) we find expressions easy for analyzing the asymmetry of individual contributions from E_e, E_μ, different angles, in energy E'_μ, E^*_μ in the form

$$(E'_\mu + E^*_\mu)/2 = E_e \cos^2(\varphi_{\mu g}) + E_\mu \cos^2(\varphi_{eg});$$
$$E'_\mu - E^*_\mu = E_{gA} \sin(2\varphi_{\mu g}) \sin(2\varphi_{eg}). \quad (24)$$

Based on the energy E_{0L} from (20) we found characteristic energies ε_{dL}, ε_{d0}, ε'_{dz} and the Higgs boson energies E_{Hd}, E'_{Hd}, E_{Hg}, E'_{Hg}, E_{HL}, E'_{HL}

$$E_{0L} = n_g \varepsilon_{0L} = n_G \varepsilon_{dL}; \quad \varepsilon_{d0} = n_{A0} \varepsilon_{dL}; \quad \varepsilon'_{dz} = z'_\mu(z'_\mu + 1)\varepsilon_{dL};$$
$$\varepsilon'_{dz} = \varepsilon_{d0} + 2\varepsilon_{0L}; \quad E^2_{Hd} = E^2_{H0} + \varepsilon^2_{dL}; \quad (E'_{Hd})^2 = E^2_{H0} - \varepsilon^2_{dL};$$
$$E^2_{Hg} = E^2_{H0} + E^2_{gA}; \quad (E'_{Hg})^2 = E^2_{H0} - E^2_{gA};$$
$$E^2_{HL} = E^2_{H0} + \varepsilon^2_{0L}; \quad (E'_{HL})^2 = E^2_{H0} - \varepsilon^2_{0L}. \quad (25)$$

Characteristic energies are $\varepsilon_{dL} = 10.04357007$ GeV (close to the energy for dark matter from [20]), $\varepsilon_{d0} = 582.9954848$ GeV, $\varepsilon'_{dz} = 590.5281624$ GeV. Energies ε_{dL}, E_{gA}, ε_{0L} describe the different nature of the Higgs field. The presence of the Higgs field leads to the appearance of active particles with energies $E_{Hd} = 125.4351201$ GeV, $E'_{Hd} = 124.6283385$ GeV, $E_{Hg} = 125.4311025$ GeV, $E'_{Hg} = 124.6323819$ GeV, $E_{HL} = 125.0890937$ GeV (corresponds to the peak for the Higgs boson decay process from [16]), $E'_{HL} = 124.9756406$ GeV. Energy differences $\delta E_{Hg} = E_{Hd} - E_{Hg} = 4.0176$ MeV, $\delta E'_{Hg} = E'_{Hg} - E'_{Hd} = 4.04343$ MeV describe the line width in the energy spectrum for the Higgs boson [16].

5 Active Microobjects

Based on the energies ε_{dL}, E_{gA}, ε_{0L} from (25) we find the radii R_{dL}, R_{gA}, R_{0L} of active microobjects associated with the different nature of the Higgs field

$$R_{dL} = A_G\, \varepsilon_{dL}; \quad R_{gA} = A_G E_{gA}; \quad R_{GL} = n_G R_{dL} = n_g R_{0L};$$
$$R'_{dz} = z'_\mu(z'_\mu + 1) R_{dL}; \quad R_{d0} = n_{A0} R_{dL}; \quad R'_{dz} - R_{d0} = 2R_{0L}. \quad (26)$$

Here $A_G = 0.960836162$ fm$(eV)^{-1}$ is the constant from [23, 24]. The gravitational radii are: $R_{dL} = 9.6502253\, \mu m$, $R_{gA} = 9.6018942\, \mu m$, $R_{0L} = 3.6188345\, \mu m$. For characteristic radii we obtain: $R_{d0} = 560.1631441\, \mu m$ (coupled to the number of quanta of the black hole n_{A0}); $R'_{dz} = 567.4008131\, \mu m$ (coupled to cosmological redshift z'_μ); $R_{GL} = 28.95067596\, \mu m$ (coupled with the number of quanta of the gravitational field in an excited state n_G, or with the number of quanta of the gluon field n_g). Next, we find the characteristic lengths l_{d0}, l'_{dz}, l_{0L} of active objects

$$l_{d0}/R_{d0} = l'_{dz}/R'_{dz} = l_{0L}/2R_{0L} = E_{\alpha u}/E_{H0} = S_{12u};$$
$$S_{12u} = S_{1u} - S_{2u}; \quad E_{\alpha u} - E_{\alpha S} = E_c;$$

$$E_{\alpha S} = S_{012} E_{H0} = \xi_{gS} E_{0g}; \quad S_{012} = S'_{01} - S'_{02}; \quad \xi_{gS} = S_{012}/n_g. \tag{27}$$

Here the parameters are: $S_{12u} = 0.013690291$, $S_{012} = 0.005451282$, $\xi_{gS} = 0.00068141$; the rest energy $E_c = 1.030142904\,\text{GeV}$, c-quark gravitational radius $R_c = A_G E_c = 0.989798554\,\mu m$; energy $E_{\alpha S} = 0.681586763\,\text{GeV}$ is determined either through the rest energy of the Higgs boson, or through the energy of the gluon, energy $E_{\alpha u} = 1.711729667$ GeV; gravitational radii $R_{\alpha S} = A_G E_{\alpha S} = 654.8932091\,\text{nm}$, $R_{\alpha u} = A_G E_{\alpha u} = 1.644691763\ \mu m$. From (27) we obtain values of characteristic lengths $l_{d0} = 7.6687965\,\mu m$, $l'_{dz} = 7.7678822\,\mu m$, $l_{0L} = 99.085795\,\text{nm}$. From (27) it follows that it is possible to describe particles and antiparticles, compound particles (hadrons), which are experimentally observed at the LHC [16], on the basis of energies $E_{\alpha u}$, $E_{\alpha S}$, E_c. As an example, consider the possibility of describing the energies E_{TQ}, E'_{TQ} of a tetraquark, a hadron by

$$\begin{aligned} &E_{TQ} = 2E_c + 2\overline{E}_c; \quad \overline{E}_c = E_c + E_{\alpha S} + \Delta^*_\mu = E_c + \xi_{gS} E_{0g} + \Delta^*_\mu = E_{\alpha u} + \Delta^*_\mu; \\ &E_{TQ} - E'_{TQ} = 2(E_\mu + E'_\mu); \quad E_{T1} = E_{TQ} - 2E'_\mu - \Delta_\mu; \\ &E_{T2} = E_{TQ} - 2E^*_\mu + \Delta^*_\mu. \end{aligned} \tag{28}$$

Here $\overline{E}_c = 1.738111117$ GeV, $E_\mu + E'_\mu = 226.4344623$ MeV are the c-antiquark, muon pair energies, respectively. Energies $E_{T1} = 6628.875515$ MeV, $E_{T2} = 6742.980837\,\text{MeV}$ determine the features of the type of local maximum, minimum on the experimental dependence of the number of events on the state of the tetraquark [18]. The base narrow peak corresponds to the tetraquark energy $E_{TQ} = 6899.681571$ MeV, and the broadened peak corresponds to the hadron energy $E'_{TQ} = 6446.812646$ MeV. Note, that the energy difference $E_c - E^*_\mu = 938.6018122$ MeV (for the c-quark and antimuon) is close to the sum of the energies $E_p + E_e = 938.7833217$ MeV (for the proton and the electron). This indicates us to the need and possibility of describing additional contributions from the neutrinos, hadron Higgs fields to the energies of active objects.

The classical decay of a neutron into a pair of proton-electron and antineutrino is described by the expressions

$$\begin{aligned} &E_n = (E_p + E_e) + n_{ra}\,\varepsilon_{\nu n}; \quad \varepsilon_{\nu n} = (\varepsilon^2_{HG} + \Delta^2_{\nu n})^{1/2}; \quad \Delta^2_{\nu n} = z_{\nu n}(z_{\nu n} + 2)\varepsilon^2_{HG}; \\ &\varepsilon_{\nu n} = \varepsilon_{HG} + z_{\nu n}\,\varepsilon_{HG} = \psi_{\nu n}\,\varepsilon_{HG}; \quad \psi_{\nu n} = 1 + z_{\nu n}. \end{aligned} \tag{29}$$

Here the rest energies are for neutrino $\varepsilon_{HG} = 280.0460475\,\text{MeV}$ [23, 24], neutron $E_n = 946.7027435\,\text{MeV}$, proton $E_p = 938.2723226\,\text{MeV}$. From (29) we find the antineutrino energy $\varepsilon_{\nu n} = 284.3344848\,\text{meV}$, energy gap $\Delta_{\nu n} = 49.1966514\,\text{meV}$, parameters of the neutrino field $z_{\nu n} = 0.015313329$, $\psi_{\nu n} = 1.015313329$.

We take into account the contribution from the hadronic Higgs field by replacing the energy of the pair $(E_p + E_e)$ in (29) with the energy of the difference $E_c - E^*_\mu$ for the c-quark and antimuon. In this case, the antineutrino energy $\varepsilon_{\nu n}$ is replaced by the renormalized antineutrino energy $\overline{\varepsilon}_{\nu n}$ and is determined from the expressions

$$
\begin{aligned}
&E_n = (E_c - E_\mu^*) + n_{ra}\overline{\varepsilon}_{vn}; \quad \overline{\varepsilon}_{vn} = (\varepsilon_{HG}^2 + \overline{\Delta}_{vn}^2)^{1/2}; \quad \overline{\Delta}_{vn}^2 = \overline{z}_{vn}(\overline{z}_{vn} + 2)\varepsilon_{HG}^2 \\
&\overline{\varepsilon}_{vn} = \varepsilon_{HG} + \overline{z}_{vn}\,\varepsilon_{HG} = \overline{\psi}_{vn}\,\varepsilon_{HG}; \quad \overline{\psi}_{vn} = 1 + \overline{z}_{vn} = \psi_{vn}^* + 0.5\,n_{vn}; \\
&n_{vn}^2 = \Omega_{\tau L}^*\ \overline{\varepsilon}_{vn} = \varepsilon_{h\,v} + \varepsilon_{vn}^*; \quad \varepsilon_{hv} = 0.5 n_{vn}\,\varepsilon_{HG}; \\
&\varepsilon_{vn}^* = (\varepsilon_{HG}^2 + (\Delta_{vn}^*)^2)^{1/2} = \psi_{vn}^*\,\varepsilon_{HG}.
\end{aligned} \tag{30}
$$

Here, the parameter $n_{vn} = 0.046008054$ from (11) describes the contribution from the hadron Higgs field to the energy $\varepsilon_{h\,v} = 6.442186838$ meV. Based on (30), we find the renormalized antineutrino energy $\overline{\varepsilon}_{vn} = 290.8512992$ meV, energy gap $\overline{\Delta}_{vn} = 78.54100538$ meV, parameters $\overline{z}_{vn} = 0.038583839$, $\overline{\psi}_{vn} = 1.038583839$. The energy $\varepsilon_{vn}^* = 284.4091123$ meV, energy gap $\Delta_{vn}^* = 49.62614656$ meV, field parameters $z_{vn}^* = \overline{z}_{vn} - 0.5n_{vn} = 0.015579812$, $\psi_{vn}^* = 1 + z_{vn}^*$ describe a different state of the antineutrino, compared to the state from (29).

Taking into account (29), we find the baryon densities of the Universe Ω_{b1} (ground state of matter), Ω_{b2} (hole state of matter) from the expressions

$$
\Omega_{b1} = (0.5 - z_{vn})\,n_{vn}; \quad \Omega_{b2} = (0.5 + z_{vn})n_{vn}; \quad \Omega_{b1} + \Omega_{b2} = n_{vn}. \tag{31}
$$

Numerical values are equal: $\Omega_{b1} = 0.022299491$, $\Omega_{b2} = 0.023708563$. At the same time $\Omega_{b1} < \Omega_{b2}$, that confirms the presence of two states of baryonic matter due to the presence of the Higgs antineutrino field z_{vn}. Replacing in (31) z_{vn} by z_{vn}^* from (30) leads to other values of the baryon density $\Omega_{b1}^* = 0.02228723$, $\Omega_{b2}^* = 0.023720824$. Hence it follows, that the baryon density of the Universe depends on the states of the antineutrino field. On the other hand, within the framework of our anisotropic model (taking into account the polarization of the CMB), the base parameter n_{vn} can be independently determined from

$$
\begin{aligned}
&n_{vn} = |\chi_{ef}|\sin(\varphi_{0g}) + \psi_{rc} + 2\Omega_{0G}; \\
&\Omega_{b1} = 0.5n_{vn} - 2n_{\tau L}\sin(\varphi_{0g}); \quad n_{\tau L}^2 = \Omega_{\tau L}.
\end{aligned} \tag{32}
$$

Here $|\chi_{ef}| = 0.2504252$, $\psi_{rc} = 0.04420725$, $\Omega_{0G} = 4.99501253 \times 10^{-6}$ from [23, 24]. The values Ω_{b1} from (32) and (31) coincide and agree with the baryon density of the Universe 0.0223 from the experimental data [19]. Note, that expressions (31) allow us to describe the inverse (at $z_{vn} < 0$) states, states with shifts Ω_{b1}', Ω_{b2}' or $\overline{\Omega}_{b1}$, $\overline{\Omega}_{b2}$ of the baryon density of the Universe

$$
\begin{aligned}
&\Omega_{b1}' = \Omega_{b1} - \Omega_{\tau L}^*; \quad \Omega_{b2}' = \Omega_{b2} + \Omega_{\tau L}^*; \\
&\overline{\Omega}_{b1} = \Omega_{b1} + \Omega_{\tau L}^*; \quad \overline{\Omega}_{b2} = \Omega_{b2} - \Omega_{\tau L}^*.
\end{aligned} \tag{33}
$$

due to the presence of a contribution from $\Omega_{\tau L}^*$ while preserving the quantum number n_{vn}. Numerical values are equal: $\Omega_{b1}' = 0.02018275$, $\overline{\Omega}_{b1} = 0.024416232$. Expressions (31)–(33) can be used to describe the effective susceptibilities χ_{vx} of active regions ($x = A, B, C, D, E$) of coronal holes on the Sun. In [15] the parameters N_{2x}, N_{1x} for these regions were measured. Based on the formulas

$$\chi_{vx} = N_{2x}/N_{1x}; \quad \chi_{bx} = \chi_{vx} + \Omega'_{b1}; \quad z_{bx} = (1+\chi_{bx}^2)^{1/2} - 1 = 1 - (1-\overline{\chi}_{bx}^2)^{1/2};$$
$$\chi_{NA}^2 = |\chi_{21} \cdot \chi_{12}| + 2\Omega_{0G}; \quad \varepsilon_{bA} = (\varepsilon_{HG}^2 + \Delta_{bA}^2)^{1/2}; \quad \overline{\varepsilon}_{bA} = (\varepsilon_{HG}^2 + \overline{\Delta}_{bA}^2)^{1/2} \tag{34}$$

estimates for each of the regions can be obtained: $\chi_{\nu A} = 0.1721131$, $\chi_{\nu B} = 0.1689743$, $\chi_{\nu C} = 0.1744639$, $\chi_{\nu D} = 0.1789925$, $\chi_{\nu E} = 0.1608336$. From (34) independently (based on the susceptibility components χ_{21}, χ_{12}, constants a_T, a_λ from our anisotropic model [23, 24]), we find χ_{NA} exactly coinciding with $\chi_{\nu A}$. For the susceptibility with a shift, we find: $\chi_{bA} = 0.1922958$, $z_{bA} = 0.0183210$, $\overline{\chi}_{bA} = 0.1905423$. These susceptibilities determine the gaps $\Delta_{bA} = \chi_{bA}\varepsilon_{HG} = 53.851680$ meV, $\overline{\Delta}_{bA} = \overline{\chi}_{bA}\varepsilon_{HG} = 53.360611$ meV (which correspond to the effective temperatures $T_{bA} = a_T\Delta_{bA} = 39.489369$ °C, $\overline{T}_{bA} = a_T\overline{\Delta}_{bA} = 36.6398$ °C), energies $\varepsilon_{bA} = 285.17677$ meV, $\overline{\varepsilon}_{bA} = 285.08445$ meV (which correspond to the wavelengths $\lambda_{bA} = a_\lambda/2\varepsilon_{bA} = 2.1734659\,\mu m$, $\overline{\lambda}_{bA} = a_\lambda/2\overline{\varepsilon}_{bA} = 2.1741698\,\mu m$) in the spectra of neutrinos with nonzero rest mass.

These active microobjects can be part of the solar and intergalactic winds and affect to various physical, chemical, biological processes on Earth and in Universe.

6 Conclusions

The relationships between the base parameters of the Higgs boson and the parameters of black holes are established. Based on the distribution density functions of the number of quanta in the ground and excited states for relic photons, a lower mass estimate for a supermassive black hole is obtained. Based on the density distribution functions of the radiation intensity, an estimate of the mass near the upper boundary is obtained. The description of the central region of a supermassive black hole is made in terms of Bose condensate from black holes. Various states for a black hole with intermediate mass are introduced. Estimates for the mass and radius of the central body, the distance from the Sun to the supermassive black hole in the center of the Milky Way galaxy, the semi-axes of the elliptical orbit of S2 (rotating around the central body) are obtained.

The model equations are used to describe the base parameters of a relativistic jet: velocities, energy, angular width of jet particles.

It is shown, that the presence of a Higgs field of different nature leads to changes in the rest energy of the Higgs boson and the energies of holes (antiparticles) for paired leptons; the appearance of active microobjects with different energies and sizes; the appearance of asymmetry of matter and antimatter. A model for the classical decay of a neutron into a proton-electron pair and an antineutrino with a nonzero rest mass is proposed. The possibility of using this model to describe tetraquarks, the baryon density of the Universe, which depends on the states of antineutrinos, is shown.

Parameter estimates are consistent with experimental data.

References

1. R. Penrose, Gravitational collapse and space-time singularities. Phys. Rev. Lett. **14**(3), 3–18 (1965)
2. S.W. Hawking, R. Penrose, The singularities of gravitational collapse and cosmology. Proc. Roy. Soc. Lond. A. **314**, 529–548 (1970)
3. A. Eckart, R. Genzel, Observations of stellar proper motions near the Galactic Centre. Nature **383**, 415–417 (1996)
4. A.M. Ghez, S. Salim, N.N. Weinberg et al., Measuring distance and properties of the Milky Way's central supermassive black hole with stellar orbits (2008). arXiv: 0808.2870v1
5. K.S. Thorne, *Black holes and time folds: Einstein's outrageous legacy* (W.W. Norton & Company, New York, London, 1994)
6. B.P. Abbott, R. Abbott, T.D. Abbott, et al., Observation of gravitational waves from a binary black hole merger. Phys. Rev. Lett., **116**, Art. no 061102, 16 (2016)
7. B.P. Abbott, R. Abbott, T.D. Abbott et al. Observation of gravitational waves from a binary neutron star inspiral. Phys. Rev. Lett. **119**, Art. no 161101, 18 (2017)
8. V.S. Abramov. gravitational waves, relic photons and higgs boson in a fractal models of the universe, in *11th Chaotic Modeling and Simulation International Conference, Springer Proceedings in Complexity*, ed. By C.H. Skiadas, I. Lubashevsky (Springer Nature, Switzerland AG, 2019), pp. 1–14
9. V.S. Abramov, Superradiance of gravitational waves and relic photons from binary black holes and neutron stars. Bull. Russ. Acad. Sci. Phys. **83**(3), 364–369 (2019)
10. V. Abramov, Super-nonradiative states in fractal quantum systems, in *XIII International Workshop on Quantum Optics (IWQO-2019). EPJ Web of Conferences* (2019)
11. V.S. Abramov, Supernonradiative states, neutrino and Higgs Boson in fractal quantum systems. Bull. Russ. Acad. Sci. Phys. **84**(3), 284–288 (2020)
12. O.P. Abramova, A.V. Abramov, Effect of ordering of displacement fields operators of separate quantum dots, elliptical cylinders on the deformation field of coupled fractal structures, in *11th Chaotic Modeling and Simulation International Conference, Springer Proceedings in Complexity*, ed. By C.H. Skiadas, I. Lubashevsky (Springer Nature, Switzerland AG, 2019), pp. 15–27
13. O.P. Abramova, A.V. Abramov, Coupled fractal structures with elements of cylindrical type, in *12th Chaotic Modeling and Simulation International Conference, Springer Proceedings in Complexity*, ed. By C.H. Skiadas, Y. Dimotikalis (Springer Nature, Switzerland AG, 2020). pp. 15–26
14. M. Espinasse, S. Corbel, Ph. Kaare, et.al., Relativistic X-ray jets from the black hole X-ray binary MAXI J1820+070. arXiv:2004.06416v1 [astro-ph.HE], 10p.
15. T. Williams, R.W. Walsh, A.R. Winebarger, et al., Is the High-resolution coronal imager resolving coronal strands? Results from AR 12712. Astrophys. J. **892**(134), 13 (2020)
16. The ATLAS Collaboration/CERN, Evidence for higgs boson decays to a low-mass dilepton system and a photon in pp collisions at $\sqrt{s} = 13$ TeV with the ATLAS detector, in *ATLAS-CONF-2021-002*, February 1, 2021
17. J. Dove, B. Kerns, R.E. McClellan et al., The asymmetry of antimatter in the proton. Nature **590**, 561–565 (2021)
18. L. An, Latest results on exotic hadrons at LHCb/CERN-LHC seminar, June 16, 2020
19. V. Mossa, K. Stöckel, F. Cavanna et al., The baryon density of the Universe from an improved rate of deuterium burning. Nature **587**, 210–213 (2020)
20. D. Hooper, The empirical case for 10-GeV dark matter. arXiv:1201.1303v1 [astro-ph.CO], 18p
21. The Event Horizon Telescope Collaboration, First M87 event horizon telescope results. VII. Polarization of the ring. Astrophys. J. Lett. **910**(L12), 48 (2021)
22. The Event Horizon Telescope Collaboration, First M87 event horizon telescope results. VIII. Magnetic field structure near the event horizon. Astrophys. J. Lett. **910**(L13), 43 (2021)

23. V.S. Abramov, Active femto- and nano-objects in relation to the solar and interstellar winds in anisotropic models. Bull. Russ. Acad. Sci. Phys. **84**(12), 1505–1510 (2020)
24. V.S. Abramov. Active Nanoobjects, Neutrino and higgs boson in a fractal models of the universe, in *12th Chaotic Modeling and Simulation International Conference, Springer Proceedings in Complexity*, ed. by C.H. Skiadas, Y. Dimotikalis (Springer Nature, Switzerland AG, 2020), pp. 1–14

Memory Cell Based on Qubit States and Its Control in a Model Fractal Coupled Structure

Olga P. Abramova and Andrii V. Abramov

Abstract Model fractal coupled structures are considered, for which a characteristic feature of the behavior of the deformation field is the presence of such superposition qubit states where there is no damping. Such states can be memory cell. The possibility of internal and external control of the structure of the memory cell, the possibility of performing the operations of write and delete information has been established. It is shown, that changes in deformation fields in a memory cell are anisotropic. External control of a memory cell in a coupled structure is performed using different fractal indices of separate structures. In this case, fractal indices do not depend on iterative processes. It is shown, that there is a critical value of the fractal index of separate structures, when passing through which effective damping occurs. This effect can be used to control the storage of information in a memory cell. When fractal indices depend on an iterative process, self-organization (internal control) occurs. By the example of the sinusoidal law of change in the fractal index of separate structures, it is shown, that structures of the following type arise: vertical, horizontal, inclined stripes; lattice structures of various orientations.

Keywords Coupled fractal structures · Memory cell · Superposition of qubit states · Deformation field · Control of memory cell structure

1 Introduction

In [1], the description of the complex deformation field of model fractal coupled structures was carried out on the basis of different qubit states of separate structures such as circular and elliptical cylinders. A distinctive feature of the behavior of the deformation field of such coupled structures is the presence of qubit states, for which there is no damping (the imaginary part of the deformation field is zero). Such states can be memory cells.

O. P. Abramova (✉) · A. V. Abramov
Donetsk National University, Donetsk, Ukraine
e-mail: oabramova@ua.fm

C. H. Skiadas and Y. Dimotikalis (eds.), *14th Chaotic Modeling and Simulation International Conference*, Springer Proceedings in Complexity,
https://doi.org/10.1007/978-3-030-96964-6_2

The relevance of the work is associated with the problem of creating quantum computers [2, 3], that encode information in qubits; with quantum cryptography, where information is recorded in a memory cell below the noise level. Physical systems, that realize qubits, can be any objects, that have two quantum states. Modern nanotechnology makes it possible to create such active objects. Various nanostructures, metamaterials [4–6], superconductors [7] can act as active objects. These active objects can be in superposition qubit states, exhibit stochastic properties, quantum entanglement, which is the basis for the creation of quantum computers. Control, storage of quantum information, the possibility of its extraction are important steps for quantum communication. Modern nanotechnologies use various periodic structures and metamaterials [4], where the amplitude and phase of the deformation field is carried out by external control. In the presence of an iterative process, quantum chaos and the phenomenon of self-organization (internal control) appear in a memory cell. Therefore, the question of preserving a memory cell for fractal coupled structures requires additional research. Random matrices are used to describe quantum chaos [8]. Elements of random matrices are formed as a result of an iterative process. In this case, it becomes necessary to describe and take into account the effect of ordering of separate operators of deformation fields in a coupled structure [9, 10], which based on various qubit states.

The aim of this work is to describe the deformation field of a memory cell in a fractal coupled structure with elements of cylindrical type, internal and external control of its structure.

2 Memory Cell of a Model Fractal Coupled Structure

To describe the deformation field of memory cells, let us consider a model coupled structure, which consists of two fractal cylinders of elliptic type ($i = 1, 2$), located in a bulk discrete lattice $N_1 \times N_2 \times N_3$, whose nodes are given by integers $n,\ m,\ j$. Nonlinear equations for the dimensionless displacement function u of the lattice node are [1, 9, 10]

$$u = \sum_{i=1}^{2} u_{Ri}; \quad u_{Ri} = R_i k_{ui}^2 (1 - 2\mathrm{sn}^2(u - u_{0i}, k'_{ui})); \quad i = 1, 2; \tag{1}$$

$$\begin{aligned} k_{ui}^2 &= (1 - \alpha_i)/Q_i; \quad k'_{ui} = (1 - k_{ui}^2)^{1/2}; \\ p_{0i} &= p_{0i} + p_{1i} n + p_{2i} m + p_{3i} j; \end{aligned} \tag{2}$$

$$Q_i = p_{0i} - b_{1i}(n - n_{0i})^2/n_{ci}^2 - b_{2i}(m - m_{0i})^2/m_{ci}^2 - b_{3i}(j - j_{0i})^2/j_{ci}^2. \tag{3}$$

Here α_i are the fractal dimensions of the deformation field u along the axis Oz; u_{0i} are the constant (critical) displacements; variable modules k_{ui}, k'_{ui} are functions of indices n, m, j nodes of the bulk discrete lattice. Different structures are characterized by parameters: p_{0i}, p_{1i}, p_{2i}, p_{3i}, b_{1i}, b_{2i}, b_{3i}, n_{0i}, n_{ci}, m_{0i}, m_{ci}, j_{0i}, j_{ci}, R_i. In our model, the choice of different states of qubits in the plane nOm is determined by the nonzero coefficients of the linear terms in the functions p_{0i}, Q_i from (2), (3). The initial state (0,0) of an separate structure is determined by zero coefficients $p_{1i} = 0$, $p_{2i} = 0$. Various basic and superposition states of qubits were considered in [1]. In this work, we will focus only on the superposition state $(-1, -1)$, in which the parameters p_{1i}, p_{2i} have the form $p_{1i} = -0.00423$, $p_{2i} = -0.00572$.

Consider a superposition state $(-1, -1)$ of two fractal coupled structures (*A*), (*B*). In structure (*A*), the operation of scalar multiplication of the complex deformation fields of separate structures (I) and (II) is realized, while the deformation field of this structure is described by a function $u_A = u_{R1} f_A(u_{R2})$ with a corresponding matrix $\mathbf{M}_A$. The elements of the matrix $\mathbf{M}_A$ are obtained by solving (1)–(3) by the method of iteration over the index n. This procedure simulates coupled (dependent) stochastic processes of the original independent stochastic processes for structures (I) and (II), which are described by the functions $u = u_{R1}$ and $u = u_{R2}$. Structure (I) is a circular cylinder with constant semi-axes $n_{c1} = m_{c1}$, and structure (II) is an elliptical cylinder with variable semi-axes n_{c2}, m_{c2}.

To take into account the ordering of separate operators of deformation fields in a coupled structure, structure (*B*) is considered, where the operation of scalar multiplication of complex deformation fields of separate structures (II) and (I) is realized. The deformation field of this structure is described by a function $u_B = u_{R2} f_B(u_{R1})$ with a corresponding matrix $\mathbf{M}_B$. In the numerical modeling, it was assumed that $N_1 = 240$, $N_2 = 240$, $u_0 = 29.537$, $p_0 = 1.0423$, $b_{1i} = b_{2i} = 1$, $n_{0i} = 121.1471$, $m_{0i} = 120.3267$, $j_{0i} = 31.5279$, $j_{ci} = 11.8247$, $b_{3i} = 0$. Values of the semi-axes of a circular cylinder (I) are $n_{c1} = m_{c1} = 57.4327$ with $R_1 = 1$. For elliptical cylinder (II) with $R_2 = 1$, we have the following dimensions of the semi-axes: variant 1 are $n_{c2} = 43.0746$, $m_{c2} = 19.1443$; variant 2 are $n_{c2} = 55.2537$, $m_{c2} = 14.9245$; variant 3 are $n_{c2} = 119.9327$, $m_{c2} = 6.8758$. Further consider only coupled structures (*A*), (*B*), in which separate structures (I) and (II), (II) and (I) have the same fractal dimensions α_i and the same superposition qubit states $(-1,-1)$, but differ in the order of the deformation field operators $u_A = u_{R1} f_A(u_{R2})$, $u_B = u_{R2} f_B(u_{R1})$.

As an example, Fig. 1 shows the behavior of the deformation field $u_A = \mathrm{Re}\,u_A$ of the structure (*A*) with the same fractal dimension $\alpha_i = 0.5$ of separate structures (I) and (II). The variable semi-axes of the elliptical cylinder of structure (II) correspond to variants 1, 2, 3. A change in the semi-axes of an elliptical cylinder of structure (II) (internal control of the structure parameters) does not lead to the appearance of an imaginary part of the displacement function, which is a characteristic feature of the behavior of the deformation field. For $u_A = \mathrm{Re}\,u_A$ the presence of a stochastic peak is characterized, for which the structure and region of localization in the plane nOm changes with depending on the semi-axes of the elliptical cylinder (II) (Fig. 1a–c). to a decrease in the semi-axis m_{c2} of the elliptical cylinder (II) (Fig. 1g–i). The cross sections $\mathrm{Re}\,u_A$ (Fig. 1g, k, l) confirm the anisotropic nature of the alteration of the

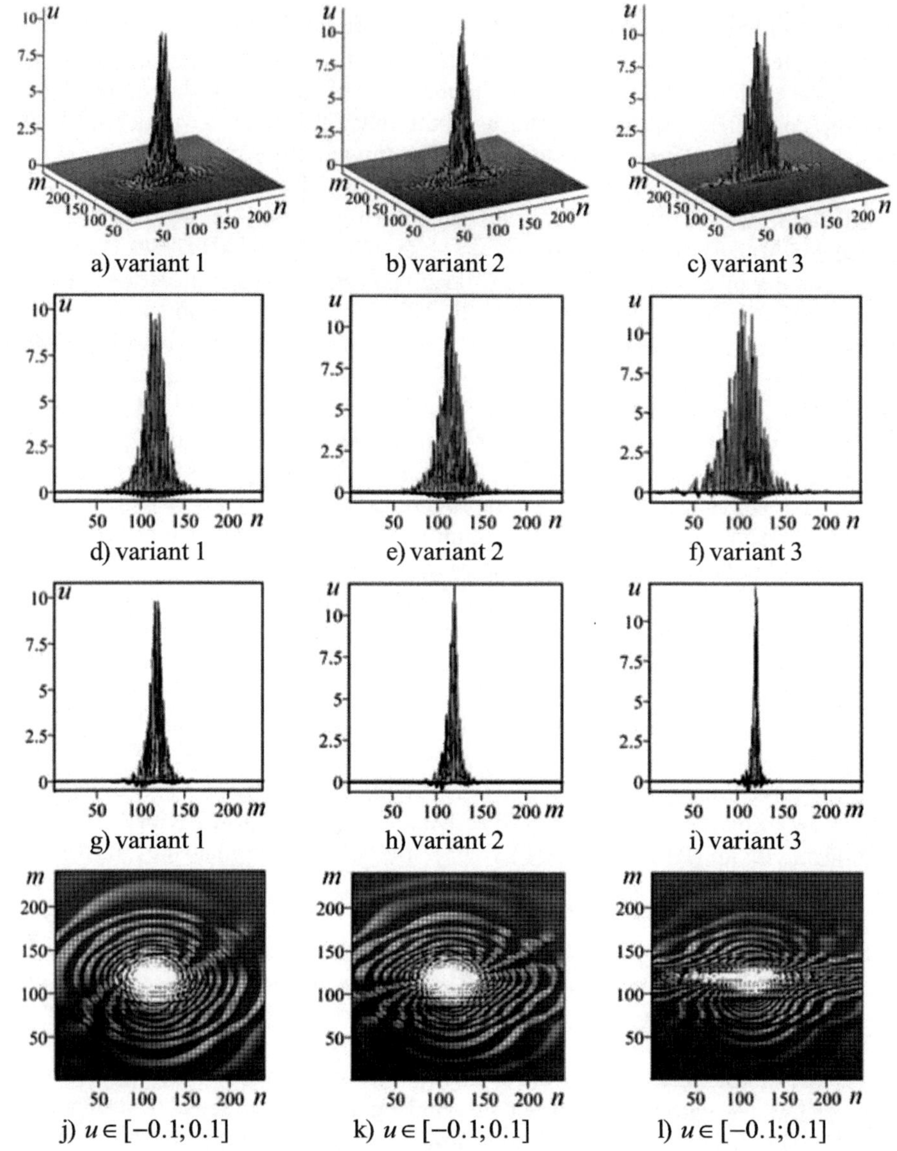

Fig. 1 Dependences of the deformation field of the structure (A) at $\alpha_i = 0.5$ on the variable semi-axes of the structure (II): $u = u_A$ (**a–c**)—general view; projections on the planes nOu (**d–f**), mOu (**g–i**); (**j–l**)—cross sections (top view)

structure of the inner region of the stochastic peak: there is a change in the shape and structure of separate elliptical rings, the effect of mixing of separate trajectories.

These changes are anisotropic. In this case, along the axis On, the peak broadens due to an increase in the semi-axis n_{c2} of the elliptical cylinder (II), the amplitudes of the peaks are of the order of 10 (Fig. 1d), 12 (Fig. 1e), 11 (Fig. 1f) dimensionless units. A narrowing of the peak occurs along the axis Om due to a decrease in the semi-axis m_{c2} of the elliptical cylinder (II) (Fig. 1g–i).

The cross sections $\mathrm{Re}u_A$ (Fig. 1g, k, l) confirm the anisotropic nature of the alteration of the structure of the inner region of the stochastic peak: there is a change in the shape and structure of separate elliptical rings, the effect of mixing of separate trajectories.

For both the structure (*A*) and the coupled structure (*B*) with the same superposition states (−1,−1) of separate structures feature of the behavior of the deformation field is the absence of effective damping in all over region ($\mathrm{Im}u_B = 0$). For $\mathrm{Re}u_B$ it is also characterized by the presence of a expanded stochastic peak with a structure close to the peak $\mathrm{Re}u_A$ (Fig. 1), but $\mathrm{Re}u_B - \mathrm{Re}u_A \neq 0$. In this case, the conditions are carried out

$$u_B - u_A = u_{R2} f_B(u_{R1}) - u_{R1} f_A(u_{R2}) \neq 0, \mathbf{M}_B - \mathbf{M}_A \neq 0, \tag{4}$$

which is related to the dependence of the considered stochastic processes. This indicates that the operators of the displacement fields of separate structures (II), (I) and (I), (II) do not commute in coupled structures (*B*) and (*A*). The results of numerical modeling for structure (*B*) are not presented in this work.

3 External Control of a Memory Cell

External control of the structure of a memory cell will be carried out due to a different choice of constant fractal dimensions α_i of separate structures (I), (II). In this case, fractal indices α_i do not depend on iterative processes. On Fig. 2 shows the behavior of the deformation fields $u_A = \mathrm{Re}u_A$ of the structure (*A*) for the same fractal dimensions of structures (I), (II): $\alpha_1 = \alpha_2 = 0.0$ (Fig. 2a, d), $\alpha_1 = \alpha_2 = 0.9$ (Fig. 2b, e), $\alpha_1 = \alpha_2 = 0.99$ (Fig. 2c, f); the semi-axes of structure (II) correspond to variant 1. When fractal dimensions of structures (I), (II) increase, then a change in the shape and structure of stochastic peaks is observed, which is accompanied by a sharp decrease in amplitudes from 46 (Fig. 2a), 0.33 (Fig. 2b) to 0.0036 (Fig. 2c) of dimensionless units. The cross sections (Fig. 2d–f) confirm a significant alteration of the structure of the inner region from a wave-like state (Fig. 2d) to an almost regular behavior (Fig. 2f). When $\alpha_1 = \alpha_2 = 1.0$ the deformation field becomes zero $u_A = \mathrm{Re}u_A = 0$. This makes it possible to interpret such a change in fractal dimensions as an operation of delete information in a memory cell. Next, consider the structure (*A*), where the fractal dimensions α_i of separate structures (I) and (II) are chosen to be different. In this case, for an elliptical cylinder (II) with $R_2 = 1$, we have the parameters of

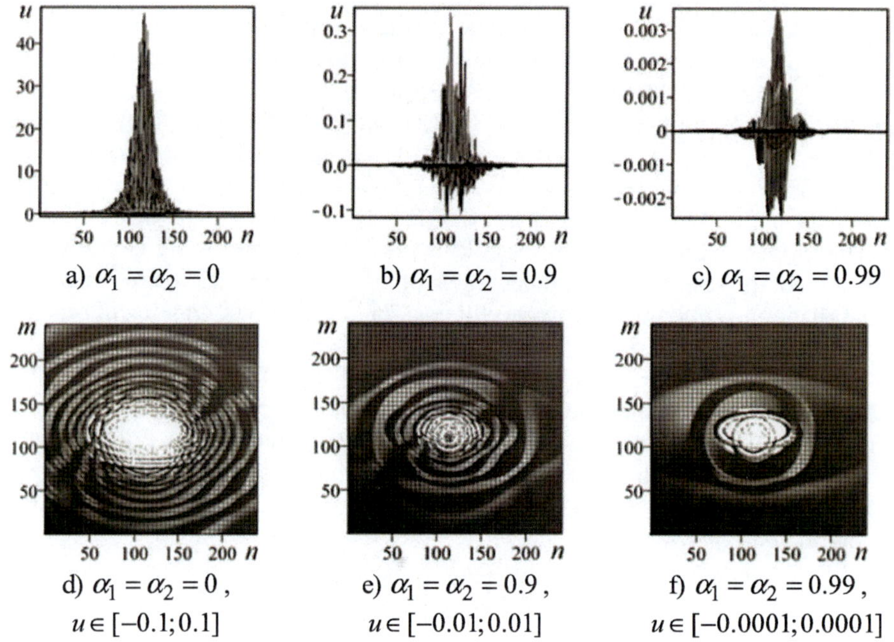

Fig. 2 Dependences of the deformation field of the structure (A) on the same fractal dimensions of the structures (I), (II): $u = u_A = \mathrm{Re} u_A$ (**a**–**c**)—projections on the plane nOu, (**d**–**f**)—cross sections (top view); $0 \le \alpha_i < 1$

variant 1. Figure 3 shows the dependences of the deformation field of the structure (A) on various joint changes in the fractal dimensions $0 \le \alpha_i < 1$ of structures (I), (II): an increase α_1 for structure (I) and a decrease α_2 for structure (II) in the segment [0; 0.99]. For projections on the plane nOu (Fig. 3a, d, g, j), mOu (Fig. 3b, e, h, k), the following characteristic features of the behavior of the deformation field are observed. The amplitudes and shapes of stochastic peaks change. Cross sections (Fig. 3c, f, i, l) $u \in [-10^{-4}; 10^{-4}]$ allow more detailed information to be extracted.

The circular cylinder of structure (I) with $\alpha_1 = 0$ defines the external regular wave-like behavior of the deformation field, and the elliptical cylinder of structure (II) with $\alpha_2 = 0.99$ defines the internal stochastic behavior of the deformation field (core) (Fig. 3c). With a further joint change in fractal dimensions, a significant change in the structure of both the core and the outer region occurs: there is an intersection (Fig. 3f), breaks (Fig. 3i) of regular and stochastic rings; the appearance of rings with superposition (Fig. 3l) of regular and stochastic behavior.

When $\alpha_1 = \alpha_2 = 1.0$ the deformation field becomes zero $u_A = \mathrm{Re} u_A = 0$, which follows from the basic Eqs. (1)–(3). This allows for the possibility of interpretation as an operation of delete information in a memory cell. With a further increase in the values of fractal dimensions $\alpha_i > 1$ of separate structures (I), (II) (Figs. 4 and 5), one should expect a significant alteration of the deformation field in the coupled structure (A).

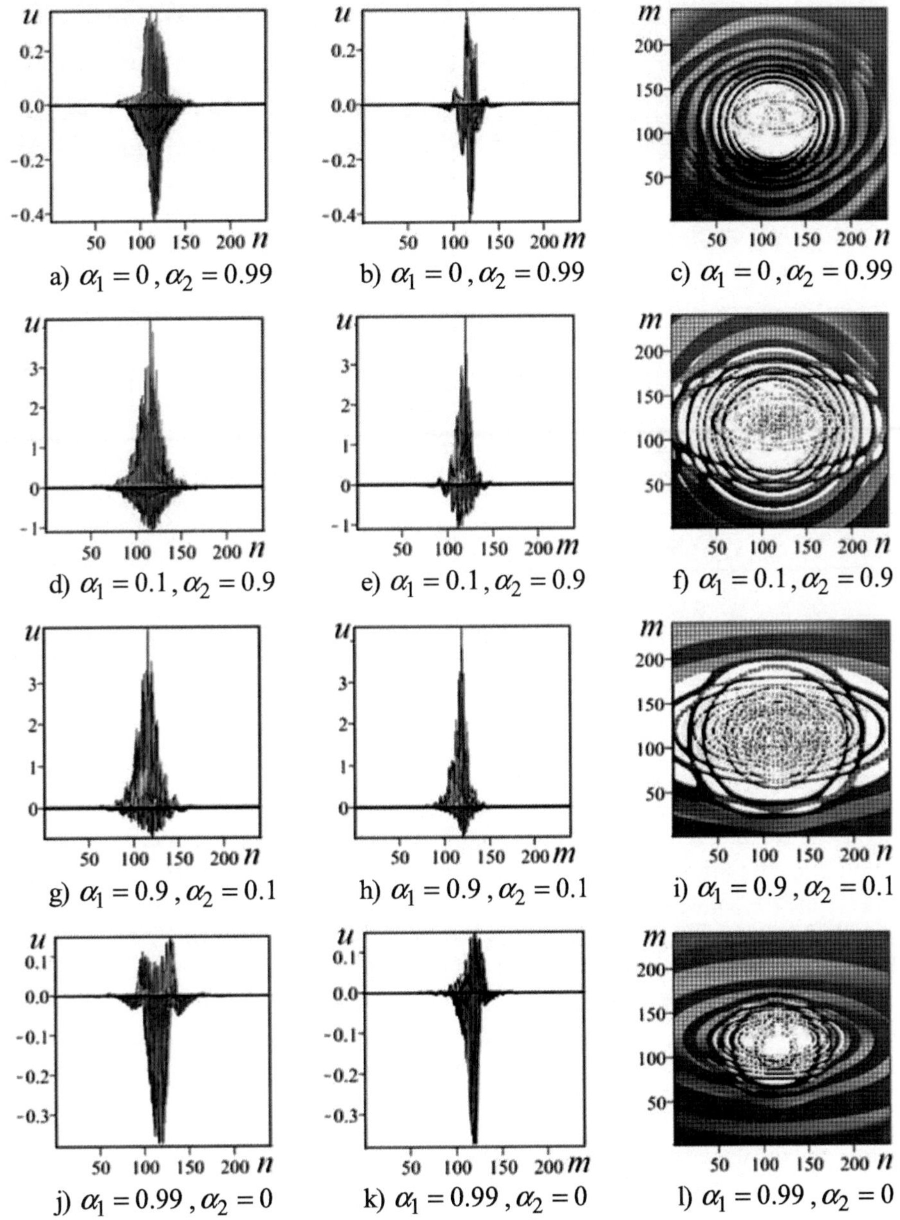

Fig. 3 Dependences of the deformation field of the structure (A) on various fractal dimensions of the structures (I), (II): projections $u = u_A = \text{Re}\, u_A$ on the planes nOu (**a**, **d**, **g**, **j**), mOu (**b**, **e**, **h**, **k**); (**c**, **f**, **i**, **l**)—$u \in [-10^{-4}; 10^{-4}]$ cross sections (top view); $0 \le \alpha_i < 1$

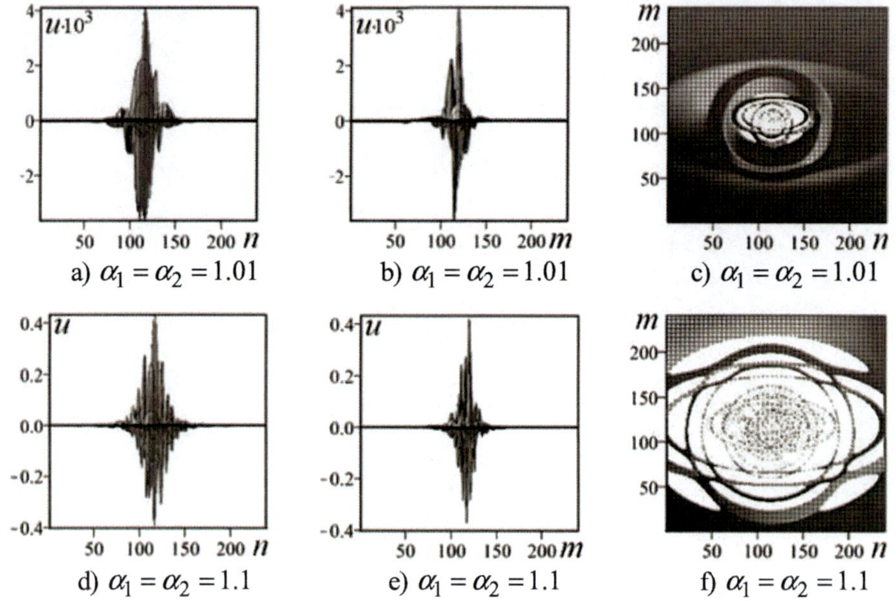

Fig. 4 Dependences of the deformation field of the structure (*A*) on the same fractal dimensions of structures (I), (II): projections $u = u_A = \mathrm{Re} u_A$ on the planes nOu (**a**, **d**), mOu (**b**, **e**); (**c**, **f**)—cross sections $u \in [-10^{-4}; 10^{-4}]$ (top view); $1 < \alpha_i \leq 1.1$

As an example, Fig. 4 shows the dependences of the deformation field of the structure (*A*) on the same fractal dimensions of structures (I), (II) for $\alpha_1 = \alpha_2 = 1.01$ and $\alpha_1 = \alpha_2 = 1.1$. Wherein, the amplitudes of the peaks increase from $4 \cdot 10^{-3}$ (Fig. 4a, b) to 0.4 (Fig. 4d, e), there are features like an inflow near the stochastic core of the peaks. The cross section (Fig. 4c) confirms the formation of an almost regular convex region (inflow) around the stochastic core. With an increase in fractal dimension, the cross section (Fig. 4f) is characterized by the formation of broadened rings of complex shape (in contrast to circular and elliptical rings from Fig. 3). Stochastic rings are present within the core (Fig. 4f).

Note, that the imaginary part of the displacement function is as before equal to zero.

A further increase in the fractal dimension leads to the appearance of the imaginary part of the displacement function $\mathrm{Im} u_A \neq 0$ (Fig. 5d–f). The amplitude of the peaks $\mathrm{Re} u_A$ (Fig. 5a–c) increases compared to (Fig. 4d–f), and the shape of the peaks becomes asymmetric. Wherein, the area of the stochastic core expands with the formation of intersecting circular and elliptical rings (Fig. 5c). The appearance of the imaginary part $\mathrm{Im} u_A \neq 0$ can be interpreted as a possible mechanism for the loss of a part of information from a memory cell. Comparison of the behavior of the deformation fields of the structure (*A*) shows, that there is a critical value of the

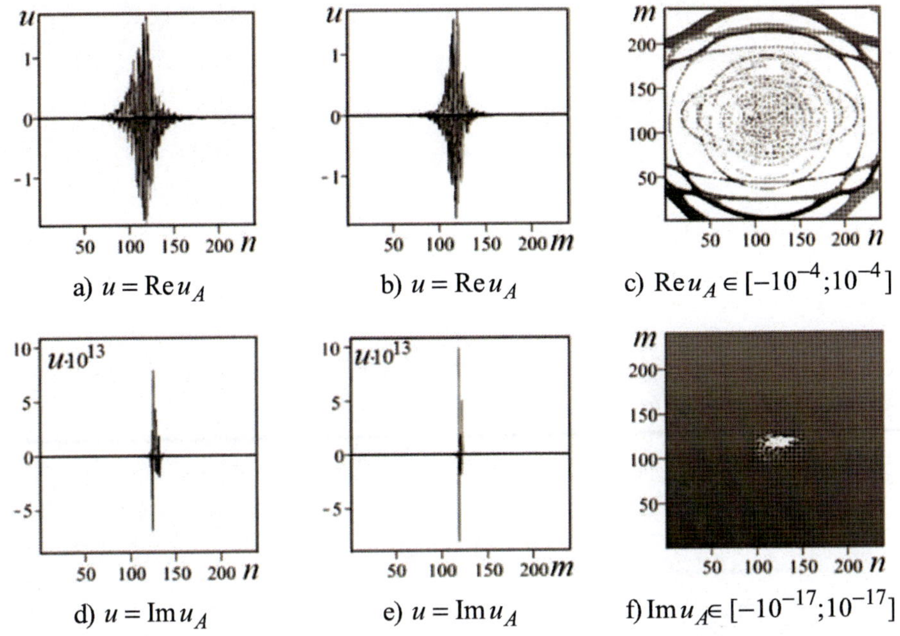

Fig. 5 Dependences of the deformation field of the structure (*A*) on the same fractal dimensions of the structures (I), (II): projection $\mathrm{Re}u_A$ and $\mathrm{Im}u_A$ on the planes nOu (**a**, **d**), mOu (**b**, **e**); cross sections (**c**, **f**) (top view); $\alpha_1 = \alpha_2 = 1.2$

fractal dimension $\alpha = \alpha_{cr} = 1.12$ (where $1.1 < \alpha < 1.2$), when passing through which, damping occurs.

4 Internal Control of a Memory Cell

Modern nanotechnology uses various periodic structures and metamaterials [4], where the amplitude and phase of the deformation field is performed by external control. The question about preservation the memory cell at the presence of an iterative process for fractal coupled periodic structures requires additional research. In this work, using examples of various sinusoidal laws of change in the fractal dimensions of separate structures (I), (II) of the coupled structure (*A*), we investigate the behavior of the deformation field, the change in the structure of the memory cell depending on the iterative process. In this case, self-organization (internal control) occurs. We realize various sinusoidal laws of change in fractal dimensions α_1 and α_2 separate structures (I) and (II) of the coupled structure (*A*) in expressions (2) as different functions of lattice indices n, m (Figs. 6 and 7). Note, that for Fig. 6, Fig. 7 the imaginary part of the deformation field are equal $\mathrm{Im}u_A = 0$.

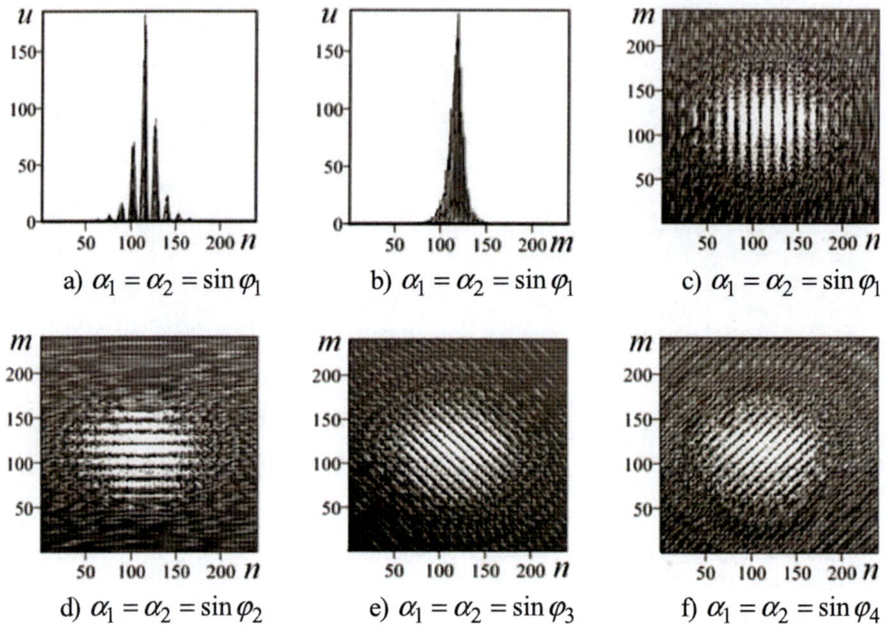

Fig. 6 Dependences of the deformation field of the structure (*A*) for the same sinusoidal laws of change in the fractal dimensions of separate structures (I), (II): projections $u = u_A = \mathrm{Re}\,u_A$ on the plane nOu (**a**), mOu (**b**); (**c–f**)—cross sections $u \in [-0.1; 0.1]$ (top view)

Figure 6 shows the dependences of the deformation field of the structure (*A*) for the same sinusoidal laws of change in the fractal dimensions of separate structures (I), (II) on the lattice indices n, m: $\alpha_1 = \alpha_2 = \sin\varphi_1, \varphi_1 = 6\pi(n-1)/39$ (Fig. 6a–c); $\alpha_1 = \alpha_2 = \sin\varphi_2, \varphi_2 = 6\pi(m-1)/39$ (Fig. 6d); $\alpha_1 = \alpha_2 = \sin\varphi_3, \varphi_3 = \varphi_1 + \varphi_2$ (Fig. 6e); $\alpha_1 = \alpha_2 = \sin\varphi_4, \varphi_4 = \varphi_1 - \varphi_2$ (Fig. 6f). Deformation field dependence $u = u_A = \mathrm{Re}\,u_A$ for projection on the plane nOu (Fig. 6a) is a zug (a sequence of peaks of different amplitude along the axis On). This is due to the presence of a sinusoidal law $\sin(6\pi(n-1)/39)$ for the fractal dimensions of separate structures (I), (II) from the lattice index n, according to which the iterative process is performed.

There is no iterative process along the axis Om, therefore a broadened stochastic peak for projection $u = u_A = \mathrm{Re}\,u_A$ on the plane mOu is observed (Fig. 6b). The cross section (Fig. 6c) shows, that the core of the coupled structure (*A*) is a sequence of broadened stochastic stripes parallel to the axis Om. Stochastic elliptical rings with internal periodicity (outer region of the coupled structure (*A*)) around the core are observed.

Note, that when choosing an iterative process along the axis Om, a zug will be observed for projection on the plane mOu, and a broadened stochastic peak will be observed for projection on the plane nOu. In this case, the core of the coupled structure (*A*) will be a sequence of broadened stochastic stripes parallel to the axis

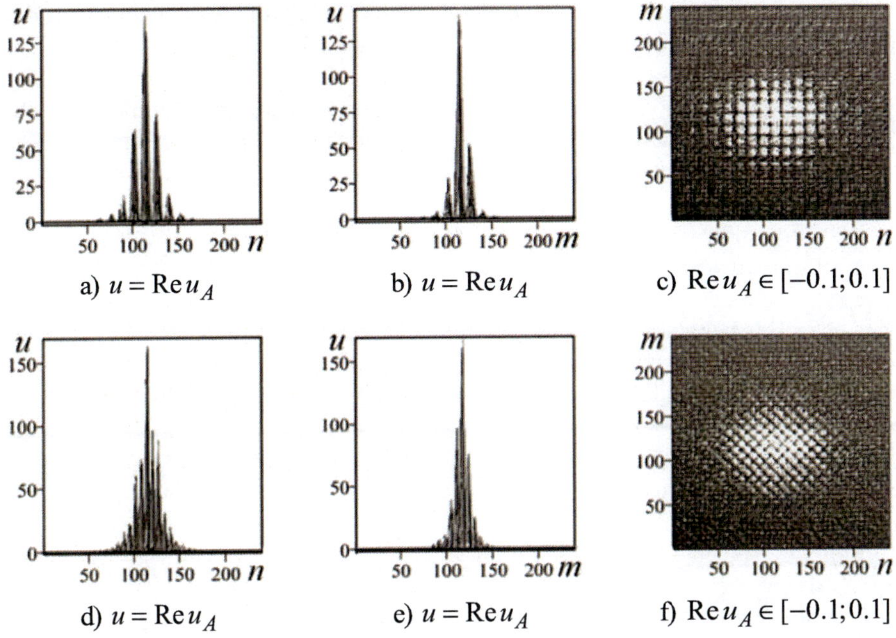

Fig. 7 Dependences of the deformation field of the structure (A) for various sinusoidal laws of change in the fractal dimensions of separate structures (I), (II): projections $u = u_A = \mathrm{Re}\,u_A$ on the plane nOu (**a**, **d**), mOu (**b**, **e**); (**c**, **f**)—cross sections $u \in [-0.1; 0.1]$ (top view)

On. Thus, the choice of the iterative process makes it possible to additionally control the deformation field of the coupled structure (A).

When the same fractal dimensions of separate structures (I), (II) $\alpha_1 = \alpha_2 = \sin\varphi_2$, $\varphi_2 = 6\pi(m-1)/39$ depend on the lattice index m (Fig. 6d), then the core of the coupled structure (A) is a sequence of broadened stochastic stripes parallel to the axis On. Other stochastic elliptic rings with internal periodicity and discontinuous trajectories around the core are observed.

When choosing fractal dimensions of separate structures (I), (II), depending on the superposition of lattice indices n, m ($\alpha_1 = \alpha_2 = \sin\varphi_3$, $\varphi_3 = \varphi_1 + \varphi_2$ (Fig. 6e); $\alpha_1 = \alpha_2 = \sin\varphi_4$, $\varphi_4 = \varphi_1 - \varphi_2$ (Fig. 6f)) inclined periodic structures appear in the core of the coupled structure (A). Note, that the angle of rotation for inclined structures of the core is anticlockwise (Fig. 6e) and clockwise (Fig. 6f). In this case, a significant difference for the deformation field of the coupled structure (A) is observed.

Figure 7 shows the dependences of the deformation field of the structure (A) for various ($\alpha_1 \neq \alpha_2$) sinusoidal laws of change in the fractal dimensions of separate structures (I), (II) on the lattice indices n, m: $\alpha_1 = sin\varphi_1$, $\alpha_2 = \sin\varphi_2$, $\varphi_1 = 6\pi(n-1)/39$, $\varphi_2 = 6\pi(m-1)/39$ (Fig. 7a–c); $\alpha_1 = sin\varphi_3$, $\alpha_2 = \sin\varphi_4$, $\varphi_3 = \varphi_1 + \varphi_2$, $\varphi_4 = \varphi_1 - \varphi_2$ (Fig. 7d–f).

Compared to Fig. 6a, b here the dependences of the deformation field $u = u_A = \mathrm{Re}u_A$ for projections on the plane nOu (Fig. 7a) and (Fig. 7b) are zugs (sequences of peaks of different amplitudes) both along the axis On and along the axis Om, respectively. This is due to the fact that the fractal dimension α_1 is a function of the lattice index n, and the fractal dimension α_2 is a function of the lattice index m. The section (Fig. 7c) shows, that the core of the coupled structure (A) is a lattice of square-shaped sub-elements. For the variant, when fractal dimensions α_1, α_2 are functions of two lattice indices n, m, instead of pronounced zugs (Fig. 7a, b), stochastic peaks with a thin structure are observed (Fig. 7d, e). The cross section (Fig. 7 f) shows, that the core of the coupled structure (A) is now a lattice of rhombic-shaped sub-elements. This behavior of the deformation field (the appearance of rhombic-shaped sub-elements) is associated with the presence of joint rotations of inclined structures both anticlockwise and clockwise, in comparison with Fig. 6e and f (where rotations are done separately).

Choosing other functions for fractal dimensions α_1, α_2 one can expect, that the core of the coupled structure (A) will be an irregular lattice of sub-elements of various shapes (such as quantum dots, curved stripes, hexagonal cells).

Modern nanotechnology makes it possible to create such structures, for example, on the surface of thin membrane.

The operators of the displacement fields of separate structures (II), (I) and (I), (II) do not commute in the coupled structures (B) and (A) in an iterative process.

Taking into account the ordering of separate operators of deformation fields in a coupled structure (B) leads to a different behavior of the deformation field depending on the functions of fractal dimensions α_2, α_1 separate structures (II), (I). The results of numerical modeling for structure (B) are not presented in this work.

5 Conclusions

It is shown, that a pronounced feature of the behavior of the deformation field of coupled structures (A), (B) with the same superposition qubit states (-1,-1) of separate structures is the absence of the imaginary part of the displacement function in the all region ($\mathrm{Im}u_A = \mathrm{Im}u_B = 0$) at $0 < \alpha_1 < 1.12$, $0 < \alpha_2 < 1.12$, which indicates to the absence effective damping. This makes it possible to interpret coupled structures (A), (B) with the same superposition states (-1,-1) of separate structures (I), (II) as memory cells. It is shown, that there is a critical value of the fractal dimension $\alpha = \alpha_{cr} = 1.12$, when passing through which, damping occurs.

The possibility of internal and external control of the parameters of the structure of the memory cell by changing the semi-axes of the elliptical cylinder of the structure (II) and the fractal dimensions α_i of separate structures (I), (II) has been established. The possibility performing operations of write, delete information in a memory cell has been established. Changes in the deformation fields of coupled structures are anisotropic.

The behavior of the deformation field of the structure (A) from constant (same and different) fractal dimensions of separate structures (I), (II) (external control of the memory cell) is investigated. It is shown, that a change in the fractal dimensions leads to alteration of the shape and structure of stochastic peaks, the core of the coupled structure (A).

Internal control of the memory cell is performed by realization various sinusoidal laws of change in fractal dimensions α_1 and α_2 separate structures (I) and (II) of the coupled structure (A), as different functions of lattice indices $n, \ m$. It is shown, that substructures of the type of vertical, horizontal, inclined strips, lattice structures with sub-elements of various shapes appear in the core of a coupled structure (A).

References

1. O.P. Abramova, Deformation field and qubit states of fractal structures with elements of cylindrical type. Bull. Donetsk National Univ. A **1**, 3–15 (2020)
2. M. Nielsen, I. Chuang, *Quantum Computation and Quantum Information.* (Cambridge University Press, New York, 2010)
3. D. Boumeister, A. Eckert, A. Zeilinger, *Physics of Quantum Information* (Springer, New York, 2001)
4. Y.S. Kivshar, N.N. Rozanov (eds.), *Nonlinearities in Periodic Structures and Metamaterials* (Fizmatlit, Moscow , 2014)
5. V.S. Abramov, Active nanoobjects, neutrinos and higgs fields in anisotropic models of fractal cosmology. Bull. Russ. Acad. Sci. Phys. **83**(12), 1516–1520 (2019)
6. V.S. Abramov. Active Nanoobjects, Neutrino and higgs boson in a fractal models of the universe, in *12th Chaotic Modeling and Simulation International Conference, Springer Proceedings in Complexity*, ed. By C.H. Skiadas, Y. Dimotikalis (Springer Nature, Switzerland AG, 2020), pp. 1–14
7. A.N. Omelyanchuk, E.V. Ilyichev, S.N. Shevchenko, *Quantum Coherent Phenomena in Josephson Qubits* (Naukova Dumka, Kiev, 2013)
8. H.-J. Stöckmann, *Quantum Chaos. An Introduction* (Philipps-Universität Marburg, Germany, 2007)
9. O.P. Abramova, A.V. Abramov. Effect of ordering of displacement fields operators of separate quantum dots, elliptical cylinders on the deformation field of coupled fractal structures, in *11th Chaotic Modeling and Simulation International Conference, Springer Proceedings in Complexity*, ed. by C.H. Skiadas, I. Lubashevsky (Springer Nature, Switzerland AG, 2019), pp. 15–26
10. O.P. Abramova, A.V. Abramov, Coupled fractal structures with elements of cylindrical type, in *12th Chaotic Modeling and Simulation International Conference, Springer Proceedings in Complexity*, ed. by C.H. Skiadas, Y. Dimotikalis (Springer Nature, Switzerland AG, 2020), pp. 15–26

To Stochastic Resonance in Homopolar Dynamo

Agalar M.-Z. Agalarov, Elena S. Alekseeva, Alexander A. Potapov, and Alexander E. Rassadin

Abstract This chapter is devoted to discussion of the behavior of one-disk dynamo under the action of harmonic and random signals. Evaluations of separated effects of harmonic and random external voltages in the framework of the linearized Bullard equations have been presented. As random signals with zero average the Gaussian delta-correlated noise and the Langevin stochastic process have been considered. In particular, as physical values characterizing these influences both autocorrelation

A. M.-Z. Agalarov
H. I. Amirkhanov Institute of Physics of Dagestan Federal Research Center, Russian Academy of Sciences, Makhachkala, Russia
e-mail: aglarow@mail.ru

Yaragskogo Str., 94, 367015 Makhachkala, Russia

E. S. Alekseeva · A. A. Potapov · A. E. Rassadin (✉)
Nizhny Novgorod Mathematical Society, Nizhny Novgorod, Russia
e-mail: brat_ras@list.ru

E. S. Alekseeva
e-mail: kometarella@mail.ru

A. A. Potapov
e-mail: potapov@cplire.ru

E. S. Alekseeva
Gagarin Ave., 23, 603600 Nizhny Novgorod, Russia

A. A. Potapov
V. A. Kotelnikov Institute of Radio Engineering and Electronics, Russian Academy of Sciences, Moscow, Russia

JNU-IREE RAS Joint Lab. of Information Technology and Fractal Processing of Signals, JiNan University, Guangzhou 510632, China

Mokhovaya Str., 11-7, 125009 Moscow, Russia

A. E. Rassadin
Higher School of Economics, Nizhny Novgorod, Russia

Pecherskaya Str, Nizhny Novgorod, 25/12, 603155 Bolshaya, Russia

C. H. Skiadas and Y. Dimotikalis (eds.), *14th Chaotic Modeling and Simulation International Conference*, Springer Proceedings in Complexity,
https://doi.org/10.1007/978-3-030-96964-6_3

functions of observables and their spectral densities have been calculated. This information is important for design and testing of homopolar dynamo layout to perform analog research of stochastic resonance in this device in nonlinear regime.

Keywords Elecromechanical system · Phase plane · Equilibrium states · Amplitude responses · The Wiener-Khinchin theorem · The Jordan's lemma · Design and testing · The Fourier transform

Abbreviations

GSSP	The Gaussian stationary stochastic process
ACF	The autocorrelation function
SD	The spectral density

1 Introduction

Stochastic resonance is known to be a cooperative effect in nonlinear systems manifesting itself in increasing of the output signal-to-noise ratio under addition of the optimal portion of noise [1].

At present great attention is paid to studying of stochastic resonance in multidimensional systems arising from physics through chemistry to biology and neuroscience [2–5]. However, in our opinion, the most correct path in investigation of stochastic resonance leading to a real understanding of the essence of this phenomenon is choosing of a fairly simple dynamic system with a relatively small dimension and a detailed study of this one. As a rule there are no analytical solutions both the nonstationary Fokker-Plank-Kolmogorov equation for such system and stochastic differential equations describing its behaviour. Numerical solution of these problems is quite hard too [6, 7]. Therefore this system ought to allow experimental investigation.

On the one hand, from the point of view of clarity, preference should be given to mechanical systems. Such systems are easily perceived and interpreted due to our daily experience. On the other hand, electrical systems are characterized by the ease of controlling of external influences. Hence it is convenient to take an electromechanical system as a model system for experimental and theoretical research of stochastic resonance.

In the framework of this approach we study one-disk dynamo (the so-called Bullard dynamo). At first this electromechanical system was suggested in article [8] in order to illustrate a number of astrophysical and geophysical effects concerning motion of electrically conducting fluid in a magnetic field (see [9] and references therein). Contrary to original article [8] we take into consideration both electrical load in parallel with the field coil and friction at the axis of the dynamo. But we restrict ourselves by investigation of the linear response of the Bullard dynamo because of our final aim is design of functioning homopolar dynamo for analog modeling of stochastic resonance in this system. We stress that in our research there is no any magnetohydrodynamic background—compare for instance with work [10].

The rest of the chapter is organized as follows: in Sect. 2, we discuss equations of motion for the Bullard dynamo and their linearization. Section 3 is devoted to calculations of influence of harmonic external voltage on the linearized Bullard system. Section 4 deals with linear responses of the system on random signals with zero average, namely, on the Gaussian delta-correlated noise and the Langevin stochastic process. Final section is devoted to discussion of results elaborated and conclusions.

2 Main Equations

Mathematical model of the homopolar dynamo is given by the following system of stochastic ordinary differential equations:

$$\begin{cases} L \cdot \frac{\mathrm{d}J}{\mathrm{d}t} + R \cdot J = M \cdot J \cdot \Omega + U(t) \\ I \cdot \frac{\mathrm{d}\Omega}{\mathrm{d}t} = K - M \cdot J^2 - 2 \cdot \gamma \cdot \Omega \end{cases}, \tag{1}$$

where

$J(t)$ is electric current via the inductance L on Fig. 1;
$\Omega(t)$ is angular speed of rotation of the disk of dynamo;
R is value of resistance in the electrical circuit on Fig. 1;
M is coefficient of mutual inductance:
$U(t)$ is an external voltage;
I is moment of inertia for the dynamo;
K is constant mechanical torque on the axis of the dynamo;
$2 \cdot \gamma$ is coefficient of mechanical friction on the dynamo axis.

To study stochastic resonance in the system on Fig. 1 one ought to choose external voltage in (1) as follows:

$$U(t) = U_0 \cdot \cos(\nu \cdot t) + V(t), \tag{2}$$

where

U_0 is amplitude of harmonic signal;
ν is circular frequency of harmonic signal;
$V(t)$ is the Gaussian stationary stochastic process (GSSP) with zero average:

$$\langle V(t) \rangle = 0, \tag{3}$$

Fig. 1 Structural scheme of the homopolar dynamo

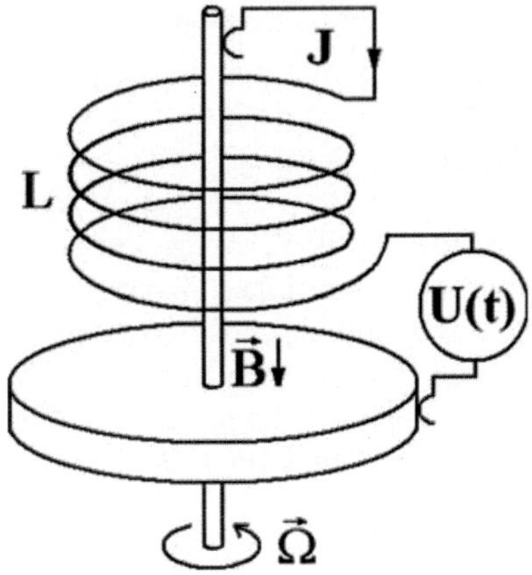

and fixed autocorrelation function (ACF):

$$\langle V(t) \cdot V(t') \rangle = B(t' - t). \tag{4}$$

We underline that our approach in (1) differs sharply from one in paper [11] because of authors of this paper apply separation of the magnetic flux on magnetic flux across disk of the dynamo and magnetic flux across the loops of inductance. This separation of magnetic flux on two parts leads to increasing of dimension of phase space of the system.

For further analysis of system (1) it is convenient to introduce the next dimensionless variables and parameters:

$$\begin{gathered} x_1 = \sqrt{\frac{M}{K}} \cdot J, \quad x_2 = \sqrt{\frac{M \cdot I}{L \cdot K}} \cdot \Omega, \quad \nu_0 = \sqrt{\frac{M \cdot K}{L \cdot I}}, \\ \mu = R \cdot \sqrt{\frac{I}{M \cdot L \cdot K}}, \delta = \gamma \cdot \sqrt{\frac{L}{I \cdot K \cdot M}}, U_m = K \cdot \sqrt{\frac{L}{I}}. \end{gathered} \tag{5}$$

After that one can rewrite system (1) in the following form:

$$\begin{cases} \dot{x}_1 = -\mu \cdot x_1 + x_1 \cdot x_2 + u(\tau) \\ \dot{x}_2 = 1 - x_1^2 - 2 \cdot \delta \cdot x_2 \end{cases}, \tag{6}$$

where

$u(\tau) = U(t) / U_m$ is dimensionless external voltage;

$\dot{x}_{1,2}$ are derivatives of dimensionless variables $x_{1,2}$ with respect to dimensionless time $\tau = \nu_0 \cdot t$.

The system (6) in the absence of external voltage is defined as:

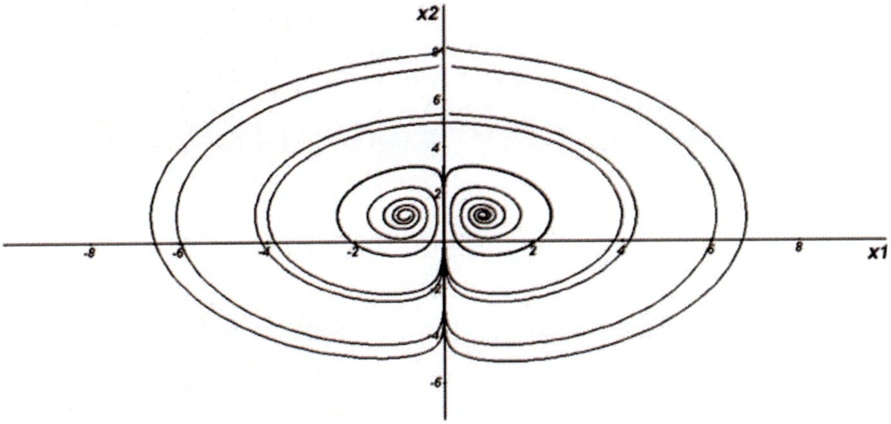

Fig. 2 Phase plane of the homopolar dynamo in the absence of external load

$$\begin{cases} \dot{x}_1 = -\mu \cdot x_1 + x_1 \cdot x_2 \\ \dot{x}_2 = 1 - x_1^2 - 2 \cdot \delta \cdot x_2 \end{cases}. \tag{7}$$

It is easy to see that if $0 < \delta < 1/2\mu$ then system (7) possesses by three equilibrium states: $O^s(0,\ 1/(2 \cdot \delta))$ and $O^{\pm}(\pm\sqrt{1 - 2 \cdot \delta \cdot \mu},\ \mu)$. It is not difficult to check that if $0 < \delta < \sqrt{2 + 4\mu^2} - 2\mu$ then points $O^{\pm}$ are stable focuses and if $\sqrt{2 + 4\mu^2} - 2\mu < \delta < 1/2\mu$ then points $O^{\pm}$ are stable nodes. Point O^s is saddle point in both cases.

We shall suppose that dimensionless damping factor δ is quite small therefore we shall deal with situation when points $O^{\pm}$ are stable focuses. Phase plane of system (7) at $\mu = 1.0$ and $\delta = 0.1$ corresponding to the case under consideration is shown on Fig. 2.

It is obvious that system (7) is invariant under transformation of variables $(x_1, x_2) \to (-x_1, x_2)$ therefore to calculate linear response of the system (6) it is enough to take into account only vicinity of the point O^+.

Introducing for system (6) new variables $y_{1,2}$ as follows:

$$x_1 = +\sqrt{1 - 2 \cdot \delta \cdot \mu} + y_1, \quad x_2 = \mu + y_2, \tag{8}$$

and rejecting terms with powers of $y_{1,2}$ greater than one we find that system (6) is reduced to this one:

$$\begin{cases} \dot{y}_1 = \sqrt{1 - 2 \cdot \delta \cdot \mu} \cdot y_2 + u(\tau) \\ \dot{y}_2 = -2 \cdot \sqrt{1 - 2 \cdot \delta \cdot \mu} \cdot y_1 - 2 \cdot \delta \cdot y_2 \end{cases}. \tag{9}$$

From system (9) it is easy to observe that variable y_2 obeys to the equation of motion for harmonic oscillator with damping factor δ and fundamental frequency

$\omega_0 = \sqrt{2 \cdot (1 - 2 \cdot \delta \cdot \mu)}$ under the action of external force:

$$\ddot{y}_2 + 2 \cdot \delta \cdot \dot{y}_2 + \omega_0^2 \cdot y_2 = -\sqrt{2} \cdot \omega_0 \cdot u(\tau), \tag{10}$$

and that the behaviour of variable y_1 is governed by the behaviour of variable y_2 as follows:

$$y_1 = -\frac{\dot{y}_2 + 2 \cdot \delta \cdot y_2}{\sqrt{2} \cdot \omega_0}. \tag{11}$$

At last for self-consistency of above presented linearization external dimensionless voltage ought to be weak: $|u(\tau)| \ll 1$.

3 Action of Harmonic Signal on the Linearized Bullard Dynamo

At first let us consider behaviour of the system (9) under the influence of external voltage:

$$u(\tau) = A_0 \cos(\omega\tau), \tag{12}$$

where in accordance with formulae (5) $A_0 = \frac{U_0}{U_m}$; $\omega = \nu/\nu_0$.

Looking at (10) with right hand side (12) one can see that in this case it describes harmonically excited linear oscillator with damping therefore we may solve it in the framework of the well-known complex amplitude method.

Seeking solution of (10) in the following form:

$$y_2(\tau) = \mathrm{Re}[A_2(\omega) \cdot \exp(i\omega\tau)], \tag{13}$$

one can easily find that complex amplitude $A_2(\omega)$ is equal to:

$$A_2(\omega) = -\frac{\sqrt{2} \cdot \omega_0}{\omega_0^2 - \omega^2 + 2i\delta\omega} \cdot A_0. \tag{14}$$

Further substituting expression (13) into (11) and using formula (14) it is not difficult to establish that

$$y_1(\tau) = \mathrm{Re}[A_1(\omega) \cdot \exp(i\omega\tau)], \tag{15}$$

complex amplitude $A_1(\omega)$ in formula (15) being equal to:

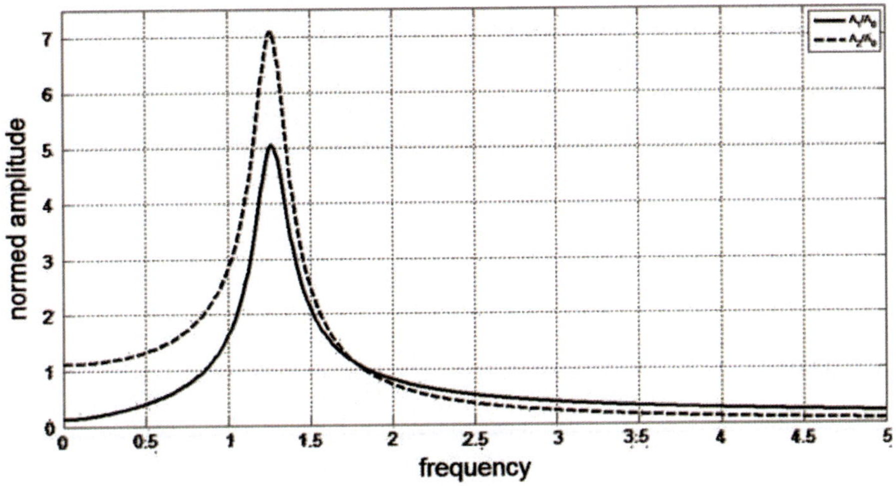

Fig. 3 Amplitude responses of the homopolar dynamo

$$A_1(\omega) = \frac{i\omega + 2\delta}{\omega_0^2 - \omega^2 + 2i\delta\omega} \cdot A_0. \tag{16}$$

Thus from formulas (14) and (16) it is easy to obtain that amplitude responses of dynamical variables of system (9) on voltage (12) are equal to:

$$\begin{array}{l} \frac{|A_1(\omega)|}{A_0} = \sqrt{\frac{\omega^2+4\cdot\delta^2}{(\omega^2-\omega_0^2)^2+4\cdot\delta^2\omega^2}} \\ \frac{|A_2(\omega)|}{A_0} = \frac{\sqrt{2}\cdot\omega_0}{\sqrt{(\omega^2-\omega_0^2)^2+4\cdot\delta^2\omega^2}} \end{array}. \tag{17}$$

Graphs of dependences (17) on dimensionless frequency ω for $\mu = 1.0$ and $\delta = 0.1$ are presented on Fig. 3. On this Figure continuous line corresponds to function $A_1(\omega)$ and dashed line corresponds to function $A_2(\omega)$. Both of them demonstrate typical resonance behavior.

4 Action of the Gaussian Delta-Correlated Noise and the Langevin Stochastic Process on the Linearized Bullard Dynamo

Let us now suppose that external voltage is GSSP purely.

In this case it is interesting to determine the following ACF:

$$B_J(t, t') =< (J(t) - J_a)(J(t') - J_a) >, \tag{18}$$

where

$$J_a = \langle J(t) \rangle \tag{19}$$

is average value of electric current in the circuit on Fig. 1.

Reducing in accordance with formulas (5) input GSSP voltage to dimensionless form:

$$u(\tau) = \frac{V(t)}{U_m} \tag{20}$$

and substituting expression (20) into formula (3) we establish that:

$$\langle u(\tau) \rangle = 0, \tag{21}$$

therefore from formulas (10) and (11) one can immediately obtain that:

$$\langle y_1(\tau) \rangle = \langle y_2(\tau) \rangle = 0. \tag{22}$$

Thus combining formulas (5), (8) and (22) it is easy to find that:

$$J_a = \sqrt{\frac{K}{M}} \cdot \sqrt{1 - \frac{2\gamma \cdot R}{K \cdot M}}, \tag{23}$$

hence

$$B_J(t, t') = \frac{K}{M} \cdot B_1(\tau, \tau'), \tag{24}$$

where

$$B_1(\tau, \tau') =< y_1(\tau) y_1(\tau') > . \tag{25}$$

On the other side in correspondence with formula (11) behavior of value $y_1(\tau)$ is controlled by value $y_2(\tau)$ therefore ACF (25) is expressed via the next ACF:

$$B_2(\tau, \tau') = \langle y_2(\tau) y_2(\tau') \rangle. \tag{26}$$

Inserting expression (11) into definition (25) and using the simplest properties of ACF [12] it is not hard to prove that:

$$B_1(\tau,\tau') = \frac{1}{2\omega_0^2}\left[\frac{\partial^2 B_2(\tau,\tau')}{\partial\tau\partial\tau'} + 2\delta\left(\frac{\partial B_2(\tau,\tau')}{\partial\tau} + \frac{\partial B_2(\tau,\tau')}{\partial\tau'}\right) + 4\delta^2 B_2(\tau,\tau')\right]. \tag{27}$$

Further after looking at formula (4) and comparing it with formula (20) it is obvious that:

$$< u(\tau)\cdot u(\tau') >= b_u(\tau'-\tau), \tag{28}$$

where

$$b_u(\tau'-\tau) = \frac{1}{U_m^2} B(t'-t). \tag{29}$$

It is clear that formulas (28) and (29) demonstrates stationary state of dimensionless input voltage therefore value $y_2(\tau)$ is GSSP too because of it obeys to linear differential equation with constant coefficients (10) [12]. It means that ACF (26) in fact depends only on variable $\theta = \tau' - \tau$:

$$B_2(\tau,\tau') \equiv B_2(\theta). \tag{30}$$

Substituting representation (30) into formula (27) one can easily derive that:

$$B_1(\theta) = \frac{1}{2\omega_0^2}\left[-\frac{d^2 B_2(\theta)}{d\theta^2} + 4\delta^2 B_2(\theta)\right], \tag{31}$$

hence $y_1(\tau)$ is also GSSP.

For further advance it is convenient in accordance with the Wiener-Khinchin theorem [12] to introduce spectral densities (SD) of ACF (30) and (31) as follows:

$$S_{1,2}(\omega) = \int_{-\infty}^{+\infty} B_{1,2}(\theta)\cdot\exp(-i\cdot\omega\cdot\theta)\cdot d\theta. \tag{32}$$

After the Fourier transform relation (31) is reduced to the next one between SD $S_1(\omega)$ and $S_2(\omega)$:

$$S_1(\omega) = \frac{\omega^2 + 4\delta^2}{2\omega_0^2}\cdot S_2(\omega). \tag{33}$$

At last it is well-known that for linear homogeneous system (10) connection between input and output SD is expressed via its amplitude response (17) [12] namely:

$$S_2(\omega) = \left| \frac{A_2(\omega)}{A_0} \right|^2 \cdot S_u(\omega), \tag{34}$$

where

$$S_u(\omega) = \int_{-\infty}^{+\infty} b_u(\theta) \cdot \exp(-i \cdot \omega \cdot \theta) \cdot d\theta \tag{35}$$

is SD for ACF (28).

Thus combining formulas (17), (33) and (34) one can obtain that:

$$S_1(\omega) = \frac{\omega^2 + 4 \cdot \delta^2}{(\omega^2 - \omega_0^2)^2 + 4 \cdot \delta^2 \cdot \omega^2} \cdot S_u(\omega). \tag{36}$$

Inverse Fourier transform of expression (36) is known to represent ACF (25):

$$B_1(\theta) = \int_{-\infty}^{+\infty} \frac{\omega^2 + 4 \cdot \delta^2}{(\omega^2 - \omega_0^2)^2 + 4 \cdot \delta^2 \omega^2} \cdot S_u(\omega) \cdot \exp(i \cdot \omega \cdot \theta) \cdot \frac{d\omega}{2\pi}. \tag{37}$$

If input voltage is the Gaussian delta-correlated noise (the white noise) then ACF (4) is equal to:

$$B(t' - t) = 2 \cdot D_V \cdot \delta(t' - t), \tag{38}$$

therefore

$$b_u(\tau' - \tau) = 2 \cdot D \cdot \delta(\tau - \tau'), \tag{39}$$

where intensity of stochastic process is renormalized in accordance with formula (29) as $D = D_V \cdot \nu_0 / U_m^2$.

Further expression (35) gives us that SD of GSSP with ACF (39) is equal to $S_u(\omega) = 2 \cdot D$. Thus integrand in formula (37) possesses by four simple poles $\pm\sqrt{\omega_0^2 - \delta^2} \pm \mathrm{i} \cdot \delta$ hence using the well-known Jordan's lemma one can calculate explicit representation of ACF (25) in this case:

$$B_1(\theta) = \frac{D \cdot \exp(-\delta|\theta|)}{2\delta\sqrt{\omega_0^2 - \delta^2}} \cdot \mathrm{Re}\left[\frac{\omega_*^2 + 4\delta^2}{\omega_*} \cdot \exp(\sqrt{\omega_0^2 - \delta^2}|\theta|) \right], \tag{40}$$

where $\omega_* = \sqrt{\omega_0^2 - \delta^2} + i \cdot \delta$.

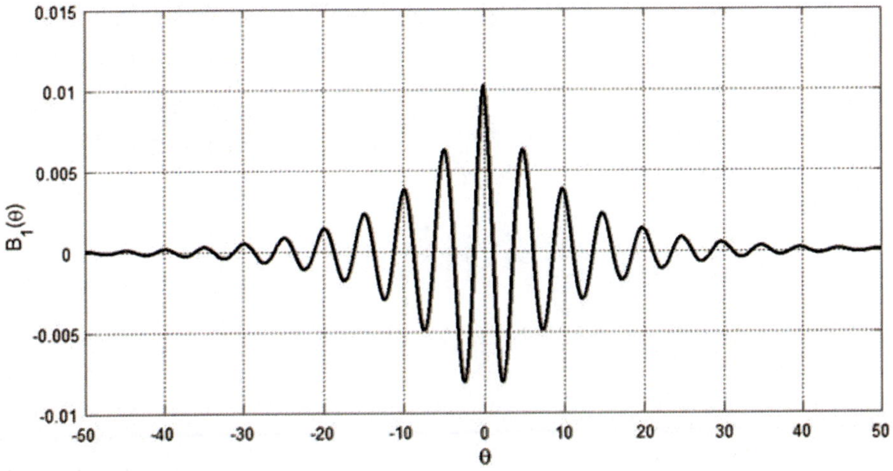

Fig. 4 Reaction of the homopolar dynamo on the Gaussian delta-correlated noise

Graph of the ACF (40) for $\mu = 1.0$, $\delta = 0.1$ and $D = 0.002$ is shown on Fig. 4.

If input voltage is the Langevin sochasic process then dimensionless ACF (29) may be chosen in the following form [13]:

$$b_u(\tau' - \tau) = \sigma^2 \cdot \exp(-\gamma|\tau' - \tau|), \quad \gamma > 0. \tag{41}$$

where σ^2 is dispersion of input GSSP $u(\tau)$.

SD corresponding to ACF (41) is equal to [13]:

$$S_u(\omega) = \frac{2 \cdot \gamma \cdot \sigma^2}{\omega^2 + \gamma^2}. \tag{42}$$

It means that in this case two additional simple poles $\pm i \cdot \gamma$ arise in integrand in formula (37).

In the same manner one can derive that for SD (42) ACF (25) is equal to the next sum:

$$B_1(\theta) = B_1^1(\theta) + B_1^2(\theta), \tag{43}$$

where

$$B_1^1(\theta) = \frac{\gamma \cdot \sigma^2 \cdot \exp(-\delta|\theta|)}{2 \cdot \delta \cdot \sqrt{\omega_0^2 - \delta^2}} \cdot \mathrm{Re}\left[\frac{\omega_*^2 + 4\delta^2}{\omega_* \cdot (\omega_*^2 + \gamma^2)} \cdot \exp(\sqrt{\omega_0^2 - \delta^2}|\theta|)\right] \tag{44}$$

and

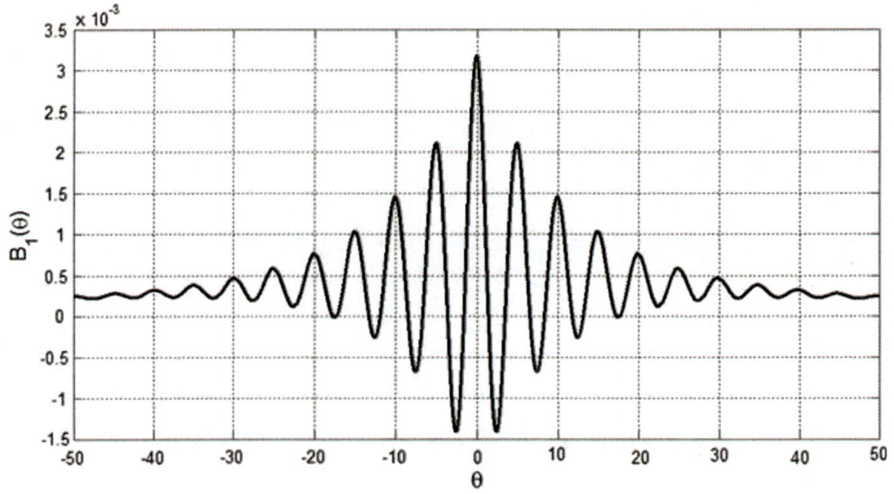

Fig. 5 Reaction of the homopolar dynamo on the Langevin stochastic process

$$B_1^2(\theta) = \sigma^2 \cdot \frac{4 \cdot \delta^2 - \gamma^2}{(\omega_0^2 + \gamma^2)^2 - 4 \cdot \delta^2 \cdot \gamma^2} \cdot \exp(-\gamma|\theta|). \tag{45}$$

Graph of the ACF (43) for $\mu = 1.0$, $\delta = 0.1$, $\gamma = 0.02$ and $\sigma = 0.2$ is presented on Fig. 5. Comparing Fig. 5 with Fig. 4 one can observe that this graph also has oscillatory character stipulated by function (44). But moreover this graph possesses by variable vertical shift caused by contribution of function (45) into expression (43).

5 Conclusion

In the chapter linear responses of the homopolar dynamo both on weak harmonic input voltage and weak GSSP input voltage have been calculated. This preliminary research gives one a possibility of investigation of stochastic resonance in the Bullard dynamo by means of analog modeling.

To realize this research program one ought to evaluate physical parameters of the system on Fig. 1 and then use them to make its layout. After that one can perform a number of tests of the operation of the layout.

The first test is an action of weak ($U_0 \ll U_m$) harmonic signal with very slowly varying circular frequency on the homopolar dynamo layout. If dimensionless circular frequency ω of this input signal gets closer to ω_0 then a sharp increase in amplitude of electric current in the circuit should be observed in accordance with formula (16) (see also Fig. 3).

The second test is an application to the layout of the weak Gaussian delta-correlated noise as an input voltage. In this case measured ACF (18) must correspond to the calculated dependence (40) (see also Fig. 4).

Moreover nonlinearity of a system is known to transform GSSP into non-Gaussian stochastic process [12], therefore, in order to control the role of nonlinearity of system (1) one should measure the following triple ACF [14]:

$$T(t_1, t_2) = \langle (J(t) - J_a)(J(t + t_1) - J_a)(J(t + t_2) - J_a) \rangle \tag{46}$$

and calculate its bispectrum [14]:

$$Q(\omega_1, \omega_2) = \int_{-\infty}^{+\infty} \int_{-\infty}^{+\infty} T(t_1, t_2) \cdot \exp(-i\omega_1 t_1 - i\omega_2 t_2) \cdot dt_1 dt_2. \tag{47}$$

If the influence of nonlinearity is small then both value (46) and value (47) must be close to zero due to the Gaussian nature of the input signal.

The third test is an action of the weak Langevin stochastic process as an input voltage. This kind of input voltage can be obtained by means of transferring of the Gaussian delta-correlated noise via four-terminal network with resistance and capacitance [12]. In this case measured ACF (18) must correspond to the calculated dependence (43) (see also Fig. 5). And it is necessary to oversee closeness to zero of values (46) and (47) too.

At last if the layout overcomes these checks successfully then one can proceed to the experimental study of stochastic resonance in the homopolar dynamo under the action of input voltage (2) in nonlinear regime.

References

1. C. Nicolis, G. Nicolis, Coupling-enhanced stochastic resonance. Phys. Rev. E. **96**(4), 042214-1-10 (2017)
2. M.A. Zaks, A. Pikovsky, Synchrony breakdown and noise-induced oscillation death in ensembles of serially connected spin-torque oscillators. Eur. Phys. J. B. **92**, 160-1-12 (2019)
3. M.I. Bolotov, L.A. Smirnov, G.V. Osipov, A. Pikovsky, Locking and regularization of chimeras by periodic forcing. Phys. Rev. E. 2020. **102**(4), 042218-1-11
4. V. Basios, Strong perturbations in nonlinear systems: the case of stochastic-like resonance and its biological relevance from a complex system's perspective. Eur. Phys. J. Spec. Top. **225**, 1219–1229 (2016)
5. E.V. Pankratova, V.N. Belykh, Consequential noise-induced synchronization of indirectly coupled self-sustained oscillators. Eur. Phys. J. Spec. Top. **222**, 2509–2515 (2013)
6. C. Floris, Numeric solution of the Fokker-Plank-Kolmogorov equation. Engineering **5**(12), 975–988
7. D. Higham, P. Kloeden, An introduction to the numerical simulation of stochastic differential equations, in *SIAM* (2021)

8. E.C. Bullard, The stability of a homopolar dynamo. Math. Proc. Cambridge Philos. Soc. **51**(4), 744–760 (1955)
9. H.K. Moffat, *Magnetic Field Generation in Electrically Conducting Fluids* (Cambridge, 1978)
10. M. Bourgoin, R. Volk, N. Plihon, P. Augier, P. Odier, J.-F. Pinton, An experimental Bullard-von Karman dynamo. New J. Phys. **8**, 329-1-14 (2006)
11. N. Leprovost, B. Dubrulle, F. Plunian, Instability of the homopolar disk-dynamo in presence of white noise. arXiv:nlin/0506050v1
12. S.M. Rytov, Yu.A. Kravtsov, V.I. Tatarskii, *Principles of Statistical Radiophysics*, vol. 1–4 (Berlin, 1987–1989)
13. A.A. Potapov, Yu.V. Gulyaev, S.A. Nikitov, A.A. Pakhomov, V.A. German, The modern methods of image processing, in *FIZMATLIT*, ed. by A.A. Potapov (2008) (in Russian)
14. A.N. Malakhov, Cumulant *Analysis of Random Non-Gaussian Processes and Their Transformations.* Sovetskoe Radio (1978) (in Russian)

Dynamic Localized Autonomous Chaotic Orbital Patterns from Rotation-Translation Sequences

Bernd Binder

Abstract Consider an ordered sequence of repeated operations given by a distance-dependent rotation followed by a translation. Operating this sequence with special parameters and initial conditions provides for characteristic spatial density patterns in the plane. In this work we introduce an additional orbital rotation and find local chaotic orbital patterns and attractors in the plane. There are two ways to form a local density from discrete long-range jumps: either a jump-back boundary condition or rotating the jump direction. We focus on real time simulations, where the chaotic evolution and vivid dynamics (the live cycles of orbitals) with characteristic numbers or stability conditions is manifest. Ring bifurcations, stable and unstable chaotic orbital patterns or solitons emerge dynamically with start conditions and without any "hard" additional boundary or radial back-jump-condition. We show some typical orbital patterns and suggest a method of categorization.

Keywords Chaos · Rotation · Bifurcation · Discrete · Translation · Reflection · Closed loop · Orbit · Soliton · Wavelet · Quantum · Attractor · Pattern · Sequence · Simulation · Commutation · Operator · Geometric phase · Magic angle · Spiral · Ray

1 Introduction

Repeating a discrete sequence of rotation—translation—reflection operations can provide for a wide range of interesting and very complex pattern emerging from chaotic jumps, see Skiadas [1–3]. Counterintuitively, discrete long-range jumps often follow a continuous type of flow pattern, e.g. in [3] very similar to v. Kármán Streets, see Fig. 1.

To get a special pattern requires adjusting the rotation strength parameter and eventually a boundary distance condition parallel to the jump direction. We found that the correspondent pattern building process can be assigned to small nonlinear

B. Binder (✉)
89182 Bernstadt, Germany
e-mail: bernd@binders.de

C. H. Skiadas and Y. Dimotikalis (eds.), *14th Chaotic Modeling and Simulation International Conference*, Springer Proceedings in Complexity,
https://doi.org/10.1007/978-3-030-96964-6_4

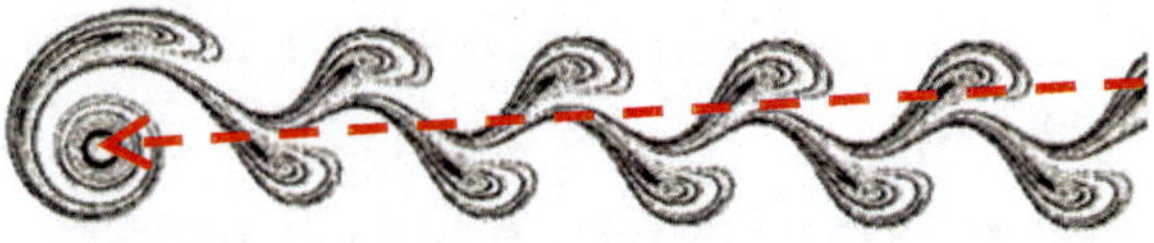

Fig. 1 Skiadas type v. Kármán Streets with rotation parameter $c_1 = \pi$, power exponent $p_1 = -2$, reflection $m = 2$, and long jump back (red arrow)

geometric (phase) shifts arising in the rotation-translation sequence on every loop Binder [4, 5]. Applying this nonlinear concept it is possible to generate a broad range of patterns, including periodic structures like waves, circles, saw-tooth or point-like discrete geometries by desktop computer simulations. The model can be generalized to multiple rotation-translation sequences on orthogonal rotation axes in higher dimensions Binder [5].

In this work we add to the basic rotation-translation-reflection model an extra orbital rotation orthogonal to the jump direction and wonder, if we can generate in these ways structures on closed orbits. This means, we add another rotation around the existing singularity (at the origin) and look for orbital structures. As a result, we find chaotic periodic structures emerging on the orbit without any additional constraint like a jump-back condition limit. These dynamical structures are more or less stable and show an internal dynamics sometimes like a vivid entity. In this paper we point to some interesting patterns/simulations and try to make a categorization according to boundary conditions and characteristic parameter.

2 The Basic Operation Sequence

In the plane the chaotic model is based on a discrete iteration sequence of the polar vector

$$\vec{r} = \begin{pmatrix} x \\ y \end{pmatrix}. \tag{1}$$

Its polar coordinates are given by the radius $r = |\vec{r}|$ and polar angle/direction φ, where

$$x = r \sin\varphi; \quad r^2 = x^2 + y^2; \quad \varphi = \arctan 2(x, y). \tag{2}$$

The iteration will be given by an ordered sequence composed by the two or three operations given by a polar **rotation R** and non-radial **translation T** and eventually a radial **inversion I**. The vector coordinate evolves in successive operations within one iteration sequence with numbering $t \to t + 1$ according to

$$\vec{r}_t \to \vec{r}_R \to \vec{r}_T \to \vec{r}_I = \vec{r}_{t+1}$$

by the following relations:

I. A radius dependent polar/central **rotation** including **reflection**

$$\vec{r}_R = \mathbf{R}(\vec{r}, \Delta\phi). \tag{3}$$

with rotation angle $\Delta\phi$ in (3) composed by the following rotation and reflection components

$$\Delta\phi = \Delta\phi_1 + \Delta\phi_2 \tag{4}$$

given by:

1. $\Delta\phi_1$ in (4) is the Skiadas type rotation angle that has a power-law radial distance dependence with exponent $p_1 < 0$

$$\Delta\phi_1 = c_1 r^{p_1}, \tag{5}$$

2. $\Delta\phi_2$ in (4) generalizes the reflection given by the difference $\sigma - \varphi$ multiplied by a reflection mode m

$$\Delta\phi_2 = m(\sigma - \varphi), \tag{6}$$

where $m = 2$ is a reflection to the opposite side with respect to the initial polar angle φ in the co-rotating frame. In our special case, the global direction σ in (6) sums up with a constant orbital rotation c_0 eventually driven by a radial power-law $c_2 r^{p_2}$

$$\sigma = \sigma_t = \sigma_{t-1} + c_0 + c_2 r^{p_2}. \tag{7}$$

The total rotation is with in (4)–(7) given by

$$\Delta\phi = c_1 r^{p_1} + m(\sigma - \varphi), \tag{8}$$

and the rotation in (3) provides for the new orientation angle

$$\varphi_R = \varphi + \Delta\phi. \tag{9}$$

II. The non-radial **translation** $\vec{r}_T = \mathbf{T}(\vec{r}_R, \varphi, \Delta r) = \mathbf{T}\left(\vec{r}_R, \overrightarrow{\Delta r}\right)$ of the rotated $\vec{r}_R$ by a translation shift $\overrightarrow{\Delta r}$

$$\vec{r}_T = \mathbf{T}\left(\vec{r}_R, \overrightarrow{\Delta r}\right) = \vec{r}_R + \overrightarrow{\Delta r} \tag{10}$$

in the initial φ—direction (and not in the actual φ_R—direction of (9)) where the polar component of the translation in (10) are given by

$$\overrightarrow{\Delta r} = \pm \begin{pmatrix} \Delta x \\ \Delta y \end{pmatrix} = \pm \left| \overrightarrow{\Delta r} \right| \begin{pmatrix} \cos \\ \sin \end{pmatrix} = c_3 r^{p_3} \begin{pmatrix} \cos \varphi \\ \sin \varphi \end{pmatrix}, \tag{11}$$

Equation (11) generalizes the usual "jump" in x—direction [1–3], where the unit jump has $\left| \overrightarrow{\Delta r} \right| = 1$ or always $c_3 = \pm 1$ with $p_3 = 0$. In this paper we will consider for simplicity only unit jumps and in Chap. 7 negative jumps with $c_3 = -1$.

III. Finally there could be an additional **inversion** operation with respect to the origin

$$\vec{r}_I = \mathbf{I}(\vec{r}_T) = \vec{r}_T / |\vec{r}_T|^2, \tag{12}$$

with invariant direction angle but inverse length to get a rotation-translation-inversion sequence. This inversion could also be used as a method to jump back near to the rotation singularity.

3 Categories of Jumping Patterns

Without orbital rotation the jumps would only go in one direction (usually the x-direction) and disappear to infinity. There are two ways to form a local density: either a jump-back boundary condition or rotating the jump direction. The v. Kármán Street pattern in Figs. 1 and 2 and the typical $2d$ wave pattern on the plane in Fig. 3 have a jump-back distance condition in the x-direction, which means, if the distance to the origin in jump direction exceeds a limit, a jump back near to the origin follows (Figs. 4 and 5).

With a new extra orbital rotation orthogonal to the jump direction we get local chaotic patterns without any orbital conditions (no jump condition) similar to the orbital solutions we know from quantum mechanics. We will call them "Localized Autonomous Chaotic Orbital Patterns" or LACOP, where we get many interesting orbital structures for integral p_1, p_2, m. It makes sense to group the jumping patterns according to the boundary conditions, where we have identified five categories given by:

1. **random starts condition**: long jump back (red arrow) randomly near to center, if distance or number of jumps exceeds a limit (see examples in Figs. 1, 2, 6 and 7).
2. **periodic boundary condition**: a defined long jump back to the back side, if distance exceeds a limit $R_{\max}$, see Fig. 3.
3. **inversion condition**: inversion operation $\vec{r}' = \vec{r}/r^2$ if distance r exceeds a limit $R_{\max}$, see (12).

Fig. 2 Long jump back with 6 arm-symmetric v. Kármán Street pattern, $c_2 = 2\pi/6$, with $m = 2$, $c_1 = \pi$, $p_1 = -2$, and $p_2 = 0$

Fig. 3 Periodic boundary jump back conditions producing waves, where $p_1 < 0, \quad p_2 = 0, \quad m = 2$

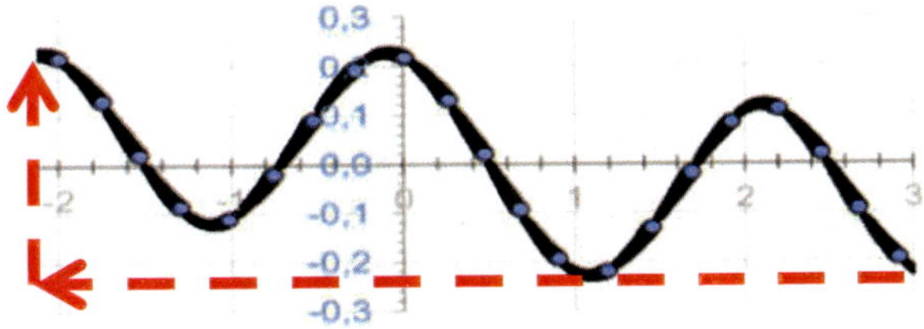

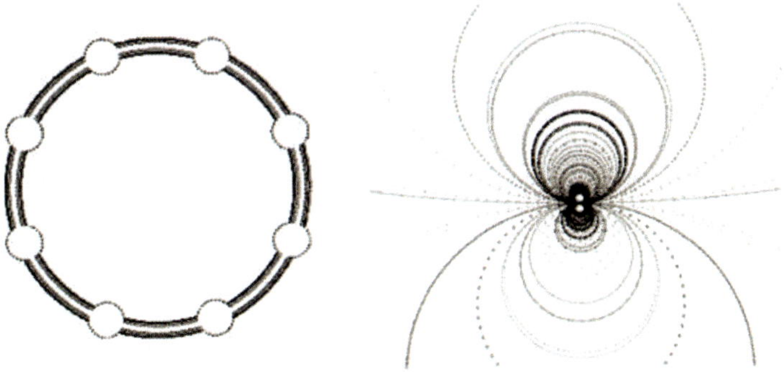

Fig. 4 $p_1 = p_2 - 1$ with $p_2 = 2$, left: LACOP with $m = 1$, right: random starts with $m = 2$ dipole

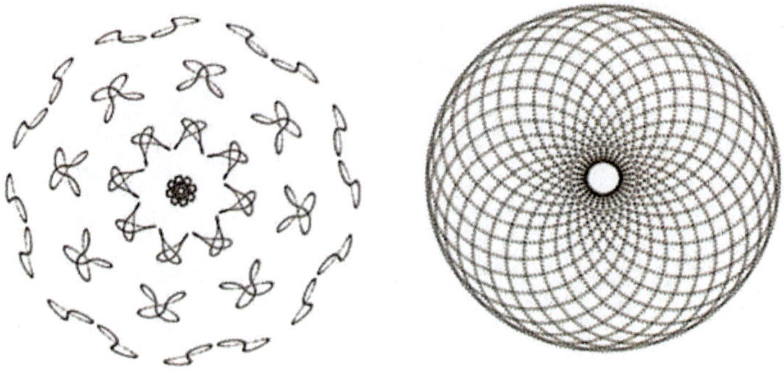

Fig. 5 LACOP with regular orbital structure, examples of patterns with $m = 0$, $p_1 = 0$, $p_2 = -1$, $p_3 = 0$.

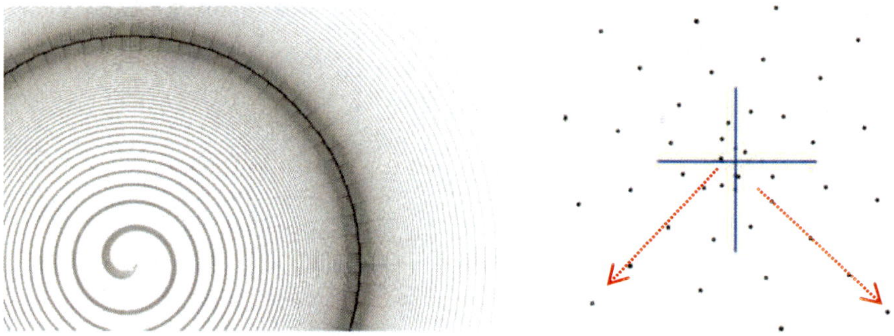

Fig. 6 Radial rays intersecting spirals according to (13) and (14), left: $m = 1$, $c_3 = -1$, $p_1 = 0$, $p_2 = 1$, $M > 100$, $J = 0.5$, right: $m = 1$, $c_3 = 1$, $p_1 = -2$, $p_2 = 0$, $M = 5$, $J = 0.5$

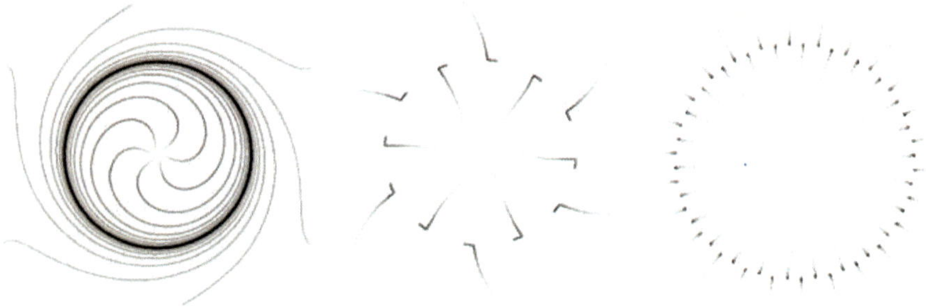

Fig. 7 Linear rays for $m = 1$, $c_3 = -1$, $p_1 = 0$, $p_2 = 1$ left and mid $M = 3$, right $M = 13$, $g = 0.001$

4. **no boundary condition**: rotated jumps around the center according to orbital quantum numbers and symmetries producing LACOP.
5. **parameter conditions**: relating the two parameter c_1, c_2 can define a family of patterns, e.g., if we define a small isotropic geometric phase shift gap $g = 1/N \ll 1$ and relating the coefficients via the geometric jumping gap g geometrically to the rotation-translation parameter according to

$$c_1 = \arccos(1 - g), \quad c_2 = 2\pi J g. \tag{13}$$

In this case we get with $= 1, 2, 3 \ldots = 1/g$, $c_3 = -1$, $p_1 = 0$, $p_2 = 1$ special radial conditions like spirals intersected radial rays, see Figs. 6 and 7 with random starts near to the center (condition 1).

4 Physics Relevance

It is interesting comparing these structures to Quantum Electrodynamic and spin symmetries. The parameter conditions categories 5 and 6 of Chap. 3 can be combined with condition categories 1, 2, 3, 4, where $p_2 = -1$ with m—pole provides for multipole type orbital ring clouds and field structures:

5 Simple $m = 0$ LACOP

Both regular and highly non-linear or chaotic are the $m = 0$ patterns in a wide parameter range, see Fig. 5.

6 Monopole $m = 1$ Negative Jumps $c_3 = -1$ with Random Start Near to Center

Both, highly chaotic and regular are the $m = 1$, $c_3 = -1$ patterns in a wide parameter range, see Figs. 6 and 7, where the geometric gap condition in (13) produces radial rays intersecting spirals with random starts. It is not surprising to get Fresnel Charge structures in Fig. 6 for $p_2 = 1$, since a Fresnel spiral has an inherent spiraling angular structure with $\varphi \propto \sum \sigma \propto r^{p_2+1} = r^2$.

With a second constraint between the two parameters c_1, c_2 in addition to (13) given by the linear relation

$$Mc_1 = 2\pi J - c_2, \tag{14}$$

we get fixed points for the parameters recovering the magic angle condition $Mc_1 = \pm J\cos(c_1)$ Binder [5] with charge J and discrete points solutions, where spirals intersect for $m = 1$ rays with center offset, see Figs. 6 and 7.

7 More Complex Vivid $m = 2$ LACOP

Very interesting and exciting is the chaotic dynamics of the $m = 2$ LACOP orbitals, see Fig. 8 and some mixed examples in Fig. 9. There is in most cases no static or stationary solution, since the orbitals often show a chaotic variation in the orientation or orbital shape. Therefore, a stable LACOP should be properly initialized; usually by an high enough orbital rotation parameter c_2 (spin, energy) while increasing the m-parameter from 1 to 2. In simulations the solutions (recorded as videos) appear to be like vivid orbitals with inherent chaotic dynamics especially in the substructure of orbital rings:

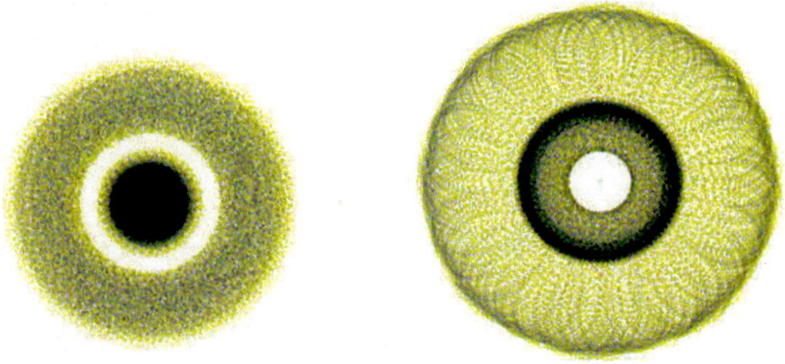

Fig. 8 Vivid pulsating LACOP with different orbital structure expanding and contracting (the two examples are snapshots from the same LACOP) with $m = 2$, $p_1 = -2$, $p_2 = 0$, $p_3 = 0$, $c_1 \approx -0.02035\pi$, $c_2 \approx 1.0021\pi$

Fig. 9 Typical chaotic LACOPs with a core and a hole in the center $p_1 < p_2, m > 0$

Fig. 10 Radial ring orbit bifurcation $p_1 = -1$, $m = 1$, left: 4th bifurcation with 16 orbits where $c_1 \cong 1.657\pi$, right: chaos starting at $c_1 \cong 1.65\pi$

8 Orbital Bifurcations for $m = 1$ LACOP

Under special conditions, the radial distribution of orbits shows a bifurcation tree, see Fig. 10.

9 Conclusion

The resulting patterns often show a localized ring shape with several mixed orbits in a kind of hydrodynamic-type orbital shelf flow and an empty region or hole at the center. m—poles or reflection modes with higher m show more complex and instable pattern. A stable form must be initialized; otherwise the pattern collapses or expands to infinity. The emerging LACOP solitons or fixed point solutions are having always characteristic.

- special parameter m, c_i, p_i, where usually $p_1 \leq 0$, p_2 and $p_3 \geq 0$, see Table 1.
- nonlinear structure, chaotic dynamics, bifurcations, and fluctuations
- radial and orbital wave numbers
- radial and orbital symmetries
- geometric phase conditions
- fixed point sets (i.e., two magic conditions (13) and (14) with numbers J, M)

Finally, we propose that this computer experiments show some relevance to quantum physical systems since we have

- a wave attractor showing quantization effects in terms of rotational units,
- a quantization of monopole and multipole charges,
- a basic non-zero quantum spin in the two main operators (rotation-translation, non-commutative) with characteristic geometric phase shifts,
- point-like local events with emerging global wave-like probabilistic patterns.

Table 1 Small LACOP exponent combinations

"Jump back" needed?	p_1	p_2	p_3	m	c_3
Yes	−2	0	0	>0	1
No	−2	0	0	>0	1
No	−1	1	0	>−1	1
No	−1	2	0	>1	1
No	−1	2	1	>0	1
No	0	1	0	1	−1
No	0	1	0	>1	1
No	0	2	0	>1	1

There will be videos of simulations available on the internet with title "Dynamic Autonomous Chaotic Orbital Patterns" or tag "#DACOPSimulation".

References

1. C.H. Skiadas, C. Skiadas, Chaos in Simple Rotation-Translation Models, nlin.CD. (2007) http://arxiv.org/abs/nlin/0701012v1
2. C.H. Skiadas, C. Skiadas, *Chaotic Modeling and Simulation: Analysis of Chaotic Models, Attractors and Forms* (Taylor and Francis/CRC, London, 2008)
3. C.H. Skiadas, Von Karman Streets chaotic simulation, in *Topics on Chaotic Systems*, ed. by C.H. Skiadas, I. Dimotikalis, C. Skiadas (World Scientific, 2009), pp. 309–313
4. B. Binder, Magic angle chaotic precession, in *Topics on Chaotic Systems: Selected Papers from CHAOS 2008 International Conference*, Singapore (World Scientific Books, 2008), pp. 31–42
5. B. Binder, Chaotic modelling and simulation: rotations-expansion-reflections chaotic modelling with singularities in higher dimensions. Chaotic Modeling and Simulation (CMSIM) **1**, 89–106 (2013)

New Discrete Chaotic Cipher Key Generation for Digital Embedded Crypto-systems

Belqassim Bouteghrine, Camel Tanougast, and Said Sadoudi

Abstract To benefit of embedded systems which are highly customized providing architectures to suit real-time computing, optimized unit size and low power consumption require highest levels of data communication security which are very useful advantage for telecommunications and networks and Internet of Things (IoT) communication applications that handle sensitive information. This paper presents an extracted new 5-dimensional (5-D) discrete time chaos system to generate a robust chaotic cipher data stream to ensure encryption application for secure communication. Dynamical behaviors and security analysis are investigated and compared to current discrete chaotic maps proving its suitability for embedded data encryption systems. Field-Programmable Gate Array (FPGA) implementation design shows better performance and good security robustness compared with previous works while proving the performance improvement of the proposed cipher block in terms of throughput, used hardware logic resources, and resistance against most cryptanalysis attacks.

Keywords Chaotic generator · Secure communication · Key space · Encryption · FPGA implementation

1 Introduction

Due to advancement in technology, thousands of devices in home, industry and health care systems are connected to each other. With advent of IoT, this number is increasing exponentially [20]. With the provision of too many features, these devices still need

B. Bouteghrine (✉) · C. Tanougast
LCOMS, Université de Lorraine, 57070 Metz, France
e-mail: belqassim.bouteghrine@univ-lorraine.fr

C. Tanougast
e-mail: Camel.Tanougast@univ-lorraine.fr

S. Sadoudi
Ecole Militaire Polytechnique, Algiers, Algeria
e-mail: sadoudi.said@gmail.com

C. H. Skiadas and Y. Dimotikalis (eds.), *14th Chaotic Modeling and Simulation International Conference*, Springer Proceedings in Complexity,
https://doi.org/10.1007/978-3-030-96964-6_5

more protection from cyber-attacks [9]. Due to the existing low security protocols, these devices are always on high threat, which makes the communication vulnerable and prone to foreign intrusion. In the literature, many schemes and protocols have been proposed to address the network security issue. In [21] authors have proposed an unclonable-based functions for authentication the protocol, for IoT devices. However, the proposed protocol lack some related computations [14]. In [16] authors have presented a simplified protocol to reduce the computations through the authentication phase But the complexity of the proposed scheme is not appropriate for the IoT devices [13]. Moreover, Random numbers are used in many cryptographic protocols, key management, identity authentication, image encryption, and so on [23]. As software generated random sequences are not truly random, fast entropy sources such as quantum systems or classically chaotic systems can be viable alternatives provided they generate high-quality random sequences sufficiently fast [3]. In [2], authors have introduced a new hardware chaos-based pseudo-random number generator, which is mainly based on the deletion of an Hamilton cycle within the N-cube plus one single permutation. Chen et al. [7] initiates a systematic methodology for real-time chaos-based video encryption and decryption communications. Based on the fundamental anti-control principles of dynamical systems, a novel 6-dimensional real domain chaotic system is designed, and then the corresponding algorithm is developed. The proposed algorithm is utilized to design a real-time chaos-based secure video communication system, with a generalized design principle derived, which is implemented on an FPGA hardware. Additionally, some other research works have been proposed which include, applications of chaotic systems for speech signal encryption [18], e-mail and database encryption [19] and image encryption [6, 8, 10, 12, 15].Unfortunately due to different technical reasons, all these schemes are not useful for different IoT applications. Because the discovery of simple chaotic systems with complex dynamics has always been an interesting research work [11], we propose through this paper an extracted low resource consumption 5-D Chaotic System for secure IoT communication. The proposed chaos-based cryptosystem is implemented by using Xilinx ZYNQ-XC7Z020 FPGA board. The rest of this paper is organised as follows. Section 2 describes the proposed 5-D map. Hardware implementation and performance analysis are presented in Sect. 3. Finally, Sect. 4 concludes this paper.

2 The Proposed 5-D Map

In [5] authors proposed a multidimensional chaotic map within good chaotic properties. From the proposed system, the extracted 5-D discrete time chaos system with nine nonlinear terms and five control parameters is described as follows:

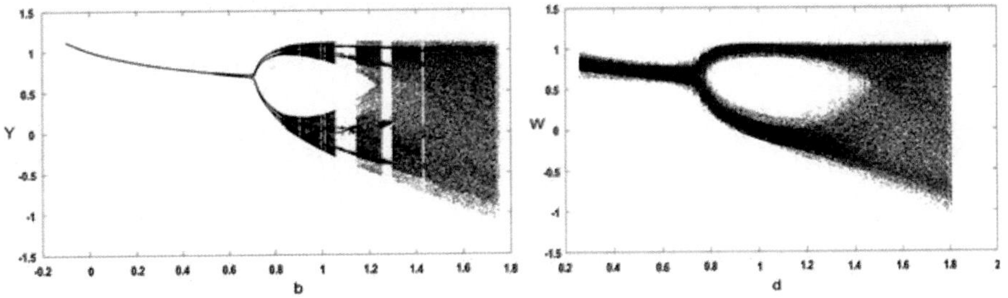

Fig. 1 The bifurcation graphs of the proposed 5-D map with parameters b, d

$$\begin{cases} X(n+1) = 1 - a * X(n)^2 + (Y(n) * Z(n) * W(n) * P(n)) \\ Y(n+1) = 1 - b * Y(n)^2 + (X(n) * Z(n) * W(n) * P(n)) \\ Z(n+1) = 1 - c * Z(n)^2 + (X(n) * Y(n) * W(n) * P(n)) \\ W(n+1) = 1 - d * W(n)^2 + (X(n) * Y(n) * Z(n) * P(n)) \\ P(n+1) = e * X(n) * Y(n) * Z(n) * W(n) \end{cases} \tag{1}$$

whereas a, b, c, d and $e \in$ R are the controllers and X, Y, Z, W, P are the state variables respectively.

2.1 Bifurcation Analysis

To investigate behaviours of the proposed system defined by the proposed 5-D map, we analyze the bifurcation diagrams related to parameters a, b, c, d and e. According to the bifurcation study, chaotic behaviour of the proposed system appears for $a \in [0.8, 1.8]$, $b \in [0.1, 1.4]$, $c \in [0.7, 1.9]$, $d \in [0.3, 1.6]$ and $e \in [0.05, 1.1]$ as shown in Fig. 1 (the bifurcation study of the parameters b, d).

2.2 Signals Analysis

To study the dynamical behaviours of the proposed model, the signal graph and the phase space trajectories defined by the state variables (X, Y, Z, W, P) can be used as an indicator to determine that the motion of that system is chaotic. In the proposed work, the technique is based on the signal output and the projection of the trajectories onto the plane, which reflects the chaotic behaviour result of the proposed system as shown in Figs. 2 and 3, respectively.

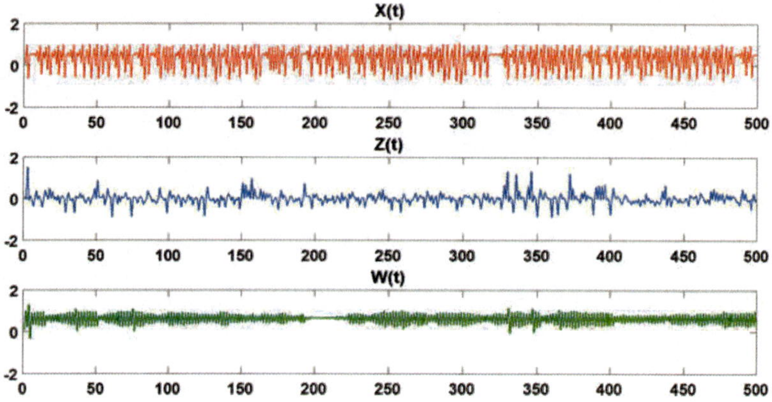

Fig. 2 Signals graphs of the proposed 5-D map

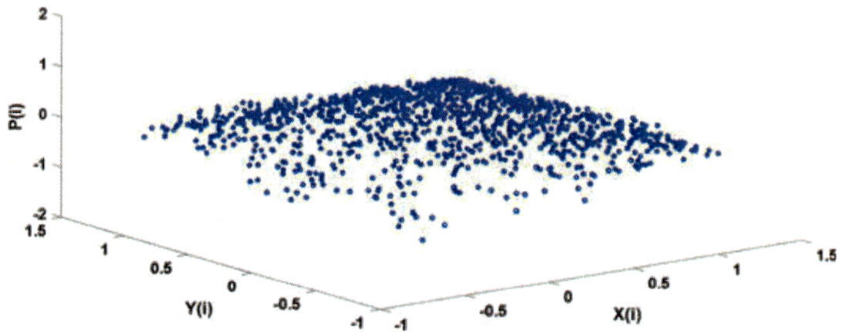

Fig. 3 Trajectory graph of the proposed 5-D map

2.3 Sensitivity Analysis

To evaluate the sensitivity to initial conditions of the proposed map, we consider a changing by 10^{-10} of the initial values $X(0)$, $Y(0)$, $Z(0)$, $W(0)$ and $P(0)$, then for the parameters a, b,c,d and e. The results shown in Figs. 4, 5, 6, 7 and 8 prove that after a few number of iteration all the signals are different from the initial ones.

3 FPGA Implementation of Secure Peer-to-Peer Communication

3.1 The Proposed Platform Test Bench

Because of flexibility, reliability, low cost, fast time-to-market, and long term maintenance, FPGA environment is considered more useful for the validation of the

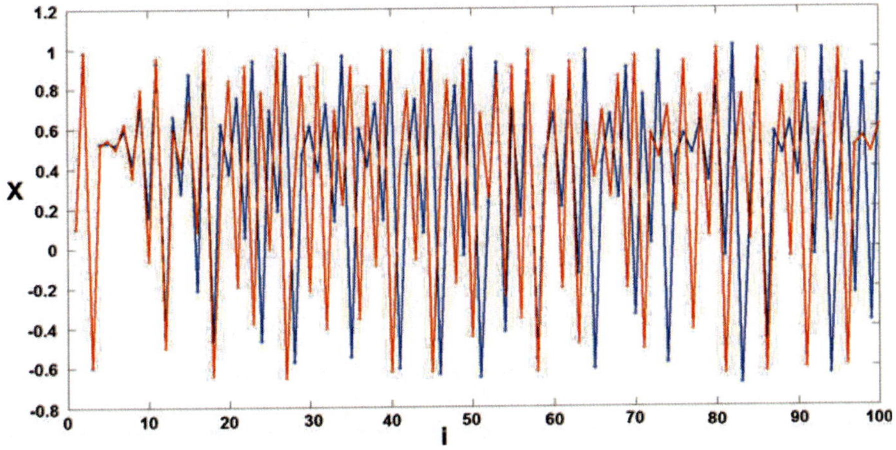

Fig. 4 Sensitivity analysis of the signal $X(n+1)$

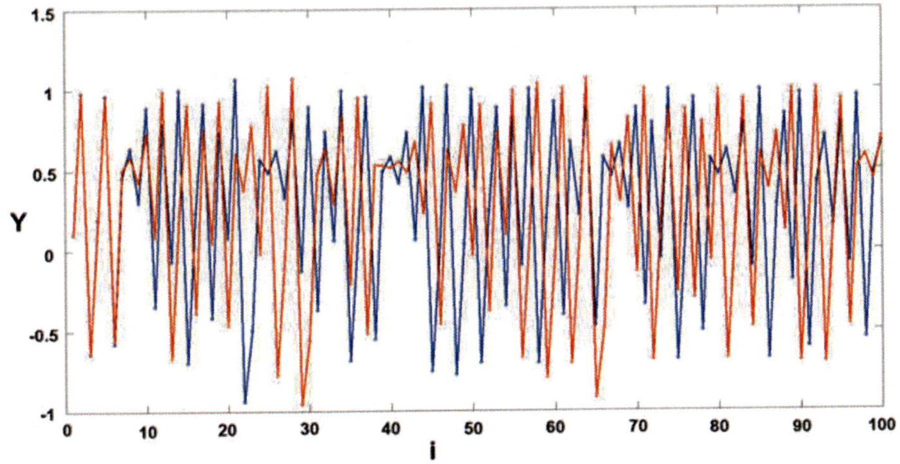

Fig. 5 Sensitivity analysis of the signal $Y(n+1)$

proposed scheme. Initially the hardware description language (VHDL) is used to implement the proposed 5-D system as a chaos-based cryptosystem (called Chaos 5-D Generator). After that, the designed is integrated as a new core or module with the other components of the Xilinx ZYNQ-XC7Z020 FPGA board as given in Fig. 9 [5].

To establish the final platform of Peer-To-Peer secure communication, we connect all the programmed FPGA boards through an Ethernet network (Fig. 10).

The proposed bench test runs as follows:

- The peers establish a new connection;
- If it is the first connection ($i = 1$), then key-Generator modules are reset;

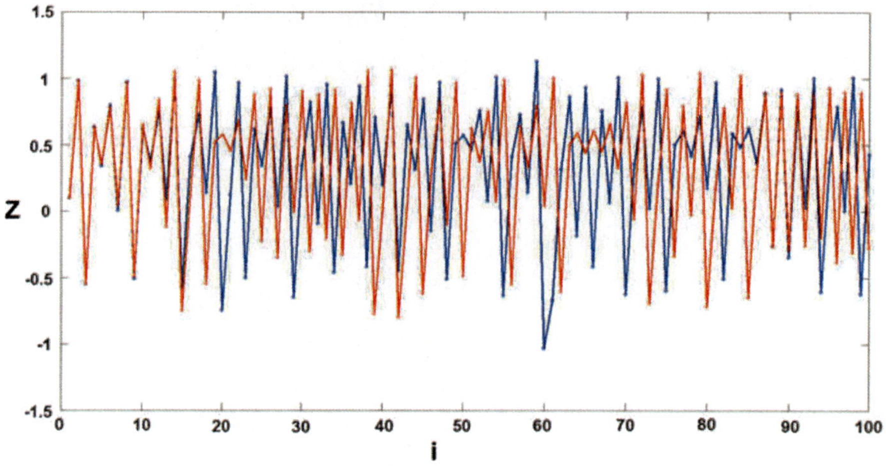

Fig. 6 Sensitivity analysis of the signal $Z(n+1)$

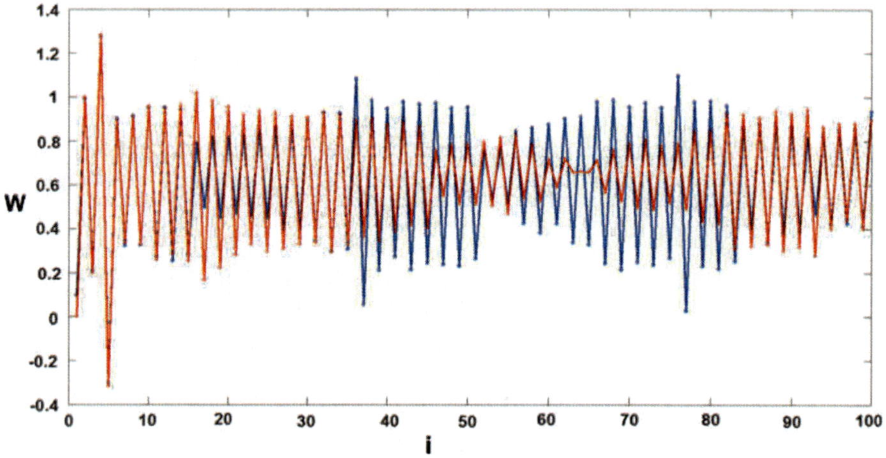

Fig. 7 Sensitivity analysis of the signal $W(n+1)$

- The key-Generator of both peers load the key corresponding to the sequence i of the implemented 5-D map;
- The generated keys are used to encrypt and send the data;
- The generated keys are used to decrypt and read the received data;

Lastly, the peers terminate the communication by closing the channel, and the key-Generator module saves the samples $(i+1)$ for the next communication.

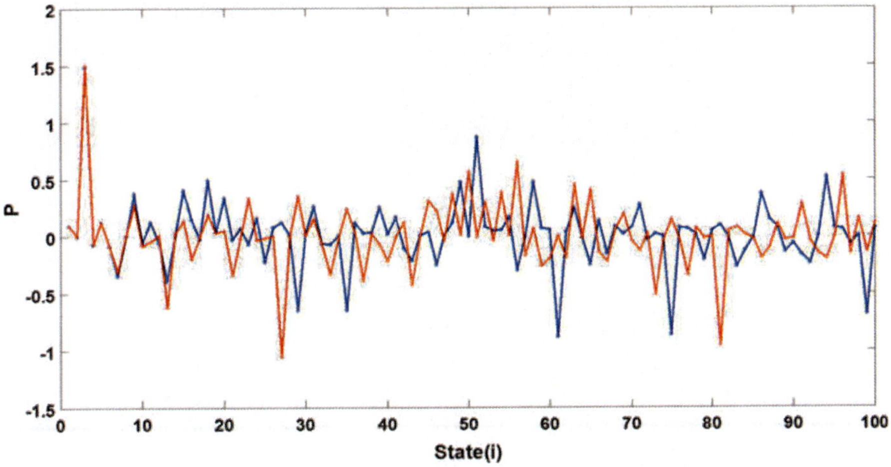

Fig. 8 Sensitivity analysis of the signal $P(n+1)$

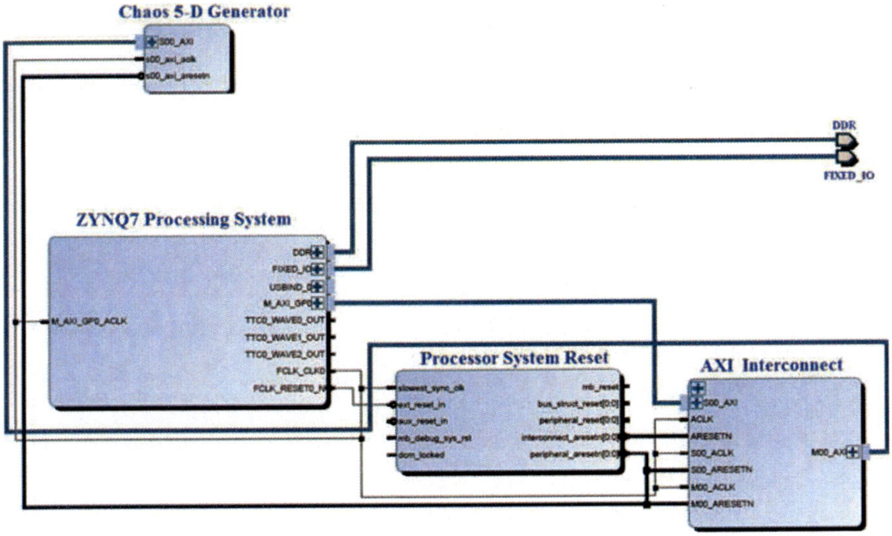

Fig. 9 FPGA block design

3.2 *Performance Analysis*

In such kind of cryptosystems, without any robust experimental solution to the chaotic synchronization issue [17], we introduce the control option for the chaos 5-D generator. Hence for each established connection i, the proposed cryptosystem generates the same encryption key which corresponds to the sample i of the proposed 5-D chaotic map.

Fig. 10 The proposed bench test platform

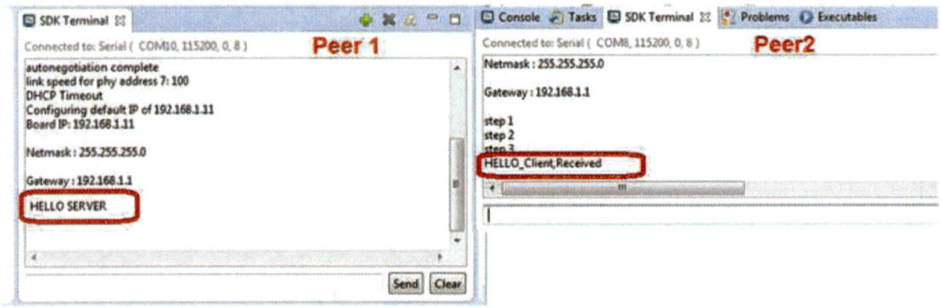

Fig. 11 Exchanged messages between the peers

Ensuring that both generators implemented on two different boards generate the same keys, thereby, the connected peers can encrypt/decrypt exchanged messages easily as shown in Fig. 11.

To show the performance of our system, the proposed model is compared with some of the state of the art works. Table 1 gives the details of the comparison. The comparison is made on the basis of LUTs, FF, DSP, maximum frequency and Slices. From table 1, it is evident that the proposed scheme has the best results with 2570 (4.83% from available) lookup tables (LUTs), 872 Slices (6.55%), 2570 flip-flops (FF), 111 DSP multipliers (50,45%) and no block RAMs, all at maximum frequency of 553.09 MHz. These results are never achieved before, which confirms the novelty and suitability of the proposed scheme.

Table 1 FPGA resource utilization comparison

Proposals	Resources avilable	LUTs 53200	FF 106400	DSP 220	Slices 13300	Fmax(MHz)
[1]	–	2986	NM	NM	26893	147.863
[24]	–	23173	1598	220	5111	125.50
[4]	–	17322	1598	220	5111	125.50
[22]	–	24836	NM	NM	27371	135.04
Proposed 5-D system	–	2570	918	111	872	553.09

4 Conclusion

With the development of chaotic theory and its applications in different domains, proposals of constructing new and higher dimensional chaotic systems become one rising trend.

In this paper, we proposed an extracted 5-D discrete chaotic map for key stream cipher generation. The chaotic behaviour of the proposed map is investigated using the bifurcation and the trajectory analysis.

Compared to some well know chaotic systems, the proposed 5-D map presents better properties in terms of resource consumption and achieved frequency. Moreover, the proposed 5-D chaotic map provides an attractive trade-off between key space, resource utilization and memory consumption proving its suitability for securing communications of resource-constrained devices such as IoTs.

References

1. A.J. Abd El-Maksoud, A.A. Abd El-Kader, B.G. Hassan, M.A. Abdelhamed, N.G. Rihan, M.F. Tolba, L.A. Said, A.G. Radwan, M.F. Abu-Elyazeed, FPGA implementation of fractional-order chua's chaotic system, in *2018 7th International Conference on Modern Circuits and Systems Technologies (MOCAST)* (IEEE, 2018), pp. 1–4
2. M. Bakiri, C. Guyeux, J.F. Couchot, L. Marangio, S. Galatolo, A hardware and secure pseudo-random generator for constrained devices. IEEE Trans. Ind. Inform. **14**(8), 3754–3765 (2018)
3. L.L. Bonilla, M. Alvaro, M. Carretero, Chaos-based true random number generators. J. Math. Ind. **7**(1), 1–17 (2016)
4. B. Bouteghrine, M. Rabiai, C. Tanougast, S. Sadoudi, Hardware implementation of secured socket communication based on chaotic cryptosystem, in *2019 International Conference on Cyber Security and Protection of Digital Services (Cyber Security)* (IEEE, 2019), pp. 1–4
5. B. Bouteghrine, C. Tanougast, S. Sadoudi, Design and FPGA implementation of new multidimensional chaotic map for secure communication. J. Circ. Syst. Computers **30**(15) (2021)
6. B. Bouteghrine, C. Tanougast, S. Sadoudi, Novel image encryption algorithm based on new 3-d chaos map. Multimedia Tools Appl. 1–23 (2021)
7. S. Chen, S. Yu, J. Lü, G. Chen, J. He, Design and FPGA-based realization of a chaotic secure video communication system. IEEE Trans. Circ. Syst. Video Technol. **28**(9), 2359–2371 (2017)

8. M. Farajallah, S. El Assad, O. Deforges, Fast and secure chaos-based cryptosystem for images. Int. J. Bifurcation Chaos **26**(02), 1650021 (2016)
9. B. Hammi, R. Khatoun, S. Zeadally, A. Fayad, L. Khoukhi, IoT technologies for smart cities. IET Networks **7**(1), 1–13 (2017)
10. P.N. Khade, M. Narnaware, 3d chaotic functions for image encryption. Int. J. Computer Sci. Issues (IJCSI) **9**(3), 323 (2012)
11. Q. Lai, P.D.K. Kuate, F. Liu, H.H.C. Iu, An extremely simple chaotic system with infinitely many coexisting attractors. IEEE Trans. Circ. Syst. II: Express Briefs **67**(6), 1129–1133 (2019)
12. J. Li, H. Liu, Colour image encryption based on advanced encryption standard algorithm with two-dimensional chaotic map. IET Inform. Secur. **7**(4), 265–270 (2013)
13. X. Li, J. Niu, M.Z.A. Bhuiyan, F. Wu, M. Karuppiah, S. Kumari, A robust ECC-based provable secure authentication protocol with privacy preserving for industrial internet of things. IEEE Trans. Ind. Inform. **14**(8), 3599–3609 (2017)
14. T. Nandy, M.Y.I.B. Idris, R.M. Noor, M.L.M. Kiah, L.S. Lun, N.B.A. Jumaat, I. Ahmedy, N.A. Ghani, S. Bhattacharyya, Review on security of internet of things authentication mechanism. IEEE Access **7**, 151054–151089 (2019)
15. A.K. Prusty, A. Pattanaik, S. Mishra, An image encryption & decryption approach based on pixel shuffling using Arnold cat map & Henon map, in *2013 International Conference on Advanced Computing and Communication Systems* (IEEE, 2013), pp. 1–6
16. S. Roy, S. Chatterjee, A.K. Das, S. Chattopadhyay, S. Kumari, M. Jo, Chaotic map-based anonymous user authentication scheme with user biometrics and fuzzy extractor for crowdsourcing internet of things. IEEE Internet Things J. **5**(4), 2884–2895 (2017)
17. S. Sadoudi, C. Tanougast, Robust hyperchaotic synchronization via analog transmission line. Eur. Phys. J. Special Topics **225**(1), 119–126 (2016)
18. S. Sheela, K. Suresh, D. Tandur, Chaos based speech encryption using modified henon map, in *2017 Second International Conference on Electrical, Computer and Communication Technologies (ICECCT)* (IEEE, 2017), pp. 1–7
19. M. Sobhy, A.E.D. Shehata, Secure e-mail and databases using chaotic encryption. Electron. Lett. **36**(10), 875–876 (2000)
20. A. Suliman, Z. Husain, M. Abououf, M. Alblooshi, K. Salah, Monetization of IoT data using smart contracts. IET Networks **8**(1), 32–37 (2018)
21. J.R. Wallrabenstein, Practical and secure IoT device authentication using physical unclonable functions, in *2016 IEEE 4th International Conference on Future Internet of Things and Cloud (FiCloud)* (IEEE, 2016), pp. 99–106
22. F. Yu, L. Li, B. He, L. Liu, S. Qian, Y. Huang, S. Cai, Y. Song, Q. Tang, Q. Wan et al., Design and FPGA implementation of a pseudorandom number generator based on a four-wing memristive hyperchaotic system and Bernoulli map. IEEE Access **7**, 181884–181898 (2019)
23. F.Yu, L.Li, Q.Tang, S. Cai, Y. Song, Q. Xu, A survey on true random number generators based on chaos. Discrete Dyn Nature Socie **2019** (2019)
24. F. Yu, L. Liu, B. He, Y. Huang, C. Shi, S. Cai, Y. Song, S. Du, Q. Wan, Analysis and FPGA realization of a novel 5d hyperchaotic four-wing memristive system, active control synchronization, and secure communication application. Complexity **2019** (2019)

A Survey on Chaos-Based Cryptosystems: Implementations and Applications

Belqassim Bouteghrine, Camel Tanougast, and Said Sadoudi

Abstract Chaos theory is considered as a tool for studying the systems that show divergence and disorder. After having used discrete mathematics to deduce non-convergence situations, these theories are modeled in the form of a dynamic system and are applied in several domains such as electronic, mechanic, network security, etc. In network security domain, the development of new cryptosystems based on chaos is a relatively new area of research and is increasingly relevant. The essence of the theoretical and practical efforts in this field derive from the fact that these cryptosystems are faster than conventional methods, while ensuring performance of security, at least similar. In this paper, we discuss several proposals about chaos-based cryptosystem and pseudo-random number generator (PRNG). Moreover, topology and architecture of the proposed chaos systems are detailed. Finally, in order to show the more suitable system for encryption and secure communication, a synthesis comparison is presented and considered.

Keywords Chaos · Network security · Crypto-systems · Communication · PRNG

1 Introduction

Nowadays, network communication is vulnerable to many threats and cyber-attacks and it becomes more important for network experts to safeguard the network access [1]. Among the available security mechanisms, chaos-based cryptosystems are considered one of the most effective solution that provides the integrity, the authentication and the confidentiality. Recently, the development of new cryptosystems based on chaos is a relatively new area of research and is increasingly relevant.

B. Bouteghrine (✉) · C. Tanougast
LCOMS, Université de Lorraine, 57070 Metz, France
e-mail: belqassim.bouteghrine@univ-lorraine.fr

C. Tanougast
e-mail: Camel.Tanougast@univ-lorraine.fr

S. Sadoudi
Ecole Militaire Polytechnique, Algiers, Algeria

C. H. Skiadas and Y. Dimotikalis (eds.), *14th Chaotic Modeling and Simulation International Conference*, Springer Proceedings in Complexity,
https://doi.org/10.1007/978-3-030-96964-6_6

In [2], an Field-Programmable-Gate-Array (FPGA) implementation of image encryption purpose using two chaotic discrete time systems. The proposed two phases algorithm is executed by using the well known Arnold Cat map and the generalized logistic map, respectively. Authors in [3] initiate a systematic methodology for securing real-time video communication. The proposed chaos-based cryptosystem have been implemented on an FPGA hardware platform via Verilog Hardware Description Language (Verilog HDL). Sreenath and Narayanan [4] presents a Hardware implementation of a Pseudo chaos signal generator using three reconfigurable discrete time systems with a linear feedback shift registers (RLFSR). The proposed technique was implemented using Verilog HDL codes, then analyzed using Xilinx Plan Ahead compiler and Model-sim software. In terms of network security protocols, [5] proposed a novel chaos-based mechanism that includes Pseudo-Random Key-Generator which can be used to secure a socket-based communication. The proposed key-generator, created by solving the Lorenz chaos-system, has the main task of delivering at each opened channel a new 32-bit key that is used for encrypting/decryption the exchanged data.

In this paper, we discuss several proposals about chaos-based cryptosystem and pseudo-random number generator (PRNG). Moreover, topology and architecture of the proposed chaos systems are detailed. Finally, in order to show the more suitable system for encryption and secure communication, a synthesis comparison is presented and considered.

The remainder of this paper is structured as follows. Section 2 describes the classification of the most used chaotic systems. Section 3 shows the hardware implementations of these chaotic systems as well as their purposes. Section 4 concludes this paper.

2 Background and Description of Chaotic Systems

Due to the sensitivity and periodicity properties, chaotic systems have been involved mainly in key generation of the recently proposed cryptography schemes. Regarding their topology and mathematical model, we can classify all existing and newly proposed chaos systems in two main categories: continuous-time systems and discrete-time systems.

2.1 Continuous Time Systems

The continuous-time systems are described by a set of linear differential equation. Moreover, in order to ensure that the dynamical systems to be chaotic, the dimensions of the system's phase space must be at least equal to three (3). In the literature, there are several well known continuous-time systems such as Lorenz [6], Chen [7], Lu [8], etc.

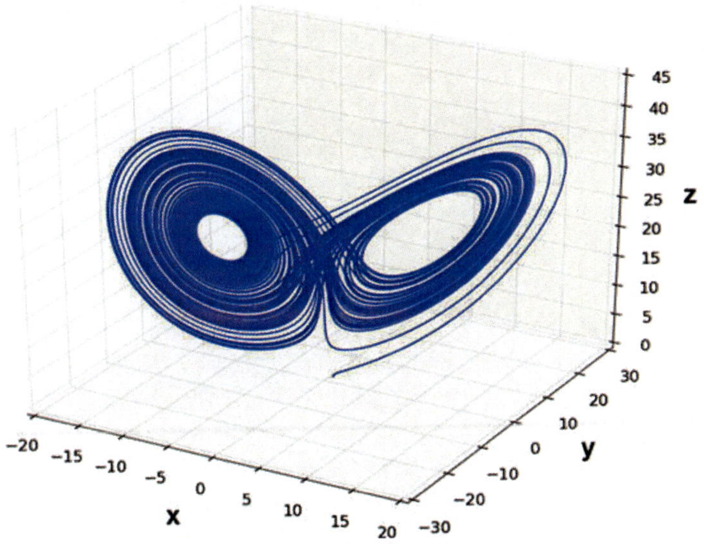

Fig. 1 Trajectory graph of the Lorenz system

Lorenz System The basic form of the Lorenz 3-D system is described by the following set of equation:

$$\dot{x} = a(y - x)$$
$$\dot{y} = y + bx - xz$$
$$\dot{z} = xy - cz$$

where x, y and z are the state variables. a, b and c are the system parameters. The chaotic behaviour (see Fig. 1) appears for a = 10, b = 28 and c = 8/3 with the initial conditions $x_0 = 0$, $y_0 = 5$ and $z_0 = 25$ [8].

Van-der-Pol System The Van-der-Pol oscillator as given in [9], is described in two dimensions as follows:

$$\dot{x} = a(x - (1/3)x^3 - y)$$
$$\dot{y} = (1/a)x$$

where x, y are the state variables, and a is the system controller. The phase portrait of the 2-D system is illustrated in Fig. 2.

Chen System Based on the 3-D Lorenz system, Chen 3-D system is proposed and described by the following set of equations:

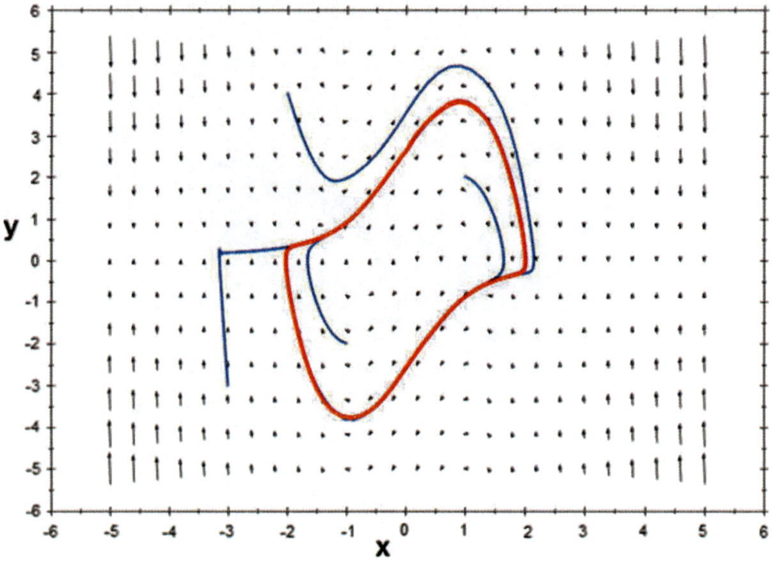

Fig. 2 Phase plan projection of the Van-der-Pol system

$$\dot{x} = a(y - x)$$
$$\dot{y} = (b - a)x + by - xz$$
$$\dot{z} = xy - cz$$

where x, y and z are the state variables. a, b and c are the system parameters. The chaotic behaviour appears for a = 35, b = 28 and c = 8/3 [10], while the phase plan projection is shown in Fig. 3.

Lu System The Lu system is known as the bridge between Lorenz system and Chen system [8]. Thereby, the mathematical model is given as follows:

$$\dot{x} = a(y - x)$$
$$\dot{y} = by - xz$$
$$\dot{z} = xy - cz$$

where x, y and z are the state variables. a, b and c are the system parameters. The trajectory graph of the proposed system is given in Fig. 4.

Linz-Sprott System Trying to simplify the formula of a chaotic system, Linz and Sprott [10] have proposed a new system which is defined as follows:

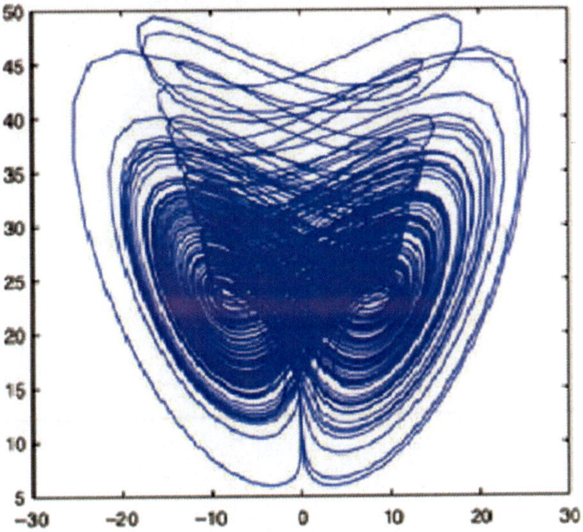

Fig. 3 Phase plan X−Y projection of the 3-D Chen system

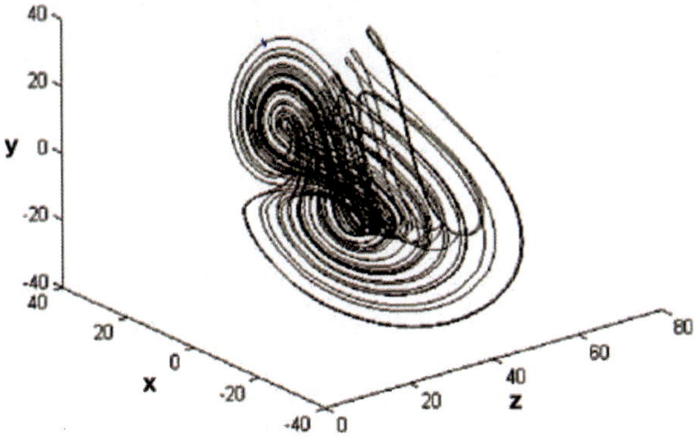

Fig. 4 Trajectory X−Y−Z of the 3-D Lu system

$$\dot{x} = y$$
$$\dot{y} = z$$
$$\dot{z} = -az - y - 1 + |x|$$

where x, y and z are the state variables and a is the system's parameter. As shown in Fig. 5, the chaotic behaviour of the proposed system is achieved for a = 0.6.

Four-Wing memristive hyperchaotic System Looking for higher dimensional chaotic system, authors in [11] have proposed a novel 4-D system which is described as follows:

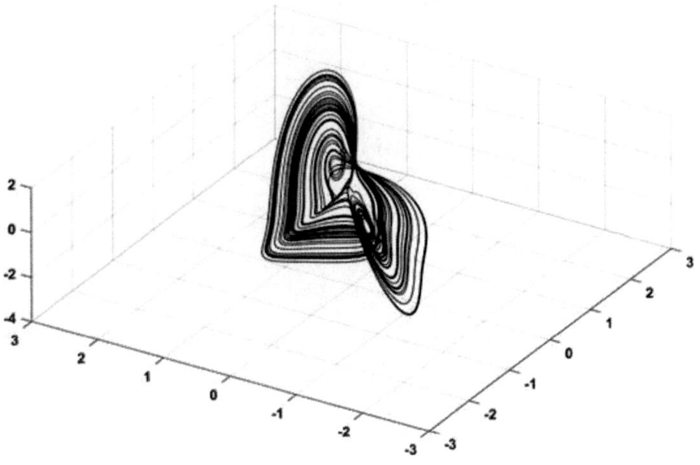

Fig. 5 Trajectory X−Y−Z of the Linz-Sprott system

$$\dot{x} = ax + byz$$
$$\dot{y} = cy + dxz - pyW(w) - Q$$
$$\dot{z} = ez + fxy + gxw$$
$$\dot{w} = -y$$
$$W(w) = m + 3nw^2$$

where x, y, z, w are the state variables. a, b, c, d, e, f, g, m, n, p, Q are the controllers of the proposed system. In order to ensure the chaotic behaviour, the controllers parameters are defined as follows: a = 0.35, b = −10, c = −0.6, d = 0.3, e = −1.6, f = 2, g = 0.1, m = 0.1, n = 0.01, p = 0.2 and Q = 0.01. The trajectory graphs corresponding to the proposed system with the associated parameters, are shown in Fig. 6.

New 3-D Continuous Time System Getting inspired from the Lorenz system [12], with only two (02) controllers, a novel 3-D system is proposed and defined as follows:

$$\dot{x} = y - x - az$$
$$\dot{y} = xz - x$$
$$\dot{z} = -xy - y + b$$

where x, y and z are the state variables. a and b are the system parameters. The chaotic behaviour of the proposed system is observed for the values $a = 0.5$ and $b = 1$ while the initial conditions are $x_0 = y_0 = z_0 = 0$ (see Fig. 7).

New 4-D Continuous Time System In [13], another new 4-D chaotic system is proposed based on the Rossler system, and defined by the following set of equations:

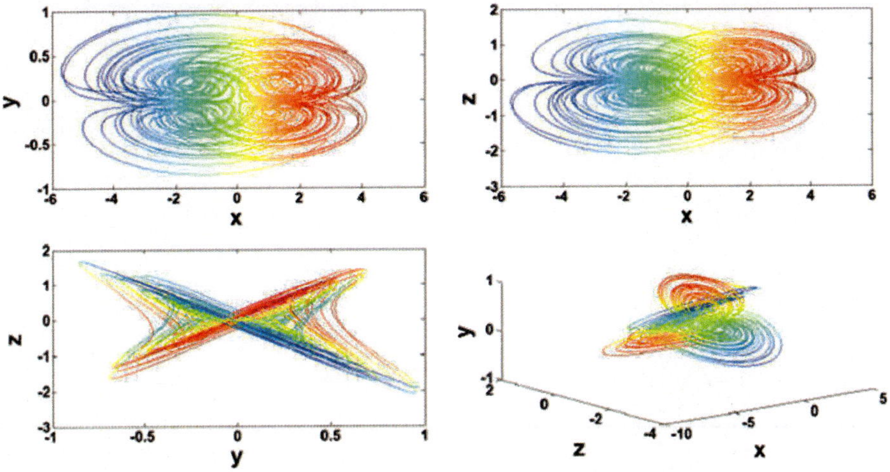

Fig. 6 Trajectory graphs of the proposed 4-Wing system

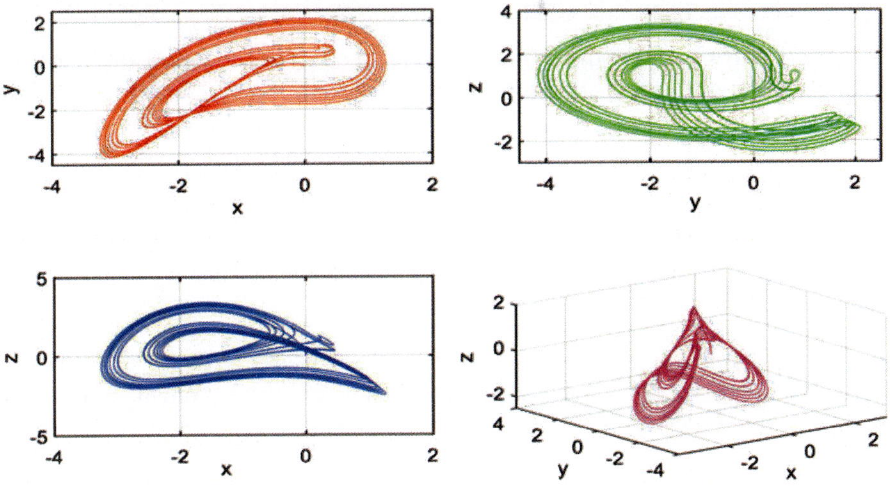

Fig. 7 Phase plan projections of the proposed 3-D system

$$
\begin{aligned}
\dot{x} &= -y - z + dw \\
\dot{y} &= x + ay \\
\dot{z} &= b + z(x - c) - a(y - w) \\
\dot{w} &= az - w
\end{aligned}
$$

where x, y, z and w are the state variables. a, b, c and d are the system parameters. By choosing a = 0.4, b = 0.6, c = 3 and d = 0.8, the chaotic behaviour of the proposed system is showed by phase plan projection (see Fig. 8).

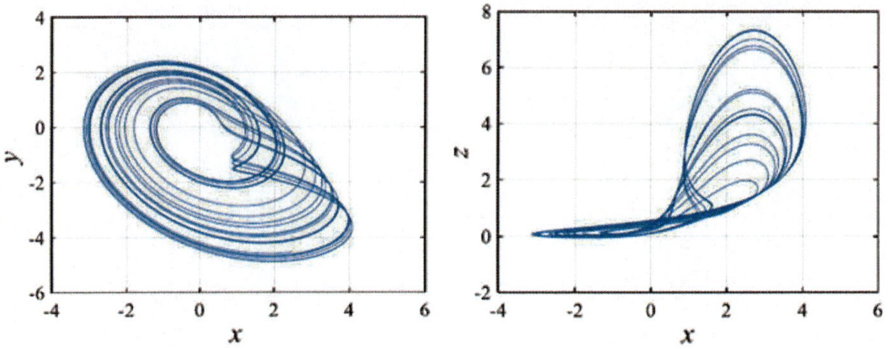

Fig. 8 Phase plan projection of the proposed system

2.2 Discrete Time Systems

The discrete time chaos system is a dynamic system which works in increments and takes the conditions at a given time t to change these conditions at a later time $t + \Delta t$. Hence, unlike the mathematical model of the continuous time systems, discrete time maps are described mathematically by an iterated function. Moreover, the dimension of the system's phase space could be only equal to one (01) to show chaos behaviour.

Logistic Map In the literature, many proposals have used the well known logistic map such as in [14] for PRNG, [2] for image encryption,[15] for chaotic signal generating, etc. The mathematical description of this map is given as follows:

$$x_{i+1} = ax_i(1 - x_i)$$

where x_i is the state variable and a is the system controller.

To ensure the chaotic behaviour (see Figs. 9 and 10) of this system, a should be in the interval $[3.57 - 4]$.

Hénon Map A simple 2-D with quadratic non-linearity, Hénon system was the first map to show strange attractor with a fractal structure [16]. The mathematical description of this map is given as follows:

$$x_{i+1} = a + y_i - x_i^2$$
$$y_{i+1} = bx_i$$

where x_i and y_i are the state variables and a, b are the system controllers.

The obtained strange attractor of this map, is shown in Fig. 11 while the controllers are a = 1.4 and b = 0.3.

Rene-Lozi Map By introducing the absolute value in the Hénon map, the Rene-Lozi map used in [17] for stream cipher purpose, is described as follows:

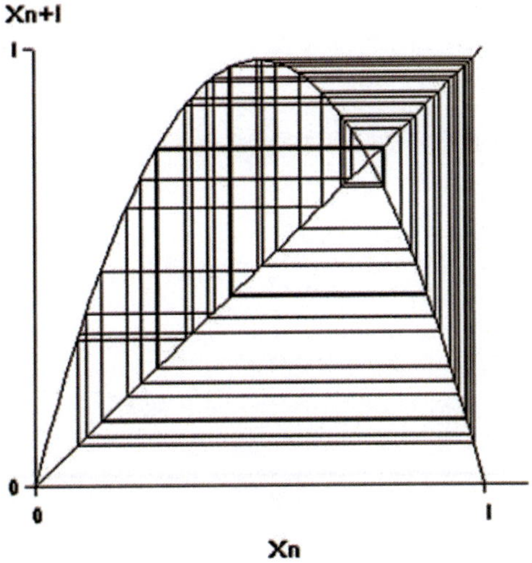

Fig. 9 Trajectory graph of the logistic map

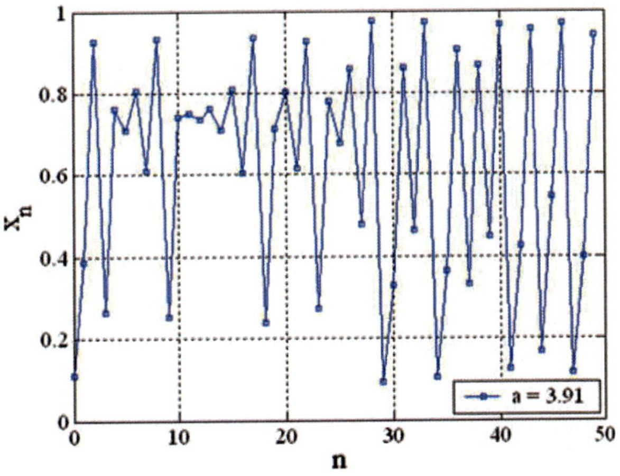

Fig. 10 Signal graph of the logistic map

$$x_{i+1} = 1 + y_i - a|x_i|$$
$$y_{i+1} = bx_i$$

where x_i and y_i are the state variables and a, b are the system controllers.

Similarly to the Hénon map, it has been shown that for a = 1.4 and b = 0.3, chaotic behaviour of this map can appear (see Fig. 12).

Bernoulli Map Unlike all the discrete time maps, Bernoulli map is composed of two piece-wise linear parts which are separated by a discontinue space of points [11] (see Fig. 13).

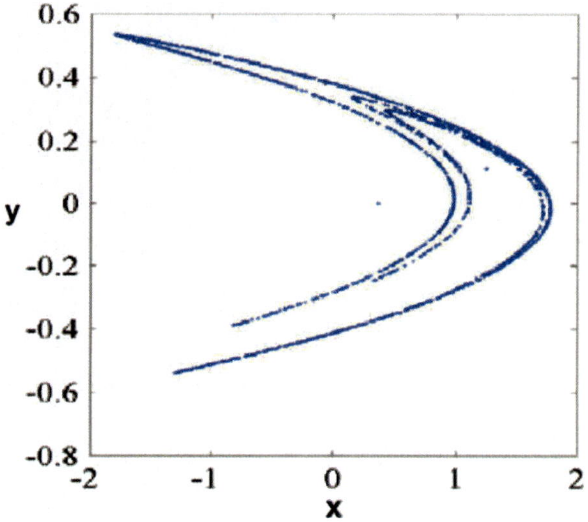

Fig. 11 Trajectory graph of the Hénon map

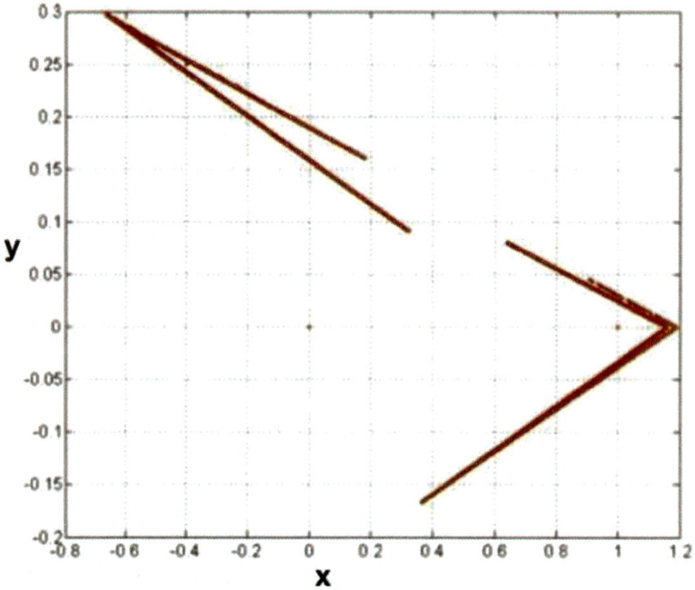

Fig. 12 Trajectory graph of the Rene-Lozi map

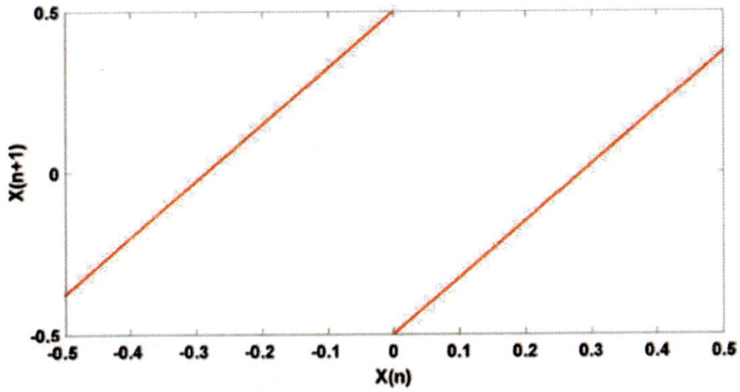

Fig. 13 Trajectory graph of the Bernoulli map

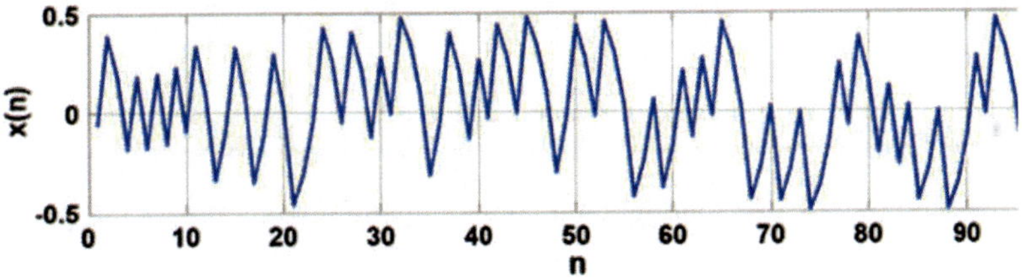

Fig. 14 Signal graph of the Bernoulli map

Mathematically, the Bernoulli map is defined as follows:

$$x_{i+1} = \begin{cases} ax_i + 0.5 & \text{if } x < 0 \\ ax_i - 0.5 & \text{if } x \geq 0, \end{cases}$$

where x_i is the state variable and a is the control parameter.
The chaotic status of this map is ensure for all the values of the parameter a inside the interval $]1.4 - 2]$ (see Fig. 14).

Sine Map The sine map is qualitatively similar to the logistic map, and the superficial similarity has resulted in a much deeper connection.

As indicated by its name, the sine map is defined by a sine function as follows:

$$x_{i+1} = a \sin(\pi x_i), \quad 0 \leq x_i \leq 1, \; a > 0$$

where x_i is the state variable and a is the system parameter. The projection graph which proves the behaviour of this map is shown in Fig. 15.

Tent Map Regarding the slope of its mathematical function, tent map with only one state variable, is considered as a slope of two (02) model. Without any control parameter, the tent map is defined as follows:

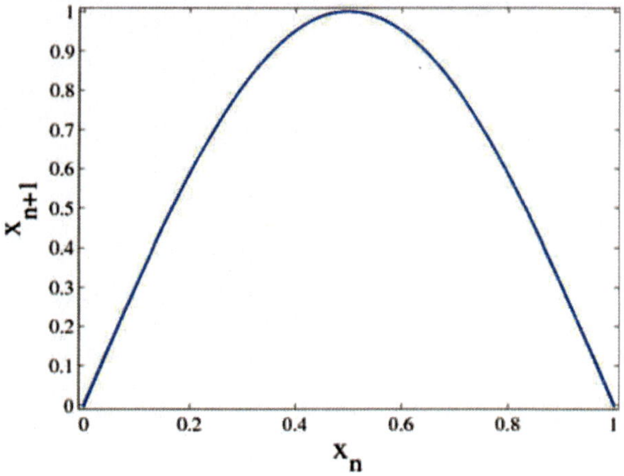

Fig. 15 Trajectory graph of the sine map

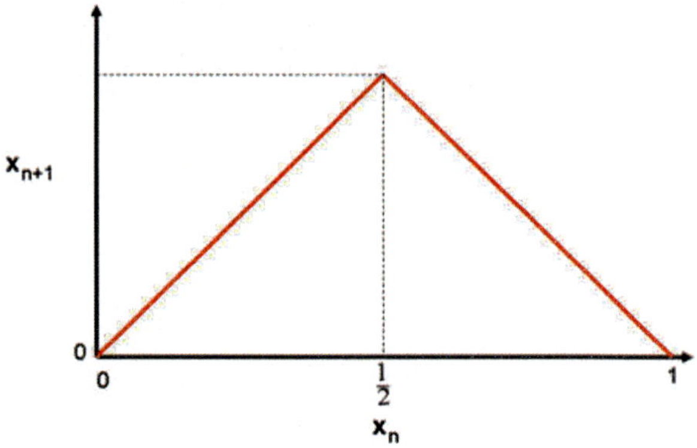

Fig. 16 Trajectory graph of the tent map

$$x_{i+1} = \begin{cases} 2x_i & \text{if } 0 \leq x_i < 1/2 \\ 2(1 - x_i) & \text{if } 1/2 \leq x_i \leq 1 \end{cases}$$

where x_i is the state variable. Moreover, the trajectory graph of the tent map is shown in Fig. 16.

All these systems have been used mainly for either generating random numbers, cipher keys or chaotic signals. They differ from each other in terms of dimension, control parameters and the purpose of use. In Table 1 we summarize all these differences obtained regarding our study.

Table 1 Summary of the chaotic systems: description and purpose of application

System	References	Type	Dimension	Controllers	Purpose
Lorenz	[8]	Continuous	3-D	3	Image encryption
Van-der-Pol	[9]	Continuous	2-D	1	Random number generator
Chen	[10]	Continuous	3-D	3	Chaos signal generator
Lu	[8]	Continuous	3-D	3	Image encryption
Linz-Sprott	[10]	Continuous	3-D	1	Chaos signal generator
4-Wing	[11]	Continuous	4-D	11	Random number generator
New 3-D	[12]	Continuous	3-D	2	Random number generator
New 4-D	[13]	Continuous	4-D	4	Image processing
Logistic	[14]	Discrete	1-D	1	Random bit Generator
Hénon	[16]	Discrete	2-D	2	Encryption
Rene-Lozi	[17]	Discrete	2-D	2	Stream cipher
Bernoulli	[11]	Discrete	1-D	1	Random number generator
Sine	[15]	Discrete	1-D	1	Chaos signal generator
Tent	[18]	Discrete	1-D	0	Chaos signal generator

3 Hardware Implementations and Applications

FPGA-based prototyping is specifically geared toward meeting the design and verification demands created by the complexities of low and constrained resources devices. Moreover, FPGA-based prototyping allows designers to develop and test their systems and provides software developers early access to a fully functioning hardware platform long before silicon is available. In order to be implemented on FPGA, the continuous time systems need to be discredited numerically using some popular methods such as Euler and Runge-Kutta (RK) methods. Euler's method is a straightforward method that estimates the next point based on the rate of change at the current point and it is easy to code [19]. It is called also a single step method. While RK methods are actually a family of schemes derived in a specific style. Higher order accurate RK methods are multi-stage because they involve slope calculations at multiple steps at or between the current and next discrete time values [20]. The next value of the dependent variable is calculated by taking a weighted average of these multiple stages based on a Taylor series approximation of the solution. The weights in this weighted average are derived by solving non-linear algebraic equations which are formed by requiring cancellation of error terms in the Taylor series. Developing higher order RK methods is tedious and difficult without using symbolic tools for computation. The most popular RK method is RK4 since it offers a good balance between order of accuracy and cost of computation. RK4 is the highest order explicit Runge-Kutta method that requires the same number of steps as the order of accuracy (i.e. RK1 = 1 stage, RK2 = 2 stages, RK3 = 3 stages, RK4 = 4 stages, RK5 = 6 stages, ...). Beyond fourth order the RK methods become relatively more expensive

Table 2 Summary of the FPGA implementations

References	FPGA technology	Resources	Chaos system	Discretization
[12]	Virtex-6	LUTs = 1070 Regs = 1196	New 3-D	Euler
[9]	Virtex-6	LUTs = 22674 Regs = 21,797	Van-der-Pol	RK4
[14]	Virtex-7	LUTs = 510 Regs = 120	Logistic	No need
[17]	Spartan-6	LUTs = 562 Regs = 386	Rene-Lozi	No Need
[11]	ZYNQ-XC7Z020	LUTs = 22,556 Regs = 264,26	Four-wing	RK4
[16]	Virtex-5	LUTs = 1496 Regs = 432	Hénon	No Need
[8]	Virtex-II	LUTs = 2490 Regs = 1316	Lorenz/Lu	RK-4
[22]	Virtex-5	LUTs = 2799 Regs = 1722	Logistic	No Need
[23]	Zynq-7000	LUTs = 856 Regs = 521	Hénon	No Need
[24]	Stratix-IV	LUTs = 49,005 Regs = 611	New 3-D	Euler

to compute. Among all the studied proposals, we have synthesised a brief comparison that includes mainly the used FPGA technology and the consumed resources. Table 2 summarizes the difference between different proposals regarding the chosen system as well as the resource consumption. However, we found that in the single-precision and the double-precision operations, there are more than 10^{-6} differences in less than 100 iterations, and the difference reaches more than one digit after 1000 iterations [21]. This is because the binary has a round-off error, so the binary cannot strictly obey the commutative law or the distribution law in floating-point operations.

4 Conclusion

In this paper, we discuss several proposals about chaos-based cryptosystem and pseudo-random number generator (PRNG). Moreover, topology and architecture of the proposed chaos systems are detailed. Finally, in order to show the more suitable system for encryption and secure communication, a synthesis comparison is presented and considered.

References

1. B. Bouteghrine, M. Rabiai, C. Tanougast, S. Sadoudi, FPGA implementation of Internet key exchange based on chaotic cryptosystem, in *10th IEEE International Conference on Intelligent Data Acquisition and Advanced Computing Systems: Technology and Applications (IDAACS)* (Metz, France, 2019), pp. 384–387
2. B. Baruah, M. Saikia, An FPGA implementation of chaos based image encryption and its performance analysis. IJCSN Int. J. Comput. Sci. Netw. **5**, 5 (2016) (Unpaginated)

3. S. Chen, S. Yu, J. Lu, G. Chen, J. He, Design and FPGA-based realization of a chaotic secure video communication system. IEEE Trans. Circuits Syst. Video Technol. **28**(9), 2359–2371 (2017)
4. H. Sreenath, G. Narayanan, FPGA implementation of pseudo chaos-signal generator for secure communication systems, in *2018 International Conference on Advances in Computing, Communications and Informatics (ICACCI)* (Bangalore, India, 2018), pp. 804–807
5. B. Bouteghrine, M. Rabiai, C. Tanougast, S. Sadoudi, Hardware implementation of secured socket communication based on chaotic cryptosystem, in *2019 International Conference on Cyber Security and Protection of Digital Services (Cyber Security)* (Oxford, UK, 2019), pp. 1–4
6. A. Qi, C. Han, G. Wang, Design and FPGA realization of a pseudo random sequence generator based on a switched chaos, in *2010 International Conference on Communications, Circuits and Systems (ICCCAS)* (Chengdu, China, 2010), pp. 417–420
7. J. Lu, X. Wu, X. Han, J. Lu, Adaptive feedback synchronization of a unified chaotic system. Phys. Lett. A **329**(4–5), 327–333 (2004)
8. M.S. Azzaz et al., FPGA implementation of new real-time image encryption based switching chaotic systems, in *IET Irish Signals and Systems Conference (ISSC 2009)*, vol. 56. (Dublin. Ireland, 2009)
9. M. Tuna, A novel secure chaos-based pseudo random number generator based on ANN-based chaotic and ring oscillator: design and its FPGA implementation. Analog Integr. Circ. Sig. Process **105**(2), 167–181 (2020)
10. A. Senouci et al., FPGA based hardware and device-independent implementation of chaotic generators. AEU-Int. J. Electron. Commun. **82**, 211–220 (2017)
11. F. Yu et al., Design and FPGA implementation of a pseudo-random number generator based on a four-wing memristive hyperchaotic system and Bernoulli map. IEEE Access **7**, 181884–181898 (2019)
12. I. Koyuncu et al., Design, FPGA implementation and statistical analysis of chaos-ring based dual entropy core true random number generator. Analog Integr. Circ. Sig. Process. **102**(2), 445–456 (2020)
13. C.H. Yang, Y.S. Chien, FPGA implementation and design of a hybrid Chaos-AES color image encryption algorithm. Symmetry **12**(2), 189 (2020)
14. M. Garcia-Bosque et al., Chaos-based bitwise dynamical pseudorandom number generator on FPGA. IEEE Trans. Instrum. Measur. **68**(1), 291–293 (2018)
15. Z. Hua, B. Zhou, Y. Zhou, Sine-transform-based chaotic system with FPGA implementation. IEEE Trans. Industr. Electron. **65**(3), 2557–2566 (2017)
16. A.M. Atteya, A.H. Madian, A hybrid Chaos-AES encryption algorithm and its implemention based on FPGA, in *IEEE 12th International New Circuits and Systems Conference (NEWCAS)*, vols. 217–220 (Quebec, Canada, 2014)
17. L. Merah et al., FPGA hardware co-simulation of new chaos-based stream cipher based on Lozi map. Int. J. Eng. Technol. **9**(5), 420–425 (2017)
18. Y. Mao, L. Cao and W. Liu, Design and FPGA implementation of a pseudo-random bit sequence generator using spatiotemporal chaos, in *2006 International Conference on Communications, Circuits and Systems* (Guangzi, China, 2006), pp. 2114–2118
19. D.I. Lanlege et al., Comparison of Euler and Range-Kutta methods in solving ordinary differential equations of order two and four. Leonardo J. Sci. **32**, 10–37 (2018)
20. S.C. Palligkinis, G. Papageorgiou, I.T. Famelis, Runge-Kutta methods for fuzzy differential equations. Appl. Math. Comput. **209**, 97–105 (2009)
21. C.H. Yang, H.C. Wu, S.F. Su, Implementation of encryption algorithm and wireless image transmission system on FPGA. IEEE Access **7**, 50513–50523 (2019)
22. P. Dabal, R. Pelka, A chaos-based pseudo-random bit generator implemented in FPGA device, in *14th IEEE International Symposium on Design and Diagnostics of Electronic Circuits and Systems* (Cottbus, Germany, 2011), pp. 151–154

23. R. Hobincu, O. Datcu, FPGA implementation of a chaos based PRNG targetting secret communication, International Symposium on Electronics and Telecommunications (ISETC), Timisoara. Romania **1–4**, 2018 (2018)
24. E. Tlelo-Cuautle et al., FPGA realization of a chaotic communication system applied to image processing. Nonlinear Dyn. **82**(4), 1879–1892 (2015)

Approximate Methods for Solving Hypersingular Integral Equations on Fractals

Ilya V. Boykov, Alla I. Boykova, Alexander A. Potapov, and Alexander E. Rassadin

Abstract The paper consists of three parts. The first one is devoted to approximate methods for evaluating Riemann integrals, singular and hypersingular integrals on closed non-rectifiable curves and fractals in the complex plane. An integral on non-rectifiable curves or fractals is defined as a double integral over a region that bounded by a non-rectifiable curve or a fractal. To evaluate double integral cubature formulas have been constructed. The second part contains methods for solving hypersingular integral equations on prefractals. Issues of solvability of singular and hypersingular integral equations with fractal in the right-hand side have been studied in the third part. Singular and hypersingular integral equations that model aerodynamics problems have been investigated. In such cases right-hand side of equations describes the gas flow which is a fractal.

Keywords Singular integrals · Hypersingular integrals · Singular integral equations · Hypersingular integral equations · Fractals · Numerical methods

---

I. V. Boykov · A. I. Boykova
Penza State University, Krasnay Str., 40, 440026 Penza, Russia
e-mail: i.v.boykov@gmail.com

A. A. Potapov (✉)
V.A. Kotelnikov Institute of Radio Engineering and Electronics, Russian Academy of Sciences, Mokhovaya Str., 11-7, Moscow 125009, Russia
e-mail: potapov@cplire.ru

JNU-IREE RAS Joint Lab. of Information Tekhnology and Fraktal Processing of Signals, JiNan University, Guangzhou 510632, China

A. E. Rassadin
Higher School of Economics, Bolshaya Pecherskaya Str., 25/12, Nizhny Novgorod 603155, Russia
e-mail: brat_ras@list.ru

C. H. Skiadas and Y. Dimotikalis (eds.), *14th Chaotic Modeling and Simulation International Conference*, Springer Proceedings in Complexity,
https://doi.org/10.1007/978-3-030-96964-6_7

Abbreviations

HI	Hypersingular integral
HIE	Hypersingular integral equation
SI	Singular integral
SIE	Singular integral equation

1 Introduction

1.1 Review of Approximate Methods for Calculating Hypersingular Integrals and Solving Hypersingular Integral Equations

Starting the middle of the last century, the methods of singular and then hypersingular integral equations have been increasingly used in the study and modeling of various problems in physics, natural science and technology: in aerodynamics, electrodynamics, elasticity theory, nuclear and atomic physics, geophysics, and mathematical physics. Analytical methods for solving singular and hypersingular integral equations are known only for very special cases. Thus numerical methods are widely employed for solving singular and hypersingular integral equations [1–7].

The development of approximate methods for solving singular integral equations (SIE) started in the 50 s of the last century. The number of publications devoted to approximate methods for solving SIE and their generalizations and related Riemann and Hilbert boundary problems has not decreased up-to-date. Main approximate methods for solving SIE are presented in [1, 2], which contain extensive bibliography.

It is interesting to note that hypersingular integrals (HI) were introduced to mathematical world around the same time as singular integrals (SI). However the development of approximate methods for solving hypersingular integral equations (HIE) started later than the development of similar methods for solving SIE. Today HIE is the fast growing field in mathematics.

An intense development of approximate method for solving SIE and HIE is caused by their numerous applications. In particular, SIE and HIE are main mathematical engine in antenna theory, composite materials theory, metamaterials.

Main approximate methods for solving HIE can be found in the publications [3–7].

Effective approximate methods for evaluating SI and HI are required to implement numerical methods for solving SIE and HIE. Analytically singular and hypersingular integrals can be evaluated pretty rare. Lack of analytical methods require the development of numerical methods for evaluating SI and HI.

There are numerous publications devoted to evaluate SI and HI over smooth curves issues. The bibliography is presented in [1, 5, 8–10]. The authors do not know about works devoted to numerical methods forcalculation SI and HI and solution SIE and HIE over fractals.

Recently the need for study of physical and technical processes on fractals has appeared. First, it should be noted synthesis and analysis of the fractal antenna problems [11], and the microwave theory and technique. It is important to know, different antenna types are modeled by SIE and HIE. Obviously, the development of approximate methods to solve SIE and HIE on non-rectifiable curves and fractals for modeling electrodynamic processes in fractal antennas will be required.

The chapter is devoted to approximate methods for calculating singular and hypersingular integrals and solving singular and hypersingular integral equations over non-rectifiable curves and fractals.

1.2 Definitions

Let L be a contour on the complex plane. Let $A = [a, b]$ or $A = L$.

Definition 1.1. Class of Holder functions $H_\alpha(M; A)(0 < \alpha \leq 1)$ consists of functions $f(x)$ given on A and satisfying at all points x' and x'' of this set the inequality $|f(x') - f(x'')| \leq M|x' - x''|^\alpha$.

Definition 2.2. The class $W^r(M; A)$ consists of functions defined on A, continuous and having continuous derivatives up to $(r-1)$-th order inclusive and piecewise continuous derivative r-th order satisfying on this set the inequality $|f^{(r)}(x)| \leq M$.

Definition 1.3. The class $W^r H_\alpha(M; A)$ consists of functions $f(x)$ belonging to the class $W^r(M; A)$ and satisfying the additional condition $f^{(r)}(x) \in H_\alpha(M)$.

Definition 1.4 [12]. Let $\varphi(t) \in W^{p-1}H_\alpha(M, A)$. The Integral $\int_a^b \frac{\varphi(\tau)d\tau}{(\tau-c)^p}$, $a < c < b$, $p = 2, 3, \ldots$, in the sense of Cauchy—Hadamard principal value is called the limit: $\int_a^b \frac{\varphi(\tau)d\tau}{(\tau-c)^p} = \lim_{v\to 0}\left[\int_a^{c-v} \frac{\varphi(\tau)d\tau}{(\tau-c)^p} + \int_{c+v}^b \frac{\varphi(\tau)d\tau}{(\tau-c)^p} + \frac{\xi(v)}{v^{p-1}}\right]$, here $\xi(v)$ is a function satisfied the conditions: (1) the limit exists; (2) $\xi(v)$ has a continuous $p-1$ degree derivative at a neighborhood of zero.

Let us give the definitions of SI and HI on a closed non-rectifiable curves and fractals.

Let γ be a simple closed curve in the complex plane forming the boundary of D, and D^+ and D^- be interior and exterior domains respectively. If $u(z)$ is continuous in $\overline{D}^+$ and has intagrable partial derivatives in в D^+, Stokes' formular occurs $\int_\gamma u(z)dz = -\int\int_{D^+} \frac{\partial u}{\partial \overline{z}} \partial z d\overline{z}$, where $\overline{z} = x - iy$, $\frac{\partial}{\partial \overline{z}} = \frac{1}{2}(\frac{\partial}{\partial x} + i\frac{\partial}{\partial y})$.

This formula allows you to enter the definition of the integral over non-rectifiable curves and fractals:

$$\int_\gamma u(z)dz = -\int\int_{D^+} \frac{\delta \widetilde{u}(z)}{\delta \overline{z}} dz d\overline{z}, \tag{1.1}$$

here $\widetilde{u}(z)$ is a continuation of $u(z)$ to the region D^+.

The [13] notes that this definition belongs to Whitney.

There are many methods of continuation. Here Whitney's continuation has been used [14], p.205.

The Whitney operator has the following properties:

(1) if the function $u(z) \in H_\lambda(\gamma)$, $z \in \gamma$, then its extension $\tilde{u}(z)$ satisfies the Holder condition in D^+;

(2) in $C\backslash\gamma$, the continuation of $\tilde{u}(z)$ satisfies the estimate $|grad\tilde{u}(z)| \leq C(dist(z,\gamma))^{\lambda-1}$.

Stein [14] shows that Definition (1.1) does not depend on Whitney's operator selection. Thus for any $\widetilde{u}_1(z)$ and $\widetilde{u}_2(z)$ Whitney's continuation appears

$$\int\int_{D^+} \frac{\partial \widetilde{u}_1(z)}{\partial \overline{z}} dz d\overline{z} = \int\int_{D^+} \frac{\partial \widetilde{u}_2(z)}{\partial \overline{z}} dz d\overline{z}.$$

It is known [15] the integral $\int\int_{D^+} \frac{\partial \widetilde{u}(z)}{\partial \overline{z}} dz d\overline{z}$ exists for $\lambda > \alpha(\gamma)-1$, where $\alpha(\gamma)$ is the cell dimension of the curve γ.

Definition 1.5. If a closed curve γ has a cell dimension $\alpha(\gamma)$, $f \in H_\lambda(\gamma)$ and $\lambda > \alpha(\gamma)-1$ occurs, then $\int_\gamma f(z)dz = -\int\int_{D^+} \frac{\partial \widetilde{f}(z)}{\partial \overline{z}} dz d\overline{z}$, where $\widetilde{f}(z)$ is a Whitney continuation for f.

In case of singularity of f, $f = f_0 v$, here $f_0 \in H_\lambda(\gamma)$, $f_0(t) = 0$, and $|v(z)| \leq c|z-t|^{-1}$, $|\partial v/\partial \overline{z}| \leq c|z-t|^{-1}$, $z \in \overline{D^+}\backslash t$, then it occurs.

Definition 1.6 [16]. If a closed curve γ has a cell dimension $\alpha(\gamma)$, the inequalities $|v(z)| \leq c|z-t|^{-1}$, $|\partial v/\partial \overline{z}| \leq c|z-t|^{-1}$, $z \in \overline{D^+}\backslash t$ and $\lambda > (\alpha(\gamma))/2$ are satisfied then $\int_\gamma f(z)dz = -\int\int_{D^+} \frac{\partial(v\widetilde{f}_0(z))}{\partial \overline{z}} dz d\overline{z}$, where $\widetilde{f}_0(z)$ is any Whitney continuation for f.

Consider the singular integral $S_\gamma f = \frac{1}{\pi i}\int_\gamma \frac{f(\tau)d\tau}{\tau - t}$, $t \in \gamma$.

If γ is a smooth curve, $S_\gamma f$ is regularized by

$$(S_\gamma f)(t) = \frac{1}{\pi i}\int_\gamma \frac{f(\tau)-f(t)}{\tau - t} d\tau + \frac{f(t)}{\pi i}\int_\gamma \frac{d\tau}{\tau - t} = \frac{1}{\pi i}\int_\gamma \frac{f(\tau)-f(t)}{\tau - t} d\tau + f(t).$$

It yields us to the following statement.

Definition 1.7 [16]. A singular integral $S_\gamma f$ over a closed non-rectifiable curve is defined by $(S_\gamma f)(t) = f(t) - \frac{1}{\pi i}\int\int_{D^+} \frac{\partial(\widetilde{f}(z))}{\partial \overline{z}} \frac{1}{z-t} dz d\overline{z}$, where $\widetilde{f}(z)$ is a Whitney continuation for f.

Consider the hypersingular integral $\frac{1}{\pi i}\int_\gamma \frac{f(\tau)}{(\tau - t)^p} d\tau$, $t \in \gamma$, $p = 2, 3, \ldots$.

In case of a smooth closed curve γ a hypersingular integral on the complex plane C is defined by

$$\frac{1}{\pi i}\int\limits_{\gamma}\frac{f(\tau)}{(\tau-t)^p}d\tau=\frac{1}{\pi i}\int\limits_{\gamma}\frac{g(\tau,t)}{\tau-t}d\tau+f(t)\frac{1}{\pi i}\int\limits_{\gamma}\frac{d\tau}{(\tau-t)^p}+$$
$$+\frac{f'(t)}{1!}\frac{1}{\pi i}\int\limits_{\gamma}\frac{d\tau}{(\tau-t)^{p-2}}+\ldots+\frac{f^{(p-1)}(t)}{(p-1)!}\frac{1}{\pi i}\int\limits_{\gamma}\frac{d\tau}{\tau-t}=\frac{1}{\pi i}\int\limits_{\gamma}\frac{g(\tau,t)}{\tau-t}d\tau+\frac{1}{(p-1)!}f^{(p-1)}(t),$$

where $g(\tau,t)=(f(\tau)-f(t)-\frac{f'(t)}{1!}(\tau-t)-\ldots-\frac{f^{(p-1)}(t)}{(p-1)!}(\tau-t)^{p-1})/(\tau-t)^{p-1}$.
Using the last formula the definition follows.

Definition 1.8. A hypersingular integral over a non-rectifiable curve γ is defined by $\frac{1}{\pi i}\int_{\gamma}\frac{f(\tau)}{(\tau-t)^p}d\tau=\frac{f^{(p-1)}(t)}{(p-1)!}-\frac{1}{\pi i}\int\int_{D^+}\frac{\partial\tilde{g}(z)}{\partial\bar{z}}\frac{1}{z-t}dzd\bar{z}$, where $\tilde{g}(z)-$ is a Whitney continuation for $g(\tau,t)$.

On a smooth closed curve γ, the hypersingular integral is also defined by the expression $\frac{1}{\pi i}\int_{\gamma}\frac{f(\tau)}{(\tau-t)^p}d\tau=\frac{1}{(p-1)!}\frac{d^{p-1}}{dt^{p-1}}\frac{1}{\pi i}\int_{\gamma}\frac{f(\tau)}{\tau-t}d\tau$.
Using this formula, we arrive at the following definition.

Definition 1.9. The hypersingular integral over the non-rectifiable contour or fractal γ is defined by the formula $\frac{1}{\pi}\int_{\gamma}\frac{f(\tau)}{(\tau-t)^p}d\tau=\frac{f^{(p-1)}(t)}{(p-1)!}-\frac{1}{(p-1)!}\frac{d^{p-1}}{dt^{p-1}}\frac{1}{\pi i}\int\int_{D^+}\frac{\partial\tilde{f}(z)}{\partial\bar{z}}\frac{1}{z-t}dzd\bar{z}$, where $\tilde{f}(z)-$ any continuation of a Whitney type function $f(z)$.

It is easy to see that Definitions 1.8 and 1.9 are equivalent.

2 Approximate Calculation of Integrals on Fractals

2.1 Riemann Integrals

Let $f(z)\in H_{\lambda}(\gamma)$,$\gamma$ be closed non-rectifiable curve.
There are two possibilities:

(1) $f(z)$ is defined in domain $\overline{D^+}$ with an intagrable partial derivative with respect to $\bar{z}$;
(2) $f(z)$ is defined only on γ.

In the first case, the Stokes formula is used directly to calculate the integral $\int_{\gamma}f(z)dz$. One can put $\int_{\gamma}f(z)dz=-\int\int_{D^+}\frac{\partial f}{\partial\bar{z}}dzd\bar{z}$ and the problem is reduced to constructing a cubature formula for calculating the integral on the right.

In the second case, it is necessary to continue the function $f(z)$, $z\in\gamma$, to the domain D^+. If $f(z)\in H_{\lambda}(\gamma)$, $0<\lambda<1$, then as the continuation operator we can take the zero Whitney operator [14], p. 204, associating the function $f(z)$ with the function $\tilde{f}(z)=\xi_0(f)$, $z\in D^+$, by formula $\int_{\gamma}f(z)dz=-\int\int_{D^+}\frac{\partial\tilde{f}(z)}{\partial\bar{z}}dzd\bar{z}$.

The specific function depends on the choice of the basic infinitely differentiable function φ^*, defined on a unit square. It is known [14] that the formula is valid for any basis functions.

Constructing the function $u(z)$, $z \in D^+$ given in [14] is rather complex. Below it is presented a numerical method for evaluating the integral $\int_\gamma f(z)dz$.

Let $G = [a, b; c, d]$, $\overline{D^+} \in G$. Let h be a grid of a cubature formula. For simplicity assume $(b-a)/h = m_1$, $(d-c)/h = m_2-$ are integer. Let $x_k = a + hk, k = 0, 1, ..., m_1;\ y_l = c + hl, l = 0, 1, ..., m_2$ be nodes. By z_{kl} denote the node $z_{kl} = x_k + iy_l, k = 0, 1, ..., m_1, l = 0, 1, ..., m_2$.

Let $\Delta_{kl} = [x_k, x_{k+1}; y_l, y_{l+1}], k = 0, 1, ..., m_1 - 1, l = 0, 1, ..., m_2 - 1$.

Fix $\varepsilon (0 < \varepsilon << h)$. Assign each point z_{kl} a point $p_{kl} \in \gamma$ attains the distance from z_{kl} to γ. Since it is rather difficult to find an accurate location of p_{kl}, it is sufficient to select any point $p'_{kl} \in \gamma$ in $B(p_{kl}, \varepsilon)$. Assume $u(z_{kl}) = f(p'_{kl}), k = 0, 1, ..., m_1 - 1, l = 0, 1, ..., m_2 - 1$.

Fix an arbitrary $l = 0, 1, ..., m_2$ and assume a sequence $u(z_{kl}), k = 0, 1, ..., m_1$. Using it we will calculate the derivative $\frac{\partial u(z)}{\partial x}|_{k,l}$. There are various methods for derivatives calculating. The method based on hypersingular integrals is used below [17].

Consider the quadrature formula [17]

$$\frac{\delta^r u(t, y_l)}{\delta t^r} = \frac{r!}{2\pi i} \sum_{k=0}^{m_1-1} u(x_k, y_l) \left[\int_{x_k}^{x_{k+1}} \frac{d\tau}{(\tau - (t + i\delta))^{r+1}} - \int_{x_k}^{x_{k+1}} \frac{d\tau}{(\tau - (t - i\delta))^{r+1}} \right] + R_{m_1}(u).$$

This formula allows you to calculate the derivatives of any finite order and has a sufficiently high accuracy and stability. The regularization parameter is δ.

Similarly using the sequence $u(z_{kl}), k = 0, 1, ..., m_2$ derivatives $\frac{\partial u(x_k, y)}{\partial y}|_{k,l}$ are calculated.

Each node z_{kl} is assigned in the complex number $\frac{\partial u(x,y)}{\partial \bar{z}}|_{x=x_k, y=y_l} = (\frac{\partial u(x,y_l)}{\partial x}|_{x=x_k} + i\frac{\partial u(x_k,y)}{\partial y}|_{y=y_l})/2, k = 0, 1, ..., m_1 - 1, l = 0, 1, ..., m_2 - 1$.

By Δ^*_{kl} denote rectangles Δ_{kl} having no intersection with γ. Let $\Omega_* = \bigcup_{k,l} \Delta^*_{kl}$. Define the function

$$w_{k,l}(z) = \begin{cases} \left(\frac{\partial u(x, y)}{\partial x}|_{x_k, y_l} + i \frac{\partial u(x, y)}{\partial y}|_{x_k, y_l} \right) / 2, z \in \Delta^*_{kl}, \\ 0, z \in G \backslash \Delta^*_{k,l}, k = 0, 1, \ldots, m_1 - 1, l = 0, 1, \ldots, m_2 - 1. \end{cases}$$

Let $w(z) = \sum_{k,l} w_{k,l}(z)$.

To evaluate the integral $\int_\gamma f(z)dz$ the following formula is used

$$\int_{\gamma} f(z)dz = -\sum_{k=0}^{m_1-1}\sum_{l=0}^{m_2-1^*} w_{k,l}\int\int_{\Delta_{k,l}} dzd\bar{z} + R_{m_1,m_2}(f),$$

where $\sum\sum^*$ means summation over the rectangles $\Delta_{k,l}$, included in Ω_*.

2.2 Singulal and Hypersingular Integrals. Consider the Integral

$$\frac{1}{\pi i}\int_{\gamma}\frac{f(\tau)}{\tau - t}d\tau = f(t) - \frac{1}{\pi i}\int\int_{\Omega^+}\frac{\partial\tilde{f}(z)}{\partial\bar{z}}\frac{d\tau}{z - t}.$$

where $\tilde{f}(z)$ is a Whitney's continuation for $f(\tau) - f(t)$.

Similarly above, construct a function $w(z)$ approximating $\frac{\partial\tilde{f}(z)}{\partial\bar{z}}$. Then a cubature formular for calculating singular integral should be constructed using results in [8]. The calculation of the HI is carried out according to the formula

$$\frac{1}{\pi i}\int_{\gamma}\frac{f(\tau)}{(\tau - t)^p}d\tau = \frac{f^{(p-1)}(t)}{(p-1)!} - \frac{1}{\pi i}\int\int_{\Omega^+}\frac{\partial\tilde{g}(z)}{\partial\bar{z}}\frac{d\tau}{z - t},$$

where $\tilde{g}(z)$ is a Whitney's continuation for $g(\tau, t)$. The function is given in Definition 1.8.

Then we construct a function $w(z)$, approximating $\frac{\partial\tilde{g}(z)}{\partial\bar{z}}$, and a cubature formula for calculating the singular integral.

3 Approximate Solution of Hypersingular Integrals Over Prefractals

Let C_n be n th prefractal of the Cantor set (in other words n th Cantor set iteration). Consider hypersingular integral equation

$$a(t)x(t) + b(t)\int_{C_n}\frac{x(\tau)}{(\tau - t)^p} + \int_{C_n} h(t,\tau)x(\tau)d\tau = f(t), t \in C_n. \tag{3.1}$$

Remark. The problem might be discussed for $t \in [0, 1]$.

Two approximation schemes for solution of the (3.1) have been introduced and justified in [18].

Let's imagine one of them. For simplicity of notation, put $h(t, \tau) = 0$.

Let $p = 2$.

An approximate solution of the (3.1) is sought in the form of a spline

$$x_n(t) = \sum_{l=1}^{2} \sum_{i_1,\ldots,i_n} \alpha_{l,i_1,\ldots,i_n} \varphi_{l,i_1,\ldots,i_n}(t),\ t \in C_n, \tag{3.2}$$

$$\varphi_{l,i_1,\ldots,i_n}(t) = \begin{cases} g_{l,i_1,\ldots,i_n}(t),\ t \in \Delta_{i_1,\ldots,i_n}, \\ 0,\ t \in [0,1]\backslash\Delta_{i_1,\ldots,i_n},\ l = 1,2,\ i_j = 0,2,\ j = 1,\ldots,n, \end{cases}$$

where functions $g_{l,i_1,\ldots,i_n}(t)$ are similar to basic functions

$$g_{1,0,\ldots,0}(t) = \begin{cases} 1,\ 0 \le t \le \frac{1}{3^{2n}}, \\ \frac{3^{2n}t - 3^n + 1}{2 - 3^n},\ \frac{1}{3^{2n}} \le t \le \frac{1}{3^n} - \frac{1}{3^{2n}}, \\ 0,\ \frac{1}{3^n} - \frac{1}{3^{2n}} \le t \le \frac{1}{3^n}, \end{cases}$$

$$g_{2,0,\ldots,0}(t) = \begin{cases} 0,\ 0 \le t \le \frac{1}{3^{2n}}, \\ \frac{3^{2n}t - 1}{3^n - 2},\ \frac{1}{3^{2n}} \le t \le \frac{1}{3^n} - \frac{1}{3^{2n}}, \\ 1,\ \frac{1}{3^n} - \frac{1}{3^{2n}} \le t \le \frac{1}{3^n}. \end{cases}$$

The coefficients $\alpha_{l,i_1,\ldots,i_l}$ are found from the system of equations:

$$a(t^l_{i_1,\ldots,i_n})\alpha_{l,j_1,\ldots,j_n} + b(t^l_{i_1,\ldots,i_n}) \int\limits_{C_n} \frac{x_n(\tau)d\tau}{(\tau - t^l_{i_1,\ldots,i_n})^2} = f(t^l_{i_1,\ldots,i_n}), \tag{3.3}$$

$l = 1, 2,\ t^1_{i_1,\ldots,i_n} = \frac{2}{3}i_1 + \frac{2}{3^2}i_2 + \ldots + \frac{2}{3^n}i_n + \frac{2}{3^{2n}}i_n,,\ t^2_{i_1,\ldots,i_n} = \frac{2}{3}i_1 + \frac{2}{3^2}i_2 + \ldots + \frac{2}{3^n}i_n + \frac{1}{3^n} - \frac{2}{3^{2n}}i_n,\ i_j = 0, 2,\ j = 1, 2, \ldots, n.$

Under hypothesis $|b(t)| \ge \gamma > 0, t \in [0, 1]$ a unique solvability and convergence of the solution of the system of (3.3) to the solution of (3.1) has been proved.

A numerical method for solution hypersingular integral equation on Hilbert's curve has been constructed and justified in [18].

In the work cited above, a spline-collocation method for solving the following equation is constructed and substantiated $a(t_1, t_2)x(t_1, t_2) + b(t_1, t_2) \iint_{\Omega_n} \frac{x(\tau_1,\tau_2)d\tau_1,d\tau_2}{((\tau_1-t_1)^2+(\tau_2-t_2)^2)^p} = f(t_1, t_2),\ (t_1, t_2) \in \Omega_n$, where Ω_n is the n-th prefractal of the Sierpinski carpet.

4 Solution of Singular Integral Equations

Consider an SIE with a Hilbert integral kernel and constant coefficients c and d

$$cx(s) + \frac{d}{2\pi}\int_0^{2\pi} x(\sigma)ctg\frac{\sigma - s}{2}d\sigma = f(s). \tag{4.1}$$

We denote by $W^c_{\alpha,\beta}(s)$ and $W^s_{\alpha,\beta}(s)$ cosinosoidal and sinusoidal Weierstrass functions $W^c_{\alpha,\beta}(s) = \sum_{k=1}^{\infty}\alpha^k\cos(\beta^k s)$, $W^s_{\alpha,\beta}(s) = \sum_{k=1}^{\infty}\alpha^k\sin(\beta^k s)$.

Hardy showed that the functions $W^c_{\alpha,\beta}(s)$ and $W^s_{\alpha,\beta}(s)$ for $0 < \alpha < 1, \beta > 1, \alpha\beta \geq 1$ are continuous, nowhere non-differentiable functions.

In the (4.1) we put $f(s) = W^c_{\alpha,\beta}(s)$. Since the right-hand side of the (4.1) is a periodic function, we put $\beta = 2$. Under the assumption that the parameters α and β satisfy the conditions $1/2 < \alpha < 1, \alpha\beta \geq 1$, the function $f(s)$ is continuous nowhere non-differentiable function.

Its fractal dimension is $D = 2 + \frac{\ln\alpha}{\ln\beta}$ [19] and varies depending on the values of the parameters α and β from 1 to 2.

We will seek an approximate solution of the (4.1) in the form of a series

$$x(t) = \sum_{k=0}^{\infty} a_k \cos kt + \sum_{k=1}^{\infty} b_k \sin kt. \tag{4.2}$$

Substituting series (4.2) into the (4.1) and using the formulas [20]

$$\frac{1}{2\pi}\int_0^{2\pi} \sin k\sigma ctg\frac{\sigma - s}{2}d\sigma = \cos ks, \quad \frac{1}{2\pi}\int_0^{2\pi} \cos k\sigma ctg\frac{\sigma - s}{2}d\sigma = -\sin ks, \tag{4.3}$$

we get

$$c\sum_{k=0}^{\infty} a_k \cos ks + c\sum_{k=1}^{\infty} b_k \sin ks - d\sum_{k=1}^{\infty} a_k \sin ks + d\sum_{k=1}^{\infty} b_k \cos ks = \sum_{k=1}^{\infty}\alpha^k\cos(\beta^k s). \tag{4.4}$$

From the (4.4) we have

$$\begin{aligned} &a = 0, \\ &a_{2^k} = \alpha^k c/(d^2 + c^2), \\ &b_{2^k} = d\alpha^k/(d^2 + c^2), k = 1, 2, 3, \ldots, \\ &a_k = b_k = 0, k = \{1, 2, ...\}\backslash 2^l, l = 1, 2, ... \end{aligned} \tag{4.5}$$

Thus, the solution of the SIE (4.1) has the form

$$x^*(s) = \sum_{k=1}^{\infty}\frac{\alpha^k c}{c^2 + d^2}\cos(2^k s) + \sum_{k=1}^{\infty}\frac{\alpha^k d}{c^2 + d^2}\sin(2^k s) = \frac{c}{c^2 + d^2}W^c_{\alpha,2}(s)$$

$$+\frac{d}{c^2+d^2}W^s_{\alpha,2}(s). \qquad (4.6)$$

Let us prove the validity of formula (4.6) under the assumption that the coefficients a_k and b_k are defined by formulas (4.5). For this, we investigate the smoothness of the function $x^*(t)$. We denote by $S_{2^n}(x^*)$ the sum

$$S_{2^n}(x^*)=\sum_{k=1}^{n}\left(\frac{\alpha^k c}{c^2+d^2}\cos(2^k s)+\frac{\alpha^k d}{c^2+d^2}\sin(2^k s)\right).$$

Then $E_{2^n}(x^*) \le |\sum_{k=n+1}^{\infty}\left(\frac{\alpha^k c}{c^2+d^2}\cos(2^k s)+\frac{\alpha^k d}{c^2+d^2}\sin(2^k s)\right)| \le C\alpha^n$. So, $E_n(x^*) \le \frac{C}{n^\gamma}, \gamma = |\log_2 \alpha|$.

Here $E_n(x^*)$ is the best uniform approximation of the function $x^*(s)$ by trigonometric polynomials of order n.

The right-hand side in the (4.1) is the Weierstrass function with exponent 2. So, one can put $\frac{1}{2} < \alpha < 1$. Hence, $0 < \gamma < 1$. From Bernstein's converse theorems of the constructive function theory [21] implies that $x \in H_\gamma$.

Let us put $R_{2^n}(s) = x^*(s) - S_{2^n}(x^*)$ and estimate the inequality

$$\left|\frac{1}{2\pi}\int_0^{2\pi} R_{2^n}(\sigma)ctg\frac{\sigma-s}{2}d\sigma\right| \le \left|\frac{1}{2\pi}\int_0^{2\pi}|R_{2^n}(\sigma)-R_{2^n}(s)|^{1-\gamma}|R_{2^n}(\sigma)\right.$$

$$-R_{2^n}(s)|^\gamma|ctg\frac{\sigma-s}{2}|d\sigma \le C(|R_{2^n}(\sigma)|^{1-\gamma}+|R_{2^n}(s)|^{1-\gamma}) \le \frac{c}{2^{n\gamma(1-\gamma)}}$$

and $\lim_{n\to\infty}\left|\frac{1}{2\pi}\int_0^{2\pi} R_{2^n}(\sigma)ctg\frac{\sigma-s}{2}d\sigma\right| = 0$. Consequently, the permutation of the operators of summation and integration is justified.

The following statement is true.

Theorem 5.1. Let $c^2+d^2 \neq 0$. Equation (4.1) has a unique solution $x^*(s)$, which is nowhere non-differentiable function.

Let us consider the singular integral equation

$$\frac{d}{2\pi}\int_0^{2\pi} x(\sigma)ctg\frac{\sigma-s}{2}d\sigma + h(s)\int_0^{2\pi} k(\sigma)x(\sigma)d\sigma = W^c_{\alpha,2}(s). \qquad (4.7)$$

We will seek an approximate solution to (4.7) in the form of series (4.2). Substituting (4.2) into (4.7), we have

$$d\left(\sum_{k=1}^{\infty} b_k \cos ks - \sum_{k=0}^{\infty} a_k \sin ks\right) + \left(\sum_{k=0}^{\infty} v_k \cos ks + \sum_{k=1}^{\infty} w_k \sin ks\right)$$
$$\times \left(\sum_{k=0}^{\infty} a_k \gamma_k + \sum_{k=1}^{\infty} b_k \delta_k) = W^c_{\alpha,2}(s)\right), \tag{4.8}$$

where $\gamma_0 = \frac{1}{2\pi}\int_0^{2\pi} k(\sigma)d\sigma$,

$$\gamma_k = \frac{1}{\pi}\int_0^{2\pi} k(\sigma)\cos k\sigma d\sigma,\ \delta_k = \frac{1}{\pi}\int_0^{2\pi} k(\sigma)\sin k\sigma d\sigma, k = 0, 1, ...$$

From (4.8) we obtain the following groups of equations

$$\begin{aligned}
&v_0\left(\sum_{l=0}^{\infty} a_l\gamma_l + \sum_{k=1}^{\infty} b_l\delta_l\right) = 0;\\
&db_k + v_k\left(\sum_{l=0}^{\infty} a_l\gamma_l + \sum_{k=1}^{\infty} b_l\delta_l\right) = 0, k = 1, 2, ...;\\
&-da_k + w_k\left(\sum_{l=0}^{\infty} a_l\gamma_l + \sum_{k=1}^{\infty} b_l\delta_l\right) = \alpha^j, k = 2^j, j = 1, 2, ...;\\
&-da_k + w_k\left(\sum_{l=0}^{\infty} a_l\gamma_l + \sum_{l=1}^{\infty} b_l\delta_l\right) = 0, k = \{1, 2, ...\}\backslash\{2^l, l = 1, 2, \ldots\}.
\end{aligned} \tag{4.9}$$

If $v_0 \neq 0$, then

$$\begin{aligned}
&\sum_{l=0}^{\infty} a_l\gamma_l + \sum_{l=0}^{\infty} b_l\delta = 0;\\
&-a_k = \alpha^j/d, k = 2^j, j = 1, 2, ...;\\
&a_k = 0, \{k = 1, 2, ...\}\backslash\{k = 2^j, j = 1, 2, ..., \}\\
&b_k = 0, k = 1, 2, ...
\end{aligned} \tag{4.10}$$

From (4.8) we find a_0 :$a_0 = -\frac{1}{\gamma_0 d}\sum_{l=1}^{\infty}\alpha^l\delta_{2^l}$.

It follows from equalities (4.10) that (4.7) has a unique solution if $\gamma_0 = \frac{1}{2\pi}\int_0^{2\pi} k(\sigma)d\sigma \neq 0$ and $v_0 = \frac{1}{2\pi}\int_0^{2\pi} h(\sigma)d\sigma \neq 0$. In other cases has a parameter-dependent solution.

5 Solution of Hypersingular Integral Equations

Consider the hypersingular integral equation

$$\frac{1}{4\pi}\int_0^{2\pi}\frac{x(\sigma)}{\sin^2\frac{\sigma-s}{2}}d\sigma = f(s), 0 \le s \le 2\pi, \tag{5.1}$$

which simulates a number of aerodynamic problems. In these cases, $f(s)$ simulates a gas flow and, therefore, is a fractal.

Let us investigate the solvability of condition (5.1) under the assumption that the right-hand side is the Weierstrass function $W^c_{\alpha,2}(s)$.

Equation (5.1) can be represented as

$$Sx \equiv \frac{d}{ds}\frac{1}{2\pi}\int_0^{2\pi} x(\sigma)ctg\frac{\sigma-s}{2}d\sigma = W^c_{\alpha,2}(s). \tag{5.2}$$

The solution of (5.2) (and, hence, (5.1)) will be sought in the form of the series $x^*(t) = \sum_{k=0}^{\infty}(a_k \cos 2^k s + b_k \sin 2^k s)$.

Acting formally, we arrive at the following equation

$$-\sum_{k=1}^{\infty}(a_k 2^k \cos 2^k s + b_k 2^k \sin 2^k s) = \sum_{k=1}^{\infty}\alpha^k \cos(2^k s). \tag{5.3}$$

It follows from formulas (4.3) that $\frac{1}{2\pi}\int_0^{2\pi} a_0 ctg\frac{\sigma-s}{2}d\sigma = 0$.

Thus, the coefficient α_0 turns out to be undefined and additional condition is required to determine it. As such condition, we can take $\int_0^{2\pi} x(s)ds = 0$. Then $a_0 = 0$. From we have $a_k = (\frac{\alpha}{2})^k$, $b_k = 0$, $k = 1, 2, \ldots$. So, $x^*(t) \in W^1 H_\alpha$.

Consider the hypersingular integral equation

$$\frac{1}{4\pi}\int_0^{2\pi}\frac{x(\sigma)}{\sin^2\frac{\sigma-s}{2}}d\sigma + h(s)\int_0^{2\pi} k(\sigma)x(\sigma)d\sigma = W^c_{\alpha,2}(s) \tag{5.4}$$

We will seek an approximate solution in the form of a series $x^*(s) = \sum_{k-0}^{\infty} a_k \cos ks + \sum_{k=1}^{\infty} b_k \sin ks$.

Substituting the series $x^*(s)$ into the (5.4) we have

$$-\sum_{k=1}^{\infty}(a_k k \cos ks + b_k k \sin ks)$$

$$+\left(\sum_{k=0}^{\infty}(v_k \cos ks + w_k \sin ks)\right)\left(\sum_{k=0}^{\infty} a_k\gamma_k + \sum_{k=1}^{\infty} b_k\delta_k\right) = W^c_{\alpha,2}(s). \tag{5.5}$$

From equality (5.5) we obtain the following groups of equations. At first we introduce the notation $A = \sum_{k=0}^{\infty} a_k\gamma_k + \sum_{k=1}^{\infty} b_k\delta_k$.

We have

$$\begin{aligned} &v_0 A = 0,\\ &-b_k k + w_k A = 0, k = 1, 2, ...\\ &-a_k k + v_k A = \alpha^l, k = 2^l, l = 1, 2, ..\\ &-a_k k + v_k A = 0, k = \{1, 2, ...\}\backslash\{2^l, l = 1, 2\}. \end{aligned} \tag{5.6}$$

If $v_0 \neq 0$, then $A = 0$ and

$$\begin{aligned} &b_k = 0, k = 1, 2, ...;\\ &a_k = -\alpha^l/k, k = 2^l, l = 1, 2, ...;\\ &a_k = 0, k = \{1, 2, ...\}\backslash\{2^l, l = 1, 2, ...; \}. \end{aligned}$$

From the condition $A = 0$, we have $a_0 = \left(\sum_{l=1}^{\infty} \frac{\alpha^l}{2^l}\gamma_{2^l}\right)\frac{1}{\gamma_0}$.

Thus, for $v \neq 0$ and $\gamma \neq 0$, (5.4) has a unique solution

$$x^*(s) = \left(\sum_{l=1}^{\infty} \frac{\alpha^l}{2^l}\gamma_{2^l}\right)\frac{1}{\gamma_0} - \sum_{l=1}^{\infty} \frac{\alpha^l}{2^l} \cos 2^l s. \tag{5.7}$$

Repeating the reasoning given above in $n^0 5$, we see that the series and the differentiated series on the right-hand side of (5.7) converge uniformly and $x^*(s) \in W^1 H_q, q = |\log_2 \alpha|$.

If $v_0 = 0$ we have a family of solutions depending on the parameter.

6 Conclusion

The approximate method for evaluating Riemann integrals, singular and hypersingular integrals on closed non-rectifiable curves and fractals has been proposed. For integrand continuation from fractal to the interior region the approximate computational scheme based on Whitney's continuation has been constructed. The approximate calculation of derivatives is based on using hypersingular integrals. It leads to two-dimensional integrals (for Riemann integrals) and to singular integrals (for singular and hypersingular integrals). Cubature formulas have been employed for evaluation of constructed integrals. The proposed method can be used to evaluate integrals over open curves.

The approximate method for solving hypersingular integral equations on the n-th prefractal of Cantor perfect setis presented. The spline-collocation method with first-order splines has been used. Justification of this method is based on theory of stability of ordinary differential equations systems. One of the main advantages of this method turns out to be its resistance to coefficients and right-hand side of equation disturbance. This method can be used for construction approximate solutions of singular and hypersingular integral equations on fractals of various types.

The solvability of singular and hypersingular integral equations with Weierstrass function in the right-hand side has been investigated.

Conflict of interest The authors declare no conflict of interest.

Funding The research received no external finding.

References

1. I.K. Lifanov, *Singular Integral Equations and Discrete Vortices* (VSP, Utrecht. The Netherlands, 1996)
2. I.V. Boykov, *Approximate Methods of Solution of Singular Integral Equations* (Penza, 2004) [in Russian]
3. M.A. Golberg, The convergence of several algorithms for solving integral equations with finite-part integrals I. J. Integral Equ. **5**(4), 329–340 (1983)
4. M.A. Golberg, The convergence of several algorithms for solving integral equations with finite-part integrals II. J. Integral Equ. **9**(3), 267–275 (1985)
5. I.K. Lifanov, L.N. Poltavskii, G.M. Vainikko, *Hypersingular Integral Equations and their Applications* (Boca Raton, Chapman Hall/CRC, CRC Press Company, London, New York, Washington, D.C., 2004)
6. I. Boykov, V. Roudnev, A. Boykova, Approximate methods for solving linear and nonlinear hypersingular integral equations. Axioms **9**(3), 2–18 (2020)
7. Boykov I.V. Approximate methods for solving hypersingular integral equations. in *Topics in Integral and Integro-Difference Equations. Theory and Applications*, Eds by H. Singh, H. Dutta, M.M. Cavalcanti (2021), pp. 63–102
8. I.V. Boykov, *Approximate Methods for Calculating Singular and Hypersingular Integrals*. Part one. Singular integrals (Penza State University Publishing House, Penza, 2005) [in Russian]
9. I.V. Boykov, *Approximate Methods for Calculating Singular and Hypersingular Integrals*. Part two. Hypersingular integrals (Penza State University Publishing House, Penza, 2009) [in Russian]
10. I.V. Boykov, E.S. Ventsel, A.I. Boykova, Accuracy optimal methods for evaluating hypersingular integrals. Appl. Numer. **59**(6), 1366–1385 (2009)
11. A.A. Potapov, Fractals in radio physics and radar, in *Sampling Topology*, (University book, 2005) [in Russian]
12. L.A. Chikin, Special cases of the Riemann boundary value problems and singular integral equations. Sci. Notes Kazan State Univ. **113**(10), 53–105 (1953). ([in Russian])
13. J. Harrison, A.A Norton, Gauss—green theorem for fractal boundaries. Duke Math. J. **67**(3), 575–586 (1992)
14. I. Stein, Singular integrals and differential properties of functions (Mir. 1973) [in Russian]
15. B.A. Kats, A jump problem and an integral over a nonrectifiable curve. Soviet Math. (Iz. VUZ) **31**(5), 65–75 (1987) [in Russian]

16. S.R. Mironova, Singular integral equations on a countable set of closed non-rectifiable and fractal curves. Russ. Math. (Iz. VUZ) **42**(5), 41–47 (1998) [in Russian]
17. I.V. Boykov, N.P. Krivulin, *Analytical and Numerical Methods for the Identification of Dynamic Systems.* (Publishing house of the Penza State University, Penza. 2016) [in Russian]
18. I. V. Boykov, A.I. Boykova, P.V. Aikashev, *Projection methods for solving hypersingular integral equations on fractals.* University Proceedings, Volga region, Physical and mathematical sciences, Mathematics, No. 1, pp. 71–86 (2016) [in Russian]
19. A.A. Potapov, Yu.V. Gulyaev, S.A. Nikitov, A.A. Pakhomov, V.A. German, in The Modern Methods of Image Processing, ed. A.A. Potapov (FIZMATLIT, 2008) [in Russian]
20. S.G. Mikhlin, Singular integral equations. Uspekhi Mat. Nauk. **3**(3), 29–112 (1948). ([in Russian])
21. I.P. Natanson, Constructive function theory. in *Uniform Approximation*, vol. I (Frederick Undar Publishing Co., N.Y., 1965)

Approximate Solution of Inverse Problems of Gravity Exploration on Fractals

Ilya V. Boykov, Alexander A. Potapov, Alexander E. Rassadin, and Vladimir A. Ryazantsev

Abstract The work is devoted to the approximate methods for solution direct and inverse problems of gravity exploration on bodies with a fractal structure. It is known that in order to construct mathematical models adequate to the geological reality, it is necessary to take into account the orderliness inherent in geological environments. One of the manifestations of orderliness is self-similarity, which remains during the transition from the microlevel to the macrolevel. Scaling of geological media can be traced in petrophysical data and in anomalous fields. It should be noted that in real structures there is no infinite self-similarity and scaling must be considered in a certain range. The work investigates analytical and numerical methods for solving inverse contact problems of the logarithmic and Newtonian potential in the generalized setting. In the case of a Newtonian potential, the problem is formulated as follows. It is required, having three independent functionals of the gravity field above the Earth's surface and additional information on the self-similarity of the disturbing body, to determine the depth, the density and the surface of the perturbing body.

Keywords Self-similarity · Fractional measure · Disturbances of the Earth's gravitational field · The logarithmic potential · The Newtonian potential · Petrophysical data

---

I. V. Boykov · V. A. Ryazantsev
Penza State University, Krasnay Str., 40, 440026 Penza, Russia
e-mail: ryazantsev@mail.ru

A. A. Potapov (✉)
V.A. Kotelnikov Institute of Radio Engineering and Electronics, Russian Academy of Sciences, Mokhovaya Str., 11-7, Moscow 125009, Russia
e-mail: potapov@cplire.ru

JNU-IREE RAS Joint Lab. of Information Technology and Fractal Processing of Signals, JiNan University, Guangzhou 510632, China

A. E. Rassadin
Higher School of Economics, Bolshaya Pecherskaya Str., 25/12, Nizhny Novgorod 603155, Russia
e-mail: brat_as@list.ru

C. H. Skiadas and Y. Dimotikalis (eds.), *14th Chaotic Modeling and Simulation International Conference*, Springer Proceedings in Complexity,
https://doi.org/10.1007/978-3-030-96964-6_8

1 Introduction

For the effective solution of direct and inverse problems of gravity prospecting, the methods of modeling bodies that perturb the potential and gravitational fields of the Earth (perturbing bodies) are of great importance. In most works, disturbing bodies are modeled by a set of the simplest geometric bodies (bar, parallelepiped, ball) [1]. In the works [2, 3], modeling is carried out with spheroids. In recent years, a large number of studies have been carried out on the fractality of individual minerals and the entire Earth as a whole [4–7]. Scaling of geological media can be traced in petrophysical data and in anomalous fields [8], etc. On the basis of the apparatus of fractional measure and fractional dimension, the processing of disturbances of the Earth's gravitational field is investigated [9].

Most minerals are porous. There are two types of porosity: the porosity of minerals and the porosity of liquids. Numerous studies have shown that in both mention cases, the porosity has a fractal structure.

In particular, the group of authors argues that sandstones have a fractal structure [4, 5, 10]. Hansen and Skjeltorp [6] investigated the fractal dimension D of a flat sandstone sample and obtained $D = 1, 73$. Brakenseik [11] determined the fractal dimension of a two-dimensional oil cut. It is equal to $D = 1, 8$. In [12], the fractal dimension of the surfaces of porous ceramic materials is investigated.

In the monograph [7] the Menger's sponge is proposed as a mathematical model of porosity, which is constructed somewhat differently from the standard construction.

In this work, when constructing fractal models of geological environments, the authors proceed not from fractals, but, following [13], from additions to fractals, since areas (volumes) of additions tend to areas (volumes) of the original body.

Taking into account the fractal components of gravitational fields makes it possible to clarify the structure of the disturbing bodies.

Methods for solving contact inverse problems of logarithmic and Newtonian potentials in a generalized setting are analyzed [14]. The problem is formulated as follows. It is required, having three independent functionals of the gravity field over the Earth's surface $z = 0$ and additional information about the self-similarity of the disturbing body, determine the depth H, the density $\sigma(x, y)$ and the surface $H - \varphi(x, y)$ of the disturbing body occupying the region $H \leq z(x, y) \leq H - \varphi(x, y)$.

Taking into account the fractal components of the gravitational and magnetic fields makes it possible to clarify the structure of the disturbing bodies.

The work is devoted to the approximate solution of direct and inverse problems of gravity prospecting on bodies with a fractal structure.

When solving inverse problems, a continuous method for solving nonlinear operator equations is used, which is presented in the next section.

2 Continuous Operator Method

Let B be a Banach space, $a, z \in B$, K be a linear operator mapping from B to B, $\Lambda(K)$ be the logarithmic norm [15] of the operator K, and I be the identity operator. We shall use the following notation: $B(a,r) = \{z \in B : \|z-a\| \le r\}$, $S(a,r) = \{z \in B : \|z-a\| = r\}$, $\mathcal{R}eK = K_R = (K+K^*)/2$, $\Lambda(K) = \lim_{h\downarrow 0}(\|I+hK\|-1)/h$.

Let a complex matrix $A = \{a_{ij}\}, i, j = 1, 2, \dots, n$, be given in $n-$ dimensional space R^n of vectors x with the norms $\|x\|_1 = \sum_{k=1}^n |x_k|$, $\|x\|_2 = \left[\sum_{k=1}^n |x_k|^2\right]^{1/2}$, and $\|x\|_3 = \max_{1\le k\le n}|x_k|$.

The corresponding logarithmic norms of the matrix A then read [16]:

$$\Lambda_1(A) = \max_j(\mathcal{R}e(a_{jj}) + \sum_{i=1, i\ne j}^{n} |a_{ij}|),$$

$$\Lambda_2(A) = \lambda_{\max}\big((A+A^T)/2\big),$$

$$\Lambda_3(A) = \max_i(\mathcal{R}e(a_{ii}) + \sum_{j=1, j\ne i}^{n} |a_{ij}|).$$

Here $\lambda_{\max}(A)$ means the largest real part of eigenvalues of the matrix A.

Consider an equation

$$A(x) - f = 0, \tag{2.1}$$

where $A(x)$ is a nonlinear operator mapping from Banach space B to B.

Let x^* be a solution of the (2.1). In [17] the connection between stability of solutions of operator differential equations in Banach spaces and resolving operator equations of the form (2.1) has been established. Here we shall summarize the results on the method.

Let us associate the (2.1) with the following Cauchy problem

$$\frac{dx(t)}{dt} = A(x(t)) - f, \tag{2.2}$$

$$x(0) = x_0. \tag{2.3}$$

Theorem 2.1 [17]. Let the (2.1) has a solution x^* and on any differentiable curve $g(t)$ in Banach space B the inequality is valid

$$\lim_{t\to\infty} \frac{1}{t}\int_0^t \Lambda(A'(g(\tau))d\tau \le -\alpha_g,\ \alpha_g > 0\ . \tag{2.4}$$

Then the solution of the Cauchy problem (2.2), (2.3) converges to the solution x^* of the (2.1) for any initial approximation.

Theorem 2.2 [17]. Let the (2.1) has a solution x^* and for any differentiable curve $g(t)$ in a ball $B(x^*, r)$ the following conditions are satisfied:

(1) for any $t (t > 0)$

$$\int_0^t \Lambda(A'(g(\tau))d\tau \leq 0; \tag{2.5}$$

(2) the inequality (2.4) is valid.

Then the solution of the Cauchy problem (2.2), (2.3) converges to a solution of the (2.1).

Note 1. In the inequality (2.4) it is assumed that the constants $\alpha_g > 0$ can differ for different curves $g(t)$.

Note 2. From inequalities (2.4), (2.5) it follows that the logarithmic norm $\Lambda(A'(g(\tau))$ can be positive for some values of τ; i.e. the Frechet derivative $A'(g(\tau))$ can degenerate into an identically zero operator along the curve.

Note 3. An example in [18] (an approximate solution of a hypersingular integral equation) has demonstrated convergence of an iterative process based on a continuous operator method when the Frechet derivative vanishes at the initial approximation.

Note 4. Logarithmic norm has the property which is very useful for numerical analysis. Let A, B be square matrices of order n with complex elements and $x = (x_1, \dots, x_n)$, $y = (y_1, \dots, y_n)$, $\xi = (\xi_1, \dots, \xi_n)$, $\eta = (\eta_1, \dots, \eta_n)$ are n-dimensional vectors with complex components. Let us consider the following systems of algebraic equations: $Ax = \xi$ and $By = \eta$. The norm of a vector and its subordinate operator norm of the matrix are fixed; the logarithmic norm $\Lambda(A)$ corresponds to the operator norm.

Theorem 2.3 [19]. If $\Lambda(A) < 0$, the matrix A is non-singular and $\|A^{-1}\| \leq 1/|\Lambda(A)|$.

Theorem 2.4 [19]. Let $Ax = \xi$, $By = \eta$ and $\Lambda(A) < 0$, $\Lambda(B) < 0$. Then

$$\|x - y\| \leq \frac{\|\xi - \eta\|}{|\Lambda(B)|} + \frac{\|A - B\|}{|\Lambda(A)\Lambda(B)|}.$$

Main properties of the logarithmic norm are given in [15].

The logarithmic norm of the operator K can have different (positive or negative) values in different spaces.

The continuous method for solving nonlinear operator equations admits the following generalization. Let us return to (2.1). Denote by $A'(x_0)$ the Gateaux (Frechet) derivative on the element x_0. We introduce the equation

$$(A'(x_0))^* A(x) - (A'(x_0))^* f = 0. \tag{2.6}$$

Equation (2.6) is associated with the Cauchy problem

$$\frac{dx}{dt} = -(A'(x_0))^* A(x) - (A'(x_0))^* f), \tag{2.7}$$

$$x(0) = x_0. \tag{2.8}$$

If $\Lambda_2(A'(x_0))^* A'(x_0)) > 0$, then in some neighborhood $B(x_0, r)$ of the element x_0 the Euclidean logarithmic norm of the operator—$A'(x_0))^* A(x)$ will be negative and $\|x(t)\| < \|x(0)\|$ for some interval $t \in (t_0, t_1]$, $t_0 = 0$.

Let the inequality $\Lambda_2(A'(x_0))^* A'(x)) > 0$ be satisfied on the segment $t \in [t_0, t_1]$, $t_0 = 0$. (Here $x(t)$ is the solution to the Cauchy problem (2.7), (2.8)).

For $t \geq t_1$, consider the Cauchy problem

$$\frac{dx_1(t)}{dt} = -(A'(x_1)^* A(x) - (A'(x_1))^* f), \tag{2.9}$$

$$x_1(t_1) = x(t_1)(3.10) \tag{2.10}$$

and define the segment $[t_1, t_2]$, in which the inequality $\Lambda_2(A'(x_1))^* A'(x_1)) > 0$ occur.

Taking $x_2(t_2) = x_1(t_2)$ as an initial value when solving the Cauchy problem

$$\frac{dx_2(t)}{dt} = -(A'(x_2))^* A(x) - (A'(x_2))^* f), \tag{2.11}$$

$$x_2(t_2) = x_1(t_2), \tag{2.12}$$

we have $\lim_{t\to\infty} \left\| \frac{dx(t)}{dt} \right\| = 0$ and therefore $\lim_{t\to\infty} x(t) = x^*$.

Assertions follow from this remark.

Theorem 2.5. Suppose that (2.6) has a solution x^* and for any differentiable curve in the Banach space B the inequality

$$\lim_{t\to\infty} \frac{1}{t} \int_0^t \Lambda((A'(g(\tau)))^* A'(g(\tau)))d\tau \leq -\alpha_g, \alpha_g > 0 \tag{2.13}$$

occur. Then the solution to the sequence of Cauchy problems ((2.7), (2.8)), ((2.9), (2.10)), ((2.11), (2.12)), etc. converges to the solution x^* of (2.6).

Theorem 2.6. Suppose that (2.6) has a solution x^* and for any differentiable curve in the sphere $B(x^*, r)$ the inequalities

$$\int_0^t \Lambda((A'(g(\tau)))^* A'(g(\tau)))d\tau < 0 \tag{2.14}$$

and (2.13) occur. Then the solution of the sequence of Cauchy problems ((2.7), (2.8)), ((2.9), (2.10)), ((2.11), (2.12)), etc. converges to the solution x^* of (2.6).

If the conditions of Theorems 2.5 and 2.6 are not satisfied, the regularization

$$\frac{dx}{dt} = -\alpha x(t) - ((A'(x_0))^* A(x) - (A'(x_0))^* f), \alpha > 0,$$

is carried out.

3 Direct Tasks

Let us consider a geological deposit represented by the uniform body D of arbitrary form. Assuming that the body has fractal dimension $D_H < 3$, we will approximate it with its complement of the Menger sponge [7]. Let the body D be situated in the cube $\Omega = [-a, a]^3$. Let us construct the n-th order prefractal (n-th iteration of the fractal) for the Menger sponge in the cube Ω. During the construction of the first iteration the cube Ω is divided into 27 equal cubes with sides $r_1 = 2a/3$, and 7 central cubes are dropped.

During the construction of the second iteration every cube from the remaining 20 cubes is divided into 27 equal cubes with the sides $2a/9$. As the result we have 729 cubes including 400 central cubes (for every initial cube with the side $2a/3$) that are dropped. Repeating the described operations n times we get the n-th Menger prefractal. As noted in the work [13], not classical fractals but their complements with respect to the initial domain should be used as the model for geological bodies. Consequently geological deposits are modeled with the set of cubes with different lengths of edges (and with different sizes).

When modeling granular and liquid media it seems that it is more efficient to model them with reduced copies of the first iteration of the Menger sponge. In that case we can construct the model using not only classical fractal but also complement to it.

Let us introduce the Cartesian three-dimensional rectangular coordinate system with down-directed z-axis and with the origin of coordinates placed at the Earth surface. Assume that the body D occurs at sufficiently great depth z = H under the Earth surface.

As the parameter H we fix the distance from the Earth surface to the average point (in vertical direction) of gravitating body.

In the introduced coordinate system the domain Ω, which the body D belongs to, rewrites as:

$$\Omega = \{(x, y, z) : -a \le x \le a, -a \le y \le a, H - a \le z \le H + a\}.$$

Let $\Delta_{i,j,k} = [x_i, x_{i+1}; y_j, y_{j+1}; z_k, z_{k+1}]$, $x_i = -a + ai/n$, $i = 0, 1, \ldots, 2n$, $y_j = -a + aj/n$, $j = 0, 1, \ldots, 2n$, $z_k = H - a + ak/n$, $k = 0, 1, \ldots, 2n$.

We refer to as marked the cubes Δ_{ijk} that have nonempty intersection with the domain D. In the marked cubes we locate the first iteration of the Menger sponge fractal with the edge length a/n. Suppose that the body is modeled by the first iteration of the fractal. Denote the constructed model of the body by D_n. For computation of the perturbed field it is sufficient to compute the vertical component of the gravity field generated by the cell Δ_{ijk} at the point $(x, y, 0)$.

The cell Δ_{ijk} consists of 20 cubes with edges having the length $a/3n$. Assuming n and H being sufficiently big we may treat $\cos(\Theta(x', y', z'))$, where $(x', y', z') \in \Delta_{ijk}$ as constant within the limits of the cell. Here $\Theta(x', y', z')$ is the angle between the radius-vector $M'P$ ($M' = (x', y', z'), P(x, y, 0)$) and the z-axis.

Let us denote by o_{ijk} the center of the cell Δ_{ijk}. Obviously,

$$o_{ijk} = (-a + a(i + 1/2)/n, -a + a(j + 1/2)/n, H - a + a(k + 1/2)/n).$$

Let us also denote by θ_{ijk} the angle between the vector $o_{ijk}P$ and the z-axis.

Thus the vertical component of the gravity force generated by the cell Δ_{ijk} at the point $P(x, y, 0)$ equals to $dV_z(i, j, k) = 20\gamma\rho(a/3n)^3 \cos(\theta_{ijk})/(r(o_{ijk}, P))^2 = \frac{20\gamma\rho a^3}{27n^3} \cdot \frac{z_k+z_{k+1}}{2}/(r(o_{ijk}, P))^3$.

Here γ is the gravitational constant, ρ is a density of body. Therefore the vertical component of the gravity force generated by the disturbing body D at the point $(x, y, 0)$ equals to $V_z(x, y, 0) = \sum_{i,j,k=0}^{2n-1} 20\gamma\rho_{ijk}a^3(\frac{z_k+z_{k+1}}{2})/27n^3(r(o_{ijk}, P))^3$, ρ_{ijk}.is a density of cell. Consider the example.

Let us se t the following parameter values: $H = 5$, $a = 1/4$, $n = 10$.

We perform calculations using the formula $dV_z(i, j, k) = \frac{20\gamma\rho a^3}{27n^3} \cdot \frac{z_k+z_{k+1}}{2}/(r(o_{ijk}, P))^3$.

Let us fix $i = j = k = n$, that corresponds to the central cell Δ_{ijk} in the domain Ω For illustrative purposes the product of the constants γ and ρ we set to 10^6.

The field $dV(i,j,k)$ of the vertical component of anomalous gravity force generated by the described cell at the Earth surface is shown in the figure (Picture 1).

For comparison we also introduce the plot of the vertical component of the anomalous gravity field generated by the continuous body occupying the domain.

The computed field is depicted in the following figure (Picture 2).

From the comparison of the computed fields it is obvious that the solution of the direct problem is essentially dependent on the chosen model for representation of the elementary cell.

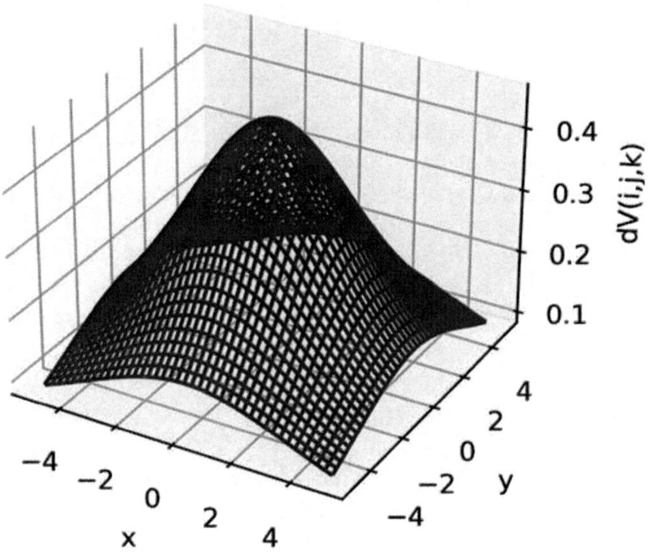

Picture 1 The vertical component of anomalous gravity force generated by the cell of the Menger sponge first order prefractal

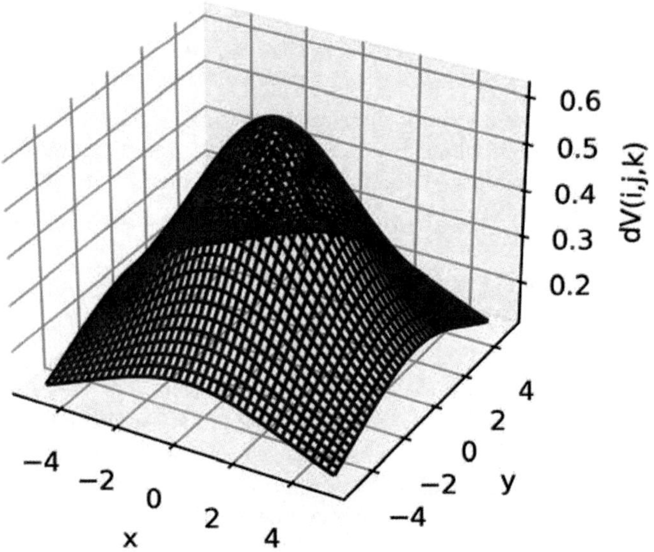

Picture 2 The vertical component of anomalous gravity force generated by the elementary cell

4 Inverce Tasks

This section examines the influence of the chosen model on the accuracy of the interpretation of the results.

Let in the domain $D\{D: -l \le x \le l, -l \le y \le l_1, H \le z \le H - \varphi(x, y)\}$ are distributed with density $\sigma(x, y, z)$ sources disturbing gravitational field of the Earth. The gravity field above the Earth's surface is determined by the equation

$$G \int_{-\infty}^{\infty} \int_{-\infty}^{\infty} \int_{H-\varphi(\zeta,\eta)}^{H} \frac{\sigma(\zeta, \eta, \xi)(\xi - z) d\zeta \, d\eta \, d\xi}{((x-\zeta)^2 + (y-\eta)^2 + (\xi - z)^2)^{3/2}} = f(x, y, z), \tag{4.1}$$

where $f(x, y, z)$ is the experimentally determined value, $G-$ gravitational constant, which for the convenience of further calculations will be set equal to $G = 1/2\pi$.

To describe the force of gravity on the Earth's surface in (4.1), one should set $z = 0$.

Having calculated the integral on the left-hand side of (4.1) by parts and assuming that the density does not depend on ξ, we have

$$\begin{aligned} &\frac{1}{2\pi} \int_{-\infty}^{\infty} \int_{-\infty}^{\infty} \sigma(\zeta, \eta) \big[((x-\zeta)^2 + (y-\eta)^2 + (H - z - \varphi(\zeta, \eta))^2)^{-1/2} \\ &-((x-\zeta)^2 + (y-\eta)^2 + (H-z)^2)^{-1/2} \big] d\zeta \, d\eta = f(x, y, z). \end{aligned} \tag{4.2}$$

We represent (4.2) in the form

$$\begin{aligned} &\frac{1}{2\pi} \int_{-\infty}^{\infty} \int_{-\infty}^{\infty} \sigma(\zeta, \eta) \big[((x-\zeta)^2 + (y-\eta)^2 + (H-z)^2)^{-1/2} (1+u)^{-1/2} \\ &-((x-\zeta)^2 + (y-\eta)^2 + (H-z)^2)^{-1/2} \big] d\zeta \, d\eta = f(x, y, z), \end{aligned} \tag{4.3}$$

where $u = \frac{\varphi^2(\zeta,\eta) - 2(H-z)\varphi(\zeta,\eta)}{(x-\zeta)^2 + (y-\eta)^2 + (H-z)^2}$. Under the assumption that $|u| < 1$, the function $\frac{1}{(1+u)^{1/2}}$ is expanded in the series

$$\frac{1}{(1+u)^{1/2}} = 1 + \sum_{n=1}^{\infty} (-1)^n \frac{(2n-1)!!}{2^n n!} u^n. \tag{4.4}$$

Substituting (4.4) into (4.3) and using the uniform convergence of series (4.4), we have

$$\frac{1}{2\pi}\sum_{n=1}^{\infty}(-1)^n\frac{(2n-1)!!}{2^n n!}\int_{-\infty}^{\infty}\int_{-\infty}^{\infty}\sigma(\zeta,\eta)\frac{(\varphi^2(\zeta,\eta)-2(H-z)\varphi(\zeta,\eta))^n\,d\zeta\,d\eta}{((x-\zeta)^2+(y-\eta)^2+(H-z)^2)^{n+1/2}}$$
$$= f(x,y,z). \tag{4.5}$$

Let us approximate (4.5), limiting ourselves to one term on the left-hand side. As a result, we obtain the equation [14]

$$-\frac{1}{4\pi}\int_{-\infty}^{\infty}\int_{-\infty}^{\infty}\sigma(\zeta,\eta)\frac{(\varphi^2(\zeta,\eta)-2(H-z)\varphi(\zeta,\eta))\,d\zeta\,d\eta}{((x-\zeta)^2+(y-\eta)^2+(H-z)^2)^{3/2}} = f(x,y,z). \tag{4.6}$$

Equation (4.6) contains three unknowns: the depth of the gravitating body H, the density of the body $\sigma(x,y)$ and the shape of the surface $H-\varphi(x,y)$. To find these unknowns, it is necessary, in addition to values of the gravity field on some surface, to have two more linearly independent sources of information. As these functionals, one can use values of the gravity field at three different levels, a combination of the values of the gravity field and its derivatives in different directions, etc.

Note. Having values of the gravity field at the same level, it is possible to restore values of the gravity field at several levels using the Poisson formula.

In the work [14], analytical and numerical methods are proposed for the simultaneous determination of the depth of the disturbing body, its density and the surface equation in contact problems of the logarithmic and Newtonian potential. In [14], the disturbing body was assumed to be solid.

Compared with iterative methods for solving (4.6), studied in [14], the preferable is the continuous operator method described in Sect. 3. In both cases, the density is interpreted as a constant function within the unit cell, which simulates the gravitating body. In the case of modeling a gravitating body with fractals, the density in elementary cells is not constant. It is of interest to study the influence of fractals chosen for modeling disturbing bodies on the accuracy of determining their densities.

In [14] the following example was analytically solved.

Let in the domain $\Omega = \{5 \le z(x,y) \le 5-\varphi(x,y), -\infty < x, y < \infty\}$, there is a perturbing body with density $\sigma(x,y)$. Let the gravity force and its first two derivatives be known on the surface $z=0$:

$$f_0(x,y,0) = \frac{24\pi}{(x^2+y^2+36)^{3/2}} - \frac{7\pi}{5(x^2+y^2+49)^{3/2}},$$
$$f_1(x,y,0) = \left.\frac{\partial f(x,y,z)}{\partial z}\right|_{z=0} = \frac{432\pi}{(x^2+y^2+36)^{5/2}} - \frac{4\pi}{(x^2+y^2+36)^{3/2}} - \frac{147\pi/5}{(x^2+y^2+49)^{5/2}} - \frac{2\pi/25}{(x^2+y^2+49)^{3/2}},$$
$$f_2(x,y,0) = \left.\frac{\partial^2 f(x,y,z)}{\partial z^2}\right|_{z=0} = \frac{12960\pi}{(x^2+y^2+36)^{7/2}} - \frac{1029\pi}{(x^2+y^2+49)^{7/2}} -$$
$$-\frac{216\pi}{(x^2+y^2+36)^{5/2}} + \frac{21\pi/25}{(x^2+y^2+49)^{5/2}} - \frac{4\pi/125}{(x^2+y^2+49)^{3/2}}.$$

It is necessary to find a depth of the gravitating body H, a density of the body $\sigma(x,y)$ and a shape of the surface $H-\varphi(x,y)$. To solve this problem, in addition to (4.6), two more equations are added

$$\int\limits_{-\infty}^{\infty}\int\limits_{-\infty}^{\infty} \frac{2Hw_1(\xi,\eta) - w_2(\xi,\eta)}{((x-\xi)^2 + (y-\eta)^2 + H^2)^{3/2}} d\xi\, d\eta = f_0(x,y),$$

$$\int\limits_{-\infty}^{\infty}\int\limits_{-\infty}^{\infty} \left\{\frac{6H^2w_1(\xi,\eta) - 3Hw_2(\zeta,\eta)}{((x-\xi)^2 + (y-\eta)^2 + H^2)^{5/2}} - \frac{2w_1(\xi,\eta)}{((x-\xi)^2 + (y-\eta)^2 + H^2)^{3/2}}\right\} d\xi d\eta$$
$$= f_1(x,y),$$

$$\int\limits_{-\infty}^{\infty}\int\limits_{-\infty}^{\infty} \left\{\frac{3w_2(\zeta,\eta) - 18Hw_1(\xi,\eta)}{((x-\xi)^2 + (y-\eta)^2 + H^2)^{5/2}} + \frac{30H^3w_1(\xi,\eta) - 15H^2w_2(\xi,\eta)}{((x-\xi)^2 + (y-\eta)^2 + H^2)^{7/2}}\right\} d\xi\, d\eta$$
$$= f_2(x,y). \tag{4.7}$$

When obtaining system (4.7), the following formulas were used $w_1(x,y) = \sigma(x,y)\varphi(x,y)$, $w_2(x,y) = \sigma(x,y)\varphi^2(x,y)$.

Its exact solution was obtained: $H = 5$, $\varphi(x,y) = \left(\frac{x^2+y^2+1}{x^2+y^2+4}\right)^{3/2}$, $\sigma(x,y) = \frac{(x^2+y^2+4)^{3/2}}{(x^2+y^2+1)^3}$.

When solving the system of (4.7) by the spline-collocation method with zero-order splines, an error is equal to $O(N^{-1})$, where $h = N^{-1}$ is a step of the computational scheme by coordinates x, y. Hence it follows that the results of the approximate solution can be interpreted as follows. In area $(x,y) : \{\frac{(x^2+y^2+4)^{3/2}}{(x^2+y^2+1)^3} \leq 1/N\}$ let us put $\sigma(x,y) = 0$. Domain G defined by the inequality $\{\frac{(x^2+y^2+4)^{3/2}}{(x^2+y^2+1)^3} \geq 1/N\}$ we will cover with elementary cells (cubes) with edges of length d/N, where d is area diameter G. Place the first-order prefractal of the Mergel sponge in the elementary cells. Then, depending on the mineral filling the addition of the Margel sponge to the unit cell, the density of the body varies from $\sigma(x,y)$ to $27\sigma(x,y)/20$. Thus, when solving inverse problems on fractals, an additional problem arises of choosing an appropriate model (fractal, multifractal) for a gravitating body.

5 Conclusions

In this work by the example of the Menger sponge approximate methods for solution of direct and inverse problems of gravity exploration using fractals are investigated. As far as inverse geophysical problems belong to the class of ill-posed problems for their solution in this work we propose the generalization of continuous operator method for solution of nonlinear equations. The proposed method allows to obtain stable solution for inverse problems which are modeled with nonlinear convolutional equations. At the core of the method there are criteria for asymptotic stability of solutions of systems of ordinary differential equations. The method can be used for solution of numerous equations of mathematical physics. In solving direct and inverse problems using fractals we show the problem of dependency of interpretation of computational results on the chosen model.

References

1. E.A. Mudretsova (ed.), Gravity Prospecting, M. Sci. (1981) [in Russian]
2. I.E. Stepanova, On one stable method for recovering ellipsoids. Phys. Earth 1, 101–106 (2001) [in Russian].
3. V.S. Sizikov, I.N. Golov, Modeling of deposits by spheroids. Phys. Earth 3, 83–96 (2009) [in Russian]
4. A.J. Katz, A.H. Thompson, Fractal sandstone pores: implications for conductivity and pore formation. Phys. Rev. Lett. **54**, 1325–1328 (1985)
5. C.E. Krohn, Fractal measurement of sandstones, shales, and carbonates. J. Geophys. Res. **93**, 3297–3305 (1988)
6. J.P. Hansen, A.T. Skjeltorp, Fractal pore space and rock permeability implications. Phys. Rev. **B38**, 2635–2638 (1988)
7. D. Turcotte, *Fractals and Chaos in Geology and Geophysics* (Cambridge University Press, Cambridge, 1992)
8. E.V. Utemov, *Flicker-noise Structure of Energy Spectra of Natural Magnetic Anomalies. Monitoring of the Geological Environment: Active Endogenous and Exogenous Processes* (Kazan University Publishing House, Kazan, 2000). ([in Russian])
9. A.A. Potapov, Yu.V. Gulyaev, S.A. Nikitov, A.A. Pakhomov, V.A. German, *The Modern Methods of Image Processing*. ed. by A.A. Potapov (FIZMATLIT, 2008) [in Russian]
10. A.H. Thompson, A.J. Katz, C.E. Krohn, The microgeometry and transport properties of sedimentary rock. Adv. Phys. 625–694 (1987)
11. D.L. Brakenseik, W.J. Rawles, S.D. Logsdon, W.M. Edwards, Fractal description of macroporosity. Soil Sci. Soc. Am. J. **56**, 1721–1723 (1992)
12. S.N. Kulkov, T. Yan, S.P. Buyakova, Fractal dimension of surfaces of porous ceramic materials. Tech. Phys. Lett. **32**(2), 51–55 (2006) [in Russian]
13. P.S. Babayants, Yu.I. Blokh, A.A. Trusov, Fundamentals of modeling potential fields of fractal geological objects. *Questions of Theory and Practice of Geological Interpretation of Gravitational, Magnetic and Electric Fields: Proceedings of the 32nd session of the international seminar named after D.G. Uspensky* (Mining Institute of the Ural Branch of the Russian Academy of Sciences, Perm, 2005), pp. 12–14 [in Russian]
14. I.V. Boikov, V.A. Ryazantsev, On Simultaneous restoration of density and surface equation in an inverse gravimetry problem for a contact surface. Numer. Anal. Appl. **13**(3), 241–257 (2020)
15. Yu.L. Daletskii, M.G. Krein, Stability of solutions of differential equations in Banach Space, Nauka, 1970 [in Russian], English transl. Transl. Math. Monographs. V. 43. Amer. Math. Soc. Providence. RI. (1974)
16. K. Dekker, J. Verver, Stability of Runge-Kutta methods for stiff nonlinear differential equations, Mir. (1988) [in Russian]
17. I.V. Boikov, On a continuous method for solving nonlinear operator equations. Differ. Equ. **48**(9), 1308–1314 (2012)
18. I.V. Boykov, V.A. Roudnev, A.I. Boykova, O.A. Baulina, New iterative method for solving linear and nonlinear hypersingular integral equations. Appl. Numer. Math. **127**, 280–305 (2018)
19. S.M. Lozinskii, Note on a paper by V.S. Godlevskii, USSR computational mathematics and mathematical physics, **13**(2), 232–234 (1973)

Exploring the Chaotic Nature of COVID-19 Pandemic: Limit Cycles and Time-Lag Around the World

Yiannis Dimotikalis and Christos H. Skiadas

Abstract In this work the Chaotic Nature of the Covid-19 Pandemic data is explored. The Total Deaths (/million) to Total Cases (% population) of Covid-19 data from https://ourworldindata.org/coronavirus website are studied. Large 28 day-delays lead to characteristic Limit Cycles while the lag between applied and adopted measures tend to form characteristic limit cycle forms. A 28-day Moving Average was tested. The data period is from January 2020 to 23 November 2021. The time period is divided to Parts based on chart data curve. Similarities but also differences are present leading to group countries accordingly. An important issue is to explore the Covid-19 spread due to variations of the virus while the vaccine measures expand considerably.

Keywords Covid-19 pandemic · Limit cycles · Chaos · Data sets · Moving average · Health state · Nonlinear analysis

1 Theory and Applications

Covid-19 pandemic and several variations of the virus passed in a new era after the introduction of several vaccines adopted. Socioeconomic, political and demographic issues influence the spread of the virus and the vaccine adoption in all countries of the World. It was clear from the beginning that the thread was very serious and radical actions should apply. The spread of the disease followed an Exponential like growth. Without adopting immediate measures the health systems would collapse. In the first period radical actions were adopted at least to reduce the speed of epidemics while waiting for the appropriate vaccine invention and perhaps new drugs or treatment methodologies.

Y. Dimotikalis
Department of Management Science and Technology, Hellenic Mediterranean University, Agios Nikolaos, Crete, Greece

C. H. Skiadas (✉)
ManLab, Technical University of Crete, Chania, Crete, Greece
e-mail: skiadas@cmsim.net

C. H. Skiadas and Y. Dimotikalis (eds.), *14th Chaotic Modeling and Simulation International Conference*, Springer Proceedings in Complexity,
https://doi.org/10.1007/978-3-030-96964-6_9

For the first time in human history our national and international systems to collect, store and analyze datasets are so-advanced. However, the analysis of so-many datasets came to be a puzzle difficult to solve. The task was to reduce the growth speed of epidemics; but how fast and what measures where the most appropriate. At least to save the socioeconomic and political system while improving the health systems as well. Now, after almost 22 months after covid-19 invasion, we have enough experience from fighting the pandemic. However, the lessons learned in the first Covid-19 period before the vaccine introduction are not enough to handle the new situation.

Some lessons of the analysis from non-linear systems theory and the related chaotic behavior may be useful (see [1–3]). Chaos or chaotic attractors appear when "Time Delays" are present. It means that as long as a delay between treatment and cure appear chaos is present. Even more longer delays appear from local Authorities measures and interventions.

Governments tend to propose measures and correct again and again after collecting appropriate data. This could act like to try to correct the Stock Exchange fluctuations by many repeated actions. The best, in this case, is to carefully study the selected actions to be effective and designed for a large time horizon. The examples presented here for several countries while the same methodology for the other countries included by introducing data from https://ourworldindata.org/coronavirus.

2 The Data and the Limit Cycles

The data are collected and stored daily for almost all World countries. Of particular interest for our study are the new death data per day. The first step of the data handling was done from the website people by smoothing to reduce the particular differences from the daily data collection. The provided Covid-19 daily data for Greece are illustrated with yellow line in the next figure. However, this smoothing is not relatively adequate to keep the important part of the virus spread by excluding parts that form local changes. Radical smoothing is done with a 28-day transformation by taking into account 14 days before and 14 days after a particular date that a ± 14 days or 28 days smoothing. This is presented by the thick blue curve in the next figure. By this transformation smoothness is clear. Of course small details are missing but the appropriate data for providing the main limit cycles remain. The ± 14-day smoothing is in accordance with the 14-day quarantine suggested for Covid-19 infected people. For a detailed Moving Average theory and applications see [4]. The method we select is also termed as Central Two-Sided Moving Average around a day x with a $\pm$ k-day summation around the point x. Accordingly the total space is $S = 2\,k + 1$ that is equal to 29 for the $k = 14$ case (Fig. 1).

Limit cycles are formed from the First difference (first derivative) of new deaths versus the new deaths per day. Of course the first derivative data is provided after a 28-day smoothing (± 14) as was done for the daily data. See next figure where

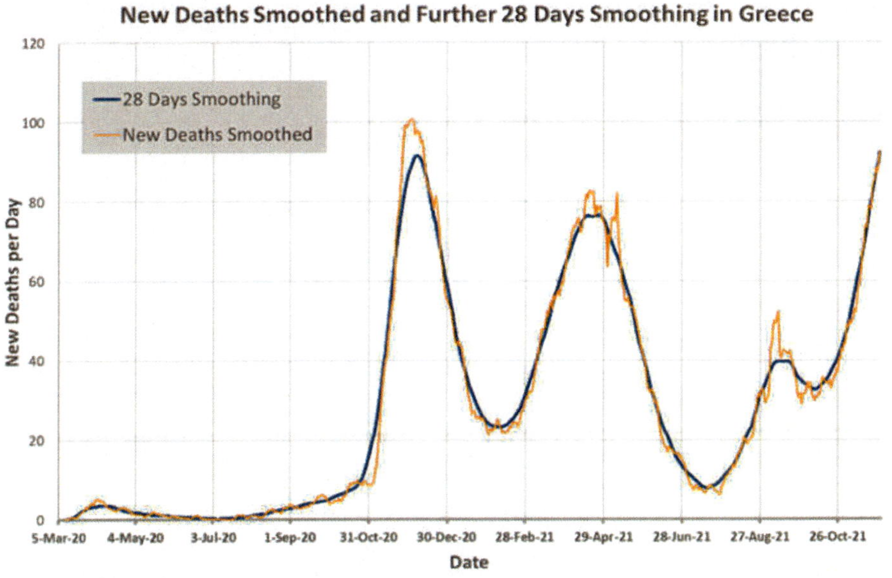

Fig. 1 Greece Covid19 data, new deaths smoothed

the data without smoothing are provided with orange whereas the blue curve is the 28-day smoothed one (Fig. 2).

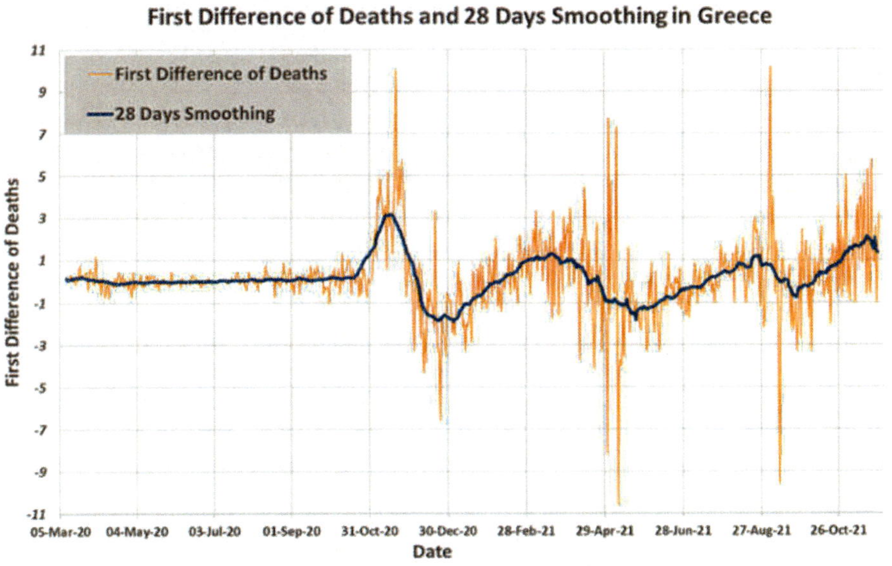

Fig. 2 Greece Covid19 data, first difference of deaths smoothed

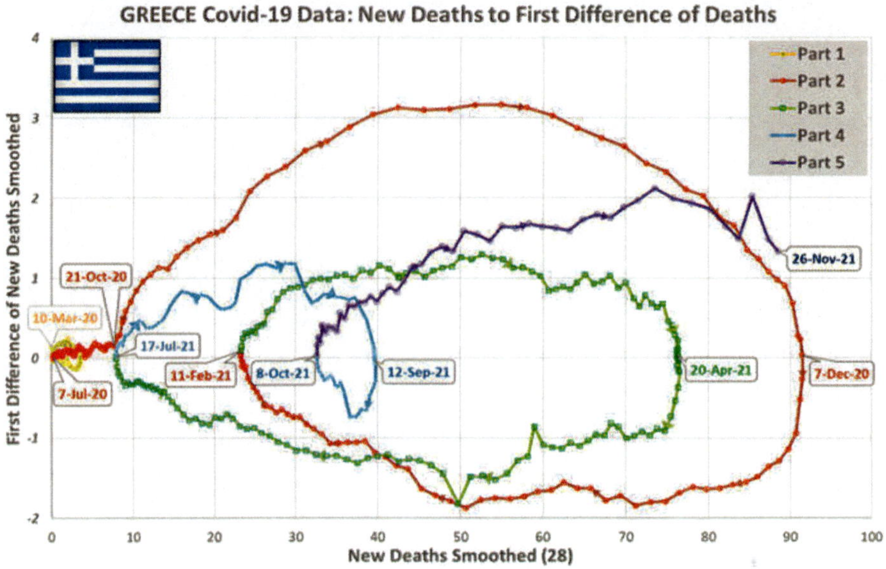

Fig. 3 Greece Covid19 data, new deaths to first difference of deaths

The chaotic like behavior is presented in the next figure for GREECE. X-Axis is for new deaths and Y-Axis represents the first difference of deaths. The 1st Part (yellow) is followed by the 2nd Part (red) and the 3rd Part (green). The final Part (cyan) is like a continuation of the 1st Part curve. The graph includes chaotic like forms with delay.

Note that a lager limit circle is followed by a smaller one (see Fig. 3). A 28-days moving average form is selected. It looks like the main delays are coming from a one-month response of the system. This is the time period that should be taken under consideration when important actions are needed. No immediate response is possible. The reaction time of a large system as a country is critical.

From the graph 3 a center of the large cycles is located at 50 deaths per day whereas the center of aa smaller cycle is at 35 deaths per day. However, after 8th October 2021 the cycle is moving outside the small period expanded to a large cycle form with a center larger to 50 deaths per day.

Following the analysis above it is clear that the chaotic circles formation need much time to appear. It could that explained the very few papers appearing in the literature. A chaotic attractor is presented for the World covid-19 data without China in the paper by e Fernandes [1] while Debbouche et al. [5] study "Chaotic dynamics in a novel COVID-19 pandemic model described by commensurate and incommensurate fractional-order derivatives."

In the following the case of Greece is presented for total deaths/million population vs total cases (pop%) with comments (Fig. 4).[1]

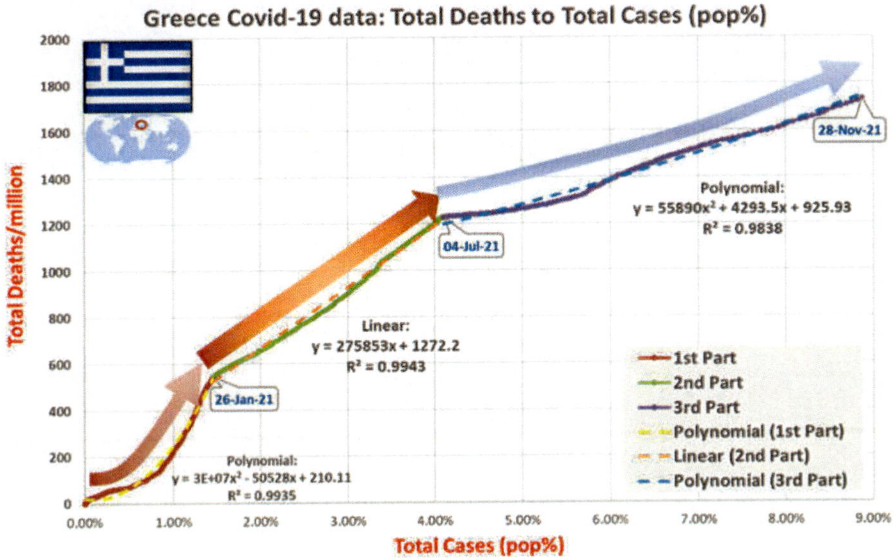

Fig. 4 Greece Covid19 data, total deaths to total cases

The slope of deaths vs cases curve declines rapidly in most countries due to: protection measures in vulnerable groups and improvement of medical treatment of cases (practically visible in the data curve of mortality from the disease).

3 Greece

The Part 4 completes the cycle on 10-Oct-21 and starts the Part 5 (5th cycle) which continues until the end of the available data (4 Nov-21) [because with Moving Average 28 days we have 4 nov + 28/2 days = 4 + 14 = 18 November. Each time we can form the circle 14 days before the end of the data].

For the initial wave propagation where part 1 at the beginning of the axes forms a small circle not clearly visible due to the graph analysis, we can see the following enlarged map forming the circle 14 days before the end of the data set (Fig. 5).

We observe that there is a cycle starting on March 4, 2020 (two weeks earlier due to MA = 28) culminating on 13 April (where the first differences = 0 ie the 1st derivative = 0 $\geq$ max). The cycle completes on July 8, 2020 where the first differences are zero again and part 2 begins.

[1] The authors invite readers feedback in YouTube® channel: *COVID-19 Data Analysis*: https://www.youtube.com/channel/UCa553hVoILqn4CJsIhiWW3w.

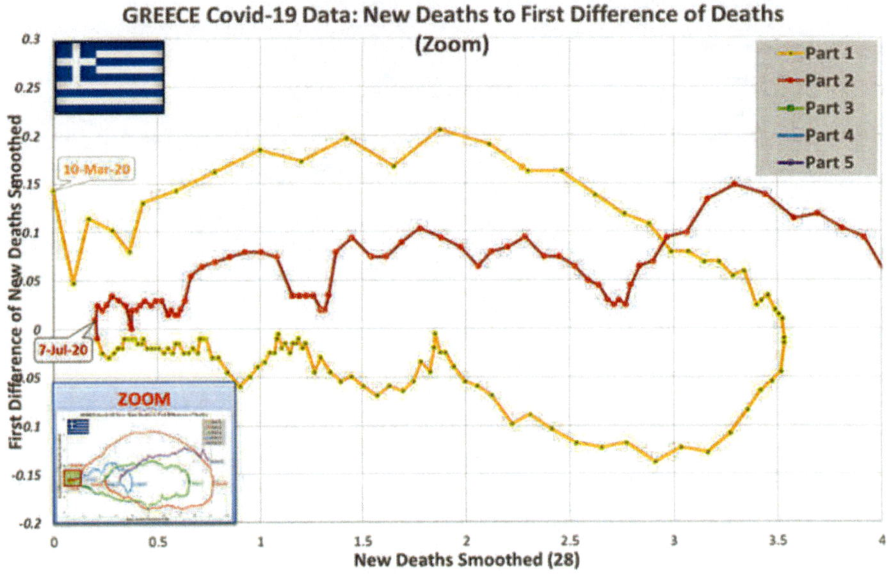

Fig. 5 Greece Covid19 data, new deaths to first difference of deaths (zoom)

We observe a small "transitional" period (from July 8, 2020 to October 21, 2020) when the cycle of part 2 (red) actually starts with the maximum reached on December 9, 2020. The cycle ends on February 11, 2021.

Then the Part 3 cycle (green) starts with a maximum on April 20, 2021 and completion on July 17, 2021. Immediately after, the part 4 cycle begins (blue) which reaches a maximum on September 12, 2021 and ends on October 10, 2021. This cycle must be due to the well-known Delta mutation that appeared in Greece in August 2021 and because it was more "contagious" it created a small cycle in the summer and with the vaccinated population of Greece over 50% (after July 2021).

The 5th cycle appear at October 10, 2021 and continues until today presenting a strong tendency to exceed the cycle of part 3 (Spring 2021).

A Greek characteristic is that the cycles move to the right in the X axis where more new deaths appear.

4 Brazil

Brazil is not known to have taken radical action against the pandemic. 28-day cycles have several periods of complex form. Furthermore, a central point of the epidemics cycles is far from the original point. Instead the point as at 1000 deaths per day (see Fig. 7). A further extended cycle has center at 1800 deaths per day (see Fig. 6). The enlargement of the graph for Brazil shows that the transition from the 2nd cycle to the

3rd cycle (at the end of Jan 21) took place immediately without a period of recession of the pandemic (semicircle in the negative region).

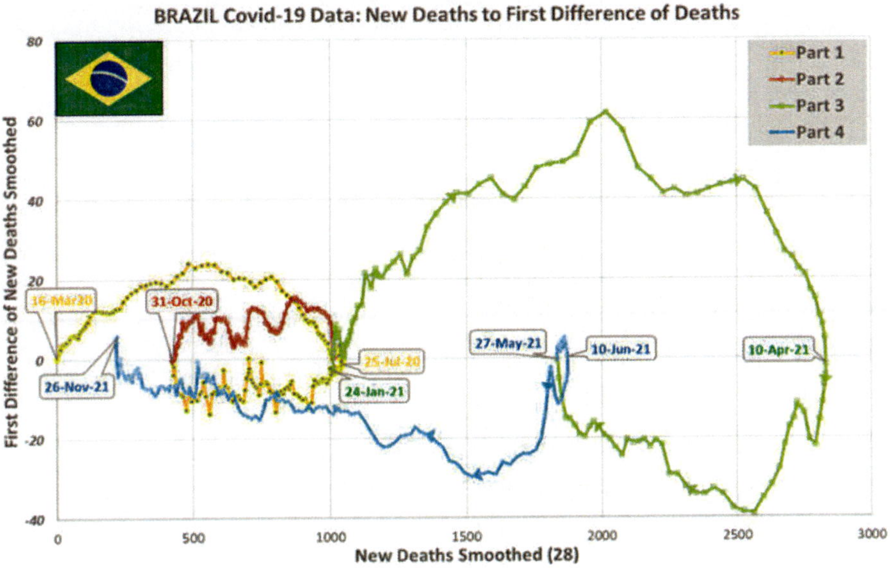

Fig. 6 Brazil Covid19 data, new deaths to first difference of deaths

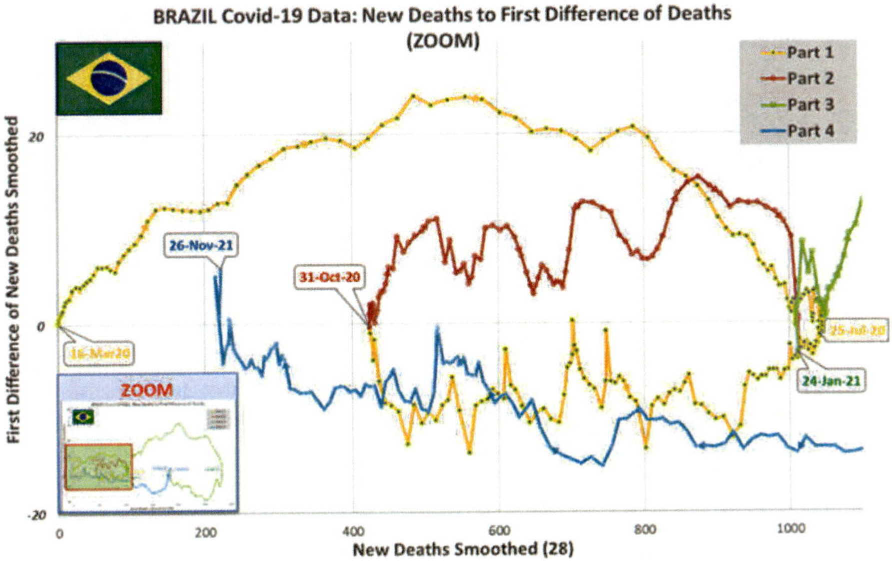

Fig. 7 Brazil Covid19 data, New Deaths to First Difference of Deaths (zoom)

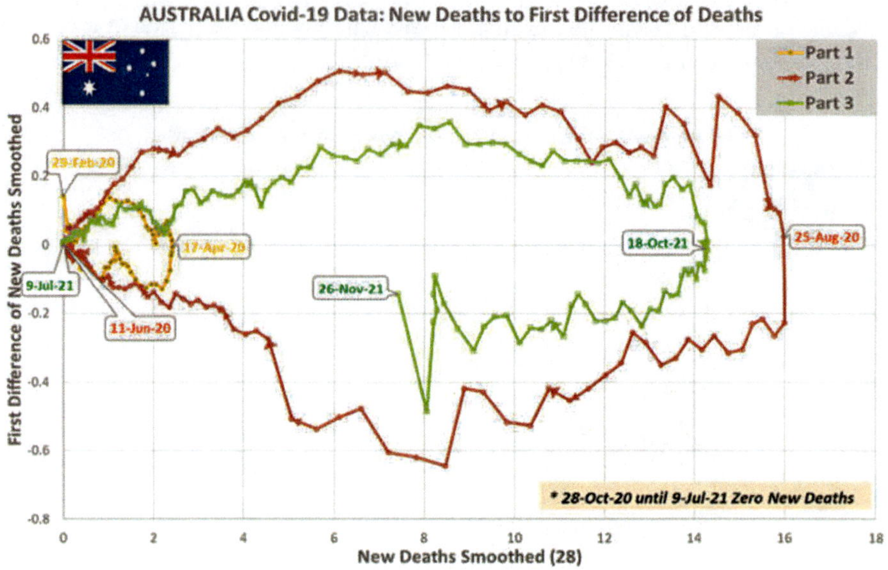

Fig. 8 Australia Covid19 data, new deaths to first difference of deaths

5 Australia

See Fig. 8.

For Australia only 3 circles are shown (the green is essentially "negligible" while the last one (blue) continuous.

Australia is the only country that managed for some time to eliminate deaths… (27-5 to 22-6/2020) and (28-10-20 to 9-7/2021). This is why we have only 3 visible circles in the graph.

6 Japan

See Fig. 9.

Japan is characterized by 5 limit cycles. The smaller 1st and 2nd (yellow and orange) appear in the first steps of the pandemic whereas the 3rd and 4th cycle periods are the larger. The 5th period shows a stabilization of cycle to a medium level with center at 30 deaths per day. Instead the 3rd and 4th cycle show a center at 60 deaths per day.

In Japan the 5 circles are visible and they are all complete. A characteristic is that the 1st and 2nd are extremely small!

The 1st, 2nd and 5th are completed with almost zero deaths in their completion.

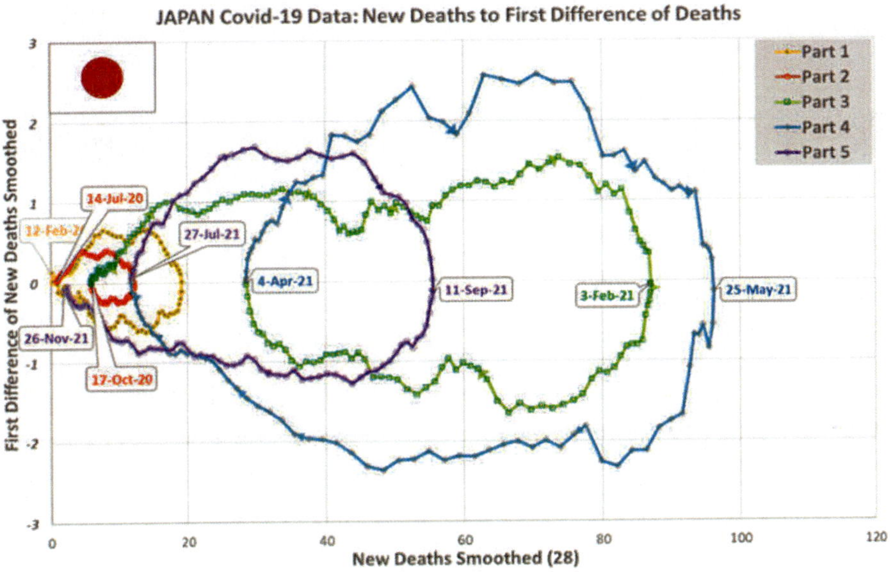

Fig. 9 Japan Covid19 data, new deaths to first difference of deaths

7 USA

See Fig. 10.

In USA the 1st, 2nd and 3rd cycles (yellow–red-green) start with a higher number of deaths than the next 4th cycle (part 4) (blue). However, the second cycle is smaller. The center for the first and second cycle are at 800 deaths per day that was expanded to 2000 deaths per day for the 3rd cycle and returned to 1000 deaths per day for the center of the 4th cycle.

8 UK

See Fig. 11.

UK is characterized by the simplicity of the cycles. The first two start and end at the origin whereas only the fourth starts after the origin and continuous as a small one until now. The center for the first cycle is at 400 death per day and at 600 deaths per day for the 2nd cycle.

The third cycle in the summer of 2021 is small and the 4th in the autumn of 2021 until today very small!

In the second cycle there is a "discontinuity" (13-Dec-20) that coincides with the period when the UK first adopted the vaccine and started mass vaccinations, while relaxing the measures.

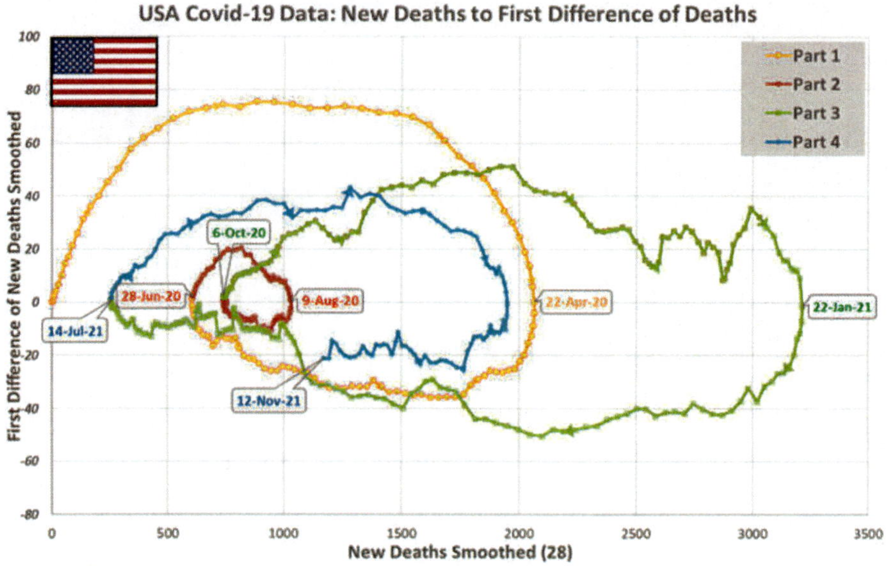

Fig. 10 USA Covid19 data, new deaths to first difference of deaths

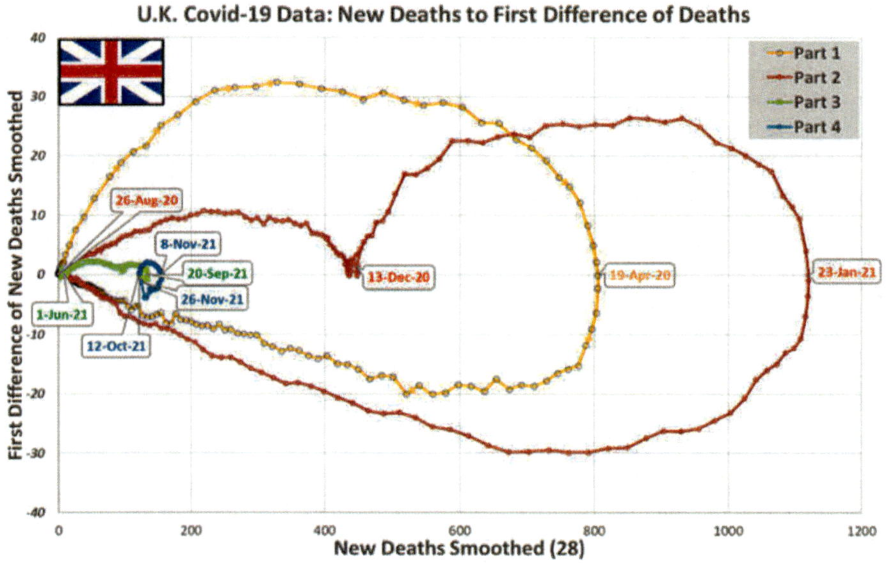

Fig. 11 UK Covid19 data, new deaths to first difference of deaths

9 Conclusions

The proposed methodology succeeds to simplify a relative complicated chaotic-like form of the Covid-19 death data set for various countries studied. Limit cycles result after a 28-day central moving average applied. The ± 14-day two sided moving average methodology accepted, compatible with usual preferred quarantine period, seems to exclude small cycles and recycles that the applied state interventions produce.

Referring to the strategies adopted from the various countries studied as can be observed by the limit cycles formed, UK provides a simple and clear strategy leading to a low number of deaths per day after the first periods with a high number of deaths per day.

Greece follows a rather complicated covid-19 spread keeping the center of the cycles relatively far from the origin. The ongoing 5th cycle covers a region with high number of deaths per day as it was for the 2nd period of the pandemic. For this and for other countries is not clear what the best strategy is. Vaccine and keeping appropriate measures are proposed while the health system waits for new medicines.

Australia and Japan cope with small numbers of deaths per day. However, they provide limit cycles indicating a systematic state intervention keeping the spread under control.

USA cycle centers are relatively far from the origin whereas these centers are very far from the origin for Brazil.

References

1. T.D.S. e Fernandes, Chaotic model for COVID-19 growth factor. Res. Biomed. Eng. (2020). https://doi.org/10.1007/s42600-020-00077-5
2. C.H. Skiadas, C. Skiadas, *Chaotic Modeling and Simulation: Analysis of Chaotic Models Attractors and Forms* (Taylor & Francis/CRC Press, London, 2009)
3. C.H. Skiadas, Chaotic Modelling. In *International Encyclopedia of Statistical Science*, ed M. Lovric (Springer, Berlin, Heidelberg, 2011) https://doi.org/10.1007/978-3-642-04898-2_166
4. R.J. Hyndman, Moving Averages. in *International Encyclopedia of Statistical Science*, ed by M. Lovric (Springer, Berlin, Heidelberg, 2011). https://doi.org/10.1007/978-3-642-04898-2_380
5. N. Debbouche, A. Ouannas, I.M. Batiha et al., Chaotic dynamics in a novel COVID-19 pandemic model described by commensurate and incommensurate fractional-order derivatives. Nonlinear Dyn. (2021). https://doi.org/10.1007/s11071-021-06867-5
6. Data Retrieved from: https://ourworldindata.org/coronavirus [Online Resource] data accessed: 23 Nov 2021

Detection of Early Warning Signals for Self-organized Criticality in Cellular Automata

Andrey Dmitriev, Anastasiia Kazmina, Victor Dmitriev, Yuriy Sanochkin, and Evgenii Gradusov

Abstract Detection the precursors of critical transitions in complex systems is one of the most difficult and still unsolved problems. This problem has not received a final solution, not only for real complex systems, but also for model systems capable to self-organize into the critical state. The presented paper is devoted to early detection of time moments of self-organized critical transitions in cellular automata as a result of the analysis of the time series they generate for a number of grains falling from the grid. It was found that cumulative moments of probability distribution and cumulative scaling exponents are quite informative indicators for early detection of critical transitions. General features of the behavior of indicators when approaching a critical point are established for the time series generated by cellular automata with different rules.

Keywords Cellular automata · Sandpile model · Self-organized criticality · Time series · Probability moments · Multifractality

1 Introduction

More than thirty years development of the theory of self-organized criticality (SOC), explaining the emergence of power law for probability density function, 1/f-noise and long-range spatial and temporal correlation in nonlinear systems far from equilibrium, has led to the emergence of the number of basic models, which have nontrivial scale-invariant dynamics under very simple local rules [1, 2]. The basic models of SOC theory are sandpile models [3]. These models have become the most important tool for studying the mechanisms of the appearance of scale-invariant properties and power statistics.

A. Dmitriev (✉) · A. Kazmina · V. Dmitriev · Y. Sanochkin · E. Gradusov
Department of Business Informatics, National Research University Higher School of Economics, Moscow, Russia
e-mail: a.dmitriev@hse.ru

A. Dmitriev
Cybersecurity Research Center, University of Bernardo O'Higgins, Santiago, Chile

C. H. Skiadas and Y. Dimotikalis (eds.), *14th Chaotic Modeling and Simulation International Conference*, Springer Proceedings in Complexity,
https://doi.org/10.1007/978-3-030-96964-6_10

The sandpile model is a conical pile of sand, on the center of which grains of sand are placed one by one. We will assume that the cohesion between grains of sand is large enough and only superficial movement of sand is possible. Then the state of the system is determined by the local slope of the surface of the sand pile (S). If S is small, then the sand is motionless. If S exceeds a certain value S_c, then there is a spontaneous flow of sand η over the surface, which increases continuously with increasing of S. This process corresponds to a second-order phase transition, in which the control parameter is S, the order parameter is η. The value S_c separates subcritical (SubC) phase and supercritical (SupC) phase. A pile of sand in these phases is resistant to small disturbances. On the contrary, the SOC state is highly volatile. Adding just one grain of sand to a pile in this state can lead to avalanches of sand of any size theoretically.

A fundamentally important property of systems, which are characterized by avalanche-like behavior, is their ability to self-organize into the critical state. In this case, it is not required to fine-tune the parameter S to the value S_c. Such systems are capable to transit to the SOC state spontaneously, which is typical for most real and model complex systems, the behavior of which is determined by the nonlinear local rules.

Scale-invariant properties and power statistics are characteristics not only for the level of the structure of the complex system and its local microscopic interactions, but also for the level of time series generated by such systems [4]. For the sandpile model, such time series are the time series (η_t), which demonstrate the stochastic dynamics of sand grains falling on the surface of the pile. When describing a pile of sand using cellular automata models, η_t is the number of grains falling from the grid. The approach to the study of self-organized critical states of complex systems based on the analysis of generated time series has, at least, one significant advantage. The approach does not require the study of detailed interactions between elements of the real systems. Information about detailed interactions is usually inaccessible for research, for example, for social networks, or inaccessible, for example, for financial networks.

Detection of early warning signals for critical transitions is a challenging task not only for the real complex systems, but also for the model systems. The overwhelming majority of the papers known to us are devoted either to detection of early warning signals associated with the critical slowing down phenomenon [5–8], or to the solution of particular problems of the early warning [9, 10]. In these studies, precursors of the critical transitions in real systems were established, which are associated with the change in the autocorrelation function, variance, skewness, and power spectral density of the observed time series when the system parameters approach their critical value.

To our knowledge, however, there is no study that investigates the search for precursors of the SOC transitions not only in real complex systems, but also in model systems, for example, in the self-organized criticality cellular automata. The detection of such precursors as a result of the analysis of the time series for the number of grains falling from the grid is the purpose of our study.

2 Methods

2.1 Time Series Generated by the Cellular Automata

It is easy to study the stochastic dynamics of the order parameter (η_t) of the sandpile models using models of cellular automata in grids of size $L \times L$. The parameter η_t is the number of sand grains falling from the grid at time t. Random integers $z_{i,j}$ are generated in the grid cells to represent the local slope of the sand pile. The cells for which $z_{i,j} \geq z_c$, where z_c is the critical value, are unstable and fall off according to the rules defined for each cellular automata. For the pile of sand, several different variants of the rules for shedding an unstable cell have been proposed. This paper considers six sandpile models, each of which belongs to one of two classes of self-organized critical models: conservative and dissipative models. Models with conservative rules are characterized by the fact that when unstable cells fall, the grains of sand removed from them are redistributed without loss and leave the grid only after reaching its edges. The boundary conditions of such systems are open. In dissipative models, after the shedding of the unstable cell, the number of grains of sand in it is zero. In the case of the supercritical number of sand grains, they are able to leave the grid also within its boundaries.

2.1.1 Conservative Systems

Let us consider in more detail the historically very first model, called the BTW-model [11]. Consideration of other models, including dissipative models, is limited to consideration of only the rules for shedding cells.

BTW-model is a cellular automaton on a square grid of size $L \times L$. A grain of sand is randomly added to a randomly selected cell (i, j), increasing the number of grains of sand ($z_{i,j}$) in the cell by one: $z_{i,j}+ = 1$. As a result, $z_{i,j} \rightarrow z_{i,j} + 1$. If $z_{i,j} \geq 4$, then one grain of sand moves to the four nearest cells: $z_{i\pm1,j\pm1}+ = 1$. In this case, the number of grains of sand in the cell (i, j) decreases by the value $z_c = 4$: $z_{i,j}- = 4$. The considered movement of sand grains can lead to loss of stability of neighboring cells, and, consequently, lead to the appearance of the avalanche with loss of stability. The introduction of the condition $z_{i,j}- = 4$ leads to the saving of the number of sand grains.

Thus, the rules of the model are as follows:

$$z_c = 4, \; z_{i,j}- = 4, \; z_{i\pm1,j\pm1}+ = 1 \tag{1}$$

In Fig. 1 the time series for the number of grains falling from the grid for a 40×40 grid of the BTW-model are presented. The rest of the time series looks the same except for the iteration number (or point in time t_c) corresponding to the SOC state. t_c depends on the grid's size: $t_c = 1656$ for 20×20 grid, $t_c = 2171$ for 30×30

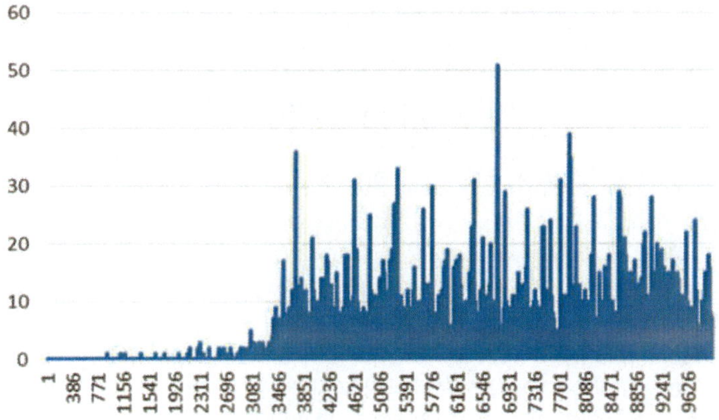

Fig. 1 Time series of the number of grains falling from the grid for the BTW-model

grid, $t_c = 3736$ for 40×40 grid, $t_c = 5491$ for 50×50 grid, and $t_c = 8234$ for 60×60 grid.

Manna model [12] is the stochastic analogue of the BTW-model. The cell (i, j) crumbles as a result of the stability loss, transferring a random number of grains of sand ($\delta_k \geq 0$) to four neighboring cells.

Formally, the rules of the model are as follows:

$$z_c = 4,\ z_{i,j}- = 4,\ z_{i\pm1,j\pm1}+ = \delta_k,\ \delta_k \geq 0,\ \sum_k \delta_k = 4 \tag{2}$$

In Fig. 2 the time series for the number of grains falling from the grid for a 40×40 grid of the Manna model are presented. The rest of the time series looks the same except for the iteration number corresponding to the SOC state. t_c depends on the

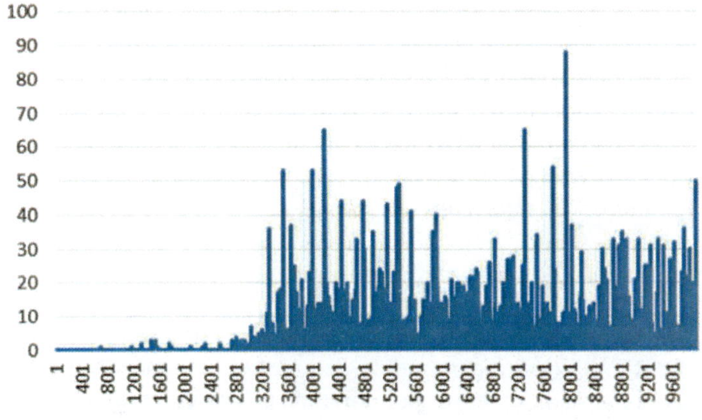

Fig. 2 Time series of the number of grains falling from the grid for the Manna model

grid's size: $t_c = 892$ for 20×20 grid, $t_c = 2510$ for 30×30 grid, $t_c = 3335$ for 40×40 grid, $t_c = 5671$ for 50×50 grid, and $t_c = 7625$ for 60×60 grid.

DR-model [13] is the cellular automaton, the rules of which are formulated on a two-dimensional hexagonal lattice. It is a cellular automaton with open boundary conditions on the lower side and periodic boundary conditions on the left and right sides. A grain of sand is randomly added to a randomly selected cell (i, j) of the top layer, increasing the number of grains of sand $(z_{i,j})$ in the cell by one: $z_{i,j}+ = 1$. When the value in any cell exceeds one, this cell loses stability and crumbles, transferring one grain of sand to the two cells lying below. It is important that the DR-model rules are anisotropic, i.e. the avalanche of sand grains, spreading from the top to the bottom, never affects the same area twice.

Formally, the rules of the model are as follows:

$$z_c = 2,\ z_{i,j}- = 2,\ z_{i\pm 1, j\pm\frac{1}{2}}+ = 1 \tag{3}$$

In Fig. 3 the time series for the number of grains falling from the grid for a 40×40 grid of the DR-model are presented. The rest of the time series looks the same except for the iteration number corresponding to the SOC state. t_c depends on the grid's size: $t_c = 1491$ for 20×20 grid, $t_c = 1821$ for 30×30 grid, $t_c = 3241$ for 40×40 grid, $t_c = 5200$ for 50×50 grid, and $t_c = 7476$ for 60×60 grid.

PSV-model [14] is the stochastic analogue of DR-model. The cell (i, j) crumbles as a result of the stability loss, transferring the random number of sand grains $(\delta_\pm \geq 0)$ to the two cells lying below.

Formally, the rules of the model are as follows:

$$z_c = 2,\ z_{i,j}- = 2,\ z_{i\pm 1, j\pm\frac{1}{2}}+ = \delta_\pm,\ \delta_\pm \geq 0,\ \delta_+ + \delta_- = 2 \tag{4}$$

In Fig. 4 the time series for the number of grains falling from the grid for a 40×40 grid of the PSV-model are presented. The rest of the time series looks the same

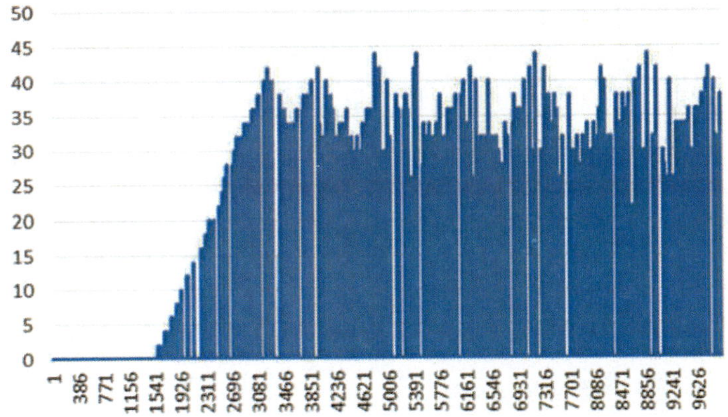

Fig. 3 Time series of the number of grains falling from the grid for the DR-model

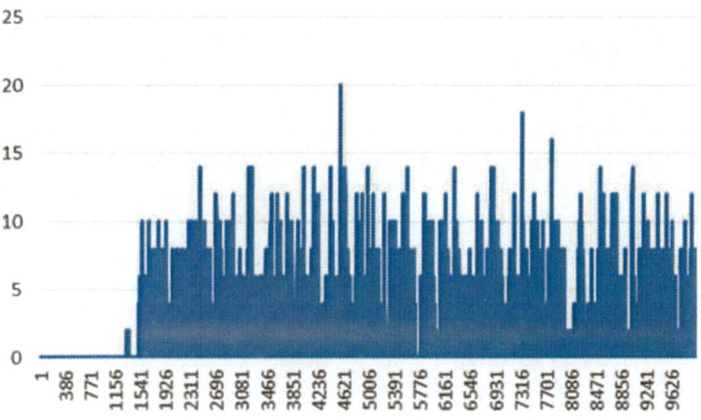

Fig. 4 Time series of the number of grains falling from the grid for the PSV-model

except for the iteration number corresponding to the SOC state. t_c depends on the grid's size: $t_c = 1$ for 20×20 grid, $t_c = 1$ for 30×30 grid, $t_c = 1$ for 40×40 grid, $t_c = 1$ for 50×50 grid, and $t_c = 1$ for 60×60 grid.

2.1.2 Dissipative Systems

The rules of the models considered above are conservative, i.e. when cells are shattered, the grains of sand removed from them are redistributed to neighboring cells without loss. The grains of sand leave the grid only when they reach its edges.

DFF-model [15] is a deterministic cellular automaton with a two-dimensional orthogonal grid of size $L \times L$. The integers in the cells $z_{i,j}$ can be interpreted as the number of grains of sand that can participate in the pouring processes. There is no designated slope direction. If $z_{i,j} \geq 4$, then the cell (i, j) is unstable and overturns. Overturn is zeroing of the number of sand grains in the cell with a simultaneous increase by 1 in the values in four cells that have a common side with this cell.

Formally, the rules of the model are as follows:

$$z_c = 4, \; z_{i,j}- = 0, \; z_{i\pm1,j\pm1}+ = 1 \tag{5}$$

In Fig. 5 the time series for the number of grains falling from the grid for a 40×40 grid of the DFF-model are presented. The rest of the time series looks the same except for the iteration number corresponding to the SOC state. t_c depends on the grid's size: $t_c = 1$ for 20×20 grid, $t_c = 1$ for 30×30 grid, $t_c = 1$ for 40×40 grid, $t_c = 1$ for 50×50 grid, and $t_c = 1$ for 60×60 grid.

The stochastic DFF-model with a random number of sand grains (δ_k) in four neighboring cells that have a common side with the cell is characterized by the following rules:

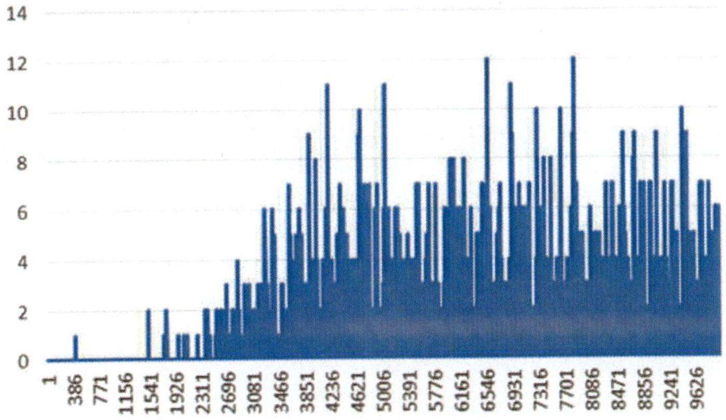

Fig. 5 Time series of the number of grains falling from the grid for the DFF-model

$$z_c = 4,\ z_{i,j} - = 0,\ z_{i\pm1,j\pm1} + = \delta_k,\ \delta_k \geq 0,\ \sum_k \delta_k = 4 \tag{6}$$

In Fig. 6 the time series for the number of grains falling from the grid for a 40 × 40 grid of the stochastic DFF-model are presented. The rest of the time series looks the same except for the iteration number corresponding to the SOC state. t_c depends on the grid's size: $t_c = 1$ for 20 × 20 grid, $t_c = 1$ for 30 × 30 grid, $t_c = 1$ for 40 × 40 grid, $t_c = 1$ for 50 × 50 grid, and $t_c = 1$ for 60 × 60 grid.

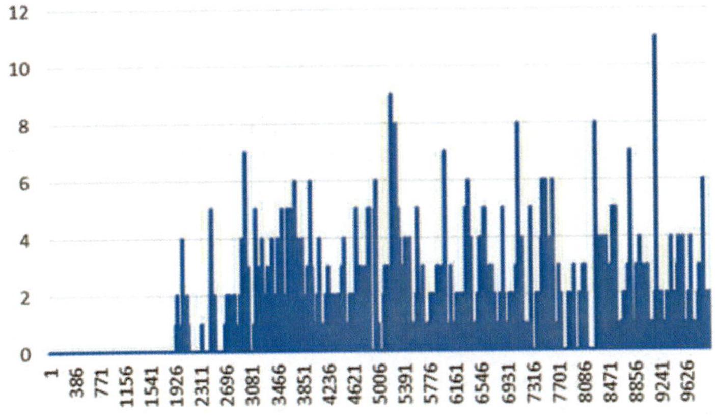

Fig. 6 Time series of the number of grains falling from the grid for the stochastic DFF-model

2.2 Moments of Probability Density Function for the Time Series

Earlier, we proposed the algorithm for detecting the self-organized critical state of the system. The algorithm is based on the analysis of scaling exponents of power laws for probability density function (α), power spectral density (β), and autocorrelation function (γ) for time series generated by SOC systems. The algorithm make it possible to identify the SubC phase and the SupC phase by belonging of α, β and γ to certain intervals. The disadvantage of the algorithm is its applicability only to the analysis of scale-invariant probability density function and the impossibility of its application to the analysis of other heavy-tailed distributions. The heavy-tailed distributions are characteristic for the time intervals of the system evolution, corresponding to its SupC phase. A famous example of a scale-invariant heavy-tailed distribution is the Pareto distribution.

Therefore, in order to go beyond the limitations of the algorithm associated only with the use of α as the only identifier of the SOC state, the SubC phase and the SupC phase, we used the main moments of probability density function as cumulative indicators.

For detection of early warning signals for self-organized criticality we used the following moments:

first raw moment (or mean) μ,
second central moment (or variance) σ^2,
standardized third moment (or skewness) γ,
standardized fourth moment (or kurtosis) κ.

2.3 Scaling Exponents for the Time Series

Even the description of model time series using the moments of their probability density function is exhaustive only for a very limited number of random processes. For example, realizations of Gaussian processes are fully described by second-order moments. Therefore, apart from the moments, other quantities should be used to describe the time series. These quantities include the scaling exponents for the time series, which determine the fractal dimensions of time series as geometric objects.

The most general approach to the study of scaling exponents of heterogeneous time series is their multifractal analysis. It is sufficient to calculate a single scaling exponent to describe the scale invariance of homogeneous model time series, since such time series demonstrate only one type of singular behavior constant in time. On the contrary, the nature of the singularity of inhomogeneous time series at different points in time may differ; therefore, the description of such time series cannot be performed using only scaling constant. Therefore, multifractal analysis, which allows to provide local analysis of heterogeneous time series, is a more informative approach.

We used multifractal detrended fluctuation analysis (MF-DFA) [16] for making of multifractal analysis of time series generated by self-organized critical cellular automata. The application of this method makes it possible to obtain estimates of the spectrum of scaling constant time series: $\{h(q)\}$.

In short, the algorithm of MF-DFA method is reduced to revealing the power law

$$F(q, s) \propto s^{h(q)} \tag{7}$$

for the fluctuation function

$$F(q, s) = \left\{ \frac{1}{2N_s} \sum_{\nu=1}^{2N_s} [\mu(\nu, s)]^{\frac{q}{2}} \right\}^{\frac{1}{q}} \tag{8}$$

To calculate function (8) from a discrete time series η_i, a fluctuation profile is formed $X_i = \sum_{k=1}^{i} (\eta_k - \overline{\eta})$, which is divided into N_s non-intersecting intervals ν containing the equal number of points s. Further, for each of the intervals, the local trend $x_{\nu,i}$ and the deviation of the fluctuation profile from the local trend $\Delta Y_{\nu,i} = X_{\nu,i} - x_{\nu,i}$ are determined. The value $\mu(\nu, s) = \max \Delta Y_{\nu,i} - \min \Delta Y_{\nu,i}$ for each split interval.

A detailed description of the algorithm of the MF-DFA method, as well as its capabilities and limitations, are presented in the paper [16]. Therefore, we will restrict ourselves by considering the main features of the time series for which the power law is satisfied (7). For multifractal time series at $q > 0$, the main contribution to function (8) is given by the partition intervals ν characterized by large values $\mu(\nu, s)$; at $q < 0$, the main contribution to function (8) comes from the partition intervals ν characterized by small values $\mu(\nu, s)$. For monofractal time series $h(q)$ does not depend on q. This is due to the fact that the behavior of function (8) when changing the scale s is the same for all intervals ν.

3 Results and Their Discussion

Cumulative mean and variance, as well as their corresponding time series, are presented in Figs. 7 and 8. These figures show moments and time series for the Manna model. The dimensions of cellular automata are 40×40. For other cellular automata and their other grid sizes, the mean and variance behavior are similar.

Cumulative mean and variance are not informative indicators characterizing the transition of cellular automata in the SOC state. Indeed, these moments are increasing functions of time and there are no significant changes in them when passing through the SOC state.

Cumulative kurtosis and skewness are quite informative precursors for the transition of cellular automata into the SOC state. The cumulative kurtosis and skewness, as well as their corresponding time series, are presented in Figs. 9 and 10. These

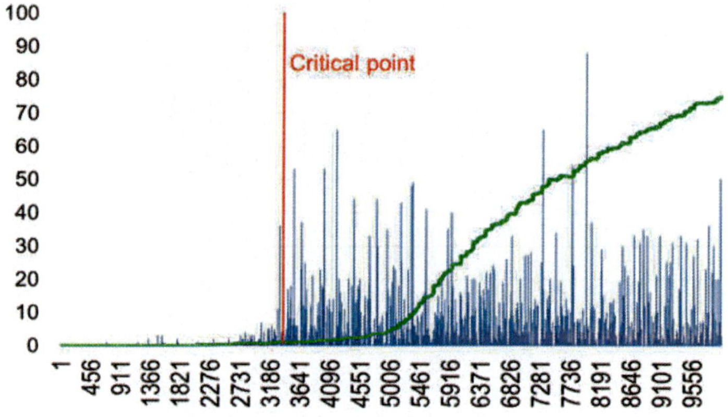

Fig. 7 Cumulative mean for the Manna model

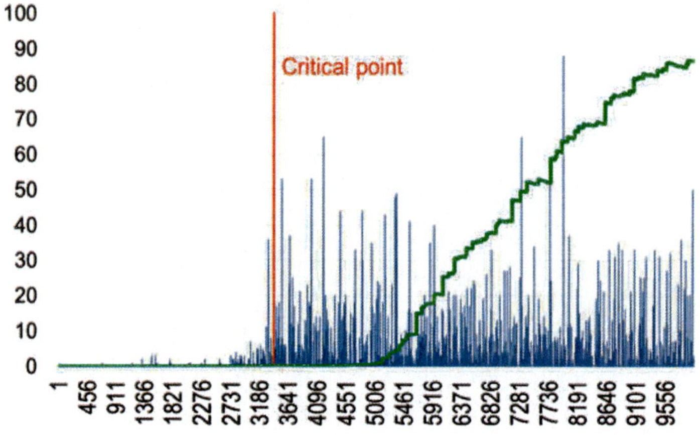

Fig. 8 Cumulative variance for the Manna model

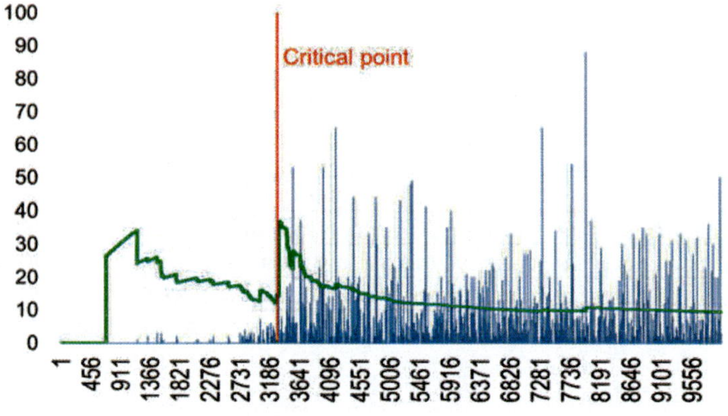

Fig. 9 Cumulative kurtosis for the Manna model

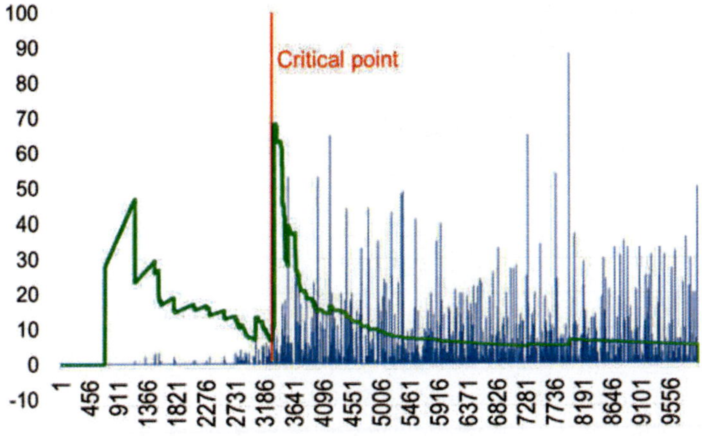

Fig. 10 Cumulative skewness for the Manna model

figures show the moments and time series for the Manna model. The dimensions of cellular automata are 50×50. For other cellular automata and their other grid sizes, the behavior of kurtosis and skewness is similar.

As the cellular automata approach to the SOC state, a noticeable decrease in kurtosis and skewness is observed up to the critical point. At the same time, a sharp increase in these cumulative moments is observed at the critical point.

The change in the cumulative moments when approaching the critical point has a simple explanation. Mean and variance increase as a result of the increase in the number of grains falling from the grid in the certain time interval $\Delta t_c \in (t, t_c)$ from the SubC phase, preceding the transition of the cellular automaton to the critical state. The decrease in the skewness in the interval Δt_c is also a consequence of the increase in the number of the grains. In this interval, a right-sided asymmetry of the distribution is still observed, characterized by an elongated right "tail," which decreases as the critical point is approached. In other words, the shortening of the right "tail" of the distribution occurs in the interval Δt_c. At the critical point, a sharp lengthening of the right "tail" of the distribution occurs as a result of the accumulation of the number of the grains from the SupC phase. An increase in the number of the grains in the interval Δt_c also leads to a decrease in kurtosis in this interval. The peak of the distribution near the mathematical expectation is sharp for the entire SubC phase, but as the critical point is approached, the peak of the distribution is smoothed out. At the critical point, there is severe increase in the sharpness of the distribution.

In Fig. 11 the cumulative scaling exponents $h(q)$ at $q = 1, 3, 5$ are shown. The behavior of scaling exponents is similar for all cellular automata and their sizes; therefore, we restrict ourselves by considering only two automata. The scaling exponents $h(1)$ are shown in blue, in orange—$h(3)$ and in gray—$h(5)$.

The time series generated by cellular automata are multifractal time series. Moreover, scale invariance in the form (7)) is characteristic only for $q > 0$, for $q < 0$ scale invariance is not observed. Therefore, there are only the scaling exponents

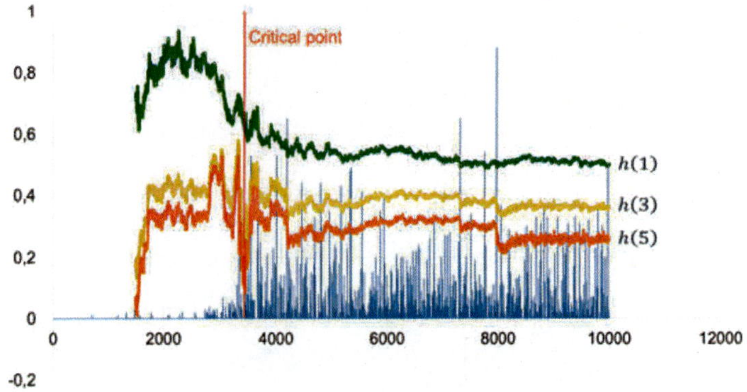

Fig. 11 Cumulative scaling exponents for the Manna model (50 × 50)

describing the intervals of time series partitioning ν with large fluctuations. Intervals ν with small fluctuations are not typical for the studied time series.

As approaching to the critical point, the distance between the points $h(q_i) = |h(1) - h(3)| + |h(3) - h(5)|$ decreases and is the smallest at the critical point. In the SupC phase, the distance between the points is almost independent of the iteration. All this is demonstrated in Fig. 11. Recall that for the Manna model (50 × 50) $t_c = 5671$.

Thus, the moments of probability density function, primarily γ and κ, as well as the scaling exponents $h(q)$, can be used as indicators of early warning for the SOC state in cellular automata.

4 Conclusions

Analysis of the behavior in time of moments and scaling exponents made it possible to provide early detection of the self-organized critical state in cellular automata. For such the early detection, it is sufficient to carry out the statistical and multifractal detrended fluctuation analysis of time series for the number of grains falling from the grid generated by cellular automata.

The results obtained allow us to make the following conclusions:

(1) Self-organized critical cellular automata generate multifractal time series, in which subcritical and supercritical phase, and self-organized criticality state can be distinguished.
(2) The most informative indicators of early detection of self-organized criticality state are cumulative skewness and kurtosis.
(3) Multifractality of time series for number of grains falling from the grid makes it possible to use cumulative scaling exponents as indicators of early detection of self-organized criticality state.

In conclusion, we briefly consider the possible practical applications of the use of the proposed indicators for early detection of critical states. If real systems are able to self-organize into the critical state, then the cumulative moments of probability distribution and the cumulative scaling exponents can be used as early warning indicators for critical states. Self-organized criticality is characteristic of phenomena and processes of a very different nature: solar flares, earthquakes, floods, forest fires, the emergence and extinction of species, demographic, ecological, economic, social, informational processes. Early detection of the critical state means predicting the critical moment in time after which the system behaves in an unpredictable manner. In this case, the system is in the supercritical phase, which is characterized by avalanche-like dynamics.

Acknowledgements The work is an output of a research project implemented as part of the Basic Research Program at the National Research University Higher School of Economics (HSE University).

References

1. P. Bak, C. Tang, K. Wiesenfeld, Self-organized criticality: An explanation of the 1/f noise. Phys. Rev. Lett. **59**, 381 (1987)
2. D. Dhar, Theoretical studies of self-organized criticality. Physica A **369**(1), 29–70 (2006)
3. M. Creutz, Self-organized criticality and cellular automata, in *Encyclopedia of Complexity and Systems Science*, ed. by R. Meyers (Springer, Berlin, 2018)
4. A. Dmitriev,V. Dmitriev, Identification of self-organized critical state on twitter based on the retweets' time series analysis. Complexity (Article ID 6612785, 2021)
5. M. Scheffer et al., Early-warning signals for critical transitions. Nature **461**, 53–59 (2009)
6. C. Diks, C. Hommes, J. Wang, Critical slowing down as an early warning signal for financial crises? Empirical Econo. **57**, 1201–1228 (2019)
7. Z. Zhang, Y. Li, L. Hu, C. Tang, H. Zheng, Predicting rock failure with the critical slowing down theory. Eng. Geol. **280**, 105960 (2021)
8. V. Dakos et al., Methods for detecting early warnings of critical transitions in time series illustrated using simulated ecological data. Plos One, **7**(7), e41010 (2012)
9. J.J. Lever et al., Foreseeing the future of mutualistic communities beyond collapse. Ecol. Lett. **23**, 2–15 (2020)
10. L. Zhao et al., Multifractality and network analysis of phase transition. Plos One **12**(1), e0170467 (2017)
11. P. Bak, C. Tang, K. Wiesenfeld, Self-organized criticality. Phys. Rev. A **38**(1), 364–374 (1988)
12. S.S. Manna, Two-state model of self-organized criticality. J. Phys. A **24**(7), 363–639 (1991)
13. D. Dhar, R. Ramaswamy, Exactly solved model of self-organized critical phenomena. Phys. Rev. Lett. **63**(16), 1659–1662 (1989)
14. R. Pastor-Satorras, A. Vespignani, Universality classes in directed sandpile models. J. Phys. A **33**(3), 33–39 (2000)
15. H.J.S. Feder, J. Feder, Self-organized criticality in a stick-slip process. Phys. Rev. Lett. **66**(20), 2669–2672 (1991)
16. E.A.F. Ihlen, Introduction to multifractal detrended fluctuation analysis in Matlab. Front. Physio. **3**, 141 (2012)

Double Symmetry and Generalized Intermittency in Transitions to Chaos in Electroelastic Systems

Serhii Donetskyi and Aleksandr Shvets

Abstract Mathematical models of a deterministic system of the type "analog generator-piezoelectric transducer" are considered. A double symmetry, atypical for dynamical systems, is found in the alternation of scenarios of transitions from regular attractors to chaotic ones. For the considered system, the symmetry inside symmetry: the described above chains of scenarios is located at the "median" point of other wider symmetric chains of transition to chaos was found. Also, for the first time for the considered system, a transition "chaotic attractor of one type-chaotic attractor of another type" through generalized intermittency was discovered. One of the distinctive features of such a transition is the appearance of coarse-grained (rough) laminar phase instead of laminar phase of usual intermittency.

Keywords Nonideal electro-elastic systems · Scenarios of transition to chaos · Generalized intermittency.

1 Introduction

Consider a cylindrical piezoceramic transducer placed in an acoustic medium. Let us assume that the oscillations of a piezoceramic transducer are excited by an analog generator. Let's also assume that the power of the generator is comparable to the power consumed by the transducer. Under these assumptions, the "generator—piezoceramic transducer" system is a typical nonideal dynamic system according to Sommerfeld-Kononenko (Sommerfeld [1, 2], Kononenko [3]). The mathematical model of such a system was described using a normal system of ordinary differential equations in Krasnopolskaya and Shvets [4].

The mathematical model of the "generator-transducer" system was derived for a real physical system based on the strict principles of the general theory of electroelastic systems in acoustic media. Subsequently, it was revealed that the "generator-

S. Donetskyi · A. Shvets (✉)
National Technical University of Ukraine "Igor Sikorsky Kyiv Polytechnic Institute", Kyiv, Ukraine
e-mail: aleksandrshvetskpi@gmail.com

C. H. Skiadas and Y. Dimotikalis (eds.), *14th Chaotic Modeling and Simulation International Conference*, Springer Proceedings in Complexity,
https://doi.org/10.1007/978-3-030-96964-6_11

transducer" system has a very wide variety of dynamic behavior. So in such a system, all the main types of regular attractors were discovered, such as equilibrium positions, limit cycles and invariant tori (Krasnopolskaya and Shvets [4], Balthazar et al. [5], Shvets and Donetskyi [6]). Chaotic attractors, including hyperchaotic ones, were also found in the "generator-transducer" system (Shvets and Krasnopolskaya [7]). Transitions to chaos (hyperchaos) through a cascade of period doubling bifurcations (Feigenbaum [8, 9] and through intermittency (Manneville and Pomeau [10]) were identified. And finally, in paper Shvets and Donetskyi [6], self-excited, hidden and rare attractors were discovered in the "generator-transducer" system.

The above allows us to assert that the "generator-converter" system has greater variety of dynamic behavior than the classical Lorentz ([11]) and Rössler ([12, 13]) systems. Such system is the "library" of regular and chaotic dynamics and can be used as a basic one in the study of the general theory of dynamical systems.

2 Mathematical Model

Using papers Krasnopolskaya and Shvets [4], Shvets and Donetskyi [6], we write the mathematical model of the "generator-converter" system in the form of a normal system of differential equations:

$$\begin{aligned}\frac{d\xi}{d\tau} &= \zeta,\\ \frac{d\zeta}{d\tau} &= -\xi + \alpha_1\zeta + \alpha_2\zeta^2 - \alpha_3\zeta^3 - \alpha_4\beta,\\ \frac{d\beta}{d\tau} &= \gamma,\\ \frac{d\gamma}{d\tau} &= -\alpha_0\beta + \alpha_5\xi + \alpha_6\zeta - \alpha_7\gamma.\end{aligned} \tag{1}$$

Here phase variables ξ, ζ describe the dynamics of piezoceramic transducer. Accordingly, phase variables β, γ describe the dynamics of analog generator. The physical meaning of these variables and parameters $\alpha_0, \alpha_1, ..., \alpha_7$ of the system (1) are described in detail in paper Krasnopolskaya and Shvets [4].

Since the system of Eq. (1) is a nonlinear system of differential equations, the study of its dynamic behavior, in the general case, can be carried out only by numerical methods. The methodology for conducting such research is described in the papers Shvets [14], Shvets and Krasnopolskaya [7].

3 Symmetry and Double Symmetry

Typical behavior for dynamical system is when, with increase(decrease) in the value of a bifurcation parameter, the following chain of transitions to chaos is observed: a cascade of bifurcations of period doubling of limit cycles, then chaos, then so called periodicity window, after which this chain repeats: cascade of period doubling →

Fig. 1 Phase-parametric characteristic

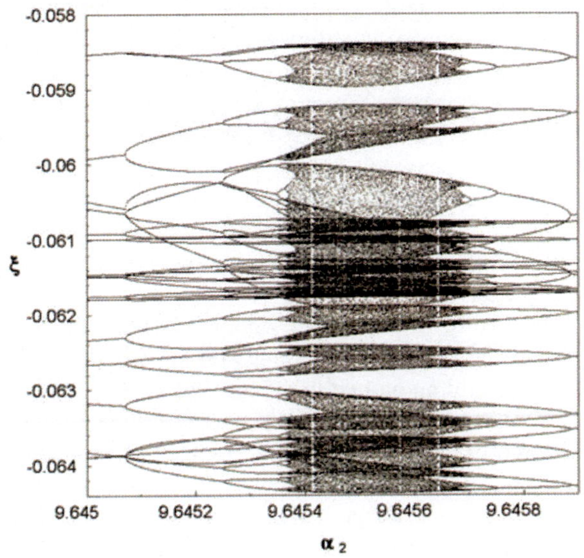

chaos → periodicity window → cascade of period doubling → ... This behavior is also known as Feigenbaum scenario. Accordingly, with a decrease(increase) in the value of a bifurcation parameter, different chain of transactions to chaos is observed. Namely limit cycle, then intermittency in chaos, then periodicity window, after which this chain repeats: limits cycle → intermittency in chaos → periodicity window → limit cycle → ... This behaviour is known as Pomeau-Manneville scenario. In this system, however, there are regions of parameters for which violation of strict chain of transactions for either scenarios is observed.

Let values of parameters be $\alpha_0 = 0.995$, $\alpha_1 = 0.0535$, $\alpha_3 = 9.95$, $\alpha_4 = -0.103$, $\alpha_5 = -0.0604$, $\alpha_6 = -0.12$, $\alpha_7 = 0.01$. And leave parameter α_2 as bifurcation one. In Fig. 1, for these values of parameters, the phase-parametric characteristic of the system (1), the so-called bifurcation tree, is constructed. Steady-state periodic regimes correspond to individual branches of this tree, and chaotic ones correspond to densely black areas. A careful study of the phase-parametric characteristics allows us to understand the bifurcations occurring in the system. As one may notice, there is some symmetry value of bifurcation parameter ($\alpha_2 \approx 9.6455$), relative to which any chain of transitions to chaos is reflected. This means that with increase in the value of the bifurcation parameter both Feigenbaum scenario and Pomeau-Manneville one occur, which is violation of strict chain of transitions to chaos. Same behavior is also true for the case of decrease in the value of the bifurcation parameter. We notice that such situation appears to be natural for this specific system, since it is not the first time when such symmetry in transition to chaos was established (Shvets and Donetskyi[6]). And as we will see further, not the last.

Consider couple more intervals of bifurcation parameter for which symmetric transition to chaos is observed along with some other interesting features.

Let us start with interval $9.075 < \alpha_2 < 9.3$. As we can see from Fig. 2, there is a double symmetry in the alternation of scenarios of transitions to chaos. One of the symmetries is clearly seen over the entire range of variation of the bifurcation parameter. Inside this symmetry, in a much smaller interval, one more symmetry is visible. Such double symmetries (symmetries within symmetries) are quite atypical for dynamical systems. Just like before, we can see violation of strict chain of transitions to chaos both with increase and with decrease in the value of the bifurcation parameter.

In the Fig. 3 you can see couple of bifurcations of Feigenbaum scenario for the Phase-parametric characteristic presented in the Fig. 2. Namely, there are three first period doubling plotted in the Fig. 3a –c. And the chaos presented in the Fig. 3d.

Another type of symmetry is realized on the interval of variation of the bifurcation parameter $9.646 < \alpha_2 < 9.64625$. The phase-parametric characteristic of the system for this interval is shown in Fig. 4. Here, the transition to chaos occurs, in one bifurcation, through the intermittency both on the left and on the right of the considered interval. Moreover, there are no periodicity windows inside the chaos. Accordingly, no other transitions to chaos are observed according to the Feigenbaum scenario.

Fig. 2 Phase-parametric characteristic

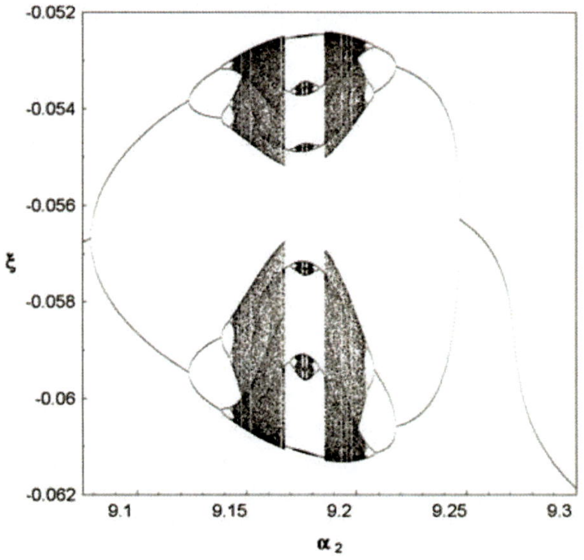

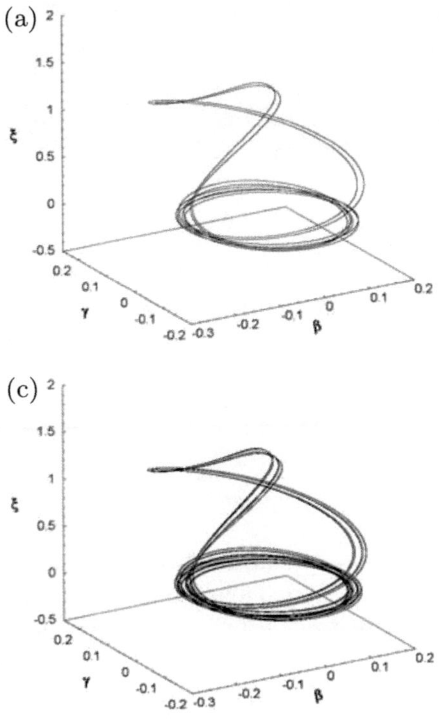

Fig. 3 Phase portrait projections: at $\alpha_2 = 9.1$ (**a**); at $\alpha_2 = 9.13$ (**b**); at $\alpha_2 = 9.14$ (**c**); at $\alpha_2 = 9.15$ (**d**).

Fig. 4 Phase-parametric characteristic

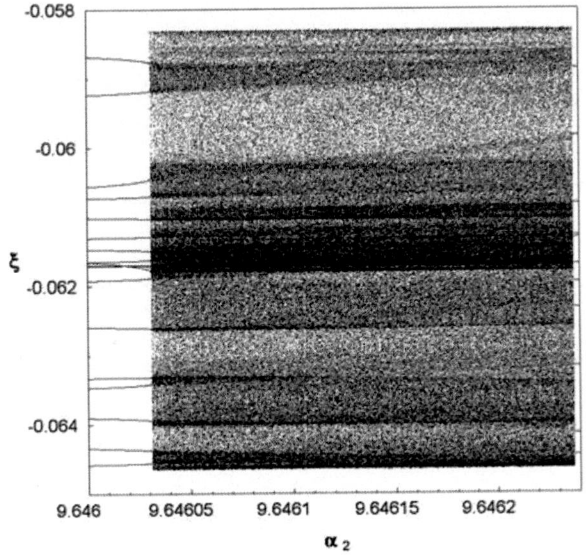

4 Generalized Intermittency and Symmetry

Finally, consider the bifurcations that occur in the system on the interval $9.64624 < \alpha_2 < 9.64665$. As the parameter α_2 increases, a cascade of bifurcations of doubling the period of limit cycles begins in the system, which leads to the appearance of a chaotic attractor. Further, as α_2 increases, the chaotic attractor is replaced with periodicity window. Then this chain of transitions is observed again: a cascade of period doubling bifurcations → chaos → a periodicity window, and so on. However, the sequence of such transitions is interrupted at $\alpha_2 \approx 9.64631$. Further, an extremely interesting transition occurs from a chaotic attractor of one type to a chaotic attractor of another type according to the scenario of generalized intermittency. This scenario is described in detail in the papers Krasnopolskaya and Shvets [15], Shvets and Sirenko [16]. One of distinctive features of such a transition is the appearance of coarse-grained (rough) laminar phase instead of laminar phase of usual intermittency Fig. 5.

We notice that all described above behavior is symmetric, i.e. exists some "median" value of bifurcation parameter α_2, such that any transition to chaos is reflected. But there is more than that, since the very first chain of transitions to chaos that happend prior the generalized intermittency is reflected too. It is worth emphasizing due to fact that regularly generalized intermittency is not a part of any other chain of transitions to chaos.

Let us illustrate the scenario of generalized intermittency using phase portraits and distributions of the invariant measure of the corresponding attractors presented in Fig. 6. In the Fig. 6a and c, the phase portrait projection and distribution of invari-

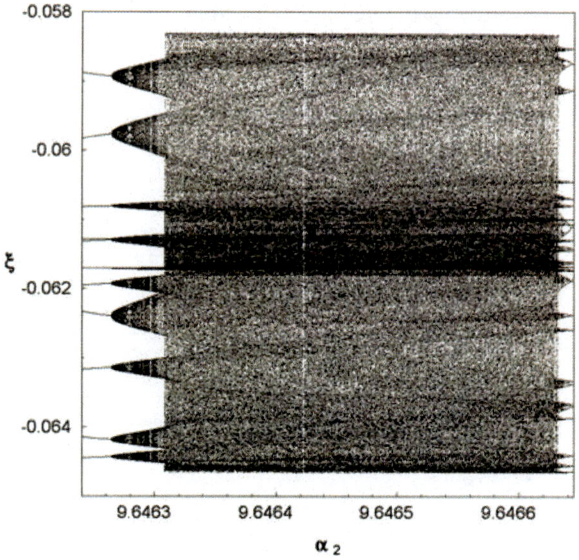

Fig. 5 Phase-parametric characteristic

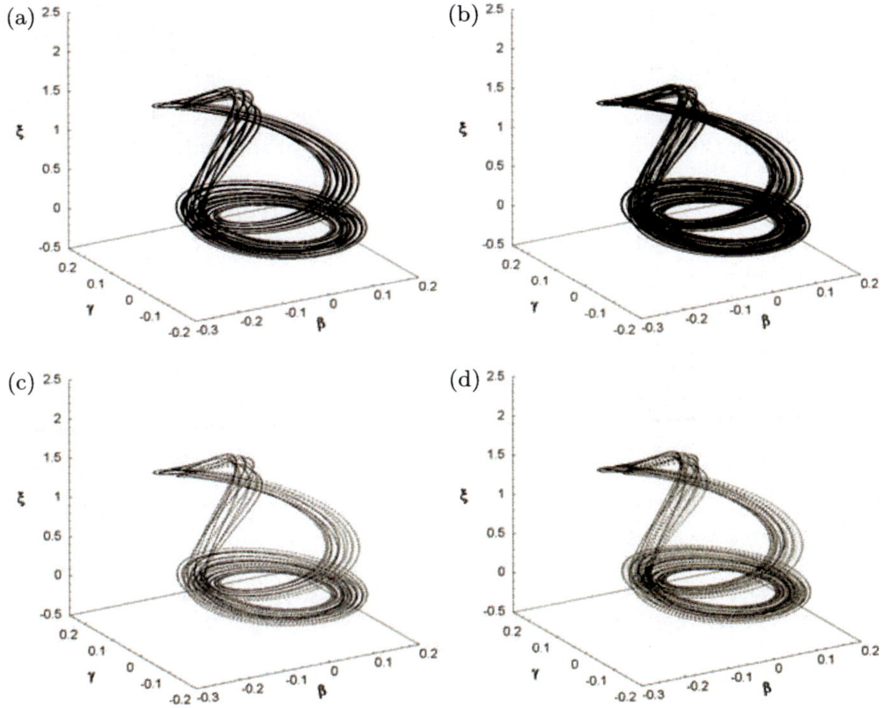

Fig. 6 Phase portrait projections: at $\alpha_2 = 9.6463$ (**a**); at $\alpha_2 = 9.64631$ (**b**). Distribution of invariant measure: at $\alpha_2 = 9.6463$ (**c**); at $\alpha_2 = 9.64631$ (**d**).

ant measure are presented respectively, prior the generalized intermittency. After the bifurcation, chaotic attractor of one type disappears, and chaotic attractor of other type borns. Phase portrait projection, as well as distribution of invarian measure for this new attractor are presented in the Fig. 6b and d respectively. Behavior of newborn chaotic attractor consists of two main phases: the rough-laminar phase and turbulent one. In the rough-laminar phase, trajectory of the caotic attractor makes chaotic movements near localization of disappeared attractor. During these movements, trajectory of newborn attractor almost coincide with trajectory of disappeared one. These correspond to the much darkened areas in Fig. 6d. Then, at the unpredictable moment of time, turbulent phase begins. During this phase, trajectory leaves localization region and moves to distant regions of the phase space. After some time, trajectory returns to rough-laminar phase. This process of switching phases is repeated infinitely many times.

5 Conclusions

Thus, the paper explored a number of symmetries in the alternation of scenarios of transitions to chaos in a nonideal dynamic system "piezoelectric converter-analog generator". The existence of double symmetry is established for such alternations of scenarios.

The possibility of transitions "chaotic attractor of one type—chaotic attractor of another type" according to the scenario of generalized intermittency was revealed for the first time.

References

1. A. Sommerfeld, Beitrage zum dynamischen Ausbau der Festigkeitslehre. Physikalische Zeitschrift **3**, 266–271 (1902)
2. A. Sommerfeld, Beitrage zum dynamischen ausbau der festigkeislehre. Zeitschrift des Vereins Deutscher Ingenieure. **46**, 391–394 (1902)
3. V.O. Kononenko, *Vibrating System with a Limited Power-Supply* (Iliffe, London, 1969)
4. T.S. Krasnopolskaya, A. Yu, Shvets Deterministic chaos in a system generator—piezoceramic transducer. Nonlinear Dyn. Syst. Theor. **6**(4), 367–387 (2006)
5. J.M. Balthazar, J.L. Palacios Felix et al., Nonlinear interactions in a piezoceramic bar transducer powered by vacuum tube generated by a nonideal source. J. Comput. Nonlinear Dyn. **4**, 011013, 1–7 (2009)
6. A. Shvets, S. Donetskyi, Transition to deterministic Chaos in some electroelastic systems, in *Springer Proceedings in Complexity* (Springer, Cham, 2019), pp. 257–264
7. A.Y. Shvets, T.S. Krasnopolskaya, Hyper-Chaos in piezoceramic systems with limited power-supply, in *IUTAM Symposium on Hamiltonian Dynamics, Vortex Structures, Turbulence* (2008), pp. 313–322
8. M.J. Feigenbaum quantative universality for a class of nonlinear transformations. J. Stat. Phys. **19**(1), 25–52 (1978)
9. M.J. Feigenbaum The universal metric properties of nonlinear transformations. J. Stat. Phys. **21**(6), 669–706 (1979)
10. P. Manneville, Y. Pomeau, Different ways to turbulence in dissipative dynamical systems. Physica D. Nonlinear Phenom. **1**(2), 219–226 (1980)
11. E.N. Lorenz, Deterministic nonperiodic flow. J. Atmos. Sci. **20**, 130–141 (1963)
12. O.E. Rössler, An equation for continuous chaos. Phys. Lett. **A57**(5), 396–397 (1976)
13. O.E. Rössler, An equation for hyperchaos. Phys. Lett. **A71**(2,3), 155–159 (1979)
14. A.Y. Shvets, Deterministic chaos of a spherical pendulum under limited excitation. Ukr. Math. J. **59**, 602–614 (2007)
15. T.S. Krasnopolskaya, A. Yu, Shvets Dynamical chaos for a limited power supply for fluid oscillations in cylindrical tanks. J. Sound Vibr. **322**(3), 532–553 (2009)
16. A.Y. Shvets, V.A. Sirenko, Scenarios of transitions to hyperchaos in nonideal oscillating systems. J. Math. Sci. **243**(2), 338–346 (2019)

Optimality Principles for Solving Nonlinear Control Problems Under Uncertainty

Tatiana F. Filippova and Oxana G. Matviychuk

Abstract The problem of estimating the reachable sets of the nonlinear dynamic control systems is investigated under the assumption of the uncertainty of the parameters of the system and its initial states. It is assumed that only bounding sets are available for some system parameters including unknown initial states, and no additional statistical information is provided on these values. Based on the previously established results of the theory control systems under uncertainty, new approaches are being developed that allow finding solutions to the problems under study and investigating the properties of the proposed solutions. Algorithms for constructing ellipsoidal estimates for trajectory tubes and their cross sections in time (reachable sets of systems) are discussed and tested. Applications to the problems of population growth models, the behavior of competing firms, environmental changes in the world, the development of some competing industries, etc. are discussed.

Keywords Control system · Nonlinearity · Set-membership uncertainty · Estimation problem · Ellipsoidal calculus · Maximum principle · HJB equation

1 Introduction

Nonlinear dynamical control systems with a special type of uncertainty related to the case of its set-membership description have attracted the attention of researchers for many years. It may be explained by several reasons, e.g. by interesting mathematical formulations of related theoretical problems and also by a close connection with the important studies of real models of various nature, where elements of nonlinearity and uncertainty are presented in their statistical and dynamical description. In this

T. F. Filippova (✉)
Ural Federal University, Ekaterinburg, Russia
e-mail: ftf@imm.uran.ru

T. F. Filippova · O. G. Matviychuk
Krasovskii Institute of Mathematics and Mechanics, Ural Branch of Russian Academy of Sciences, Ekaterinburg, Russia
e-mail: vog@imm.uran.ru

C. H. Skiadas and Y. Dimotikalis (eds.), *14th Chaotic Modeling and Simulation International Conference*, Springer Proceedings in Complexity,
https://doi.org/10.1007/978-3-030-96964-6_12

regard, it is important to specially highlight fundamental and important research that gave rise to further branches of development in this direction and in related areas, namely Kurzhanski and Varaiya [19], Kurzhanski and Valyi [18], Chernousko [5], Polyak et al. [21], Schweppe [22] and a number of other interesting publications in this area.

It should also be emphasized that the study of nonlinear problems of control theory is very important not only for theoretical specialists. Such developments are in demand because of the practical formulations of real control problems, for a more accurate account of the characteristics of dynamic models, which in reality can be very complex, contain nonlinearity and uncertainty of an unexplained nature, when it is possible to determine only the boundaries of any unpredictable effects on the object under study.

The important issue in these studies is to develop related techniques, which may help to construct external or internal estimates for unknown system characteristics. In this context, it is possible to point out not only the theoretical academic interest in such a study, but also the possibility of practical application which may be based on the theoretical results and may be used as a basis for algorithmic support for a number of applied control problems with both uncertainty and nonlinearity. It would be worth mentioning in this context, for example, researches by Allgower and Zheng [1], August et al. [2], Boscain et al. [3], Ceccarelli et al. [4].

In this paper we use the modified state estimation approaches and investigate control systems having the nonlinearity of a special kind. In the general analytical schemes we use also the advantages of ellipsoidal calculus. Namely, we deal here with dynamical control systems with a special structure of nonlinearity in the description of their dynamics, when unknown inaccuracies are present simultaneously in the initial data and in the matrix of linear components of the phase velocities of the system. The presented modified approaches to solving problems of estimating the unknown states of the system take into account the specifics of the problem under study and present a new modified approach to solve it.

The techniques presented here relate to the theory of trajectory tubes of differential control systems with uncertainty and are based on the following basic principles:

- set-membership estimation approach to deal with system uncertainty,
- optimality principles to analyze properties of reachable sets of control problems,
- Hamilton-Jacobi-Bellman (HJB) approach applied for finding the external set-valued estimates of uncertain solutions of dynamical systems.

The studies presented in this paper are based on the previous results of the authors (in particular, Filippova [7–9] and Filippova and Matviychuk [12, 13]) and develop them for new classes of estimation problems in the control theory for dynamical systems under conditions of uncertainty.

2 Statement of the Problem and Basic Notation

Introduce first a number of standard, but necessary notations. Here we denote as R^n the Euclidean space of dimension n, also let comp R^n be the set of all compact subsets of R^n and $R^{n\times m}$ be the set of all real $n \times m$-matrices.

We use further the standard notation $x'y = (x, y) = \sum_{i=1}^{n} x_i y_i$ for the inner product of $x, y \in R^n$, two specific norms of $x \in R^n$ are denoted as

$$\|x\| = \|x\|_2 = (x'x)^{1/2}, \quad \|x\|_\infty = \max_{1\le i\le n} |x_i|.$$

We use also symbols $I \in R^{n\times n}$ for the identity matrix and tr (A) for the trace of $n \times n$-matrix A (the sum of its diagonal elements), $|A|$ means the determinant of a square matrix A.

Denote as

$$B(a, r) = \{x \in R^n : \|x - a\| \le r\}$$

a ball in R^n with a center $a \in R^n$ and with a radius $r > 0$. Also let

$$E(a, Q) = \{x \in R^n : \ (Q^{-1}(x - a), (x - a)) \le 1\}$$

be an ellipsoid in R^n with a center $a \in R^n$ and with a symmetric positive definite $n \times n$-matrix Q.

We study further the nonlinear control system described by the system of differential equations with uncertain terms

$$\dot{x} = A(t)x + \tilde{f}(x)d + u(t), \quad x_0 \in X_0, \quad u(t) \in U, \quad t \in [t_0, T], \tag{1}$$

where $x, d \in R^n$. Here $\tilde{f}(x)$ is the nonlinear function, which assumed to be quadratic in x, that is $\tilde{f}(x) = x'Bx$ and B is a given symmetric and positive definite $n \times n$-matrix.

In the general setting of the problem, we only assume that both sets X_0 and U in (1) are convex and compact in R^n. Further, the structure of the sets X_0, U will be refined: namely, the main results obtained here will refer to the case when both these sets are ellipsoids with given parameters.

We assume here that the matrix $A(\cdot)$ in (1) contains uncertain elements, namely we have

$$A(t) = A^0 + A^1 \in \mathbf{A} = A^0 + \mathbf{A}^1, \quad t \in [t_0, T]. \tag{2}$$

Here A^0 is a given $n \times n$-matrix, but the second matrix term A^1 in (2) is assumed to be unknown and we know only that A^1 is containing in the set $\mathbf{A}^1$, where

$$\begin{gathered}\mathbf{A}^1 = \{\{a_{ij}\} \in R^{n\times n} : a_{ij} = 0 \ \text{ if } i \ne j, \ \text{ and} \\ a_{ii} = a_i, \ i = 1, \dots, n, \ a = (a_1, \dots, a_n), \ \text{ with } (a, Da) \le 1\},\end{gathered} \tag{3}$$

here D is a given symmetric and positive definite $n \times n$-matrix.

The assumption in (1) that the initial state x_0 of the dynamical system (1) belongs to a compact set X_0 corresponds to the case when we do not know a true value of x_0, we know it only with some inaccuracy of the above mentioned type.

Let the absolutely continuous function $x(t) = x\big(t; u(\cdot), A(\cdot), x_0\big)$ be a solution to the nonlinear dynamical system (1) and (2) with initial state $x_0 \in X_0$, control $u(\cdot)$ and with a matrix $A(\cdot)$ satisfying (2) and (3). We assume here that the solutions $\{x(t)\}$ are extendable up to the instant T and are bounded $\|x(t)\| \leq K$ (with some $K > 0$) (in this regard, it is useful to read, for example, the discussion and results presented in Filippova and Berezina [10]).

We define the reachable set $X(t)$ at time t $(t_0 < t \leq T)$ of system (1)–(3) as follows

$$X(t) = \{x \in R^n : \exists\, x_0 \in X_0, \exists\, u(\cdot) \in U, \exists\, A(\cdot) \in A,\ x = x\big(t; u(\cdot), A(\cdot), x_0\big)\}. \quad (4)$$

Recently in Filippova [6–8] and Filippova and Matviychuk [11–13] several approaches have been proposed for solving problems of estimating unknown states of dynamical systems of this kind.

Here we propose a modified approach and describe new techniques which allow us to find the external ellipsoidal estimate $E(a^+(t), Q^+(t))$ of the reachable set $X(t)$ $(t_0 < t \leq T)$ in more complicated case of dynamical systems when unknown values are present simultaneously both in the initial data and in the state velocities of the control system. Moreover, in the general initial formulation of the problem, it is not assumed here that the initial set of the studied system should be necessarily an ellipsoid.

In general, the main problem studied here is related to the questions how to determine the reachable sets of nonlinear dynamical systems of the class specified above. It is well known that it is very difficult to construct the exact reachable set of the control system with nonlinearities and uncertainty and it takes a very long time to construct it even using computer simulations and numerical approximations, so this research and presented results may well turn out to be useful, especially in cases when approximate solutions of optimization problems are sufficient.

3 Hamilton-Jacobi-Bellman Inequalities in State Estimation for Uncertain Systems

We develop here in some features the notion of generalized solutions of Hamilton-Jacobi-Bellman inequalities which allows to find the external set-valued estimates of reachable sets as level sets of a related cost functional. In this regard, it is worth highlighting recent and important studies in this direction, namely Kurzhanski [17], Gurman [14], Gusev and Kurzhanski [16], Gusev [15] and also results Filippova [6, 8].

The problems of state estimation and also related problems of control synthesis for systems described by ordinary differential equations or differential inclusions with unknown but bounded disturbances can be transformed into the studies of first order PDEs of the Hamilton-Jacobi-Bellman (HJB) type and related modifications.

To investigate this possibility for uncertain systems studied here, let us consider the following control system

$$\dot{x} = f(t, x, u(t)), \quad t \in [t_0, T] \tag{5}$$

$$u(\cdot) \in U = \{u(\cdot) : u(t) \in U_0 \in \text{comp}R^m, \; t \in [t_0, T]\}, \tag{6}$$

$$x(t_0) = x_0 \in X_0 \tag{7}$$

with a solution $x(t)=x(t, u(\cdot), t_0, x_0)$ and with the reachable set $X(t) = X(t;\ t_0, X_0)$ generated by the trajectory tube

$$X(\cdot) = X(\cdot;\ t_0, X_0) = \bigcup \{\, x(\cdot) = x(\cdot, u(\cdot), t_0, x_0) \mid x_0 \in X_0,\ u(\cdot) \in U \,\}. \tag{8}$$

We need further the important auxiliary result formulated first in Kurzhanski [17].

Lemma 1 (Kurzhanski [17]) *Suppose that there exists a function $\mu(t)$ which is integrable on $[t_0, T]$ and such that we have*

$$V_t(t, x) + \max_{u \in U} (V_x, f(t, x, u)) \le \mu(t), \; t_0 \le t \le T. \tag{9}$$

Then the external estimate of the reachable set $X(t) = X(t; t_0, X_0)$ is valid

$$X(t) \subseteq \{\, x : V(t, x) \;\le \int_{t_0}^{t} \mu(s)ds + \max_{x \in X_0} V(t_0, x) \,\}, \; t_0 \le t \le T. \tag{10}$$

Remark 1 The inequality of a more general type (Gurman [14], Gusev [15])

$$V_t(t, x) + \max_{u \in U} (V_x, f(t, x, u)) \le g(t, V(t, x)) \tag{11}$$

is true also with a function $g(t, V)$ which is integrable in $t \in [t_0, T]$ and continuously differentiable in V.

Consider the following ordinary differential equation

$$\dot{U}(t) = g(t, U), \quad U(t_0) = U_0, \tag{12}$$

which is called a comparison equation for (5)–(7).

Theorem 1 (Kurzhanski [17], Gusev [15]) *Assume that (11) and (12) are fulfilled. Assume also that*

$$\max_{x \in X_0} V(t_0, x) \leq U_0. \tag{13}$$

Then the following upper estimate is valid

$$X(t) \subseteq \{x : V(t, x) \leq U(t)\}, \ \ t_0 \leq t \leq T. \tag{14}$$

4 Main Results

4.1 Ellipsoidal Estimates Through HJB Inequalities

Consider the dynamical control system

$$\dot{x}(t) = Ax(t) + \tilde{f}(x(t))d + u(t), \ \ x \in R^n, \ \ t_0 \leq t \leq T, \tag{15}$$

where we have

$$\tilde{f}(x) = x'Bx, \ \ x \in R^n,$$

with a $n \times n$-matrix B being positive definite and with a vector $d \in R^n$ fixed and known. We assume that the following constraints are valid now and

$$x(t_0) = x_0 \in X_0, \tag{16}$$

and the set X_0 is compact and convex, but not necessarily it is an ellipsoid. Bearing this fact in mind, we can nevertheless find its outer (upper) ellipsoidal estimate. Moreover, there can be found several such external ellipsoidal estimates, including those with different centers a_0^i and different ellipsoid matrices Q_0^i. Some of these estimates can provide the mathematical accuracy of the estimation for a number of direction vectors, while the other part can be caused by the empirically found or expected accuracy of approximations of the applied models under study.

So we may assume that the following upper estimate is given for the unknown initial state x_0 of the system, namely we have the inclusion

$$X_0 \subseteq \bigcup_{i=1}^{k} E(a_0^i, Q_0^i) \tag{17}$$

where centers a_0^i of constructed new ellipsoids and their matrices Q_0^i are fixed now.

The control constraint on variable u is still the same as before, that is

$$u(t) \in U = E(\hat{a}, \hat{Q}).$$

Here Q_0^i, $\hat{Q}$, B are symmetric positive definite matrices. Let us denote $k_{0,i}^+$ a positive number such that the inclusion

$$E(a_0, Q_0^i) \subseteq E(a_0, (k_{0,i}^+)^2 B^{-1}), \quad i = 1\ldots, r,$$

is true. The parameters similar to $k_{0,i}^+$ were used also for state estimation schemes in Filippova [7, 8]. It is important to underline that because of a special type of constraints (17) on initial states we have to apply here more complicated approximation constructions.

Define also two types of functions $a_i^+(t)$ and $r_i^+(t)$ $(i = 1, \ldots, r)$ to be the solutions of the following differential equations,

$$\dot{a}_i^+(t) = A^0 a_i^+(t) + ((a_i^+(t))' B a_i^+(t) + r_i^+(t))d + \hat{a}, \tag{18}$$

$$\begin{aligned} \dot{r}_i^+(t) = \max_{\|l\|=1} \Big\{ l' \big(2r_i^+(t) B^{1/2} (A^0 + 2d(a_i^+(t))' B) B^{-1/2} \\ + q^{-1}(r_i^+(t)) B^{1/2} \hat{Q}^* B^{1/2} \big) l \Big\} + q(r_i^+(t)) r_i^+(t), \end{aligned} \tag{19}$$

$$q_i(r_i) = ((nr_i)^{-1} Tr(B\hat{Q}_i^*))^{1/2}, \quad t_0 \leq t \leq T, \tag{20}$$

where a positive definite matrix $\hat{Q}^*$ is such that

$$\mathcal{A}^1 a_0 + E(0, \hat{Q}) + k_i D^{1/2} B^{1/2} B(0, 1) \subset E(0, \hat{Q}_i^*), \tag{21}$$

and where also related initial values are defined as

$$a_i^+(t_0) = a_{0,i}, \quad r_i^+(t_0) = k_i^2, \; i = 1\ldots, r. \tag{22}$$

Now we may formulate a new result which provides the upper estimate for reachable sets of the considered nonlinear control system with uncertainty.

Theorem 2 *Let*

$$V_i(t, x) = (x - a_i^+(t))'(r_i^+(t))^{-1} B (x - a_i^+(t)) - 1 \tag{23}$$

with $a_i^+(t)$ and $r_i^+(t)$ defined in Eqs. (18)–(20). Then $V(t, x) = V_i(t, x)$ satisfies the HJB inequality Eq. (9) with the boundary condition

$$V(t_0, x) = (x - a_0)'(k_0^+)^{-2} B\,(x - a_0) - 1 \leq 0. \tag{24}$$

Moreover, the related upper estimate

$$X(t) \subseteq \bigcup \{x : V(t, x) \leq 0\}, \quad t_0 \leq t \leq T \tag{25}$$

is true.

Proof The proof of the theorem is carried out by direct calculations, taking into account relations (18)–(22) and a rather complicated structure of constraints on initial data and parameters.

We observe that Theorem 2 allows us to find the solution of HJB inequality explicitly and to use it in the estimation problem for a class of nonlinear dynamical control systems.

We would like to underline that the analysis of appropriate numerical schemes and approximations including presented here gives the way to establish a useful connection between the techniques of ellipsoidal calculus [5, 18, 21, 22] for dynamic control systems with uncertainties and results based on the theory of Hamilton-Jacobi-Bellman equations and related inequalities.

4.2 Example

Consider an example which show that in nonlinear case the reachable sets of the dynamical system of the studied type (with simultaneously presenting nonlinearity and uncertainty) may lose the convexity property with increasing time $t > t_0$. Nevertheless the related external estimates calculated on the basis of above ideas and results are ellipsoids (and therefore convex) and these ellipsoids contain the true reachable sets of the studied nonlinear system. The ellipsoidal estimates in some directions are tight that is, they cannot be further reduced, otherwise they will stop evaluating the reachable set and to give the guaranteed upper estimate. In this sense, the proposed estimates are accurate.

Example. Consider the following control system in the space R^3

$$\begin{cases} \dot{x}_1 = -a_1 x_1 + x_1^2 + x_2^2 + x_3^2 + u_1, \\ \dot{x}_2 = a_2 x_2 + u_2, \\ \dot{x}_3 = a_3 x_3 + u_3. \end{cases} \tag{26}$$

Here we take $x_0 \in X_0 = B(0, 1)$, $0 \le t \le 0.4$ and $U = B(0, 0.1)$. System coefficients $\{a_1, a_2, a_3\}$ are unknown but bounded,

$$a_1^2 + a_2^2 + a_3^2 \le 1.$$

Applying above constructions and using the numerical algorithm similar to those described in Filippova and Matviychuk [12, 13] we can construct the upper ellipsoidal tube $E^+(t)$ (it is shown in blue colour in Fig. 1) which estimate the real set-valued solution of the system $X(t)$ (shown in black in Fig. 1).

Figures 2, 3 and 4 show the dynamics in time of projections of reachable sets onto the spaces of pairs of state coordinates, respectively.

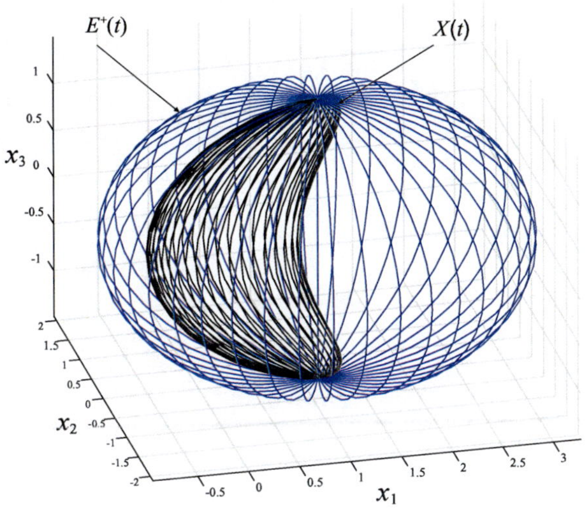

Fig. 1 The reachable set (black colour) and its upper estimate (blue colour)

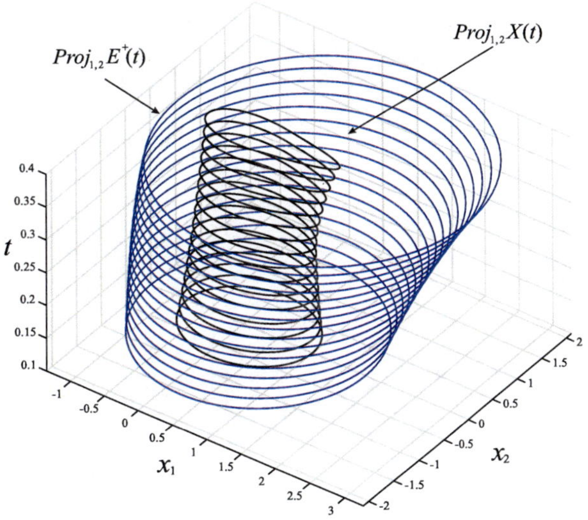

Fig. 2 The projections of reachable set (black colour) and of its upper estimate (blue colour) to the space $\{x_1, x_2, t\}$

This example confirms that the upper ellipsoidal estimates in some directions are tight, so they cannot be further reduced, otherwise they will stop evaluating the reachable set and will not give the guaranteed upper estimate. In this sense, the proposed estimates are accurate. In solving the problems of numerical modeling, carried out in this example, previously developed and officially registered computer programs Filippova and Matviychuk [11] and Matviychuk and Matviychuk [20] were used. Further discussions and other numerical examples may be found also in Filippova and Matviychuk [12, 13].

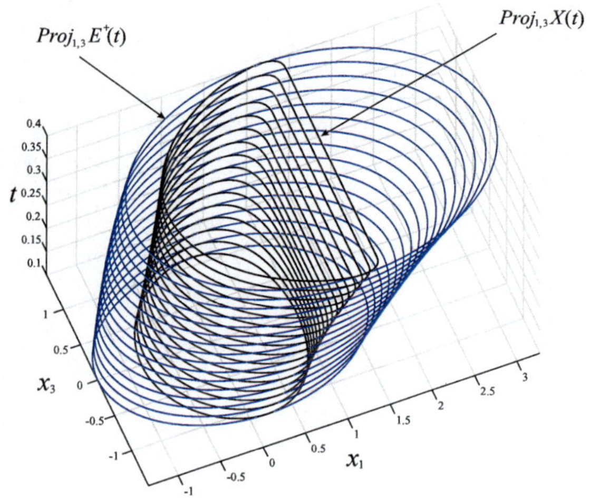

Fig. 3 The projections of reachable set (black colour) and its upper estimate (blue colour) to the space $\{x_1, x_3, t\}$

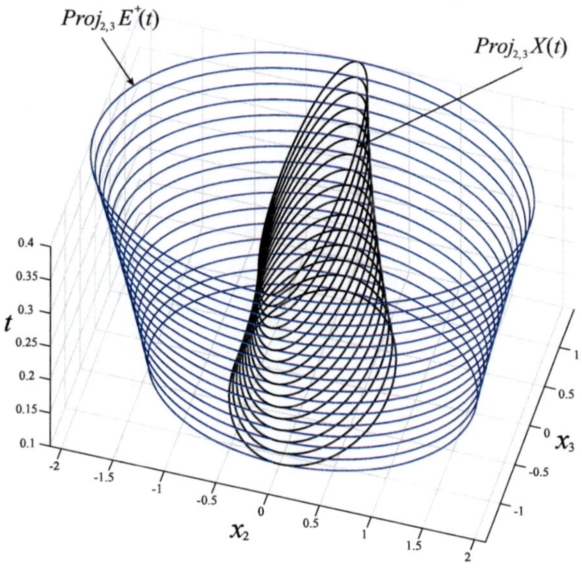

Fig. 4 The projections of reachable set (black colour) and its upper estimate (blue colour) to the space $\{x_2, x_3, t\}$

5 Conclusion

The problems of state estimation for nonlinear dynamical control systems with unknown but bounded initial state were considered.

The solution was studied through the techniques of trajectory tubes with their cross-sections $X(t)$ being the reachable sets at instant t to control system.

We presented the modified state estimation approach which uses the special structure of the control system and the techniques of generalized solutions of Hamilton-Jacobi-Bellman equations and inequalities and is based on the comparison method for analogies to related Lyapunov functions.

References

1. F. Allgower, A. Zheng (eds.), *Nonlinear Model Predictive Control* (Birkhäuser, Basel, 2000)
2. E. August, J. Lu, H. Koeppl, Trajectory enclosures for nonlinear systems with uncertain initial conditions and parameters, in *Proceedings—American Control Conference*, Montréal, Canada, Fairmont Queen Elizabeth, 2012, pp. 1488–1493
3. U. Boscain, T. Chambrion, M. Sigalotti, On some open questions in bilinear quantum control, in *European Control Conference (ECC)*, Zurich, Switzerland, July 2013, pp. 2080–2085
4. N. Ceccarelli, M. Di Marco, A. Garulli, A. Giannitrapani, A set theoretic approach to path planning for mobile robots, in *Proceedings of the 43rd IEEE Conference on Decision and Control, Atlantis, Bahamas*, Dec 2004, pp. 147–152
5. F.L. Chernousko, *State Estimation for Dynamic Systems* (Nauka, Moscow, 1988)
6. T.F. Filippova, State estimation for a class of nonlinear dynamic systems with uncertainty through dynamic programming technique, in *Proceedings of the PHYSCON-2013*, San Luis Potosi, Mexico, 2013, pp. 1–6
7. T. F. Filippova. Estimation of star-shaped reachable sets of nonlinear control systems, *Lecture Notes in Computer Science*, vol. 10665, ed. by I. Lirkov, S. Margenov (Springer, Cham, 2018), pp. 210–218
8. T.F. Filippova, The HJB approach and state estimation for control systems with uncertainty. IFAC-PapersOnLine **51**(13), 7–123 (2018)
9. T.F. Filippova, Reachable sets of nonlinear control systems: estimation approaches, in *Large-Scale Scientific Computing, LSSC 2019, Proceedings*, ed. by I. Lirkov, S. Margenov, Lecture Notes in Computer Science, vol. 11958 (Springer, Cham, 2020), pp. 510–517
10. T.F. Filippova, E.V. Berezina, On state estimation approaches for uncertain dynamical systems with quadratic nonlinearity: theory and computer simulations, *Large-Scale Scientific Computing*, Lecture Notes in Computer Science, vol. 4818 (2008), pp. 326–333
11. T.F. Filippova, O.G. Matviychuk, Calculation of ellipsoidal estimates for reachable sets of dynamical systems with bilinear-quadratic nonlinearity and uncertainty (a software package). Certificate of state registration of a computer program. Reg. no. 18-618122990008-2 (ROSRID) dated December 29, 2018, https://www.rosrid.ru/rid/detail/I0OIOXAHZREYLWVPQJSHB2UJ
12. T.F. Filippova, O.G. Matviychuk, Control problems for set-valued motions of systems with uncertainty and nonlinearity, in *Lecture Notes in Control and Information Sciences, Proceedings of the International Conference "Stability, Control, Differential Games" (SCDG2019)*, ed. by A. Tarasyev, V. Maksimov, and T. Filippova (Springer, Cham, 2020), pp. 379–389
13. T.F. Filippova, O.G. Matviychuk, Approaches to estimating the dynamics of interacting populations with impulse effects and uncertainty, in *Springer Proceedings in Complexity*, ed. by C. Skiadas, Y. Dimotikalis (Springer, Cham, 2020), pp. 85–99
14. V.I. Gurman, *The Extension Principle in Problems of Control* (Fizmatlit, Moscow, 1997)
15. M.I. Gusev, Application of penalty function method to computation of reachable sets for control systems with state constraints. AIP Conf. Proc. **1773**(050003), 1–8 (2016)
16. M.I. Gusev, A.B. Kurzhanski, On the Hamiltonian techniques for designing nonlinear observers under set-membership uncertainty, in *Preprints of the 7th IFAC Symposium on Nonlinear Control Systems*, South Africa, Pretoria, 2007, pp. 343–348
17. A.B. Kurzhanski, Comparison principle for equations of the Hamilton-Jacobi type in control theory. Proc. Steklov Inst. Math. **253**, S185–S195 (2006)

18. A.B. Kurzhanski, I. Valyi, *Ellipsoidal Calculus for Estimation and Control* (Birkhäuser, Boston, 1997)
19. A.B. Kurzhanski, P. Varaiya, *Dynamics and Control of Trajectory Tubes, Theory and Computation, Systems & Control*. Foundations & Applications, vol. 85 (Birkhäuser, Basel, 2014)
20. O.G. Matviychuk, A.R. Matviychuk, Program for constructing and three-dimensional visualization of ellipsoidal estimates of reachable sets of the bilinear control systems with uncertainty (a software package). Certificate of state registration of a computer program. Reg. no. 2020618808 (ROSPATENT) dated August 4, 2020, https://new.fips.ru/registers-doc-view/fips_servlet?DB=EVM&DocNumber=2020618808&TypeFile=html
21. B.T. Polyak, S.A. Nazin, C. Durieu, E. Walter, Ellipsoidal parameter or state estimation under model uncertainty. Automatica **40**, 1171–1179 (2004)
22. F.C. Schweppe, *Uncertain Dynamical Systems* (Prentice-Hall, Englewood Cliffs, NJ, 1973)

Piezo Spintronic Effect in DNA Molecular Chains

Masumeh Garagozi, Samira Fathizadeh, and Fatemeh Nemati

Abstract Recent efforts have been focused on producing nanoscale spintronic systems based on molecular materials. Molecular spintronics is an exciting concept for spin-based quantum computing. Spintronics combines the electronic with electron spin, which is an attractive field for processing and transferring the information. On the other hand, creating the pure spin currents in response to strain can be studied in the content of the piezo spintronic effect. In this regard, we have tried to design a DNA-based piezo spintronic device. We have proposed a theoretical model for controlling the spin current in DNA based on coupling between mechanical distortions and spin degrees of freedom. We have used the chaos theory tools to study the spin transport properties in the system. The obtained results determine that different DNA sequences show distinct behavior with respect to the mechanical tension. Also, the regions in the parameter values in which the maximum spin current flows through the system can be investigated. The mechanical tension can adjust the spin current flowing through the system. Therefore, one can design and control a novel piezo spintronic nanodevice based on DNA sequences.

Keywords Piezo spintronic · Mechanical tension · DNA chain · Spin current · Chaos theory tools

1 Introduction

The spintronic field can detect, inject and manipulate electron spins into solid-state systems [1]. Researchers have recently shown, using experiments and theoretical work, that they can perform similar and even better functions in making spinning devices than inorganic metals and semiconductors [2]. This phenomenon, known as molecular spintronics, has grown exponentially over the past few decades for practical applications [3]. The molecular spintronic field uses the spin state of organic molecules to produce electromagnetic devices widely used in sensors, memory, and

M. Garagozi · S. Fathizadeh (✉) · F. Nemati
Department of Physics, Urmia University of Technology, Urmia, Iran
e-mail: s.fathizadeh@sci.uut.ac.ir

C. H. Skiadas and Y. Dimotikalis (eds.), *14th Chaotic Modeling and Simulation International Conference*, Springer Proceedings in Complexity,
https://doi.org/10.1007/978-3-030-96964-6_13

quantum computing [4]. Molecular spintronic devices can produce future spin valves and quantum computing devices [5]. In these devices, polar spin currents are transmitted through molecules [1]. In 2011, scientists discovered that the transfer of electrons through chiral molecules depends on the direction of the electron spin. It has recently been shown that the transmission of charge in these molecules is spin polarized [6].

In this work, we use a new effect, called piezo spintronic (piezo in Greek means stress), to generate spin current in DNA nanowires, which is based on mechanical connection and the degree of spin release [7]. This effect, unlike the effects of piezo magnetism and piezoelectricity, is a phenomenon limited to the simultaneous presence of systems with time-reversal (T), inversion (I), and symmetry failure [8]. This mechanism opens the way to obtain and measure net spin currents. In essence, a crystal may exhibit piezoelectric, piezomagnetic, and piezo spintronic effects simultaneously. The piezo voltage is one of the most effective methods for controlling magnetic switching, in which the deformation of the crystal structure of the magnetic material changes the crystal magnetic anisotropy, which is directly related to a spin–orbit interaction in the crystal [9]. On the other hand, the DNA molecule is widely used as a complex nanostructure with high flexibility in nanotechnology [10]. The double-stranded DNA molecule is a piezoelectric material. Piezoelectric materials are a class of dielectrics that can be polarized by an electric field and mechanical stress [11]. This particular property of piezoelectric materials is due to the crystal structure of the material [12]. Piezoelectric materials are used in converters and devices that convert electrical energy into mechanical energy or vice versa. Piezoelectric materials have many applications in diodes, switches, memories, transistors, sensors, energy storage devices, etc. [13–15]. DNA molecule is a chiral molecule due to its asymmetric crystal structure (mirror asymmetry) that can exhibit conductivity, insulation, semiconductor, and superconductivity. On the other hand, chiral organic molecules are a good candidate for transmitting information encoded in spin and spin-polarized current sources [16]. The piezo spintronic effect is very similar to the polarization of charge currents caused by pressure-induced spin–orbit interactions. In this work, we show that many polar spin currents can be produced by applying external mechanical stress to the molecular junctions of the DNA chain. We also show that spin-dependent charge transport can be observed in DNA nanowires by applying the mechanical stress and in the presence of an external magnetic field. For this purpose, we designed a piezo spintronic nanostructure based on the DNA sequence according Fig. 1.

2 Model and Methods

In the current work, we have studied the spin currents along DNA nanowires through the piezo spintronic effect using the Peyrard–Bishop–Holstein (PBH) model modified for the spin degree of freedom. PBH model considers the pairing of bases in the direction of hydrogen bonding and plots the DNA molecule as a one-dimensional

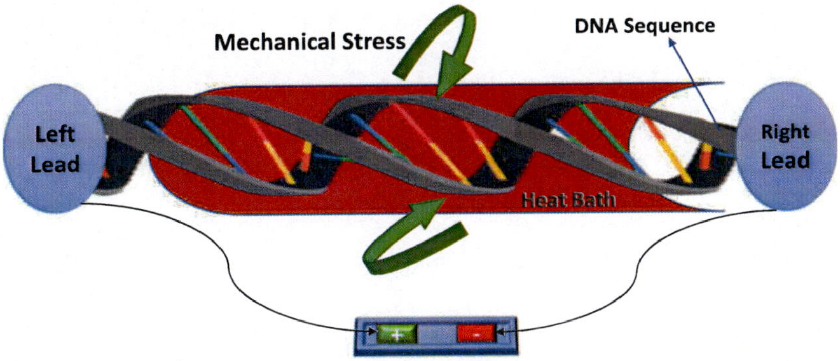

Fig. 1 A schematic illustration that shows the DNA nanowires immersed in a thermal bath and connected at both ends to the metal leads in the presence of an external mechanical stress

network [17]. The Hamiltonian of the system can be presented as follows:

$$\mathrm{H} = \mathrm{H}_{\mathrm{DNA}} + \mathrm{H}_{\mathrm{so}} + \mathrm{H}_{\mathrm{lead}} + \mathrm{H}_{\mathrm{DNA-lead}} + \mathrm{H}_{\mathrm{Bath}} + \mathrm{H}_{\mathrm{DNA-Bath}} + \mathrm{H}_{\mathrm{fields}} \tag{1}$$

where, the first term is Hamiltonian related to DNA molecule written as follows:

$$\begin{aligned} H_{DNA} = & \sum_{i,j=1,2}\sum_{\sigma=\uparrow\downarrow}\left[\varepsilon_{i,j}c_{i,j}^{+\sigma}c_{i,j}^{\sigma} + V_{i,i+1,j}c_{i+1,j}^{+\sigma}c_{i,j}^{\sigma}\right] + \sum_{i,\sigma}\lambda_i c_{1,i}^{+\sigma}c_{2,i}^{\sigma} \\ & + \sum_{i,j}[2it_{so}\cos\theta(c_{i,j}^{+\uparrow}c_{i+1,j}^{\uparrow} - c_{i,j}^{+\uparrow}c_{i-1,j}^{\uparrow} - c_{i,j}^{+\downarrow}c_{i+1,j}^{\downarrow} \\ & + c_{i,j}^{+\downarrow}c_{i-1,j}^{\downarrow} + D_{i,i+1}c_{i,j}^{+\uparrow}c_{i+1,j}^{\downarrow} - D_{i,i+1}^{+}c_{i,j}^{*\downarrow}c_{i+1,j}^{\uparrow} \\ & + D_{i-1,i}^{*}c_{i,j}^{+\downarrow}c_{i-1,j}^{\uparrow} - D_{i-1,i}c_{i,j}^{+\uparrow}c_{i-1,j}^{\downarrow})] + H.c. \end{aligned} \tag{2}$$

where, t_{so} is a spin–orbit coupling constant, θ is the helix angle, and i, j indicate the number of sites and strings, respectively. Also, ε is electron energy and $c_{i,j}^{\dagger}$, $c_{i,j}$ are the electron creation and annihilation operators at the site (i, j), respectively. λ_i is the interaction coupling between the DNA chains and

$$D_{n,n+1} = it_{so}\sin\theta\{\sin[n\Delta\varphi] + \sin[(n+1)\Delta\varphi] + i\cos[n\Delta\varphi] + i\cos[(n+1)\Delta\varphi]\} \tag{3}$$

where $\varphi = n\Delta\varphi$ is the angle in the cylindrical coordinate and $\Delta\varphi$ is defines the twist angle. To maintain inverse symmetry, we have

$$D_{\mathrm{n,n-1}} = D_{n-1,n}^{*}$$

Also, $V_{i,i+1,j}$ shows the mutation between the nearest neighbors, which is written as follows:

$$V_{i,i+1} = V_o e^{-\beta_i (y_{i+1} - y_i)}$$

where V_o is the constant of the hopping integral and the β_i indicates the intensity of the coupling.

H_{so} is a spin–orbit coupling Hamiltonian written as follows:

$$\begin{aligned} H_{SO} = \sum_n [& 2it_{so} \cos\theta (c_i^{\dagger\uparrow} c_{i+1}^{\uparrow} - c_i^{\dagger\uparrow} c_{i-1}^{\uparrow} - c_i^{\dagger\downarrow} c_{i+1}^{\downarrow} + c_i^{\dagger\downarrow} c_{i-1}^{\downarrow}) \\ & + D_{i,i+1} c_n^{\dagger\uparrow} c_{i+1}^{\downarrow} - D_{i,i+1}^{*} c_i^{\dagger\downarrow} c_{i+1}^{\uparrow} + D_{i-1,i}^{*} c_i^{\dagger\downarrow} c_{i-1}^{\uparrow} \\ & - D_{i-1,i} c_i^{\dagger\uparrow} c_{i-1}^{\downarrow}] \end{aligned} \tag{4}$$

H_{lead} is the Hamiltonian related to the electrodes expressed as follows:

$$H_{lead} = \sum_{j=1,2} \sum_{k,\sigma} \left(\varepsilon_{L_{j,k}} + \frac{eV_b}{2} \right) a_{L_{j,k}}^{+\sigma} a_{L_{j,k}} + \sum_{j=1,2} \sum_{k,\sigma} \left(\varepsilon_{R_{j,k}} - \frac{eV_b}{2} \right) a_{R_{j,k}}^{+\sigma} a_{R_{j,k}} \tag{5}$$

where V_b is the bias voltage applied to the system and $a_{\beta_{j,k}}^{+}$, $a_{\beta_{j,k}}$ are the operators of electron creation and annihilation in the electrode $\beta = R, L$, respectively. $H_{DNA-lead}$ is the Hamiltonian relating to the interaction of the DNA molecule with the electrodes written as follows:

$$H_{DNA-lead} = \sum_{j=1,2} \sum_{k,\sigma=\uparrow\downarrow} \left(t_L a_{L_{j,k}}^{+\sigma} c_{j,1}^{\sigma} + t_R a_{R_{j,k}}^{+\sigma} a_{j,N}^{\sigma} + H.c. \right) \tag{6}$$

H_{Bath} is the Hamiltonian of thermal bath defined as follows [18]:

$$H_{Bath} = \sum_{i=1}^{N} \hbar\omega_i b_i^{+} b_i + 2 \sum_{i=1}^{N-1} \hbar\Omega_i (b_i^{+} b_{i+1} + b_{i+1}^{+} b_i) + H.c. \tag{7}$$

where b_i^{+} and b_i are the oscillator creation and annihilation operators at the i site, respectively. Ω is the reciprocal coupling constant and ω is the oscillator frequency at the site. The Hamiltonian of the interaction of the thermal bath with the DNA molecule is written as follows:

$$H_{DNA-lead} = \sum_{j=1,2} \sum_{k,\sigma=\uparrow\downarrow} \left(t_L a_{L_{j,k}}^{+\sigma} c_{j,1}^{\sigma} + t_R a_{R_{j,k}}^{+\sigma} a_{j,N}^{\sigma} + H.c. \right) \tag{8}$$

here, t_i is related to the elements of the tunneling matrix. In recent equations, the phrase $H.c.$ is entered for the effect of a Hermitian conjugate.

Finally, H_{fields} is related to the external electric and magnetic field Hamiltonian written as follows:

$$\mathrm{H_{fields}} = \mathrm{H_E} + \mathrm{H_B} \tag{9}$$

The Hamiltonian of an electric and a magnetic field is written as follows:

$$H_{\mathrm{E}} = -e \sum_{i,\sigma=\uparrow,\downarrow} Ed\cos[(i-1)\Delta\varphi] c_i^{\sigma\dagger} c_i^{\sigma} \tag{10}$$

$$H_B = \sum_{i,j} \left(-\mu_B B c_{i,j}^{\uparrow\dagger} c_{i,j}^{\uparrow} + \mu_B B c_{i,j}^{\downarrow\dagger} c_{i,j}^{\downarrow}\right) \tag{11}$$

where d is the radius of DNA and $\mu_B = \frac{e\hbar}{2mc} = 5/78838 \uparrow \times 10^{-5}$ is the constant of magneton Bohr.

In the current study, to apply a mechanical stress to the system, we corrected the electron onsite and electron hopping constants at site n through the stress parameter as follows [19]:

$$\varepsilon_{n'} = \varepsilon_n^0 + \frac{er\left(1 - \sigma\frac{s_1}{100}\right)}{L\left(1 + \frac{s_1}{100}\right)} \tan\alpha V_{sd} \cos\left(\frac{2\pi i}{10} + \varphi_0\right) \tag{12}$$

$$V_{n'+1} = V_{n+1} e^{-\left(1 + \frac{s_1}{100}\right)} \tag{13}$$

where $r = 10$ is the radius of DNA, α is the angle of DNA rotation, V_{sd} is the source voltage, $\sigma = 0.5$ is the Poisson rate, and S_1 is the longitudinal stress applied to DNA [20].

Here, the evolution equations of our dynamical system can be derived through the Heisenberg equation $\dot{c}_n^{\sigma} = -\frac{i}{\hbar}\left[c_n^{\sigma}, H\right]$ for up spin and down spin operators, respectively [21]. On the other hand, the spin current corresponding to up and down spins can be extracted via the continuity equation as follows:

$$I^{\uparrow}(t) = \frac{-ie}{\hbar} \sum_n \left\{ \begin{array}{l} W_{n,n+1} c_n^{\dagger\uparrow} c_{n+1}^{\uparrow} + W_{n-1,n}^{*} c_{n-1}^{\dagger\uparrow} c_n^{\uparrow} + \\ D_{n,n+1} c_n^{\downarrow\dagger} c_{n+1}^{\downarrow} - D_{n-1,n} c_{n-1}^{\dagger\downarrow} c_n^{\downarrow} \end{array} \right\} \tag{14}$$

$$I^{\downarrow}(t) = \frac{-ie}{\hbar} \sum_n \left\{ \begin{array}{l} W_{n,n+1}^{*} c_n^{\dagger\downarrow} c_{n+1}^{\downarrow} + W_{n-1,n} c_{n-1}^{\dagger\downarrow} c_n^{\downarrow} - \\ D_{n,n+1}^{*} c_n^{\uparrow\dagger} c_{n+1}^{\uparrow} + D_{n-1,n}^{*} c_{n-1}^{\dagger\uparrow} c_n^{\uparrow} \end{array} \right\} \tag{15}$$

Therefore, the net charge current I_C and the net spin current I_s can be defined as follows:

$$I_c = I^{\uparrow} + I^{\downarrow}$$
$$I_s = I^{\uparrow} - I^{\downarrow} \quad (16)$$

3 Results and Discussion

In this study, we have tried to investigate the spin transfer and generation of pure spin currents in different sequences of DNA in the presence and absence of mechanical stress and magnetic field. The system shows high sensitivity to the initial conditions since the dynamics behavior is nonlinear.

3.1 Mechanical Stress Effect

To investigate the effect of mechanical stress on the system in the absence of a magnetic field, we have applied micro-positive and negative mechanical stress to the system (negative stress means DNA compression and positive stress means DNA elongation). According to Fig. 2, in $S_1 = 3$, the maximum spin current and in $S_1 = -3$, the minimum spin current flow through the system.

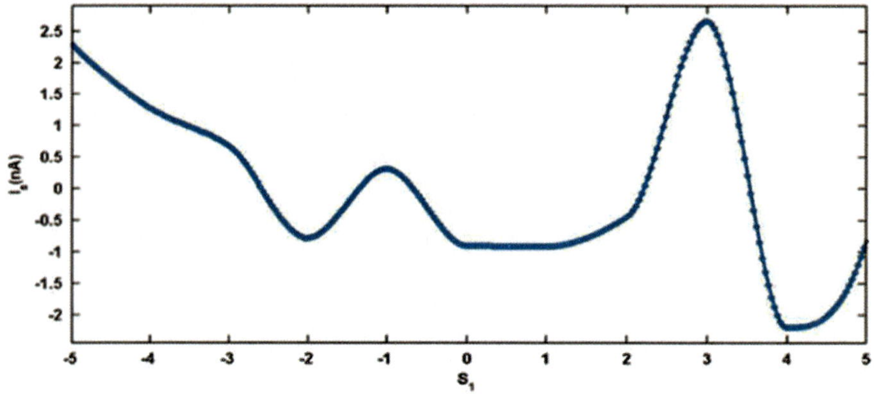

Fig. 2 The spin current with respect to the mechanical stress parameter (B = 0)

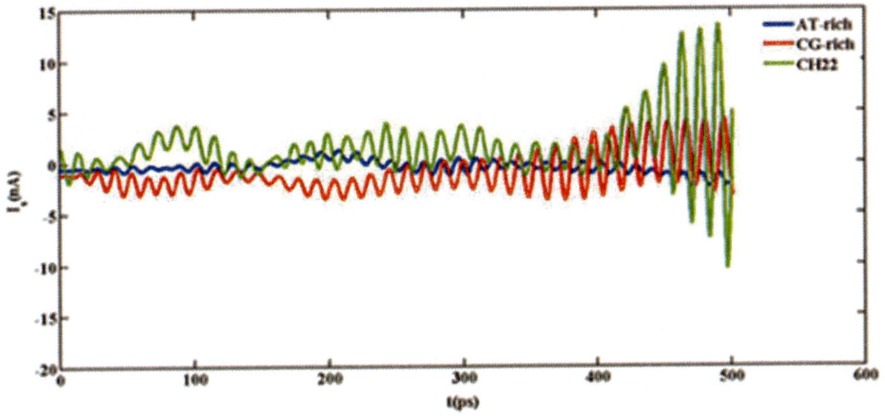

Fig. 3 The time-series of spin currents for CH22, AT-rich and CG-rich sequences in the presence of tension $S_1 = 3$ and $B = 4.5$ (mT)

3.2 *Effect of Different Sequences on the Spin Current*

The conductivity dependence of the DNA molecule on the type and length of its sequence are studied, previously [22]. One of the effective parameters on the electrical properties of DNA molecule is the variation the sequence type since different arrangement of adjacent pairs in the molecule changes the coupling and the energy of the pair [23].

In this study, we have chosen three types of sequences: CH22, AT-rich, and CG-rich, with a length of 60 bp. Therefore, we have studied the spin transport yn system by applying stress $S_1 = 3$ and in the presence of a magnetic field $B = 4.5$ (mT). According to Fig. 3, at $t = 500$ (ps), the maximum spin current flows in CH22 sequence. A moderate spin current value flows in the CG-rich sequence, and a minimum spin current flows in the AT-rich sequence. The result indicates the effect of type the sequence in the spin current flows through the molecule chain.

3.3 *Voltage Effect*

The external electric field, or in other words the gate voltage, is an influential factor on the spin current flowing through the DNA molecule. DNA molecule behaves distinctly against variable voltage [24]. According to Fig. 4, the spin current in terms of voltage shows an increase in spin current with increasing voltage in some regions can be called quasi-ohmic regions. In some regions, a decreasing spin current is observed by increasing the voltage which can be expressed as spin-polarized negative differential resistance (SPNDR) regions. In the interval 7–8 (mV), system shows the

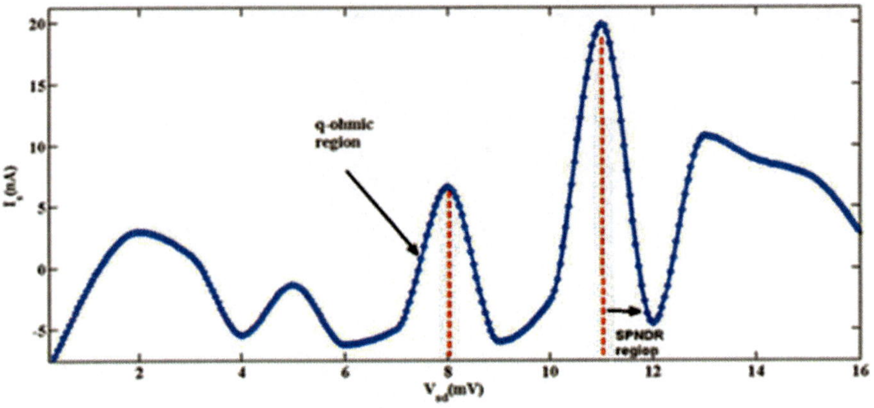

Fig. 4 The I-V characteristic diagram for the spin current in the presence of a stress $S_1 = 3$ ($B = 0$)

quasi-ohmic behavior while in the interval 11–12 (mV), the SPNDR behavior is observed.

3.4 *Spin Current in the Presence of Simultaneous Variation of Mechanical Stress and DNA Twist Angle*

We have tried to examine the simultaneous effect of mechanical stress and DNA twist angle as the most effective parameters in piezo spintronic effect on the spin current flowing through the CH22 sequence (Fig. 5). Figure 5 shows the creation of island-like areas in different parameter values. The simultaneous effect of the mechanical

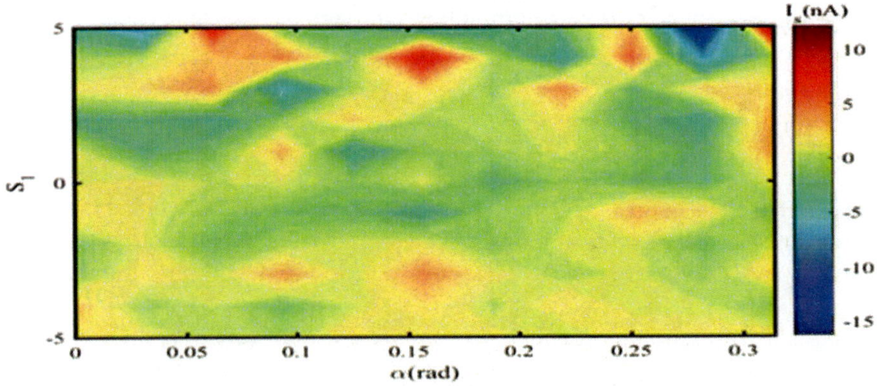

Fig. 5 The simultaneous effect of mechanical stress and DNA twist angle on spin current

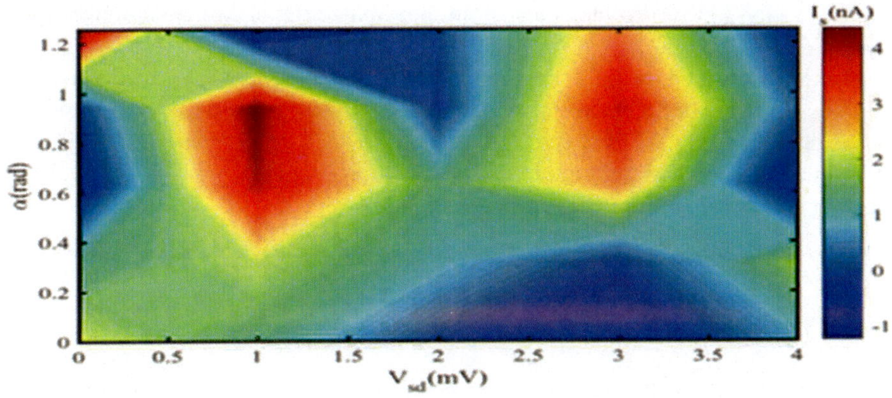

Fig. 6 The simultaneous effect of voltage and DNA twist angle on spin current

stress parameter and twist angle leads to the maximum and the minimum spin current flowing regions. It is clear in Fig. 5 that no significant current flows through the system as long as the mechanical tension have the zero or negative value, but by increasing the twist angle and applying mechanical stress $S_1 = 3$ and higher, an increase in net spin current is observed, so that at $S_1 = 5$ and $\alpha = 0.27$ (rad), the maximum spin current flows through the system.

3.5 Spin Current in the Presence of Simultaneous Variation of the External Electrical Field and DNA Twist Angle

Figure 6 shows the simultaneous effect of DNA twist angle and applied voltage on the spin current flowing through the CH22 sequence. We have observed island-like regions in which the simultaneous effect of the molecule's twist angle and voltage creates areas with a maximum and minimum spin current. It is clear that by increasing the applied voltage to the system in the presence of mechanical tension, the islands with maximum spin current increases. The result indicates the positive effect of simultaneous application of voltage and stress on the spin current in the system.

4 Conclusions

In this study, we have studied the piezo spintronic effect led to creating the net spin current in response to stress. We discussed the piezo spintronic response of DNA nanowires to create net spin currents. By studying the spin current in terms of voltage, we have observed quasi-ohmic and SPNDR regions. Finally, by examining the simultaneous variation of the DNA twist angle and the applied voltage to the system, we have observed the islands that represented the maximum and minimum spin current.

The results expand the field for spin mechanical systems since it provides a direct coupling between the spin current and the tension. In this work, we used a simplified model that considers the structure of DNA as a ladder. To continue the work, it is suggested that the natural structure of DNA, which is considered as a double helix with degrees of freedom of angle be considered.

References

1. Sanvito, Stefano, A. Reily Rocha, Molecular-spintronics: The art of driving spin through molecules. J. Comput. Theor. Nanosci. **3**(5), 624–642 (2006)
2. S. Sanvito, Molecular spintronics. Chem. Soc. Rev. **40**(6), 3336–3355 (2011)
3. A. Grizzle, C. D'Angelo, P. Tyagi, Monte Carlo simulation to study the effect of molecular spin state on the spatio-temporal evolution of equilibrium magnetic properties of magnetic tunnel junction based molecular spintronics devices. AIP Adv. **11**(1), 015340 (2021)
4. K.V. Raman, Interface-assisted molecular spintronics. Appl. Phys. Rev. **1**(3), 031101 (2014)
5. H. Brown et al., Impact of direct exchange coupling via the insulator on the magnetic tunnel junction based molecular spintronics devices with competing molecule induced inter-electrode coupling. AIP Adv. **11**(1), 015228 (2021)
6. R. Naaman, D.H. Waldeck, Y. Paltiel, Chiral molecules-ferromagnetic interfaces, an approach towards spin controlled interactions. Appl. Phys. Lett. **115**(13), 133701 (2019)
7. C. Ulloa et al., Piezospintronic effect in honeycomb antiferromagnets. Phys. Rev. B **96**(10), 104419 (2017)
8. I. Samathrakis, H. Zhang, Piezospintronic effect in antiperovskite Mn_3GaN. arXiv preprint arXiv:1905.11798 (2019).
9. Á.S. Núñez, Theory of the piezo-spintronic effect. Solid State Commun. **198**, 18–21 (2014)
10. D. Winogradoff et al., Chiral systems made from DNA. Adv. Sci. **8**(5), 2003113 (2021)
11. C. Cui et al., Two-dimensional materials with piezoelectric and ferroelectric functionalities. npj 2D Mater. Appl. **2**(1), 1–14 (2018)
12. U. Sundar et al., Dielectric and piezoelectric properties of percolative three-phase piezoelectric polymer composites. J. Vac. Sci. Technol. B Nanotechnol. Microelectron.: Mater. Process. Meas. Phenom. **34**(4), 041232 (2016)
13. I. Díez-Pérez et al., Rectification and stability of a single molecular diode with controlled orientation. Nat. Chem. **1**(8), 635–641 (2009)
14. S.J. van der Molen, P. Liljeroth, Charge transport through molecular switches. J. Phys.: Condens. Matter **22**(13), 133001 (2010)
15. S. Fathizadeh, S. Behnia, Control of a DNA based piezoelectric biosensor. J. Phys. Soc. Jpn. **89**(2), 024004 (2020)
16. S. Behnia, S. Fathizadeh, Spintronics in nano scales: an approach from DNA spin polarization. Sci. Iran. **24**(6), 3448–3451 (2017)
17. T. Holstein, Studies of polaron motion: Part I. The molecular-crystal model. Ann. Phys. **281**(1–2), 706–724 (2000)
18. S. Fathizadeh, S. Behnia, J. Ziaei, Engineering DNA molecule bridge between metal electrodes for high-performance molecular transistor: an environmental dependent approach. J. Phys. Chem. B **122**(9), 2487–2494 (2018)
19. Y.S. Joe, S. Malakooti, E.R. Hedin, Controllable negative differential resistance on charge transport through strained and tilted DNA molecules. Int. J. Mod. Phys. B **33**(11), 1950099 (2019)

20. M. Tassi et al., RT-TDDFT study of hole oscillations in B-DNA monomers and dimers. Cogent Phys. **4**(1), 1361077 (2017)
21. A.-M. Guo, Q.-f. Sun, Spin-selective transport of electrons in DNA double helix. Phys. Rev. Lett. **108**(21), 218102 (2012)
22. Y. Matsuura, F. Kato, Structure-dependent spin transport in a DNA molecule. Comput. Theor. Chem. **1188**, 112943 (2020)
23. O. Vaughan, Single molecules: piezoresistivity in DNA. Nat. Nanotechnol. 1–1 (2015)
24. S. Behnia, S. Fathizadeh, A. Akhshani, DNA spintronics: charge and spin dynamics in DNA wires. J. Phys. Chem. C **120**(5), 2973–2983 (2016)

The Atom, from a Mathematical-Physical Perspective

Alina Gavrilut and Maricel Agop

Abstract In this paper, an exhaustive study on the problem of atomicity with respect to set functions is provided. Different types of atoms are discussed, the relationships among them are studied and several examples and physical possible implications and applications are obtained.

Keywords Atom · Pseudo-atom · Minimal atom · Set function · Self-similarity

1 Introduction

Using different notions, concepts and results, in this paper we shall try to answer the question "What is the atom?" from a mathematical-physical perspective, offering at the same time a series of possible interpretations and meanings that exceed its strict limits. We shall see that the mathematical perspective preserves the intimate, defining property of the atom, in its various forms and mathematical meanings of being, in a sense, the essential indestructible, indivisible, irreducible, minimal and self-similar unity. We emphasize that an atom is a mathematical object (an entity) that, in essence, has no other subobjects (subentities) than the object itself or the null subobject. The idea is also found in computer science, for example. In partially ordered sets, atoms are generalizations of the singletons (that is, sets containing only one element) of the sets theory. Moreover, in this sense, atomicity (the property of a mathematical object of being atomic), provides a generalization in an algebraic context of the possibility of selecting an element from a nonempty set. In mathematical logic, *an atomic formula* is a formula without a deep propositional structure, that is, a formula that does not contain logical connections, or, equivalently, a formula that does not have strict subformulas. Atoms are thus the simplest well-formed formulas of logic,

A. Gavrilut (✉)
Department of Mathematics, Alexandru Ioan Cuza University from Iasi, Iasi, Romania
e-mail: gavrilut@uaic.ro

M. Agop
Department of Physics, Gheorghe Asachi Technical University of Iasi, Iasi, Romania

C. H. Skiadas and Y. Dimotikalis (eds.), *14th Chaotic Modeling and Simulation International Conference*, Springer Proceedings in Complexity,
https://doi.org/10.1007/978-3-030-96964-6_14

the compound formulas being formed by combining atomic formulas using logical connections. Also, also in logic, an atomic sentence is a type of declarative sentence that is either true or false and that cannot be broken down into other simpler sentences. In some models of set theory, an atom is an entity (a mathematical object) that can be an element of a set but does not itself contain elements with similar properties (hence the "ultimate" character of an atom). In mathematical analysis, a set's property of being an atom is defined in relation to another mathematical object, namely, with respect to a set (multi)function.

2 The Mathematical-Physical Perspective

2.1 Set Functions

Let $\mathcal{C}$ be a ring of subsets of a non-empty abstract set T and $m : \mathcal{C} \to \mathbb{R}_+$ be a set function which satisfies the condition $m(\emptyset) = 0$. The following notions generalize the notion of a measure in its classic sense (as a foundation of measure theory). In mathematical analysis, a measure (in classic sense) is a function which "measures", assigning to certain sets of a class (family) of sets, a positive real number or $+\infty$. In this sense, a measure is a generalization of the concepts of length, area or volume. One particularly important example is the Lebesgue measure on a Euclidean space, which assigns the conventional length, area and volume of Euclidean geometry to appropriate subsets of the Euclidean space $\mathbb{R}^n$. For instance, the Lebesgue measure of the interval $[0, 1]$ is its length in the ordinary sense of the word, namely, 1 [1, 9]. A measure must be additive, which means that the measure of a set representing the union of a finite (or countable) number of smaller sets that are pairwise disjoint is equal to the sum of the measures of these smaller subsets.

The notions that we shall introduce next have contributed to the development in recent years of the theory of non-additive measures, sometimes known as the fuzzy measures theory [7, 8]. These notions prove their utility due to the necessity to model phenomena from the real world, in circumstances in which the condition of additivity (either finite or countable), as an immediate property of a measure, is much too restrictive.

The set function m is called:

(i) *null-additive* if $m(A \cup B) = m(A)$, for every sets $A, B \in \mathcal{C}$, satisfying the condition $m(B) = 0$;
(ii) *null-null-additive* if $m(A \cup B) = 0$, for every sets $A, B \in \mathcal{C}$, satisfying the condition $m(A) = m(B) = 0$;
(iii) *diffused* if $m(\{t\}) = 0$, whenever $\{t\} \in \mathcal{C}$;
(iv) *monotone* if $m(A) \leq m(B)$, for every sets $A, B \in \mathcal{C}$, so that $A \subseteq B$;
(v) *null-monotone* if for every two sets $A, B \in \mathcal{C}$, having the property that $A \subseteq B$, if $m(B) = 0$ holds, then one necessarily has also $m(A) = 0$;

(vi) *finitely additive* if $m(A \cup B) = m(A) + \nu(B)$, for every disjoint sets $A, B \in \mathcal{C}$;

(vii) *subbaditive* if $m(A \cup B) \leq m(A) + \nu(B)$, for every (disjoint or not) $A, B \in \mathcal{C}$.

Example. (i) Let us suppose that $T = \{t_1, t_2, ..., t_n\}$, where for every $i \in \{1, 2, ..., n\}$, t_i represents a particle, and $m : \mathcal{P}(\mathrm{T}) \rightarrow \mathbb{R}_+$ is a set function representing the mass of the particle. In the macrosopic world, m is a finitely additive set function. At quantum scale, however, this statement no longer remains valid due to the phenomena of annihilation. For instance, if t_1 and t_2 represents an electron and a positron, respectively, then $m(\{t_1\}) = m(\{t_2\}) = 9,11 \times 10^{-31}$ kg, but $m(\{t_1, t_2\}) = m(\{t_1\} \cup \{t_2\}) = 0$;

(ii) Entropy in Shannon's sense is a subadditive set function, taking real values [2, 3].

2.2 *Types of Atoms*

In the following, we shall present several types of atoms in their mathematical meaning, we shall establish some relationships among these types of atoms and we shall also highlight several possible interpretations. Unless stated otherwise, $\mathcal{C}$ will represent a ring of subsets of an arbitrary nonvoid set T and $m : \mathcal{C} \rightarrow \mathbb{R}_+$, an arbitrary set function satisfying the condition $m(\emptyset) = 0$. This abstract set function represents the generalization of the classic notion of a *measure* used in measure theory and it is the mathematical object through which the process of so-called "measurement" is performed.

Atoms and pseudo-atoms

These are the main types of atoms from the mathematical perspective:

I. A set $A \in \mathcal{C}$ is called an *atom* of m if $m(A) > 0$ and for every $B \in \mathcal{C}$, with $B \subseteq A$, it holds either $m(B) = 0$ or $m(A \backslash B) = 0$.

We observe that, in a certain sense, an atom is a special set, of strictly positive "measure", having additionally the property that any of its subsets either has zero "measure", or the difference set between the initial set and its subset we refer to has zero "measure". An atom can be interpreted, from a physics viewpoint, as the correspondent of a black hole.

II. The set function m is said to be *non-atomic* if it has no atoms, that is, for every set $A \in \mathcal{C}$ with $m(A) > 0$, there exists a subset $B \in \mathcal{C}(\mathrm{B} \subseteq \mathrm{A})$ so that $m(B) > 0$ and $m(A \backslash B) > 0$.

III. A set $A \in \mathcal{C}$ is called a *pseudo-atom* of m if $m(A) > 0$ and for every subset $B \in \mathcal{C}$ $(B \subseteq A)$ one has either $m(B) = 0$ or $m(B) = m(A)$.

In other words, a pseudo-atom is a special set, of strictly positive "measure", for which any of its subsets either has null "measure", or has the same "measure" as the set itself. Thus, it can be stated that a pseudo-atom possesses the property that any of its subsets either has null "measure" (that *is, it is negligible during the "measurement" process), or it entirely "covers" the set (during the same "measurement" process). In other words, assuming that the set function m is monotone, then a pseudo-atom is a set of strictly positive "measure" and which does not contain any proper subset of strictly smaller and strictly positive "measure".*

IV. The set function m is said to be *non-pseudo-atomic* if it does not have pseudo-atoms, that is, for any set $A \in \mathcal{C}$ with $m(A) > 0$, there exists a subset $B \in \mathcal{C}(\mathrm{B} \subseteq \mathrm{A})$ so that $m(B) > 0$ and $m(B) \neq m(A)$.

For instance, the Lebesgue measure on the real line is a measure (in the classic sense) which is non-pseudo-atomic [9], and therefore it does not have any pseudo-atom. The non-pseudo-atomic measures satisfy the following remarkable property, which we owe to Sierpinski, a property which states that if m is a non-pseudo-atomic measure (in classic sense), defined on a σ-algebra $\mathcal{A}$ (of subsets of an abstract space T), and $A \in \mathcal{A}$ is an arbitrary set so that $m(A) > 0$, then for every element $b \in [0, m(A)]$, there exists a set $B \in \mathcal{A}$, so that $B \subseteq A$ and $m(B) = b$ (in other words, the set function m takes a continuum of values, and thus it does not omit any intermediate value).

V. A set function m is called *purely-atomic* if the space T can be represented as a finite or countable union of atoms of m.

Examples. (i) Let be the set $T = \{1, 2, ..., 9\}$. We define the set function $m : \mathcal{P}(\mathrm{T}) \to \mathbb{R}_+$ as follows: $\forall A \subseteq T, m(A) = card A$. Then $\forall i \in \{1, 2, \ldots, 9\}$, the singleton $\{i\}$ is an atom of m: $\forall i \in \{1, 2, \ldots, 9\}$, $m(\{i\}) = 1 > 0$ and $\forall B \subseteq \{i\}$, we have either $B = \emptyset$, in which case $m(B) = 0$, or $B = \{i\}$, in which case $m(\{i\} \backslash B) = m(\emptyset) = 0$. So, in this case, any singleton is an atom.

(ii) Generally, *there is no relationship between the notion of an atom and that of a pseudo-atom*: Let us consider an abstract set $T = \{t_1, t_2\}$ and let also be the set function $m : \mathcal{P}(\mathrm{T}) \to \mathbb{R}_+$ defined for every $A \subset T$ by

$$m(A) = \begin{cases} 2, \text{ if } A = T \\ 1, \text{ if } A = \{t_1\} \\ 0, \text{ if } A = \{t_2\} \text{ or } A = \emptyset. \end{cases}$$

Then T *is an atom and it is not a pseudo-atom for* m. Indeed, $m(T) = 2 > 0$. Let be an arbitrary subset B of T. If $B = \emptyset$, then $m(B) = 0$;

If $B = \{t_1\}$, then, by the definition, $m(T \backslash B) = m(\{t_2\}) = 0$;

If $B = \{t_2\}$, then, by the definition, $m(B) = 0$;

If $B = \{t_1, t_2\} (= T)$, then $m(T \backslash B) = m(\emptyset) = 0$.

Therefore, T is indeed an atom of m. On the other hand, let us note that there exists the singleton $\{t_1\}$ for which $m(\{t_1\}) = 1 \neq 0$ and $m(\{t_1\}) = 1 \neq 2 = m(T)$. Consequently, T is not a pseudo-atom of m.

However, we note that, *if the set function m is null-additive, then any atom of m is a pseudo-atom* (*). Indeed, let us assume that $m : \mathcal{C} \to \mathbb{R}_+$ is a null-additive set function, and that the set $A \in \mathcal{C}$ is an atom of m. We shall prove that A is also a pseudo-atom of m: Obviously, since A is an atom, then $m(A) > 0$. If we consider an arbitrary set $B \in \mathcal{C}$, with $B \subseteq A$, from the fact that A is an atom it follows that either $m(B) = 0$ or $m(A \backslash B) = 0$. In the latter case, since m is null-additive, it follows that $m(A) = m((A\backslash B) \cup B) = m(B)$. Consequently, A is a pseudo-atom of m. Conversely, *if the set function* $m : \mathcal{C} \to \mathbb{R}_+$ is, moreover, finitely additive, then *any pseudo-atom* $A \in \mathcal{C}$ of m *is an atom*, too, and this immediately yields based on the equality $m(A) = m((A\backslash B) \cup B) = m(A\backslash B) + m(B) = m(B)$, which implies $m(A\backslash B) = 0$.

That is why, *in the framework of the classic measure theory* (a measure always possesses the null-additive property), *the notions of an atom and that of a pseudo-atom coincide*. The converse of the above statement (*) does not generally hold since *there exist pseudo-atoms which are not atoms*:

(ii) Let $T = \{t_1, t_2\}$ be an abstract set, containing two arbitrary elements, and let us consider the set function $m : \mathcal{P}(\mathrm{T}) \to \mathbb{R}_+$, defined for every set $A \subseteq T$, by
$$m(A) = \begin{cases} 1, \text{if } A \neq \emptyset \\ 0, \text{ if} A = \emptyset. \end{cases}$$

Then m is null-additive and $T = \{t_1, t_2\}$ is a *pseudo-atom of* m*, but it is not an atom of* m. Let $A, B \subseteq T$ be so that $m(B) = 0$. By the definition of m we note that we must necessarily have $B = \emptyset$, whence $m(A \cup B) = m(A)$, and this proves that the set function m is null-additive.

We prove now that $T = \{t_1, t_2\}$ is a pseudo-atom of m. Indeed, we have $m(T) = 1 > 0$ and let $B \subseteq T$ an arbitrary subset. If $B = \emptyset$, then $m(B) = 0$. If $B \neq \emptyset$, then the set B either is a singleton, or is the set T, itself consisting of two elements. In both situations, one has $m(T) = 1 = m(B)$, which proves that $T = \{t_1, t_2\}$ is a pseudo-atom of m.

Let us prove now that $T = \{t_1, t_2\}$ is not an atom of m. Indeed, $m(T) = 1 > 0$ and there exists the singleton $\{t_1\}$ for which we have $m(\{t_1\}) = 1 \neq 0$ and $m(T\backslash\{t_1\}) = m(\{t_2\}) = 1 \neq 0$. Therefore, $T = \{t_1, t_2\}$ is not an atom of m.

(iii) The Dirac measure (or, the unit mass measure) (or, the δ-measure) δ_t concentrated in an arbitrary fixed point t of an abstract set T, is an example of a measure (in the classical sense) which is purely-atomic [5]. The Dirac measure is defined as follows: If $\mathcal{A}$ is a σ-algebra of subsets of T, then $\delta_t(A) = \begin{cases} 1, & t \in A \\ 0, & t \notin A \end{cases}, \quad \forall A \in \mathcal{A}$.

Obviously, T is an atom of δ_t (because $\delta_t(T) = 1 > 0$ and $\forall A \in \mathcal{A}$, it holds either $\delta_t(A) = 0$ or $\delta_t(cA) = 0$, as $t \notin A$ or $t \in A$, that is, $t \notin cA$).

Let us recall now the following:

If $\mathcal{C}$ is a ring of subsets of an abstract space T and if $m : \mathcal{C} \to \mathbb{R}_+$ is a set function satisfying the condition $m(\emptyset) = 0$, two *sets* A_1, A_2 are said to be *equivalent* if $m(A_1 \Delta A_2) = 0$.

We note that if the set function m is additionally null-monotone and null-additive, then $m(A_1) = m(A_2)$ (which justifies the terminology, since the equivalence of the sets takes place in the sense of the "measurement" process). Indeed, since $m(A_1 \Delta A_2) = m((A_1\backslash A_2) \cup (A_2\backslash A_1)) = 0$ and m is null-monotone, it follows that $m(A_1\backslash A_2) = 0$ and $m(A_2\backslash A_1) = 0$, whence, because m is null-additive and $m(A_1) = m((A_1\backslash A_2) \cup (A_1 \cap A_2)) = m(A_1 \cap A_2)$, $m(A_2) = m((A_2\backslash A_1) \cup (A_1 \cap A_2)) = m(A_1 \cap A_2)$ it follows that $m(A_1) = m(A_2)$. We note that, *with respect to the Dirac measure* δ_t, *the atom* T *(the space itself, unreduced to a single point) is equivalent to the singleton* $\{t\}, t \in T$ [5]. Indeed, we have $m(T\Delta\{t\}) = 0$ (so, *with respect to the Dirac measure, the space "collapses" into a single point*).

We shall prove in the following that, *with respect to a monotone and null-additive set function, any set which is equivalent to an atom is itself an atom:* Let us assume that the set A_1 is an atom and we prove that the set A_2, which is equivalent to the set A_1, possesses the same property. Indeed, according to the above statements, we have $m(A_2) = m(A_1) > 0$ and let $B \in \mathcal{C}$, B $\subseteq$ A, be arbitrary. If $m(B) = 0$, then the proof ends. If $m(A_1 \backslash B) = 0$, then, since m is monotone and $m(A_1 \Delta A_2) = 0$, it follows that $m(A_2 \backslash A_1) = 0$.

On the other hand, again from the monotonicity of m we have $m(A_2 \backslash B) \leq m((A_2 \backslash A_1) \cup (A_1 \backslash B)) = m(A_1 \backslash B) = 0$, based also on the fact that m is null-additive and $m(A_2 \backslash A_1) = 0$. Consequently, $m(A_2 \backslash B) = 0$, and this finally proves that A_2 is an atom of m, too.

Let us also note that, with respect to a monotone and null-additive set function, *any set which is equivalent to a pseudo-atom is, itself, a pseudo-atom*:

We assume that A_1 is a pseudo-atom and we prove that the set A_2, which is equivalent to the set A_1, possesses the same property. Indeed, from the above statements, we have $m(A_2) = m(A_1) > 0$ and let $B \in \mathcal{C}$, B $\subseteq$ A, be arbitrary. If $m(B) = 0$, then the proof ends. If $m(A_1) = m(B)$, then, since $m(A_2) = m(A_1) = m(B)$, it follows that A_2 is also a pseudo-atom of m.

Atoms and fractality

Next, we shall underline the fact that *both the notion of atom and that of pseudo-atom* (in the mathematical sense*) possess a remarkable property, namely that of self-similarity* (every part reflects the whole), a property which is a characteristic to fractals, both from a mathematical point of view and from the perspective of modern physics. This finding, among others, justifies the extension we illustrate in the last section, in which we address the necessity to introduce the notion of a fractal atom [3, 4].

The self-similarity property of the atoms (pseudo-atoms, respectively).

(i) If $m : \mathcal{C} \to \mathbb{R}_+$ is a null-monotone set function, with $m(\emptyset) = 0$, $A \in \mathcal{C}$ is an atom of m and $B \in \mathcal{C}$ is a subset of A having the property $m(B) > 0$, then B is also an atom of m and, moreover, $m(A \backslash B) = 0$. (which means that the "measure" of what remains when the set B is removed from the set A is null). Indeed, one has $m(B) > 0$ and if we consider an arbitrary set $C \in \mathcal{C}$, with $C \subseteq B$, then, since $B \subseteq A$, it follows that $C \subseteq A$. If $m(C) = 0$, the proof ends. Let us assume now that $m(C) \neq 0$. Because $A \in \mathcal{C}$ is an atom of m, it follows that $m(A \backslash C) = 0$. Since $B \backslash C \subseteq A \backslash C$ and m is null-monotone it gets that $m(B \backslash C) = 0$ and, therefore, B is an atom of m. Moreover, since $A \in \mathcal{C}$ is an atom of m and $B \in \mathcal{C}$ is a subset satisfying the property $m(B) > 0$, then we must necessarily have $m(A \backslash B) = 0$.

(ii) If $A \in \mathcal{C}$ is a pseudo-atom of m and the set $B \in \mathcal{C}$ satisfies $B \subseteq A$ and $m(B) > 0$, then B is also a pseudo-atom of m and, moreover, $m(B) = m(A)$

(the sets A are B are "identical" with respect to the "measure" m). Indeed, we have $m(B) > 0$ and, if we consider an arbitrary set $C \in \mathcal{C}$, with $C \subseteq B$, then, since $B \subseteq A$, it follows that $C \subseteq A$. If $m(C) = 0$, the proof ends. Let us assume now that $m(C) \neq 0$. Since $A \in \mathcal{C}$ is a pseudo-atom of m, it follows that $m(A) = m(C)$. On the other hand, since $A \in \mathcal{C}$ is a pseudo-atom of m, the set $B \in \mathcal{C}$ satisfies $B \subseteq A$ and $m(B) > 0$, then $m(B) = m(A)$. In consequence, $m(B) = m(C)$, and this finally proves that B is also a pseudo-atom of m.

Let us make, at the end of this section, the following observation: Assuming that a set function $m : \mathcal{C} \to \mathbb{R}_+$ is monotone, null-additive and regular (meaning that, roughly speaking, we can, through it, approximate sets about which we have little information, with sets about which we have more information), one can prove that for each atom A of m (if it exists), there exists a unique element $a \in A$ so that $m(A) = m(\{a\})$ [7, 8] (this means that the *"measure" of the atom is equal to the measure of each "point" it contains, and this reflects the holographic perspective, according to which the information is concentrated in a single point.*

Minimal atoms

We shall now introduce a very special category of atoms, which we show to reflect the property of indivisibility (non-decomposability). Let $\mathcal{C}$ be an arbitrary ring of subsets of an abstract space T and let $m : \mathcal{C} \to \mathbb{R}_+$ be a set function so that $m(\emptyset) = 0$. A set $A \in \mathcal{C}$ is called a *minimal atom* of m if $m(A) > 0$ and for every subset $B \in \mathcal{C}$ ($B \subseteq A$) it holds either $m(B) = 0$, or $B = A$ [6].

In other words, a minimal atom is a special set, of strictly positive "measure", so that any of its subsets has either zero "measure", or identifies with the set itself. Thus, a minimal atom has the property that any of its subsets has either zero "measure" (that is, it is negligible during the "measurement" process), or identifies with the initial set (without the need of a "measurement" process). The terminology is justified. Indeed, if $A \in \mathcal{C}$ is a minimal atom of m, then for m there cannot exist other minimal atom $A_1 \in \mathcal{C}$, which is different from A and satisfies $A_1 \subset A$. Indeed, if we assume, on the contrary, that there exists another minimal atom $A_1 \in \mathcal{C}$ which is different from A and satisfies $A_1 \subset A$, then, since A_1 is a minimal atom, we get that $m(A_1) > 0$. Because $A_1 \subsetneq A$, then $A_1 = A$, and this is false due to our assumption.

Example. Let $T = \{a, b, c, d\}$ be an abstract set, constituted of four distinct elements and let also be the set function $m : \mathcal{P}(T) \to \mathbb{R}_+$, defined for every $A \subseteq T$ by $m(A) =$

$$\begin{cases} 5, & \text{if } A = T \\ 2, & \text{if } A \neq T \\ 0, & \text{if } A = \emptyset. \end{cases}$$

We note that any singleton (i.e., a set containing only one element) is a minimal atom of m. Indeed, the "measure" m of any singleton is, according to the definition, 2, so it is strictly positive and any subset is either void and hence has zero measure, or is the set itself.

In general, *any minimal atom is, particularly, an atom and also a pseudo-atom.* Indeed, if $A \in \mathcal{C}$ is a minimal atom of m, then $m(A) > 0$ and for any of its subset

$B \in \mathcal{C}$ $(B \subseteq A)$ it holds either $m(B) = 0$, or $B = A$. The latter possibility yields $m(A \backslash B) = 0$ and $m(B) = m(A)$, so A is both an atom and a pseudo-atom of m.

The following examples show that there is generally no relationship between the notions of atom/pseudo-atom and that of minimal atom:

Examples. (i) Let $T = \{a, b\}$ be an abstract set constituted of two distinct elements and let also be the set function $m : \mathcal{P}(\mathrm{T}) \to \mathbb{R}_+$ defined as follows:

$$\forall A \subseteq T, m(A) = \begin{cases} 1, & \text{if } A = \{a\} \text{ or } A = T \\ 0, & \text{otherwise.} \end{cases}$$

Then T *is an atom of* m: Obviously, $m(T) = 1 > 0$. Let $B \subseteq T$ be an arbitrary set. If $B = \emptyset$, then $m(B) = 0$. If $B = \{a\}$, then $m(T \backslash B) = m(\{b\}) = 0$. If $B = \{b\}$, then $m(B) = 0$. If $B = T = \{a, b\}$, then $m(T \backslash B) = m(\emptyset) = 0$.

But T *is not a minimal atom of* m : Obviously, one has $m(T) = 1 > 0$ and let $B \subseteq T$ be an arbitrary set. We observe that there exists the set $B = \{a\} \neq T$ for which $m(B) = 1 \neq 0$. We also note that *the set* $\{a\}$ *is an atom* (we have $m(\{a\}) = 1 > 0$ and any subset $B \subseteq \{a\}$ either is void, so $m(B) = 0$, or is the set $\{a\}$ itself, so $m(\{a\} \backslash \{a\}) = 0$). The set $\{a\}$ *is also a minimal atom* of m since $m(\{a\}) = 1 > 0$ and any subset $B \subseteq \{a\}$ either is void, so $m(B) = 0$, or is $\{a\}$ itself.

(ii) Let $T = \{a, b, c, d\}$ be an abstract set and let also be the set function m : $\mathcal{P}(\mathrm{T}) \to \mathbb{R}_+$, defined as follows: $\forall A \subseteq T$,

$$m(A) = \begin{cases} 5, & \text{if } A = T \\ 3, & \text{if } A = \{a, b, c\} \text{ or } A = \{a, b, d\} \text{ or } A = \{a, c, d\} \\ 2, & \text{if } A = \{a, b\} \text{ or } A = \{a, c\} \\ 0, & \text{otherwise.} \end{cases}$$

Then $\{a, b\}$ and $\{a, c\}$ *are minimal atoms of* m. We shall prove the statement, for instance, for the set $\{a, b\}$: Indeed, we have $m(\{a, b\}) = 2 > 0$ and let B be an arbitrary subset. If $B = \{a, b\}$, the statement is verified. If $B = \{a\}$ or $B = \{b\}$, then, according to the definition, we have $m(\{a\}) = m(\{b\}) = 0$, so the statement is again verified. If $B = \emptyset$, then $m(B) = 0$.

In the following, we note that if $m : \mathcal{C} \to \mathbb{R}_+$ is a null-null-additive set function and $A, B \in \mathcal{C}$ are two different minimal atoms of m, then they must be necessarily disjoint, that is, $A \cap B = \emptyset$. Indeed, let us assume that, on the contrary, $A \cap B \neq \emptyset$. Since $A, B \in \mathcal{C}$ are two minimal atoms of m, $A \backslash (A \cap B) = A \backslash B \subseteq A$ and $A \cap B \subseteq B$, it follows that $[m(A \backslash B) = 0$ or $A \backslash B = A]$ and $[m(A \cap B) = 0$ or $A \cap B = B]$.

(i) If $A \backslash B = A$, then $A \cap B = \emptyset$, which is false since, according to our assumption, we have $A \cap B \neq \emptyset$.
(ii) If $m(A \backslash B) = 0$ and $m(A \cap B) = 0$, then, since m is null-null-additive, one gets that $m(A) = m((A \backslash B) \cup (A \cap B)) = 0$, which is false, since $m(A) > 0$, the set A being a minimal atom of m.

(iii) If $m(A\backslash B) = 0$ and $A \cap B = B$, then $B \subseteq A$, so, since A is a minimal atom of m, one gets from the above observation that $B = A$, which is false.

Consequently, $A \cap B = \emptyset$.

The property we shall demonstrate next reflects *the non-decomposability (non-partitionability) of the minimal atoms:* A minimal atom $A \in \mathcal{C}$ of a null-null-additive set function m cannot be partitioned in sets that are elements of $\mathcal{C}$. Indeed, if we suppose, on the contrary, that there exists a partition of a lui A, this means that there exists a family $\{A_i\}_{i\in\{1,2,\ldots,p\}}$ of nonvoid sets of $\mathcal{C}$ so that $\bigcup_{i=1}^{p} A_i = A$ and the sets A_i are pairwise disjoint.

Referring to the first set A_1, since $A \in \mathcal{C}$ is a minimal atom, it follows that we cannot have the situation $A_1 = A$. Therefore, $m(A_1) = 0$. Analogously, for the second set,A_2, we get that $m(A_2) = 0$. Recurrently, it gets that $m(A_3) = \cdots = m(A_p) = 0$. Since m is null-null-additive, it follows that $m(A) = m\left(\bigcup_{i=1}^{p} A_i\right) = 0$, which is obviously false.

Consequently, any minimal atom is non- decomposable.

In the following, we shall prove that the converse of this statement also holds, namely, we shall demonstrate that *any non-decomposable atom $A \in \mathcal{C}$ is necessarily a minimal atom.* Indeed, since the set A is an atom, then $m(A) > 0$. Since the set A is not partitionable, there cannot exist two nonvoid disjoint subsets $A_1, A_2 \in \mathcal{C}$ of A so that $A = A_1 \cup A_2$. Let be an arbitrary set $B \in \mathcal{C}$, with $B \subseteq A$. If $m(B) = 0$, then the proof ends. If $m(B) > 0$, since $B \subseteq A$, one gets that $B = A$ (otherwise, the family $\{A\backslash B, B\}$ is a partition of A: $A\backslash B$, $B \in \mathcal{C}$, $(A\backslash B) \cap B = \emptyset$, $(A\backslash B) \cup B = A$, which is false). Consequently, A is a minimal atom. From the two statements above, one arrives at the following conclusion: *an atom is minimal if and only if it is not partitionable (it is non-decomposable).*

In the following, we shall highlight that, *in the case when the abstract set T is finite*, then *any set $A \in \mathcal{C}$, satisfying the condition $m(A) > 0$ possesses at least one set $B \in \mathcal{C}, B \subseteq A$, which is a minimal atom minimal of m.* Moreover, in the particular case when A is an atom of m and the set function m is null-additive, one gets that $m(A) = m(B)$ and the set B is unique. Indeed, let us consider the family of sets $\mathcal{M} = \{M \in \mathcal{C}, M \subseteq A, m(M) > 0\}$. Obviously, since $A \in \mathcal{C}$, then $\mathcal{M} \neq \emptyset$. We note that any minimal element $M \in \mathcal{M}$ of $\mathcal{M}$ is a minimal atom of m. Indeed, since M is a minimal element, there cannot exist another set $D \in \mathcal{M}$ so that $D \subsetneq M (**)$.

Since $M \in \mathcal{M}$, this means that $M \in \mathcal{C}, M \subseteq A$ and $m(M) > 0$.

We shall prove that M is a minimal atom of m. Indeed, for any set $S \subseteq M$, $S \in \mathcal{C}$, we have either $m(S) = 0$ or $m(S) > 0$. In the latter case, we have either $S = M$ (which is suitable) or $S \neq M$, which contradicts the statement $(**)$.

Let us assume, moreover, that the set A is an atom of m and m is null-additive. According to the considerations proved above, there exists at least one set $B \in \mathcal{C}, B \subseteq A$, which is a minimal atom of m. This means that $m(B) > 0$ and, because A is an atom, we must necessarily have $m(A\backslash B) = 0$. Since m is null-additive, this yields $m(A) = m((A\backslash B) \cup B) = m(B)$.

It only remains to prove that the set B is unique. Indeed, if we suppose, on the contrary, that there exist two different minimal atoms B_1 and B_2 of m, this would imply, as before, that $m(A\backslash B_1) = m(A\backslash B_2) = 0$. If $m(B_1 \cap B_2) = 0$, then $m(A) = m(A\backslash(B_1 \cap B_2) \cup (B_1 \cap B_2)) = m(A\backslash(B_1 \cap B_2)) = m((A\backslash B_1) \cup (A\backslash B_2))$, which is false. If $m(B_1 \cap B_2) > 0$, since B_1 and B_2 are minimal atoms of m, it results that $B_1 = B_1 \cap B_2 = B_2$, which is again false. Finally, we shall prove that, *if the set* T *is finite*, the set function m is null-additive, and $\{A_i\}_{i\in\{1,2,\ldots,p\}}$ is the family of all different minimal atoms which are contained in a set $A \in \mathcal{C}$, satisfying $m(A) > 0$ (we proved in the above considerations that such atoms exist), then $m(A) = m\left(\bigcup_{i=1}^{p} A_i\right)$.

(This means that *the set A identifies itself, from the "measure" m viewpoint, with the union of all different minimal atoms which it contains*, therefore the *minimal atoms are the only ones that matter from the "measurement" point of view.*)

Let us note that $m\left(A\backslash \bigcup_{i=1}^{p} A_i\right) = 0$ (if, on the contrary, one has $m\left(A\backslash \bigcup_{i=1}^{p} A_i\right) > 0$, from the statement proved above it would follow that there exists at least one set $B \in \mathcal{C}, B \subseteq A\backslash \bigcup_{i=1}^{p} A_i \subseteq A$, which is a minimal atom of m, and this is false since $A_1, \ldots, A_p$ are the only different minimal atoms contained in A). Since $m\left(A\backslash \bigcup_{i=1}^{p} A_i\right) = 0$ and m is null-additive, it follows that $m(A) = m\left(\left(A\backslash \bigcup_{i=1}^{p} A_i\right) \cup \left(\bigcup_{i=1}^{p} A_i\right)\right) = m\left(\bigcup_{i=1}^{p} A_i\right)$.

We finally note the following:

1. Any minimal atom is also an atom and a pseudo-atom (which justifies the terminology);
2. If the set function is null-additive, then any of its atoms is a pseudo-atom, too;
3. If, moreover, the set function is finitely additive, then the converse of the above statement is also valid, therefore any pseudo-atom is particularly an atom.

Consequently, for a finitely additive set function (which is automatically null-additive), the notion of atom and that of pseudo-atom coincide.

2.3 Extensions of the Notions of Atom

Generalizations of the mathematical notion of an atom have been made, so far, in two major directions. A first direction is given by the fact that, instead of set functions, which are indispensable to the process of the so-called "measurement", one could generally operate with set multifunctions (that is, functions that associate a set to another set). Thus, results with a higher degree of generalization and abstraction can be obtained. The second direction is given by the correlation that can be made by placing the notion of (minimal) atom within the fractal sets theory, thus resulting in the notion of *fractal (minimal) atom* [2–4].

The first direction: Set-valued approach

Let be an abstract nonvoid set T, $\mathcal{C}$ a ring of subsets of T, X a real linear normed space with the origin θ and $\mathcal{P}_0(X)$, the family of all nonvoid subsets of X. By a *set multifunction* we mean a function (or, application) which associates a set to another set, in contrast with the notion of a function, which associates a point to another point. So, in what follows, let $\mu : \mathcal{C} \to \mathcal{P}_0(X)$ be an arbitrary set multifunction, with $\mu(\emptyset) = \theta$.

The notions of atom, pseudo-atom, minimal atom introduced with respect to a set function m can be generalized in this context, with respect to the set multifunction μ, as follows. We say that a set $A \in \mathcal{C}$ is:

(i) an *atom* of μ if $\mu(A) \supsetneq \{\theta\}$ and for every set $B \in \mathcal{C}$, with $B \subseteq A$, we have either $\mu(B) = \{\theta\}$ or $\mu(A \backslash B) = \{\theta\}$;
(ii) a *pseudo-atom* of μ if $(A) \supsetneq \{\theta\}$ and for every set $B \in \mathcal{C}$, with $B \subseteq A$, it holds either $\mu(B) = \{\theta\}$, or $\mu(A) = \mu(B)$;
(iii) a *minimal atom of* μ if $\mu(A) \supsetneq \{\theta\}$ and for every set $B \in \mathcal{C}$, with $B \subseteq A$, one has either $\mu(B) = \{\theta\}$ or $A = B$.

Detailed considerations on atomicity with respect to set multifunctions can be found in Gavriluţ and Agop [2] and also in Gavriluţ et al. [3, 4].

The second direction: Towards a fractal theory of atomicity

The main idea in the quantum theory of measure and in generalized quantum mechanics is to provide a description of the world in terms of histories. A history is a classical description of the system considered for a certain period of time, which may be finite or infinite.

If one tries to describe a particle system, then a history will be given by classical trajectories. If one deals with a field theory, then a history corresponds to the spatial configuration of the field as a function of time. In both cases, the quantum theory of measure tries to provide a way to describe the world through classical histories, extending the notion of probability theory, which is obviously not enough to shape our universe.

On the other hand, ordinary structures, self-similar structures etc. of nature can be assimilated to complex systems, if one considers both their structure and functionality. The models used to study the complex systems dynamics are built on the assumption that the physical quantities that describe it (such as density, momentum, and energy) are differentiable. Unfortunately, differentiable methods fail when reporting to physical reality, due to instabilities in the case of complex systems dynamics, instabilities that can generate both chaos and patterns.

In order to describe such dynamics of the complex systems, one should introduce the scale resolution in the expressions of the physical variables describing such dynamics, as well as in the fundamental equations of the evolution (density, kinetic moment and equations of the energy). This way, any dynamic variable which is dependent, in a classical sense, both on the space and time coordinates, becomes, in this new context, dependent on scale resolution as well. Therefore, instead of working with a dynamic variable, we can deal with different approximations of a mathematical function that is strictly non-differentiable. Consequently, any dynamic variable acts as the limit of a family of functions. Any function is non-differentiable at a zero resolution scale and it is differentiable at a non-zero resolution scale. This approach, well adapted for applications in the field of complex systems dynamics, in which any real determination is made at a finite resolution scale, clearly involves the development of both a new geometric structure and a physical theory (applied to the complex systems dynamics) for which the motion laws, that are invariant to the transformations of spatial and temporal coordinates, are integrated with scale laws, which are invariant to transformations of scale. Such a theory that includes the geometric structure based on the assumptions presented above was developed in the scale relativity theory and, more recently, in the scale relativity theory with constant arbitrary fractal dimension. Both theories define the class of fractal physics models. In this model, it is assumed that, in the complex systems dynamics, the complexity of interactions is replaced by non-differentiability. Also, the motions forced to take place on continuous, differentiable curves in a Euclidean space are replaced by free motions, without constraints, that take place on continuous, non-differentiable curves (fractal curves) in a fractal space. In other words, for a time resolution scale that seems large when compared to the inverse of the largest Lyapunov exponent, deterministic trajectories can be replaced by a set of potential trajectories, so that the notion of "defined positions" is replaced by the concept of a set of positions that have a definite probability density. In such a conjecture, quantum mechanics becomes a particular case of fractal mechanics (for the structural units motions of a complex system on Peano curves at Compton scale resolution). Therefore, the quantum theory of the measure could become a particular case of a fractal measure theory. One of the concepts that needs to be defined is that of a fractal minimal atom, as a generalization of the concept of a minimal atom [3, 4].

3 Conclusions

An exhaustive study on the problem of atomicity with respect to set functions is provided. Different types of atoms are discussed, the relationships among them are studied and several examples and physical possible implications and applications are obtained.

References

1. D.H. Fremlin, *Measure Theory*, vol. 5 (2000)
2. A. Gavriluţ, M. Agop, *An Introduction to the Mathematical World of Atomicity through a Physical Approach* (ArsLonga Publishing House, Iaşi, 2016)
3. A. Gavriluţ, I. Mercheş, M. Agop, *Atomicity Through Fractal Measure Theory: Mathematical and Physical Fundamentals with Applications* (Springer, 2019). https://doi.org/10.1007/978-3-030-29593-6. ISBN: 978-3-030-29592-9
4. G. Gavriluţ, A. Gavriluţ, M. Agop, Extended minimal atomicity through non-differentiability: a mathematical-physical approach. Adv. Math. Phys. **160**, 1–16 (2019). https://doi.org/10.1155/2019/8298691
5. V. Kadets, *A Course in Functional Analysis and Measure Theory* (Springer, 2018)
6. Y. Ouyang, J. Li, R. Mesiar, Relationship between the concave integrals and the pan-integrals on finite spaces. J. Math. Anal. Appl. **424**(2), 975–987 (2015)

7. E. Pap, *Null-Additive Set Functions* (Kluwer Academic Publishers, Dordrecht, 1995)
8. E. Pap, Regular null-additive monotone set functions, Univ. u Novom Sadu, Zb. Rad. Prirod. Mat. Fak. Ser. Mat. **25**(2), 93–101 (1995)
9. H.L. Royden, *Real Analysis*, 3rd edn. (Macmillan, New York, 1988)

Nonlinear Phenomena in the Dynamics of a Class of Rolling Pendulums: A Trigger of Coupled Singularities

Plenary Review Lecture

Katica R. (Stevanović) Hedrih

Abstract In the introductory part of the plenary lecture, an overview of nonlinear differential equations of heavy ball and heavy thin dick rolling along curvilinear paths and surfaces of different shapes were presented. This is reason that this content is omitted from present review paper. In the introductory part of this article, we will present nonlinear phenomena of motion of a heavy material point moving along a rotating, smooth circle around a vertical, central or eccentric axis, as well as around an eccentric oblique axis relative to the vertical, at a constant angular velocity. Using linear and nonlinear approximations of the nonlinear differential equation in the vicinity of singular points of the observed dynamics, the analysis of the local dynamics of the heavy material point system along the rotating circle around the oblique axis is given. A mathematical analogy between this model and the model of the dynamics of a thin heavy disk rolling in a rotating circle around an eccentric-centric oblique axis is pointed out. Using linear and nonlinear approximations of the nonlinear differential equation in the vicinity of the singular points of the observed dynamics, the analysis of the local dynamics of the heavy material point system along the rotating circle around the oblique axisat a constant angular velocity is given. A mathematical analogy between this model and the dynamics model of a thin heavy disk rolling in a rotating circle around an eccentric-centric vertical-oblique axis is pointed out. The central and main subject of the paper is the identification and presentation of nonlinear phenomena in the nonlinear dynamics of a class of generalized rolling pendulums, whose heavy bodies roll along curvilinear paths, lying in a vertical plane, rotating around a vertical axis, at a constant angular velocity. The bifurcation parameter of coupled rotations is identified. The bifurcation of the position of stable equilibrium of the generalized rolling pendulum and the corresponding representative singular points of the type of the stable center is described, as well as the stratification and transformation of phase trajectories in the phase portrait of nonlinear dynamics of the generalized rolling pendulum in the Earth's gravitational

K. R. (Stevanović) Hedrih (✉)
Department of Mechanics, Mathematical Institute of Serbian Academy of Science and Arts, Belgrade, Serbia
e-mail: khedrih@sbb.rs; katicah@mi.sanu.ac.rs

Faculty of Mechanical Engineering, University of Niš, Nis, Serbia

C. H. Skiadas and Y. Dimotikalis (eds.), *14th Chaotic Modeling and Simulation International Conference*, Springer Proceedings in Complexity,
https://doi.org/10.1007/978-3-030-96964-6_15

field, and along curvilinear route in rotate vertical plane around vertical axis at a constant angular velocity. Additionally a theorem of trigger of coupled singularities and a homoclinic orbit in the form of the number "eight" is graphically proofed. A series of graphs of characteristic equation oh nonlinear dynamics as well as series of phase portraits for different coefficients of curvilinear paths described by parabola, bi-quadratic parabola or polynomials of the eighth degree is presented and sets of transformed phase trajectories and homoclinic orbits in the form of the number "eight" are presented, which include one or more triggers of coupled singular points in nonlinear dynamics of relative rolling thin heavy disk along these curvilinear trace in rotate vertical plane around vertical axis at a constant angular velocity.

Keywords Generalized rolling pendulum · Bifurcation · Trigger of coupled singularities · Curvilinear rolling path · Phase trajectory portrait · Homoclinic orbit · Theorem · Mathematical analogy

1 Introduction

This review paper presents the main content of the Plenary Lecture, which Author held at the international conference "CHAOS 2021 Conference", traditionally organized by Professor Christos H Skiadas, CHAOS Conference Chair. He is the "spiritus movens" of the high scientific level program of the series of these good conferences and the accompanying series of publications of papers presented at them.

In the introductory part of the Plenary lecture, an overview of nonlinear differential equations of heavy ball [1–3] and heavy thin dick rolling along curvilinear paths [4–7] and surfaces of different shapes [3] were presented.

Also, an overview of nonlinear differential equations and nonlinear equations of phase trajectories were given for a number of special rolling points on spherical surfaces, on a cone and on a torus (see [7]).

For a number of nonlinear dynamics of ball and thin disk rolling along curvilinear paths, phase portraits were presented with the definition of the term generalized rolling pendulum (see [5]).

This review paper of mine contains my author's original scientific results, one part of which has already been presented or published, and some of which have now been shown for the first time and have not been published before. The presentation begins with an introduction to the nonlinear dynamics of rheonomic discrete systems with coupled rotations, both with two degrees of mobility and one degree of freedom of movement. Such systems are described by one rheonomic coordinate and one independent generalized coordinate. The rheonomic coordinate introduces a kinematic constrain, i.e. kinematic excitation, into the system, and with the help of an independent generalized coordinate, which describes a degree of freedom of movement of the rheonomic system, a nonlinear differential equation of nonlinear dynamics of the rheonomic system is formed.

1.1 Models of Fascinating Nonlinear Dynamics of Abstraction of Real Systems and Phenomenological Mapping of Local Dynamics Around Stationary States

We first present several models of abstraction of nonlinear dynamics of real rheonomic mechanical systems, which simultaneously represent models of motion of a heavy meterail point in a circle, which rotates with a constant angular velocity around the centric/eccentric vertical or oblique axis (see [8–16]), but in terms of mathematical analogy and models of abstraction of real systems of rigid bodies (rolling without sliding a thin heavy disk), which perform dynamics with coupled rotations around two or three passing (no intersecting) axes [17, 18]. In the conditions of rotation around the hair, in relation to the vertical, the axis, there is another member that explicitly depends on both the generalized coordinate and the time.

In this part, we focus on determining the linearization and the linear and nonlinear approximation of the nonlinear differential equation around stationary singular points. The stability as well as the instability of the system dynamics around stationary singular points-positions of relative rest (equilibrium) of the material point on the rotating circle were, also, analyzed.

The appearance of triggers of coupled singularities [13, 19–25] is especially pointed out, as well as the fact that the sources of chaos dynamics in such systems are. It is indicated how the character of local properties of linear and nonlinear dynamics around stationary singular points is examined.

Figure 1 shows three models of structurally the same system, with one axis around which the circle rotates at a constant angular velocity and along which a heavy metal point moves.

The difference between these three models is in the different position of the fixed axis, around which the circular line rotates, in relation to the horizon, as well as the position of the axis in relation to the center of the circular line. Depending on the position of the axis around which the circular line rotates at a constant angular velocity, the nonlinear dynamics of the relative motion of a material point along a circular, ideally smooth line is described by the following nonlinear differential equations [8, 16]:

$$\ddot{\varphi} + \Omega^2(\lambda - \cos\varphi)\sin\varphi = \Omega^2\lambda ctg\alpha \cos\varphi \cos\Omega t \tag{1}$$

(Model in Fig. 1a.1*, за $\varepsilon = 0$)

$$\ddot{\varphi} + \Omega^2(\lambda - \cos\varphi)\sin\varphi = 0 \tag{2}$$

(Model in Fig. 1a.2*, за $\varepsilon = 0$)

$$\ddot{\varphi}_2 + \Omega^2\langle\lambda - \cos\varphi_2\rangle\sin\varphi_2 - \Omega^2\varepsilon\cos\varphi_2 = 0 \tag{3}$$

(Model in Fig. 1a.3*, за $\varepsilon \neq 0$).

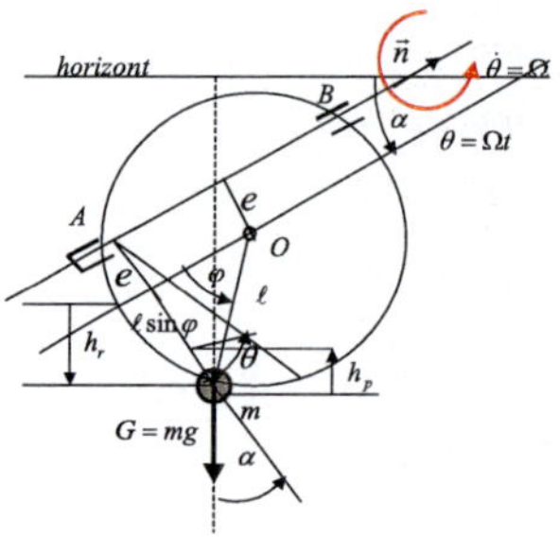

a.1*

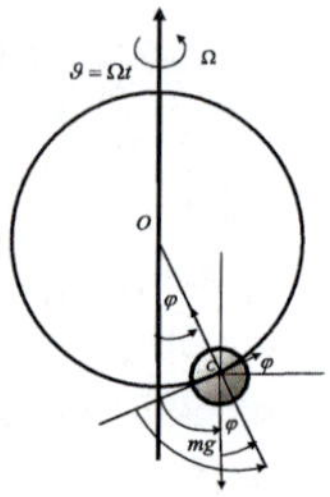

a.2*

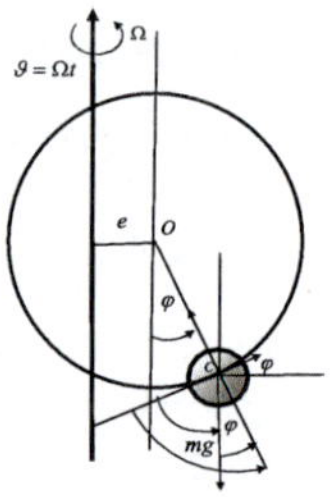

a.3*

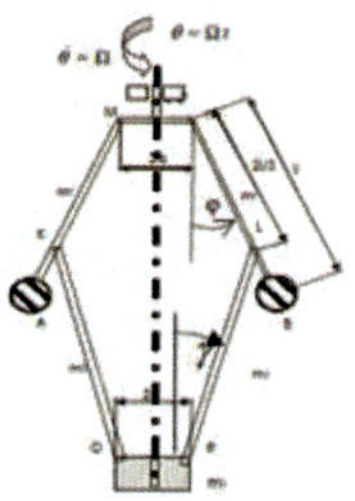

a.4*

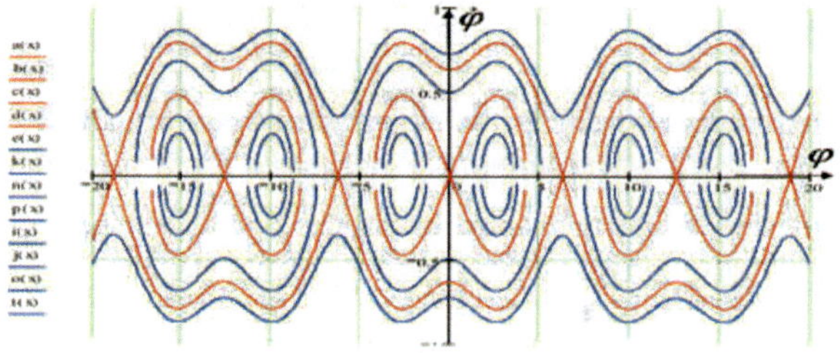

$$\dot{\varphi}^2 - \dot{\varphi}_0^2 + \frac{2\Omega^2}{\kappa}\left\langle \lambda(\cos\varphi_0 - \cos\varphi) - \frac{1}{2}\left(\cos^2\varphi_0 - \cos^2\varphi\right)\right\rangle = 0$$

$$\lambda = \frac{g}{(R-r)\Omega^2}$$

b.2*

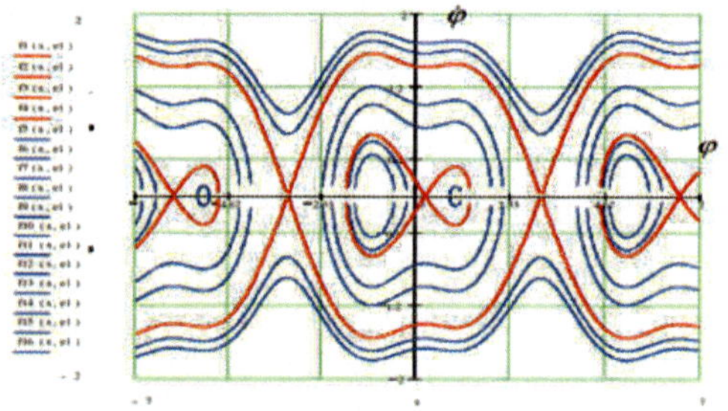

$$\ddot{\varphi} + \frac{\Omega^2}{\kappa}(\lambda - \cos\varphi)\sin\varphi - \varepsilon\frac{\Omega^2}{\kappa}\cos\varphi = 0$$

$$\varepsilon = \frac{e}{(R-r)}$$

b.3*

Fig. 1 Models of a heavy material point moving in a rotating circle, at a constant angular velocity, around an eccentric axis obliquely positioned with respect to the horizon (**a.1***), or a vertical centric (**a.2***) or vertical eccentric axis (**a.3***); **b*** Phase portrait of the nonlinear dynamics of a difficult material point, moving in a rotating circle, at a constant angular velocity, around a centric (**b.2***) or eccentric vertical axis (**b.3***)

These differential equations (1)–(3) in Petrović's terminology would be phenomenological differential equations (see [8, 16]).

Figure 1 shows models of a heavy material point moving in a rotating circle, at a constant angular velocity, around an eccentric axis obliquely positioned relative to the horizon (a.1*), or a vertical centric (a.2*) or vertical eccentric axis (a.3*); Fig. 1b.2* shows the phase portraits of the nonlinear dynamics of a heavy material point moving along a rotating circle, at a constant angular velocity, around a centric (b.2*) or eccentric vertical axis (b.3*). Both phase portraits are in the case of the appearance of triggers of coupled singularities. Each of the triggers of coupled singularities contains a homoclinic phase trajectory in the shape of the number "eight" which intersects at a singular point of the unstable saddle type, and surrounds a stable center-type singular point on the left and on the right side of the unstable saddle type singular point. The trigger of coupled singularities in the phase portrait in Fig. 1(b.2*), i.e. (b.3*), was created by bifurcation of a singular point of the stable center type. At certain values of the bifurcation parameter of the system, this singular point of the stable center type was transformed into a singular point of the unstable saddle type, and two new singular points of the stable center type appeared around it and a new separatrix-homoclinic phase trajectory appeared in the form of number "eight" with self-intersection at a singular point of the unstable saddle type and surround singular points of stable centre types. At certain values of the bifurcation parameter of the system, in the phase portrait of the material point nonlinear dynamics along the rotating circle there is no trigger structure of coupled singularities and there is no transformation of the singular point of the stable center type into the singular point of the unstable saddle type. These cases of phase portraits, which do not contain in their structure a substructure of triggers of coupled singularities, were not considered to be presented here.

1.2 Linearization of a Nonlinear Differential Equation Around Singular Points and Local Properties of System Dynamics

We start with the linearization of the nonlinear differential equations (1) around the singular points, obtained for the differential equations (2) obtained from the previous one for $\alpha = \frac{\pi}{2}$. In particular, we will consider cases of linearizations [8, 16] around singularity: 1 * for, $\lambda > 1$ around singularity $\varphi = 0$ and 2 * for $\lambda < 1$, around singularity $\varphi = \pm \arccos \lambda$.

1* In the case that $\lambda > 1$, and we examine small forced oscillations around a singular point $\varphi = 0$ of the center type, by linearization around it we obtain a linearized differential in the following form

$$\ddot{\varphi} + \Omega^2(\lambda - 1)\varphi = \Omega^2 \lambda ctg\alpha \cos \Omega t \tag{4}$$

in which we performed linearization by the following approximations $\sin\varphi \approx \varphi + \cdots$ and $\cos\varphi \approx 1 + \cdots$, and measured the coordinate from the singularity, around when we performed linearization, and introduced the assumption that the coordinate is small.

From the linearized differential equations (4) we can conclude that for the case $\lambda > 1$, around a singular point $\varphi = 0$ for small force amplitudes $\Omega^2\lambda ctg\alpha$ and small values of initial angle coordinate $\varphi_0 = \varphi(0)$ and small initial angular velocity $\dot{\varphi}_0 = \dot{\varphi}(0)$, that the circular frequency of natural oscillations is $\omega \approx \Omega\sqrt{\lambda - 1}$, around that singular point, while the frequency Ω of forced oscillations corresponds to angular velocity of a rotating circle in a rheonomic system.

We can also conclude that when $\Omega = \Omega_{rez}$ and $\lambda = 2$ when there is a resonant state for small oscillations around the singular point $\varphi = 0$ for $\lambda = 1$, so the linearization of the differential equation can be accepted only in a very short time interval, while the amplitude of forced oscillation, which increases with time in which the system is exposed coercion, does not go beyond the assumed limits, which allow linearization.

In the resonant case, the linearization around a singular point $\varphi = 0$, when Ω^2_{rez} and $\lambda = 2$, for given initial conditions, the solution of the linearized differential equation is of the form:

$$\varphi(t) \approx \varphi_0 \cos\frac{\omega_0 t\sqrt{2}}{2} + \frac{\sqrt{2}\dot{\varphi}_0}{\omega_0}\sin\frac{\omega_0 t\sqrt{2}}{2} + \left[\frac{\omega_0 t\sqrt{2}}{2}\sin\frac{\omega_0 t\sqrt{2}}{2}\right] ctg\alpha \tag{5}$$

where $\Omega^2_{rez} = \frac{g\sin\alpha}{2\ell} = \frac{\omega_0^2}{2}$, $\omega_0^2 = \frac{g\sin\alpha}{\ell}$. $\lambda = \frac{\omega_0^2}{\Omega^2}$, $\lambda_{rez} = 2$.

2* For the case when $\lambda < 1$, we will report the linearization around the singular points $\varphi_s = \pm\arccos\lambda$, so instead of the coordinate φ in the nonlinear differential equations (1), we enter $\varphi_s + \varphi$ by writing:

$$\ddot{\varphi} + \Omega^2[\lambda - \cos(\varphi_s + \varphi)]\sin(\varphi_s + \varphi) = \Omega^2\lambda ctg\alpha\cos(\varphi_s + \varphi)\cos\Omega t \tag{6}$$

now the coordinate φ is measured from these singular points $\varphi_s = \pm\arccos\lambda$, as the beginnings in which that coordinate is zero. After linearization of the nonlinear differential equations (6) around singular points $\varphi_s = \pm\arccos\lambda$, the linearized differential equation takes the form:

$$\ddot{\varphi} + \Omega^2(1 - \lambda^2)\left[1 + \frac{\lambda ctg\alpha}{\sqrt{1-\lambda^2}}\cos\Omega t\right]\varphi \approx \Omega^2\lambda ctg\alpha\cos\Omega t \tag{7}$$

Obtained by linearization, nonlinear differential equations around singular points $\varphi_s = \pm\arccos\lambda$, the previous differential equation is rheolinear and Mathieu-Hill type:

$$\frac{d^2\varphi}{d\tau^2} + (\tilde{\lambda} + \tilde{\gamma}\cos\tau)\varphi = h\cos\tau \tag{8}$$

whose shape coefficients are:

$$\omega = \Omega\sqrt{(1-\lambda^2)}, \quad \tilde{\lambda} = 1 - \lambda^2, \quad \text{and} \quad \tilde{\gamma} = \lambda\sqrt{1-\lambda^2}ctg\alpha = h\sqrt{\tilde{\lambda}}, \quad \tau = \Omega t.$$

A general solution to the Mathieu-Hill differential equation can be found in References to Mathieu's differential equation [26, 27, 28] or Floquet, Annales de l'Ecole Normale, 1883.

$$\varphi(t) = Ae^{\mu t}p_1(t) + Be^{-\mu t}p_2(t) \tag{9}$$

in which A and B there are also integration constants; μ is a characteristic exponent, and $p_i(t), i = 1, 2$ are periodic functions of the period 2π, which depend on the parameters $\tilde{\lambda}$ and $\tilde{\gamma}$.

The main problem in studying the stability around these singular points, which in the basic autonomous nonlinear system described by the autonomous nonlinear differential equations (2) represent stable centers (stable relative positions of the relative equilibrium of the rheonomic system, when the axis of rotation is vertical), and in the nonautonomous system described by nonlinear differential equations (1) and approximation (7) and (8), respectively, open more complex questions of motion stability testing around these same singular points. References to the solution of the solution of the Mathieu-Hill differential equation of the form (8) can be found in [29, 26]. Using the Ince-Strutt stability map, we can determine the parameter areas by $\tilde{\lambda}$ and $\tilde{\gamma}$, in which the solutions are stable or unstable, and then conclude about the character of the stability of the system motion around the relative equilibrium position or singular points of the nonlinear differential equations (1).

1.3 Phenomenological Approximate Mapping of Nonlinear Dynamics

In this previously presented example, linearization of nonlinear differential equations, we have shown that the analysis of nonlinear dynamics of a nonlinear system can be performed by decomposing the analysis to the analysis of local dynamics in the vicinity of singular points by linear mappings of dynamic phenomena into approximate local dynamics or linear or rheolinear and in depending on the initial conditions to study the properties of these dynamics. Knowing the properties of local dynamics or stable singular or unstable singular points, we can assemble-compose a whole of global motion and rare nonlinear phenomena of nonlinear system dynamics.

This is a *phenomenological mapping* [29–31] of partial, linear and rheolinear phenomena from local dynamics to global nonlinear system dynamics by analyzing the phenomenon of dynamics in local domains and synthesizing the results of these analyzes into global ones.

We see that by approximations around some singularities, harmonic natural or forced oscillations occur with the possibility of the occurrence of the basic resonant state of local dynamics, i.e. in rheolinary natural or forced dynamics, with the possibility of the appearance of the parametric resonant state. Both the basic resonant state and the parametric resonant state, depending on the initial conditions, as well as the length of the time interval in which the nonlinear dynamical system is subjected to that local state of dynamics, come out of that state of dynamics, when the descriptions do not applicable, and study must to continue in global.

The ideas of Mihailo Petrović and his mathematical or qualitative phenomenology [29–31] can also be recognized in these approaches. And if we take into account that these models of nonlinear dynamics of a heavy material point have the same mathematical description as the motion of a rolling rigid body (homogeneous heavy thin disk), which performs coupled rotations around the passing (no intersecting) axes, then Mathematical phenomenology and phenomenological mapping become the right tool for searching for models of abstraction of nonlinear dynamics of real mechanical systems, which are described by one or more nonlinear differential equations.

2 Bifurcation and Trigger of Coupled Singularities in the Dynamics of Generalized Rolling Pendulums Along Curvilinear Route in a Rotating Vertical Plane at a Constant Angular Velocity About a Vertical Axis

In a series of [4, 5, 6, 7] of the author of this paper, the results of research of nonlinear dynamics of special cases of generalized rolling pendulums on curvilinear line in a rotating vertical plane, at a constant angular velocity Ω around the vertical axis are presented, and a given series of phase trajectory portraits in phase planes. Each of these phase portraits contains at least one trigger of coupled singularities, consisting of a singular point of the unstable saddle type, and two singular points of the stable center type, surrounded by a single-separator phase trajectory in the form of number "eight", which intersects at a singular point of the unstable saddle type. The angular velocity Ω of rotation of the vertical plane around the vertical axis appears as a bifurcation parameter, whose change can achieve the disappearance of the trigger of coupled singularities, or the appearance of that trigger in the phase portrait, or the appearance of bifurcation of a stable type singular position, and two new singular points of the stable center type appear around it, and in the phase portrait a separatrix phase trajectory in the form of a number of "eight" that surrounds them and self-intersect at singular point which has lost stability and bifurcated into unstable saddle-type singular point.

In such a system, there is now a phenomenon of a bifurcation [8, 19, 32, 33], because the trigger of coupled singularities is now in results caused by the property of nonlinearity in the form of bifurcation and nonlinear dynamics of such a system.

And with the existence of a bifurcation parameter with the change of which the trigger of coupled singularities appears or disappears.

In such a system, there is at least one parameter with the change of which such a trigger of coupled singularities would disappear or appear, which is caused by the properties of both, the curvilinear path and the existence of extremums in a set of one maximum and two minimums of the curvilinear trajectory in the rotate vertical plane at constant angular velocity Ω around vertical axis. Bifurcation and trigger of coupled singularities are nonlinear properties of nonlinear dynamics of generalized rolling pendulum.

For example, the ordinary nonlinear differential equation of non-linear rolling dynamics, non-slip, heavy homogeneous thin disk, radius r, in a circle, radius R, in a rotating vertical plane at a constant angular velocity Ω about the vertical central axis, is [6]:

$$\ddot{\varphi} + \frac{\Omega^2}{\kappa}(\lambda - \cos\varphi)\sin\varphi = 0 \tag{10}$$

which φ is generalized independent coordinate, $\kappa = 1 + \frac{\mathrm{i}_C^2}{\mathrm{r}^2}$ is the coefficient of disk rolling and $\lambda = \frac{g}{(R-r)\Omega^2}$ is the bifurcation parameter (see [12, 23] for details).

In Fig. 2, a* and b*, shown graphic representation of the transformations, by changing the bifurcation parameter $\lambda = \frac{g}{(R-r)\Omega^2}$, of the separatrix phase trajectories of the phase portrait of the non-linear rolling dynamics of a heavy homogeneous thin disk, radius r, in a circle, radius R, in a rotating vertical plane at a constant angular velocity Ω around the vertical central axis. These graphs are also presentation continuous process of bifurcations followed by change of bifurcation parameter $\lambda = \frac{g}{(R-r)\Omega^2}$ depending of angular velocity Ω of vertical plane rotation around vertical axis, and also of difference between radiuses of rolling disk r and circle trace R.

The nonlinear differential equation of rolling dynamics of a heavy homogeneous disk in a circle in a rotating vertical plane of the central vertical axis at a constant angular velocity is thematically analogous to the nonlinear differential equation of motion of a heavy material point along a smooth circular line in a rotating vertical plane around a vertical axis at a constant angular velocity. We have shown a phase portrait in Fig. 1b.2*, for the case when a tiger of conjugate singularities is visible.

Figure 2 shows the transformation of a singular point of the stable center type into an unstable singular point of the unstable saddle type by bifurcation into a trigger of coupled singularities with a homoclinic phase orbit in the shape of the number "eight". Figure 2a* shows the bifurcation of the singular point of the stable center type into the trigger of coupled singularities, and Fig. 2b* shows the stratification of phase trajectories by changing the bifurcation parameter (see also [19, 32, 33]).

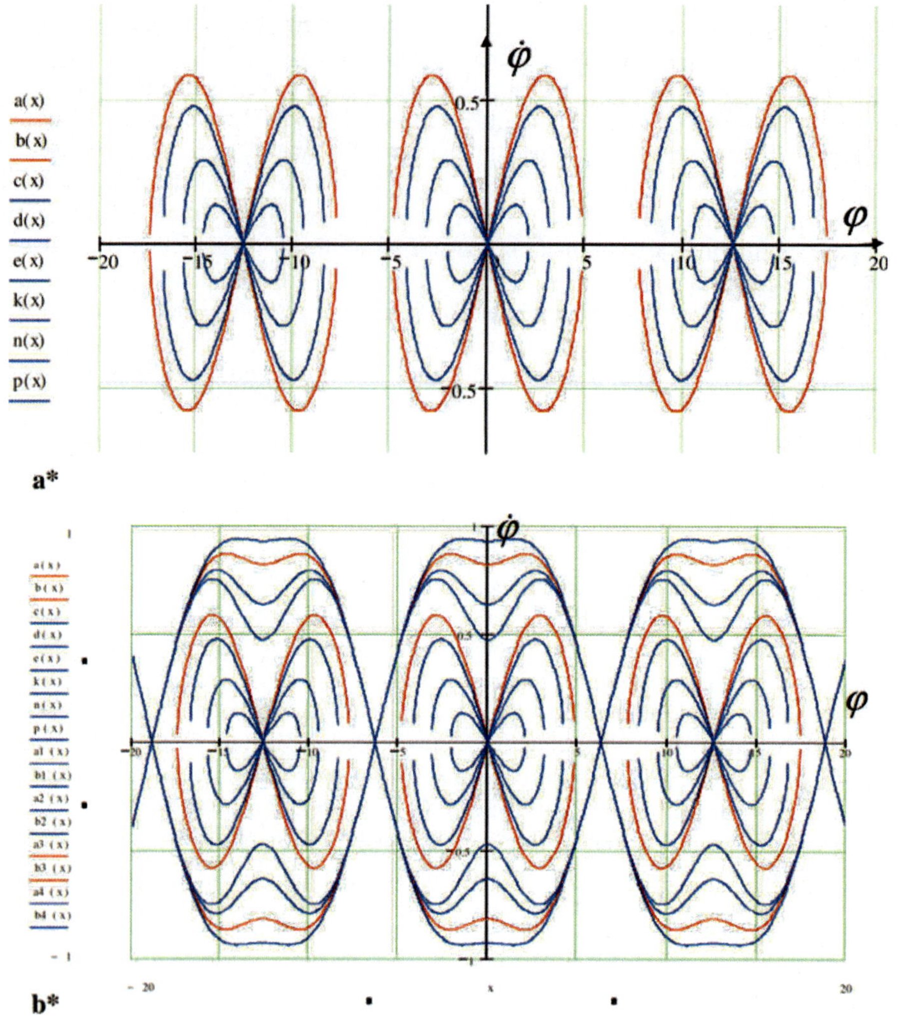

Fig. 2 Transformation of a singular point of the stable center type into an unstable singular point of the unstable saddle type by bifurcation into a trigger of coupled singularities with a homoclinic phase orbit in the form of the number "eight"; **a*** bifurcation of a stable center-type singular point into a trigger of coupled singularities and **b*** stratification of phase trajectories by changing the bifurcation parameter

3 Nonlinear Differential Equations and Phase Trajectory Equation of Nonlinear Dynamics of a Class of Rolling Pendulums

Reference [5] by Hedrih (Stevanović) K. R., titled by "Generalized rolling pendulum along curvilinear trace: Phase portrait, singular points and total mechanical energy surface", is the full paper of Plenary Lecture given in Mini-symposium on Computational Aspects of Classical and Celestial Mechanics, Stability and Motion Control at CASTR (Computer Algebra Systems in Teaching and Research—CASTR'2017. Thus paper contains description of a generalized rolling pendulum along curvilinear trace consisting by three circle arches in vertical plane. Sets of three non-linear differential equations of dynamics of described generalized rolling pendulum along each of three circle arches, is presented. Three integrals of previous three nonlinear differential equations present a set of three equations of each of three phase trajectory branches which correspond to dynamics of described generalized rolling pendulum along each of three circle arches. Phase portrait, set of singular points and total mechanical energy surface are graphically presented for particular case of geometrical parameters of the system. Paper contains basic elements of the methodology for investigation of the vibro-impact dynamics of the system with two rolling bodies along defined curvilinear trace in successive collisions.

Reference [6] by Hedrih (Stevanović) K. P. titled by "Rolling heavy disk along rotating circle with constant angular velocity" is a paper of Plenary lecture in section of Mini-symposium on Computational Aspects of Classical and Celestial Mechanics, Stability and Motion Control included in the Conference Program Computer Algebra Systems in Teaching and Research CASTR'2015. In Abstract of this reference we read following: "Non-linear differential equation of non-linear dynamics of a rolling heavy disk along rotating circle trace, with constant angular velocity, about central axis in vertical direction is derived. For this case, corresponding equation of phase trajectory portraits depending on kinetic parameters of the system are obtained. Existence of trigger (see [5, 6]) of coupled three singular points and homoclinic orbit in the form of number "eight" depending on system kinetic parameters and appearance of the bifurcation of relative equilibrium positions are investigated. Functional dependence between angle of disk relative arbitrary position on rotating circle trace and time of motion duration is derived. For obtaining this solution, an elliptic integral is derived. For solving elliptic integral, series of transformations are introduced and functions under the elliptic integral are expanded in three series along angle of disk relative arbitrary position on rotate circle trace. By use obtained functional dependence between time of disk rolling and angle of disk relative position, discussion of different period duration of rolling disk oscillations along rotating circle trace about vertical central axis is done depending of initial conditions and constant angular velocity of the circle rotation".

Next [7] titled by "Vibro-impact dynamics of two rolling heavy thin disks along rotate curvilinear line and energy analysis" written by Hedrih (Stevanović) K. P. is published as original article in Journal Nonlinear Dynamics, Springer Nature.

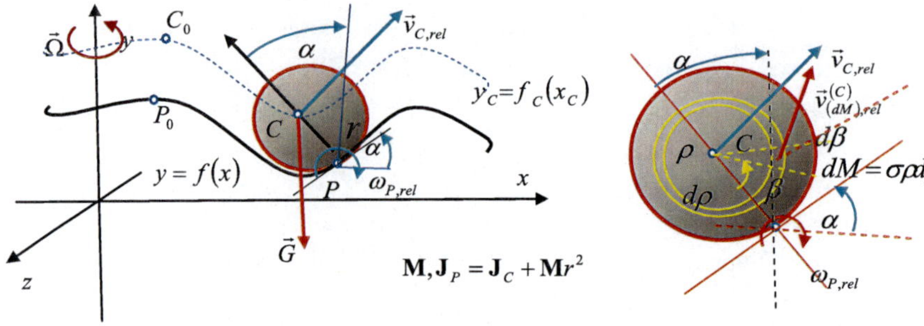

Fig. 3 Geometric parameters of a rolling, without slipping, of heavy thin rigid discs on a rotating curvilinear trace in a vertical plane around a vertical axis with constant angular velocity

Reference [34] by Hedrih (Stevanović) K. P. titled by "Dynamics of a rolling heavy thin disk along rotate curvilinear trace on vertical plane about vertical axis" is an extended abstract of first presentation lecture in session of General Mechanics at Congress of Serbian Society of Mechanics helped in Sremski Karlovci 2019. In the lecture Nonlinear differential equation of dynamics of a rolling, without slipping, heavy thin disk along rotate general curvilinear trace, in vertical plane, around vertical axis with constant angular velocity, is presented. First integral of this nonlinear differential equation is presented. First integral present the nonlinear equation of the phase trajectory in phase plane of a rolling, without slipping, heavy thin disk along rotate general curvilinear trace, in vertical plane, around vertical axis with constant angular velocity.

Based on the new author's authentic research and new results, which represent new contributions by generalizing previous results, without the article being clear and we think that there is no need to historically present the results of other authors, which are mostly in the field of mathematics.

We observe an axially symmetric heavy rigid body with one central plane of symmetry, which is in the case of a thin disk of radius r, mass M, the axial mass inertia moment $\mathbf{J}_C$ for the central axis parallel to the rolling axis. See Fig. 3.

Suppose there is a curvilinear trace, determined by $y = f(x)$, such that the radius of curvature of each of its concave arch is larger than the radius of the contour of the disc circle in the plane of symmetry, by which the disk rolls, without slipping, along the curvilinear trace, rotating, around the vertical axis with constant angular velocity Ω, in the rotating vertical plane.

If we introduce the coefficient of disk rolling, without slipping, in the form $\kappa = \frac{\mathbf{J}_P}{Mr^2} = \frac{\mathbf{i}_P^2}{r^2} = \frac{\mathbf{i}_C^2}{r^2} + 1 = \kappa$ [26, 27, 35, 28], which for thin disk is $\kappa = \frac{3}{2}$, then the nonlinear ordinary differential equation of rolling, without sliding, a heavy rigid thin disk along a curvilinear line route in rotate vertical plane around vertical axis at a constant angular velocity $\dot{\vartheta} = \Omega$, is in the following form:

$$\ddot{x}F(x,r) + \frac{1}{2}\dot{x}^2 F'(x,r) - \frac{r^2}{2}\Omega^2 \frac{\mathbf{J}_z'(x,M,r)}{\mathbf{J}_P} + \frac{g}{\kappa} f_C'(x) = 0 \qquad (11)$$

where $\vartheta = \Omega t$ is rheonomic coordinate, as a kinematic excitation, x independent generalized coordinate, correspond to one degree of freedom of this rheonomic system with two degrees of mobility. Next, $f_C(x)$ and $F(x, r)$ are expressed by following expressions (for details see [8] by Hedrih (Stevanović) K. R.):

$$F(x,r) = \left\langle 1 + [f'(x)]^2 \right\rangle \left\{ 1 - \frac{rf''(x)}{[1 + [f'(x)]^2]^{\frac{3}{2}}} \right\}^2 \tag{12}$$

$$f_C(x) = y + r\frac{1}{\sqrt{1 + [f'(x)]^2}} \tag{13}$$

$$\mathbf{J}_z(x, M, r) = \mathbf{J}_{Cz} + M[x_C]^2 = M\frac{r^2}{4} + M\left[x - r\frac{f'(x)}{\sqrt{1 + [f'(x)]^2}}\right]^2 \tag{14}$$

Expression (14) presents axial mass inertia moment of thin disk for vertical axis of vertical plane rotation in which curvilinear trace of disk rolling lies.

Then, previous integral of nonlinear differential equations (11) finally take the following form:

$$[\dot{x}(x)] = \mp\sqrt{[\dot{x}(x_0)]^2 + \Omega^2\frac{r^2[\mathbf{J}_z(x, M, r) - \mathbf{J}_z(x_0, M, r)]}{\mathbf{J}_P F(x, r)} - \frac{2g}{\kappa F(x, r)}[f_C(x, r) - f_C(x_0, r)]} \tag{15}$$

where $f_C(x)$ is expressed by (13) and $F(x, r)$ is expressed by (12) and $\mathbf{J}_z(x, M, r)$ is expressed by (14).

We can now write an expression for the relative angular velocity $\omega_P(x, \dot{x})$ of thin rigid disk relative rolling, without slipping, along a curved line route $y = f(x)$ in rotate vertical plane around vertical axis with constant angular velocity Ω, based on a expression: $[\omega_P(x, \dot{x})]^2 = \frac{1}{r^2}\dot{x}^2 F(x, r)$ and previous obtained integral (15) in the following form: $[\omega_P(x, \dot{x})] = \frac{1}{r}\dot{x}\sqrt{F(x, r)}$, and finally in the form:

$$[\omega_P(x)] = \pm\frac{1}{r}\sqrt{\left[\dot{x}_0^2 F(x_0, r)\right] + \Omega^2\frac{r^2[\mathbf{J}_z(x, M, r) - \mathbf{J}_z(x_0, M, r)]}{\mathbf{J}_P} - \frac{2g}{\kappa}[f_C(x_0 r) - f_C(x, r)]} \tag{16}$$

where $f_C(x)$ is expressed by (13) and $F(x, r)$ is expressed by (12) and $\mathbf{J}_z(x, M, r)$ is expressed by (14).

4 Characteristic Equations of Nonlinear Dynamics of a Class of Rolling Pendulums and Bifurcation of Relative Stable Equilibrium Positions

In this paper, main attention is paid to a more detailed analysis of the characteristic equation $K(x)$ of dynamics of the generalized rolling pendulum, along trajectory in rotate vertical plane at a constant angular velocity Ω about vertical axis, which was performed in [7] in the form:

$$K(x) = f'(x)\left\{1 - r\frac{f''(x)}{\left[1+[f'(x)]^2\right]^{\frac{3}{2}}}\right\}$$
$$-\frac{2\kappa}{3g}\Omega^2\left\langle x - \frac{rf'(x)}{\sqrt{1+[f'(x)]^2}}\right\rangle\left\langle 1 - \frac{rf''(x)}{\left[1+[f'(x)]^2\right]\sqrt{1+[f'(x)]^2}}\right\rangle = 0 \tag{17}$$

and in which: $y = f(x)$ in general, or in particular cases $y = f(x) = kx^2$ or $y = f(x) = kx^2\left(x^2 - a^2\right)^2$ or $y = f(x) = kx^2\left(x^2 - a^2\right)^2\left(x^2 - b^2\right)$ or $f(x) = -kx^2\left(x^2 - a^2\right)\left[c^4 - \left(x^2 - b^2\right)^2\right]$ is equation of the curvilinear path, where a, b, c and k are known constants, and with the following relation $a < b$.

For various changes of values of the angular velocity Ω of rotation of the vertical plane, in which the curvilinear path along which the generalized rolling pendulum rolls is located, the numerical analysis shows the obtaining of the zero (roots) of characteristic equation (17), or singular points and triggers, each of coupled three singular points.

5 Nonlinear Phenomena in the Dynamics of a Class of Rolling Pendulums: Bifurcations and Trigger of Coupled Singular Points

In this part of our paper, using the equation of a curvilinear trajectory in a rotating vertical plane, around a vertical axis at a constant angular velocity Ω: $y = f(x)$ in general, for in particular cases $y = f(x) = kx^2$ or $y = f(x) = kx^2\left(x^2 - a^2\right)^2$ or $y = f(x) = kx^2\left(x^2 - a^2\right)^2\left(x^2 - b^2\right)$ or $f(x) = -kx^2\left(x^2 - a^2\right)\left[c^4 - \left(x^2 - b^2\right)^2\right]$, where a, b, c and k are known constants, and with the following relation $a < b$, we will analyze the zeros of the graph of the characteristic equation $K(x) = 0$, defined by expression (17).

Figures 4, 5, 6, 7, 8, 9 and 10 show, in pairs, the characteristic curves of the shape of the curvilinear trajectory, $y = f(x) = kx^2$ or $y = f(x) = kx^2\left(x^2 - a^2\right)^2$ or

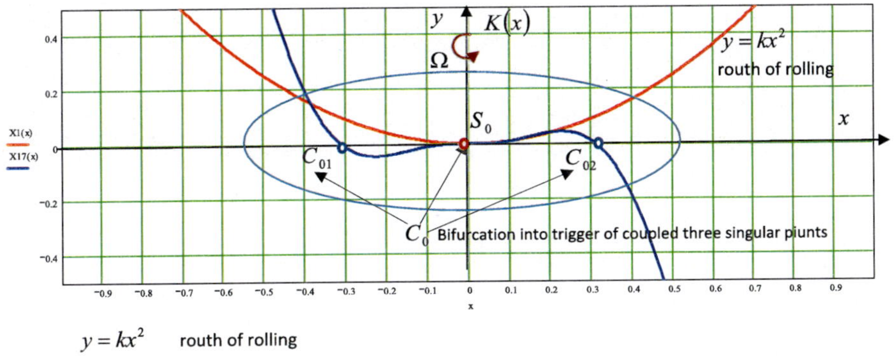

$$K(x)=f'(x)\left\{1-r\frac{f''(x)}{\left[1+[f'(x)]^2\right]^{\frac{3}{2}}}\right\}-\frac{2\kappa}{3g}\Omega^2\left\langle x-\frac{rf'(x)}{\sqrt{1+[f'(x)]^2}}\right\rangle\left\langle 1-\frac{rf''(x)}{\left[1+[f'(x)]^2\right]\sqrt{1+[f'(x)]^2}}\right\rangle=0$$

Fig. 4 Graphs of the curvilinear route, defined by parabola in the form $y = f(x) = kx^2$ as well as the frequency function $K(x)$ of the nonlinear rolling dynamics of a rigid heavy thin disk, in a rotating vertical plane with a constant angular velocity Ω around the vertical axis defined by (17): case with a trigger of coupled three singular points

$y = f(x) = kx^2(x^2 - a^2)^2(x^2 - b^2)$ or $f(x) = -kx^2(x^2 - a^2)\left[c^4 - (x^2 - b^2)^2\right]$, with the corresponding extreme points, the minimum and maximum, and the graphs of the corresponding characteristic equations $K(x) = 0$ with the corresponding zeros, for the cases of the indicated curvilinear paths the chosen angular velocity Ω of rotation of the system around the vertical axis.

Im Figs. 4 and 5, graphs of the curvilinear route, defined by parabola in the form $y = f(x) = kx^2$ as well as the frequency function $K(x)$ of the nonlinear rolling dynamics of a rigid heavy thin disk, in a rotating vertical plane with a constant angular velocity Ω around the vertical axis defined by (17) are presented for different geometrical and kinetic parameters. We can see different cases of characteristic function graphs depending of the value of angular velocity Ω, which appears as a bifurcation parameter. In Fig. 4 a graph of characteristic equation poses thee zero points around minima of curvilinear trace of disk rolling and dynamics system is with a trigger of coupled three singular points. Analogous graph of characteristic equation $K(x)$ is presented in Fig. 5b* for different values of angular velocity Ω, also a trigger of coupled three singular points exists. The graph of characteristic equation $K(x)$ presented in Fig. 5a* for zero values of angular velocity $\Omega = 0$, is without any trigger of coupled three singular points and in minima of curvilinear parabolic trace of disk rolling no bifurcation.

In Figs. 6 and 8, graphs of the curvilinear route, defined by be-quadratic parabola in the form $y = f(x) = kx^2(x^2 - a^2)$ as well as the characteristic frequency function $K(x)$ of the nonlinear rolling dynamics of a rigid heavy thin disk, in a rotating vertical plane with a constant angular velocity Ω around the vertical axis defined by (17) are presented for different geometrical and kinetic parameters. We can see different cases

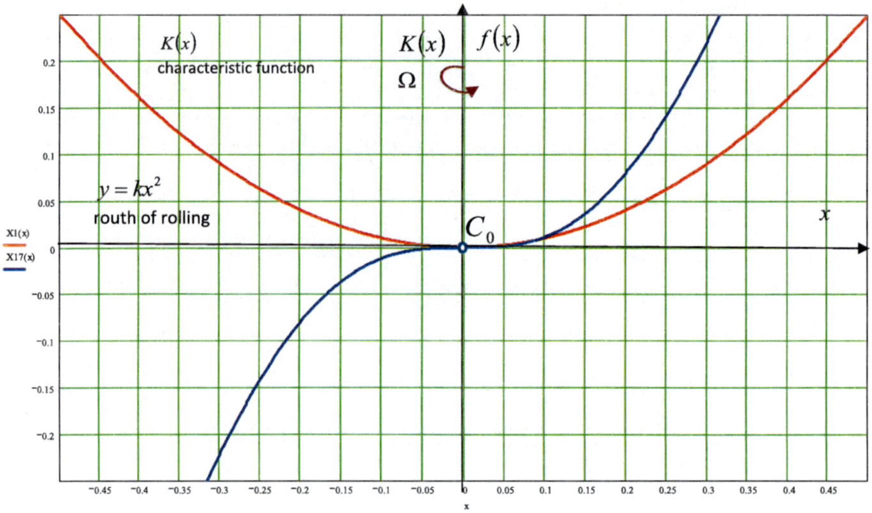

a* $y = kx^2$ routh of rolling

$$K(x)= f'(x)\left\{1-r\frac{f''(x)}{\left[1+[f'(x)]^2\right]^{\frac{3}{2}}}\right\}-\frac{2\kappa}{3g}\Omega^2\left\langle x-\frac{rf'(x)}{\sqrt{1+[f'(x)]^2}}\right\rangle\left\langle 1-\frac{rf''(x)}{\left[1+[f'(x)]^2\right]\sqrt{1+[f'(x)]^2}}\right\rangle=0$$

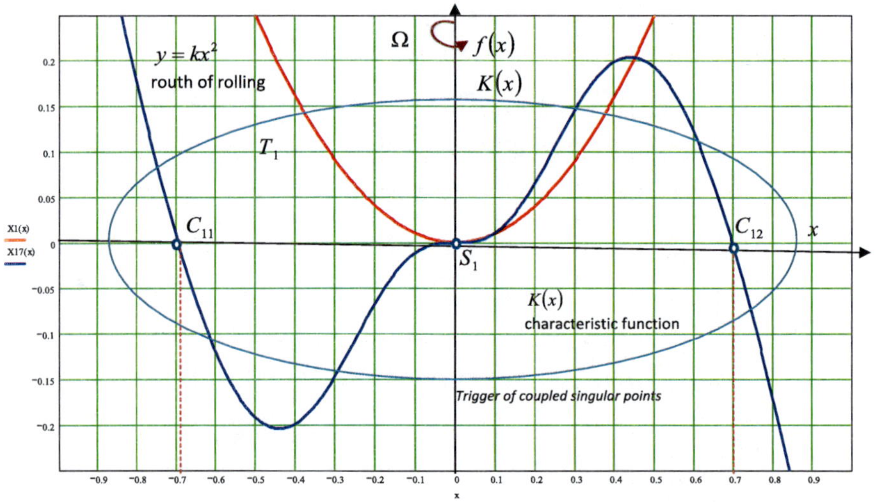

$y = kx^2$ routh of rolling

$$K(x)= f'(x)\left\{1-r\frac{f''(x)}{\left[1+[f'(x)]^2\right]^{\frac{3}{2}}}\right\}-\frac{2\kappa}{3g}\Omega^2\left\langle x-\frac{rf'(x)}{\sqrt{1+[f'(x)]^2}}\right\rangle\left\langle 1-\frac{rf''(x)}{\left[1+[f'(x)]^2\right]\sqrt{1+[f'(x)]^2}}\right\rangle=0$$

b*

Fig. 5 Graphs of the curvilinear route, defined by parabola in the form $y = f(x) = kx^2$ as well as the frequency function $K(x)$ of the nonlinear rolling dynamics of a rigid heavy thin disk, in a rotating vertical plane with a constant angular velocity Ω around the vertical axis defined by (17): a* case without a trigger of couple three singular points and b* case with a trigger of coupled three singular points

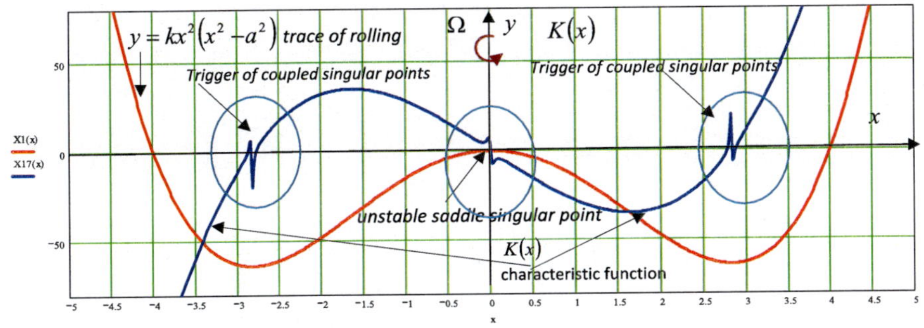

$$K(x) = f'(x)\left\{1 - r\frac{f''(x)}{\left[1+[f'(x)]^2\right]^{\frac{3}{2}}}\right\} - \frac{2\kappa}{3g}\Omega^2\left\langle x - \frac{rf'(x)}{\sqrt{1+[f'(x)]^2}}\right\rangle\left\langle 1 - \frac{rf''(x)}{\left[1+[f'(x)]^2\right]\sqrt{1+[f'(x)]^2}}\right\rangle = 0$$

Fig. 6 Graphs of the curvilinear route, defined by be-quadratic parabola in the form $y = f(x) = kx^2(x^2 - a^2)$ as well as the frequency functions $K(x)$ of the nonlinear rolling dynamics of a rigid heavy thin disk, in a rotating vertical plane with a constant angular velocity Ω around the vertical axis defined by (17): case with a trigger of coupled three singular points

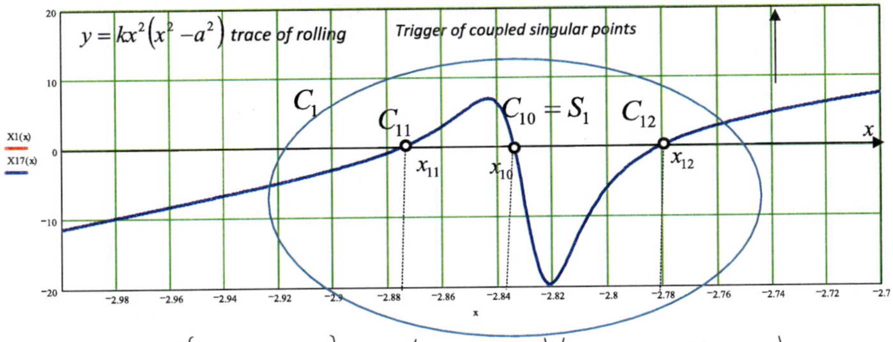

$$K(x) = f'(x)\left\{1 - r\frac{f''(x)}{\left[1+[f'(x)]^2\right]^{\frac{3}{2}}}\right\} - \frac{2\kappa}{3g}\Omega^2\left\langle x - \frac{rf'(x)}{\sqrt{1+[f'(x)]^2}}\right\rangle\left\langle 1 - \frac{rf''(x)}{\left[1+[f'(x)]^2\right]\sqrt{1+[f'(x)]^2}}\right\rangle = 0$$

Fig. 7 Detail of the graph of the frequency functtion $K(x)$ of nonlinear rolling dynamics of a rigid heavy thin disk, along a curvilinear path in a rotating vertical plane with constant angular velocity Ω around the vertical axis, defined by the equation $y = f(x) = kx^2(x^2 - a^2)$: detail shows the phenomenon of bifurcation of a stable singular point centre type into unstable saddle-type brush and basket of two new stable singular points center type around—appearance of a trigger of coupled three singular points

of characteristic function graphs depending of the value of angular velocity Ω, in which appears two bifurcation in each of two minima of curvilinear trace of disk rolling.

In Fig. 6, graphs of the curvilinear route, defined by equation $y = f(x) = kx^2(x^2 - a^2)$, as well as the frequency function expressed by expression $h(x) = K(x)$ of the nonlinear rolling dynamics of a rigid heavy thin disk, in a rotating

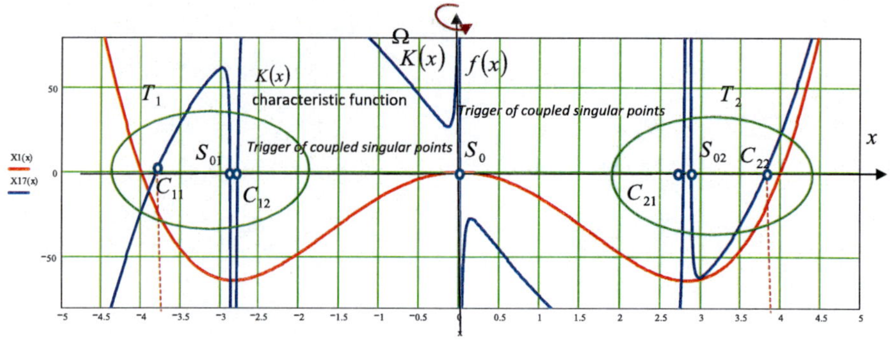

$$y = kx^2\left(x^2 - a^2\right) \text{ trace of rolling}$$

$$K(x) = f'(x)\left\{1 - r\frac{f''(x)}{\left[1+[f'(x)]^2\right]^{\frac{3}{2}}}\right\} - \frac{2\kappa}{3g}\Omega^2\left\langle x - \frac{rf'(x)}{\sqrt{1+[f'(x)]^2}}\right\rangle\left\langle 1 - \frac{rf''(x)}{\left[1+[f'(x)]^2\right]\sqrt{1+[f'(x)]^2}}\right\rangle = 0$$

Fig. 8 Graphs of the curvilinear route, defined by be-quadratic parabola in the form $y = f(x) = kx^2(x^2 - a^2)$ as well as the frequency functions $K(x)$ of the nonlinear rolling dynamics of a rigid heavy thin disk, in a rotating vertical plane with a constant angular velocity Ω around the vertical axis defined by (17): case with a trigger of coupled three singular points

vertical plane with a constant angular velocity around the vertical axis defined by (17) are presented. In Fig. 6 graph of characteristic equation poses thee zero points around each of two minima of curvilinear trace of disk rolling and dynamics system is with two triggers, each of coupled three singular points. Analogous graph of characteristic equation $K(x)$ is presented in Fig. 7 for different values of angular velocity Ω, also two triggers, each of coupled three singular points. The graph of characteristic equation $K(x)$, for zero values of angular velocity $\Omega = 0$, is without any trigger of coupled three singular points and in each of two minima of curvilinear be-parabolic trace of disk rolling no bifurcation.

In Fig. 7, detail of the graph of the frequency functtion $h(x) = K(x)$, from Fig. 6, of nonlinear rolling dynamics of a rigid heavy thin disk, along a curvilinear path in a rotating vertical plane with constant angular velocity Ω around the vertical axis, defined by the equation $y = f(x) = kx^2(x^2 - a^2)$: detail shows the phenomenon of bifurcation of a stable singular point centre type into unstable saddle-type brush and basket of two new stable singular points center type around—appearance of a trigger of coupled singularities, are visible around a of two minima in trace of rolling in the form of be-quadratic parabola.

In Fig. 9, two graphs of the curvilinear route, defined by polynomial expression $y = f(x) = kx^2(x^2 - a^2)^2(x^2 - b^2)$, as well as the frequency function $h(x) = K(x)$, defined by (17), of the nonlinear rolling dynamics of a rigid heavy thin disk, in a rotating vertical plane with a constant angular velocity Ω around the vertical axis are presented. Series of triggers each of three coupled singular points are visible around each of four minimum C_s, $s = 1, 2, 3, 4$ of curvilinear route of disk rolling.

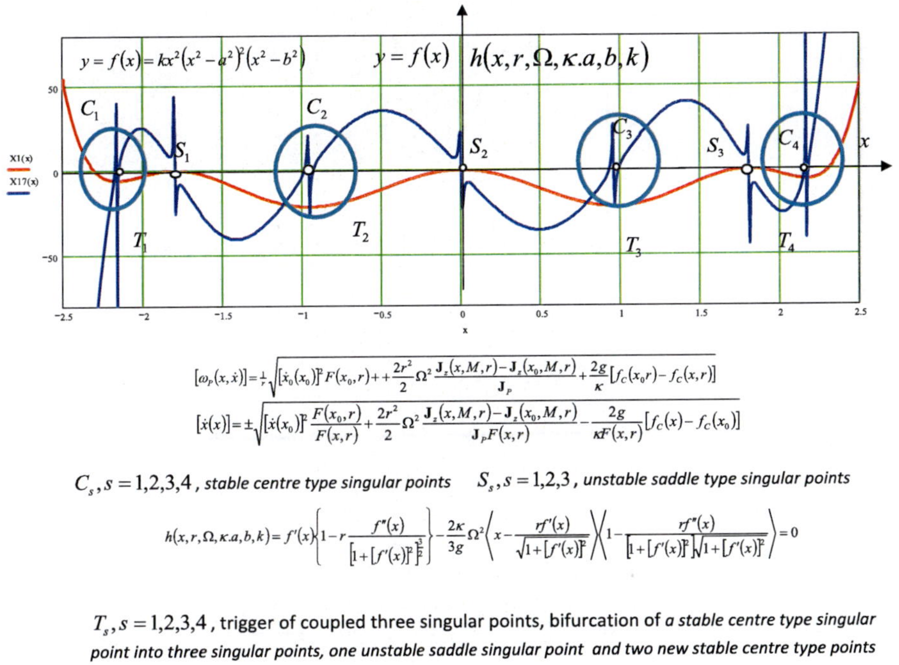

Fig. 9 Two graphs of the curvilinear route, defined by polynomial equation $y = f(x) = kx^2(x^2 - a^2)^2(x^2 - b^2)$, as well as the frequency function $h(x) = K(x)$ of the nonlinear rolling dynamics of a rigid heavy thin disk, in a rotating vertical plane with a constant angular velocity Ω around the vertical axis defined by (17): case with a series of the four triggers, each of coupled three singular points

There are four minima C_s, $s = 1, 2, 3, 4$ in which and around which appear in total four bifurcations and four triggers of coupled singularities, and three maxima S_s, $s = 1, 2, 3$ around which no appearing bifurcations. Four bifurcations and four triggers of coupled singularities occur $C_s \to T_s$, $s = 1, 2, 3, 4$ in each point of the four minimums C_s, $s = 1, 2, 3, 4$ are visible in Fig. 9.

In Fig. 10, detail, of Fig. 9, of the graph of the frequency functtion $h(x) = K(x)$ of nonlinear rolling dynamics of a rigid heavy thin disk, along a curvilinear path in a rotating vertical plane with constant angular velocity around the vertical axis, defined by the equation $y = f(x) = kx^2(x^2 - a^2)^2(x^2 - b^2)$ is presented. Detail shows the phenomenon of bifurcation of two stable singular points each of a stable centre type into unstable saddle-type brush and basket of two new stable singular points center type around each - appearance of two triggers each of coupled three singular points.

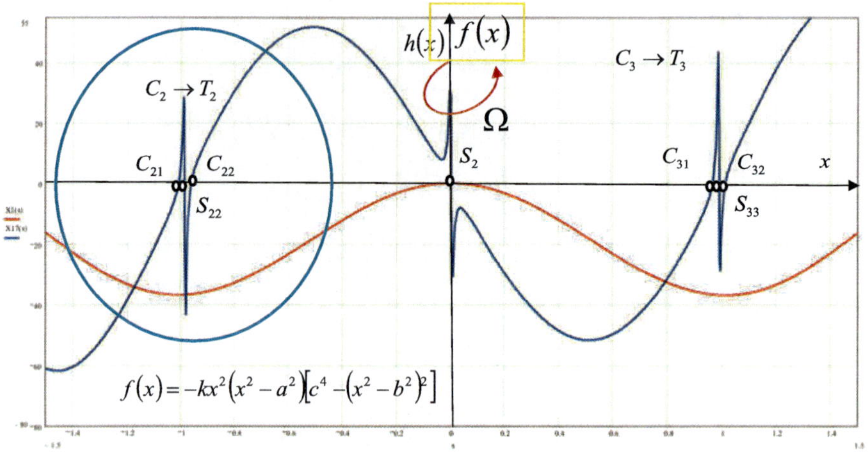

$$h(x) = f'(x)\left\{1 - r\frac{f''(x)}{\left[1+[f'(x)]^2\right]^{\frac{3}{2}}}\right\} - \frac{2\kappa}{3g}\Omega^2\left\langle x - \frac{rf'(x)}{\sqrt{1+[f'(x)]^2}}\right\rangle\left\langle 1 - \frac{rf''(x)}{\left[1+[f'(x)]^2\right]\sqrt{1+[f'(x)]^2}}\right\rangle = 0$$

$$k = 1; \quad a = 1{,}8; \quad b = 2{,}3; \quad c = 1{,}4; \quad \frac{g}{\kappa} \approx 10; \quad r = 0{,}5; \quad \Omega = 1$$

$$[\omega_P(x,\dot{x})] = \pm\frac{1}{r}\sqrt{[\dot{x}_0(x_0)]^2 F(x_0,r) + + \frac{2r^2}{2}\Omega^2\frac{\mathbf{J}_z(x,M,r) - \mathbf{J}_z(x_0,M,r)}{\mathbf{J}_P} + \frac{2g}{\kappa}[f_C(x_0,r) - f_C(x,r)]}$$

$$[\dot{x}(x)] = \pm\sqrt{[\dot{x}(x_0)]^2\frac{F(x_0,r)}{F(x,r)} + \frac{2r^2}{2}\Omega^2\frac{\mathbf{J}_z(x,M,r) - \mathbf{J}_z(x_0,M,r)}{\mathbf{J}_P F(x,r)} - \frac{2g}{\kappa F(x,r)}[f_C(x) - f_C(x_0)]}$$

$C_s, s = 1,2,3,4$, *stable centre type singular points* $S_s, s = 1,2,3$, *unstable saddle type singular points*

$T_s, s = 1,2,3,4$, trigger of coupled three singular points, bifurcation of *a stable centre type singular point into three singular points, one unstable saddle singular point and two new stable centre type points*

Fig. 10 Detail, of Fig. 9, of the graph of the frequency functtion $h(x) = K(x)$ of nonlinear rolling dynamics of a rigid heavy thin disk, along a curvilinear path in a rotating vertical plane with constant angular velocity Ω around the vertical axis, defined by the equation $y = f(x) = kx^2(x^2-a^2)^2(x^2-b^2)$: detail shows the phenomenon of bifurcation of two stable singular points each of a stable centre type into unstable saddle-type brush and basket of two new stable singular points center type around each—appearance of two triggers each of coupled three singular points

6 Phase Trajectory Portraits in the Nonlinear Dynamics of a Class of Rolling Pendulums and Structural Analysis: Bifurcations, Layering of Phase Trajectories, Trigger of Coupled Singularities

In this part of our paper, using the equation of a curvilinear trajectory in a rotating vertical plane, around a vertical axis at a constant angular velocity Ω: $y = f(x)$ in general, for in particular cases $y = f(x) = kx^2$ or $y = f(x) = kx^2(x^2-a^2)^2$ or $y = f(x) = kx^2(x^2-a^2)^2(x^2-b^2)$ or $f(x) = -kx^2(x^2-a^2)[c^4-(x^2-b^2)^2]$, where a, b, c and k are known constants, and with the following relation $a < b$, we will analyze the structure of phase portraits using the equation of phase trajectories in the form (16), in the phase plan—the relative angular velocity $\omega_P(x, \dot{x})$ of thin rigid disk relative rolling, without slipping, along a curved line route $y = f(x)$ and independent generalized coordinate x, i.e. in the form (15), in the phase plane a derivative $\dot{x} = \dot{x}(x)$ of the independent generalized coordinate and the independent generalized coordinate x.

Figures 11, 12, 13, 14 and 15 show the characteristic phase portraits of the nonlinear rolling dynamics of a heavy thin, rigid disk along curvilinear paths of curvilinear path shape, $y = f(x) = kx^2$ or $y = f(x) = kx^2\left(x^2 - a^2\right)^2$ or $y = f(x) = kx^2\left(x^2 - a^2\right)^2\left(x^2 - b^2\right)$ or $f(x) = -kx^2\left(x^2 - a^2\right)\left[c^4 - \left(x^2 - b^2\right)^2\right]$, with corresponding extreme points, minimum and maximum. We use the findings from the analysis of the number of zeros and the existence of triggers of coupled singular points, which we conducted in the previous chapter of this paper, by analyzing the number of zeros of the characteristic equation $h(x) = K(x)$ for a certain shape of the curvilinear trajectory, and for the corresponding value of the

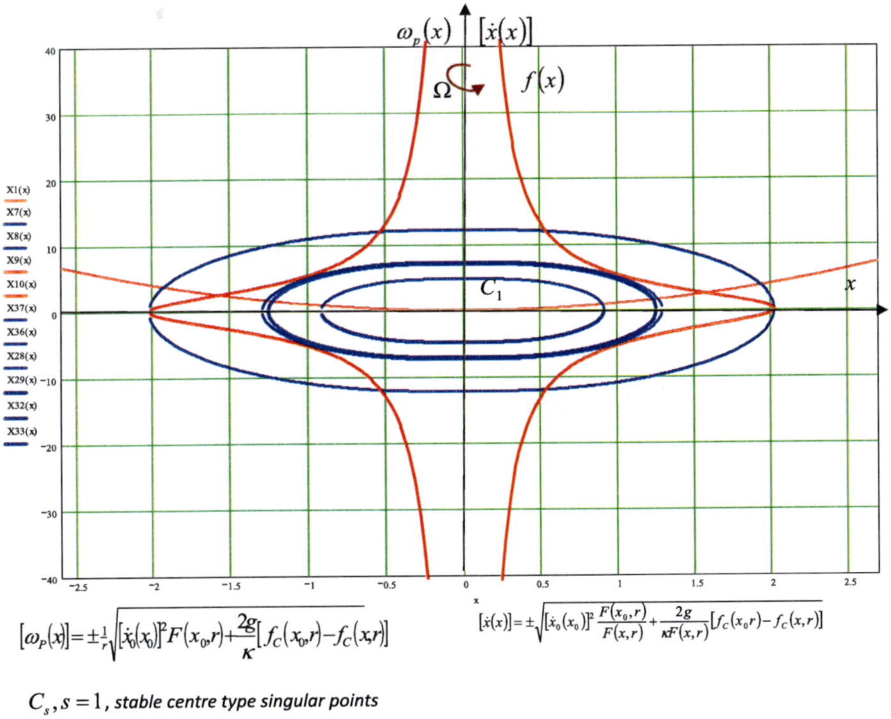

Fig. 11 A phase portrait of the dynamics of a generalized rolling pendulum, which rolls, without slipping, along a curvilinear path defined by a parabola in the form $y = f(x) = kx^2$, where k, is a known constant, in a stationary vertical plane, $\Omega = 0$, and in Earth's field of gravity, in phase coordinates $\omega_P(x)$, x and a phase trajectory in phase plane $\dot{x} = \dot{x}(x)$, x

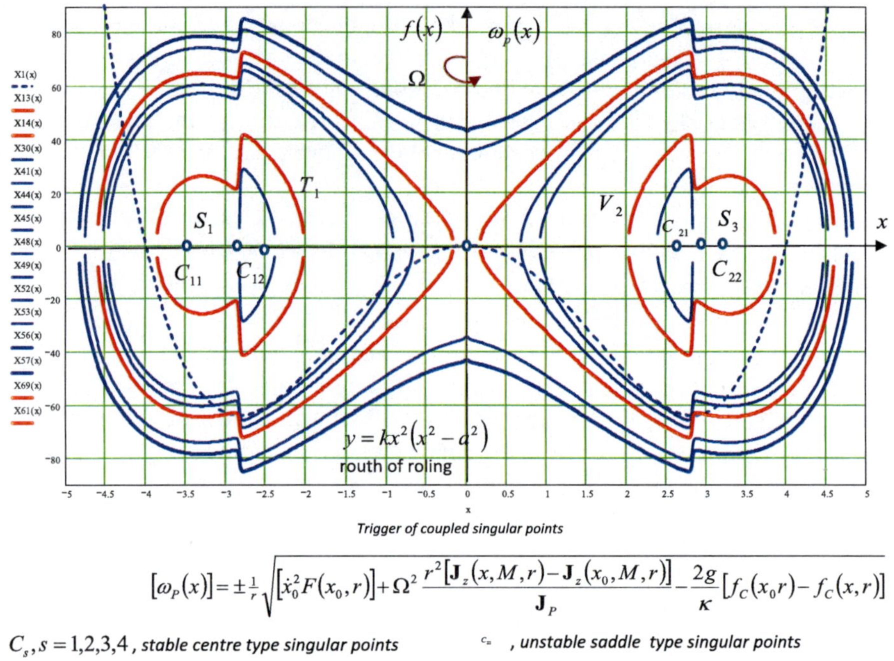

Fig. 12 A phase portrait of the dynamics of a generalized rolling pendulum, which rolls, without slipping, along a curvilinear path defined by a polynomial of the four degree in the form $y = f(x) = kx^2(x^2 - a^2)$, where a and k, are known constants, in a rotate vertical plane around the vertical axis at a constant angular velocity Ω, and in Earth's field of gravity, in phase coordinates $\omega_P(x)$,x

bifurcation parameter—the angular velocity Ω of rotation of the vertical plane around the vertical axis.

In order to obtain one of the phase portraits, it is necessary to draw a series of phase trajectories for different values of the initial conditions, using the equation of phase trajectories in the form (16), in the phase plane- the relative angular velocity $\omega_P(x, \dot{x})$ of thin rigid disk relative rolling, without slipping, along a curved line route $y = f(x)$ and independent generalized coordinates x, i.e. in the form (15), in the phase plane a derivative $\dot{x} = \dot{x}(x)$ of the independent generalized coordinate x and the independent generalized coordinate x. From an infinite set of phase trajectories, we choose characteristic series, as well as separatrix phase trajectories-homoclinic phase trajectories, which separated individual series of phase trajectories, which describe similar properties of the dynamics of the studied nonlinear dynamics.

In Fig. 11 a phase portrait of the dynamics of a generalized rolling pendulum, which rolls, without slipping, along a curvilinear path defined by a parabola in the form $y = f(x) = kx^2$, where k, is a known constant, in a stationary vertical plane, $\Omega = 0$, and in Earth's field of gravity, in phase coordinates $\omega_P(x)$, x and a phase trajectory in phase plane $\dot{x} = \dot{x}(x)$, x are presented.

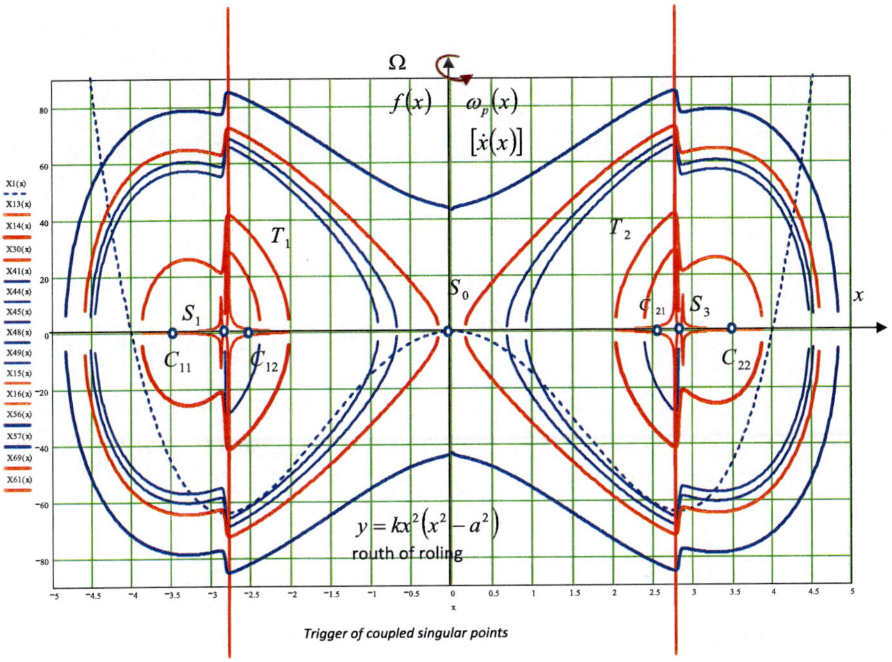

$$[\omega_P(x)] = \pm\frac{1}{r}\sqrt{[\dot{x}_0^2 F(x_0,r)] + \Omega^2\frac{r^2[\mathbf{J}_z(x,M,r)-\mathbf{J}_z(x_0,M,r)]}{\mathbf{J}_P} - \frac{2g}{\kappa}[f_C(x_0,r)-f_C(x,r)]}$$

$$[\dot{x}(x)] = \mp\sqrt{[\dot{x}(x_0)]^2 + \Omega^2\frac{r^2[\mathbf{J}_z(x,M,r)-\mathbf{J}_z(x_0,M,r)]}{\mathbf{J}_P F(x,r)} - \frac{2g}{\kappa F(x,r)}[f_C(x,r)-f_C(x_0,r)]}$$

$C_s, s = 1,2,3,4$, *stable centre type singular points* $S_s, s = 1,2,3$, *unstable saddle type singular points*

Fig. 13 A phase portrait of the dynamics of a generalized rolling pendulum, which rolls, without slipping, along a curvilinear path defined by a polynomial of the four degree in the form $y = f(x) = kx^2(x^2 - a^2)$, be-quadratic parabola, where a and k, are known constants, in a rotate vertical plane around the vertical axis at a constant angular velocity Ω, and in Earth's field of gravity, in phase coordinates $\omega_P(x)$, x and a phase trajectory in phase plane $\dot{x} = \dot{x}(x)$, x

In Fig. 12 a phase portrait of the nonlinear dynamics of a generalized rolling pendulum, which rolls, without slipping, along a curvilinear path defined by a polynomial of the four degree in the form $y = f(x) = kx^2(x^2 - a^2)$, as a be-quadratic parabola, where a and k, are known constants, in a rotate vertical plane around the vertical axis at a constant angular velocity Ω, and in Earth's field of gravity, in phase coordinates $\omega_P(x)$, x is presented.

The same Fig. 12, also, shows the curvilinear path of the shape of the square parabola of the equation $y = f(x) = kx^2(x^2 - a^2)$ along which the disk rolls. This path has two minimums and one maximum. At the selected value of the angular velocity Ω of rotation around the vertical axis of the vertical plane in which the curvilinear trajectory is, bifurcation and trigger of coupled singular points occur

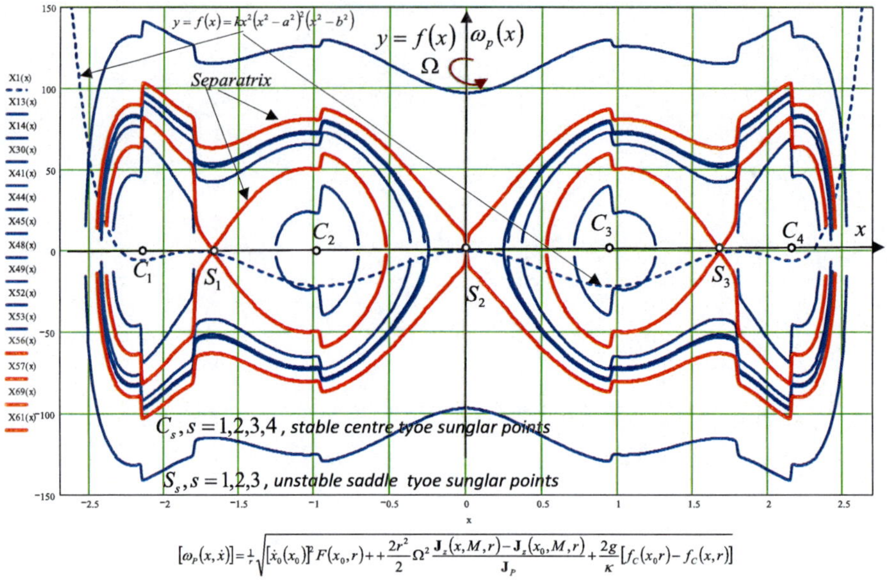

Fig. 14 A phase portrait of the dynamics of a generalized rolling pendulum, which rolls, without slipping, along a curvilinear path defined by a polynomial of the eighth degree in the form $y = f(x) = kx^2(x^2 - a^2)^2(x^2 - b^2)$, where a, b and k, are known constants, in a rotate vertical plane around the vertical axis at a constant angular velocity Ω, and in Earth's field of gravity, in phase coordinates $\omega_P(x)$, x

around each minimum trajectory, as we determined by analyzing the zeros of the characteristic equation $h(x) = K(x)$.

From the structure of the phase portrait from Fig. 12, we see that it contains two types of separatrix trajectories - homoclinic orbits in the shape of the number "eight". One of these homoclinic orbits self-intersects at a singular point of the unstable saddle type, which corresponds to the maximum of the curvilinear trajectory and exists in the phase portrait and when the disk rolls along a stationary path and then surrounds two singular points of stable center type on each side of the singular point of stabile center corresponding to minima of trace of rolling.

When the curvilinear trajectory is in a rotating vertical plane, around the vertical axis at a constant angular velocity, and when bifurcation of each of the singular points of the stable center type occurs at the minimum of the curvilinear trajectory, then this phase trajectory surrounds on each side of the self-intersection unstable saddle type point. Each of these triggers was created by bifurcation and contains two singular points of the stable center type and one singular point between them of the unstable saddle type.

The other two separatrix phase trajectories in the shape of the number "eight" intersect at singular points of the unstable saddle type of each of the formed triggers of coupled singularities, about two minimum curvilinear paths along which a thin

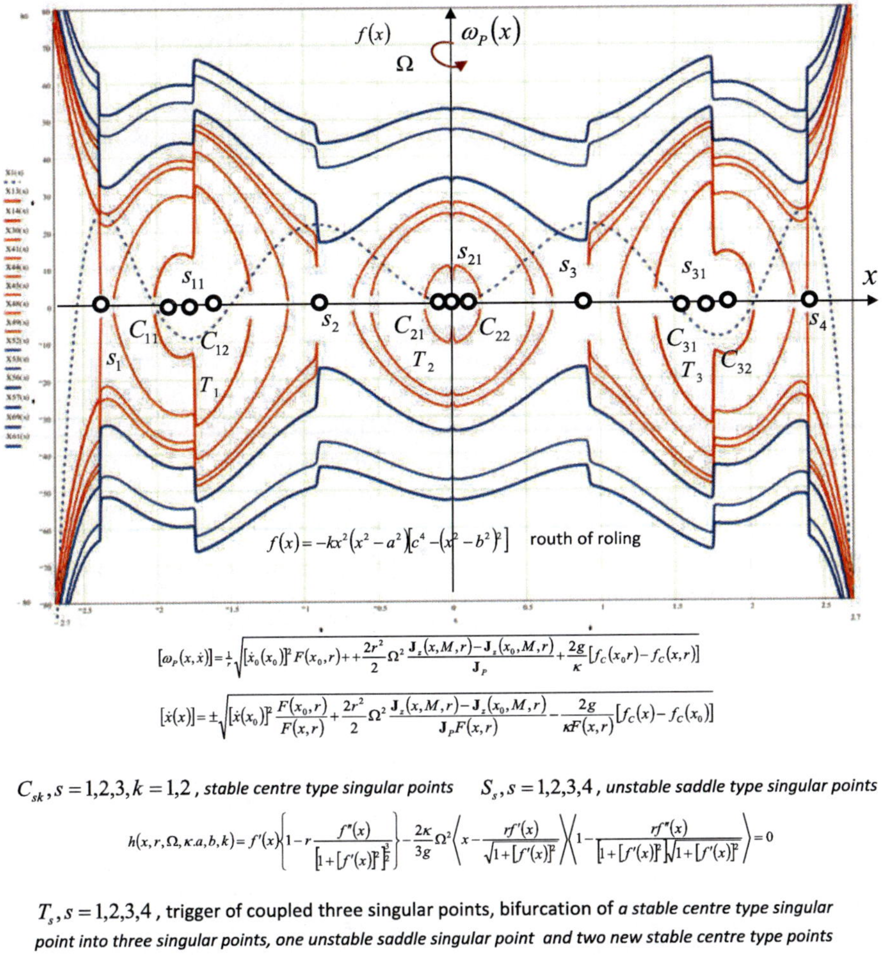

Fig. 15 A phase portrait of the dynamics of a generalized rolling pendulum, which rolls, without slipping, along a curvilinear path defined by a polynomial of the eighth degree in the form $y = f(x) = kx^2(x^2 - a^2)^2(x^2 - b^2)$, where a, b and k, are known constants, in a rotate vertical plane around the vertical axis at a constant angular velocity Ω, and in Earth's field of gravity, in phase coordinates $\omega_P(x)$, x

disk rolls. They mean that in the observed phase portrait, two types of triggers of conjugated singularities appeared with bifurcation.

One homoclinic orbit in the form of number "eight" contains two coupled triggers of coupled singular points, and two homoclinic orbits in the form of number "eight" contains each one trigger of coupled singular points.

The term trigger of coupled singular points contains three singular points, one type of unstable saddle and two types of stable centers, and was created by bifurcation of a singular point of the stable center type. The term trigger of coupled singularities includes a trigger of three singular points and a trajectory separatrix-homoclinic orbit in the form of number "eight" with a self-intersect at a no stable saddle type singular point and sound to singular points stable centre types.

We can conclude that this phase portrait contains two triggers of coupled singularities and one trigger of coupled triggers of coupled singularities, because it surrounds two triggers of coupled singularities.

In Fig. 13 a phase portrait of the dynamics of a generalized rolling pendulum, which rolls, without slipping, along a curvilinear path defined by a polynomial of the four degree in the form $y = f(x) = kx^2(x^2 - a^2)$, be-quadratic parabola, where a and k, are known constants, in a rotate vertical plane around the vertical axis at a constant angular velocity Ω, and in Earth's field of gravity, in phase coordinates $\omega_P(x)$, x and a phase trajectory in phase plane $\dot{x} = \dot{x}(x)$, x are presented.

In Fig. 14 a phase portrait of the dynamics of a generalized rolling pendulum, which rolls, without slipping, along a curvilinear path defined by a polynomial of the eighth degree in the form $y = f(x) = kx^2(x^2 - a^2)^2(x^2 - b^2)$, where a,b and k, are known constants, in a rotate vertical plane around the vertical axis at a constant angular velocity Ω, and in Earth's field of gravity, in phase coordinates $\omega_P(x)$, x is presented.

We can conclude that, in the observed case, in the phase trajectory portrait, from Fig. 14, three types of separatist phase trajectories—homoclinic orbits in the shape of the number "eight" are observed:

The first type of separatrix phase trajectories surrounds only three coupled singular points, two types of stable center and one type of unstable saddle, which is intersect, and all these elements represent a trigger of first order coupled singularities. There are as four triggers of coupled singularities as there are four minimum of the rolling paths and in this case.

The second type of separatrix phase trajectories surrounds two triggers of coupled singularities and only intersects at one singular point of the unstable saddle type between them. There are two such separatrix phase trajectories in the studied phase portrait.

The third type of third-order homoclinic orbits surrounds one second-order homoclinic orbit, as well as two first-order homoclinic orbits. Here, in the observed case, in the phase trajectory portrait, in Fig. 14, there is only one such homoclinic orbit—the separatrix phase trajectory, and it surrounds all four triggers of coupled singularities, each of which is about one of the four minimum positions on the generalized rolling pendulum rolling path.

Between these separatrix phase trajectories in phase trajectory portrait, are regular closed phase trajectories corresponding to periodic rolls of the generalized rolling pendulum with corresponding periods of oscillatory rolling which depend on the initial conditions and the value of total mechanical energy which achieves conservative nonlinear dynamics, and the number of equilibrium positions on the path through which the body passes for a period of one rolling oscillation.

In Fig. 15 a phase portrait of the dynamics of a generalized rolling pendulum, which rolls, without slipping, along a curvilinear path defined by a polynomial of the eighth degree in the form $y = f(x) = kx^2(x^2 - a^2)^2(x^2 - b^2)$, where a, b and k, are known constants, in a rotate vertical plane around the vertical axis at a constant angular velocity Ω, and in Earth's field of gravity, in phase coordinates $\omega_P(x)$, x is presented.

By analyzing the shape of the paths along which the body of the generalized rolling pendulum rolls, without slipping, as well as by analyzing a series of phase portraits and the structure of sangular points in them, as well as structural stability and sensitivity to changes in the system's bifurcation parameters, bused on series published author's [11, 12, 23, 31], as well as a large number of numerical experiments and obtained different graphs of nonlinear phenomena in nonlinear dynamics of generalized rolling pendulum, *a new theorem of bifurcation and of trigger of coupled singularities* can be defined in the following formulation:

Theorem on bifurcation and on the trigger of coupled singularities in the dynamics of generalized rolling pendulums along curvilinear routes in a rotating vertical plane around a vertical axis at a constant angular velocity on bifurcation and on the trigger of coupled singularities in the dynamics of generalized rolling pendulums along curvilinear routes in a rotating vertical plane around a vertical axis at a constant angular velocity: *Let the curved line, given with $y = f(x)$, for which it is valid $f(x) = f(-x)$, and which has at the points for extreme values $EX_s(x_s, y_s = f(x_s))$ for $f'(x_s) = 0$, the minimums $C_s(x_s, y_s = f(x_s))$ for $f'(x_s) = 0$, $f''(x_s) > 0$, and the maxima $S_s(x_s, y_s = f(x_s))$ for $f'(x_s) = 0$, $f''(x_s) < 0$, the curvilinear route, along which rolls, without slipping, a heavy homogeneous thin disk, of radius $r > 0$, and let it located in the Earth's gravitational field, and in the vertical plane, which rotates around the vertical axis, at a constant angular velocity $\Omega > 0$. The characteristic equation for determining the singular points, as well as the position of the relative equilibrium of the disk on the curvilinear path, in the vertical rotating plane around the vertical axis at a constant angular velocity $\Omega > 0$, is of the form:*

$$h(x) = K(x) = f'(x)\left\{1 - r\frac{f''(x)}{\left[1 + [f'(x)]^2\right]^{\frac{3}{2}}}\right\} - \frac{2\kappa}{3g}\Omega^2\left\langle x - \frac{rf'(x)}{\sqrt{1 + [f'(x)]^2}}\right\rangle\left\langle 1 - \frac{rf''(x)}{[1 + [f'(x)]^2]\sqrt{1 + [f'(x)]^2}}\right\rangle = 0 \quad (18)$$

in which it is $\kappa = \frac{\mathbf{J}_P}{Mr^2} = \frac{\mathbf{i}_P^2}{r^2} = \frac{\mathbf{i}_C^2}{r^2} + 1 = \kappa$, that is $\kappa = \frac{3}{2}$, the rolling coefficient of the disk, because is $\mathbf{J}_{Cz} = \sigma\frac{r^4}{4}\pi = M\frac{r^2}{4}$ and $\mathbf{J}_P = \mathbf{J}_C + Mr^2 = \frac{3}{2}Mr^2$, and g the acceleration of the Earth is heavier. Around each extremum of the curvilinear trajectory, which is the minimum defined by $C_s(x_s, y_s = f(x_s))$ for $f'(x_s) = 0$, $f''(x_s) > 0$, in the dynamics of thin dick rolling, triggers of conjoined singularities appear, and around each extremum, which is maximum defined with $S_s(x_s, y_s = f(x_s))$ for

$f'(x_s) = 0$, $f''(x_s) < 0$, *there is no trigger of coupled singularities (see* Figs. 11, 12, 13, and 14*).*

7 Concluding Remarks

The paper presents an analogy [29–31, 36] of the nonlinear dynamics of a heavy material point along curvilinear paths in a rotating vertical plane, at a constant angular velocity, around a vertical axis and the nonlinear dynamics of a generalized rolling thin heavy disk pendulum along the same curvilinear paths in both these cases. One theorem are presented and additionally graphically proofed. The theorem describes the process of bifurcation and occurrence and disappearance of triggers of coupled three singular points in the local area of the minimum of curvilinear paths in rotating vertical planes, at a constant angular velocity around the vertical axis, caused by the angular velocity of rotation as a bifurcation parameter.

Based on a numerical experiment with various curvilinear rolling routes, a large number of graphs of the characteristic function of nonlinear dynamics of generalized rolling thin heavy disk pendulum, were obtained, such as phase trajectory portraits of nonlinear dynamics of a generalized rolling thin heavy disk pendulum along curvilinear paths in a rotating vertical plane at different values of constant angular velocity about a vertical axis, also are presented.

From a large number of obtained graphics, the most characteristic examples were selected and presented in the paper. The results of previous published author's references for particular examples of the shape of curvilinear paths along which the thin heavy disk of a generalized rolling pendulum rolls were also used as initial ideas for research continuations.

The observed bifurcation and triggers of coupled singularities is a property of the nonlinear dynamics of generalized rolling thin heavy disk pendulums along rotating curvilinear route about vertical axis at a constant angular velocity. Identification of the triggers of coupled singularities in the coupled rotations in system dynamics is very important for explanation of numerous phenomena in real engineering system dynamics.

References

1. K.R. Hedrih (Stevanović), *Vibro-impact dynamics in systems with trigger of coupled three singular points: Collision of two rolling bodies*, in *The 24th International Congress of Theoretical and Applied Mechanics (IUTAM ICTAM 2016)*, Montreal, Canada, 21–26 Aug 2016, Book of Papers, pp. 212–213. IUTAM permanent site. ISBN: NR16-127/2016E-EPUB; Catalogue Number: 978-0-660-05459-9
2. K.R. Hedrih (Stevanović), Non-linear phenomena in vibro-impact dynamics: central collision and energy jumps between two rolling bodies. Dedicated to memory of Professor and important

scientist Ali Nayfeh (December 21, 1933–March 27, 2017). Nonlinear Dyn. **91**(3), 1885–1907 (2018). https://doi.org/10.1007/s11071-017-3988-x
3. K.R. Hedrih (Stevanović), Rolling heavy ball over the sphere in real Rn3 space. Nonlinear Dyn. **97**, 63–82 (2019). https://doi.org/10.1007/s11071-019-04947-1
4. K.R. Hedrih (Stevanović), Vibro-impact dynamics of two rolling heavy thin disks along rotate curvilinear line and energy analysis. J. Nonlinear Dyn. **98**(4), 2551–2579 (2019). https://doi.org/10.1007/s11071-019-04988-6. ISSN 0924-090X
5. K.R. Hedrih (Stevanović), Generalized rolling pendulum along curvilinear trace: phase portrait, singular points and total mechanical energy surface. J. Comput. Algebra Syst. Teach. Res. **VI**, ed. by A. Prokopenya, A. Gil-Swiderska (Siedlce University of Natural Sciences and Humanities (Siedlce, Poland), 2017), pp. 204–216. ISSN 2300-7397. http://www.castr.uph.edu.pl
6. K.R. Hedrih (Stevanović), Rolling heavy disk along rotating circle with constant angular velocity, in *Computer Algebra Systems, in Teaching and Research*, Chapter 2. Problems of Classical Mechanics, vol. V, ed. by A.N. Prokopenya, M. Jakubiak (Siedlce University of Natural Sciences and Humanities, Siedlce, 2015), pp. 293–304. ISSN 2300-7397; ISBN 978-83-7051-779-3
7. K.R. Hedrih (Stevanović), Vibro-impact dynamics of two rolling heavy thin disks along rotate curvilinear line and energy analysis. J. Nonlinear Dyn. **98**(4), 2551–2579 (2019). https://doi.org/10.1007/s11071-019-04988-6. http://www.springer.com/home?SGWID=0-0-1003-0-0&aqId=3763390&download=1&checkval=27b55b07bee6baeb0e6c0ca9c045ba38
8. K.R. Hedrih (Stevanović), Beseda o Mihajlu Petroviću, Legende Beogradskog Univerziteta, Univerzitet u Beogradu, Univerzitetska biblioteka „Svetozar Markovic" u Beogradu, 2005, pp. 37–48
9. K.R. Hedrih (Stevanović), Nonlinear dynamics of a gyro-rotor, and sensitive dependence on initial conditions of a heavy gyro-rotor forced vibration/rotation motion, semi-plenary invited lecture, in *Proceedings: COC 2000*, ed. by F.L. Chernousko, A.I. Fradkov, IEEE, CSS, IUTAM, SPICS, St. Petersburg, Inst. for Problems of Mech. Eng. of RAS, 2000, vol. 2 of 3, pp. 259–266
10. K.R. Hedrih (Stevanović), The optimal control in nonlinear mechanical systems with trigger of the coupled singularities, in *Advances in Mechanics: Dynamics and Control: Proceedings of the 14th International Workshop on Dynamics and Control*, ed. by F.L. Chernousko, G.V. Kostin, V.V. Saurin (A.Yu. Ishlinsky Institute for Problems in Mechanics RAS, Nauka, Moscow, 2008), pp. 174–182, ISBN 978-5-02-036667-1
11. K.R. Hedrih (Stevanović), On rheonomic systems with equivalent holonomic conservative system, in *International Conference on ICNM-IV, 2002*, ed. by W. Chien et al. (T. Nonlinear Dynamics, Shanghai, pp. 1046–1054
12. K.R. Hedrih (Stevanović), On rheonomic systems with equivalent holonomic conservative systems applied to the nonlinear dynamics of the Watt's regulator, in *Proceedings, vol. 2: The Eleventh World Congress in Mechanism and Machine Sciences, IFToMM*, China Machine Press, Tianjin, China, April 1–4 April 2004 (China Machine Press, Tianjin, China), pp. 1475–1479. ISBN 7-111-14073-7/TH-1438. http://www.iftomm2003.com
13. K.R. Hedrih (Stevanović), Discontinuity of kinetic parameter properties in nonlinear dynamics of mechanical systems, Keynote Invited Lecture, in *9º Congresso Temático de Dinâmica, Controle e Aplicaçõesm*, 07–11 June 2010. UneSP, Sao Paolo (Serra negra), Brazil (Proceedings of the 9th Brazilian Conference on Dynamics Control and their Applications, Serra Negra, 2010), pp. 8–40. SP-ISSN 2178-3667
14. K.R. Hedrih (Stevanović), Energy and nonlinear dynamics of hybrid systems, in *Dynamical Systems and Methods*, Part 1, ed. by A. Luo (Springer, 2012), pp. 29–83. https://doi.org/10.1007/978-1-4614-0454-5_2
15. K.R. Hedrih (Stevanović), Vibrations of a heavy mass particle moving along a rough line with friction of Coulomb type. Int. J. Nonlinear Sci. Numer. Simul. **11**(3), 203–210 (2010). http://www.reference-global.com/toc/ijnsns/11/3
16. K.R. Hedrih (Stevanović), Nonlinear dynamics of a heavy material particle along a circle which rotates and optimal control, in *IUTAM Symposium on Chaotic Dynamics and Control*

of Systems and Processes in Mechanics, ed. by G. Rega, F. Vestroni, Solid Mechanics and Its Applications, vol. 122 (Kluwer and Springer, 2003), pp. 37–45. ISBN: 978-1-4020-3267-7 (Print) 978-1-4020-3268-4 (Online). http://springerlink.com/content/v3426kp5483w3v00/?p=af7754855fef4487bf3c169778ad50e0&pi=3
17. K.R. Hedrih (Stevanović), G. Janevski, Nonlinear dynamics of a gyro-disc-rotor and structural dependence of a phase portrait on the initial conditions, in *Proceedings of Dynamics of Machines 2000*, Institute of Thermomechanics, Czech Committee of the European Mechanics Society, Prague, 8–9 Feb 2000, pp. 81–88
18. L.J. Veljović, Nelinearne oscilacije giro-rotora (Non-linear oscillations of Gyro-rotors), Doctor's Degree Thesis, Faculty of Mechanical Engineering in Niš, 2011. Supervisor K. Hedrih (Stevanović) [in Serbian]
19. A.A. Andronov, A.A. Vitt, S. Haykin, *Teoriya kolebaniy* (Theory of oscillations). (Nauka, Moskva, 1981), p. 568 (in Russian)
20. K.R. Hedrih (Stevanović), *Trigger of Coupled Singularities (invited plenary lecture)*, Dynamical Systems—Theory and Applications, ed. by J. Awrejcewicz et al. (Lodz, 2001), pp. 51–78
21. K.R. Hedrih (Stevanović), a trigger of coupled singularities. MECCANICA **39**(3), 295–314 (2004). https://doi.org/10.1023/B:MECC.0000022994.81090.5f
22. K.R. Hedrih (Stevanović), Dynamics of coupled systems. Nonlinear Anal.: Hybrid Syst. **2**(2), 310–334 (2008)
23. K.R. Hedrih (Stevanović), Existence of trigger of coupled singularities in nonlinear dynamics of mechanical system with coupled rotations. Sci. Tech. Rev. **62**(1), 48–56 (2012). ISSN 1820-0206. http://www.vti.mod.gov.rs/ntp/entp.htm
24. K.R. Hedrih (Stevanović), Dynamics of a rolling heavy thin disk along rotate curvilinear trace on vertical plane about vertical axis, in *Proceedings of the 7th International Congress of the Serbian Society of Mechanics (32th Yugoslav Congress)*, 24–16 June 2019, Sremski Karlovci, Serbia, pp. 96–97. ISBN 978-86-909973-7-4. http://www.ssm.org.rsm
25. D.P. Rašković, *Analitička mehanika* (Analytical Mechanics). (Mašinski fakultet Kragujevac, 1974)
26. M. Petrović, *Sur l' équation différentielle de Riccati et applications chimiques*. Sitzungsberichte der Kőnigl, Bőhmischen Gesellschaft der Wissenschaften, Praha, 1896, pp. 39, 1–25
27. D.P. Rašković, *Teorija oscilacija* (Theory of oscillations). Naučna knjiga, 1st edn., 1952, 2nd ed., 1965. http://elibrary.matf.bg.ac.rs/search. http://elibrary.matf.bg.ac.rs/handle/123456789/3778
28. D.P. Rašković, *Tablice iz Otporbosti matematijala, Dodatak iz Matematike* (Tables from Steinght of Material, Appendix in Mathematics), XIth issue (University of Belrade, Publisher „Gradjevinska Knjiga", Beograd, 1979), p. 240
29. M. Petrović, *Elementi matematičke fenomenologije* (Elements of mathematical phenomenology). (Srpska kraljevska akademija, Beograd, 1911), p. 89. http://elibrary.matf.bg.ac.rs/handle/123456789/476?locale-attribute=sr
30. H. Kauderer, *Nichlineare Mechanik* (Nook, Berlin, 1958) (there are Russian translation Nauka, Moskva)
31. M. Petrović, *Fenomenološko preslikavanje* (Phenomenological mapp). (Srpska kraljevska akademija, Beograd, 1933), p. 33
32. J. Guckenheimer, Ph. Holmes, *Nonlinear Oscillations, Dynamical Systems, and Bifurcations of Fields* (Springer, 1983), p. 461
33. J. Gerard, J.J. Daniel, *Elementary Stability and Bifurcation Theory* (Springer, 1980)
34. K.R. Hedrih (Stevanović), Discontinuity of kinetic parameter properties in nonlinear dynamics of mechanical systems, Keynote Invited Lecture, in *9th Brazilian Conference on Dynamics, Control and Their Applications—9º Congresso Temático de Dinâmica, Controle e Aplicaçõesm*, 07–11 June 2010. UneSP, Sao Paolo (Serra Negra), Brazil (Postconference Proceedings of the 9th Brazilian Conference on Dynamics Control and their Applications, Serra Negra, 2010), pp. 8–40. SP—ISSN 2178–3667. http://www.rc.unesp.br/igce/demac/dincon2010/instrucao.php

35. D.P. Rašković, *Mehanika III—Dinamika* (Mechanics IIII—Dynamics), 4th edn. (Naučna knjiga, 1972). http://elibrary.matf.bg.ac.rs/handle/123456789/3777; http://elibrary.matf.bg.ac.rs/search
36. K.R. Hedrih (Stevanović), Analogy between models of stress state, strain state and state of the body mass inertia moments. Facta Univ. Ser. Mech. Autom. Control Robotics Niš **1**(1), 105–120 (1991)

A Quantum Dynamical Map in the Creation of Optimized Chaotic S-Boxes

Nafiseh Hematpour, Sodeif Ahadpour, and Sohrab Behnia

Abstract The substitution boxes are an open challenge due to not meeting the theoretical criteria of a good S-box. Recently, the use of chaos in the design of efficient S-boxes was proposed. In this article, after introducing a new quantum system, we examine its effect on the formation of chaotic S-boxes. We compare the proposed quantum chaotic map with previous results. Also, in the previous work, the PSO algorithm was improved with the help of the classical map and then used in the optimization of chaotic S-boxes. We are using and improving the performance of PSO in generating the S-box, by the introduced quantum chaotic map. Then, by changing the type of optimization, we examine its effects. For the first time, the harmony search algorithm is improved by the said quantum map, and then we use it to optimize the produced chaotic S-box. By examining the performance of generated S-boxes by common attacks such as nonlinearity, BIC, SAC, LP, and DP. The results for the improved harmony search algorithm is better.

Keywords Quantum dynamical map · Substitution box (S-box) · Harmony search algorithm · Particle swarm optimization (PSO) · Nonlinearity

1 Introduction

Many researchers in recent decades, to achieve higher security, have combined the two fields of chaos and cryptography under the heading of chaotic-based cryptography [1–4]. Due to their many applications, quantum dots are one of the favorite topics of researchers. So far, quantum dots have been used in solar cells [5], diodes

N. Hematpour (✉) · S. Ahadpour
University of Mohaghegh Ardabili, Ardabil, Iran
e-mail: n_hematpour@uma.ac.ir

S. Ahadpour
e-mail: ahadpour@uma.ac.ir

S. Behnia
Department of Physics, Urmia University of Technology, Urmia, Iran
e-mail: s.behnia@sci.uut.ac.ir

C. H. Skiadas and Y. Dimotikalis (eds.), *14th Chaotic Modeling and Simulation International Conference*, Springer Proceedings in Complexity,
https://doi.org/10.1007/978-3-030-96964-6_16

[6], medical imaging [7], and quantum computing [8]. When quantum dots are paired with other quantum dots or external fields, They have a long periodicity, making them suitable for use in cryptography. The National Institute of Standards and Technology (NIST) proposed the Data Encryption Standard (DES) for the encryption and decryption process in 1977 [9], which was replaced by the AES symmetric-key algorithm in 2001 [10]. S-box, which performs confusion, has been widely employed in traditional cryptographic standards such as DES and AES. Making efficient boxes is a major issue for security experts. Recently, some S-box algorithms based on the chaotic map have been proposed [11–14]. Then optimization algorithms are used to improve the performance of chaotic S-boxes [11, 15, 16]. All optimizers require a fitness function, which [11] shown to use nonlinearity fitness for better results. In this reference, classical maps are proposed to improve the performance of the PSO algorithm. Considering the theoretical criteria of a good S-box, there is a need to form new S-boxes.

In this work, a quantum map is replaced by a classical map. Also, the harmony search algorithm is replaced with the PSO algorithm to investigate the effect of the type of optimization.

The paper continues as follows: In Sect. 2, preliminary is proposed that includes the introduction of quantum dots and the study of their behavior. In Sect. 3, S-box criteria are presented. Sections 4 includes the algorithm for creating improved PSO and optimized S-box. Improved HS and optimized S-box is offered in Sect. 5. Section 6 provides an analysis of the performance of the S-boxes. Finally, a conclusion is proposed.

2 Preliminary

We introduce a generalized Dicke model presenting a new quantum chaotic map. It also investigates the chaos of this created system.

2.1 The Maps of Generalized Dicke Model

The dynamical system governed by a generalized Dicke Hamiltonian form is constructed as follows:

$$H = a^\dagger a + \omega_A J_z + \frac{\gamma}{\sqrt{N}}(a^\dagger + a)(J_- + J_+) + \frac{\gamma}{\sqrt{N}} V(J_-, J_+)$$

$$\Sigma_n \delta(t - nT). \tag{1}$$

In fact, we consider delta function added to Dicke Hamiltonian. where, a and $a^\dagger$ are respectively bosonic annihilation and creation operators. The parameter $\hbar\tilde{\omega}_A$ denotes the energy separation of N two-level atoms [17]. We assume that $\hbar = 1, \omega_A =$

$\tilde{\omega}_A/\tilde{\omega}_f \geq 0$, and $\tilde{\omega}_f$ is the field of frequency. $\gamma = \tilde{\gamma}/\tilde{\omega}_f$ is the coupling parameter. Furthermore, J_z and $J_{\pm}$ are the atomic relative population operator and the atomic transition operator, respectively [18]. In [12], we introduced a chaotic mapping based on this Hamiltonian.

$$< J_{+(n+1)} >= \alpha \left(< J_{+(n)} > - < J_{-(n)} J_{+(n)} >\right). \tag{2}$$

According to previous studies for quantum systems [19], here in the same way map with quantum corrections for a system of coupled quantum dots is extracted. To appear the effect of the quantum correlations using $J_+ =< J_+ > + \delta J_+$ and $J_- =< J_- > + \delta J_-$, we have:

$$\begin{aligned} < J_{+(n+1)} > &= r\left(< J_{+(n)} > - < J_{-(n)} >< J_{+(n)} >\right) \\ &\quad - r < \delta J_- \delta J_+ > . \end{aligned} \tag{3}$$

Taking time derivation of $(\delta J_+ \delta J_-)$ implies

$$\frac{d}{dt}(\delta J_+ \delta J_-) = \delta \dot{J}_+ \delta J_- + \delta J_+ \delta \dot{J}_-. \tag{4}$$

Next, by applying $< \delta J_+(nT) \delta J_-(nT) >= y_n$, $< \delta J_+ \delta J_+ >= z_n$, $< J_+(nT) >= x_n$, we obtain (see Appendix 1):

$$\begin{cases} X_{n+1} = r(x_n - x_n^2) - r y_n \\ Y_{n+1} = -y_n + r e^{-\beta}((1 - x_n + e^{2\beta} - x_n e^{2\beta}) y_n - z_n x_n - e^{2\beta} z_n x_n) \\ Z_{n+1} = -z_n e^{2\beta} + r e^{\beta}(2z_n - 2x_n z_n - x_n y_n - y_n x_n) \end{cases}. \tag{5}$$

Equation (5) gives the lowest-order quantum corrections. For convenience, we consider that $\beta = i w_A T$ [20]. The sensitivity of the map to its initial values are shown in Fig. 1. We plotted Fig. 1 for constant parameter $r = 9$, $b = 0.5$, $y_0 = 0.435$ and $z_0 = 0.777$ as well as variable initial condition $x_0 = 0.423$ and $x_0 = 0.424$. Lyapunov exponent curve are seen in Fig. 2.

3 S-Box Criteria

An $n * m$ S-box is a nonlinear mapping $S : V_n \rightarrow V_m$, where V_n and V_m represent the vector spaces of n, m elements from $GF(2)$. Important tests to check the performance of S-box are nonlinearity (NL), strict avalanche criterion (SAC), bit independence criterion (BIC), linear approximation probability (LP), and differential approximation probability (DP).

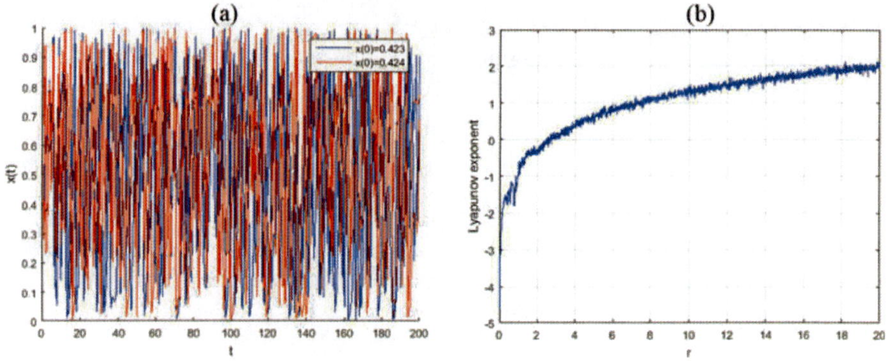

Fig. 1 **a** The sensitivity of the chaotic map to initial values the maps of generalized Dicke model for $x_0 = 0.423$ and $x_0 = 0.424$ where the control parameter $r = 9$, $b = 0.5$, $y_0 = 0.435$ and $z_0 = 0.777$. **b** The variation of the Lyapunov exponent the maps of generalized Dicke model in term of parameters r

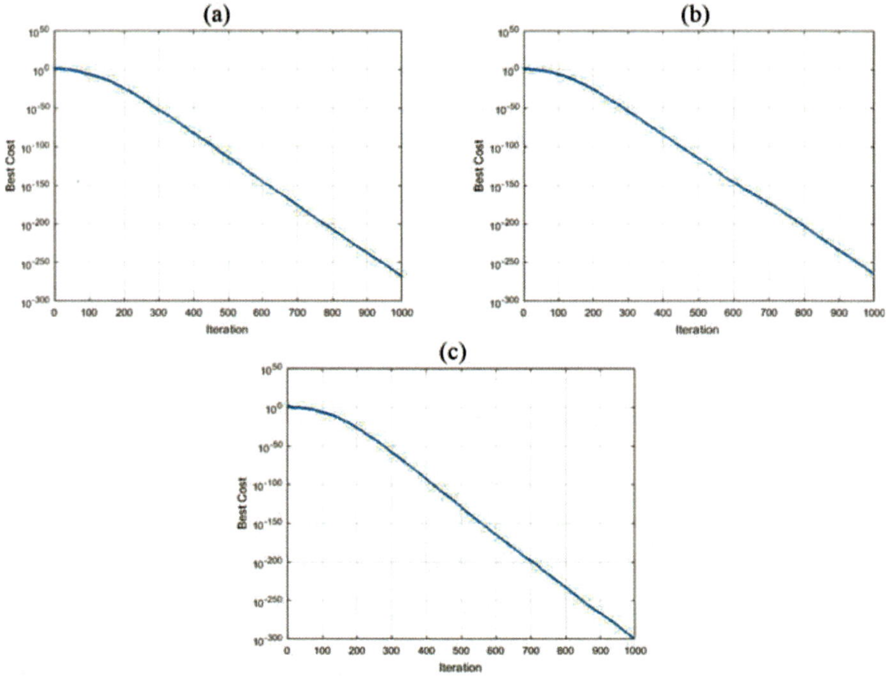

Fig. 2 The variation of the cost function (sphere) for **a** PSO algorithm with unifrnd and rand functions **b** improved PSO algorithm with quantum maps **c** improved PSO algorithm with quantum maps and hierarchy of rational-order chaotic maps

3.1 Nonlinearity

The nonlinearity value is calculated from the following equation:

$$N = 2^{n-1} - \frac{1}{2} \max_{a \in B^n} \left| \sum_{x \in B^n} (-1)^{f(x)+a \cdot x} \right|$$

where $B = \{0, 1\}$, $f : B^n \to B$, $a \in B^n$ and $a \cdot x$ is the dot product between a and x (see [21] for example). Since the affine functions are weak in terms of cryptography, the similarity of the Boolean function variable of S-box is measured with the affine variable.

3.2 Strict Avalanche Criterion (SAC)

Webster and Tavares introduced SAC. If one bit in the input of Boolean function changed, half of the output bits should be changed [22]. The value of SAC = 0.5 is necessary for passing this test.

3.3 Bit Independence Criterion (BIC)

BIC, which calculate the independence of the avalanche vectors sets, is a desirable feature for any encryption transformation for S-box analysis (Webster and Tavares defined this test in [22]). If one changes the inverse of input single bits, these sets are created [23].

3.4 Linear Approximation Probability (LP)

LP [24] is:

$$LP = \max_{a,b \neq 0} \left| \frac{\#\{x | x \cdot a = f(x) \cdot b\}}{2^n} - 0.5 \right|$$

where a, b are the input and output masks, and the set x contains all the possible inputs, and 2^n is the number of its elements. The maximum value of imbalance in the event between input and output bits is called LP. Low LP is necessary for resistance against linear attacks.

3.5 Differential Approximation Probability (DP)

DP is:

$$DP = \max_{\Delta_x \neq 0, \Delta_y} (\#x \in X, f_x \oplus f(x + \Delta_x) = \Delta_y / 2^n)$$

where X shows the set of all possible input values, and 2^n is the number of its elements. DP which calculate XOR the distribution between input and output bits of S-Box is introduced by Biham and Shamir [25]. The closer distribution between the input and output bits is necessary for resistance against differential attacks.

4 Improved PSO and Optimized S-Box

In PSO, the swarm consists of particles, each one representing a potential solution in the optimization problem. These particles have position and velocity. The PSO algorithm uses the unified function for the initial population and the rand function to update the speed and position. In this paper, we use a quantum map for the initial population. Instead of the rand function, once we use the quantum map and for the second time, the classical hierarchy of rational-order chaotic maps (the best result of [11]) (see Fig. 1). As can be seen, the best results are obtained for improved PSO with quantum maps and the hierarchy of rational-order chaotic maps (see Fig. 2). Now we use this optimization algorithm to get the best S-box based on the highest nonlinear value (see Appendix 2). The best S-box is seen in Table 2 (Table 1). The highest obtained nonlinearity value is 106.

5 Improved HS and Optimized S-Box

Zong Woo Geem et al. in 2001 developed Harmony search which is a music-based metaheuristic algorithm [26]. It used to solve many optimization problems such as function optimization, engineering optimization [27], water distribution networks [28]. To enhance the global convergence and to prevent to stick on a local solution, different HS methods based on chaotic maps have been proposed [29]. The improved HS (see Fig. 3) steps and its application for optimizing the designed chaotic S-box are discussed. The steps of the algorithm are as follows:

Step 1 Enter improved HS parameters (number of decision variables, decision variables matrix size, Maximum number of iterations, Harmony Memory size, number of new Harmonies, Harmony Memory consideration rate, Pitch Adjustment rate, Fret width(Band width), Fret width Damp ratio) and $r = 5.5$, $\beta = 0.5$ for (5).

Table 1 New S-box for the map of (5)

99	206	2	73	228	88	191	176	6	101	211	98	231	153	62	207
164	179	49	195	108	31	141	8	185	57	27	249	91	128	209	154
252	201	138	205	247	76	60	165	14	55	5	56	12	238	139	240
149	125	192	54	188	183	39	229	193	117	180	13	233	146	30	150
214	97	106	82	35	109	131	230	173	152	127	182	41	25	47	236
92	196	160	122	242	111	34	220	212	81	175	170	77	118	132	4
26	145	119	168	15	187	63	136	7	148	181	123	17	221	241	53
254	250	255	67	1	239	93	103	46	226	157	90	167	51	184	105
72	219	140	133	194	203	59	115	232	70	246	243	199	112	142	224
19	42	213	186	177	66	94	68	129	79	21	256	234	80	172	223
171	58	74	156	126	38	16	33	48	178	78	52	114	143	104	23
200	32	251	151	216	237	65	89	28	190	75	202	83	159	69	245
20	96	45	225	9	50	174	113	137	95	198	44	162	244	18	87
210	130	102	61	107	85	215	147	248	43	71	29	64	24	121	100
116	134	22	155	124	135	217	235	189	163	11	253	144	3	84	218
204	110	86	208	158	10	197	161	120	222	37	169	40	36	227	166

Table 2 New optimized S-box for the map of (5) with improved PSO algorithm

185	241	245	54	115	154	198	63	190	228	29	94	177	213	186	240
192	191	28	200	208	193	194	238	34	244	188	132	254	164	107	2
151	239	125	128	171	231	181	96	220	71	21	204	43	101	39	95
256	33	41	218	127	141	137	230	207	201	44	4	102	124	70	10
248	153	212	13	158	119	69	1	143	167	14	3	195	121	206	6
81	17	152	82	111	210	109	113	199	27	140	211	131	148	233	112
48	221	92	253	187	57	243	60	217	78	234	130	116	173	216	120
31	227	246	179	83	7	162	196	232	23	182	47	45	126	72	91
90	76	62	215	30	169	88	222	99	172	176	237	136	189	139	100
197	235	64	156	229	77	87	142	157	98	166	105	51	183	61	59
106	38	68	67	144	155	202	247	40	123	104	174	147	122	163	117
79	36	255	22	37	236	20	74	32	138	223	165	35	86	97	226
58	19	110	209	108	114	103	118	25	9	50	5	160	12	129	252
65	24	149	16	249	52	224	184	55	66	178	225	219	150	242	93
11	53	49	84	175	146	205	15	26	56	89	18	250	159	180	8
170	214	42	133	46	161	75	145	134	85	203	80	251	73	168	135

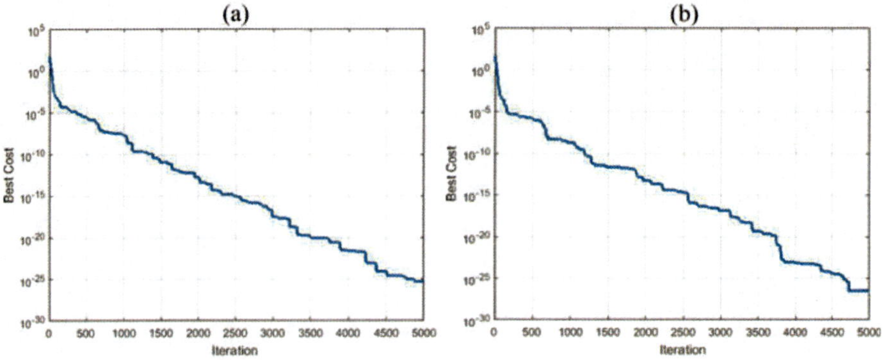

Fig. 3 The variation of the cost function (sphere) for **a** HS algorithm with unifrnd **b** improved HS algorithm with quantum maps

Step 2 Initialize Harmony Memory using liana function(liana function produce random number between 100 and 120 by using (5)).

Step 3 Creation of S-box based on quantum map (5):

1. Enter $r = 5.5$, $\beta = 0.5$ for (5) (consider Fig. 1).
2. Pass the transition state by repeating the map (5).
3. We create empty $16 * 16$ box.
4. Repeat the map (5) and select $x(f)$.
5. The S-box numbers are obtained:

$$S(i) = x(f) * 10^5 \bmod 256$$

6. The process continues from 4 and select different $S(i)$.

Step 4 Calculation of nonlinearity for all Harmony Memory positions.

Step 5 Sort Harmony Memory from MAX to MIN.

Step 6 Update Best solution ever found.

Step 7 Create new Harmony position using liana function.

Step 8 Pitch Adjustment using nafis function(nafis function produce random number between -1 and 1 by using (5)).

Step 9 If Nonlinearity(new position) > best solution save S-box.

Step 10 Merge Harmony memory and new Harmonies.

Step 11 Sort Harmony Memory from MAX to MIN.

Step 12 Update Best solution ever found.

Step 13 Save Best Nonlinearity.

Step 14 If iteration finished, print Best Nonlinearity.

Optimized S-box creation algorithm using improved HS algorithm with quantum maps is presented in Fig. 4. The created S-box are seen in Table 3. Figure 5 shows

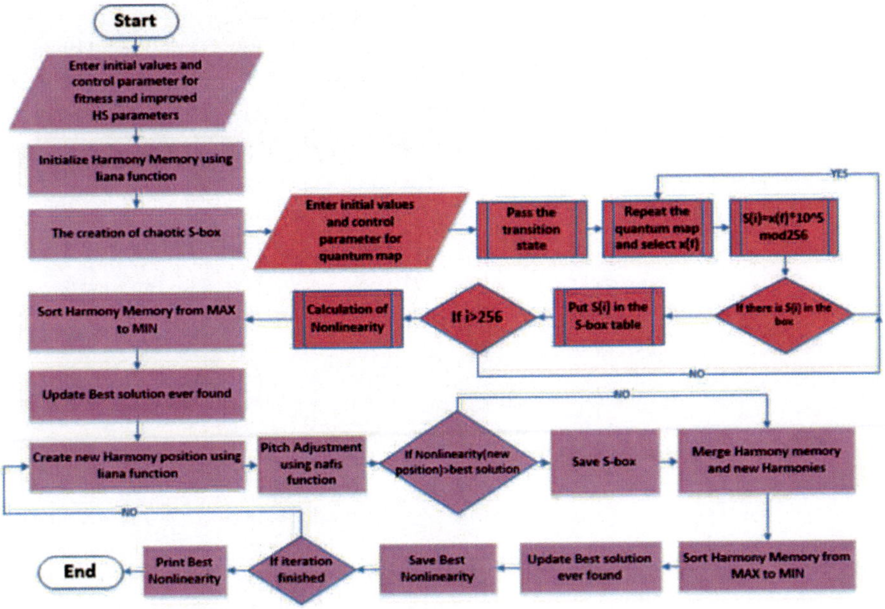

Fig. 4 Optimized S-box creation algorithm using improved HS algorithm with quantum maps

best nonlinearity of optimized S-box with nonlinearity fitness function for improved PSO algorithm with quantum maps and hierarchy of rational-order chaotic maps and improved HS algorithm with quantum maps.

6 Security Analysis

The security of any encryption is measured by its key(the keyspace size more than 2^{100} [30, 31]). We prob the keyspace of a quantum map to create the S-box. The order of complexity for decoding is:

$$T(r, \beta, x_0, y_0, z_0) = \theta(r \times \beta \times x_0 \times y_0 \times z_0)$$

If the computer's analysis power is 10^{16} decimal, the accuracy for each variable is 10^{-16}. The number of these parameters for the quantum map in (5), is 5. So the keyspace for each is $10^{80}(2^{265})$. These spaces could resist all types of brute-force attacks.

Table 4 represent nonlinearity, SAC, BIC, LP and DP results for new S-boxes and compares with the other results.

Table 3 New optimized S-box for the map of (5) with improved HS algorithm

206	4	51	105	57	121	73	247	36	152	101	109	18	134	119	173
25	222	43	122	78	242	30	110	83	114	12	65	23	185	58	138
141	96	1	64	209	135	116	126	156	226	212	84	237	238	160	128
47	255	103	253	40	67	98	229	153	225	14	8	66	29	99	217
21	155	146	219	37	246	181	227	108	17	171	220	7	52	256	94
89	130	211	20	77	133	82	190	24	10	50	44	62	120	136	234
224	208	80	3	163	251	245	195	148	143	203	235	113	72	216	117
144	115	16	142	162	111	70	193	191	38	177	174	213	165	194	86
145	42	34	45	202	204	22	158	139	31	157	75	92	180	241	198
11	188	61	26	151	132	197	39	233	207	97	170	184	68	214	104
149	182	35	49	112	189	60	140	107	239	56	100	199	150	87	186
250	231	196	187	33	19	168	161	46	183	249	76	221	2	93	95
9	201	240	91	13	90	192	236	223	125	28	5	147	131	244	129
230	41	71	210	254	167	69	200	27	205	48	54	228	85	172	218
166	176	248	55	159	106	88	102	15	243	59	164	6	53	124	179
81	178	252	169	154	32	123	118	63	74	79	232	137	175	127	215

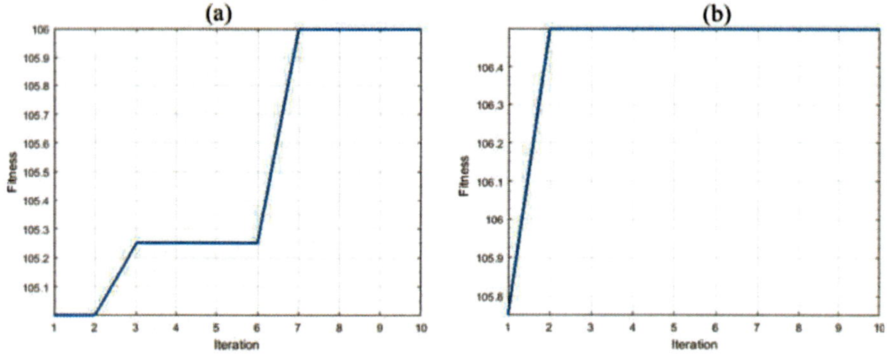

Fig. 5 Best nonlinearity of optimized S-box for **a** improved PSO algorithm with quantum maps and hierarchy of rational-order chaotic maps **b** improved HS algorithm with quantum maps

7 Conclusion

We are using the introduced quantum map based on quantum dots to generate chaotic S-boxes. The proposed map results, improving in performance of introduced PSO and HS optimization algorithms. In comparing the with classic ones, it is effectively acting on generation the S-box. The obtained results show the importance of optimization algorithms in generating the S-box. The Harmony search algorithm for the known sphere function has a weaker answer than the PSO algorithm. In optimizing chaotic S-boxes, the use of Harmony search algorithms produces better results. The introduced

Table 4 Nonlinearity, SAC, BIC, LP, and DP results for new S-boxes and compare with the other results

	Nonlinearity	SAC	BIC-nonlinearity	BIC-SAC	LP	DP
New S-box	105.5	0.512939	103.714		0.140625	10
New optimized S-box with PSO	106	0.499512	103.5	0.500837	0.132813	10
New optimized S-box with HS	106.5	0.501465	104.071	0.498047	0.132813	10
[11]	106.5	0.503662	102.857	0.499512	0.140625	10
[12]	105.25	0.495605	104.571	0.504325	0.140625	12
[13]	104.2	0.4931	103.3	0.4988	0.1563	12
[14]	106	0.52881	100	–	–	10
[10]	112	0.5048	112	–	–	4

S-boxes can be used in all image encryption, steganography, watermarking, and quantum digital signatures to increase security.

Appendix 1

In this appendix we derive Eq. (5). By inserting expressions $J_+ =< J_+ > +\delta J_+$ and $J_- =< J_- > +\delta J_-$ into force equation ([12]) we get

$$\begin{aligned} f(J_+, J_-) &= - < J_+ > -\delta J_+ \\ &+ e^{-iw_A T} r(< J_+ > +\delta J_+ - < J_- >< J_+ > -\delta J_- \delta J_+ \\ &- < J_- > \delta J_+ - \delta J_- < J_+ >). \end{aligned} \tag{6}$$

By considering $\dot{J}_+ = \delta \dot{J}_+$, $\dot{J}_- = \delta \dot{J}_-$, and due to

$$\delta \dot{J}_- = \delta \dot{J}_+^{\dagger}, \tag{7}$$

we use [12] for obtaining:

$$\begin{aligned} &\frac{d}{dt}(\delta J_+ \delta J_-) = [i w_A (< J_+ > +\delta J_+) - i \frac{\gamma}{\sqrt{N}} a^{\dagger}(0) e^{it} (\delta J_+ \\ &\delta J_- - \delta J_- \delta J_+) - i \frac{\gamma}{\sqrt{N}} a(0) e^{-it} (\delta J_+ \delta J_- - \delta J_- \delta J_+) \\ &+ [- < J_+ > -\delta J_+ + e^{i w_A T} r(< J_+ > +\delta J_+ - < J_- >< J_+ > \\ &- \delta J_- \delta J_+ - < J_- > \delta J_+ - \delta J_- < J_+ >)] \, \Sigma_n \delta(t - nT)] \, \delta J_- \end{aligned}$$

$$+\delta J_+(-iw_A(< J_- > +\delta J_-) + i\frac{\gamma}{\sqrt{N}}a(0)e^{-it}[\delta J_-\delta J_+$$
$$-\delta J_+\delta J_-] + i\frac{\gamma}{\sqrt{N}}a^\dagger(0)e^{it}[\delta J_-\delta J_+ - \delta J_+\delta J_-] + [- < J_- >$$
$$-\delta J_- + e^{iw_AT}r(< J_- > +\delta J_- - < J_+ >< J_- > -\delta J_+$$
$$\delta J_- - < J_+ > \delta J_- - \delta J_+ < J_- >)]\ \Sigma_n\delta(t - nT)]\,. \tag{8}$$

By integrating Eq. (8), from nT to $(n+1)T$, and take the expectation value, by taking into account $< \delta J_-(nT) >=< \delta J_+(nT) >= 0$, $< a^\dagger(0) >=< a(0) >= 0$ we obtain:

$$< \delta J_+((n+1)T)\delta J_-((n+1)T) >= - < \delta J_+(nT)\delta J_-(nT)$$
$$> +re^{-iw_AT} < \delta J_+(nT)\delta J_-(nT) > -re^{-iw_AT} < J_- >$$
$$< \delta J_+(nT)\delta J_-(nT) > -re^{-iw_AT}r < \delta J_-(nT)\delta J_-(nT) >$$
$$< J_+ > +re^{iw_AT} < \delta J_+(nT)\delta J_-(nT) > -re^{iw_AT}$$
$$< \delta J_+(nT)\delta J_-(nT) >< J_+ > -re^{iw_AT} < \delta J_+(nT)$$
$$\delta J_+(nT) >< J_- >\,. \tag{9}$$

The calculation of $< \delta J_+\delta J_+ >$ goes as follows:

$$\frac{d}{dt}(\delta J_+\delta J_+) = \delta\dot{J}_+\delta J_+ + \delta J_+\delta\dot{J}_+. \tag{10}$$

We end up with:

$$\frac{d}{dt}(\delta J_+\delta J_+) = [iw_A(< J_+ > +\delta J_+) - i\frac{\gamma}{\sqrt{N}}a^\dagger(0)e^{it}[\delta J_+$$
$$\delta J_- - \delta J_-\delta J_+] - i\frac{\gamma}{\sqrt{N}}a(0)e^{-it}[\delta J_+\delta J_- - \delta J_-\delta J_+]$$
$$+ [- < J_+ > -\delta J_+ + e^{-iw_AT}r(< J_+ > +\delta J_+ - < J_- >< J_+ >$$
$$-\delta J_-\delta J_+ - < J_- > \delta J_+ - \delta J_- < J_+ >)] \times \Sigma_n\delta(t - nT)]$$
$$\delta J_+ + \delta J_+[iw_A(< J_+ > +\delta J_+) - i\frac{\gamma}{\sqrt{N}}a^\dagger(0)e^{it}[\delta J_+\delta J_-$$
$$-\delta J_-\delta J_+] - i\frac{\gamma}{\sqrt{N}}a(0)e^{-it}[\delta J_+\delta J_- - \delta J_-\delta J_+] + [- < J_+ >$$
$$-\delta J_+ + e^{-iw_AT}r(< J_+ > +\delta J_+ - < J_- >< J_+ > -\delta J_-\delta J_+$$
$$- < J_- > \delta J_+ - \delta J_- < J_+ >)]\Sigma_n\delta(t - nT)]. \tag{11}$$

By integrating form Eq. (11), from nT to $(n+1)T$, and by assuming $< \delta J_-(nT) >=< \delta J_+(nT) >= 0$, $< a^\dagger(0) >=< a(0) >= 0$ we obtain:

$$
\begin{aligned}
&< \delta J_+((n+1)T)\delta J_+((n+1)T) > e^{-2i\omega_A(n+1)T} - < \delta J_+(nT) \\
&\delta J_+(nT) > e^{-2i\omega_A nT} = e^{-2i\omega_A nT}(- < \delta J_+(nT)\delta J_+(nT) > \\
&+ e^{-i\omega_A T} r(< \delta J_+(nT)\delta J_+(nT) > - < J_-(nT) > \\
&< \delta J_+(nT)\delta J_+(nT) > - < J_+(nT) >< \delta J_-(nT)\delta J_+(nT) >)) \\
&+ e^{2i\omega_A nT}(- < \delta J_+(nT)\delta J_+(nT) > + e^{-i\omega_A T} r(\\
&< \delta J_+(nT)\delta J_+(nT) > - < J_-(nT) >< \delta J_+(nT)\delta J_+(nT) > \\
&- \delta J_+(nT)\delta J_-(nT) >< J_+(nT) >)).
\end{aligned} \tag{12}
$$

Appendix 2

This appendix describes the improved PSO steps and its application for optimizing the designed chaotic S-box. The steps of the algorithm are as follows:

Step 1 Enter improved PSO parameters (number of decision variables, size of decision variables matrix, Maximum number of iterations, population size, inertia weight, inertia weight damping ratio, personal learning coefficient, global learning coefficient) and $a_1 = 2.61$, $a_2 = 3.168$ for the Hierarchy of rational order chaotic maps [11].

Step 2 Initial population production using chaotic map (5).

Step 3 Creation of S-box based on quantum map (5):

1. Enter $r = 5.5$, $\beta = 0.5$ for (5) (consider Fig. 1).
2. Pass the transition state by repeating the map (5).
3. We create empty $16 * 16$ box.
4. Repeat the map (5) and select $x(f)$.
5. The S-box numbers are obtained:

$$S(i) = x(f) * 10^5 \bmod 256$$

6. The process continues from 4 and select different $S(i)$.

Step 4 Calculate nonlinearity of all primary particles and search personal and global best for this population.

Step 5 Update the speed and position (consider jth dimension at iteration t of each particle i):

$$
\begin{aligned}
V_{i,j}(t+1) = wV_{i,j}(t) &+ (c1)(r1)(BestX_{i,j}(t) - X_{i,j}(t)) \\
&+ (c2)(r2)(GlobalBest(t) - X_{i,j}(t))
\end{aligned} \tag{13}
$$

$$X_{i,j}(t+1) = X_{i,j}(t) + V_{i,j}(t+1) \tag{14}$$

where $V_{i,j}(t)$ is a velocity of particle i at iteration t; $X_{i,j}(t)$ it is a position of i particle at iteration t; r1 and r2 are two random number between (0, 1) provided by the Hierarchy of rational order chaotic maps [11]; $BestX_{i,j}(t)$ is the local best particle i in all swarm and $GlobalBest(t)$ is the leader of the swarm or global best position of all population.

Step 6 Local and global search and save the best nonlinearity and related S-box.

References

1. S. Behnia, A. Akhshani, S. Ahadpour, A. Akhavan, H. Mahmodi, Cryptography based on chaotic random maps with position dependent weighting probabilities. Chaos, Solitons and Fractals **40**, 362–369 (2009)
2. S. Behnia, S. Ahadpour, P. Ayubi, Design and implementation of coupled chaotic maps in watermarking. Applied Soft Computing **21**, 481–490 (2014)
3. S. Behnia, M. Yahyavi, R. Habibpourbisafar, Watermarking based on discrete wavelet transform and q-deformed chaotic map. Chaos, Solitons and Fractals **104**, 6–17 (2017)
4. N. Hematpour, S. Ahadpour, S. Behnia, Digital signature: Quantum chaos approach and bell states, Chapter 9 Springer Science and Business Media LLC, 2019
5. T. K. Das, S. P Ilaiyaraja C, Whispering gallery mode enabled efficiency enhancement: Defect and size controlled cdse quantum dot sensitized whisperonic solar cells, *Scientific Reports*, 8, 9709, 2018
6. Q. Li, X.Wang, Z. Zhang, H. Chen, Y. Huang, C. Hou, J.Wang, R. Zhang, J. Ning, Z. J Min C, Development of modulation p-doped 1310 nm inas/gaas quantum dot laser materials and ultrashort cavity fabry-perot and distributed-feedback laser diodes, *ACS Photonics*, 5, 1084-1093, 2018
7. K. J. McHugh, L. Jin, A. M. Behrens, S. Jayawardena, W. Tang, M. Gao, J. R Langer A, Biocompatible semiconductor quantum dots as cancer imaging agents, *Advanced Materials*, 30, 1706356, 2018
8. M. Wilson, Silicon-based quantum dots have a path to scalable quantum computings. Physics Today **71**, 17–20 (2018)
9. National Institute of Standards and Technology, FIPS PUB 46-3: Data Encryption Standard (DES), super-sedes FIPS, 46-2, 1999
10. Advanced encryption standard (aes), Federal Information Processing Standards Publication 197 Std
11. N. Hematpour, and S. Ahadpour, Execution examination of chaotic S-box dependent on improved PSO algorithm, *Neural Computing and Applications*, 1-23, 2020
12. N. Hematpour, S. Ahadpour, and S. Behnia, Presence of dynamics of quantum dots in the digital signature using DNA alphabet and chaotic S-box, *Multimedia Tools and Applications*, 1-23, 2020
13. F. Ozkaynak, S. Yavuz, Designing chaotic S-boxes based on time-delay chaotic system. Nonlinear Dynamics **74**, 551–557 (2013)
14. D. Lambic, A new discrete-space chaotic map based on the multiplication of integer numbers and its application in s-box design, *Nonlinear Dynamics*, 1-13, 2020
15. H.A. Ahmed, M.F. Zolkipli, M. Ahmad, A novel efficient substitution-box design based on firefly algorithm and discrete chaotic map. Neural Computing and Applications **31**, 7201–7210 (2019)
16. T. Farah, R. Rhouma, S. Belghith, A novel method for designing S-box based on chaotic map and Teaching-Learning-Based Optimization. Nonlinear Dynamics **88**, 1059–1074 (2017)
17. C. Emary, T. Brandes, Chaos and phase transitions in quantum dots coupled to bosons. Physical review E **67**, 0662031–06620322 (2003)

18. J.G. Hirsch, O. Castaños, R.L. Pena, E.N. Achar, Mean field description of the Dicke Model. Found. Probab. Phys. 6 AIP Conf. Proc. **1424**, 144–148 (2012)
19. M.E. Goggin, B. Sundaram, P.W. Milonni, Quantum logistic map. Physical review A **41**, 5705 (1990)
20. S. Ahadpour, N. Hematpour, *Quantum chaos in quantum dots coupled to bosons*, arXiv:1207.5590v1, 2012
21. T. Cusick, P. Stanica, *Cryptographic boolean functions and applications* (Elsevier, Amsterdam, 2017)
22. A. Webster, S. Tavares, *On the design of s-boxes*, in: Conference on the theory and application of cryptographic techniques, Springer, 523-34, 1985
23. H. Zhang, T. Ma, G. Huang, Z. Wang, Robust global exponential syn- chronization of uncertain chaotic delayed neural networks via dual-stage impulsive control, IEEE Trans. Syst. Man. Part B Cybern **40**, 831–844 (2009)
24. M. Matsui, *Linear cryptanalysis method for des cipher, Workshop on the theory and application of cryptographic techniques*, Springer, Berlin, Heidelberg, 386-397, 1993
25. E. Biham, A. Shamir, Differential cryptanalysis of des like cryptosystems. J. Cryptol. **4**, 3–72 (1991)
26. Z.W. Geem, J.H. Kim, G.V. Loganathan, A new heuristic optimization algorithm: Harmony search. Simulation **76**, 60–68 (2001)
27. K.S. Lee, Z.W. Geem, A new meta-heuristic algorithm for continuous engineering optimization: harmony search theory and practice. Comput. Methods Appl. Mech. Engrg. **194**, 3902–3933 (2005)
28. Z.W. Geem, Optimal cost design of water distribution networks using harmony search. Engineering Optimization **38**, 259–280 (2006)
29. B. Alatas, Chaotic harmony search algorithms. Applied Mathematics and Computation **216**, 2687–2699 (2010)
30. B. Schneier, *Applied cryptography: Protocols, algorithms, and source code* in c, John Wiley and Sons, 2007
31. M. Mollaeefar, A. Sharif, M. Nazari, A novel encryption scheme for colored image based on high level chaotic maps. Multimedia Tools and Applications **76**, 607–629 (2017)

The Interaction of Memristor in Cellular Nonlinear Network for Image and Signal Processing

Aliyu Isah, A. S. Tchakoutio Nguetcho, Stéphane Binczak, and Jean-Marie Bilbault

Abstract In this paper, we describe the application of memristor in the neighborhood connections of 2D cellular nonlinear networks (CNN) essentially for image and signal processing. We focused particularly on the interaction of memristors between two cells allowing us to study the contribution of the memristor qualitatively and quantitatively. The dynamics and the steady state response of each cell is described. The resistance of a memristor is not fixed, hence the study takes into account the initial state of the memristance characterized by the previous amount of charge passed through the memristor. We show that the system transition and the steady state response depend strongly on the history of the memristor.

Keywords Memristor · 2D cellular nonlinear networks · System of two cells · System dynamic · Steady state response

1 Introduction

Memristor is a nanoscale two terminals solid state device whose conductivity is controlled by the time-integral of the current flowing through it or the time-integral of the voltage across its terminals [1, 2]. The dynamic conductance modulation and

A. Isah (✉) · S. Binczak · J.-M. Bilbault
ImViA, Université de Bourgogne, BP 47870-21078, Dijon Cedex, France
e-mail: aliyuisahbabanta@gmail.com

S. Binczak
e-mail: stbinc@u-bourgogne.fr

J.-M. Bilbault
e-mail: Jean-Marie.Bilbault@u-bourgogne.fr

A. Isah
ELE-FAENG, Kano University of Science and Technology, BP 3244, Kano, Nigeria

A. S. Tchakoutio Nguetcho
LISSAS, Faculté des Sciences, Université de Maroua, BP 814, Maroua, Cameroun

C. H. Skiadas and Y. Dimotikalis (eds.), *14th Chaotic Modeling and Simulation International Conference*, Springer Proceedings in Complexity,
https://doi.org/10.1007/978-3-030-96964-6_17

connection compatibility with complementary metal-oxide semiconductor neurons, are essential features of memristor affirming its potentiality as synaptic function and memristive gird network [3–5].

This paper describes a memristor based 2D cellular nonlinear network using memristor in the coupling mode. The network is essentially for signal and image processing applications. Figure 1 shows the conventional 2D cellular nonlinear network with each cell constituting one linear capacitor in parallel with one nonlinear resistance, and a series resistance coupling [6]. Figure 2 shows the equivalent network using memristors in the coupling mode.

For any cell at a node n and voltage V_n, the nonlinear current function through the nonlinear resistance is given by:

$$I_{NL_n} = \frac{V_n(V_n - V_a)(V_n - V_b)}{R_o V_a V_b}, \tag{1}$$

and the nonlinear resistance, R_{NL_n}:

$$R_{NL_n} = \frac{R_o V_a V_b}{V_n(V_n - V_a)(V_n - V_b)}.$$

The characteristic roots of the cubic resistance are 0, V_a and V_b, meanwhile R_o is its linear approximation. The corresponding potential energy $W(V_n)$ is obtained from (1) as:

$$W(V_n) = \frac{1}{4}V_n^4 - \frac{V_a + V_b}{3}V_n^3 + \frac{V_a V_b}{2}V_n^2 + \kappa,$$

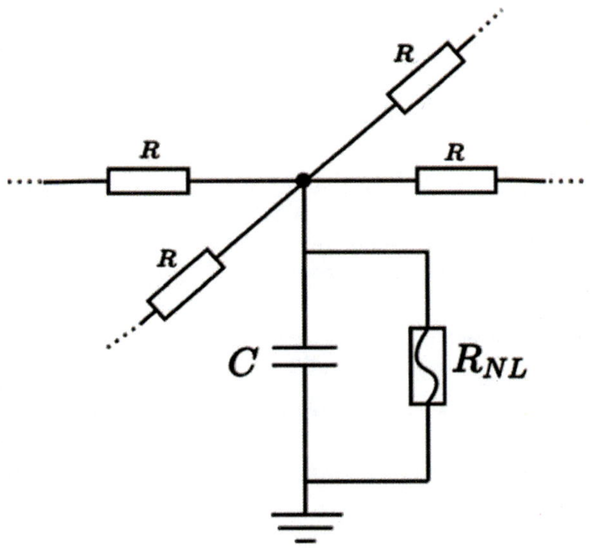

Fig. 1 2D CNN using series resistance coupling [6]

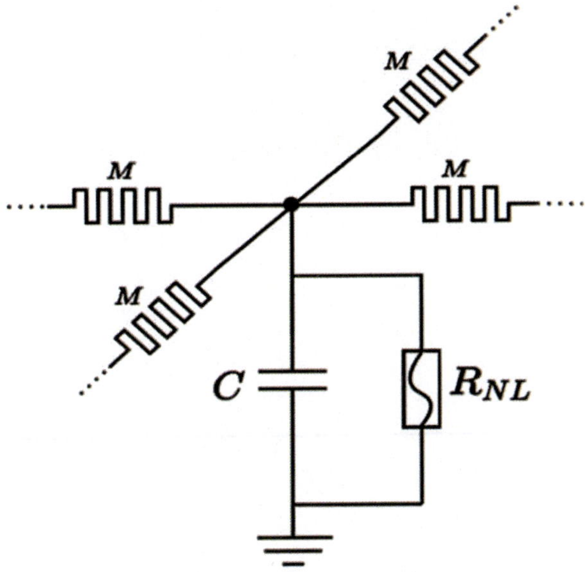

Fig. 2 2D CNN using memristors in the coupling mode

where κ is a constant. Figure 3 shows the response of the cubic resistance and the corresponding potential energy showing the possible equilibrium state. The lower potential energy state are at 0 and V_b marked by numbers 1 and 2 respectively, meanwhile V_a is the unstable state.

We focus on the study of memristor response based on the interaction of one cell with its neighbouring cells. Therefore, using the system of two cells coupled by a memristor, allows us to observe the quantitative and qualitative interaction of memristors in the network.

2 System of Two Cells

Figure 4 shows the network of two cells coupled by a memristor, thus the cells communicate together bidirectionally. One of the cells acts as the master while the other one as slave so that the direction of the flowing charge through the memristor becomes specific. The switchs s_1 and s_2 are closed simultaneously and the network gives the following set of bidirectional coupled equations:

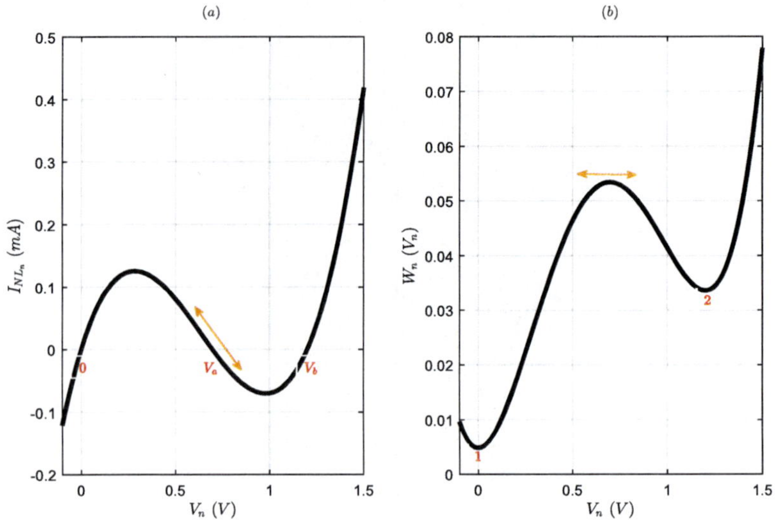

Fig. 3 Response of the nonlinear resistance in the cells. $V_a = 0.7$ V, $V_b = 1.2$ V and $R_o = 1023\,\Omega$. **a** I-V characteristic and **b** the corresponding potential energy. Labels 1 and 2 show the two possible equilibrium states corresponding to $V_n = 0$ and $V_n = V_b$

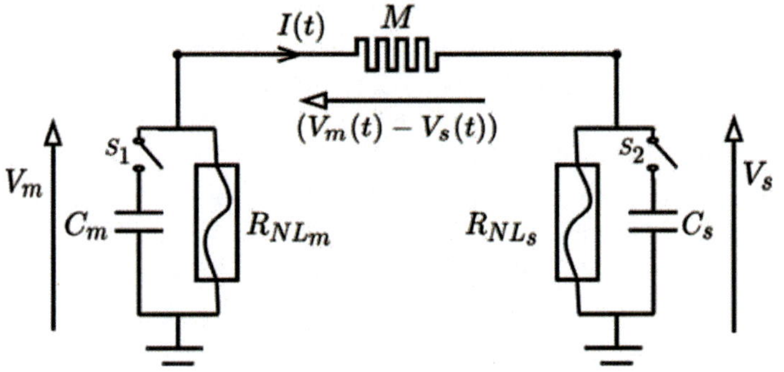

Fig. 4 System of two cells coupled by a memristor. The master and slave cells with their elements given by the subscripts letters m and s respectively

$$\frac{dq}{dt} = -C\frac{dV_m}{dt} - \frac{V_m(V_m - V_a)(V_m - V_b)}{R_o V_a V_b}, \tag{2a}$$

$$\frac{dq}{dt} = C\frac{dV_s}{dt} + \frac{V_s(V_s - V_a)(V_s - V_b)}{R_o V_a V_b}, \tag{2b}$$

$$\frac{dq}{dt} = \frac{V_m - V_s}{M(q)}, \tag{2c}$$

where: $I(t) = \dfrac{dq}{dt}$ is the flowing current through the memristor and $M(q)$ is the memristance function that has a desirable continuity for all the flowing charge [7]:

$$M(q) = \begin{cases} R_{off}, & \text{if } q(t) \leq 0 \\ R_{off} - \frac{3\,\delta R}{q_d^2}\, q^2 + \frac{2\,\delta R}{q_d^3}\, q^3, & \text{if } 0 \leq q(t) \leq q_d \\ R_{on}, & \text{if } q(t) \geq q_d \end{cases} \tag{3}$$

$q_d = \dfrac{D^2}{\mu_v R_{on}}$ is a charge scaling factor depending on the technology parameters [2, 8] and $\delta R = R_{off} - R_{on}$ is the difference between the two limiting resistance values of the memristor, that is, the ON and OFF states, represented here by R_{on} and R_{off} respectively. Charge flows from the master cell to the slave one through the memristor until $V_m(t) = V_s(t)$ and that is when the network is stabilized. Equation (2) is reformulated as:

$$\frac{dV_m}{dt} = -\frac{V_m - V_s}{C.M(q)} - \frac{V_m(V_m - V_a)(V_m - V_b)}{R_o V_a V_b C}, \tag{4a}$$

$$\frac{dV_s}{dt} = \frac{V_m - V_s}{C.M(q)} - \frac{V_s(V_s - V_a)(V_s - V_b)}{R_o V_a V_b C}, \tag{4b}$$

$$\frac{dq}{dt} = \frac{V_m - V_s}{M(q)}. \tag{4c}$$

As illustrated in Fig. 3b, 0 and V_b are the only two possible equilibrium states. The stability of the cells at 0 or V_b is determined by V_a. It can be seen that if $V_b - 2V_a > 0$ the cell stabilizes at V_b and if $V_b - 2V_a < 0$ the cell stabilizes at 0.

The initial conditions of V_m, V_s and q are V_{m_0}, V_{s_0} and q_0 respectively. Figure 5 shows the time evolution of the 2 cells network for $V_{m_0} = 2\,\text{V}$, $V_{s_0} = 0\,\text{V}$, $V_b = 1.5\,\text{V}$, $V_a = 0.7\,\text{V}$, $R_o = 10\,\text{k}\Omega$ and $C_m = C_s = 1\,\mu\text{F}$, hence $V_b - 2V_a > 0$. Initially, the voltage $V_m(t)$ and $V_s(t)$ evolve in the differential mode and thereafter the common mode when $V_m(t) = V_s(t)$ which continues to evolve until the steady state V_b. The charge $q(t)$ flows through the memristor until $V_m(t) = V_s(t)$. When $V_m(t) = V_s(t)$, the voltage across the memristor is 0 (i.e $V_d(t) = 0\,\text{V}$).

3 Discussion

Variation of the system parameters, such as V_a, R_o and q_0 affects the steady state response of the system. The results of Fig. 6 show the variation of V_a with respect to V_b, for example $V_a = \Upsilon V_b$. The results are obtained for $R_0 = 2833\,\Omega$, $V_b = 1.5\,\text{V}$, $V_{m_0} = 2\,\text{V}$, $V_{s_0} = 0\,\text{V}$ and $C_m = C_s = 1\,\mu\text{F}$. Hence, V_a varies according to $\Upsilon = [0.25, 0.45, 0.49, 0.5, 0.51, 0.55, 0.75, 0.9]$ with the corresponding results given by Fig. 6a–h, respectively. Furthermore, the difference $V_b - 2V_a$ is calculated and

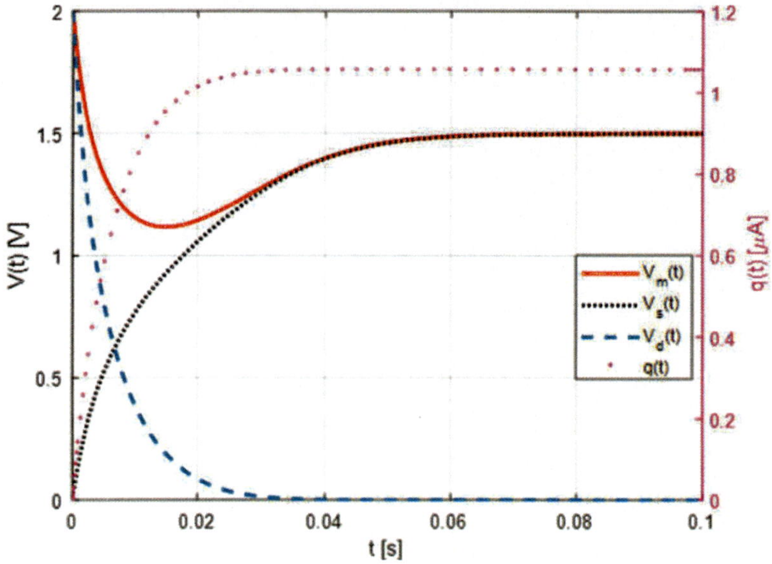

Fig. 5 The evolution of $V_m(t)$, $V_s(t)$, $V_d(t) = V_m(t) - V_s(t)$ and $q(t)$, for $V_{m_0} = 2\,\text{V}$, $V_{s_0} = 0\,\text{V}$, $V_b = 1.5\,\text{V}$, $V_a = 0.7\,\text{V}$, $R_o = 10\,\text{k}\Omega$ and $C_m = C_s = 1\,\mu\text{F}$. The charge $q(t)$ flows through the memristor until $V_d(t) = 0$ and at this time, the combined evolution of $V_m(t)$ and $V_s(t)$ is the common mode which evolves further to stabilizes at the steady state V_b

tabulated in Table 1. The results show two noticeable effects on the evolution of $V_m(t)$ and $V_s(t)$ based on the variation of V_a. The results show different timing at which $V_m(t) = V_s(t)$ and the change in the steady state at V_b or 0 for $V_a < \frac{V_b}{2}$ or $V_a > \frac{V_b}{2}$ respectively.

The initial charge q_0 characterizes the initial memristance of a memristor. The initial condition of a memristor has strong effect on its circuit functionality [9]. Figure 7 shows the effect of changing initial condition of the memristor on the system evolution and stability. It also takes into account the variations of R_0. The initial memristance of the memristor is given by the initial charge q_0. Four different initial charges are considered as: $q_{0_1} = 20\,\mu\text{C}, q_{0_2} = 40\,\mu\text{C}, q_{0_3} = 60\,\mu\text{C}$ and $q_{0_4} = 80\,\mu\text{C}$, as indicated respectively, by the subscripts numbers **1–4** in Fig. 7a, b. Notice that only one parameter is varied at a time. Figure 7a: $R_0 = 2833\Omega$ while q_0 varied and Fig. 7b: $R_0 = 10\,\text{k}\Omega$ while q_0 varied. In each case, $V_a = 0.7\,\text{V}$, $V_b = 1.3\,\text{V}$, $V_{m_0} = 1.5\,\text{V}$ and $V_{s_0} = 0\,\text{V}$. Even though V_a is the main parameter that plays significant role on the system steady state, the results show that other parameters (e.g. R_o and q_0) affect the dynamics and steady state of the system.

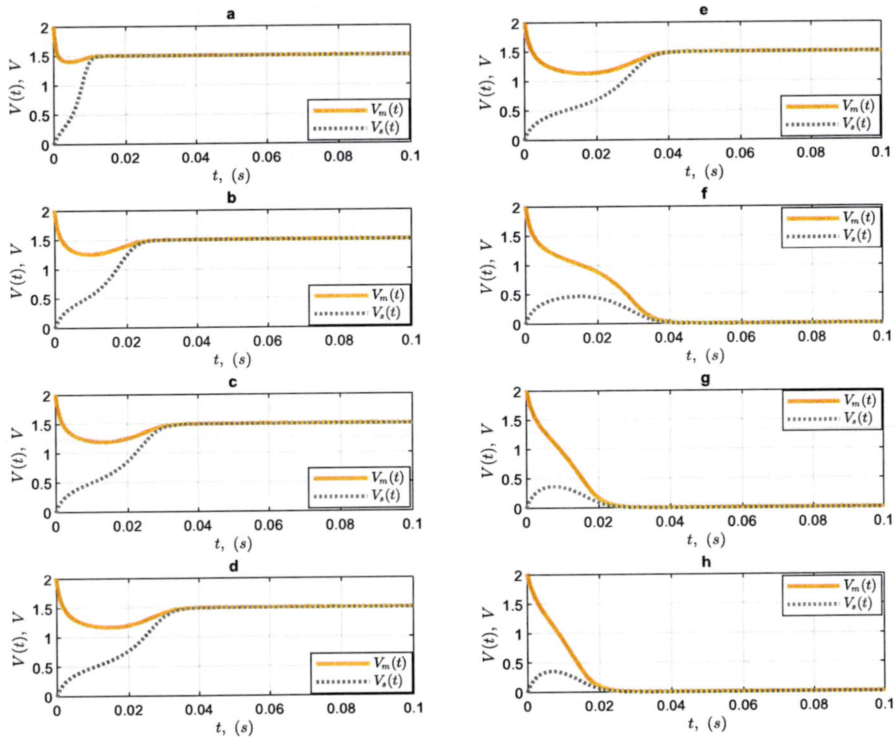

Fig. 6 Evolution of $V_m(t)$ and $V_s(t)$ showing the variations of $V_a \in [0, V_b]$: $R_0 = 2833\,\Omega$, $V_{s0} = 0\,\text{V}$, $V_{m0} = 2\,\text{V}$, $V_b = 1.5\,\text{V}$ and $\Upsilon = [0.25, 0.45, 0.49, 0.5, 0.51, 0.55, 0.75, 0.9]$ as shown respectively by figures (**a**–**h**). Variation of V_a affects the time at which $V_m(t) = V_s(t)$ and the steady state at V_b or 0 depending on whether $V_a < \frac{V_b}{2}$ or $V_a > \frac{V_b}{2}$ respectively

Table 1 Table of $V_b - 2V_a$ for Fig. 6. $V_{s0} = 0\,\text{V}$, $V_a = \Upsilon V_b$, $V_b = 1.5\,\text{V}$ and $V_{m0} = 2\,\text{V}$

Fig. 6	a	b	c	d	e	f	g	h
Υ	0.25	0.45	0.49	0.5	0.51	0.55	0.75	0.9
$V_b - 2V_a$ (V)	0.75	0.15	0.03	0	−0.03	−0.15	−0.75	−1.20

4 Conclusion

Memristor based 2D cellular nonlinear network is introduced using memristors in the coupling mode. The cells correspond to pixels in image processing applications. Each elemental cell consists of one linear capacitor in parallel with one nonlinear resistance such as Fitzhugh Nagumo. Using the system of two cells coupled by a memristor, the dynamics and the steady state of each cell are observed, with mainly a competition between the role of cubic resistance on one hand, and the role of

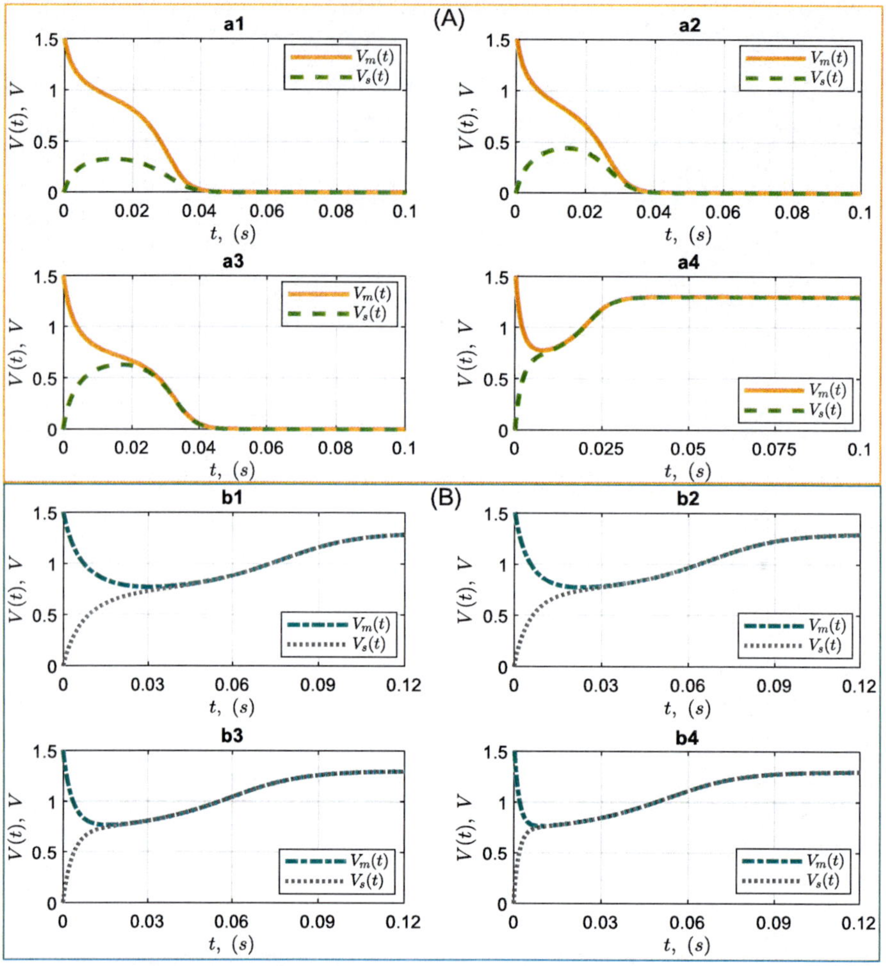

Fig. 7 Results showing the variation effect of q_0 and R_o on the system evolution and the steady state. Four different initial charges are considered as: $q_{0_1} = 20\,\mu\text{C}$, $q_{0_2} = 40\,\mu\text{C}$, $q_{0_3} = 60\,\mu\text{C}$ and $q_{0_4} = 80\,\mu\text{C}$, as indicated respectively by the subscripts numbers **1–4** in figures (**a** and **b**). In each case, $V_a = 0.7\,\text{V}$, $V_b = 1.3\,\text{V}$, $V_{m_0} = 1.5\,\text{V}$, $V_{s_0} = 0\,\text{V}$ and $C_m = C_s = 1\,\mu\text{F}$. **a** $R_0 = 2833\,\Omega$ and **b** $R_0 = 10\,\text{k}\Omega$. It shows that values of q_0 and R_0 have an effect on the evolution and steady state of the system

memristor on the other hand. The parameter V_a predominantly decides the system steady state, however other parameters (e.g R_o, q_0 etc...) affect the system steady state. The results show that the network can be used to realize a binarization circuit, for example, to generate different gray scaling. The ongoing study focuses on the implementation of the generalized 2D network for processing any number cells.

References

1. L. Chua, Memristor-the missing circuit element. IEEE Trans. Circuit Theory **18**(5), 507–519 (1971)
2. D.B. Strukov, G.S. Snider, D.R. Stewart, R.S. Williams, The missing memristor found. Nature **453**(7191), 80–83 (2008)
3. S.H. Jo, T. Chang, I. Ebong, B.B. Bhadviya, P. Mazumder, W. Lu, Nanoscale memristor device as synapse in neuromorphic systems. Nano Lett. **10**(4), 1297–1301 (2010)
4. A. Chowdhury, A. Ayman, S. Dey, M. Sarkar, A.I. Arka, Simulations of threshold logic unit problems using memristor based synapses and CMOS neuron. In: *2017 3rd International Conference on Electrical Information and Communication Technology (EICT)* (IEEE, 2017), pp. 1–4
5. S.P. Adhikari, C. Yang, H. Kim, L.O. Chua, Memristor bridge synapse-based neural network and its learning. IEEE Trans. Neural Netw. Learn. Syst. **23**(9), 1426–1435 (2012)
6. J.C. Comte, P. Marquié, J.M. Bilbault, Contour detection based on nonlinear discrete diffusion in a cellular nonlinear network. Int. J. Bifurcat. Chaos **11**(01), 179–183 (2001)
7. A. Isah, A.S.T. Nguetcho, S. Binczak, J.M. Bilbault, Memristor dynamics involved in cells communication for a 2D non-linear network. IET Signal Process. **14**(7), 427–434 (2020)
8. Y.N. Joglekar, S.J. Wolf, The elusive memristor: properties of basic electrical circuits. Eur. J. Phys. **30**(4), 661 (2009)
9. L. Chua, Everything you wish to know about memristors but are afraid to ask. Radioengineering **24**(2), 319 (2015)

The Turing Model and Discrete Limit Cycles with Eddy and Convection

Shunji Kawamoto

Abstract The Turing model and discrete limit cycles are considered, from the viewpoint of chaos functions. Firstly, the Turing model is explained as a reaction–diffusion system of chemical substances, and a three-dimensional (3-D) time-dependent solvable chaos map corresponding to the model is derived on a basis of chaos function solutions. Secondly, a 2-D chaos map for the 2-D Turing model is proposed for simplicity to present chaotic dynamics, and the 2-D map is shown to have symmetric bifurcation diagrams, a ring of cells and limit cycles with different patterns, depending on the system parameter. In particular, the limit cycles appear in pairs, and are discussed on left-handed and right-handed eddies, which generate convections, as nonlinear dynamics of non-equilibrium open systems.

Keywords Turing model · Turing pattern · Chaos function · Bifurcation diagram · Symmetry · Fixed point · Limit cycle · Eddy · Convection · Non-equilibrium open system

1 Introduction

For the study of nonlinear phenomena, nonlinear dynamics such as soliton, chaos and fractals, have been considered in the field of physics, chemistry, biology and engineering, as nonlinear science [1–4]. It is well known that one-dimensional (1-D) nonlinear difference equations possess a rich spectrum of dynamical behavior as chaos, and the chaotic modeling and simulation have been extended to medicine, optics, living systems, neuro- science and life science, as interdisciplinary fields of science [5–8]. At the same time, large scale systems, such as atmosphere, climate, human brain, power grid, information system and communication network have been studied as non-equilibrium open systems and/or complex systems [9–14].

On the other hand, the Fisher equation has been proposed as a model for the propagation of gene [15]. After that, travelling wave solutions to reaction–diffusion

S. Kawamoto (✉)
Osaka Prefecture University, Sakai, Osaka, Japan
e-mail: kawamoto@eis.osakafu-u.ac.jp

C. H. Skiadas and Y. Dimotikalis (eds.), *14th Chaotic Modeling and Simulation International Conference*, Springer Proceedings in Complexity,
https://doi.org/10.1007/978-3-030-96964-6_18

systems are discussed widely, and the effect of boundaries on convection in a shallow layer of fluid heated from below has been considered in the field of fluid mechanics [16, 17]. Then, the propagation of waves observed in a chemical reaction system has been reported and considered as a non-equilibrium open system [18, 19]. Later, the data obtained in the experiment on the Belousov–Zhabotinsky (BZ) reaction have been analyzed [20, 21]. Moreover, a 2-D model of nonlinear differential equations is derived for the interaction between local reaction and diffusion as chaotic dynamics [22, 23]. In the meantime, a reaction–diffusion system called the Turing model has been presented as a chemical basis of morphogenesis [24]. The chemical pattern is considered as the Turing pattern, and stripe patterns on the skin of marine angelfish have been discussed for understanding biological pattern formation [25–27].

Recently, 1-D, 2-D and 3-D time-dependent solvable chaos maps and a nonlinear time series expansion method have been proposed [28, 29]. Then, the 2-D maps corresponding to the FitzHugh-Nagumo (FHN) model, the BZ reaction and reaction–diffusion systems are derived, and the bifurcation diagrams and the discrete limit cycles have been studied for population growth, neural cell and chemical cell, respectively [30–33]. In addition, a limit cycle analysis and the interaction of limit cycles are presented for the 2-D logistic maps, as non-equilibrium open systems [34–36].

In this paper, firstly the Turing model is explained in Sect. 2 as a reaction–diffusion system of chemical substances, and a 3-D time-dependent solvable chaos map, which corresponds to the model, is derived on a basis of chaos function solutions. Then, in Sect. 3, a 2-D solvable chaos map for the 2-D Turing model is proposed for simplicity to find chaotic properties. The 2-D map is shown numerically to possess symmetric bifurcation diagrams, a ring of cells and discrete limit cycles with different patterns, depending on the system parameter. In particular, the limit cycles appear in pairs, and are discussed on left-handed and right-handed eddies of cells, which generate convections, as nonlinear dynamics of non-equilibrium open systems. Finally, Conclusions are summarized in the last section.

2 The Turing Model and 3-D Chaotic Maps

As is known, a reaction–diffusion system called the Turing model has been presented as a chemical basis of morphogenesis [24], in where a mathematical model of the growing embryo in biology is described, and the chemical reaction and diffusion are explained, under the assumption of reaction rates. In addition, the spherical symmetry of embryo is introduced, and the system is assumed to be far from homogeneous, in left-handed and right-handed organisms. Finally, from the mathematics of the ring, a set of nonlinear differential equations are formulated as the model equations:

$$\frac{dz_0}{dt} = p_0 z_0 + A z_1^2 + B z_2^2, \tag{1}$$

$$\frac{dz_1}{dt} = p_1 z_1 + C z_1 z_2 + D z_0 z_1, \tag{2}$$

$$\frac{dz_2}{dt} = p_2 z_2 + E z_1^2 + F z_0 z_2 \tag{3}$$

with system parameters $\{p_0, p_1, p_2, A, B, C, D, E, F\}$, where $z_0(t)$, $z_1(t)$ and $z_2(t)$ are dimensionless variables [24]. Then, the model equations can be rewritten into the following nonlinear difference equations by the difference method for numerical calculation as

$$z_{0,n+1} = (1 + p_0(\Delta t))z_{0,n} + (\Delta t)\big(A z_{1,n}^2 + B z_{2,n}^2\big), \tag{4}$$

$$z_{1,n+1} = (1 + p_1(\Delta t))z_{1,n} + (\Delta t)(C z_{1,n} z_{2,n} + D z_{0,n} z_{1,n}), \tag{5}$$

$$z_{2,n+1} = (1 + p_2(\Delta t))z_{2,n} + (\Delta t)\big(E z_{1,n}^2 + F z_{0,n} z_{2,n}\big), \tag{6}$$

where the passage from a point $(z_{0,n} \equiv z_{0,n}(t_i), z_{1,n} \equiv z_{1,n}(t_i), z_{2,n} \equiv z_{2,n}(t_i))$ to the next one $(z_{0,n+1} \equiv z_{0,n+1}(t_{i+1}), z_{1,n+1} \equiv z_{1,n+1}(t_{i+1}), z_{2,n+1} \equiv z_{2,n+1}(t_{i+1}))$ is considered as a 3-D mapping with the discrete time t_i and the time step $\Delta t = t_{i+1} - t_i$ [30].

On the other hand, we introduce time-dependent chaos functions

$$x_n(t_i) = a_1 \cos\big(2^n t_i\big) + b_1, \tag{7}$$

$$y_n(t_i) = a_2 \sin(2^n t_i), \tag{8}$$

$$z_n(t_i) = a_3 \sin(2^n t_i) \tag{9}$$

with real parameters $\{a_1, a_2, a_3, b_1\}$, and find a 3-D solvable chaos map with system parameters $\{\beta_1, \beta_2, \beta_3\}$ [28], corresponding to the 3-D map (4)–(6), as

$$\begin{aligned} x_{n+1} = {} & \left(\frac{1}{a_1}\right)(x_n^2 - 2b_1 x_n) - a_1(1 - \beta_1)\left(\frac{1}{a_2}\right)^2 y_n^2 \\ & - a_1\beta_1\left(\frac{1}{a_3}\right)^2 z_n^2 + \left(1 + \frac{b_1}{a_1}\right)b_1, \end{aligned} \tag{10}$$

$$y_{n+1} = -2\left(\frac{b_1}{a_1}\right)y_n + 2\left(\frac{a_2}{a_1 a_3}\right)\beta_2 x_n z_n + 2\left(\frac{1}{a_1}\right)(1 - \beta_2)x_n y_n, \tag{11}$$

$$z_{n+1} = -2\left(\frac{b_1}{a_1}\right)z_n + 2\left(\frac{a_3}{a_1 a_2}\right)\beta_3 x_n y_n + 2\left(\frac{1}{a_1}\right)(1-\beta_3)x_n z_n, \tag{12}$$

where $x_n \equiv x_n(t_i)$, $y_n \equiv y_n(t_i)$ and $z_n \equiv z_n(t_i)$. It is notable that the time step $\Delta t, 0 < \Delta t \ll 1.0$ is not included in the coefficients of the 3-D map (10)–(12) [31].

3 2-D Chaotic Maps and Discrete Limit Cycles

In this Section, we find a 2-D Turing map for simplicity, by setting $z_{2,n} = 0$ in the 3-D Turing map (4)–(6), as

$$z_{0,n+1} = (1 + p_0(\Delta t))z_{0,n} + (\Delta t)Az_{1,n}^2, \tag{13}$$

$$z_{1,n+1} = (1 + p_1(\Delta t))z_{1,n} + (\Delta t)Dz_{0,n}z_{1,n}, \tag{14}$$

and then, from the chaos function solutions (7), (8) and the condition given by

$$\left(\frac{1}{a_1}\right)^2 (x_n - b_1)^2 + \left(\frac{1}{a_2}\right)^2 y_n^2 = 1, \tag{15}$$

we have a 2-D chaotic map:

$$x_{n+1} = (a_1 + b_1) - 2a_1\left(\frac{1}{a_2}\right)^2 y_n^2, \tag{16}$$

$$y_{n+1} = \alpha\left(\frac{1}{a_1}\right)(x_n - b_1)y_n \tag{17}$$

with a system parameter α, $0 < \alpha \leq 2.0$, where the 2-D chaotic map (16) and (17) has chaos function solutions at $\alpha = 2.0$. Here, it is important to note that the time step $\Delta t, 0 < \Delta t \ll 1.0$ is not included in the 2-D map (16) and (17), and the solutions $x_n(t_i)$ and $y_n(t_i)$ are restricted by the condition (15) for the generation of discrete limit cycles [34].

Bifurcation diagrams for the 2-D map (16) and (17) are firstly illustrated in Fig. 1a and b, where we carry out 200 iterations of the 2-D map at each value of α, $\Delta\alpha = 0.01$ with an initial point $(x_0, y_0) = (0.2, 0.1)$ for Fig. 1a and $(x_0, y_0) = (0.2, -0.1)$ for Fig. 1b, respectively. Here, the bifurcation diagram depends on the initial point (x_0, y_0) for iterations at each value of α. Then, bifurcation diagrams of Fig. 1a and b are presented in Fig. 1c, and it is found that the diagrams y_{n1} and y_{n2} are symmetric with respect to the α-axis on the $\alpha - x_n, y_{n1}, y_{n2}$ plane. Thus, the bifurcation diagrams

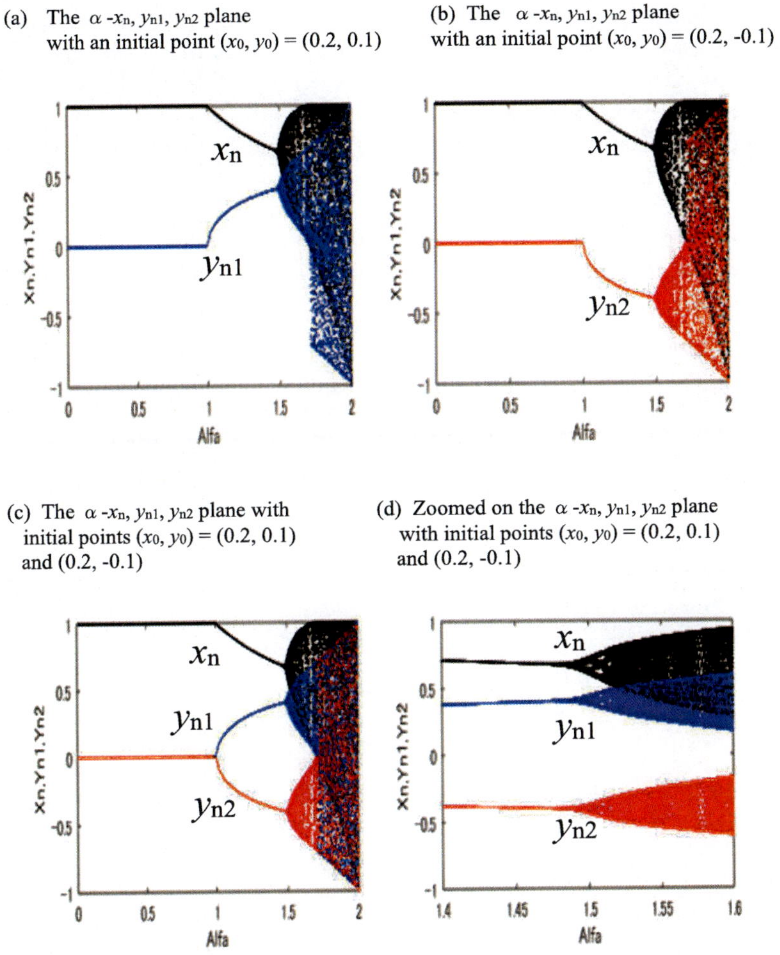

Fig. 1 Bifurcation diagrams: **a** An initial point $(x_o, y_O) = (0.2, 0.1)$, **b** $(x_o, y_O) = (0.2, -0.1)$ and $\Delta\alpha = 0.01$, **c** Initial points $(x_O, y_O) = (0.2, 0.1)$ and $(0.2, -0.1)$, and **d** A zooming diagram of **c** and $\Delta\alpha = 0.005$, at $a_1 = 1.0$, $a_2 = 1.0$ and $b_1 = 0.0$ for the 2-D map (16) and (17)

are enlarged for the interval $1.4 \leq \alpha \leq 1.6$ with $\Delta\alpha = 0.005$, and are illustrated in Fig. 1d to show the pitchfork bifurcation [37]. In addition, the 2-D map (16) and (17) has fixed points, which are defined by $x_n^* = F(x_n^*, y_n^*)$ and $y_n^* = G(x_n^*, y_n^*; \alpha)$, and are given as

$$(x_n^*, y_n^*) = \begin{cases} (a_1 + b_1, 0), 0 < \alpha \leq 2.0, \\ \left(\dfrac{a_1}{\alpha} + b_1, \pm a_2\sqrt{\dfrac{\alpha - 1}{2\alpha}}\right), 1.0 < \alpha \leq 2.0, \end{cases} \tag{18}$$

where the 2-D map is found to have pitchfork bifurcations with three fixed points in the interval $1.0 \leq \alpha \leq 2.0$. Thus, the numerical result at $a_1 = 1.0, a_2 = 1.0$ and $b_1 = 0.0$ of the 2-D map (16) and (17) is presented in Fig. 1a–d. The MATLAB program for Fig. 1c is shown in Appendix 1.

On the other hand, the 2-D chaotic map (16) and (17) corresponding to the 2-D Turing map (13) and (14) is derived from the chaos function solutions (7) and (8), and as the numerical result, the chaotic solutions $x_n(t_i)$ and $y_n(t_i)$, orbit solutions on the $x_n - y_n$ plane and a ring of sequential points are illustrated in Fig. 2a–d, on the basis of the solutions (7) and (8), at $a_1 = 1.0, a_2 = 1.0$ and $b_1 = 0.0$ with an

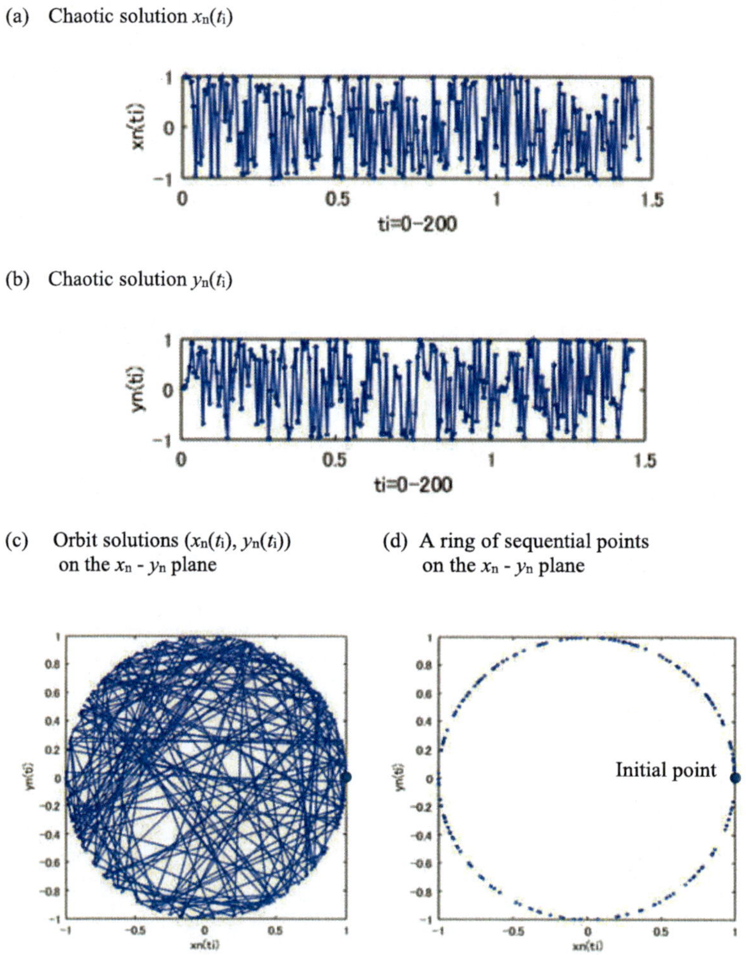

Fig. 2 Chaotic time series: **a** Chaotic solution $x_n(t_i)$ (7), **b** Chaotic solution $y_n(t_i)$ (8), **c** Orbit solutions $(x_n(t_i), y_n(t_i))$ on the x_n–y_n plane, **d** A ring of sequential points on the x_n–y_n plane, at $a_1 = 1.0, a_2 = 1.0$ and $b_1 = 0.0$ with an initial point $(x_0, y_0) = (1.0, 0.0)$

initial point $(x_0, y_0) = (1.0, 0.0)$. The chaotic time series shown in Fig. 2a–d are calculated without the accumulation of round-off error caused by numerical iterations of nonlinear equations [30]. Then, the numerical result of the 2-D map (16) and (17) is presented in Fig. 3a–d: (a) Symmetric orbit solutions on the $x_n - y_n$ plane, (b) A ring of sequential points at $\alpha = 2.0$ with initial points $(x_0, y_0) = (1.0, \pm 0.000001)$, (c) Symmetric orbit solutions and (d) Symmetric limit cycles in pairs at $\alpha = 1.69$ with inside initial points $(x_0, y_0) = (0.5, \pm 0.5)$, as a numerical example of 'dappled' pattern shown for the Turing model in [24].

Moreover, we find symmetric limit cycles in pairs with inside initial points $(x_0, y_0) = (0.5, \pm 0.5)$ for the 2-D map (16) and (17) as shown in Fig. 4a–d. Then,

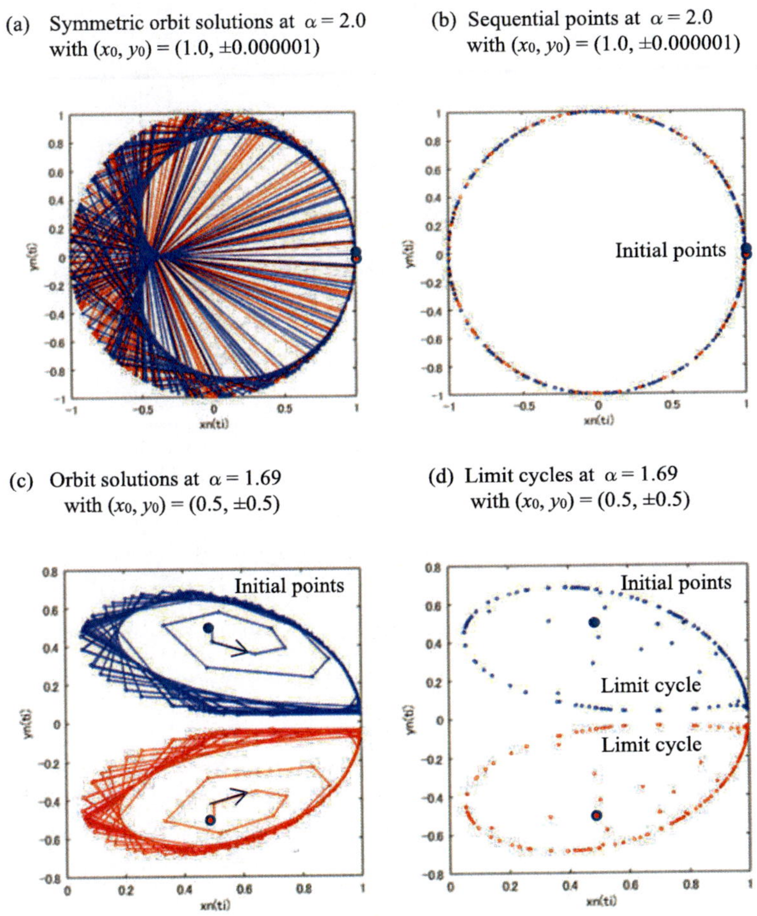

Fig. 3 Symmetric orbits and limit cycles of the 2-D map (16) and (17): **a** Symmetric orbit solutions and **b** Sequential points at $\alpha = 2.0$ with $(x_0, y_0) = (1.0, \pm 0.000001)$, **c** Symmetric orbit solutions and **d** Symmetric limit cycles at $\alpha = 1.69$ with $(x_0, y_0) = (0.5, \pm 0.5)$, at $a_1 = 1.0$, $a_2 = 1.0$ and $b_1 = 0.0$

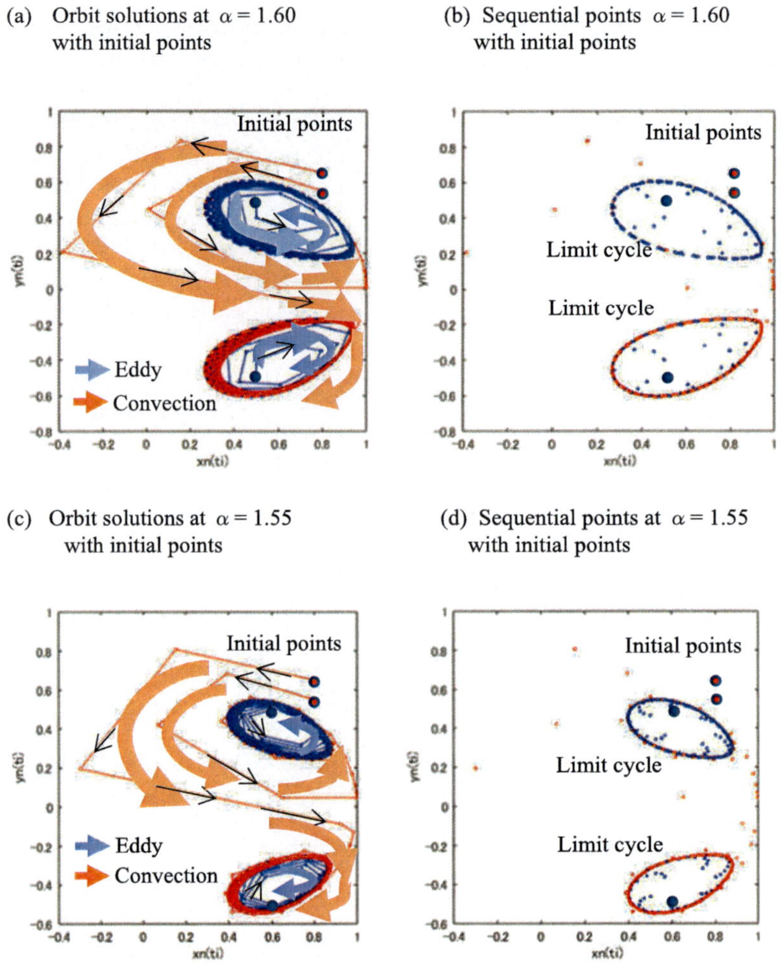

Fig. 4 Limit cycles in pairs of the 2-D map (16) and (17): **a** Orbit solutions with left-handed and right-handed eddies and convections, **b** Sequential points at α = 1.60, **c** Orbit solutions with eddies and convections, **d** Sequential points at α = 1.55, with inside (blue) and outside (red) initial points

it is found that inside initial points (blue) converge to the limit cycles and form left-handed and right-handed eddies of cells, and one of outside initial points (red) converges to the opposite limit cycle, that is, the red initial points generate convections, as shown for $\alpha = 1.60$ and $\alpha = 1.55$ in Fig. 4a and c, respectively. The MATLAB program for Fig. 4a and b is presented in Appendix 2.

Here, it is interesting to emphasize that the 2-D logistic maps derived from chaos function solutions $x_n(t_i) = \sin^2(2^n t_i)$ and $y_n(t_i) = \cos(2^n t_i)$ have discrete limit cycles, corresponding to the FHN model, the F-KPP equation, the BZ reaction and the reaction–diffusion systems, as presented in [30–33], and the 2-D chaotic map (16) and (17) derived as a 2-D Turing map (13) and (14) has chaos function solutions

$x_n(t_i) = \cos(2^n t_i)$ and $y_n(t_i) = \sin(2^n t_i)$, which corresponds to the Lorenz system for atmospheric convection [28], the reaction–diffusion systems for fluid convections [16, 17] and the equation of motion derived from the Hénon-Heiles Hamiltonian for the third integral with chaotic properties [38].

4 Conclusions

The 3-D solvable chaos map corresponding to the Turing model was firstly derived in Sect. 2, and it is explained that the Turing model has nonlinear dynamics, such as spherical symmetry, ring of cells and numerical patterns on the *x*–*y* plane. In Sect. 3, the 2-D chaotic map for the 2-D Turing model is proposed in order to show the nonlinear dynamics, and the 2-D map (16) and (17) is shown to have symmetric bifurcation diagrams, the ring of cells and limit cycles in pairs with different patterns. From the numerical result, the limit cycles are presented to generate discrete eddies and convections, as chaotic dynamics of non-equilibrium open systems. Therefore, nonlinear dynamics of the Turing model may correspond essentially to fluid dynamics with chaotic properties of the Lorenz system for atmospheric convection and the reaction–diffusion systems for convection cells.

Appendix 1

```
% MATLAB program for Fig. 1c by S. Kawamoto.
   % initial conditions.
   ALFA = zeros(1, 400);
   X1 = zeros(1, 200);
   Y1 = zeros(1, 200);
   XX1 = zeros(1, 200);
   YY1 = zeros(1, 200);
   X2 = zeros(1, 200);
   Y2 = zeros(1, 200);
   XX2 = zeros(1, 200);
   YY2 = zeros(1, 200);
   X10 = 0.2;
   Y10 = 0.1;
   X20 = 0.2;
   Y20 = -0.1;
   A1 = 1.0;
   A2 = 1.0;
   B1 = 0.0;
   % system parameter ALFA and bifurcation diagrams.
   figure('Position', [100 100 350 200]).
```

```
for I=1:400
   ALFA(I)=0.005*I;
   for J=1
      X1(I, J)=(A1+B1)-2*A1*((Y10)^2)/A2^2;
      Y1(I, J)=ALFA(I)*(X10-B1)*Y10/A1;
      X2(I, J)=(A1+B1)-2*A1*((Y20)^2)/A2^2;
      Y2(I, J)=ALFA(I)*(X20-B1)*Y20/A1;
   end
   for J=2:200
       X1(I, J)=(A1+B1)-2*A1*((Y1(I, J-1))^2)/A2^2;
       Y1(I, J)=ALFA(I)*(X1(I, J-1)-B1)*Y1(I, J-1)/A1;
       X2(I, J)=(A1+B1)-2*A1*((Y2(I, J-1))^2)/A2^2;
       Y2(I, J)=ALFA(I)*(X2(I, J-1)-B1)*Y2(I, J-1)/A1;
   end
   for J=150:200
       XX1(J)=X1(I, J);
       YY1(J)=Y1(I, J);
       XX2(J)=X2(I, J);
       YY2(J)=Y2(I, J);

       plot(ALFA(I), XX1(J), ‘k.’,’MarkerFaceColor’,’k’,’MakerSize’,4); hold on
       plot(ALFA(I), YY1(J), ‘b.’,’MarkerFaceColor’,’b’,’MakerSize’,4); hold on
       plot(ALFA(I), YY2(J), ‘r.’,’MarkerFaceColor’,’r’,’MakerSize’,4); hold on
   end
end
xlabel(‘Alfa’); ylabel(‘Xn, Yn1, Yn2’)
```

Appendix 2

```
% MATLAB program for Fig. 4a–b by S. Kawamoto.
   % initial conditions.
   T = zeros(1, 200);
   TT = zeros(1, 200);
   X1 = zeros(200, 200);
   Y1 = zeros(200, 200);
   XX1 = zeros(1, 200);
   YY1 = zeros(1, 200);
   X2 = zeros(200, 200);
   Y2 = zeros(200, 200);
   XX2 = zeros(1, 200);
   YY2 = zeros(1, 200);
   X3 = zeros(200, 200);
   Y3 = zeros(200, 200);
   XX3 = zeros(1, 200);
   YY3 = zeros(1, 200);
   X4 = zeros(200, 200);
   Y4 = zeros(200, 200);
   XX4 = zeros(1, 200);
   YY4 = zeros(1, 200);
   L0 = 1;
   PR = 431;
   T0 = 0.0;
   X01 = 0.5;
   Y01 = 0.5;
   X02 = 0.8;
   Y02 = 0.65;
   X03 = 0.8;
   Y03 = 0.55;
   X04 = 0.5;
   Y04 = -0.5;
   ALFA = 1.6;
   A1 = 1.0;
   A2 = 1.0;
   B1 = 0.0;
   % limit cycles in pairs with initial points.
```

```
for I=1:200, T(I)=T0+I*L0*pi/PR; end
for I=1:200
    for N=1
        X1(I, N)=(A1+B1)-2*A1*(Y01^2)/(A2^2);
        Y1(I, N)=ALFA*(X01-B1)*Y01/A1;
        X2(I, N)=(A1+B1)-2*A1*(Y02^2)/(A2^2);
        Y2(I, N)=ALFA*(X02-B1)*Y02/A1;
        X3(I, N)=(A1+B1)-2*A1*(Y03^2)/(A2^2);
        Y3(I, N)=ALFA*(X03-B1)*Y03/A1;
        X4(I, N)=(A1+B1)-2*A1*(Y04^2)/(A2^2);
        Y4(I, N)=ALFA*(X04-B1)*Y04/A1;
    end
    for N=2:I
        X1(I, N)=(A1+B1)-2*A1*(Y1(I, N-1)^2)/(A2^2);
        Y1(I, N)=ALFA*(X1(I, N-1)-B1)*Y1(I, N-1)/A1;
        X2(I, N)=(A1+B1)-2*A1*(Y2(I, N-1)^2)/(A2^2);
        Y2(I, N)=ALFA*(X2(I, N-1)-B1)*Y2(I, N-1)/A1;
        X3(I, N)=(A1+B1)-2*A1*(Y3(I, N-1)^2)/(A2^2);
        Y3(I, N)=ALFA*(X3(I, N-1)-B1)*Y3(I, N-1)/A1;
        X4(I, N)=(A1+B1)-2*A1*(Y4(I, N-1)^2)/(A2^2);
        Y4(I, N)=ALFA*(X4(I, N-1)-B1)*Y4(I, N-1)/A1;
    end
end
for I=1
     TT(I)=T0;
end
for I=2:200
     TT(I)=T(I-1);
end
for I=1
    XX1(I)=X01;
    YY1(I)=Y01;
    XX2(I)=X02;
    YY2(I)=Y02;
    XX3(I)=X03;
    YY3(I)=Y03;
    XX4(I)=X04;
    YY4(I)=Y04;
end
for I=2:200
    XX1(I)=X1(I-1, I-1);
    YY1(I)=Y1(I-1, I-1);
    XX2(I)=X2(I-1, I-1);
```

```
    YY2(I)=Y2(I-1, I-1);
    XX3(I)=X3(I-1, I-1);
    YY3(I)=Y3(I-1, I-1);
    XX4(I)=X4(I-1, I-1);
    YY4(I)=Y4(I-1, I-1);
end
% figures (a)-(b)
figure('Position', [100 100 350 350])
plot(XX4, YY4, '-b.','MarkerFaceColor','b','MarkerSize', 7); hold on
plot(XX3, YY3, '-r.','MarkerFaceColor','r','MarkerSize', 7); hold on
plot(XX2, YY2, '-r.','MarkerFaceColor','r','MarkerSize', 7); hold on
plot(XX1, YY1, '-b.','MarkerFaceColor','b','MarkerSize', 7); hold off
xlabel('xn(ti)'); ylabel('yn(ti)')

figure('Position', [100 100 350 350])
plot(XX4, YY4, 'b.','MarkerFaceColor','b','MarkerSize', 7); hold on
plot(XX3, YY3, 'r.','MarkerFaceColor','r','MarkerSize', 7); hold on
plot(XX2, YY2, 'r.','MarkerFaceColor','r','MarkerSize', 7); hold on
plot(XX1, YY1, 'b.','MarkerFaceColor','b','MarkerSize', 7); hold off
xlabel('xn(ti)'); ylabel('yn(ti)')
```

References

1. M.J. Ablowitz, H. Segur, *Solitons and the Inverse Scattering Transform* (SIAM, Philadelphia, 1981)
2. F.C. Moon, *Chaotic and Fractal Dynamics* (Wiley, New York, 1992)
3. H. Peitgen, H. Jurgens, D. Saupe, *Chaos and Fractals—New Frontiers of Science* (Springer, New York, 1992)
4. A. Scott, *Nonlinear Science* (Routledge, London, 2005)
5. R.J. Bird, *Chaos and Life: Complexity and Order in Evolution and Thought* (Columbia University Press, New York, 2003)
6. E.M. Izhikevich, *Dynamical Systems in Neuroscience: The Geometry of Excitability and Bursting* (The MIT Press, London, 2007)
7. G.B. Ermentrout, D.H. Terman, *Mathematical Foundations of Neuroscience* (Springer, London, 2010)
8. Ed. by C. H. Skiadas and C. Skiadas. *Handbook of Application of Chaos Theory* (Chapman and Hall/CRC Press, 2016)
9. M.M. Waldrop, *Complexity: The Emerging Science at the Edge of Order and Chaos* (Simon and Schuster Paperbacks, New York, 1993)
10. G. Nicolis, C. Rouvas-Nicolis, Complex systems. Scholarpedia **2**(11), 1473 (2007)
11. K. Aihara, T. Takabe, M. Toyoda, Chaotic neural networks. Phys. Lett. **A144**, 333–340 (1990)
12. D.J. Watts, S.H. Strogatz, Collective dynamics of 'small-world' networks. Nature **393**, 440–442 (1998)
13. K. Kaneko, I. Tsuda, *Complex Systems: Chaos and Beyond: A Constructive Approach with Applications in Life Sciences* (Springer, Berlin, 2000)
14. S.H. Strogatz, Exploring complex networks. Nature **410**, 268–276 (2001)
15. R.A. Fisher, The wave of advance of advantageous genes. Ann. Eugenics **7**, 355–369 (1937)
16. A.C. Newell, J.A. Whitehead, Finite bandwidth, finite amplitude convection. J. Fluid Mech. **38**, 279–303 (1969)

17. L.A. Segel, Distant side-walls cause slow amplitude modulation of cellular convection. J. Fluid Mech. **38**, 203–224 (1969)
18. R.J. Field, R.M. Noyes, Explanation of spatial band propagation in the Belousov reaction. Nature **237**, 390–392 (1972)
19. G. Nicolis, I. Prigogine, *Self-Organization in Non-Equilibrium Systems* (John Wiley & Sons, New York, 1977)
20. J.-C. Roux, R.H. Simoyi, H.L. Swinney, Observation of a strange attractor. Physica **8D**, 257–266 (1983)
21. F. Argoul, A. Arneodo, P. Richetti, J.C. Roux, H.L. Swinney, Chemical chaos: from hints to confirmation. Acc. Chem. Res. **20**, 436–442 (1987)
22. J.J. Tyson, P.C. Fife, Target patterns in a realistic model of the Belousov–Zhabotinskii reaction. J. Chem. Phys. **73**, 2224–2237 (1980)
23. J.A. Vastano, T. Russo, H.L. Swinney, Bifurcation to spatially induced chaos in a reaction-diffusion system. Physica D **46**, 23–42 (1990)
24. A.M. Turing, The chemical basis of morphogenesis. Philos. Trans. Roy. Soc. London **B327**, 37–72 (1952)
25. S. Kondo, R. Asai, A reaction-diffusion wave on the skin of the marine angelfish Pomacanthus. Nature **376**, 765–768 (1995)
26. S. Kondo, T. Miura, Reaction-diffusion model as a framework for understanding biological pattern formation. Science **329**, 1616–1620 (2010)
27. S. Kondo, An updated kernel-based Turing model for studying the mechanisms of biological pattern formation. J. Theor. Biol. **414**, 120–127 (2017)
28. S. Kawamoto, 2-D and 3-D solvable chaos maps. Chaotic Model. Simulation (CMSIM) **1**, 107–118 (2017)
29. S. Kawamoto, Chaotic time series by time-discretization of periodic functions and its application to engineering. Chaotic Model. Simulation (CMSIM) **2**, 193–204 (2017)
30. S. Kawamoto, The FitzHugh-Nagumo model and 2-D solvable chaos maps. Chaotic Model. Simulation (CMSIM) **3**, 269–283 (2018)
31. S. Kawamoto, in *The FitzHugh-Nagumo Model and Spatiotemporal Fractal Sets Based On Time-Dependent Chaos Functions. 11th Chaotic Modeling and Simulation International Conference, Springer Proceedings in Complexity* ed. by C. H. Skiadas, I. Lubashevsky (Springer Nature, Switzerland AG, 2019), pp. 159–173
32. S. Kawamoto, in *Intermittency of Chaos Functions and the Belousov–Zhabotinsky Reaction. 12th Chaotic Modeling and Simulation International Conference, Springer Proceedings in Complexity*, ed. by C. H. Skiadas, Y. Dimotikalis (Springer Nature Switzerland AG, 2020), pp. 123–134
33. S. Kawamoto, in *Reaction-Diffusion Systems and Propagation of Limit Cycles with Chaotic Dynamics. 12th Chaotic Modeling and Simulation International Conference*, ed. by C. H. Skiadas, Y. Dimotikalis (Springer Nature, Switzerland AG, 2020), pp. 135–149
34. S. Kawamoto, Limit cycle analysis for 2-D time-dependent logistic maps. (Accepted)
35. S. Kawamoto, in *Interaction of limit cycles for the FitzHugh-Nagumo model. 13th Chaotic Modeling and Simulation International Conference*, ed. by C. H. Skiadas, Y. Dimotikalis (Springer Nature, Switzerland AG, 2021), pp. 345–359
36. S. Kawamoto, in *Pattern formation of limit cycles for 2-D generalized logistic maps. 13th Chaotic Modeling and Simulation International Conference*, ed. by C. H. Skiadas, Y. Dimotikalis (Springer Nature, Switzerland AG, 2021), pp. 333–344
37. S.H. Strogatz, *Nonlinear Dynamics and Chaos*, 2nd edn. (Westview Press, Boulder, 2015)
38. M. Hénon, C. Heiles, The applicability of the third integral of motion: some numerical experiments. Astron. J. **69**(1), 73–79 (1964)

Cloud Electrification as a Source of Ignition for Hydrogen Lift-Gas Airships Disasters

V. J. Law and D. P. Dowling

Abstract The first half of the twentieth century became the Golden Age of the dirigible airship. After the Hindenburg disaster, (1937), the dirigible use fell into rapid decline leaving the non-rigid airships to serve in maritime roles until the mid 1960s. Throughout dirigible and non-rigid use, violent storm systems have been associated with in-flight airship disasters. In particular, the popular press at time instilled into the public perception that lightning strikes were the guilty ignition source of the disasters. Over the past 25 years, Saint Elmo's Fire has come forward as an alternative ignition source for in-flight airship disasters. Understanding the role of low energy discharges events is important for the emerging hydrogen economy that is intended to reduce the world's energy consumption and greenhouse emissions. This paper reviews $2H_2 + O_2 = 2H_2O$ combustion chemistry, the role of heterogeneous graupel chemistry within electrification of Cumulonimbus, and how the empirical mathematical construct of Peek's Law which attempts to identify the visual inception voltage in terms of the minimum electrical field stress required for the generation of Saint Elmo's Fire. Using this electrochemical knowledge, in-flight airship disasters associated with nearby cloud electrification, or violent storms systems, are correlated and reviewed. This study is supported by firsthand accounts (from survivors), including radio messages prior to an airship disaster, ground eyewitness accounts, along with the structural design of the airship. The hydrogen lift-gas airships reviewed here are four dirigibles (LZ-4 (*L-10*), SL-9, Dixmude and Hindenburg) and one non-rigid airship (NS.11). As a comparative control, this paper reviews the worst airship disaster, that of the helium lift-gas flying aircraft carrier, USS Akron (ZRS-4), which led to the loss of 73 lives. In addition to that of the sister airship, USS Macon (ZRS-5) disaster where two lives were also lost.

Keywords Saint Elmo's fire · Plasmoid · Dirigible · Non-rigid airship · Lightning

V. J. Law (✉) · D. P. Dowling
School of Mechanical and Materials Engineering, University College Dublin, Belfield, Dublin 4, Ireland
e-mail: viclaw66@gmail.com

C. H. Skiadas and Y. Dimotikalis (eds.), *14th Chaotic Modeling and Simulation International Conference*, Springer Proceedings in Complexity,
https://doi.org/10.1007/978-3-030-96964-6_19

1 Introduction

In 2020, the authors of this paper presented two talks at the 13th virtual CHAOS2020 conference Florence-Italy on how Peek's formula [1] may be used to estimate the visual inception voltage stress point on natural and artificial structures [2] and microwave oven plasma processing of nanomaterials [3]. In the former paper, the dirigible airship was one of the artificial structures examined. In the follow-up question and answer section, the main question was 'if not lightning, how does Saint Elmo Fire (SEF, also sometimes-called brush discharge, corona, or partial discharge) become a lethal threat to a hydrogen (H_2) lift-gas airship. The answer to this question involves a complexity of factors, including a detailed knowledge of the airship construction, the prevailing metrological conditions at the time of the disaster and if the airship's captain was 'Extraordinary Good', or 'Lucky'.

This paper consolidates published information regarding five H_2 lift-gas airships (four dirigibles and one non-rigid airship) disasters that are associated with cloud electrification surrounding or near-by the airships. In these disasters, the airships are in-flight (not tied to a mooring-mast or their ground handling ropes secured to the ground, i.e. the airships obtain a quasi-equilibration to the local weather electrical field conditions. In this context amongst the factors and accounts considered are the prevailing metrological electrical and chemical environment, radio messages prior to the airship disaster, firsthand accounts (from survivors) and ground eyewitness accounts, along with the structural design of the airship. References from the '*first rough draft of history*' (newspapers and movie-reels) are used along with board of inquiries, contemporary and current aeronautical journals, metrological, physical chemistry and electrical engineering journals. The chronology of these articles reveals the complex processes (physical, commercial and political) were not interlinked, but evolved overtime. To clarify these complex issues, the paper is organized as follows: Sect. 2 gives a historical view of SEF. Section 3 describes the process of airship disasters selection and classification of the selected dirigible and non-rigid airships (Sect. 3.1). Section 4 looks at the airship construction, gasbag (Sect. 4.1), non-rigid envelopes (Sect. 4.2) and dirigible airframes (Sect. 4.3). Section 5 provides an anatomy of H_2 lift-gas fires. Section 6 revisits Peek's formula for a single metal electrode. Section 7 lists the airship under consideration in this paper. Finally, Sect. 8 provides summary of this review.

2 Saint Elmo's Fire (SEF)

Since classical Greek and Roman times Ermus of Fomia has been the patron Saint of Mediterranean sailors, to whom he appeared as SEF on the masts and spars of sailing ships as an electrical storm began to dissipate in electrical intensity. These good omens being manifest as characteristic cracking or hissing sound with a blue/violet flame-like glow. Between the years 1610–1611, art emulates real life when William

Table 1 WW1 dirigible non-lethal airship-SEF encounters

Airship	Date	Location	Weather observations	Reference [11]
LZ-41 (*L-11*)	Aug 10, 1915	Dogger bank	Thunderstorm, cloud-to-cloud and cloud-to-sea lighting. 1,000–4,600 feet	Page 121–122
LZ 41 (*L-11*)	Mar 5, 1916	Spurn head	Squall, cloud, snow and ice Wind speeds 55 m.p.h Temperature + 3F	Page 148
LZ-53 (*L-17*)	May 3, 1916	North Sea	Thunderstorm, rain and hail	Page 160
LZ-91 (*L-42*)	May 23–24, 1917	Over London	Squall, hail, solid cloud	Page 244
LZ-104 (*L-59*)	Nov 21, 1917	Eastern Crete	Thunderstorm, black clouds and flashes of lightning close-by	Page 310

Strachey's account of the ill-fated '*Sea adventure*' voyage from the new world in 1610, is retold by William Shakespeare within the play '*The Tempest*' [4]. In this play, SEF takes on a more sinister role as the spirit '*Arial*' who manipulates the mariners off the ship. By 1886, this atmospheric phenomenon started to be systematically complied and reported as SEF, Ball lightning (BL) [5–8] and fireball (FB) [9, 10]. The latter two types proving to be more life threatening when compared to SEF. In addition, it has become clear that BL has the ability to interfere with radio broadcasts and to transfer part of its information through a glass windowpane with and without damage to the glass [5–7]. During world war one (WW1; 1914–1918) reports of non-lethal SEF encounters, both inside and outside of the airship airframe accumulates as airships flew though bad weather on their bombing raids and reconnaissance, due to the necessity of war, see for example Douglas W. Robinson's book '*The Zeppelin in Combat*' [11]. Table 1 provides five examples of non-lethal SEF encounters, in each case the prevailing metrological conditions being a squall or thunderstorm containing lighting with a mixtures of rain, hail or snow.

Since the early 1900s, metrology has shown that cloud-to-cloud and intra-cloud electrification has its origins in the Earth's troposphere (0–12 km) [5–10]. Fair weather dc electric fields are modulated by ac and RF fields due to thunder and lightning activity. Moreover, the appearance of SEF around conducting tips and protrusions being due to the geometric field enhancement where equal-potential lines become bunched [1]. By 1928, the term for this electrical phenomenon began to be classified as 'plasma' (Greek: meaning mouldable substance), which considers an assembly of gas molecules that has some of its atoms or molecules temporally ionized or excited [12]. In 1952, Bostick [13] added the subclass 'plasmoid' that defines a separate plasma-magnetic entity that may be ejected from the parent plasma.

3 Airship Heuristic Selection Criteria and Classification

This section lists the H_2 lift-gas airship heuristic elimination criteria used to identify the airships destroyed by violent weather conditions. Figure 1, provides a chorological (1895–1960) time-stamp of forty airship accidents against the number of deaths per accident. In this period, two airship design generations appeared, the dirigible (1905–1937), and blimp (Russian (ca, 1920–1947) and USA (1930–1960)). Within this period, five-hundred and five lives were lost to airship disasters. In addition, during this period there were twenty-two nonlethal airships accidents (not shown) where airships were either lost or written-off. Note airships lost to enemy action are excluded from these tallies. The data in Fig. 1, is given as a function disaster type (fire/explosion, midair and ground collision, pilot error, structural failure, lost, H_2 and helium (He) lift-gas airships destroyed in storm conditions. Where multiple disasters occurred in one year (i.e. 1902, 1912, 1913, 1915...), the total loss of life is denoted with a + sign.

3.1 Heuristic Elimination Criteria

Using the forty airships disasters listed in Fig. 1 as a starting point, the heuristic elimination criteria (removal of airships from the list) is given as follows.

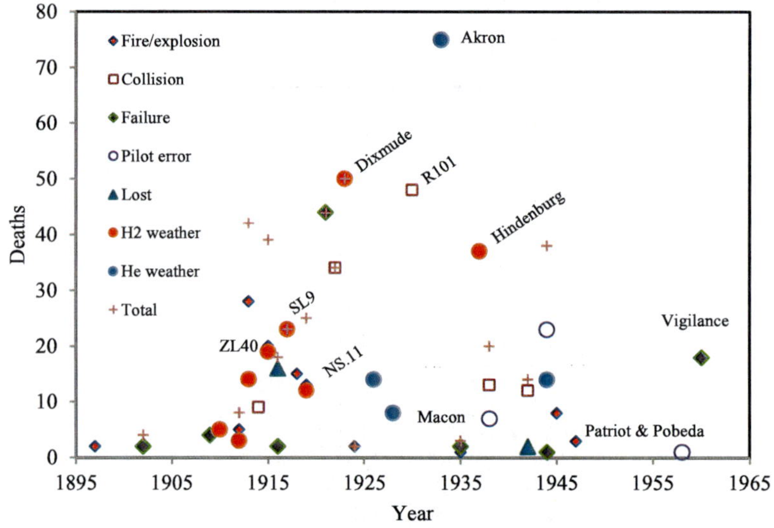

Fig. 1 Lethal airships disasters between 1897 and 1960 as a function of related potential cause. The + symbol denotes the total number of in each year. For reference purposes only, ten airships are named here

1. Remove all airship decommissioned (for example SL-8, 1918 [11], page 281)
2. Remove all airships set alight during inflation of gasbags within their hangar (for example, L-6 and L-9, Fuhlabüttel air field, 1916 [11], page 199)
3. Elimination of airships involved in a collision (for example, high voltage power lines (USS Roma TR-4, 1922 [14]), crash-landed on ice near the North Pole (Italia, 1928 [16] and grounded on a hillside (R101, 1930 [15], and SSSR-V6 1938 [17]).
4. Remove all airships destroyed whist flying in good weather (for example, LZ-104 (*L-59*), 1918 [11], page 315)
5. Remove due to pilot error (for example, SSSR-V10 1938 [17]
6. Remove all He lift-gas airships excluding the USS Akron and the USS Macon.

3.2 The Airship Classification

The airships disasters examined in this work are the LZ-40 (*L-10*) and SL-9 (*Type E*), the NS.11, Dixmude (formerly the LZ-72) and the Hindenburg (LZ-129). For comparison purpose, the He lift-gas airships, (USS Akron (ZRS-4) and USS Macon (ZRS-5)) are used. Note in the American airship number system, Z refers to Zeppelin mode of construction, R refers to ridged airframe and, S refers to flying aircraft carrier. As all of the airships have aero-engines as a means of propulsion, the airships are classified as either non-rigid or dirigible.

3.3 Zeppelin Production Number and Tactical Number Classification

The Zeppelin company gave their airship a production number (LZ-xxx) whereas the German military gave their airships a tactical number (L-xxx). This dual number system has led to some confusion. In this work, the Zeppelin production number is used. The Zeppelin tactical number is given in *italics* and is only used in Sects. 1, 2 and 3 to provide a link between the airships, after which only the production number is used.

3.4 Non-rigid (Pressure) Airship

The non-rigid airship uses a H_2, or He, lift-gas envelope that is pressurized with air-filled ballonets (air-filled compartments) to control lift and pitch, plus envelope shape and structural integrity. To achieve the weigh off (initial static-lift) the envelope is filled with H_2 until the airship's volume equalizes with ground-level air volume. In equilibrium flight, the effect of slow varying updrafts, temperature changes and loss of fuel weight requires the airship to be maintained by blowing air into the ballonets or releasing air from the ballonets. Whereas dynamic lift is achieved by altering the elevator position with aero-engine power). When a rapid and violent updraft occurs, automatic spring-loaded lift-gas valves open to prevent the airship pressure ceiling being exceeded, resulting in a corresponding rapid loss of lift.

3.5 *Dirigible Airship*

Unlike non-rigid airships, the dirigibles LZ-40 and SL-9, the NS.11, Dixmude, and the Hindenburg have multiple H_2 lift gasbags located within a metal or wood airframe. In the case of the USS Akron and USS Macon, the lift gas is He. The Dirigible design ensues the airframe provides structural protection to the gasbags and greater shape protection from aerodynamic forces and moments. In the event of one of the gasbags is compromised, buoyancy is maintained by discharging ballast at the location of the compromised gasbag. See for example the USS Shenandoah (ZR-1) which was torn from its mooring-mast in 1924, and crashed in 1925 [18] and the R-33 30-h unscheduled flight in 1925 [19]. Again when a rapid and violent updraft occurs, the automatic spring-loaded lift-gas valve opens, resulting in a rapid corresponding loss of lift.

4 Airship Construction

This section reviews WW1, lighter-than-air flight. Section 4.1 looks the development of the gasbag (sometimes called cell), the non-rigid envelope (Sect. 4.2), and the dirigible airframe structure (Sect. 4.3). The airships in question are the LZ-40 and the SL-9, the Dixmude, the USS Akron and USS Macon and the Hindenburg, plus the non-rigid airship NS.11. Ladislas D'Orcy's 'International register and compendium of airships (built between 1873 and 1917)' [20], and Robinson's 'The Zeppelin in combat' [11] provides information on the techniques used in the manufacture of LZ-40, SL-9 and LZ-72 (latter to be named the Dixmude). In addition, written articles in the 'Journal Dirigible' are extensively used.

4.1 *Gasbags*

Since 1782 in Paris-France, 18-inch diameter balloons made from goldbeater skins filled with H_2 were flown for recreational purposes [21]. The goldbeater skins originally obtained from cow intestines (*cecum*, or, *caecum*). This very lightweight material was found to exhibit a high inherent strength and is almost impervious to H_2 gas. When cleaned and stretched having an approximate area of 20 cm in length and 25 cm in width. In 1883 move from making toy balloons to manufacturing for the 10,000 cubic foot balloon 'Heron' was performed by the Weinling family under direction of Captain Templar at Chatham, England [22]. The Weinling family tried very hard to keep their gasbag manufacturing process at secret, but there is a suggestion of industrial espionage between Templar and the Italian government [23]. McKechnie's (Vickers. Ltd.) 1919 patent details the manufacture of a lightweight and gastight 4-layer gasbag for airships and balloons [24]. The layers comprise; a single ply of linen coated with unvulcanized rubber followed by Goldbeater skin and vanishes. The patent, also states *'This represents about one ton increase of lift for a million cubic feet capacity'* The quantity of Goldbeater skins for a standard WW1 German Navy dirigible airship is in the order of 20,000. It is no wonder that the Zeppelin Company had to recycle old gasbag material with greater outward (H_2) and inward (air) permeability properties that may lead to the loss of a dirigible and its crew [25]. After WW1, the number of skins used for dirigibles grew considerably. For example, the USS Shenandoah used over 750,000 goldbeater skins [18].

The early Zeppelins had a multitude of gasbags within a metal airframe covered with a waterproof, non-gas-tight skin. This construction allowed leakage of lift-gas to mix with the natural airflow up and round the gasbags and eventually permeate through the outer skin, see Fig. 2a. However, within certain H_2-air mixing ratios the gas mixture is flammable and liable to explode given a source of ignition. To counter act this problem, Schütte-Lanz airships improved on the design by adding forced ventilation which expelled the gas mixture via ducting from the bottom skin to the upper outer skin. In addition, a gas-tight coating to the bottom skin is added

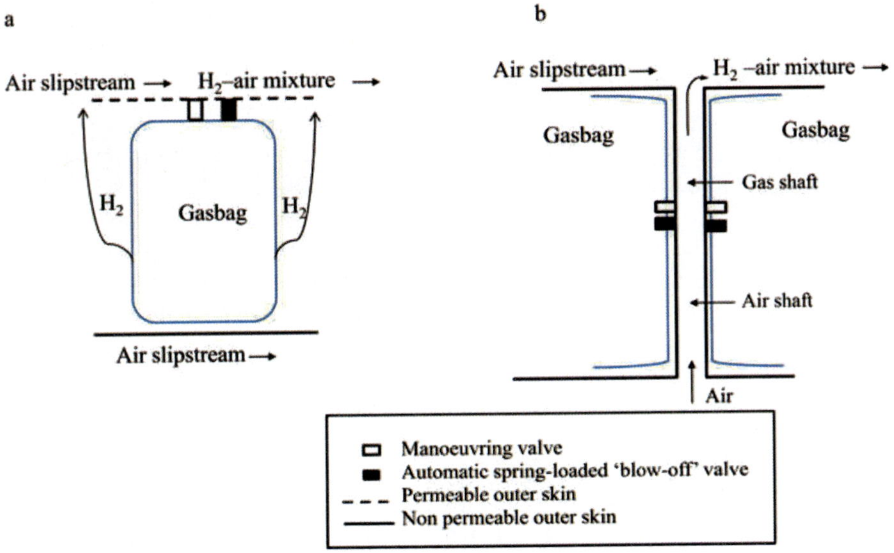

Fig. 2 **a** Schematic of early zeppelin lift-gas venting arrangement: **b**, Schütte-Lanz H_2 and latter Zeppelin venting valve arrangement (schematic redrawn from Thomas [26]).

to prevent leaking lift-gas reaching the aero-engine exhausts. Both of these Schütte-Lanz designs were taken up by the Zeppelin Company during WW1, and ultimately, in a modified form for Hindenburg, see Fig. 2b.

4.2 Non-rigid Envelopes

Goldbeater skins, although having excellence gas-tightness, exhibited relatively low tensile strength and proved less than satisfactory against water. To cope with the stress encountered in non-rigid airship envelopes rubberized fabric of high tensile strength is used. Typical 2–3 layers are used where the threads of each layer is diagonal opposed. The envelope fabric, however, when subjected to an electrostatic fields may become electrified; and under certain conditions (such as when the envelope is deflated (less taut) whilst the H_2 lift-gas is being released) a fire may be ignited.

By 1917, Britain's answer to Germanys U-boat threat in the North Sea was the North Sea (NS) class H_2 filled non-rigid airship. Using a tri-lobe lift-gas envelope based on the Astra-Torres design [27], fourteen of these airships were built. Within the envelope, there were six ballonets fitted with air-blowers for buoyancy control: the control car and engine gondola being slung under the envelope. Initially designed for 24-h flight endurance, on February 9–13, 1919 the NS.11 smashed the non-rigid flight endurance specification by a record-breaking endurance flight of 400 miles in 100 h and 50 min [28]. Table 2 lists the gasbag/envelope details of the six dirigibles and the one non-rigid airship discussed here.

4.3 Dirigible Airframe Structure

The Pre-WW1 Zeppelin designs and Schütte-Lanz dirigible airframe designs are notable for their very different materials and methods of construction [11, 20]. The early Zeppelin designs on the Dave Schwarz of Zagreb patents using zinc aluminum alloy airframes. The general appearance of a Zeppelin is one of cigar shaped, with a parallel mid-section built from many transverse polygon rings of the same form. While the short (with respect to the mid-section) front and rear sections use similar reducing polygon rings apart from the aft section that has four tails fins built-in using a cruciform girder construction. The overall design allows mass production techniques to be used. By the start of WW1, aged-harden aluminum alloy (duralumin) containing copper (3.5–4%) and manganese (0.5–1%) began to be used for the airship airframes in Germany [29].

In the case of the twenty-four airships built by Schütte-Lanz, the airframe was one of the first successful geodesic latticework constructions. All, but two [SL-23 and SL-24] used wood and laminated wood all boned together with minimal metal fixings. Due to the large number of individual parts used, the construction time of the airframe was considerably greater than a comparable Zeppelin. However, the

Table 2 Airship gasbag and envelope information

Airship	Classification	Gasbag construction	Gas	Number of gasbags/ballonets	Gas capacity (m^3)
LZ-40	Dirigible	3 layers of goldbeater on cotton	H_2	12	31,900
SL-9	Dirigible	3 layers of goldbeater on cotton	H_2	12	38,750
NS.11	Non-rigid	Rubberized cotton	H_2	6 ballonets in one envelope	10,194
Dixmude	Dirigible	3 layers of goldbeater on cotton	H_2	16	68,470
USS Akron	Dirigible	Rubberized cotton, and cotton impregnated with gelatin-latex	H_e	12	194,000
USS Macon	Dirigible	Cotton impregnated with gelatin-latex	H_e	12	194,000
Hindenburg	Dirigible	2 cotton fabric layer with celluloid in between which was the impregnated with a gelatin-latex applied	H_2	16	200,000

airframe tensgrity (tension and integrity) was flawed, as the laminated wood was prone to delaminate under moist conditions encounter in maritime roles leading the German Imperial Navy to mistrust these airships. Towards the end of WW1, a Schütte-Lanz engineering manager Hermann Müller, (Swiss by birth) defected to Britain and gave his knowledge of building wood airframe airships to the Short Brothers [30]. The outcome of which was R-31 and R-32 airships, which proved to have the same delaminating problem as the Schütte-Lanz airships. Later in 1928, Barnes Wallis patented the geodesic construction method using tubular metal for the contiguous transverse space frame design in the R101 [31].

During WW1 Britain, France and USA studied the construction of shot-down German airships, in particular the duralumin airframes. With final terms of the WW1 armistice signed on June 28, 1919 Germany was mandated to handover its airships (and High Sea Fleet) as war reparations. The political and revolutionary feelings within Germany at the time resulted in the scuttling of the High Sea Fleet at Scapa Flow and after seven airships were destroyed on the ground. Among the reparation demands following this act of destruction, Germany had to make-good the lost airships and handover all their airship technology. This forced reparation process meant that the allies received the *'Height Climber'* class of Zeppelin that where designed for high-altitude (such as the Dixmude) at the expense of airframe structural integrality whist maneuvering at low altitude, a design feature that would plague the allied countries development of commercial airships for years to come.

In 1925, flying aircraft carrier proof-of-concept trails using the British R-33 airship that involved the launching and recapture of parasitic fighters. By 1929, the USA experimented (under land-based conditions) the concept of the flying aircraft carrier airship using the USS Los Angeles (ZR-3) as the mother ship. With completion of the British and American trials, the Goodyear-Zeppelin Corporation was formed for the design and construction of the first purpose built He lift-gas flying aircraft carrier: USS Akron and USS Macon. The hull design incorporated improvements in transverse frames for rigid airships as lay out by Richmond and Scott [32], which later appears to morph from the Barnes Wallis's space frame design, 1928 [31]. The airships used twelve He-lift gasbags using the Goodyear Tire and Rubber's rubberized cotton as the outer skin.

The original ship design used eight Maybach VL 11 aero-engines placed inside the hull (four each side) for driving propellers located in-line outside of the hull. In this configuration the engines disturbed air (wash) to the next inline propeller resulting sever airframe vibration and loss in available aero-engine power. To reduce vibration to the airframe, the propellers had to be operated in contra-rotation to the next in-line aero-engine. This also provided greater engine thrust. In addition, during the design stage, the Navy requested for the bottom of the lower fin to be visible from the control car. To achieve this goal the goal car was moved 2.4 m aft and all the fins were shortened and deepened. The alteration meant that the leading edge root of the fins no longer coincided with an original main transverse frame fixing; instead, the attachment was now to a weaker intermediate traverses frame. The contra-rotating propeller preference combined with the weak tail fine attachment points have been the subject of much speculation of the USS Akron's many crashes and its final demise of the USS Akron (Sect. 7.5) along her sister airship (USS Macon (Sect. 7.6).

5 Anatomy of a H_2 Lift-Gas Fire

To prevent an airship exceeding its safe pressure ceiling under rapid and violent updraft conditions, automatic spring-loaded lift-gas valves blow-off gas from the gasbags. Early Zeppelin airships (pre 1920s) the valves where located at the bottom of the gasbags to enable contaminated H_2 gas to be blown off. Manually venting of H_2 life-gas in storm conditions was prohibited in German airships from late 1915 (Sect. 7.1). During this automatic process, the released H_2 gas mixed with air within the airship's volume then diffuses through the outer airship fabric to mix with the airship's slipstream.

Unlike H_2-air mixtures, pure H_2 is difficult to ignite as many aircraft pilots firing solid metal bullets into the WWI Zeppelin and Schütte-Lanz airships found, see for example LZ-76 first and last raid on London [11]. When the metal bullets did hit the airship gasbags, they simply went through leaving small holes were H_2 would slowly escaped and become quickly diluted by the surrounding air. It was not until the autumn of 1916 when the newly developed explosive bullet (Pomperoy, containing nitro-glycerin) and the incendiary bullet (Brock, containing potassium chlorate) were fired in combination, the gasbags become blown apart when hit. As large quantities of released H_2 mixes with atmospheric air, the incendiary bullet [33] ignites the flammable oxyhydrogen gas mixture.

Pure H_2 gas burns with low radiant heat, almost without color, and becomes red-yellow depending upon the amount and variety low molecular weight carbide and carbon monoxide impurities). In early 1900_s, the German preferred method of H_2 production was to pass steam over hot iron at high temperature to produce Knallgas (bang gas). Today's hydrogen economy the process is known as Steam Methane Reforming and the product H_2 gas termed as gray H_2 (a mixture H_2 and CO_2) or Blue H_2 if the CO_2 is removed. However, as Zeppelin warfare increased the production of blue H_2 could not keep-up with lift-gas demand leading to greater CO_2 impurities in the supplied life-gas. In addition, production accidents (Seddin gas plant, June 7, 1917) and train supply problems between the North Sea and Baltic bases [11], page 271–273 affected continuity of life-gas supply to the airships. In the inter war years, Britain faced a smaller but similar problem which was overcome by using mobile batch process units containing sodium hydroxide, ferrosilicon, and water that generated sodium metasilicate and H_2 gas (99.3–99.6 pure), see (1) [34].

$$2NaOH(s) + 2HO_2(l) + Si(s) \overset{>150C}{\longrightarrow} 2NaSiO_3(l) + 2H_2(g) \tag{1}$$

Depending on pressure (p) and temperature (T), the flammability limit of H_2 in air is generally between 4 and 75% H_2 by volume, and the explosive limit of H_2 in air is 18.3–59% by volume. It only requires spark or electrical discharge of sufficient energy to crack both the H–H bond (432 $kJ.mol^{-1}$) and the O–O bond (146 $kJ.mol^{-1}$) to ignite the mixture and burn until the H_2 fuel is consumed. Equation 2 depicts an almost physically impossible exothermic stoichiometric equation for these reactants to form water vapor (H_2O) along with the associated $-482\Delta H$ value per two molecules of H_2 fuel.

$$2H_2(g) + O_2(g) \rightarrow 2H_2O(g), \Delta H = -482\,kJ \tag{2}$$

At atmospheric pressure, the stoichiometric mixture auto ignition temperature is in the order of 570 C (843.15 K) with a calculated minimum spark energy of the order of 0.02×10^{-3} J [35]. However, this simple thermodynamic equation greatly misrepresents the electrical breakdown process of the gases, as both pressure and temperature; electric field stress, ignition frequency (dc, ac, or radio frequency), relative gas buoyancy, and the liquid–gas interface at the airship outer skin surface in storm conditions all have a role in the breakdown process.

5.1 H_2–O_2 Gas Vapor Chain Reaction Mechanism

This section postulates a limited series of reaction steps within the H_2–O_2 gas vapor reaction. The steps proceed by initiation (3), branching (4, 5) and propagation (6).

$$Initiation : H_2 + O_2 \rightarrow HO_2 \cdot + H\cdot \tag{3}$$

where, the initiation step (2) proceeds with the dissociation of some amount of molecular gas (H_2 and O_2) by a spark, flame, or electric discharge.

The resulting hydrogen radical (H·) attacks the reactants O_2 through (4)

$$Branching : H \cdot + O_2 \rightarrow HO \cdot + O\cdot \tag{4}$$

Followed by the products above steps attack the H_2 fuel (5, 6)

$$Branching : O \cdot + H_2 \rightarrow HO \cdot + H\cdot \tag{5}$$

$$Propagation : HO \cdot + H_2 + \cdot \rightarrow H \cdot + H_2O + heat \tag{6}$$

In these oxygen–hydrogen reactions, the chain cycle starts with one H· atom product (4), then the cycle generates additional H· atoms (5, 6). Steps (4 and 5) are named branching steps because they produce OH· radicals which further attacks the H_2 fuel to generate two further radicals. The branching steps therefore promotes the rate of heat release which may increase exponentially, to the point that the heat generated cannot be removed faster enough from the vapor at which an explosion occurs. In addition, in each chain cycle, the propagation step (6) produces a water molecule with an associated release of energy, which in turn promotes steps (4 and 5) along with the energy kick from the Pomperoy and Brock bullets. When all of the H_2 is consumed, the cycle process is terminated. In this context, reaction steps 3 to 6 redefine the role of H_2 from a lift-gas to an energy source.

In chemistry textbooks [36], the H_2–O_2 reaction is shown to have a complex dependence on pressure and temperature, specifically a zigzag curve that separates the non-explosive (p, T) regimes from the explosive (p, T) regimes. The free branches of the curve are called the first, the second and the third explosion limit. Early airships could gain altitudes of 2.5 km [20] and later WW1 version 'Height Climbers' reaching altitudes of 6 km [11]. These altitudes equate a standard pressure range of 1013–47.1 kPa along with ambient temperature variation of approximately 15 to −24 C (~288 to ~249 K). This p, T range places the airship flight altitude is well within the first and second explosive limits branching steps (7, 8) are explosively efficient.

On YouTube there are many slow motion photography sequences of balloon detonations filled with stoichiometric mixtures of H_2–O_2, see for example [37]. The slow motion films reveal that the initial shockwave ruptures the balloon, followed by the oxyhydrogen mixture burning with a typically yellow-orange that expands out from center of where the balloon once was. When a balloon filed with pure H_2 is ignited, the reaction with the surrounding air is less rapid and the sound is less loud. From these demonstrations, the explosion is caused by a sudden pressure effect through the action of heat.

5.2 *Fairweather Electric Field Between Cloud and Ground/Sea*

In this section, the heterogeneous chemistry within Cumulonimbus cloud is consisted as a source of ignition for the H_2 filled airships. First consider the convection of warmed air (mainly a mixture of N_2 and O_2) from the earth as it expands adiabatically as it rises through the troposphere until it reaches the stratosphere, where the sun's energy reheats the circulated air. This natural convection process allows the cloud to capture positive charged particles resulting in a initial electrification of the cloud. With increasing electrification a negative charge begins to be formed on the upper cloud boundary which then flows down outside to the base of the cloud. The accumulation of negative charge at the base of the cloud, now by convention called 'Cumulonimbus' reinforces the cloud-ground/sea electric field. The electric field in this region is of the order of 1–3 kV cm^{-1} that is not sufficient to overcome the dielectric strength of air. To achieve the required field strength an inductive charge process within clouds has been considered by Saunders [38] and Prevenslik [39] where water moisture (H_2O) is propelled to high altitudes by updroughts and cools to form graupel (a mixture of water and ice particles) that undergoes a continuous dissociation_recombination process forming hydronium ions (H_3O^+) and hydroxyl ions (OH^-) intermediate products. This reversible reaction process is given in (7) where approximately 20% of the intermediate product ions are available for electrification.

$$2H_2O \overset{graupel}{\longleftrightarrow} H_3O^+ + \mathrm{OH}^- \qquad (7)$$

Under natural background acidic conditions, charge separation of the available ions then follows, where the H_3O^+ ions move into the vapor phase, and due to their buoyancy are lifted by updroughts to the top of the cloud leaving the larger and denser OH^- charged graupel to fall under gravity to bottom of the cloud. This dynamic process generates a potential difference between the top and bottom cloud boundaries. With increasing gravitational separation, the negative charged graupel forms a negative space-charge that enhances the pre-existing fairweather electric field between the cloud and ground/sea. When the charge attraction between the cloud bottom boundary and ground strengthens, electrons and negative charged ions shoot down from the cloud as stepped leaders to meet upward positive charged streamers to produce a lighting channel. As the enhanced electric field subsides, sufficient energy remains to partially ionized nitrogen molecules (N_2) at the enhanced electrical fields at metal protrusions, at ground level, or in airships flying through, or near, the cloud [9] to produce the characteristic blue/violet. Westcox, using optical emission spectroscopy (OES) of SEF on aircraft measurement reveals a blue/violet emission that can be attributed to the 2nd positive system of N_2 $\left(C^3\Pi_u^+ - B^3\Pi_g^+\right)$: < 18 eV [40].

Prevenslik [39] proposed that where a cluster (10 or more) of charged graupel particles fall to the ground in the absence of a metal protrusion SEF does not occur but a collective discharge action occurs breaking down the surrounding atmospheric air causing the production of buoyant BL. The characteristic optical emission of which ranges from yellow, through orange, to red $\lambda = 550$–780 nm) as indicated by [41]. The associated atomic and molecular ion spectra are: atomic-H-Balmer-α line ($\lambda = 656$ nm), the 1st positive system nitrogen ($\lambda = 580$ and 654 nm), the O ($3p^5P - 3s^5S$) ($\lambda = 777$ nm) and the excited $NO_2{}^*$ molecule continuum ($\lambda = 450$–800 nm) [42]. Plus metastable neutral molecular oxygen (O_2; $\lambda = 557.7$ nm) [43]. The emission lines and bands quenching as the graupel finally melts.

5.3 Precipitation Static

Wireless equipment having a range of 300 km started to be installed in airships as early as 1910 [11, 20] followed by their installation in aircraft. From the outset, the performance of the wireless communication degraded when flying through rain, mist and snow and it is thought that this precipitation caused an accumulation of electrical charge on the wireless antenna and other surfaces. To prevent electrical arcing and flashovers the standard approach was to bond all electrical equipment along with the airship's outer surface to central Earth point so that the airship has 'theoretically' an equal-potential throughout. Operationally it becomes standard practice to reel-in all wireless antennas when passing through a thunderstorm [9, 44], see Sect. 7.4 (Dixmude).

Marriot reports one of the first investigations of electrostatic interference in 1914 [45]. By 1937, this electrostatic interference became known as precipitation static, or P-static [46]. With the advent of aircraft, high-speed flight the flux of charged particle due to increased antenna drag became a major problem and aerodynamic shielding measures where sought [47]. From the early 1990s it was shown, based on space born X-ray measurements that lightning produces high-energy radiation, in form of an abundance of electrons (up to 10 s MeV) and gamma-ray glow / flash that drives the thundercloud electrostatic interference [48].

6 Peek's Formula for a Single Metal Electrode

It is well known that increasing the electrical stress around a single metal electrode tip (or protrusions) ultimately results in local air breakdown around the electrode.

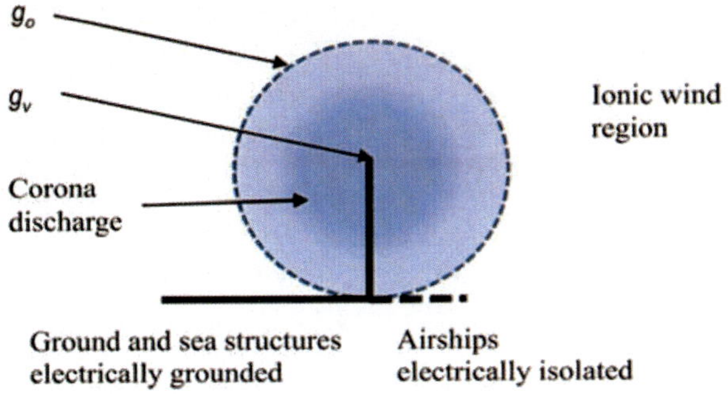

Fig. 3 Schematic of corona discharge boundary limits (g_v and g_o) for a single electrode. The ionic wind region is where unipolar charge carriers (for dry air, N_2^+ and O_2^+ ions) drift away for the corona region with insufficient energy to generate further reactions and/or ion creation

At this level of stress, the local air volume is weakly ionized followed by a rapid electron recombination back to the ground state discharge. On the milliseconds time scale the outer boundary of this volume, the ionization frequency (ν_i) just balances the electron loss frequency (ν_a) by attachment [49], see (8), and Fig. 3. Under these condition a static corona discharge, or SEF, appears attached to the electrode with the visual inception voltage being higher than the visual extinction voltage because, once started there are always electrons to ionize gas molecules.

$$V_i - V_a \approx 0 \tag{8}$$

Upon increasing the voltage stress level further (~5 kV cm^{-1}), the discharge extends outward to form multiple streamers flowing from the electrode, where breakdown is enhanced by the production of electrons at the head of streamer. If the voltage becomes large or a counter electrode is close by (1–10 cm) a conducting trail or channel may form producing a flashover discharge. If the applied voltage is maintained sparks may be also formed. Further increasing the voltage stress creates bidirectional leaders are formed, which involve; space-charge and a gas heating ($\geq$500 K) mechanisms, rather than corona onset alone.

From this sequence of increasing discharge energy states, it is reasonable to assume that SEF influences streamer and leader production. Table 3 (adopted from Gibson [50]) provides a guide to the inception voltage for the three different discharge types. The data shows that although the corona inception voltage for lighting rods has the lowest value for the three discharges (where the variation in the values is due to physical structure orientation of the rods [51]).

Peek's formula was originally proposed as an empirical formula for coaxial cylindrical configurations, parallel wires and spheres in the 1920s [1]. Peek's empirical formula utilizes the local atmospheric condition and the surface condition of a conductor to estimate the corona visual inception voltage at local gas breakdown. For a manmade ac voltage source, see (9).

Table 3 Corona inception electrical field values and characteristic temperature (K) for atmospheric discharge at ground/sea level

Parameter	Corona discharge 1–2 cm diameter grounded lighting rods [50, 51]	Streamer discharge [51]	Leader discharge [51]
Electric field (kV cm^{-1})	0.2–2.7	~5	~1–5
Gas temperature (K)	~300	~300	≥3000 K

$$g_v = g_o \delta m \left(1 + \frac{k}{\sqrt{\delta r_0}}\right) \text{Measured in units of kV cm}^{-1} \tag{9}$$

In (9), g_v is the voltage gradient (kV cm^{-1}) at the visual corona inception voltage; g_o is the disruptive electric gradient, for an ac voltage the value varies from: 27.2 k.cm^{-1} for a sphere, 30 kV.cm^{-1} for parallel wires, and 31 kV cm^{-1} for coaxial geometries. The parameter δ is the local relative air density (at sea level, $\delta =$ 1 under fair weather conditions and 0.9–0.8 for storm conditions), m is the surface roughness factor ($m = 1$ represents dry and smooth clean surface under laboratory conditions). For wet conditions, Peek found that the g_v fell sharply and considered this as a special case for m by substituting it with $g_o = 9$ kV cm^{-1}. The parameter k is an empirical dimension factor (0.301–0.308) and r_o is the tip geometry radius (cm). As energy is required to start a corona discharge the single electrode surface-to-space boundary limits requires that the surface electrical stress be raised to g_v so that at a finite distance away in space where $k\sqrt{r_o}$ is g_o air breakdown occurs. For dry air, the conducting carriers are typically N_2^+ and O_2^+ ions within a background of neutral gas molecules that drifts away from the corona discharge [52].

Natural occurring disruptive electric gradients formed by thunderstorms may also have a direct current voltage component [1], therefore (9) may be rewritten as follows:

$$g_v = 21.9 \delta m \left(1 + \frac{k}{\sqrt{\delta r_0}}\right) \text{Measured in units of kV cm}^{-1} \tag{10}$$

where 21.9 is the route mean square (RMS) of the ac disruptive electric gradient for air (g_o). The parameters: g_v, k and r_0 having the same meaning as in (3).

Given that Peek's Law, in its different forms (9 and 10), is an empirical mathematical construct, parameters δ m k and r_o may be varied to fit the scenario of an airship entering a Cumulonimbus vertical cloud formation where the fair weather electric field is enhanced. In this scenario, the electrically isolated airship may become negatively charged with respect to the cloud were the amount of negative charge is determined by the competing effects of the rate of positive ions pulled to the charge surface as compared to the rate of electron generation by photoemission at the surface under ion bombardment conditions. Under this negative corona condition, the initial visual inception voltage generates discrete discharge points, or tufts, on airship sharp edges where the electrical stress is the greatest. These discrete discharges only grow in number to produce a uniform discharge as the voltage is increased. In addition, beyond the corona boundary (g_o) electrons are propelled away from the discharge with sufficient number and energy to drive electron-impact reactions with neutral molecules [52, 53]. The presence of a visible SEF glow on flying aircraft surface also appears to be dependent on its airspeed. Researchers at MIT have recently demonstrated that high airspeed the SEF become detached leaving the electrical stress level to raised to it pre-visible inception voltage level [53, 54]. Which of these two corona mechanisms (positive or negative) has the greater potential as a H_2-air ignition source is of interest when considering the destruction of H_2 lift-gas airships?

7 Storm and Thunderstorm Activity Leading to Airship Disasters

This section considers six notable dirigible and one non-rigid airship disaster attributed to storm and thunderstorm activity. These are LZ-40 and SL-9; The Royal Air Force (RAF) North Sea class NS.11; the French Navy Dixmude, originally built by as LZ-114 for the Imperial German Navy; the USS Akron and USS Macon; and the Hindenburg that ended the dirigible airship adventure.

7.1 LZ-40 (1915)

Two of the earliest known military H_2 lift-gas dirigible disasters caused by natural atmosphere electrostatic disturbance were the LZ-40 and the SL-9 during WW1.

The LZ-40 took part in a number of bombing raids on England between June and September 1915 [11, 54 and 55]. On the LZ-40 last reconnaissance (commanded by *Kapitänleunant* Klaus Hirsch), the airship encountered a thunderstorm whilst returning to base on September 3, 1915. Robinsons [11], page 124–125, provides details of the disaster. The following text is therefore complied from Robinson's account. In the afternoon of September 3, a radio message from LZ-40 informs Nordholz airship base that they would be returning at 3.30 pm. The metrology conditions in the local area were thunder and lighting, and at 2.30 pm, a number of eyewitness at the base saw in the direction of the town of Cuxhaven a *'large flash of flame like that of an explosion'*. There accounts detailed how the explosion was red in color and the LZ-40 smothered in flames falling into the tidal region between the island of Neuwerk and Cuxhaven. Immediately rescue attempts were underway, but it was not until the following day that the salvage teams were able to recover eleven bodies out of a total of twenty on board the airship, along with the airships recording barograph.

Posted on the www.wrecksite [55] is a photograph of the LZ-40 salvage operation with the airship's airframe laying in the shallow tidal sea.

In the report of the disaster (written by *Kapitän* Peter. Strasser; Chief of German Imperial Naval Airship Division) it was detailed that the airship was above its pressure ceiling, venting H_2 lift-gas at the time of the disaster and ventures to relate the disaster to lighting even though there was no eyewitness to verify this. Strasser goes on to not: *'airships should in all circumstances try go around thunderstorms. If this is not possible, they should go through as far as possible under the pressure height as the squalls will allow. The Airships of the Division now have such orders; also in thunderstorms they are ordered to reel in antennas'*.

On a final note, D'Orcy mistakenly lists the LZ-40 destroyed at Ostend harbor on August 10, 1915 by the Royal Navy Air Service (RNAS) [20], in reality it was LZ-43 that was destroyed [11].

7.2 Sl-9 (1917)

Commissioned into the Germany imperial Navy at Seddin in Pomerania, the SL-9 was the ninth in the series of twenty-four airships built by Schütte-Lanz. Although it wood and laminated wood airframe caused concerns to the German Imperial Navy, in particular Strasser [11], page 56. However, thirteen reconnaissance flights were made by SL-9, and in the summer of 1916, the airship bombed the port of Mariehimn, Finland (July 25, 1916) [56] and later took part in joint Army-Navy bombing raids over the South-East coast of England where airframe damage was sustained that required a month of repairs. After these raids, the SL-9 fell in flames into the Baltic Sea near Pillau on March 30, 1917 with the loss of twenty-three lives. Robinson [11], page 393 mentions that SL-9 burnt in a thunderstorm, while the Wikipedia website [57] claims the possible cause of the crash was a lightning strike. As with the LZ-40, the ignition source has not been determined. A possible explanation is that SL-9 rose above its pressure ceiling by a violent updraft causing an automatic blow off of H_2 lift-gas which was then ignited by cloud electrification.

7.3 NS.11 Non-rigid Airship (1919)

The NS.11 entered service with the recently formed RAF in 1918. Based at RN airship station Longside Aberdeenshire, the airship made its record-breaking endurance flight of 400 miles in 100 h and 50 min on February 9–13, 1919 [28]. The R-34 broke this endurance record some months later when flying East-to-West transatlantic flight from East Fortune, England to Mineola, long island, USA in 108 h on July 2–6, 1919. Following this new endurance record, Captain W.K.F.G. Warneford of the NS.11 filed a circular 48-h flight plan over the North Sea from Pulham airship station. Thus, the NS.11, with Warneford and eight crewmembers rather than the usual nine, commenced its last journey at 9 pm on July 13, 1919 from Pulham.

Some 8 min past midnight, a routine radio message revealed no problem with the flight, but some 15 min later, a Mr. E.T. Elwin from the hamlet of Newgate heard the NS.11 aero-engines making 'a lot of noise' and thought the NS.11 was in trouble. A few minutes later in the town of Cley, Mr. A.E Stangroom heard the NS.11 pass overhead, again making 'a 'tremendous noise'. By 12.45 am a number, people heard the NS.11 pass over Blakeney. At approximately 1.45 am, a violent explosion out to sea was heard, with the noise being carried as far as Wells and Cromer. The NS.11 underwent a midair explosion and fell burning and exploding again before crashing into the shallow North Sea some 5–6 miles of the Norfolk coast with the loss of nine lives [58, 59]. Unlike the ZL-40 and SL-9, there were many ground-based eyewitnesses to the unfolding NS.11 disaster. Several unsuccessful, rescue attempts were made. Most of the eyewitness heard the noise of an explosion then turned to look at where the noise came from, at which point they described what they saw. Two eyewitnesses saw the explosion; an old seaman saw the airship turning under the cloud before the explosion and then 'turned on end', whilst the other stated that, 'she 'took a header'. Both the witnesses inferring that the airship aft tail fins went up as the ship took fire. The staff at Pulham airship station unaware of the unfolding disaster until someone from Easter Daily Press (Norwich) phoned to ask if they could comment on the disaster. As for the recovery of NS.11 crew, only the body of the second Coxswain (Sgt. C.H. Lewry) was found, it was washed up on the beach at Salthouse 2 weeks later on July 31.

In all of the accounts, the most notable metrological feature of the unfolding disaster was the isolated 'greasy black cloud', which the NS.11 was approaching and then turned away when the initial explosion occurred. Importantly there is no mention of lightning. Peake [60] has reconstructed a plausible account of NS.11 destruction that starts with the eyewitnesses seeing the NS.11 turning thereby presenting its rear gasbag and tail fins to the cloud and the second explosion occurring as the remaining gasbags rupture on impact with the sea. Figure 4 graphically shows the account. In this account the appearance of the 'greasy black cloud' is characteristic of an advancing 'cold front', Where the front is formed dry denser cold air pushes under moisture-laden clear air which is forced up where upon moisture is condenses out as water droplets to form the greasy appearance of the cloud. The condensation process also releases heat that causes a self-sustaining warm updraft leading to the formation of cumulonimbus and ultimately thunderclouds. Under these metrological conditions, the fair weather electric field is enhanced as the 'greasy black cloud' grows with a potential to induce SEF on the outer fabric of the NS.11 envelope. This scenario in itself does not explain the explosion. However, factoring in that the caption and coxswain anticipated the updraft would force the NS.11 above its pressure ceiling, and vent H_2 lift-gas to counteract the uplift thereby creating the very distinct possibility that SEF would ionize the escaping H_2 gas particularly if the H_2 gas was of poor purity and within the flammable limit. This scenario has credibility if the reports were true that Captain Warneford was attempting to break his own endurance record, by leavening Pulham with maximum fuel and minimum

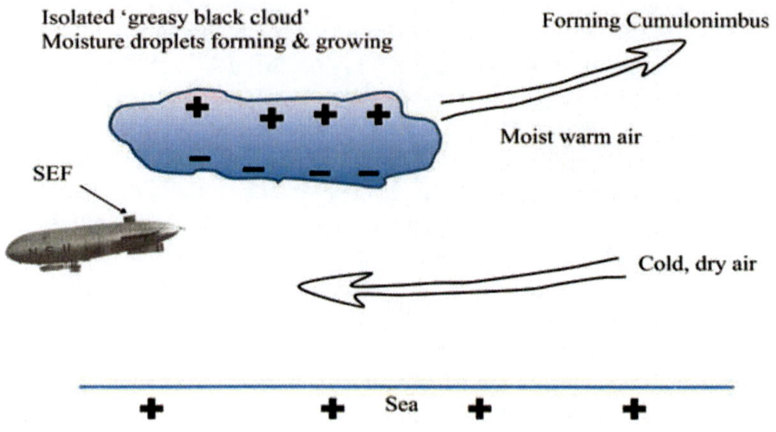

Fig. 4 A schematic depiction of the NS.11 near what the eyewitness described as a 'greasy black cloud'

crew. In which case there would be no air in airship bonnets and her pressure ceiling would be much reduced. The official court of enquiry findings was inconclusive, but lighting was considered as a possible cause, despite no forthcoming evidence.

7.4 Dixmude Dirigible (1923)

The L-72 was the third and final 1918 'Height Climber' X-class Zeppelin, designed to have a working altitude 6,000–6,400 m within a bombing raid duration of two-days. These airships required a significant increase in length (addition of one gasbag) and a reduction in weight. The achieved weight loss through the removal of parts of the original airframe, along with one of the original seven Maybach IVa aero-engines from the rear gondola, a reduction in fuel and water ballast capacity, as well as a reduction in machine gun armaments.

As part of the war reparations, in July 1920, the LZ-72 was turned over to France in '*perfect condition*' and renamed the Dixmude. At the French naval air base Cuers-Pierrefeu near Toulon, the airship came under the command of the charismatic twenty-eight year-old naval officer: Lt. Cdr. Jean Du Plessis du Grenedan. Du Plessis supervised a three-year rebuild program of the Dixmude for extended flight duration (4–5 days) at low altitude (2,000 m). To achieved this goal, new goldbeater's skin gasbags supplied by the newly formed Astra Company, rather than the original German Company and the airframe strengthen to carry the increased fuel and water ballast plus crew and passengers.

After a number of trial flights, the Dixmude began its last flight on December 18, 1923, a planned return flight from Cuers-Pierrefeu air field-Toulon to the Algerian oasis of Ain-Salah (Fig. 5). At 8.00 am on Thursday (some 50 h of flying time) the Dixmude turned north, bound for the Algerian coast, the airship encountered strong

Fig. 5 The last flight of the Dixmude, December 1923. Route taken from Ridley-Kitts (2010) [64] and overlaid on to a modern map of the western Mediterranean Sea, Algeria and Tunisia. Red solid-line: known flight path. Red dashed-line: assumed 'free-balloon' flight path

impeding winds and as the Dixmude fought against the winds, radio messages were sent reporting that fuel was running low and two aero-engines had broken down. The Dixmude was now being force east to Tunisia as a 'free-balloon' and at the mercy of the winds. The airships last radio message (02:08 am Saturday morning December 21) reported that they were following standard operating procedures to reel-in its radio antenna due to thunderstorm activity. Soon afterwards (02:30 am) railway workers and a hunter near Sciacca—Sicily reported a red flash in the Western night sky followed by burning objects falling in to the sea. On the morning of December 26, 1923, burnt wreckage of the Dixmude was found along with the charged corpse of her Commandant and the radio operator. As the news of the crash spread around the world, many newspapers speculated that lightning struck the Dixmude and was the cause of death of the fifty crew and passengers [61, 62].

Throughout this period, French newspapers reported the Dixmude voyage up to the last radio massage, as for the 4–5 days the Dixmude was missing reports emerge that the airship was lost in the Tunisian desert, and French, Italian and British naval ships searched for the airship in the Mediterranean Sea. Confusion reigned in the French

Ministry of Marine and international newspapers. Later the French commission of enquiry confirmed the newspaper speculation that the Dixmude was destroyed by lightning.

In 1924, Dr. Hugo Eckener (Manger of Luftschiffbau Zeppelin and later Commandant of the Graf Zeppelin) wrote in the '*Luftfahrt*' on the Dixmude disaster [44]. In the article, the known German construction details of the L-72 and the subsequent conversion to the Dixmude are analyzed and the probable last 4–5 days timeline of the disaster presented. The following text provides a summary of his analysis. Firstly, repurposing the L-72 airframe from one of a high altitude bomber to one intended for extended flight duration by altering the distributed payload would cause excess stress on the airframe at low altitude. Secondly, the six Maybach aero-engines were pushed well beyond their military specification maintenance schedule of 1–2 days. Indeed, Maybach refused to guarantee more than 48 h continuous use, especially for the crankshafts. Thirdly, as for a lightning strike being the energy source of the disaster, Dr. Eckener comments that duralumin airframe are designed to withstand routine lighting strikes by dissipating the electrical charge throughout the metal airframe, particularly at the nose and rear of the airship. [*N.B. The Grafe Zeppelin and the Hindenburg are a case in-point, as both were struck by lightning many time as they voyaged between Europe and the Americas*]. Fourthly, the burnt condition of the wreckage and the body parts found were consistent with a gasoline fire rather than a H_2 fire that is less destructive to immediate surroundings. Fifth, automatic opening of the pressure ceiling valves due to violent updrafts may have been a contributing factor. Finally, even the radio message sent by the Dixmude build a picture of the storms it encountered; it is most likely that we will never know true cause of the airframe sudden and catastrophic failure.

A contemporary in-depth analysis of the Dixmude may be read in Ridley-Kitts three-part history of the Dixmude: published in Dirigible (2010 and 2011) [63–65].

7.5 USS Akron Dirigible (1933)

In 1929, the USS Akron (the first purpose built flying aircraft carrier airship) was laid down and took her maiden voyage on November 2, 1931. After two ground-handling accidents, both captured on newsreels February 22, 1932 [66] and May 8, 1932 [67], a third ground handling occurred at the Lakehurst hangar 1 on August 22, 1932. On April 4, 1933, the worst airship disaster unfolded as the USS Akron crashed at sea off the coast of New Jersey with the loss of seventy-six crewmembers. The high death toll being due to drowning or hypothermia a factor being that there were no life jackets onboard the airship [68]. On this occasion, the surviving crewmembers were able to give a firsthand account of the disaster. The disaster happened whilst the airship was navigating at low altitude through a thunderstorm when her lower tail section hit the water. As with the first three accidents, the fourth and final accident provides a real-life and death example of the dangers of violent crosswinds and vertical winds to airships at or close to ground/sea level.

7.6 *USS Macon (1935)*

The USS Macron airship took to the sky on April 21, 1933, two week after the lost of her sister airship, USS Akron. In April 1934, whilst maneuvering through mountains in Arizona the USS Macron was forced to exceed its pressure ceiling height (910 m) and climb to 1,800 m to past the mountain range which required 7,300 kg of ballast and fuel to be jettisoned. To gain a safe altitude it became necessary to jettison H_2 lift-gas leaving the airship's ability to compensate for further changes in buoyancy greatly reduced. That is the USS Macon began take on balloon flying characteristics. With minimal ballast, and fuel, she pasted through the next mountain range in Texas, where violent up- and downdrafts could not be compensated for, resulting large aerodynamic pressures buckling the leading tail fin girder ring (#17.5). Subsequent repairs where made to the lower and lateral fins, but where not finished on the upper tail before her next flight on February 12, 1935. In this flight, the USS Macon encountered a storm off Point Sur-California where aerodynamic pressure at the rear of the airship caused the upper fin to shear off. The tail fin structural failure caused the USS Macon to climb above it pressure ceiling where the He lift-gas was automatically released, subsequently the USS Macon slowly glided in to the sea. Unlike the USS Akron, life jackets and rafts where on board and SOS messages sent, resulted in only two lives being lost with the remaining eighty-one crewmembers rescued [69]. Here again the tail fins attachment appears to be a contributing factor under storm conditions. A similar tail cone problem was to plague the British R-100 airship test flights and on the airship's maiden transatlantic crossing to Canada (July 29–August 1, 1930) where on a rival the outer fabric of the starboard elevator became ripped [70].

7.7 *Hindenburg Dirigible (1937)*

At 18.00 local time on May 6, 1937 the second worst dirigible airship disaster, with the lost 36 lives, unfolded at Lakehurst, New Jersey when the Hindenburg commenced its tethering procedure at the airship mooring-mast. The airship had been delayed by poor weather and nearby thunderstorms as portrayed by British Pathé newsreel of the unfolding disaster [71]. Out of the sixty-two survivors, many gave testament to the disaster along with many ground witnesses. The disaster has evoked many books, journal papers [26, 72–77] and aired TV programs [78]. This section considers the H_2 lift-gas ignition theory based around four eyewitness accounts (Broadcaster: Herbert Morrison [79]: history Professor Mark Healed [76, 77]: photographer Arthur Cofod Jr [80]; and Helmsman Helmut Lau [81]. To aid the reader with these accounts, Fig. 6 shows a sketch of the Lakehurst airfield and a schematic of the final 30 min of the Hindenburg flight. In addition, the approximate location of the external eyewitness is as given: H. Lau position is within the Hindenburg's lower tail fin auxiliary control room.

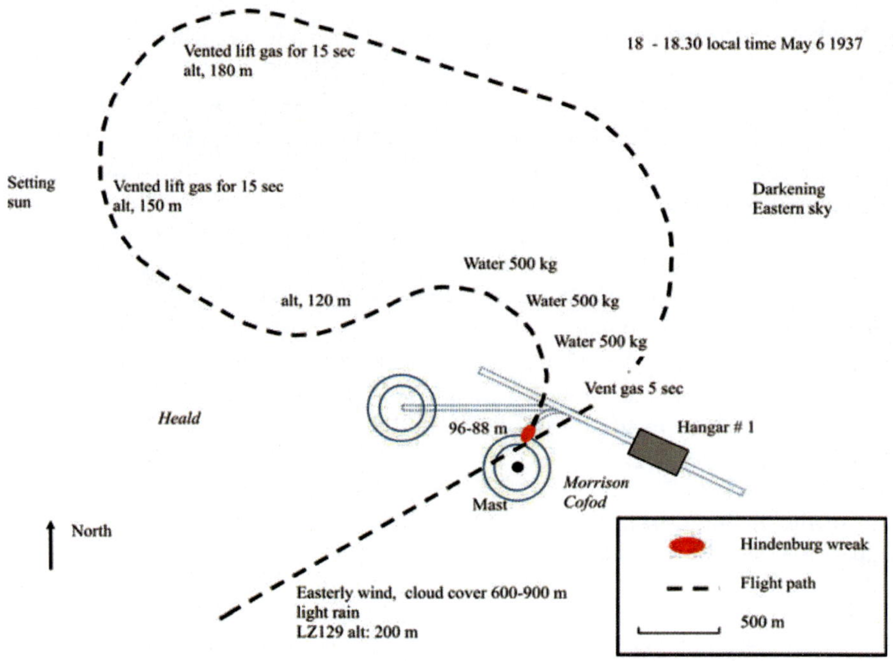

Fig. 6 Schematic of the Lakehurst airfield and the final flight of the Hindenburg

Positioned between Hangar # 1 and the mooring-mast, Herbert Morrison and Charles Nehlson's [79] sound recording of the Hindenburg disaster, transmitted on the following day of the disaster, imprinted such public reflective memories to give rises the '*Hindenburg syndrome*' [72]. Morrison and Nehlson's account would not allow H_2 gas to be used in public transport for many decades. The German Board of Inquiry into the Hindenburg disaster (picked out from many plausible reasons) two H_2 gas ignition theories: (a) and (b).

Theory (a), proposed that due to atmospheric electric disturbances at the time of landing of the airship a corona discharge, otherwise known, as SEF or brush discharge, was the ignition source.

Theory (b), after dropping of the landing ropes, the airships outer fabric became less well grounded than the framework of the airship due to the lower conductivity of the outer fabric. Under these conditions, a spark possibly caused ignition of a H_2-air mixture present over the gasbags four and five.

Professor M. Heald with his wife and son were located outside the main gate of the naval base on a trip to see the Hindenburg. From the car park lot, he records seeing a dim blue flame flickering along the Hindenburg's top ridge minutes before the fire started [76, 77]. The Heald's account from outside the airfield that gave a starboard side view of the Hindenburg against the backdrop of darkening eastern sky rather than the view from port side of the airship as told by Morrison and Nehlson.

Given this information, it is generally thought that the blue color of the dim flicker might be SEF, thus supporting theory (a).

Closet to the initial ignition of the fire was Helmsman H. Lau who was (stationed with three over crewmembers) at the auxiliary control room within in the lower tail fin. In his testimony to the board of inquiry, he states the first time he notices anything wrong is when he '*hears a muffed detonation and looked up and saw from the starboard side down inside the gas cell a bright reflection on the front bulkhead of cell No. 4*'. He goes on to stay: "*The bright reflection in the cell was inside. I saw it through the cell. It was at first red and yellow and there was smoke in it. The cell did not burst on the lower side. The cell suddenly disappeared by the heat.... The fire proceeded further down and then it got air. The flame became very bright and the fire rose up to the side, more to the starboard side, as I remember seeing it, and I saw that with the flame aluminum parts and fabric parts were thrown up. In that same moment the forward cell and the back cell of cell 4 also caught fire [cell 3 and cell 5]. At that time parts of girders, molten aluminum and fabric parts started to tumble down from the top. The whole thing only lasted a fraction of a second.*

Helmut Lau's testimony (translated by Willy von Mesiter) uses the word 'aluminum', which is assumed a simple transcription mistake as in the US Navy the names are interchangeable [29]. Given this, the Hindenburg's airframe would be expected to exhibit pronounced airframe deformation at approximately 471 °C and produce molted duralumin (aluminum-copper alloy) 630 °C [82].

Arthur Cofod Jr (AC) took a series of black and white photographs of the Hindenburg from a location to the north of Hangar # 1. His most memorable photograph (Fig. 7) shows the starboard and aft section of the Hindenburg 10 s of seconds before crashing to the ground. With the back of the dark cloudy sky, the image photograph graphically details how the fire progresses forward with the keel just buckling aft of the rear two aero-engines, suggesting that the temperature generated by the fire is >471 °C. Some 250 m above the Hindenburg, the updraft from the fire forms as a pyrocumulus mushroom-shaped cloud: where the upper bright region is normally associated with condensation of water vapor and the lower dark region contains burning debris of the airship with the most heaviest parts falling back down under the force of gravity. The moving H_2-air flame-front is said to create a mantle effect between the patches of un-burnt outer fabric [74, 75]. It is also clear that at this stage of the fire the lower tail fin with its auxiliary control room is horizontal and still intact. Presumable, it is this aspect of the fire that enables H. Lau, (along with three other crewmembers (H. Freund, R. Kollmer and R. Sautar)) to escape the inferno when the intact lower tail fin crashes to the ground.

Alan Thomas writing in the Dirigible [26], advances the plausible theory of how the last of the three vented H_2 lift-gas volumes may have be ignited. This theory may be dived into both what is known and speculation as to what may have happened. What is known is that, in the final minutes before mooring the maneuvering valves are operated to vent H_2 gas to stabiles the airships neutral bouncy at the mooring-mast. In addition, with all of the Hindenburg aero-engines reversed the airship comes to a stop at the mooring-mast. At this point, the vent shaft theory may come into play that the airship slows down and the aerodynamic extraction force at the top of the vent shaft

Fig. 7 Hindenburg seconds before dropping to the Lakehurst airfield. The photo is downloaded from Wikimedia and attributed to Arthur Cofod Jr/Public domain [80]

is corresponding reduced. Thus, H_2 gas slowly builds-up in the vertical vent shaft (Fig. 2), with little of the H_2-air mixture diluted in the airships slowing slipstream. With H_2-air mixture, exiting the top of vent shaft becomes partially ionized by SEF. In addition, the ionized H_2-air mixture flows-back down the vent shaft to combust the concentrated H_2 gas.

8 Discussion

Since the beginning of recorded history, St Elmo's Fire (SEF) has been widely observed at the closing stages of thunderstorm activity: both at sea level and in mountain regions. The systematic study of these naturally occurring atmospheric weather disturbances has proved difficult due to verifiable eyewitness accounts and real-time high-voltage air breakdown measurements. However, at least five non-lethal airship-SEF encounters are known to have occurred in WW1, see Table 1. The five H_2 lift-gas airships (5 dirigibles and one non-rigid) disasters presented here represent the most notable storm weather related airships disasters. In contrast to these 5 airships disasters the He-lift-gas USS Akron disaster and its sister ship USS Macon had similar tail fin structural and aero-engine design faults, both of which played a significant part in their encounter with violent storms systems. Lightning, SEF or another form of static discharge did not have a role in these two airship disasters.

The boards of enquiries in to each disaster indicate a combination of Pilot error and structural failures where the primary contributing factors in the destruction of the two airships.

This review has looked at H_2 lift-gas airship disasters where blue H_2 gas is the main source of lift. In the first two discussed (LZ-40 and SL-9), all the crew of both the airship died, thus providing no firsthand evidence to the cause of either crash. However, Nordholz airfield ground crew did see the LZ-40 burst into flames as the airship prepared to land. The recovered barograph from the wreck indicted that the airship was a height of 2,400 feet and valving H_2 lift-gas at the time of the disaster. As for the cause of SL-9 disaster, there can only be speculation.

This work has reviewed the anatomy of a H_2 lift-gas airship fire along with cloud electrification as the ignition source using Peeks formula to describe the point of ignition. From the forgoing line of reasoning, it is hypothesized that a potential ignition source in weather related airship disasters, in part, might be due to cloud electrification and the production SEF. The mechanism of positive and negative corona discharge along the airship airspeed may also have a role in the production SEF. Notwithstanding this observation, SEF is most likely to be prevalent at high electrical stress points on the airship external surface coupled with automatic H_2 blow off, or the manual operations of maneuvering valves to blow off H_2 gas. Out of five H_2 lift-gas airship disasters reviewed here, three airships (NS.11, Dixmude and the Hindenburg) are likely candidates as the means of the airship destruction. The scenario in which the disasters occurred is as follows.

The NS.11 disaster was witnessed by many people and recorded in newspapers of the time where lighting strike was portrayed as the guilty party was even though a thunderstorm was not present at the time. Without clear evidence, the board of enquiry found that lightning was the most probable cause of the disaster thereby deflecting blame form unauthorized flight endurance attempts by the captain of the NS.11.

As Dixmude turned home on its endurance flight from the Algerian oasis of Ain-Salah, there was no eyewitness of unfolding disaster. The disaster was pieced together in French national newspapers from radio massages and the discovery of the airship wreckage some four to six days after the event. One year later, a detailed forensic analysis of the Dixmude disaster (by Dr. Hugo Eckener) highlights the failings of the airship's aero-engines and modification (strengthening) to the original L-72 airframe as being a major contributing factors to the lost of the airship.

In the case of the Hindenburg, Professor Heald's family provided visual evidence of SEF flickering along the upper ridge close to the tail fin of the Hindenburg moments before the disaster. This however was not given at the board of inquiry. Although late, this new evidence gives weight to the first option (a), where atmospheric electric disturbances at the time of landing of the airship, a corona discharge, otherwise known as SEF or brush discharge was the ignition source of a H_2-air gas mixture. In this case as manual maneuvering valves where operated to vent the H_2 lift-gas as the airship approached the mooring mast.

It may be concluded that the NS.11, Dixmude and the Hindenburg fell victim to the '*first rough draft of history*' as portrayed in the newspapers where a lightning strike was stated to the likely guilty party. However, the Hindenburg disaster was the first airship disaster to be captured using real-time sound-recordings, black-and-white movie-reels and photos. Herbert Morrison's recorded radio broadcast of the Hindenburg disaster was the final death blow to Germanys dirigible travel, but in reality the Pan American Airways M-130 China Clipper scheduled flight across the Pacific on November 22, 1935, (some three months before the Hindenburg first took to the air) was the first blow. In Russia, the end of H_2 lift-gas dirigible service did not end until the SSSR-V6 and SSSR-10 crashed in 1948 with a combined lost of twenty lives. Non-rigid airship service continued throughout the '*Great Patriotic War*' and beyond as unpressurised H_2 bulk transporters. The lost of the Patriot and Pobeda (*Victory*) in 1947 may be considered as the end of the H_2 lift-gas airship golden age.

Acknowledgements The Authors declare that there is no conflict of interest regarding the publication of this paper. The Authors dedicate this work to all the people who perished in airship disasters.

References

1. F.W. Peek, *Dielectric Phenomena in High Voltage Engineering*, 2nd edn. (McGraw Book Company, Inc, New York, USA, 1920)
2. V.J. Law, D.P. Dowling, The St Elmo's fire: its formation and measurement on both natural and artificial structures, in *Springer Proceedings in Complexity, 13th Chaotic Modeling Simulation International Conference*, eds. C. Skiadas, Y. Diomtiakis, Chap 34 (2021). ISBN: 978-3-030-70794-1
3. V.J. Law, D.P. Dowling, Application of microwave oven plasma reactors for the formation of carbon-based nanomaterials, in *Springer Proceedings in Complexity, 13th Chaotic Modeling Simulation International Conference,* eds. C. Skiadas, Y. Diomtiakis, Chap 35 (2021). ISBN 978-3-030-70794-1
4. W. Shakespeare, The Tempest. (Act 1, Scene 2). *First Folio*, eds. E. Blount, I. Jaggard (London, 1623)
5. Meessen, Ball lightning: bubbles of electronic plasma oscillations. J. Unconvent. Electromag Plasmas. **4**, 163–179 (2012)
6. A.I. Grigor'ev, D. Grigor'eva, S.O. Shiryaeva, Ball lightning penetration into closed rooms: 43 eyewitness accounts. J. Scient. Explor. **6**(3), 261–279 (1992)
7. M. Donoso, J.L. Trueba, A.F. Rañada, The riddle of ball lightning: a review. Scient World J **6**, 254–278 (2006)
8. J.J. Lowke, D. Smith, K.E. Nelson, R.W. Crompton, A.B. Murph, Birth of ball lightning. J. Geophys. Res. **117**, D19107 (2012)
9. F. Arago, Sur le tonnerre, Annuaire au Roi par le Bureau des Longitudes. Notices Scientifiques. **221** (1838)
10. G. Hartwig, *The Aerial World.* (London. 1886). p 310.
11. D. Robinson, The Zeppelin in combat: A history of german navel airships division 1912–1918. Schiffer Publishing Ltd US (1997). ISBN: 9780887405105
12. Langmuir, Oscillations in ionized gases. Proc. Nat. Acad. Sci. U. S. A. **14**(8), 627–637 (1928)
13. W.H. Bostick, Experimental study of ionized matter projected across a magnetic field. Phys. Rev. **104**(2), 292–299 (1956)
14. C.P. Hall, Rigid verses semi-rigid. Dirigible: J Airship Heritage Trust. **83**, 21 (2018)
15. Report of the R.101 inquery. (His Majesty's Stationary Office, London, 1931).

16. S.B. Nelson, Airships in the arctic. Arctic **46**(3), 278–283 (1993)
17. Russian airships of the soviet era 1921–1950. (first printed in http://dolgoprud.org). Reprinted (additional information by N. Caley, H. Bantock and M. Rentell) in Dirigible. J Airship Heritage Trust. **65**, 15–17, (2012)
18. USS Shenandoah. https://www.airships.net/us-navy-rigid-airships/uss-shenandoah/. Accessed May 2021
19. S. Snelling (Features editor), R-33. Dirigible: J Airship Balloon Museum **V(1)**, 18–20 (1995)
20. D'Orcy L, *D'Orcy's airship manual: an international register of airships with compendium of the airships elementary mechanics*. (Century, Co. New York, 1917)
21. L. Chollet, Balloon fabric made of goldbeater skins. L' Aéronautique, August 1922
22. Goldbeaters skins for airships. Dirigible: The Journal of the Airship Heritage Trust. IX(3–4), 20, (1998–99). The original article appeared in the Yorkshire Post on 28 March 1919
23. G. Camplin, Stealing the weinling family secerts. J Airship Heritage Trust. **66**, 10–11 (2012)
24. J. Mckechnie, Fabric for gasbags for aircraft and method of manufacturing the same. US. Patent 1,301,955 (issued April 29, 1919)
25. M. Steadman, The cow and the airship, MuseumPosten and Post & Tele (2006)
26. A. Thomas, Politics and propaganda: a theory to explain the loss of the hindenbug. Dirigible: J. Airship Heritage Trust. **73**, 26–27, (2014)
27. F.A.G. Rendondo, Leonardo Torres Quevedo 1902–1908: the foundation for 100 years of airship design, in *The 7th international airship convention* (2008). Document ID 71192
28. M.J. Dunn, H.M.A, NS11 and the world endurance record. Dirigible: J. Airship Heritage Trust. **71**, 12–16 (2014)
29. G.L. Faurote, Feasibility study of modern airships (phase I) vol III- historical overview (task I). Goodyear Aerospace Corporation. (1975). https://ntrs.nasa.gov/citations/19750023961. Accessed May 2021
30. G. Camplin. The wooden ones. Dirigible: J Airship Heritage Trust. **84**, 17, (2018)
31. B.N. Wallis. Rigid airship. US. Patent 1,675,009 (issued June 26, 1928)
32. C.P. Hall. Two conflicting patents: a question of an origin of invention. Dirigible: J Airship Heritage Trust. **84**, 18–20 (2018)
33. P. Garwood, The vortex gun and other air defence weapons. Dirigible: J Airship Heritage Trust. **67**, 10–12 (2012)
34. J. Frith, The a2 mobile silicol plant mk1. Dirigible: J. Airship Heritage Trust. **79**, 14–15, (2016)
35. U. Mass, J. Warnatz, Ignition process in hydrogen-oxygen mixtures. Combust. Flame **74**, 54–69 (1988)
36. P.W. Aktins, *Physical Chemistry*, 2nd edn. (Oxford University Press, UK, 1982)
37. Hydrogen explosions (slow motion)—Periodic Table of Videos. https://www.youtube.com/watch?v=qOTgeeTB_kA. Accessed March 2021
38. C. Saunders, Charge separation mechanisms in clouds. J. Appl. Meteorol., 335–353 (1993)
39. T.V. Prevenslik. A unified theory for sprites, st. elmo's fire, and ball lighting. J. Metrol. **26**(260), 204–211
40. E.M. Wescott, D.D. Sentman, M.J. Heavner, T.J. Hallinan, D.L. Hampton, D.L. Osborn, The optical spectrum of aircraft St. elmo's fire. Geophys. Res. Lett. **23**(25), 3687–3690 (1996)
41. A.I. Grigor'ev, I.D. Grigor'eva and S.O. Shiryava, Ball lightning and St. elmo's fire as forms of thunderstorm activity. J. Scient. Explor. **5**(2), 163–190 (1991)
42. D.P. Dowling, F.T. O'Neill, V. Milosavljević, V.J. Law, DC pulsed atmospheric pressure plasma jet image information. IEEE Trans Plasma Sci. **39**(11), 2326–2327 (2011)
43. S. Hughes, Green fireballs and ball lightning. Proc. R. Soc. A. **467**, 1427–1448 (2011)
44. H. Eckener. Technical Memorandum No. 256. Lost of the Dixmude. Translated from the *Luftfahrt*. January 23, pp 1–3 (1924). https://digital.library.unt.edu/ark:/67531/metadc56906/m1/2/. Accessed May 2021
45. R.H. Marriott, Radio range variations. Proc. IRE. **2**(3), 37–52 (1914)
46. R.A. Perala. A critical review of precipitation static research since the 1930's and comparison with aircraft charging by dust (2009). https://www.ema3d.com/wp-content/uploads/downloads/AEP_3.pdf. Accessed May 2021

47. R.H. Bryan, T.L.J. Wilmette, Loop antenna shield. US. Patent 2,419,480 (issued April 22, 1947)
48. M. Marsaldi, D.M. Smith, S. Brandt, M.S. Briggs, C.Budtz-JØrgenson, R. Campana, et al., High-energy radiation from thunderstorms and lightning with LOFT. arXiv:1501.02775v1 [astro-ph.HE] 12 Jan 2015
49. M. Goldman, A. Goldman, R.S. Sigmond, The corona discharge, its properties and specific uses. Pure Appl. Chem. **57**(9), 1353–1362 (1985)
50. A.S. Gibson, Jeremy A. Rioussety, V.P. Pasko. Minimum breakdown voltage for corona discharge in cylindrical and spherical geometries. Ann. Res. J. Electr. Eng. Res. Exp. Undergrad. Penn State. **7**, 1–17 (2009)
51. N.L. Aleksandrov, E.M. Bazelyan, F. D'Alessandro, Y.P. Raizer, Numerical simulations of thunderstorm-induced corona processes near lightning rods installed on grounded structures. J. Electrostat. **64**(12), 802–816 (2006)
52. W. Grassi, D. Testi, Induction of waves on a horizontal water film by an impinging corona wind. IEEE Trans. Dielectr. Electr. Insul. **16**(2), 337–385 (2009)
53. C. Guerra-Garcia, N.C. Nguyen, T. Mouratidis, M. Martinez-Sanchez, Corona discharge in wind for electrically isolated electrodes. JGR: Atmos. **125**(6), e2020JD032908 (2020)
54. P. Carradice. The zeppelin: an illustrated history. (Fonhill Media, 2017). https://www.airships.net/hydrogen-airship-accidents/. Accessed May 2021
55. https://www.wrecksite.eu/wreck.aspx?259426. Accessed May 2021
56. J. Passivirta, *Finland and Europe: the early years of independence 1917–1939* (Finish historical Society, Helsinki, Finland, 1988)
57. https://en.wikipedia.org/wiki/List_of_Sch%C3%BCtte-Lanz_airships. Accessed May 2021
58. N. Walmsley, The loss of N.S.11—a local view. Dirigible: J. Airship Balloon Mus. **V**(2), 5–10 (1994)
59. N. Peake, The loss of NS11. Dirigible: J. Airship Balloon Mus. **VI**(2), 5 (1995)
60. N.B. Peake, NS11 and the greasy cloud. Dirigible: J. Airship Balloon Mus. **V**(2), 10–11 (1994). Reprinted in Dirigible: J. Airship Heritage Trust. **71**, 16–17 (2014)
61. Dixmude (airship) https://en.wikipedia.org/wiki/Dixmude_(airship). Accessed May 2021
62. Dixmude (LZ114) (+1923) https://www.wrecksite.eu/wreck.aspx?230834. Accessed May 2021
63. D.G. Ridley-Kitts, Dixmude (part 1): the biggest aerial cruiser of the french navel air arm -1920. Dirigible: J. Airship Heritage Trust. **60**, 12–13 (2010)
64. D.G. Ridley-Kitts, Dixmude (part 2): the biggest aerial cruiser of the french navel air arm -1920. Dirigible: J. Airship Heritage Trust. **61**, 12–13 (2010)
65. D.G. Ridley-Kitts, Dixmude (part 3- conclusion): the biggest aerial cruiser of the french navel air arm -1920. Dirigible: J. Airship Heritage Trust. **62**, 13–14 (2011)
66. USS akron crash. https://www.youtube.com/watch?v=cLFLAj-9-vU. Accessed May 2021
67. British Movietone, Akron Test Ends In Tragedy. https://www.youtube.com/watch?v=iJKF3RHKDtY. Accessed May 2021
68. Loss of the akron, Nature **131**, 499–500 (1933)
69. https://www.airships.net/us-navy-rigid-airships/uss-akron-macon/. Accessed May 2021
70. C.P. Hall, Two tales of R100 failed tail. Dirigible: J. Airship Heritage Trust **60**, 18–21 (2010)
71. Hindenburg Disaster—Real Footage (1937) | British Pathé. https://www.youtube.com/watch?v=fURATK5Yt30. Accessed March 2021
72. A. Bain, W.D. Van Vorst, The hindenburg tragedy revisited: the fatal flaw found. Int. J. Hydrogen Energy **24**, 392–403 (1999)
73. R.G. van Treuren, What happen to the Hindenburg…indeed. Dirigible: J. Airship Heritage Trust. **X1**(3), 4–6, (2000)
74. A.J. Dressler, The hindenburg hydrogen fire: fatal flaws in the addison bain incendiary-paint theory, p. 21 (2004) https://spot.colorado.edu/~dziadeck/zf/LZ129fire.pdf. Accessed May 2021
75. A.J. Dressler, D.E. Overs, W.H. Appleby, The hindenburg fire: hydrogen or incendiary-paint. Buoyant Flight. **52**(2–3), p. 11 (2005)
76. D.H. Robinson, *LZ-129 Hindenburg*, 1st edn. (Arco Publishing Co, New York, 1964)

77. T. Terrel. LZ-129 hindenburg. Curious dragonfly LLC. (2002). https://curiousdragonflycom.files.wordpress.com/2017/12/the-hindenburg.pdf. Accessed May 2021
78. D. Grossman, What destroyed the hindenburg. Channel4.com (2013)
79. C. O'Dell, Crash of the hindenburg (Herbert Morrison, reporting) (6 May, 1937). National Registry (2002)
80. A. Codfod. (https://commons.wikimedia.org/wiki/File:Hindenburg_disaster_(1).jpg). Accessed May 2021
81. H. Lau, Testimony to board of inquiry. http://facesofthehindenburg.blogspot.com/search/label/Helmut%20Lau. Accessed May 2021
82. W.M. Nelson, Heat treatment of duralumin. Aviation 362–365 (1927). National Advisory Committee for Aeronautics. Technical Memorandum. No 401. (1927). https://ntrs.nasa.gov/search.jsp?R=19930090764. Accessed May 2021

'Dubro' Resophonic Guitar: Glissando Gestures

V. J. Law and D. P. Dowling

Abstract Whether in the Hawaiian, Bluegrass, Rock 'n' roll, film sound track or animated cartoon genre, the swoop (glissando) sound made on a slide-guitar is one the most instantly recognizable in western music. This paper reports on the complex acoustical and perceptual glissando of the opening few seconds of Warner Brothers '*Looney Tunes*' ascending glissando, and its counterpart (descending glissando), both played on a '*Dubro*' resophonic guitar. The aim is to analyze these guitar themes in an attempt to provide both a historical development, as well as a technical understanding of the generated sound. With the resophonic guitar tuned to open G (D-G-D-G-B-D), the radiated sounds, includes the guitarist gestures and the glissando sound of steel and glass bottleneck**,** Using the toolbox within Audacity software (time-domain, standard autocorrelation, spectrogram and noise reduction), the recorded tracks are transcribed for tempo, consonant, dissonant, string squeaks, and incoherent/coherent noise. This study also attempts to map the complex psychoacoustic tonal quality of a resophonic guitar, which has been demonstrated to impact emotionally on the listener. It is found that dynamic slide movement divides the string scale length into two coupled longitudinal vibrating segments, each producing a coherent continuous mirrored exponential varying pitch that extends to the guitar brilliance region (4.5–20 kHz). Incoherent or 'hiss-like' noise is found within the lower psychoacoustic warm region (0–0.5 kHz). This incoherent noise is linked to a slip-stick friction process between the slide and string. Slide material and slide direction varies the intensity of the noise that has a Voss-Clarke *1/f*-like response with a Brownian ~ −7 dB/10 Hz roll-off. It is proposed that the guitarists fretting arm musculoskeletal system plays a role in the generation incoherent or hiss-like noise.

Keywords Dubro · Resophonic guitar · Bottleneck · Glass · Steel · Glissando · Incoherent noise

V. J. Law (✉) · D. P. Dowling
School of Mechanical and Materials Engineering, University College Dublin, Belfield, Dublin 4, Ireland
e-mail: viclaw66@gmail.com

C. H. Skiadas and Y. Dimotikalis (eds.), *14th Chaotic Modeling and Simulation International Conference*, Springer Proceedings in Complexity,
https://doi.org/10.1007/978-3-030-96964-6_20

1 Introduction

Chordophones have an important cognitive and emotional role in the development of world music [1], and western music, where the guitar is the main instrument in this classification of stringed instruments. When playing a string instrument, the listener's perceptual experience can invoke a strong psycho physiological response (chills and tears [2], change in heartbeat and respiration rate [3], pupil dilation response [4], and dance [5]. Over riding these emotions is whether the music is played in the major or minor cord: where a major cord instills joy and happiness, and calmness and sadness is found in minor cords [6]. The average tempo of a composition also influences the listener's emotions. For example, an Allegro composition of 140 beat per minute (BPM) has been found to increase listener's blood pressure and heart rate, whereas Andante of 80–82 BPM produces calmness [7]. For a human body mechanism to produce these responses, Voss and Clarke [8] proposed that the natural chemical oscillations within nerve membranes are a likely candidate. Arguably the resonant guitar (or, resophonic guitar [9, 10] played with a bottleneck slide provides one of the most distinctive glissando [11] sounds within the family of string instruments. For example, Freddie Travares's crystal-clear attention grabbing opening 2 s for the Warner Brothers instrumental theme *'Looney Tunes'* (based on the song Merry-go-round broke down) [12, 13]. Travares's credited guitar work on Elvis Presley's *'Blue Hawaii'* [14] is another, if not well known, example of the guitar glissando. It is no surprise then that *'Loony Tunes'* is associated with comedy and happiness and *'Blue Hawaii'* is associated with mellow emotions. Ry Cooders reworking of Blind Willies Johnson slide guitar chords in the film *'Paris, Texas'*, goes one further by introducing vibrato at the end of each fading cord to evoke the feeling of doubt, sadness and yearning of the American dry desert landscape [15].

Beyond the resophonic guitar patents [9, 10], online commentary of partitioning the resophonic guitar psychoacoustic pitch/frequency bands [16], mechanical modal analysis of the resophonic guitar [17, 18] and the development of virtual slide guitar software [19], detailed mapping of the resophonic guitar psychoacoustic space has not be documented as played by a guitarist. The aim of this work is to analysis the radiated sound of a *'Dubro'* resophonic guitar with the guitarist playing the instantaneously recognizable opening seconds of the ascending glissando of the *'Looney Tunes'* instrumental theme and its counterpart associated with the instrumental theme music to the film *'Paris, Texas'*. The capture and analysis of these guitar themes and guitarist gestures maps the perceived sound by the musician and nonmusicians alike thus providing a greater insight to the resophonic guitar psychoacoustics space.

This paper is organized as follows: Sect. 2 reviews the origins of the resophonic guitar and open G tuning. Section 3 provides the experimental. Section 4 describes Benchmarking of the guitar under steel- and glass-bottleneck at a fixed fret position when strumming a using a plastic plectrum [20]. The frequency range of the benchmark extends through three psychoacoustic regions: warm (0–2 kHz), bright (2–4.5 kHz) and the lower brilliance region (4.5–8 kHz) [16]. Section 5 explores the ascending and descending glissando in these three psychoacoustic regions. In Sect. 6

the extended brilliance range up to the human audible threshold limit (20–22 kHz) is examined for ascending and descending glissandos. In Sect. 7, the background detailed in Sects. 4, 5 and 6, informs and identifies the incoherent noise produced by the guitarist gestures. Finally, Sect. 8 provides a discussion on this work.

2 Development of Resophonic Guitar

2.1 Pythagoras String Instrument Theory

The employment of music in the treatment of disease dates back to the earliest times including when David strummed his harp before Saul [21]. Later Pythagoras [ca, 570–495 BC] became interested in understanding the notes and scales used in Greek music for the healing of disease. In particular, the use of the stringed instrument, called the lyre. It is from this time the use of a mathematical approach to help achieve a greater understanding of western music became established. Pythagoras studies found that when two strings with the same length, tension, and thickness, sounded the same when they were plucked, or picked. This means they have a unison sound to the human ear (or consonant), when played together. He also found that if the strings have different lengths (keeping the tension and thickness the same); the strings have a different sound and generally sounds bad (or dissonant) when played together. He also noted that strings having different lengths produce sounds but were consonant rather than dissonant. Pythagoras called the relationship between two notes an interval. Since these discoveries, music containing consonant tones has treated disorders of the ear and epilepsy, sciatic gout and a range of mental disorders [21]. Today, when two strings of the same length are plucked, or picked we say they have the same pitch and, if one string is plucked, or, picked at exactly one-half of the length of the other string, the pitch is doubled and are consonant when played together. This interval is called an octave (harmonic). Furthermore, if one string has a length that is two-thirds the length of the other, the strings again sound consonant when played together and this interval is called a Perfect Fifth. Finally, if one string has a length that is three-quarters the length of the other, the strings again sound consonant, when played together and this interval is a Perfect forth. Hence, the length of the strings being a certain ratio defines interval. Musically speaking the intervals discussed have ratios of: unison (1:1), octave (2:1), a perfect fifth (3:2) and perfect forth (4:3) and so on. The frequency response of the human ear however can only spatially differentiate a limited number of tones within an octave, which are 12 half tones or semitones.

2.2 Vincenzo Galilei's Fret Fingerboard

In the late Renaissance period, the composer, experimentalist, mathematician, and father to Galileo Galilei, Vincenzo Galilei [ca, 1520–1591], developed Pythagoras linear ideas for string instrument to perhaps the first non-linear theory of stringed instruments. From his work we get the rule of eighteen i.e., the division of the active vibrational length of the string [string length (*SL*)] by 18, to obtain the first fret position ($fret_1$) on the string fingerboard (Eq. 1) and dividing the remaining string length by 18 again to get the second fret, and so on. The distance from front of the nut at the headstock to the bridge is defined as *SL*. Today we use the more precise calculation of 17.817, although the rule of eighteen is still commonly used. This rule places the string octave at the twelfth fret', thus providing an equal temperament between each fret. However, the exact overall length from nut to bridge varies slightly with each string, due to the different mass of each string. In this case, the bridge is orientated at an angle to make a slightly longer sounding length for the lower strings and a shorter one for the high strings, thereby, altering each string scale length minutely to improve intonation across all strings in relation to each other for more accurate tuning when playing up the neck. Equations 1 and 2 help to demonstrate this relationship [22].

$$fret_1 = \frac{SL}{17.817} \tag{1}$$

Equation 2 computes the *nth* fret position from the front of nut at the headstock.

$$fret_n = SL - \left(\frac{SL}{(2^{n/12})}\right) \tag{2}$$

The posture of the guitarist is in the seated position with the finger board held by the left hand at about 45° to the horizontal. The area between the thumb Interphalangeal joint and the Metacarpophalangela joints of the left hand are warped around the neck two allow the bottleneck (placed on the ring finger) to act a mobile fret. In this position the left hand is moved up- and -down the fingerboard using the musculoskeletal arm system (with minimal wrist flexion). To produce the rich and complex Delta and bluegrass sound, the index, second and fourth finger do not mute (dampen) the strings. In addition, the guitarist uses a 0.5 mm thick plastic plectrum held in the right-hand to down-stroke the cord string while the palm and lower fingers mute (dampen) the remaining strings.

Using Eq. 2, the fret-offset distance to the nut (fret = 0) can be plotted as log to the base 10 on the horizontal axis against the Fret number, as shown in Fig. 1. This example is for a Dubro guitar that has a 19 fret fingerboard with a fixed *SL* value of 61.2 cm, see experimental section. The exponential trend-line fitted to the data points is associated with the fitting parameter. The trend-line deviation towards fret = 0 and fret = 19, indicates that equal temperament is not directly achieved. In

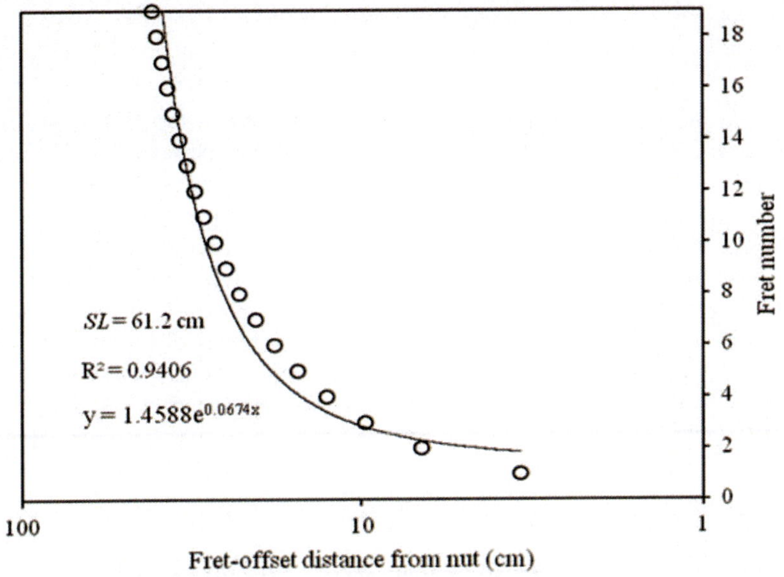

Fig. 1 Fret distance from nut plotted on a $\log_{10}$-linear scale with the data points represented by open-circles, fitted with a Microsoft Excel exponential trend-line

practice however, equal temperament is achieved by altering the bridge orientation and string tension as mentioned above.

Unlike non-fretted instruments (violin), the guitar fretted fingerboard allows people with limited musical knowledge to know when to stop at a given target pitch. This is because the additional tactile and visual cues add to the audible cues to provide an all round cognitive feedback system between the guitarist and the sound of the guitar strings when plucked or picked.

2.3 Origins of the Six-String Acoustic Guitar

The six-string acoustic guitar as we know it today has its origins in post Braque Europe, in particular in Spain where Antonio de Torres Jurado [1817–1892] developed the classical hour glass look and the introduction of the evolutionary "fan" bracing pattern within the guitar's body. Using a circular aperture (hole) in the top plate as the principle mode of acoustic amplification and sound projection (see Helmholtz Eq. 3) [23], his design improved the volume and tone of the guitar when using the rapidly accepted standard guitar tuning of (lowest pitch, thickest string) E-A-D-G-B- E (highest pitch, thinnest string).

$$f_0 \approx \frac{v}{2\pi}\sqrt{\frac{A}{V_{oL_{eq}}}} \quad (3)$$

where f_o is the resonant frequency of aperture in the guitar top plat, v is the speed of sound (at 20 °C, $v \approx 343$ m s^{-1}), A is the area of the aperture, V_o is the volume of the guitar body and L_{eg} is the equivalent length of the neck plus end correction.

f-holes were originally developed for the violin in the Braque period and Antonio Stradivari [1644–1737], is widely regarded as having produced the best design in sound projection and pleasing appearance. Later the physicist Félix Savart [1791–1841] brought this innovation to the guitar, thereby helping to separate the guitar from its classical roots and gain a new audience in the form of country and jazz. In 2015, a study of f-hole sound projection revealed that the axial-length of the f-hole rather than its area that determines the acoustic power projection [23].

2.4 The Steel-String Acoustic Guitar

The first steel-strings for the banjo and guitar are generally considered to have been offered by Christian F Martin [1796–1867] in the mid 1920s, when Hawaiian music became popular in the USA. The union of the steel-strings with the guitar produced a brighter and louder sound that could complete with horns, pianos and drums at mid west American barn dances. Here it's worth noting that a direct and contemporary comparison between the 5 steel-string banjo and the 6 steel-string guitar can be found in the 1972 film Deliverance [24]. The emerging expressive music (Cajon, country, Folk and Bluegrass music) also meant that standard guitar tuning had to change to an open G (lowest pitch, thickest string first) D-G-D-G-B-D (highest pitch, thinnest pitch last) to enable the G major chord (G-B-D) to be strummed on all six strings without the use of the guitarists fret hand, or a capo.

As open G tuning only requires the re-tensioning of only three strings, this new tuning style was readily adopted in bands with a wide spread of music genre. Open G tuning requires the sixth and five strings pitch to be lowered in to D_2 and G_2 respectively. The next three strings (4, 3, and 2) remain the same while the first string (1) with the highest pitch and thinnest string is lowered in pitch from E_4 to D_4. Table 1 tabulates this process, where the last row provides the comparative frequency compression (brightness) of open G tuning with respect to standard tuning.

2.5 Lap-Steel-Guitar and Slide-Guitar

It is said 'that in the 1890s, Joseph Kekuku [1873–1932], accidently strummed a Spanish guitar with a discarded bolt and from that day Kekuku become the inventor of the Hawaiian '*lap-steel-guitar*'. This guitar music requires the guitar to be played

Table 1 Standard and open G tuning of a guitar with pitch values rounded to the nearest whole number

	Standard		Open G	
String	SPN[a]	Pitch (Hz)	SPN[a]	Pitch (Hz)
6	E_2	82	D_2	73
5	A_2	110	G_2	98
4	D_3	*147*	D_3	*147*
3	G_3	*196*	G_3	*196*
2	B_3	*247*	B_3	*247*
1	E_4	330	D_4	294
Frequency range (center point)		248 (128)		221 (110.5)

The rows emphasized with italic (string 4, 3, and 2) have no change of tuning
[a] Scientific pitch notation (based on 400 Hz), subscript denotes the octave in which the note is played

in a flat and horizontal position across the guitarist's knee. Bolts, nails, back of a pocketknife and steel combs all give a pleasing descending—glissando sound that invokes a vision of Hawaiian palm beaches and rolling surf. Around the turn of the nineteenth century, the Steel guitar began to be held against the body as in the Spanish style with the guitarist using a metal, or glass cylindrical object worn on the fretting finger. These fretting techniques, known as '*Slide-guitar*' in the Mississippi Delta: where in the Deep South, Blind Willie Johnson [25], Elmore James [26] and others developed and popularized Gospel Delta blues and Bluegrass. By the early 1920s, the term bottleneck came in use, due to a common idea that the remnants of broken glass bottles left over from bar room fights were picked-up and played on the guitar frets, and if not up to the task than another bottleneck could be picked-up from the floor and used.

In practices, the bottleneck divides the guitar string into two coupled vibrating string-lengths, with the extreme ends of the two sting lengths fixed and the opposing ends coupled through the damping action point of the bottleneck. When it comes to the sound quality '*slide-guitar*' guitarists consider that glass slides offer a smoother playing feel, and produces a warmer and thicker sound that emphasizes the low to mid overtones within the harmonic series compared to metal slides that give a longer sustain that is also brighter and harsher [16].

3 Experimental

This study firstly investigates the sound generated by the Dubro DM-33 Hawaiian resonator guitar (Fig. 2a). The name '*Dubro*' is a portmanteau of '*Dopyera Brothers*' who invented this type of resonant guitar. The guitar has a chrome-plated brass

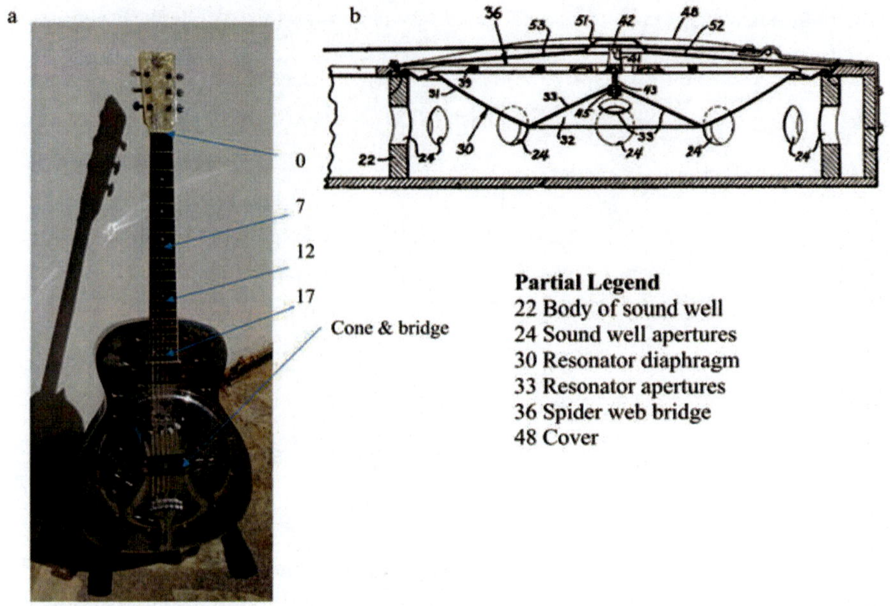

Fig. 2 a–b Photo of the Dubro DM-33 resonator guitar (**a**). Cross-section of Dubro cone/resonator diaphragm based on R. Dopyera's 1932 US patent (**b**)

metal body with sandblasted palm trees, two rolled *f*-holes (axial length 112 mm × 15 mm at their widest point), and a 19-fret rosewood fingerboard with pearl dot inlays (Fig. 2a). The average scale length of the strings is 61.2 cm, and the string action is 3.5 mm, to minimize accidental fret notes. The Dubro is a relatively complex string instrument, compared to the classical acoustic guitar, where the primary mechanical sound amplification is produced by a 26.7 cm diameter outward facing resonator cone/diaphragm, at the top of which is attached a biscuit bridge (Fig. 2b and R. Dopyera US patterns [9, 10]). The purpose of the cone is twofold, (1), to project the string vibrational sound out and away from the guitar and (2), send part of the sound in to sound-well and out via body ports. The two main ports being rolled *f*-holes that are **set** symmetrically **set** either side of the strings. Using this arrangement, the cone produces a harsh mid-high frequency range (1 kHz and above) while the *f*-holes project sound energy in the low to mid frequency range 70–150 Hz. The mechanical complexity of the Dubro does mean regular carful maintenance and regular tune-up is required. A range of short videos of resophonic guitar tuning can be found on YouTube, see for example [27, 28].

3.1 Recording of the Dubro Guitar

For this study, the digital recordings of the acoustic resophonic guitar were made during a performance in Kastollos, Crete in August 2020. The guitar radiated sound is recorded using a Zoom H4n handy recorder (frequency response ~ 30 Hz to 22 kHz) positioned one meter in front of the guitar. The sound levels were set using an alto ZMX122FX mixer and the track recordings saved in waveform audio files on a SD card. The choice of a microphone rather than an electric pick-up is deliberate as this gives both acoustical and perceptual information of the guitar sound as played by the guitarist.

The posture of the guitarist is in the seated position with the fingerboard held by the left hand at about 45 degrees to the horizontal. The area between the thumb Interphalangeal joint and the Metacarpophalangela joints of the left hand are warped around the neck two allow the bottleneck (placed on the ring finger) to act a mobile as the left hand is moved up- and -down the fingerboard using the musculoskeletal arm system (with minimal wrist flexion). The bottleneck divides the strings into two vibrating portions that are designated as string bridge (S_b) and string nut (S_n) respectively. The guitarist may choose to mute (dampen) S_n to generate a crystal-clear tone as in the case of Freddie's swoop in the opening seconds of '*Looney Tunes*' (Fig. 3a) or un-mute (Fig. 3b) to provide a rich and complex sound that is character of Delta blues. Steel and glass bottleneck slides are used in the recordings. The steel has a length = 51 mm, inside diameter = 19 mm and outside diameter = 26.5 mm and glass has a length = 70 mm, inside diameter = 20 mm and outside diameter = 2.5 mm). In the following text the slides are designates as s-slide and g-slide. In addition, the guitarist uses a 0.5 mm thick plastic plectrum held in the right-hand to down-stroke the cord string while the palm and lower fingers mute (dampen) the remaining strings.

Track transcription is performed within a Lenovo laptop running Microsoft Windows 10, therefore the xxx.wav files are fully combatable with Microsoft's Resource Interchange File format (RIFF) specification. Audacity® version 2.4.2 (a

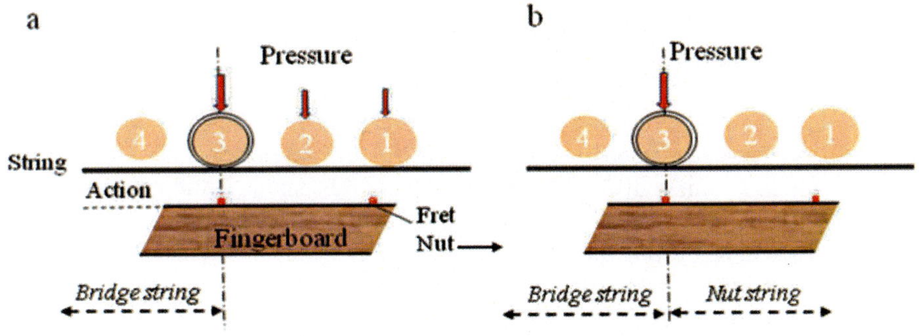

Fig. 3 **a–b** Cross-section schematic of slide and fingers in the S_n muted position (**a**). Cross-section schematic of slide and fingers in the S_n un-muted position (**b**)

Table 2 Audacity software project rate and display information

	Lower Frequency (Hz)	Upper Frequency (kHz)	Sampling rate ($s\ s^{-1}$)	Display video bandwidth (Hz)	Screen shot
Waveform	N/A	N/A	44,100 32-float	N/A	Yes
Spectrum analyzer	30	22	44,100 32-float	N/A	
Spectrogram default maximum	30 30	8 22	44,100 32-float	50	Yes

free, open-source, cross platform audio software) is used to transcribe the guitar sound recordings [29]. The software uses a sampling rate of 44,100 Hz with a dynamic range of 32-bit float to provide a coupled time-domain and spectrogram (3-D plot of sound intensity (color) as a function of frequency and time) of the selected audio track recording. Frequency spectrum analyzer is also available. Table 2 provides the basic metadata for these displays.

Generally, open-access spectrometer software is limited in its ability to provide real-time frequency analysis due to the latency within the software. The latency is because of the lack of processing power to handle the large amount of time-series data that is needed to be converted into the frequency-domain using a Fast Fourier Transform (FFT) algorithm. To overcome this problem Audacity toolbox contains a set autocorrelation algorithms used to identify the SPN frequencies. This option measures how many times SPNs are repeated within the selected waveform record length. This is achieved by taking two copies of the waveform data set, and moving one waveform data set piecewise ($n = 1$) followed by multiplying the two waveform data sets together. The piecewise process is repeated, up to the selected size option. This mathematical noise reduction tool is one of many embedded delay time-series analysis tools used in chaos theory to extract periodic signals (overtones, octave, and harmonics) out of incoherent noise [30].

In the case of the spectrogram, a noise reduction algorithm (NRA) uses the FFT with a Hann window to sample the local neighborhood noise to obtain an incoherent noise profile. Subtracting the noise profile from the whole of spectrogram leaves the coherent acoustic signature of the resophonic guitar. Three operator parameters (amplitude, sensitivity, and frequency smoothing bands) settings determine the impact upon the guitars acoustic signature and the surrounding noise floor. The RNA is used here to estimate of the specific incoherent noise contribution for ascending and descending glissandos, rather than to clean-up the musical signature of the guitar. In mathematical terms this noise reduction technique is called spectral noise gating [31] and is used the compare the SPN and glissando modes of the bottlenecks. Other pixel thresholding methods may be applied using different software platforms, such as LabVIEW [32].

4 Benchmarking

Viewing the stereo sound tracks from the recordings, revealed that there was no different in X- and Y-tracks presumably this because the closeness of the microphones to each other (0.01 m), with respect to the guitar position (1.5 m), even though microphones have an XY orientation. Given this, only the X-tracks are used. The purpose of the Benchmarking the plectrum down-stroke is to establish both the acoustic signature of the guitar and the guitarist's gesture. Two Benchmarks are made, (1) strumming open G and (2) the first-string triad (strings: 3, 2, 1). The first-string triad is frequently used in the Rock 'n' roll genre [33] and therefore is included in this study. The knowledge gained from the Benchmarking informs the identification process of ascending and descending glissandos and incoherent noise.

4.1 Open G Tuning Benchmark

To establish the plastic plectrum down-stroke Benchmark, the guitar is strummed, and recorded, for 35 s. An initial analysis of the total waveform record-length yielded an average BMP of 144. A more detailed standard autocorrelation of a 2.2-s period encompassing both down and up cords yields the tones and overtones. To produce the greatest definition, the autocorrelation algorithm is set with a Hann window and sample size of 2048.

Figure 4 provides the computation where the correlation delay time is on the horizontal axis and SPN level on the vertical axis. In this representation and Figs. 7 and 8, frequency decreases to the right, therefore the root tones are to the right and the higher overtones progress to the left. Note, the delay time 0.01 and 0.025 corresponds to the at-rest human heart beat range (60–100 BPM). Using this representation, the tones G_1, B_1, and overtones C_2, D_2 and G_2 fall within the at-rest heat beat range, and the higher overtones (B_2, and D_3) are in the + 38 BPM elevated/stressed human heart beat range.

4.2 First-String Triad Benchmark

The plastic plectrum down-stroke of the first-string triad (strings: 3, 2 and 1), with the bottleneck slides damping the fifth fret produces corresponding values of s_b ~ 45 cm and S_n ~ 15.3 cm, respectively. In this style of strumming, the second-string triad (strings: 6, 5 and 4) are damped by the guitarist palm. This procedure changes the open G cord by five semitones without changing the original open G tuning.

Figure 5a–d shows two typical strumming cord acoustic waveforms (5a–b) and their associated default spectrogram (5c–d) for both s-slide and g-slide positioned on the fifth fret. In the case of the waveforms, there are two features of note. First is the

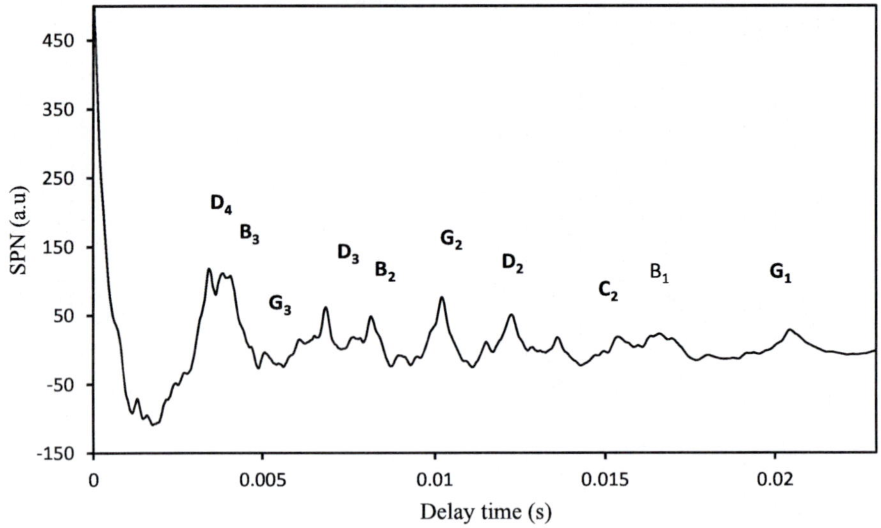

Fig. 4 Standard autocorrelation of the 2.2 s Benchmark. The root tones are to the left and the higher overtones to the rights. The human heart beat range is between G_1 and D_3

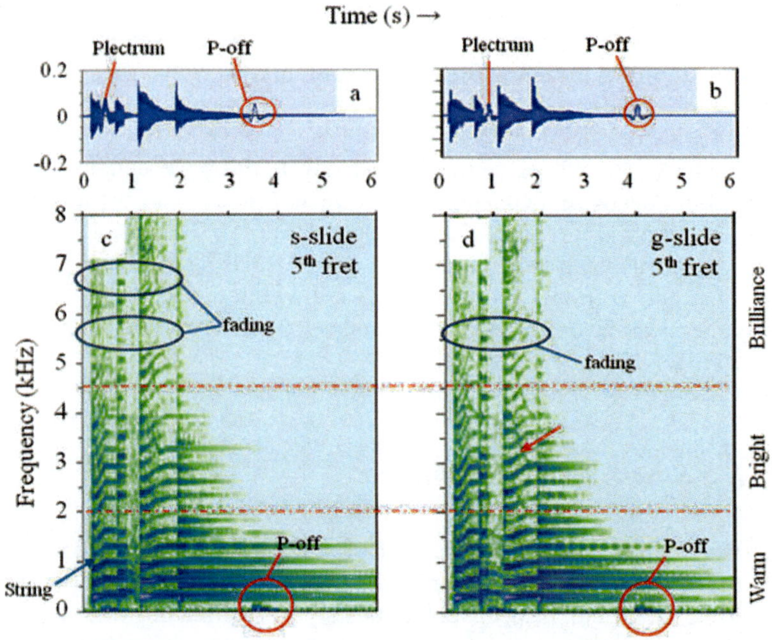

Fig. 5 a–d Waveforms and spectrograms obtained from the first-string triad with bottleneck damping on the fifth fret: s-slide 4a–c, g-side 4b–d

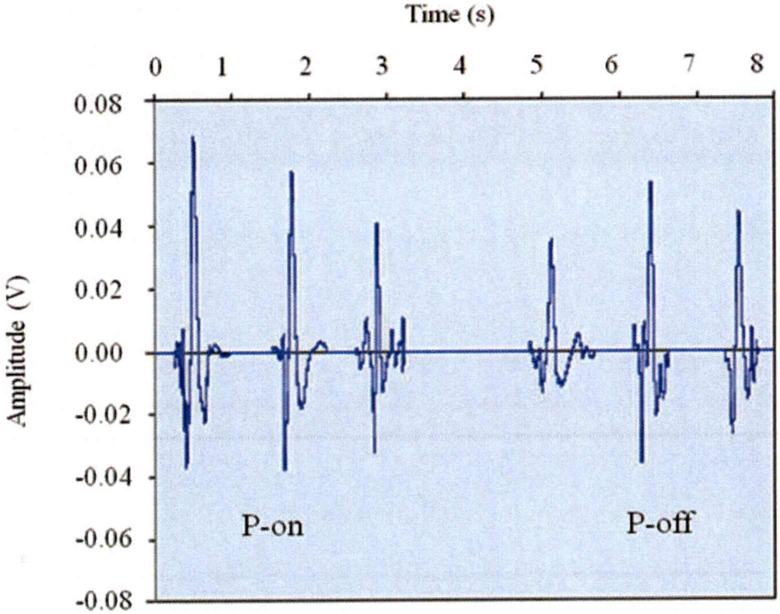

Fig. 6 Typical bottleneck pressure-on and pull-off signatures

truncation of the first and second envelopes by the third envelope that fades out to completion. Note also and that the s-slide fades by an additional 0.5 s compared to the g-side. Second, for both s-slide and g-slide, a string speak caused by an involuntary guitarist gesture is present in the first envelope (plectrum annotation in Fig. 5a–b). Note also, string squeaks are not found in the second or third envelope. In addition, at the end of the completed cord, a 10–20 Hz non-complex resonance is present. Not shown in these figures, but shown in Fig. 6, are further examples of complex resonances that appear at the start of additional recorded cords. The resonances have similar timestamps to the guitarist applied bottleneck pressure-on (p-on) and release (here called pull-off), thus bracketing the cord.

The two default spectrograms in Fig. 5c–d provide a greater insight to the fifth fret bottleneck position. To aid the reader's eye, the psychoacoustic terms: warm (0–2 kHz), bright (2–4.5 kHz) and brilliance (4.5–8 k Hz, which extend to 22 kHz, see Sect. 6) are located on the right-hand frequency axis of the spectrograms. Within the two spectrograms there are five features of note, these are listed as follows:

1. The bottleneck p-off points at the end of the cord are located at the lower-end of the warm region.
2. An intermodulation of tones are observed in the psychoacoustic brilliance region that are caused by energy being transferred up- and down- in frequency range where addition and subtraction of consonant and dissonant tones result in fading in-and-out in the higher frequency range. Unlike electromagnetic signals, the origin of the acoustic energy (in this case the strings and body of the guitar) is

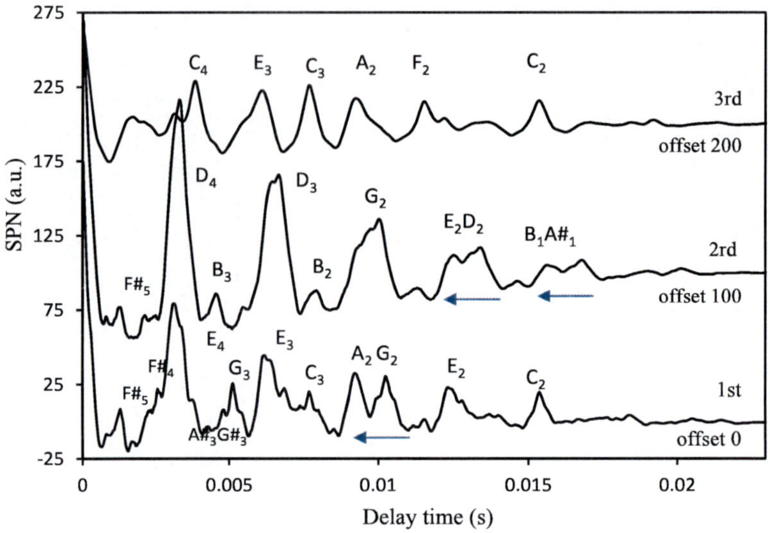

Fig. 7 First-string triad (3, 2, and 1) with s-slide damping the fifth fret

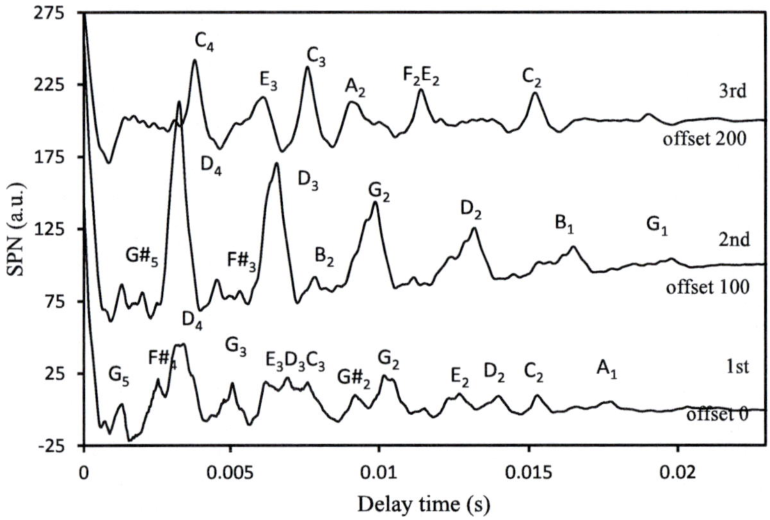

Fig. 8 First-string triad (3, 2, and 1) with g-slide damping the fifth fret

directly altered by the vibration mode of the strings and body, and the medium that the sound is traveling through. Thus, each pitch has a non-zero bandwidth [34, 35] that periodically fads when subtraction occurs.

3. A series short rising tones of approximately + 0.5 kHz (blue arrow annotation) that have an initial timestamp corresponding the picking of the strings. A second

set of descending glissandos of approximately—2 kHz (red arrow annotation) are launched after the raised tones are also present.

4. There is a marked difference in the s-slide and the g-slide sustain periods within the warm and bright regions. In the warm region, the s-slide produces a stronger spectral density compared to g-slide. However both slide produce similar short sustain periods.
5. The string squeaks caused by an involuntary guitarist gesture appear mixed and muddled within the warm region.

Figures 7 and 8 shows the standard autocorrelation (Hann window and 2084-sample size) of the waveform for the s-slide and g-slide damping the fifth fret. Note for clarity, the sound level of the second and third envelopes are offset by + 100, and + 200, respectively. For each envelope, the major overtones and the root along with prominent flat (dissonant) overtones are labeled.

In Fig. 7 (s-slide), the first envelope overtones $E_{3\text{-}4}$ and $G_{3\text{-}2}$ and $C_{3\text{-}2}$ are clearly defined, along with the flat (dissonant) $F\#_{4\text{–}5}$ and $A\#_3$. Note also G_2 transposes up in pitch to A_2 within the duration of the envelope period. In the second envelop the $D_{4\text{-}2}$ and the $B_{3\text{-}2}$ are present, in addition $A\#_1$ transposes up in pitch to the root B_1, and D_2 transposes up in pitch to E_2 within the envelope period. The F#5 is also present. In contrast, the third envelope contains the major overtones $C_{4\text{-}2}$, E_3, A_2 and F_2, with the flat (dissonant) tones less prominent.

Figure 8 reveals that the first envelope contains the overtones $G_{5\text{-}2}$, $E_{2\text{-}3}$, and the roots B_1 and A_1, along with the flat (dissonant) tones $F\#_4$ and $G\#_2$. In the second envelope, $D_{4\text{-}2}$ and $B_{3\text{-}1}$ are present along with $G\#_5$ and $F\#_3$, but at reduced amplitude compared to the s-slide. Again, in contrast, the first two envelopes, the envelope exhibit the major tones $C_{4\text{-}2}$, E_3, A_2 and F_2), with the flat (dissonant) tones less prominent.

5 Ascending and Descending Glissandos

This section looks at the glissando sound production between the seventh and twelfth fret for open G tuning and different bottleneck material (steel and glass). Using Eq. 2, S_b therefore varies between approximately 30 and 40 cm, and S_n varies between approximately 20 and 30 cm. For ease of comparison, spectrograms of a first-string triad ascending glissando using the s-slide is presented followed by two pairs of comparative '*Looney Tunes*' and it counterpart tracks.

Figure 9a–b, depicts the default spectrograms for first-string-triad with steel and glass bottleneck for the descending glissando (twelfth to seventh fret). Annotated on the right-hand vertical axis is the warm, bright and brilliance regions and the horizontal dashed-lines (at 2 and 4.5 kHz) delineate the regions. Within these two spectrograms, three contrasting features are observed and are listed as:

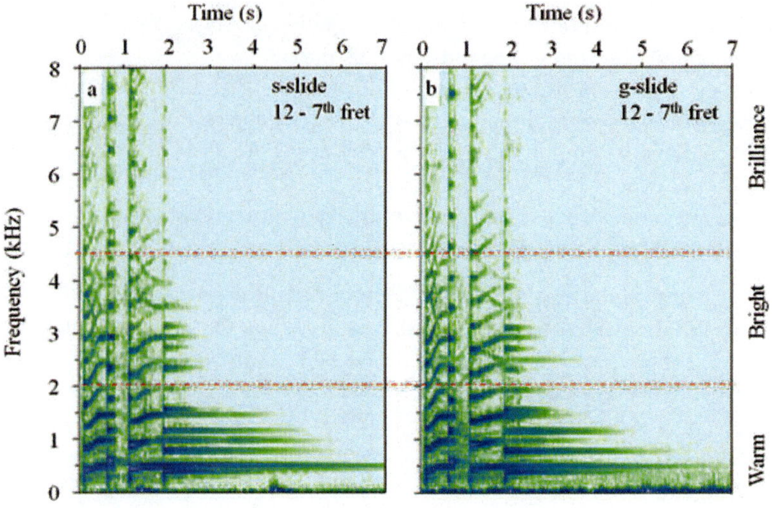

Fig. 9 a–b First-string triad: s-slide descending glissando (twelfth to seventh fret) for s-slide 9a, and g-slide 9b

1. The root and overtones within the warm psychoacoustic region have differing sustain lengths, where the s-slide produce longer and stronger tones compared to the g-side.
2. As the slides physically moves perpendicular across the strings (at rate of approximately 50 mm s^{-1}) a mirrored bifurcation occurs where the glissandos have an exponential trajectory with a frequency shift of approximately 2.2 kHz with time. These mirrored glissandos extend through the bright region and fades into the brilliance region.
3. There is a marked and contrasting noise floor between the two bottlenecks? In the case of the g-slide, a greater incoherent (hiss-like [19]) noise is present at the lower end of the warm region (0–0.5 kHz) as compared to the s-slide bottleneck. Section 6 further quantifies these noise features.

Figure 10a–b depicts the default spectrogram for the '*Looney Tunes*' ascending glissando (seventh to twelfth fret, 10a) and its counterpart descending glissando (twelfth to the seventh fret, 10b). Both spectrograms are for s-slide. Again, the psychoacoustic regions are annotated on the right-hand vertical axis. The spectrograms reveal a number of contrasting features.

1. The inclusion of the thicker strings (4, 5, and 6) generates a high-frequency content that ranges to the top of the bright psychoacoustic region.
2. Now as the slides physically moves perpendicular across the strings at a rate of approximately 50 mm s^{-1} mirrored bifurcation of the glissando occurs. As in Fig. 9a–b, the frequency shift is some 2.2 kHz, however in this case the glissando extend through the bright and well in to the brilliance region. To separate apart these mirrored glissandos it is reasonable to assigned the string

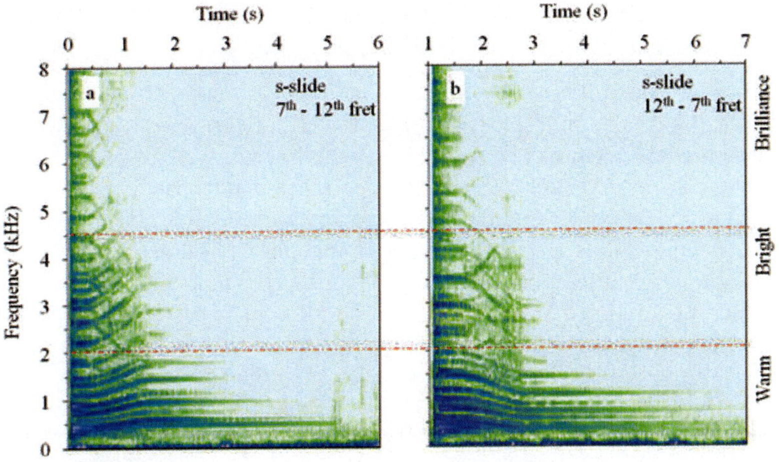

Fig. 10 a–b '*Looney Tunes*'s-slide ascending glissando (seventh to twelfth fret) 10a; and s-slide descending glissando (twelfth to seventh fret) 10b

ascending glissando with increasing fret number, hence the mirrored glissando is assigned to the slip-stick friction process of the slide.

3. The noise floor at the lower-end of the warm (0–0.5 kHz) region is raised with incoherent, or hiss-like, noise. For comparison, see Fig. 9a.

Figure 11a–b provides the default spectrogram for the '*Looney Tunes*' ascending glissando (seventh to twelfth fret, 10a), and its counterpart (twelfth to the seventh

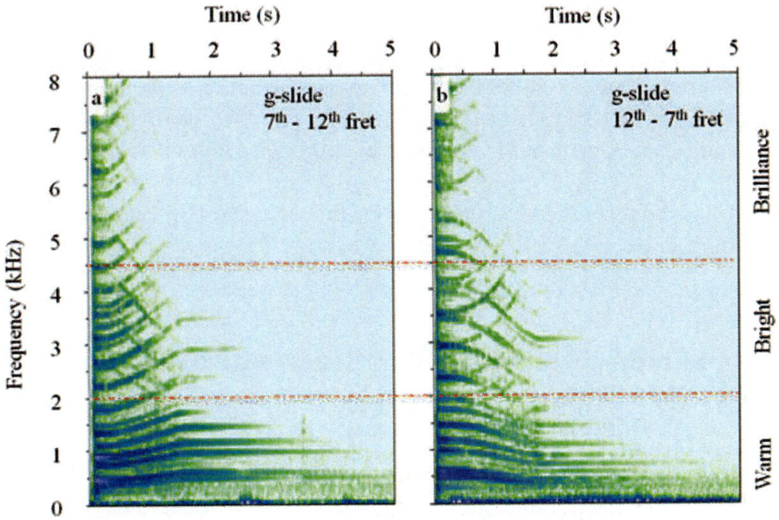

Fig. 11 a–b '*Looney Tunes*' g-slide ascending glissando (seventh to twelfth fret) 11a; and g-slide descending glissando (twelfth to seventh fret) 11b

fret, 10b). Both spectrograms are for the g-slide. Again the right-hand vertical axis depicts the psychoacoustic regions. Main features of note are:

1. As in Fig. 10a–b, mirrored bifurcation of the glissandos produce exponential trajectories as the slide moves perpendicularly across the strings at a rate of approximately 50 mm s^{-1}.
2. Incoherent, or hiss-like, noise is prominent has marked increase in lower-end of the warm region (0–0.5 kHz) as compared to the s-slide (Fig. 10a–b).
3. Taking Figs. 9a–b, 10a–b and 11a–b together, psychoacoustic feature of ascending and descending seventh to twelfth glissando may be summarized. Firstly, the sound of the first-string triad slide extends to the bright region, whereas the thicker strings extend the guitar response in to brilliance region. Secondly, pronounced mirrored glissandos are produced when all six strings are played with the slides. Third, incoherent, or hiss-like, noise in the lower-end of the warm region is produced by the g-slide first-string triad mode, and when all strings are played using either the s-slide or g-slide.
4. For a slide acoustic guitar, Pakarinen et al. [19] has demonstrated that slide divides the damped string into two longitudinal excited string segments, where the main sound originates from the slide to bridge string segment (here labeled S_b) and second excitation originates from the slide to nut (here labeled S_n) segment. This excitation process appear to hold in the resophonic guitar, although a string portion from bridge to the tailpiece must vibrate due to the up-and-down motion of bridge, albeit a smaller bandwidth of S_b and S_n. Given this scenario, vibrational energy is continuously flowing between the string segments as slide moves across the string. Following this, it is reasonable to assign the origin of the mirrored exponentially varying pitch glissandos. Hence, an ascending glissando associated with increasing fret number (7_12) originates in S_b, whilst the mirrored descending glissando has it origin in S_n.

Pakarinen et al. [19] has also identified incoherent, or hiss-like, noise in the steel-string acoustic guitar and assigned this noise to contact points as the side moves across the string. When they synthesized this form of noise they used a noise pulse train thereby evoking an impact and friction modal, otherwise known as slip-stick friction between the surface of the string and slide/Bow [36]. The low-frequency nature of the noise also suggests there is Voss-Clarke flicker noise ($1/f^{\alpha}$ noise) content [8]. Section 6 further explores this psychoacoustic noise for the resophonic guitar.

6 Resophonic Guitar Upper Psychoacoustic Brilliance (0–22 kHz) Region

Given the lack of full range psychoacoustic data for the resophonic guitar, this section looks at the *'Dubro'* resophonic guitar's radiated sound in the 0 to 22 kHz frequency range to understand the interaction and delineation of each psychoacoustic region.

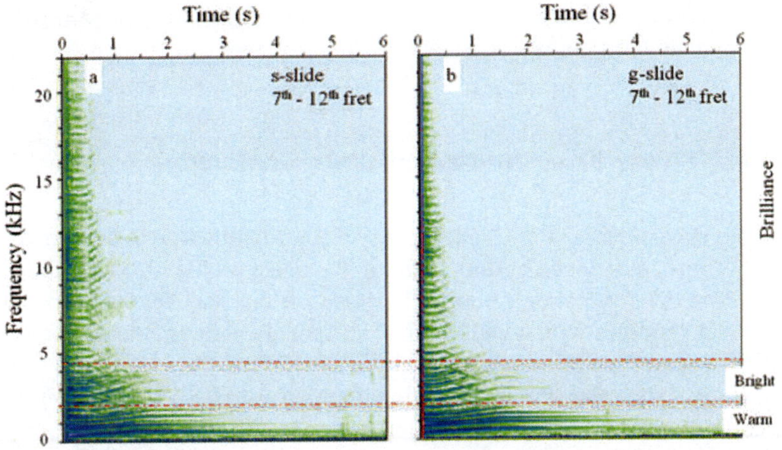

Fig. 12 a–b. Extended frequency range of '*Looney Tunes*' g-slide ascending glissando (seventh to twelfth fret) 12a, and g-slide ascending glissando (twelfth to seventh fret) 12b

This is achieved by using the Audacity spectrogram with a selected full frequency range (22 kHz) for '*Looney Tunes*' ascending s-slide (11a) and g-slide glissando (11b).

Figure 12a–b provides the comparison between the s-slide and g-slide ascending glissando. In the case of s-slide (12a), the ascending glissando overtones extend through the warm region with typical sustain periods of 3 s and to a lesser extent (0–2 s) in the bright region. Whereas the initial plastic plectrum attack overtones have sustain periods of typically 0.5 s throughout the 4.5–22 kHz brilliance region. In addition there is some evidence of weak glissando overtones with typical sustain periods of 1 s. In comparison, the g-slide (12b) produces weaker sustain periods in all three psychoacoustic regions. The least marked being in the brilliance region where the initial plectrum attack overtones have sustain periods decaying from 1 s at 4.5 kHz to 0.25 s at 22 kHz. Within the decay process, the ascending glissandos overtones also become less pronounced.

7 Noise Reduction

Figure 9a–b has revealed, that for a first-string triad ascending glissando, the g-slide induces more incoherent (or hiss-like) noise at the lower-end of the warm region, when compared the s-slide. The aim of this section is therefore threefold: First to isolate and remove the incoherent noise to, or below, the noise floor of s-side glissando, thereby providing an estimate of the noise contribution. The second is to extend the noise reduction knowledge to the '*Looney Tunes*' (Figs. 10a and 11a) and the counterpart descending glissandos (Figs. 10b and 11b). Third, identify and map the characteristic morphology of the noise [8].

7.1 *First-String Triad Bottleneck Noise Reduction*

The first step in estimating the incoherent noise contribution is to identify and isolate the noise. This is performed by first defining the noise profile (np) within the spectrogram (13b). The selection criterion is based on that incoherent noise contains random pixel variables with a well-defined statistical characteristic as compared to the coherence pixel regions of glissando.

The removal step uses three parameters to control the noise reduction process. These parameters are noise reduction level, sensitivity, and frequency-smoothing band. The noise reduction controls the volume reduction (in dB) applied to the noise. The sensitivity controls the amount the signal to be considered as noise (using a scale of 0 to 24). In addition, the frequency-smoothing band controls the spread of the smoothing in neighboring bands, therefore altering the original sampling rate (using a scale of 0–12 and is set to zero so that direct comparison between the original and modified dataset is made) (*N. B. Further details on how the three parameters are used, see Audacity software* [29–31]). A series of iteration processes follows, where the noise reduction value and sensitively value is changed, with the aim of reducing the incoherent noise with minimal damage to the coherent glissando feature within the spectrogram. Figure 13a–d shows the overall process in spectrogram format where Fig. 13a is the first-string triad for the s-slide (taken from Fig. 9a). Figure 13b is the first-string triad for the g-slide along with the incoherent noise profile region selected. Figure 13c is the noise-reduced image using a noise reduction value of 12 dB and a sensitivity value of eight. A comparison of Fig. 13c with the s-slide (13a) reveals similar coherent features and the incoherent spectral densities for both slides is in the 0–0.4 kHz range. Thus the indicating the g-slide incoherent noise contribution is in the order of 12 dB.

Figure 13d, depicts the removed residue noise spectrogram in the low-frequency region of the acoustic spectrogram. It is noted that the isolated noise inevitably captures part of the overtone structure, and therefore some of overtone in Fig. 13c is lost. The overall discrimination process may not be perfect, but it is more beneficial in this case when compared to a low-pass filter that would remove higher frequency noise in the bright and brilliance regions [29–32].

7.2 *Incoherent Noise Reduction*

To quantify the visible incoherent noise in the audio spectrogram Figs. 10a–b and 11a–b, the same attenuation process as in Fig. 13b–d is undertaken. To allow a direct comparison throughout, only the noise attenuation (dB) is altered, whist

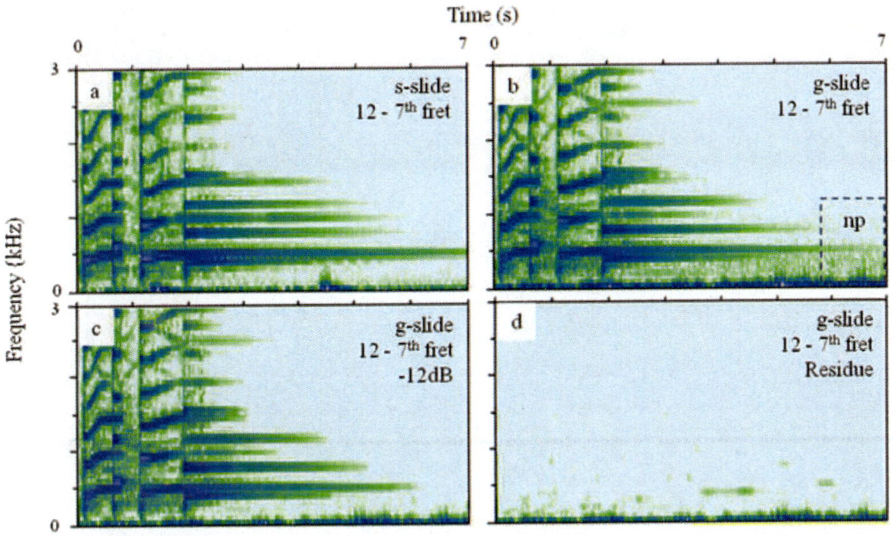

Fig. 13 a–d First-string triad ascending glissando spectrogram for s-slide 13a. First-string triad ascending glissando for g-slide 13b. The g-slide reduced noise profile after noise reduction (−12 dB) 13c. The g-slide residue noise spectrogram 13d

the sensitivity or frequency-smoothing band values are fixed at eight and zero, respectively.

Table 3 depicts the required incoherent noise reduction values to achieve the desired noise floors for each spectrogram. The results support the general concept that a g-slide produces more (3 dB) incoherent, hiss-like, noise than an s-slide. In addition, a descending glissando (twelfth to seventh fret) also produces more incoherent. This finding suggests that the guitarist gesture whether emotional or and musculoskeletal (movement of the upper extremity as the left moves away from the body when play a descending glissando) may also have a role in the production of slide noise.

Table 3 Incoherent noise reduction algorithm variable parameter values

Spectrogram figure	Noise reduction (dB)	Sensitivity level	Frequency-smoothing band
10a (s-slide: 7–12 fret)	9	8	0
10b (s-slide: 12–7 fret)	12	8	0
11a (g-slide 7–12 fret)	9	8	0
11b (g-slide:12–7 fret)	12	8	0

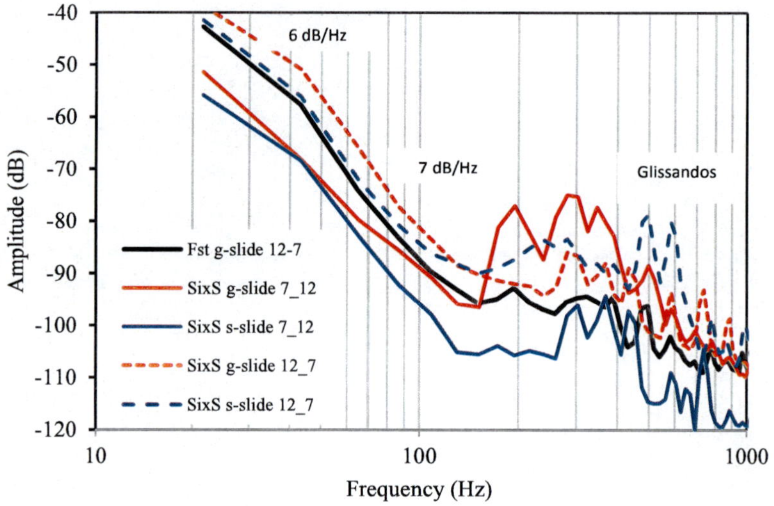

Fig. 14 Frequency spectra of all five residue datasets (see Fig. 13d and Table 3)

7.3 Characteristic Noise Morphology

Using the Audacity FFT algorithm, the five-residue noise datasets obtained in Sects. 7.1 and 7.2 are analyzed for their spectral morphology (color). Figure 14 depicts the FFT results as log–log plot, where frequency (Hz) plotted on the horizontal-axis and the sound amplitude (dB) plotted on the vertical-axis. In this representation, all five datasets exhibit a $1/f$ -like response: e.g. − 6 dB per 10 Hz in the 10–20 Hz frequency band and − 7 dB per 10 Hz in the 30–150 Hz frequency band. Note the 6 dB roll-off in 10–20 Hz band is most likely an artifact of the microphone cut-off frequency. In addition, the structures above 150 Hz are the captured coherent portions of the glissandos and are not considered further as they are not the primary interest here.

Using the first-string triad (Fst, solid black line) as a comparative control, remaining four six-string triads are partitioned around the control. Where the descending glissando for both steel and glass produce the greatest residue noise and therefore are above the control. The opposing ascending glissandos produce the least noise and therefore are positioned below the control. The limited measurements present here appear to indicate that the direction of the slide movement along the fingerboard determines the relative residue noise level, also. One possible cause for this differentiation in the musculoskeletal locomotion force required to extend and retract the guitarist fret arm [37, 38], as similarly observed in violinist [39]. In the case of guitar descending glissando, the guitarist musculoskeletal systems extends the left arm, hand and hence bottleneck from the twelfth to seventh fret so altering the body's center of gravity from the seated position (and vice-versa for the ascending glissando). These varying locomotion conditions are known to induce ulnar nerve entrapment, and therefore merit further investigation.

8 Discussion

This paper has presented a study of the *'Dubro'* resophonic guitar psychoacoustic response to a plastic plectrum applied in the down-stroke for ascending and descending glissando where both steel and glass slide are used. The *'Dubro'* is chosen for its enhanced mechanical sound amplification as compared the classic acoustic and electrical guitar. The slide is placed on the ring finger on of the left hand with the index, second and fourth finger not used to mute (dampen) the strings. This style of slide play provides the rich and complex guitar Delta bluegrass sound. The work has focused on a guitarist's gesture (rather than a mechanical modal based method [17, 18]) to help provide the human psychoacoustic perception of the *'Dubro'*.

The measured radiated sound recordings extend through the psychoacoustic warm (2–4.5 kHz), bright (4.5–8 kHz) and brilliance region up to a frequency of 22 kHz where the initial attack, rather than chord overtones are present. It is worth noting that online commentary declares that the resophonic guitar brilliance may extend to 20 kHz [16]. As the guitarist musculoskeletal system physically moves the slide perpendicular across the full six strings (at approximately 50 mm s^{-1}) between the seventh to twelfth fret, glissando overtones are generated that instantly undergo mirrored bifurcation forming two exponential trajectories: one decreasing in pitch and the other increasing in pitch.

The glissando overtones extend throughout the bright psychoacoustic region and the lower brilliance region for both s-slide and g-slide. In the case of the s-slide, the overtones are weakly present in the mid brilliance region (8–15 kHz). The exponential trajectory of the glissandos as the slide traverses perpendicularly across the strings demonstrates *Vincenzo Galilei's* non-linear theory of fretted string instruments (Eqs. 1 and 2 and graph 1).

Fading in the brilliance psychoacoustic region is also indentified, and is attributed to intermodulation (or, addition and subtraction) of overlaying consonant and dissonant tones. Due to the inner ear's inability to separate high pitch overtones, a listener may perceive the fading process as roughness or timbre in the guitar overall sound. Incoherent, hiss-like, noise is identified and shown to be associated with the slip-stick friction processes between the moving slides and the vibrating strings, where the intensity of the noise is more pronounced on the thicker (wound) steel strings. The g-slides produce a higher intensity (some 3 dB) incoherent, hiss-like, noise than the s-slide.

The direction of slide movement is observed to produce a variation in the amplitude of the incoherent, hiss-like, noise. Slide movements away from the gustiest body centre of gravity (associated with a descending glissando) produce an increase in noise amplitude. As musculoskeletal pain and stress in string instrument players is common [37–39] the measured noise may be a significant finding. This aspect of the work requires further research both in guitarist gesture and in mechanical based modals.

The time-domain and frequency-domain information presented here provides control-data (slide contact gestures) for improved slide-music and slide-noise synthesis within virtual slide guitar systems as reported by Pakarinen, Puputti and Välimäki 2008 [19]. In their work an Omni-directional contact-noise building block is used that did not differentiate the direction of slide movement. Our new work (this paper) indicates that a bidirectional contact-noise building block should be used to synthesize possible differences in musculoskeletal induce noise.

To conclude this work, the opening seconds of the original Warner Brothers instrumental theme *'Looney Tunes'* as played by F. Travares is used as a control. In the original Travares recording, muting (damping) of the strings is performed to make the non-complex (crystal-clear) sound. Our findings reveal that the mirrored bifurcation is present when the strings are not muted. This finding supports the two vibrating string portion mechanism when the slide employ with muting and damping.

Acknowledgements With grateful thanks to Carl Axon for playing the *'Dubro'* resophonic guitar and Nick Dutton for recording the *'Dubro'* sound.

References

1. C.L. Krumhansl, Music: a link between cognition and emotion. Curr. Dir. Psychol. Sci. **11**(2), 45–50 (2002)
2. K. Mori, M. Iwanaga, Two types of peak emotional response to music: the psychophysiological of chills and tears. Sci. Rep. **7**, 460631 (2017)
3. K. Wantanabe, Y. Ooishi, M. Kashino, Heart beat response induced by acoustic tempo and it interaction with basal heart rate. Sci. Rep. **7**, 43856 (2017)
4. R. Jagiello, U Pomper, M. Yoneya, S. Zhao, M. Chait, Rapid brain responses to familiar vs. unfamiliar music—an EEG and pupillometry study. Sci. Rep. **9**, 15570 (2019)
5. V.J. Law, M. Donegan, B. Creaven, Acoustic metrology: from atmospheric plasma to solo percussive irish dance. Chaotic Model. Simul. **4**, 663–670 (2012)
6. P. Virtala, M. Tervaniemi, Neurocognition of major-minor and consonance-dissonance. Music. Percept. **34**(4), 387–404 (2017)
7. B. Bora, M. Krishna, K.D. Phukan, The effects of tempo of music on heart rate, blood pressure and respiratory rate- a study in Gauhati medical college. Indian J. Physiol. Pharmacol. **61**(4), 445–448 (2017)
8. R.E. Voss, J. Clark, 1/f noise in music and speech. Nature **258**(5533), 317–318 (1973)
9. R. Dopyera, Stringed Musical Instrument. US Patent 1,872,633 (issued 16 Aug 1932)
10. R. Dopyera, Stringed Musical Instrument. US Patent 3,931,753. (issued 13 Jan 1976)
11. J.L. Snyder, Evolution and notation of glissando in string music. Indiana Theor. Rev. **1**(2), 35–49 (1978)
12. F. Travares. https://en.wikipedia.org/wiki/Freddie_Tavares and https://youtu.be/qv8WpfTCU9o
13. Looney Tunes. https://www.youtube.com/watch?v=0jTHNBKjMBU
14. E. Presley, Blue Hawaii. RCA Victor Records. (Recorded 20–23 Mar 1961
15. Nebudkadnezar. Ry Cooder. Paris, Texas. https://www.youtube.com/watch?v=X6ymVaq3Fqk&list=PLOKMpbyL5AcsiMt6gDk_gY9WvmAnICS1S

16. The acoustic guitar (2020). https://www.theacousticguitarist.com/best-acoustic-guitar-strings-for-a-warm-sound/
17. M. Rau, J.M. Smith, Measurement and modeling of a resonator guitar, in *Proceedings of the ISMA* (2019), pp. 269–276
18. C.M. Lavallee, Spectral response of acoustic guitars, in *Thinking Matter Symposium, poster session* (2020), p. 10. https://digitalcommons.usm.maine.edu/thinking-matters-symposium/2020/poster-sessions/10/
19. J. Pakarinen, V. Puputti, V. Välimäki, Virtual slide guitar. Capture Music. J. **32**(3), 43–54 (2008)
20. S. Carral, M. Paset, The influence of plectrum thickness on the radiated sound of a guitar. J. Acoust. Soc. Am. **123**, 3380 (2008)
21. E.A. Vescelius, Music and health. Music. Q. **4**(3), 376–401 (1918)
22. M. Atre, S. Apte, Mathematical analysis of acoustic guitar notes. Int. J. Signal Process. **2**, 21–22 (2017)
23. H.T. Nia, A.D. Jain, Y. Liu, M. Alam, R. Barnas, N.C. Makris, The evolution of air resonance power efficiency in the violin and its ancestors, in *Proceedings of the Royal Society A: Mathematical, Physical and Engineering Sciences*, vol. 471, p. 26 (2015)
24. A. Smith. Dueling banjos. Deliverance. https://www.youtube.com/watch?v=pDlZLsJJkVA
25. B.W. Johnson, You're gonna need somebody on your bond. Columbia Records, New Orleans, Louisiana (Recorded 11 Dec 1929)
26. J. Elmore, Dust my broom. Trumpet Records (Recorded 5 Aug 1951)
27. Rev. R. Jones. Blues guitar lesson-slide, resonator, and open tunings. https://www.youtube.com/watch?v=Af57gMSYoXE
28. K.L. Burgus, B. Helferich, Blues harp & bottleneck guitar duet # 2 B.W. Johnson. https://www.youtube.com/watch?v=IO8vytUTO1E
29. Audacity software, version 2.4.2. https://www.audacityteam.org/
30. V.J. Law, D.P. Dowling, Time-series and moments analysis of atmospheric pressure plasma jet treatment of composite surfaces. CMSIM J. January issue, 85–95 (2016)
31. C.E. Deforest, Noise-gating to clean astrophysical image data. Astrophys. J. **838** (155), 10 (2017)
32. V.J. Law, D.P. Dowling, Steganalysis of a pulsed plasma jet ICCD camera image using LabVIEW. Math. Comput. Sci. Indus. **50**, 15–20 (2015). ISBN: 978-1-61804-327-6
33. The Rolling Stones. I wanna be your man. Decca UK (Recorded 7 Oct 1963)
34. V.J. Law, Plasma harmonic and overtone coupling, in *Handbook of Applications of Chaos Theory*, ed. by C.H. Skiadas, C Skiadas, Chap 20, pp. 405–421 (CRC Press, Taylor and Frances 2016)
35. P. Virtala, M. Houtilainen, E. Eilia, Distortion and western music chord processing: an ERP study of musicians and nonmusicians. Music. Percept. **35**(3), 315–331 (2017)
36. R.T. Schumacher, S. Garoff, J. Woodhouse, Probing the physics of slip-stick friction using a bowed string. J. Adhes. **81**, 723–750 (2005)
37. G. Genani, M.C. Dekker, J. Molenbroek, Design of an ergonomic electric guitar. Tijdschrift voor Ergonomie **38**(2), 43–49 (2013)
38. S. Sd, T. Acar, V.K. Gannamaneni, Disability and severity of playing related soft tissue injuries among professional guitar players: a cross sectional survey. Eur. J. Mol. Clin. Med. **7**(2), 4767–4771 (2020)
39. H-S. Lee, H.Y. Park, J.O. Yoon, S. Kim, J.M. Chun, I.W. Aminata, W-J. Cho, I-H, Jean, Musicians' medicine: musculoskeletal problems in string players. Clin. Orthop. Surg. **5** (3), 155–160 (2013)

On the Origin of the Universe: Chaos or Cosmos?

Wiesław M. Macek

Abstract I would like to consider the Universe according to the standard Big Bang model, including various quantum models of its origin. In addition, using the theory of nonlinear dynamics, deterministic chaos, fractals, and multifractals I have proposed a new hypothesis, Macek (The Origin of the World: Cosmos or Chaos? Cardinal Stefan Wyszyński University (UKSW) Scientific Editions, 2020). Namely, I have argued that a simple but possibly nonlinear law is important for the creation of the Cosmos at the extremely small Planck scale at which space and time originated. It is shown that by looking for order and harmony in the complex real world these modern studies give new insight into the most important philosophical issues beyond classical ontological principles, e.g., by providing a deeper understanding of the age-old philosophical dilemma (Leibniz, 1714): *why does something exist instead of nothing?* We also argue that this exciting question is a philosophical basis of matters that influence the meaning of human life in the vast Universe.

Keywords Chaos · Cosmos · Universe · Creation

> *Chaos* is the score on which reality is written.
> Henry Miller (1891–1980)

1 Introduction

In science the evolution the Universe is based on the Big Bang model, which has now become a standard scenario. However, very little is known about the early stages of this evolution, where we should rely on some models, because the required quantum gravity theory is still missing. On the other hand, creation of the Universe is usually

W. M. Macek (✉)
Faculty of Mathematics and Natural Sciences, Institute of Physical Sciences, Cardinal Stefan Wyszyński University, Wóycickiego 1/3, 01-938 Warsaw, Poland
e-mail: macek@uksw.edu.pl
URL: http://www.cbk.waw.pl/~macek

Space Research Centre, Polish Academy of Sciences, Bartycka 18 A, 00-716 Warsaw, Poland

C. H. Skiadas and Y. Dimotikalis (eds.), *14th Chaotic Modeling and Simulation International Conference*, Springer Proceedings in Complexity,
https://doi.org/10.1007/978-3-030-96964-6_21

an important issue of philosophy. Hence, one should return to great philosophers starting from the Greeks asking the questions about the origin of existence of the world [7], including

- Plato's creation: a Demiurg transformed an initial *chaotic* stuff into the ordered *Cosmos*.
- Aristotle's universe is *eternal*: the world always existed, but needed the possibly atemporal Prime Mover or the *First Cause*.

In this paper, we would like to consider the origin of the Universe in view of the modern science, including quantum models of creation, and the recent theory of nonlinear dynamics, deterministic chaos, and fractals, see [12]. We hope that these modern studies give also new insight into the most important philosophical issues exceeding the classical ontological principles, e.g., providing a deeper understanding of the age-old philosophical question:

> *Why does something exist instead of nothing?*
> Gottfried Wilhelm von Leibniz (1646–1716)

2 The Universe in Modern Science

Here we discuss the Standard Model of the Evolution of the Universe based on the Standard Model of Forces together with selected models of the creation of the world based on quantum theory and modern mathematics [12, ch. 2].

A veritable revolution in understanding of the evolution of the Universe was achieved only a century ago owing to the foundation of general relativity by Albert Einstein in 1916. This theory is based on the principle of relativity insisting that physical laws should be independent of the observer, even in the case of a noninertial frame of references (i.e., moving with acceleration).

2.1 The Geometry of Spacetime

According to general relativity, gravitation is revealed by the curvature of local spacetime, as schematically shown in Fig. 1. Instead of the flat four-dimensional Minkowski spacetime we should involve a non-Euclidean spacetime with positive (elliptic type) or negative (hyperbolic) curvatures, respectively, as formulated by Georg F. B. Riemann (1826–1866). Minkowski geometry (corresponding to four-dimensional Euclidean pseudo-space) is only a special case of Riemannian geometry. General theory of relativity can well be applied even in the case of strong gravitational fields. Therefore, one should conclude that spacetime and matter cannot be independent. We may briefly state that mass (energy) tells spacetime geometry about its curvature, but curved spacetime tells the mass how to move.

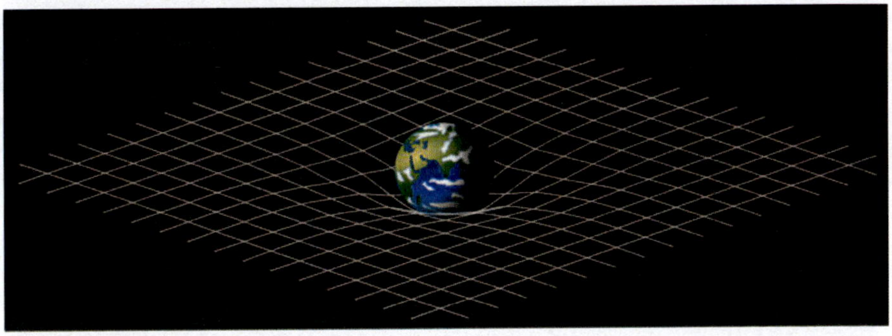

Fig. 1 Gravitation and geometry

2.2 Gravitational Waves

Since the formulation of the theory of general relativity, it was expected that strong gravitational waves, which are actually distortions of spacetime, can arise during the merger of two massive black holes. Figure 2 shows computer simulations of a possible generation mechanism of gravitational waves in the vicinity of black holes. On the one-hundredth anniversary of this theory, we can now confirm its important implications. In fact, the measurements of experimental signals by two independent detectors of the Laser Interferometer Gravitational-Wave Observatory (LIGO) in Hanford and Livingston (separated by ~3000 km) are consistent with observations of a gravitational-wave strain, which is of the order of the amplitude of a gravity

Fig. 2 The generation of gravitational waves (LIGO)

wave, with a relative amplitude of $\sim 10^{-21}$) [1]. For the first time this proves that the international experiment LIGO directly detected gravitational waves originating several billions years ago from the merging of two black holes (of masses about 30 times larger than the mass of the Sun) in the rotating binary system GW150914. Therefore, a large fraction of energy ($\sim$5%, corresponding to three solar masses) has been released in this process in form of gravitational waves. In 2017 the Nobel Prize in Physics was awarded to the American experimental and theoretical physicists Rainer Weiss, Kip Thorne, and Barry Barish for their role in the detection of gravitational waves.

2.3 *The Big Bang Model*

According to the Big Bang model, the Universe expanded from an extremely dense and hot state and continues to expand today. It is worth noting that space itself is expanding, carrying galaxies with it. A representation of the Universe's evolution is schematically shown in Fig. 3, based on the best available measurements of the Wilkinson Microwave Anisotropy Probe (WMAP) operating from 2001 to 2010. The far left depicts the earliest moment we can now probe: size is depicted by the vertical extent of the grid in this graphic. The original state of the Universe began around 13.8 billion years ago, when the Big Bang occurred. This was possibly followed

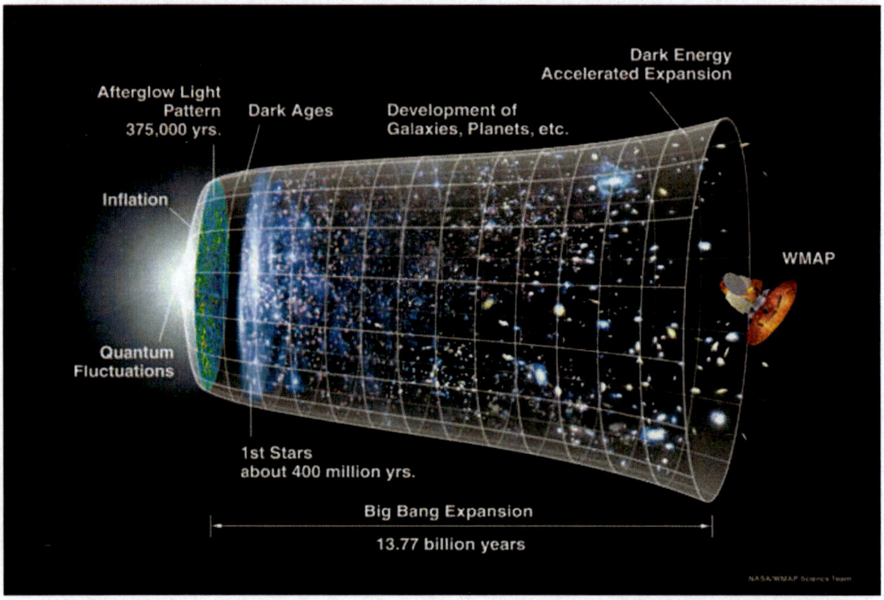

Fig. 3 Schematic of the evolution of the universe. *Credit* NASA/WMAP Science Team

by ‘inflation’, producing a burst of exponential growth in the size of the Universe. The first microsecond, consisting of electroweak, quark, and hadron epochs, together with the lepton epoch (until 3 min of its existence) was decisive for further evolution, leading to the nucleosynthesis of helium from hydrogen. Only after 70 thousand years was light separated from matter. The afterglow light seen by WMAP was emitted about 400 thousand years after the beginning (when the electrons and nucleons were combined into atoms, mainly hydrogen) and has traversed the Universe largely unimpeded since then. The conditions of earlier times are imprinted on this light; it also forms a backlight for later developments of the Universe. The first stars appeared about 400 million years later.

Also the Planck mission launched in 2009 (deactivated in 2013) has become the most important source of information about the early Universe by providing unique data at microwave and infra-red frequencies with high sensitivity and small angular resolution. The Planck data suggest that the Dark Ages (before the first star appeared) ended somewhat later, i.e., 550 million years after the Big Bang. This mission has also provided a new catalog of more than 1500 clusters of galaxies observed in the Universe. More than 400 of these galaxy clusters have large masses ranging between 100 and 1000 times that of our Milky Way galaxy.

After the formation of galaxies, and finally, our solar system, about 4.5 billion years ago, for the next several billion years the expansion of the Universe gradually slowed down as the matter in the Universe pulled on itself by gravity. One can ask whether the present expansion will continue forever or if it might eventually stop, thereby allowing a subsequent contraction. Even though we cannot give a definitive answer to this question, recently it has appeared that the expansion has begun to speed up again, as the repulsive effects of mysterious *dark energy* have come to dominate the expansion of the Universe. The Planck data also support the idea of dark energy acting against gravity. At present this accounts for about 70% of the entire mass of the Universe, and it will presumably increase in the future.

2.4 *The Birth and Evolution of the Universe*

The role of the elementary interactions during the evolution[1] is depicted in Fig. 4. One can see that the splitting of one force after the Big Bang into the four kinds of forces that we know today, after 1.38×10^{10} years of the evolution, happened in a very tiny fraction of the first second. Strong forces should be limited only to the scales (nucleon size of $\sim 10^{-15}$ m) in the microworld, while general relativity models long-range gravitational interactions on very large scales of up to the size ($\sim 10^{27}$ m) of the observed Universe. It is interesting that timescales are from 10^{-24} s in atomic nuclei to nearly 10^{18} s of the experimentally confirmed age of the Universe. This means a range of 42 orders of magnitude is the same as for spacescales; the masses span the range of about 83 orders of magnitude, between 10^{-30} kg for the electron

[1] From http://web.williams.edu/Astronomy/Course-Pages/330/images/forces.jpg.

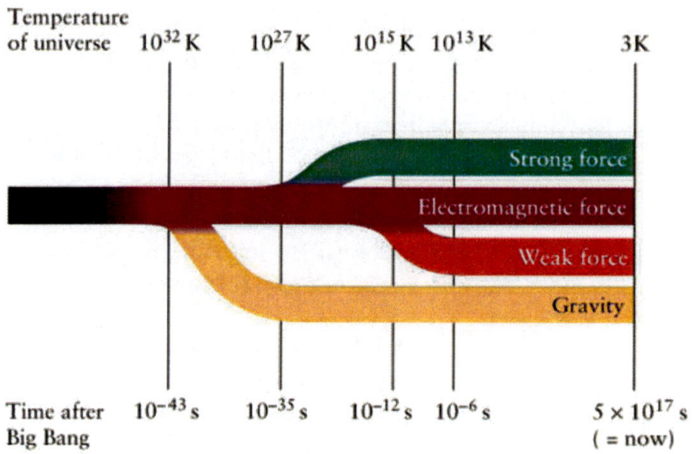

Fig. 4 The grand unification theory (GUT) for the universe

mass and about 10^{53} kg for the of mass of the whole world ($\sim 10^{80}$ baryons, mainly nucleons: protons and neutrons with mass of $\sim 10^{-27}$ kg); this range is roughly twice as large as the time or space scale range.

Because the Universe has already expanded to that extremely huge size, gravitational forces (basically about 40 orders of magnitude weaker than strong nuclear forces) dominate the evolution of the Universe at present. However, at early stages of its evolution both forces resulted from an unknown simple law and could have been of a similar strength. The other long-range electromagnetic interactions between charged particles have already been unified with the short-range weak interactions responsible for the decay of nuclei (electroweak forces). Of course, the Grand Unification Theory (GUT) in Fig. 4 describing the unknown primordial force responsible for the creation of the Universe at a Planck scale of 10^{-43} s will facilitate a better understanding of the physical processes at very early stages of the history of our world.

2.5 *Quantum Models for the Creation of the Universe*

Using the three available universal physical constants—namely the gravitational constant G, the speed of light c, and the Planck constant h, we can construct a quantity called a Planck length $l_P = \sqrt{G\hbar/c^3}$, where $\hbar = h/(2\pi)$. Another quantity l_P/c is the respective Planck time scale, t_P. Because we do not have a quantum theory of gravitation quantum gravity a number of models for the creation creation of the Universe with the following characteristics have been proposed, including:

- The quantum model [2]
  creation from 'nothing', ex nihilo
- Noncommutative geometry [4]
  beginning is everywhere
- String theory, M-theory [30]
  collision of branes
- Cyclic (*ekpyrotic*) model [26, 27]
  big bangs and crunches
- Eternal chaotic inflation [5]
  bubble of universes.

The concept of the quantum wave function of the primordial Universe was put forward in [2]. This point of view was illustrated in a simple minisuperspace model with an invariant scalar field as the only gravitational degree of freedom. The authors of this model focus on the ground state with minimum excitation of an initial Universe on extremely small scales. Providing that the time is changed to imaginary values it, spacetime with a four-dimensional geometry becomes positive-defined. This allows us to obtain the path integral of the respective Euclidean action. In this way, the authors obtained finite nonzero probabilities of propagating from the ground (vacuum) state to the spectrum of possible excited states.

It is worth noting that below the Planck threshold $l_P = 1.6 \times 10^{-35}$ m $\sim 10^{-35}$ m and $t_P = 5.4 \times 10^{-44}$ s $\sim 10^{-43}$ s, in space and time, respectively, any time could be formally eliminated in the quantum model. In this scenario the Universe interpreted without any boundary conditions [2]. Moreover, because one can obtain the excited state from the vacuum state, they argue for the creation out of *nothing*, even *ex nihilo*. However, one should bear in mind that a quantum vacuum state is not actually 'nothingness'—indeed it could be interpreted as a 'sea' of various possibilities [3].

An alternative interesting solution for the origin of spacetime on extremely small scales has been proposed in [4], where it was suggested that these critical values would correspond to a phase transition from a smooth commutative geometry to a rather singular noncommutative régime, with no space points and no time instances. Hence, noncommutative algebra is the other quantum gravity counterpart of the observable in the standard quantum theory, which can help in the application of quantization methods to the origin of the primordial Universe. Therefore, as one can paradoxically put it: *the beginning is everywhere.*

Following the M theory [30], in the context of an initial universe resulting from a *collision of branes*, another interesting non-standard cosmological scenario has been proposed in [26, 27]. According to their proposed model, the Universe undergoes a sequence of cosmic epochs each of which begins with a created world with a standard big bang event, followed by a slowly accelerating expansion with radiation and matter domination periods, but ends by contraction with a crunch. This model is called *ekpyrotic*, because in ancient Greece's Stoic philosophy *ecpirosi* means 'escape from fire'. This endless cycle of *big bangs and crunches* would avoid any particular singularity, but is able to explain the approximate homogeneity of distribution of mass, instead of a hypothetical inflation following the Planck epoch. It is worth noting that the model produces the recently observed flatness of spacetime geometry, providing the energy needed to restore the Universe from the same vacuum state in the next cycle. These authors also assure us that, owing to acceleration, this continuously repeating cyclic solution is an attractor [27].

Taking the wave function of the Universe [2], it can be shown that the large scale fluctuations of the quantum scalar field can generate an infinite process of self-reproducing primordial mini-universes. Therefore, one can suggest an eternally existing chaotic inflationary scenario, describing the Universe as a self-generating fractal that springs up from the multiverse [5]. Because it seems improbable that only one such Universe is chosen in reality by compactification during the expansion, it is argued that there exists a *bubble* of all possible *universes* that is always growing until a new universe is created by chaotic inflation in the bubble [5]. Therefore, there should exist an exponentially large number of causally disconnected mini-universes corresponding to all possible vacuum states followed by inflations. Admittedly, in the last two models time is *eternal*, but it is difficult to verify these models according to the criterion of falsifiability required for any scientific theory of Popper [24].

3 Nonlinear Dynamics and Fractals

In the second part of this paper we focus on nonlinear chaotic dynamics and fractals in a search for implications of an unknown nonlinear law related to a hidden order responsible for the creation of the Cosmos at the Planck epoch, see [12, ch. 3].

3.1 Deterministic Chaos

CHAOS ($\chi\acute{\alpha}o\varsigma$) according to [29] is (see the excellent popular book by Stewart [28]):

- NON- PERIODIC long-term behavior
- in a DETERMINISTIC system
- that exhibits SENSITIVITY TO INITIAL CONDITIONS.

More precisely, we say that a bounded solution $\mathbf{x}(t)$ of a given dynamical system, $\dot{\mathbf{x}} = \mathbf{F}(\mathbf{x})$, is SENSITIVE TO INITIAL CONDITIONS if there is a finite fixed distance $d > 0$ such that for any neighborhood $\|\Delta\mathbf{x}(0)\| < \delta$, where $\delta > 0$, there exists (at least some) other solutions $\mathbf{x}(t) + \Delta\mathbf{x}(t)$ for which for some time $t \geq 0$ we have $\|\Delta\mathbf{x}(t)\| \geq d$. This means that there is a fixed distance d such that, no matter how precisely one specifies an initial state, there exists a solution of a dynamical system starting from a nearby state (at least one) that gets a distance d away.

Given $\mathbf{x(t)} = \{x_1(t), \ldots, x_N(t)\}$, any positive finite value of Lyapunov exponents (or equivalently metric entropy)

$$\lambda_k = \lim_{t \to \infty} \frac{1}{t} \ln \left| \frac{\Delta x_k(t)}{\Delta x_k(0)} \right|, \tag{1}$$

where $k = 1, \ldots N$, implies chaos.

One example comes from the dynamics of irregular flow in viscous fluids, which is still not sufficiently well understood. It appears that the behavior of such systems can be rather complex: from equilibrium or regular (periodic) motion, through intermittency (where irregular and regular motions are intertwined) to nonperiodic behavior. Two types of such nonperiodic flows are possible, namely chaotic and hyperchaotic motions. As discovered by Lorenz (1963) deterministic chaos exhibits sensitivity to initial conditions leading to the unpredictability of the long-term behavior of the system (the 'butterfly effect') [6].

3.2 Hyperchaos

Hyperchaos is a more complex nonperiodic flow, which was discovered by Macek and Strumik (2010) [15] in the generalized Lorenz system previously proposed in [14]. Mathematical and physical aspects of this new low-dimensional model of hydromagnetic convection together with the detailed derivation from the basic partial differential equations, including the magnetic diffusion equations and naturally the anisotropic tension of the magnetic field lines, has been addressed in detail in [11].

Within the theory of dynamical systems transitions from fixed points to periodic or nonperiodic flows often occur in a given system through bifurcations, intermittency, resulting in a turbulent irregular behavior of the nonlinear system. In fact, we have identified type I and III intermittency [23] in the generalized Lorenz model of hydromagnetic convection, as also discussed in the papers [10, 14, 15]. It would be interesting to look for the remaining basic type II intermittency and the respective Hopf bifurcation in this model.

The following ordinary differential equations are obtained in the generalized Lorenz system [14]:

$$\left.\begin{aligned}
\dot{X} &= -\sigma X + \sigma Y - \omega_0 W \\
\dot{Y} &= -XZ + rX - Y \\
\dot{Z} &= XY - bZ \\
\dot{W} &= \omega_0 X - \sigma_{\mathrm{m}} W.
\end{aligned}\right\} \tag{2}$$

In this simplified system, $X(t)$ denotes a time amplitude of the potential of the velocity of a viscous horizontal fluid layer in the vertical gravitational field heated from below, with the normalized (dimensionless) Rayleigh number r, proportional to an initial temperature gradient δT_0, which is a control parameter of the system. Similarly, $Y(t)$ and $Z(t)$ correspond to the two lowest-order amplitudes of the deviation from the linear temperature profile of the layer (of height h) during the convection. The other parameter $\sigma = \nu/\kappa$ is the ratio of the kinematic viscosity ν to thermal conductivity κ (the Prandtl number) characterizing the fluid and $b = 4/(1 + a^2)$ is a geometric factor related to the aspect ratio a of the convected cells.

Admittedly, Lorenz (1963) only took three of several coefficients appearing in the lowest-order of the bispectral Fourier expansion, cf. [25]. In addition to the standard Lorenz system [6], a new time dependent variable W in (2) describes the profile of the magnetic field induced in the convected magnetized fluid. We have also introduced the second control parameter proportional to an initial horizontal magnetic field strength B_0 applied to the system, more precisely defined here as a basic dimensionless magnetic frequency $\omega_0 = \upsilon_{A0}/\upsilon_0$, which is the ratio of the Alfvén velocity $\upsilon_{A0} = B_0/(\mu_0\rho)^{1/2}$, with a constant magnetic permeability μ_0 and mass density ρ, to a characteristic speed $\upsilon_0 = 4\pi\kappa/(abh)$. Naturally, besides $\sigma = \nu/\kappa$, the magnetized viscous fluid is characterized by an analogue parameter $\sigma_m = \eta/\kappa$, defined as the ratio of the magnetic resistivity η to the thermal conductivity κ (related to the magnetic Prandtl number, $Pr_m = \sigma/\sigma_m$).

The results of our another paper illustrate how all these complex motions can be studied by analyzing this simple model [15, Fig. 1]. For example, for a chosen value of $\sigma_m = 3$ (other parameters have the same values as for the classical Lorenz model, $\sigma = 10, b = 8/3$), Fig. 5 plots the largest Lyapunov exponent, calculated according to (1), depending on the control parameters ω_0 and r. Convergence of the asymptotic solutions of (2) to equilibria described by fixed points ($\lambda_1 < 0$) is shown in black, to periodic (limit cycles) solutions ($\lambda_1 = 0$) – in violet/blue color (see the color bar for $\lambda_1 = 0$), to chaotic (nonperiodic) solutions ($\lambda_1 > 0$)—in a color, consistently with the color bar scale, from violet/blue to yellow. For the panel an enlargement of the region bounded by black lines is shown in the right-bottom part of plots. Fine structures are shown in the inset. This proves that various kinds of complex behavior are closely neighbored in the space of control parameters ω_0 and r.

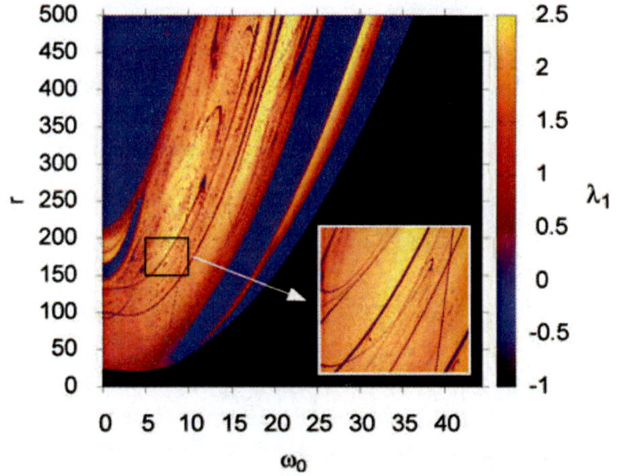

Fig. 5 Color-coded dependence of the long-term asymptotic solutions of the generalized Lorenz system on the control parameters ω_0 and r parameters (for $\sigma_m = 3$). Equilibria (fixed points) (with a negative largest Lyapunov exponent, $\lambda_1 < 0$) are shown in black, periodic solutions ($\lambda_1 = 0$)—in violet/blue, and (nonperiodic) chaotic solutions ($\lambda_1 > 0$)—in a color, on the color bar scale, from violet to yellow. Fine structures are shown in the inset, as taken from Macek and Strumik (2014)

Convection appears naturally in plasmas, where electrically charged particles interact with the magnetic field. Therefore, the obtained results could be important for explaining dynamical processes in solar sunspots, planetary and stellar fluid interiors, and possibly for plasmas in nuclear fusion devices. Generally speaking, nonlinear differential equations or iterated discrete maps are useful models of some phenomena appearing naturally in the contexts in biology (e.g., animal population), economics, including finance theory, e.g., [22], and social sciences.

4 Fractals and Multifractals

Let us now move on to a basic concept of a *fractal* coined from the Latin adjective *fractus* and the corresponding verb *frangere*, which means 'to break into irregular fragments', see p. 4 of [20]; Mandelbrot (1982) always argued that fractal geometry is important for understanding the structure of nature describing, for example clouds, mountains, and coastlines, e.g. p. 1 of [20]. We can say that a **fractal** is a rough or fragmented geometrical object that can be subdivided in parts, each of which is (at least approximately) a reduced-size copy of the whole. Fractals are generally *self-similar* and independent of scale, described by a fractal dimension.

Namely, fractal structure is obtained recursively using a simple rule. The initial stages of the construction of two typical fractals in one-dimensional and two-dimensional space are schematically illustrated in Fig. 6 for a middle Cantor (a) and a Koch triangle (b) sets, respectively, which are also discussed in many textbooks, e.g. [21, 29]. First, as proposed by the German mathematician Georg Cantor in 1883, let us take a unit closed interval on a one-dimensional line and remove its open middle third, but necessarily leaving the endpoints behind. Second, we remove the open middle thirds of both closed smaller intervals, and in each of the following k-th step this produces 2^k closed (more and more narrower) intervals of length $(2/3)^k$, where $k = 1, \ldots, n$. Now imagine that the repetitions never end, one obtains the limiting set that consists of the intersection of all such closed intervals. Provided that $n \to \infty$, the resulting set has structure at arbitrarily small scales; the remaining elements during the construction are separated by various gaps. Surprisingly enough, two paradoxically opposite topological properties of the Cantor set (called also a dust) can be reconciled: the set itself is totally disconnected (without any closed intervals), but arbitrarily close to each elements one can always find another neighboring element (there are no isolated points).

Further, it is worth noting that each element of this set is specified by its location at successive steps, in the left (denoted by zero) or right (marked by one) fragment. One now sees that elements of the Cantor set are equivalent to various infinite sequences of zeros and ones, and can be put into one-to-one correspondence with the elements of the entire initial interval (in binary representation). Because common sense has some difficulty in comparing countable with uncountable infinity, this is somewhat strange that the Cantor set is uncountable, notwithstanding its total length equal to zero (the length of all the removed parts is equal one). Mainly because of this paradox, such sets are commonly called strange fractals, even though one can also construct fractals with length or in general volume (strictly a Lebesgue measure) different than

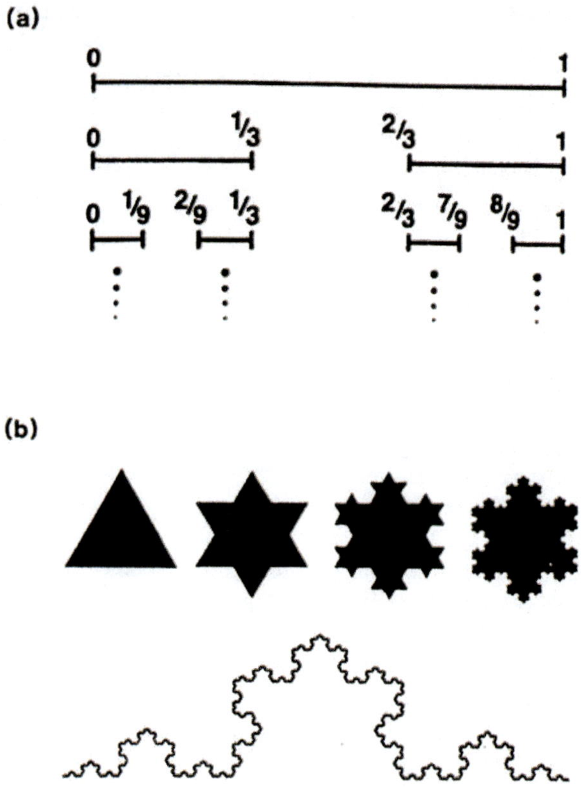

Fig. 6 Self-similar fractals of the Cantor (**a**) and Koch (**b**) sets

zero. Similar fractal sets with zero Lebesgue measures constructed starting from a triangle or a full square on a two-dimensional plane were proposed by the Polish mathematician Wacław Sierpiński (1882–1969) in 1916.

Figure 6b shows another interesting snowflake curve obtained on a plane by adding onto sides of an initial equilateral triangle additional triangles that are three times smaller, after removing as before open middle thirds of any side. Blowing up this van Koch curve by a factor of three results in its length four times as large, and hence the length of perimeter of the triadic Koch island increases and becomes ultimately infinite, despite the fact that the area of course remains finite. Surprisingly, the arc length between any two elements of such a Koch set is also infinite. Therefore, because every element of this set is located infinitely far from any other element, the length cannot be used to identify the elements of such a strange fractal.

Mandelbrot (1982) noted that a fractal (Hausdorff) dimension,[2] which plays a central roles in case of fractal sets, exceeds the topological dimension, D_T [20]. Anyway,

[2] Strictly speaking, the Hausdorff dimension is more involved that a usual fractal capacity dimension. The boxes needed to cover a set may vary in sizes and one needs to take a supremum of the cover of the set.

the concept of dimension should be modified as compared with a standard topological dimension useful in the Euclidean linear geometry. However, a somewhat different definition of a fractal set is generally accepted. The capacity dimension D_F, which takes into account how many elements (cubes) of size l in phase space is needed to cover the set, is defined by

$$D_F = \lim_{l \to 0} \frac{\ln N(l)}{\ln 1/l}. \tag{3}$$

This means that fractal dimension is calculated by taking the limit of the quotient of the logarithm change in object size and the logarithm in scale as the limiting scale approaches zero. For example, the fractal dimensions of the Cantor and the Koch sets are $D_F = \ln 2/\ln 3 \approx 0.63$ (this means $D_F > 0$) and $D_F = \ln 4/\ln 3 \approx 1.26$ (> 1) i.e., greater than the respective topological dimensions, $D_T = 0$ and 1. As is known, the later non-integer dimension describes sufficiently well the length of the rocky western coast of Great Britain as a function of diminishing scale size, see [19]; in reality the lowest scale is admittedly limited.

4.1 Multifractal Models for Turbulence

A deviation from a strict self-similarity is also called INTERMITTENCY, and that is why a generalized two-scale weighted Cantor set has been applied for modeling intermittent turbulence in fluids [8, 9].

In fact, this complex process can be described by the generalized weighted Cantor set, as illustrated in Fig. 7 taken from [8]. In the first step of the two-scale model construction, we have two eddies of sizes l_1 and l_2 satisfying $p_1/l_1 + p_2/l_2 = 1$. Therefore, the initial energy flux ε_0 is transferred to these eddies with the different proportions: $\varepsilon_0 p_1/l_1$ and $\varepsilon_0 p_2/l_2$. In the next step the kinetic or magnetic energy flux is divided between four eddies in the following way: $\varepsilon_0 (p_1/l_1)^2$, $\varepsilon_0 p_1 p_2/(l_1 l_2)$, $\varepsilon_0 p_2 p_1/(l_2 l_1)$, and $\varepsilon_0 (p_2/l_2)^2$. At nth step we have $N = 2^n$ eddies and partition of energy ε can be described by the binomial formula, e.g., [9]:

$$\varepsilon = \sum_{i=1}^{N} \varepsilon_i = \varepsilon_0 \sum_{k=0}^{n} \binom{n}{k} \left(\frac{p_1}{l_1}\right)^{(n-k)} \left(\frac{p_2}{l_2}\right)^{k}. \tag{4}$$

For any real number $-\infty < q < +\infty$, one obtains the generalized dimension defined by $D_q = \tau(q)/(q-1)$ by solving numerically the transcendental equation, e.g., [21],

$$\frac{p_1^q}{l_1^{\tau(q)}} + \frac{p_2^q}{l_2^{\tau(q)}} = 1, \tag{5}$$

which is only somewhat more general than the analytical solution. In particular, for the one-scale multifractal model with $l_1 = l_2 = \lambda$, we have $D_q = -\ln(p_1^q + p_2^q)/\ln \lambda$, and a special case for λ = 1/2 is called P-model, as classified on the right

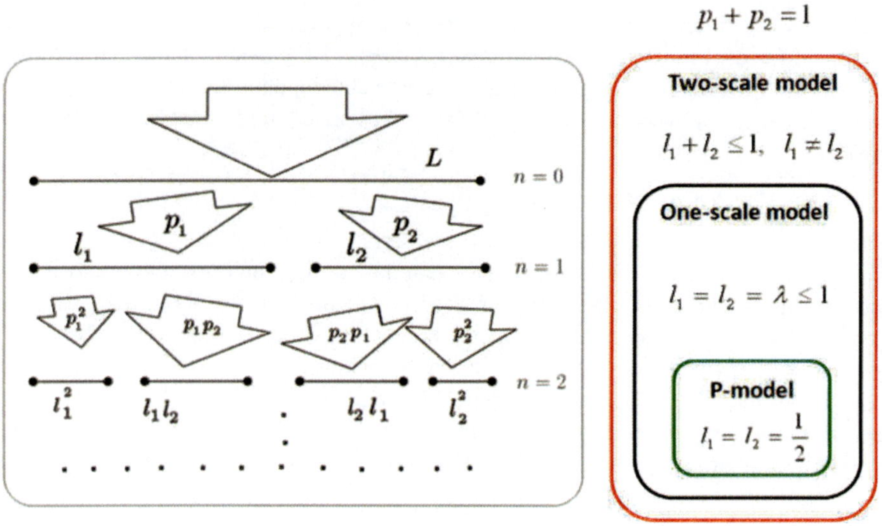

Fig. 7 The generalized two-scale weighted Cantor set model for turbulence

side of Fig. 7. We see that only for equal scales together with equal weights ($p_1 = p_2 = 1/2$) there is no multifractality, and we have a monofractal with a fractal dimension given by (3).

5 Implications for Cosmology and the Creation of the Universe

This method was extensively used in various situations in solar wind magnetized plasmas based on space missions penetrating various regions of the solar system, see [9, 13, 16, 17]. In this way, based on a wealth of data acquired from Helios in the inner heliosphere and especially from deep space Voyager 1 and 2 spacecraft in the outer heliosphere, we have shown that turbulence is intermittent in the entire heliospheric system, even at the heliospheric boundaries [18]. However, it appears that the heliosphere is immersed in a relatively quiet very local interstellar medium. Therefore, after crossing the heliopause (on 25 August 2012), which is the ultimate boundary separating the heliospheric and interstellar plasmas, Voyager 1 only detected smoothly varying magnetic fields. As expected this change in the behavior of plasma parameters (with a frozen-in magnetic field) was confirmed by the crossing of the heliopause by Voyager 2 in 2018.

Moreover, based on scientific experience, I have argued that a simple but possibly **nonlinear law** [7], within the theory of *chaos* and (multi-)fractals, can describe a hidden ORDER for the creation of the COSMOS, at the Planck epoch, when space (at a scale of 10^{-35} m) and time (10^{-43} s) originated, see [12, p. 3.4].

6 Conclusions

To summarize, based on space, astrophysical, and even cosmological applications, one can say that

- **Nonlinear** systems exhibit complex phenomena, including bifurcation, intermittency, and *chaos*.
- Fractals can describe complex shapes in the real word.
- Strange *chaotic* attractors have fractal structure and are sensitive to initial conditions.
- Within the complex dynamics of the fluctuating intermittent parameters of turbulent media there is a detectable, hidden *order* described by a generalized Cantor set that exhibits a multifractal structure.

Acknowledgements The research studies on chaos and turbulence in space plasmas have been funded in part by the National Science Centre (NCN), Poland, through grants 2014/15/B/ST9/04782 and 2021/41/B/ST10/00823. I would like to thank the NASA/WMAP Science and ESA/NASA Planck Teams for providing the schematic of the Evolution of the Universe and the illustrations of gravitation and geometry.

References

1. P.B. Abbot et al., Observation of gravitational waves from a binary black hole merger. Phys. Rev. Lett. **116**, 061102 (2016). https://doi.org/10.1103/PhysRevLett.116.061102
2. J.B. Hartle, S.W. Hawking, Wave function of the Universe. Phys. Rev. D **28**, 2960–2975 (1983). https://doi.org/10.1103/PhysRevD.28.2960
3. M. Heller, *Ultimate Explanations of the Universe* (Springer, Berlin, Heidelberg, 2009). https://doi.org/10.1007/978-3-642-02103-9
4. M. Heller, W. Sasin, Noncommutative structure of singularities in general relativity. J. Math. Phys. **37**, 5665–5671 (1996). https://doi.org/10.1063/1.531733
5. A.D. Linde, Eternally existing self-reproducing chaotic inflationary universe. Phys. Lett. B **175**, 395–400 (1986). https://doi.org/10.1016/0370-2693(86)90611-8
6. E.N. Lorenz, Deterministic nonperiodic flow. J. Atmos. Sci. **20**, 130–141 (1963). https://journals.ametsoc.org/view/journals/atsc/20/2/1520-0469_1963_020_0130_dnf_2_0_co_2.xml
7. W.M. Macek, On being and non-being in science, philosophy, and theology, in *Interpretazioni del reale. Teologia, filosofia e scienze in dialogo* (Interpretations of Reality: a Dialogue among Theology, Philosophy, and Sciences). Pontificia Università Lateranense (Pontifical Lateran University), Rome, Italy, ed. by P. Coda, R. Presilla. Quaderni Sefir, vol. 1 (2000), pp. 119–132
8. W.M. Macek, Multifractality and intermittency in the solar wind. Nonlinear Process. Geophys. **14**(6), 695–700 (2007). https://doi.org/10.5194/npg-14-695-2007
9. W.M. Macek, Multifractal turbulence in the heliosphere, in *Exploring the Solar Wind*. ed. by M. Lazar (Intech, Croatia, 2012), pp. 143–168
10. W.M. Macek, Intermittency in the generalized Lorenz model, in *Chaotic Modeling and Simulation*, ed. by C. Skiadas. Int. J. Nonlinear Sci. **4**, 323–328 (2015). http://www.cmsim.eu/
11. W.M. Macek, Nonlinear dynamics and complexity in the generalized Lorenz system. Nonlinear Dyn. **94**(4), 2957–2968 (2018). https://doi.org/10.1007/s11071-018-4536-z

12. W.M. Macek, *The Origin of the World: Cosmos or Chaos?* (Cardinal Stefan Wyszyński University (UKSW) Scientific Editions, 2020). ISBN 978-83-8090-686-0, e-ISBN 978-83-8090-687-7
13. W.M. Macek, A. Wawrzaszek, Evolution of asymmetric multifractal scaling of solar wind turbulence in the outer heliosphere. J. Geophys. Res. **114**(A3), A03108 (2009). https://doi.org/10.1029/2008JA013795
14. W.M. Macek, M. Strumik, Model for hydromagnetic convection in a magnetized fluid. Phys. Rev. E **82**(2), 027301 (2010). https://doi.org/10.1103/PhysRevE.82.027301
15. W.M. Macek, M. Strumik, Hyperchaotic intermittent convection in a magnetized viscous fluid. Phys. Rev. Lett. **112**(7), 074502 (2014). https://doi.org/10.1103/PhysRevLett.112.074502
16. W.M. Macek, A. Wawrzaszek, V. Carbone, Observation of the multifractal spectrum at the termination shock by Voyager 1. Geophys. Res. Lett. **38**, L19103 (2011). https://doi.org/10.1029/2011GL049261
17. W.M. Macek, A. Wawrzaszek, V. Carbone, Observation of the multifractal spectrum in the heliosphere and the heliosheath by Voyager 1 and 2. J. Geophys. Res. **117**, A12101 (2012). https://doi.org/10.1029/2012JA018129
18. W.M. Macek, A. Wawrzaszek, L.F. Burlaga, Multifractal structures detected by Voyager 1 at the heliospheric boundaries. Astrophys. J. Lett. **793**, L30 (2014). https://doi.org/10.1088/2041-8205/793/2/L30
19. B.B. Mandelbrot, How long is the coast of Britain? Statistical self-similarity and fractional dimension. Science **156**(3775), 636–638 (1967). https://doi.org/10.1126/science.156.3775.636
20. B.B. Mandelbrot, *The Fractal Geometry of Nature* (W. H. Freeman, New York, 1982)
21. E. Ott, *Chaos in Dynamical Systems* (Cambridge Univ. Press, Cambridge, 1993)
22. E.E. Peters, *Chaos and Order in the Capital Markets: A New View of Cycles, Prices, and Market Volatility*, 2nd edn. (Addison-Wesley, Reading, MA, 1996). Wiley, 978-0-471-13938-6
23. Y. Pomeau, P. Manneville, Intermittent transition to turbulence in dissipative dynamical systems. Commun. Math. Phys. **74**, 189–197 (1980). https://doi.org/10.1007/BF01197757
24. K.R. Popper, *The Logic of Scientific Discovery* (Routledge, London, New York, 1959)
25. B. Saltzman, Finite amplitude free convection as an initial value problem—I. J. Atmos. Sci. **19**, 329–341 (1962). https://journals.ametsoc.org/view/journals/atsc/19/4/1520-0469_1962_019_0329_fafcaa_2_0_co_2.xml
26. P.J. Steinhardt, N. Turok, A cyclic model of the universe. Science **296**, 1436–1439 (2002). https://doi.org/10.1126/science.1070462
27. P.J. Steinhardt, N. Turok, Cosmic evolution in a cyclic universe. Phys. Rev. D **65**(12), 126003 (2002). https://doi.org/10.1103/PhysRevD.65.126003
28. I. Stewart, *Does God Play Dice?: The New Mathematics of Chaos* (Blackwell Publishers, 1990)
29. S.H. Strogatz, *Nonlinear Dynamics and Chaos: With Applications to Physics, Biology, Chemistry, and Engineering* (Addison-Wesley, Reading, MA, 1994)
30. E. Witten, String theory dynamics in various dimensions. Nucl. Phys. B **443**, 85–126 (1995). https://doi.org/10.1016/0550-3213(95)00158-O

Maximum Lyapunov Exponent Calculation

Vasiliy D. Pechuk, Tatyana S. Krasnopolskaya, and Evgeniy D. Pechuk

Abstract The maximum Lyapunov exponent characterizes the degree of exponential divergence of close trajectories. The presence of a positive Lyapunov exponent in the system indicates a rapid divergence over time of any two close trajectories and sensitivity to the values of the initial conditions. Therefore, the determination of the Lyapunov exponent makes it possible to identify a dynamical system as a system with chaotic dynamics in it. When studying the output signals of dynamic systems, it is often necessary to quantify the degree of randomness of the output signal when equations of the system are unknown. In this paper, the possibilities of accuracy improvement of the numerical algorithms of Benettin and Wolf for estimating the maximum Lyapunov exponents of an attractor of a dissipative dynamical system are shown. Under these procedures a system itself can be specified both analytically (by a system of differential equations) and only by output signal.

Keywords Maximum Lyapunov exponent · Algorithms of Benettin · Algorithms of Wolf · Initial conditions · Chaotic modeling

1 Introduction

As is known, one of the necessary conditions for the randomness of the dynamic behavior of the system is sensitivity to the values of the initial conditions [1–7]. As a quantitative criterion for this concept, the senior Lyapunov exponent (Lyapunov exponent) is usually used. Let a dynamical system be given analytically by a system

---

V. D. Pechuk
Kyiv National University of Construction and Architecture, Kyiv, Ukraine
e-mail: pechuk.vd@knuba.edu.ua

T. S. Krasnopolskaya (✉) · E. D. Pechuk
Institute of Hydromechanics NASU, Kyiv, Ukraine
e-mail: t.krasnopolskaya@tue.nl

E. D. Pechuk
e-mail: evgdmp@gmail.com

C. H. Skiadas and Y. Dimotikalis (eds.), *14th Chaotic Modeling and Simulation International Conference*, Springer Proceedings in Complexity,
https://doi.org/10.1007/978-3-030-96964-6_22

of differential equations in the Cauchy form:

$$\frac{dx(t)}{dt} = f(x(t), t), \tag{1}$$

where $x \in \mathbb{R}^n$, $\mathbb{R}^n$—is the phase space of the system, the initial conditions $x(t_0) = x_0$, $f : \mathbb{R}^{n+1} \to \mathbb{R}^n$—are a continuous vector—function satisfying the Lipschitz conditions for all arguments, except for time t. Then a solution to system (1) exists and is unique. Let us denote by x_i a point in the phase space $\mathbb{R}^n$ of the dynamical system corresponding to the radius vector of the state $\mathrm{x}(t_i)$. If in the course of time the solution of system (1) approaches a certain manifold A when time goes to infinity, then we call A the attractor of system (1). There can be one or several attractors in the phase space, hidden, large, etc., depending on the properties of the system itself (1). In this case, the most important characteristics of the system are [4, 6, 8, 9]:

- chaos or regularity (presence or absence of chaotic dynamics);
- dissipation or conservatism (presence or absence of energy dissipation);
- topological invariants of the attractor (for example, fractal dimension).

Depending on these characteristics, the behavior of the solution of the system in the phase space changes significantly. Let the solution of the system be obtained for some initial conditions and, after the transient process, a certain attractor is found. To find out whether the attractor has a sensitivity to the values of the initial conditions, which is characteristic of a strange (chaotic) attractor, let us calculate the maximum Lyapunov exponent (MLE).

Consider a point x_0 on the attractor of a dynamical system (Fig. 1) at the initial moment of time t_0. Having given some small value ε_0, we choose one more point $\tilde{x}_0$ on the attractor that satisfies the condition.

$$|\widetilde{x_0} - x_0| = \varepsilon_0.$$

It should be noted here that for correct operation it is a point on the attractor that is needed, and not close to the attractor [4, 6, 8, 9]. Otherwise, the obtained result will characterize not the behavior of the trajectory on the attractor, but near it, which is the main source of errors in calculating the MLE for non-conservative systems.

Fig. 1 Evolution of two close points on the attractor dynamic system

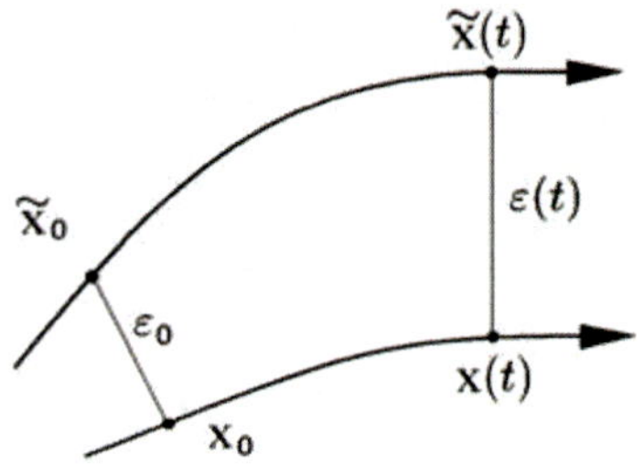

Through a period of timeΔt, the points x_0 and $\tilde{x}_0$ evolve into points $x(t)$ and $\tilde{x}(t)$. The Distance between them is denoted by $\varepsilon(x)$, here $t = t_0 + \Delta t$. The value $\varepsilon(t)$ depends on the initial position of the points x_0 and $\tilde{x}_0$, as well as the time interval Δt and the dynamic system as a whole. However, approximately, we can assume that

$$\varepsilon(t) \approx \varepsilon_0 e^{\lambda \Delta t},$$

where λ is the MLE. Thus, the parameter characterizing the dynamics of the representing point on the attractor is [1–4]

$$\lambda \approx \frac{1}{\Delta t} \ln \frac{\varepsilon(t)}{\varepsilon_0}. \tag{2}$$

Here it is necessary to take into account the fact that the boundedness of the attractor implies boundedness $\varepsilon(t)$ and, therefore, Δt should increase until $\varepsilon(t)$ is significantly less than the size of the attractor, otherwise λ will be equal to zero at $\Delta t \to \infty$. The value λ obtained in accordance with (2) should be considered as averaged over all the initial points x_0 of the attractor. Therefore,

$$\lambda = \overline{\lim_{\Delta t \to \infty} \frac{1}{\Delta t} \ln \frac{\varepsilon(t)}{\varepsilon_0}},$$

where $\varepsilon(t)$ is much smaller as the size of the attractor. This approach is based on the Oseledts ergodic theorem, see Oseledets [10], according to which the exponential divergence of two randomly selected points on the attractor characterizes the maximum Lyapunov exponent with probability equals 1.

In practice, the Benettin algorithm is used to find the values of the maximum Lyapunov exponent. Proposed by a team of authors in 1976 for the conservative Hénon-Heiles system [1], this calculation method works well and is suitable, first of all, for conservative systems. In their 1980 work Benettin et al. [2], the same authors used their results for smooth Hamiltonian dynamical systems. Due to the lack of alternatives, it is often used for dissipative systems, but in this case errors inevitably arise due to the fact that the dimension of the attractor of such a system is lower than the dimension of the phase space.

And the application of the classical Benettin's algorithm in the case of the existence of several attractors in the system and, moreover, hidden attractors, is generally incorrect. In this paper, we propose methods to improve the accuracy of this algorithm for the case of a dissipative dynamical system.

2 Procedure for Improving the Accuracy of Benettin's Algorithm

Let us denote by x_0 a point $x(t_0)$ on the attractor of the dynamical system (1) at the initial moment t_0, after the transient process. We choose a positive value ε—much less than the linear dimensions of the attractor and a point $\widetilde{x_0}$ satisfying the equality

$$|\widetilde{x_0} - x_0| = \varepsilon. \tag{3}$$

Let's track the evolution of points x_0 and $\widetilde{x_0}$ after a short period of time T. The resulting values will be denoted by x_1 and $\widetilde{x_1}$ (Fig. 2).

The vector $\Delta x_1 = \widetilde{x_1} - x_1$ is called the vector of disturbance, and its absolute value $|\Delta x_1|$ is called the amplitude of the disturbance. The MLE value for this stage is estimated by the formula:

$$\lambda_1 = \frac{1}{T} \ln \frac{|\widetilde{x_1} - x_1|}{\varepsilon} = \frac{1}{T} \ln \frac{|\Delta x_1|}{\varepsilon}.$$

Then the following renormalization is performed:

$$\Delta x_1^{'} = \frac{\Delta x_1}{|\Delta x_1|} \varepsilon,$$

and the described procedure is repeated for points $\widetilde{x_1^{'}} = x_1 + \Delta x_1^{'}$ and x_1, instead of $\widetilde{x_0}$ and x_0 (Fig. 2). After M repetitions, we find the MLE as the arithmetic mean of the values obtained at each stage

$$\lambda \approx \frac{1}{M} \sum_{k=0}^{M} \lambda_k = \frac{1}{M} \sum_{k=0}^{M} \frac{1}{T} \ln \frac{|\Delta x_k|}{\varepsilon} = \frac{1}{MT} \sum_{k=0}^{M} \ln \frac{|\Delta x_k|}{\varepsilon}.$$

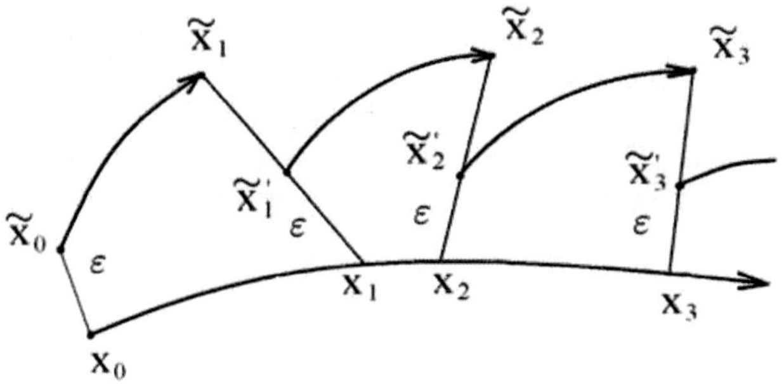

Fig. 2 Benettin's algorithm for computation the maximum Lyapunov exponent

It is clear that the algorithm works well in conservative models, where the volume of the phase space does not change over time. For example, in problems of mixing liquids in a tank, where the result obtained in this way directly characterizes the degree of mixing of the points of the medium [11, 12], where, however, it is shown that the maximum value of the maximum Lyapunov exponent does not guarantee the best mixing quality.

The main problem of this method in the case of a dissipative system is the fact that the points $\widetilde{x_1'}$, $\widetilde{x_2'}$, ..., generally speaking, do not lie on the attractor. The dimension of the attractor is lower than the dimension of the phase space; therefore, the probability of a randomly taken point hitting it is zero. So, for example, if the only attractor of the system is the limit cycle, then as a result of the application of Benettin's algorithm, the MLE will be negative, but should be zero. The following hitting options for a point are possible x_k':

- on the investigated attractor (probability zero);
- on another, possibly hidden attractor (probability zero);
- into the basin of attraction of the investigated attractor (the probability is positive);
- into the basin of attraction of another attractor (the probability is positive).

In the last two most probable cases, the value λ_k will characterize not the MLE value of the attractor, but the behavior of the trajectory close to it and introduce an error into the calculation result. To avoid this and increase the calculation accuracy, you need to select points $\widetilde{x_0}$, $\widetilde{x_1'}$, $\widetilde{x_2'}$,... on the investigated attractor. With the known right-hand sides of system (1), the solution can be obtained numerically, and the algorithm can look as follows. After the transient process, we select points x_0 and $\widetilde{x_0}$ on the solution of the system of Eq. (1) so that condition (3) is satisfied for some small one ε. Solving the system further, after T we get points x_1 and $\widetilde{x_1}$, respectively. Find the first value

$$\lambda_1 = \frac{1}{T} \ln \frac{|\widetilde{x_1} - x_1|}{\varepsilon}.$$

Next, we fix one of the obtained points, for example x_1, and we find $\widetilde{x_1'}$ by solving the system further from the point $\widetilde{x_1}$ until the condition $0 < \left| |\widetilde{x_1'} - x_1 \right| \leq \varepsilon.$ is fulfilled. Repeating this procedure M once, we find the MLE as the arithmetic mean of the values obtained at each stage:

$$\lambda \approx \frac{1}{M} \sum_{k=0}^{M} \lambda_k.$$

The obtained value, according to the ergodic theorem of Oseledts (see Oseledets [10]), characterizes the maximum Lyapunov exponent of the attractor under study with a probability of 1.

3 The Method of Applying the Wolf's Algorithm

When studying the output signals of dynamic systems, it is often necessary to quantify the degree of randomness of the output signal, with unknown equations of the system. In this case, the method proposed by Wolf et al. [3] in 1985 for calculating the Lyapunov exponent from the chaotic time realization of the system under study is usually used. It is based on the classical Benettin's algorithm and Takens theorem, see Takens [13]. According to this theorem, having an one-dimensional time realization $a(t)$ of a dynamical system belonging to a smooth manifold—an attractor of dimension d, the delay method can be used to reconstruct the original attractor as a n-dimensional set of state vectors $x(t) \in \mathbb{R}^n$, for $n \geq 2d + 1$

$$x(t) = (a(t), a(t + \tau), \ldots, a(t + (n - 1)\tau)).$$

The method for calculating the maximum Lyapunov exponent is as follows. Let the time realization $a(t)$ be given over a finite time interval at the moments

$$t_i = i\Delta t, \ i = 0, \ldots, N.$$

Let us denote by a x_i point in space $\mathbb{R}^n$ corresponding to the radius – vector of the state $x(t_i)$. Then, as a result of reconstruction, we obtain the attractor of the system as a sequence of points in space $\mathbb{R}^n$ (Fig. 3):

$$x_i = (a(i\Delta t), a(i\Delta t + \tau), \ldots, a(i\Delta t + (n - 1)\tau)), \tag{4}$$

here $\tau = m\Delta t$,$i = 0, \ldots, N - (n - 1)m$.

Take an arbitrary point from the sequence (4) and denote it an x_0. Further, passing through the sequence (4), we find on it a point $\widetilde{x_0}$, satisfying the inequality

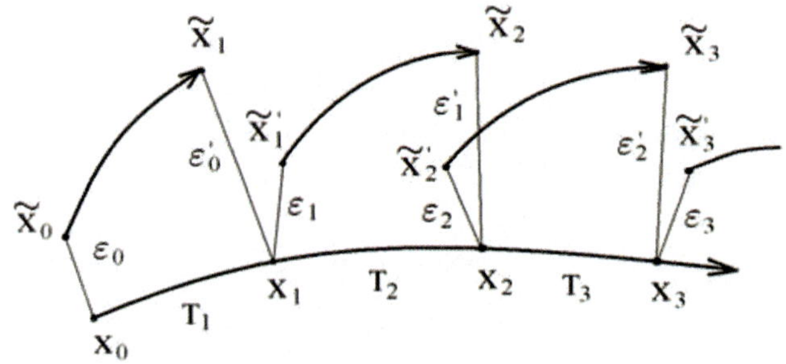

Fig. 3 Wolf's algorithm for computing the maximum Lyapunov exponent from the output signals set

$$|\widetilde{x_0} - x_0| = \varepsilon_0 \leq \varepsilon,$$

where ε is some constant significantly less than the linear dimensions of the reconstructed attractor. In this case, the points x_0 and $\widetilde{x_0}$ must be separated in time. After that, we track their evolution in time on the attractor until the distance between them exceeds a given value ε_{max}. Let's designate this time interval as T_1. Then, again going through the sequence (4), we find a point $\widetilde{x_1'}$ close to x_1, that the inequality

$$|\widetilde{x_1'} - x_1| = \varepsilon_1 \leq \varepsilon,$$

realized and vectors $\widetilde{x_1} - x_1$ and $\widetilde{x_1'} - x_1$ have the closest direction. The procedure is repeated again, only instead of points x_0 and $\widetilde{x_0}$ are taken already x_1 and $\widetilde{x_1'}$. After M repetitions, the maximum Lyapunov exponent is estimated as follows:

$$\lambda \approx \frac{1}{M} \sum_{k=0}^{M-1} \frac{\ln\left(\varepsilon_k'/\varepsilon_k\right)}{T_{k+1}}.$$

The considered Wolf algorithm is applicable only for a chaotic output signals set implementation (with a positive Lyapunov exponent), which somewhat reduces the universality of its practical application. In this paper, a modification of this algorithm is considered, which makes it possible to obtain both positive and negative values of MLE. The difference is as follows. The evolution of the selected starting points x_0 and $\widetilde{x_0}$ is tracked over an interval T, of a fixed length, and the distance ε_0' between the points x_1 and $\widetilde{x_1}$ obtained in this case is used to evaluate the MLE. The same is repeated for points x_1 and $\widetilde{x_1'}$. After M such length T steps, the value of the Lyapunov exponent can be estimated:

$$\lambda \approx \frac{1}{MT} \sum_{k=0}^{M-1} \ln\left(\varepsilon_k'/\varepsilon_k\right).$$

In the study the interaction in the human cardiorespiratory system Konovalyuk et al. [14], chaotic modes generated by the interaction of the respiratory and cardio subsystems were found. The classical Benettin's algorithm did not allow identifying the differences between quasiperiodic and chaotic dynamics. The use of the above upgrades made it possible to increase the accuracy of calculation the maximum Lyapunov exponent by an order of magnitude. For example, in the system of discrete maps describing the cardiorespiratory system, quasiperiodic and chaotic modes were found (Figs. 4 and 5). After the application of the classical Benettin's algorithm for the quasiperiodic regime $\lambda_1 \approx 0.008$, and for the chaotic regime were obtained $\lambda_2 \approx 0.01$. As a result of the application of the modified algorithm, the values were

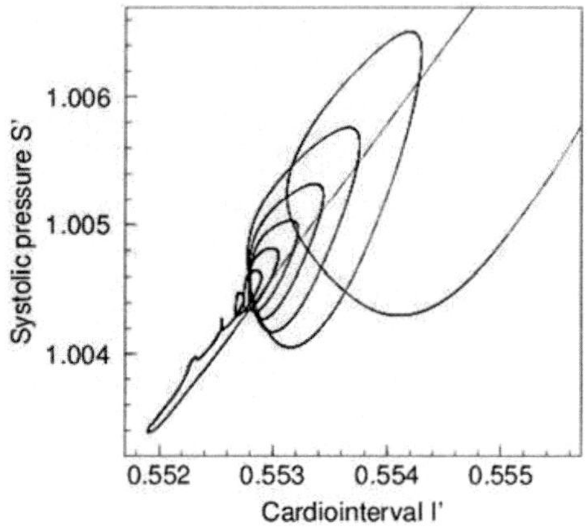

Fig. 4 Quasiperiodic mode in the system of discrete maps of the cardiorespiratory system model

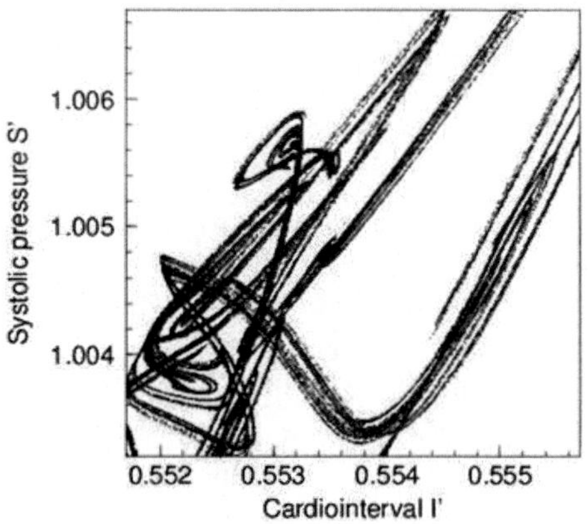

Fig. 5 Chaotic mode in the system of discrete maps of the cardiorespiratory system model

obtained as following $\lambda_1 \approx 0.001$ and $\lambda_2 \approx 0.01$, that made it possible to more accurately characterize the dynamics of the model.

	Quasiperiodic regime	Chaotic regime
Benettin's algorithm	0.008365	0.009623
Modified algorithm	0.001224	0.010334

4 Conclusions

Numerical calculation of the maximum Lyapunov exponent based on the classical Benettin's algorithm does not always give a good result if there is energy dissipation in the dynamic system. The article discusses modifications that improve the accuracy of calculations and expand the scope of the well-known algorithms of Benettin and Wolf. The use of the proposed modifications for the model of the human cardiorespiratory system made it possible to more accurately identify the differences between quasiperiodic and chaotic dynamics generated by the interaction of the respiratory and cardio subsystems.

References

1. G. Benettin, L. Galgani, J.M. Strelcyn, Kolmogorov entropy and numerical experiments. Phys. Rev. A **4**, 2338–2342 (1976)
2. G. Benettin L. Galgani, A. Giorgilli, J.M. Strelcyn, Lyapunov characteristic exponents for smooth dynamical systems and for Hamiltonian systems; a method for computing all of them. Meccanica 9–30 (1980)
3. A. Wolf, J.B. Swift, H.L. Swinney, J.A. Vastano, Determining Lyapunov exponents from a time series. Physica **16**(D), 285–317 (1985)
4. S.P. Kuznetsov, *Dynamic Chaos* (Fizmatlit, Moscow, 2001)
5. J. Laskar, K. Froscchle, A. Celletti, Measuring chaos using the numerical analysis of fundamental frequencies, appendix to the standard map (Translation from English—A.G. Arzamastseva, 1992)
6. V.A. Golovko, *Neural Network Methods for Processing Chaotic Processes* (Scientific session of MEPhI, MEPhI, Moscow, 2005), pp. 43–91
7. F. Moon, *Chaotic Oscillations* (Translation from English Yu.A. Danilova, A. M. Shukurova, Mir, Moscow, 1990)
8. P. Berger, I. Pomo, K. Vidal. *Order in Chaos*. On the deterministic approach to turbulence (Translated from French. Yu.A. Danilova, Mir, Moscow, 1991)
9. D.P. Crutchfield, Chaos. The World Sci. **2**, 18–28 (1987)
10. V.I. Oseledets, Multiplicative ergodic theorem. Lyapunov characteristic exponents of dynamical systems. Proc. Moscow Math. Soc. **19**, 179–210 (1968)
11. T.S. Krasnopolskaya, V.V. Meleshko, G.W.M. Peters, H.E.H. Meijer, Mixing in Stokes flow in an annular wedge cavity. Eur. J. Mech. B_Fluids **18**, 793–822 (1999)
12. V.V. Meleshko, T.S. Krasnopolskaya, G.W.M. Peters, H.E.H. Meijer, Coherent structures and scales of Lagrangian turbulence. (Advances in Turbulence_VI,, Kluwer, Dordrecht, 1996), pp. 601–604
13. F. Takens, in *Detecting Strange Attractors in Turbulence*, ed. by D. Rand, L.S. Young, Dynamical System in Turbulence, Warlock 1980. Lecture Notes in Mathematics, vol. 898 (Springer, Berlin, 1980), pp. 366–381
14. T.P. Konovalyuk, E.D. Pechuk, T.V. Sobol, T.S. Krasnopolskaya, *Influence of the Heart Rate on Dynamics of Cardiorespiratory System* (Springer Proceedings in Complexity, 2020), pp 211–216

New Fractal Features for Textural Morphologic Analysis

Alexander A. Potapov, Viktor A. Kuznetsov, and Anton N. Pototskiy

Abstract Significantly increased resolution of image formation systems (down to a few centimetres) causes a possibility of more effective using of objects textural features and signs in case of thematic processing of radar and optical images. The existing methods of image fractal features measurement allows to evaluate numerically the following topological characteristics of image texture: fractal dimension (FD); directional FD in the analysis directions (DFD); multifractal dimension (MFD) (a widespread case — the spectrum of Renyi dimensions (SRD)); morphological multifractal exponent (MME); fractal signature (FS) and directional FS (DFS); morphological MFS (MMFS) and lacunarity. However today there are no complex methods allowing to measure at the same time parameters of the scaling, multifractal and anisotropic properties of a texture possessing reciprocal relationships. In this work the specificities of new Directional Multifractal Blanket method (morphological) ($DMBM_M$) for fractal features measurement of an image textures synthesized on the basis of two best ABRG and MBM_M methods in the groups, are considered. Simultaneous accounting of multifractal, singular and anisotropic properties of the image texture with limited scaling character allowed to increase measuring accuracy both FD, and FS at each analysis scale. This feature is the most representative on comparing with all features considered in this work as the functional correlation of the derived features. The increased informativeness of the developed feature in case of image processing is caused by additional determination, along with multifractal and singular properties, anisotropic properties and their joint account and implied the possibility of its using for the properties description of different images textures and also in images clustering and segmentation tasks.

A. A. Potapov (✉)
V. A. Kotelnikov Institute of Radio Engineering and Electronics, Russian Academy of Sciences, Moscow, Russia
e-mail: potapov@cplire.ru

JNU-IREE RAS Joint Lab. of Information Technology and Fractal Processing of Signals, JiNan University, Guangzhou 510632, China

V. A. Kuznetsov · A. N. Pototskiy
Air Force Academy, Voronezh, Russia

C. H. Skiadas and Y. Dimotikalis (eds.), *14th Chaotic Modeling and Simulation International Conference*, Springer Proceedings in Complexity,
https://doi.org/10.1007/978-3-030-96964-6_23

Keywords Fractal dimension · Multifractal signature · Directional features · Anisotropic textures

Abbreviations

ABRG	Augmented iterative covering blanket method with rotating grid
AFS	Anisotropic fractal surface
DFD	Directional fractal dimension
DFS	Directional fractal signature
$DMBM_M$	Directional multifractal blanket method (morphological)
DMFS	Directional multifractal signature
DMMFS	Directional morphological multifractal signature
FD	Fractal dimension
FS	Fractal signature
L	Lacunarity
LFD	Local fractal dimension
LMME	Local morphological multifractal exponent
MBM_M	Morphological multifractal iterative covering blanket method
MFD	Multifractal dimension
MFS	Multifractal signature
MME	Morphological multifractal exponent
MMFS	Morphological multifractal signature
SRD	Spectrum of Renyi dimensions

1 Introduction

The modern air- and space-based monitoring systems of Earth provides formation of high resolution optical and radar images for the wide range solutions of remote sensing, reconnaissance and special tasks. Significantly increased resolution of image formation systems (down to a few centimeters) causes a possibility of more effective using of objects textural features and signs in case of thematic processing of radar and optical images [1, 2]. Textural approach [3] is based on the fact, that in most cases spatial configurations of high resolution images brightness units within boundaries of heterogeneous classes of objects have essential differences. Specific numerical values of textural features decide on the help of different mathematical apparatuses among which it is possible to select the wavelet-analysis, Fourier analysis, variance analysis and also a number of the modern methods, based on the fractal theory [1, 2].

In fundamental research in the fractal theory field [1, 4], and also related practical applications of digital image processing [5–10], formulated and proved statements about presence of fractal properties and the characteristic features corresponding to them at images of natural objects. The most distinctive properties of fractal sets are the scale invariance (scaling), the continuity and nondifferentiability described within the mathematical theory of the fractional integro-differential equations [1, 4].

Fractal processing implies receiving numerical evaluations of scale invariance indices of the image texture, by means of a research of local and global topological features of spatial structure of its intensity field, and the subsequent image differentiation on homogeneous areas on the basis of the measured values [1, 10]. Now the methods of the automatic analysis and thematic processing of images based on separate using of the estimated values of different fractal features are developed. Such widely used features include the fractal and the multifractal dimensions (FD and MFD), the fractal and the multifractal signatures (FS and MFS), allowing to find areas of textural homogeneity on images with different efficiency.

Both in foreign, and in domestic scientific publications there was a many results describing researches of new methods of measurement of scaling, spatial, statistical and other parameters of fractal sets taking into account textural images formation features (see, for example, [1, 2, 10–14]). At the same time such properties of images as multifractality, singularity (in a broad sense — the local non-uniformity), limited scaling feature and anisotropic are taken into account. However today there are no complex methods allowing to measure at the same time parameters of the scaling, multifractal and anisotropic properties of a texture possessing reciprocal relationships.

The purpose of this work is to synthesize of a new directional morphological multifractal signature computation method for image texture, a research of its functional capabilities, assessment of measuring accuracy of some fractal features by the developed method when processing of the test synthesized and real images both fitting, and not meeting a self-similarity condition and also determination of new textural-fractal features informativeness in tasks of the images textural analysis.

2 Description of Images Texture Fractal Features

The existing methods of the image fractal features measurement al-lows to evaluate numerically the following topological characteristics of image texture: the FD; the directional FD in the analysis directions (DFD); the MFD (in the wide sense — the spectrum of Renyi dimensions (SRD)); the morphological multifractal exponent (MME); the FS and the directional FS (DFS); the morphological MFS (MMFS) and lacunarity (L) [1, 2, 15]. These fractal features are systematized in Table 1. The main properties of these features are characterized as follows.

In geometry terms, the FD D characterizes a level of the isotropic surface texture «roughness» (in case of $D = 2$ the surface formed by a combination of the image brightness units is absolutely smooth, and in case of $D = 2$ — infinitely «wrinkled»).

The Renyi dimensions D_q are sensitive to the inhomogeneities of the analyzable surface which is characterized by combining of areas with different the FD's and allow to describe the global and local topological features of the texture. The DFD

Table 1 Fractal features of an images texture

Sign	Symbol	Mathematical formulations
FD	D	$D = \lim_{\varepsilon\to 0} \frac{\log N(\varepsilon)}{\log(1/\varepsilon)}$
MFD	$[D_q]$	$D_q = (1-q)^{-1} \lim_{\varepsilon\to 0} \frac{\log(I(\varepsilon))}{\log(1/\varepsilon)}$
DFD	$[D(\varphi_n)]$	$D(\varphi_n) = \lim_{\varepsilon\to 0} \frac{\log N(\varepsilon,\varphi_n)}{\log(1/\varepsilon)}$
FS	$\mathbf{S} = [D(\varepsilon)]$	$D(\varepsilon) = \left(\log \frac{\varepsilon}{\varepsilon+1}\right)^{-1} \log \frac{A(\varepsilon+1)}{A(\varepsilon)}$
DFS	$\mathbf{S}_n = [D(\varepsilon, \varphi_n)]$	$D(\varepsilon, \varphi_n) = \left(\log \frac{\varepsilon}{\varepsilon+1}\right)^{-1} \log \frac{A(\varepsilon+1,\varphi_n)}{A(\varepsilon,\varphi_n)}$
MME	$[L_q]$	$L_q = \frac{1}{\lvert q\rvert} \lim_{\varepsilon\to 0} \frac{\log(Z(q,\varepsilon))}{\log(1/\varepsilon)}$
MFS	$\mathbf{S}_q = [L_q(\varepsilon)]$	$L_q(\varepsilon) = \left(\log \frac{\varepsilon}{\varepsilon+1}\right)^{-1} \log \frac{Z(q,\varepsilon+1)}{Z(q,\varepsilon)}$
L	$[\Lambda(\varepsilon)]$	$\Lambda(\varepsilon) = \left(M^2(\varepsilon) - (M(\varepsilon))^2\right)/(M(\varepsilon))^2$
	$[C(\varepsilon)]$	$C(\varepsilon) = (M(\varepsilon) - N(\varepsilon))/(M(\varepsilon) + N(\varepsilon))$
DMFS	$\mathbf{S}_{q,n} = [L_q(\varepsilon, \varphi_n)]$	$L_q(\varepsilon, \varphi_n) = \left(\log \frac{\varepsilon}{\varepsilon+1}\right)^{-1} \log \frac{Z(q,\varepsilon+1,\varphi_n)}{Z(q,\varepsilon)}$
	$\mathbf{S}_{q,n}^{\text{opt}} = [L_q(\varepsilon(\varphi_{n_{opt}}))]$	$\exists L_q(\varepsilon(\varphi_{n_{opt}})) \in [L_q(\varepsilon, \varphi_n)]$, $n_{\text{opt}} = \lfloor \Delta\varphi^{-1}(\pi/2 + \psi(q, \varepsilon)) \rfloor$

components collection $D(\varphi_n)$ allows to define correctly a level of the anisotropic fractal surface (AFS) «roughness» which is characterized by the different values of the FD along the analysis different angular directions $n_{\max}$.

In [2, 16, 17] it is marked that in case the FD computation D of the images texture loss of information on its singularities can happen by the known methods. This negative aspect is caused by the fact that the image **I** self-similarity remains only within some limited range of degree dependence of the evaluated image measure on the analysis scale ε constructed in double logarithmic scale and approximated by linear dependence. At the same time the great value is acquired by the «personal» topological features of the image texture, but not average implementations having often absolutely other character [1, 2, 9, 14, 18].

For successful permission of this mismatch in the above-mentioned articles the signature approach consisting in finding of the local fractal dimensions (LFD) $D(\varepsilon)$ calculated for the adjacent analysis scales ε and $\varepsilon + 1$ is offered and reasonable and with their subsequent combining in an ordered set — the FS $S = \{D(\varepsilon)\}$, where $\varepsilon = \overline{1, \varepsilon_{\max} - 1}$, and $\varepsilon_{\max}$ is the maximum number of analyzable scales.

The FS distinctive feature of the images having strictly scale and invariant properties is the persistence of the LFD values in all range of the analyzable scales. Computation of the FS allows to reveal the images texture singularities even in that case when they have no the fractal properties. Thus, the FS **S** characterizes the scaling ratio of a measure variation of the researched surface and existence of the scale singularities of its texture.

It is necessary to mark a number of articles [19–26] devoted to the description of the anisotropic properties of a texture in the case of measurement of the DFS Sn in the tasks of automated processing of X-ray images of a human bone tissue and microsamples of the constructional materials. This sign is closely related to the parameterization of the texture directional properties taking into account scale singularities.

The sign $\left[L_q\right]$ is the SRD $\left[D_q\right]$ alternative in the case of the MME spectrum calculation with using the image morphological processing.

The applying of the MMFS $\mathbf{S}_q$ is intended for assessment of the local MME variation measure (LMME) $L_q(\varepsilon)$ and allows to consider at the same time both the singular and multifractal texture properties.

The lacunarities $[\Lambda(\varepsilon)]$ and $[C(\varepsilon)]$ respectively allow to obtain the average information on (filling) of the fractal surface «mass» distribution at the big and small analysis scales.

Arguments of the features shown in Table 1 are provided by the following parameters: $N(\varepsilon)$ is the number of cubes with the side (scale) ε in the case of cellular partition completely covering the researched image surface; $\mathbf{I}(q,\varepsilon)$ is the generalized statistical amount (a probability multifractal measure of the image surface area distribution) with number of the scaling moments orders q; $Z(q,\varepsilon)$ is the generalized statistical amount evaluated in the case of morphological computation $\mathbf{I}(q,\varepsilon)$ with use of the structural elements set which value w corresponds to the analyzable scale, where $w = 2\varepsilon + 1$; $A(\varepsilon)$ is the image surface area evaluated at the scale ε; $M^2(\varepsilon)$ is the fractal set mass; $(M(\varepsilon))^2$ is the fractal set expected mass; n is the analysis angular direction number; $\Delta\varphi$ is the elementary angular direction; $\mathbb{Z}$ is the set of integer numbers; $[\cdot]$ is the array of values; $\exists$ is the existential quantifier at least the one element from the definition range.

The high resolution optical and radar images with the significantly heterogeneous texture are characterized by the both scale singularities, and the anisotropic and multifractal properties therefore the productive applying of the known fractal features in the case of image processing of the similar character encounters a restrictions number. This is because, on the one hand, the fractal features describes preferentially separated aspects of the texture properties, without its complex, integrative character, and on the other hand, in the case of their sharing, the correlations existence accounting between the texture elements components is not carried out, i.e. the signs components connected, in essence, in fact are calculated «separately» (independent) from each other. Finally, the characteristics measured in this way lose the image texture specified properties description integrity.

The peculiarities discussed in terms of the high resolution images fractal theory and the also limited information pithiness of the listed fractal features point to need for the complex morphological method development allowing to derive at the same time scaling, multifractal and anisotropic properties of the image texture with measurement of the most relevant of them that is taking into account their reciprocal relationships.

3 Synthesis of the Directional Morphological Multifractal Signature Measurement Method

Basis of the new method allowing to reveal not only multifractal and singular, but also anisotropic properties of processed images, the morphological implementation of the augmented iterative covering blanket method with rotating grid (ABRG) [24] with the modified choice procedures of the rotating grid size and formation of the horizontally oriented structural elements set makes. The choice in favor of morphological processing is because the morphological extension (dilatation) and the morphological corrupting (erosion) of the processed image surface by the structural element allows to consider all surface irregularities which extent we will commensurate with the size of the structural element, and, thus, to make the image analysis at the given scale ε. Besides, the FD and the FS measurement methods by means of the morphological implementation have the high accuracy. For accounting of the multifractal properties of an image texture in this work the morphological multifractal iterative covering blanket method ($\mathrm{MBM_M}$) is used [17].

The entity of the developed method consists in computation of the upper and lower coverings set by means of the ABRG method [24] rotating grid modified by authors, the LMME $L_q(\varepsilon,\varphi_n)$ used in the computation case (see a line 9 of Table 1) for the required number of the analysis angular directions of the processed image by the $\mathrm{MBM_M}$ method and formation of the directional morphological multifractal signature (DMMFS) in the «direction-scale» coordinates for each order q of the scaling moment.

The procedure of the image **I** texture DMMFS $S_{q,n}$ computation can be presented in a general view by the linear and non-linear operators set.

$$\mathbf{I} \xrightarrow{\mathcal{L}_R} \{\mathbf{I}_{\varphi_n}\} \longrightarrow \begin{matrix} \nearrow \xrightarrow{\mathcal{L}_U} \mathbf{U}_{\varepsilon,\varphi_n} \longrightarrow \searrow \\ \searrow \xrightarrow{\mathcal{L}_B} \mathbf{U}_{\varepsilon,\varphi_n} \longrightarrow \nearrow \end{matrix} \xrightarrow{\mathcal{L}_Z}$$

$$\xrightarrow{\mathcal{L}_Z} Z(q,\varepsilon,\varphi_n) \xrightarrow{\mathcal{L}_L} L_q(\varepsilon,\varphi_n) \xrightarrow{\mathcal{L}_{1S}} \mathbf{S}_{q,n} \xrightarrow{\mathcal{L}_{2S}} \mathbf{S}_{q,n}^{\mathrm{opt}},$$

where $\mathcal{L}_R$ is the set $\{\mathbf{I}_{\varphi_n}\}$ formation function of turned on the required number angular provisions of source image I copies; $\mathcal{L}_U$ and $\mathcal{L}_B$ is the calculation functions according to the dilatation and the erosion $\{\mathbf{I}_{\varphi_n}\}$ with use of the horizontally oriented structural elements set $\Upsilon\varepsilon$; $\mathcal{L}_Z$ is the calculation function of the generalized statistical amount $Z(q,\varepsilon,\varphi_n)$; $\mathcal{L}_L$ is the LMME $[L_q(\varepsilon,\ \varphi_n)]$ array calculation function; $\mathcal{L}_{1S}$ is the $\mathbf{S}_{q,n}$ formation operator; $\mathcal{L}_{2S}$ is the realizing the LMME $L_q(\varepsilon(\varphi_{n_{\mathrm{opt}}}))$ choice operator which is corresponding to the prevailing direction of a texture elements orientation for each scale of the analysis from a set $[L_q(\varepsilon,\ \varphi_n)]$ and the $\mathbf{S}_{q,n}^{\mathrm{opt}}$ formation, where nopt is the image turn number (the analysis direction) corresponding to the prevailing direction of a texture elements orientation.

As the input, intermediate and output variables serve: $\mathbf{I} = [I(i,\ j)]$, $I_{i,j} \in \overline{0,\ 2^{v-1}}$, $i = \overline{1,\ M}$, $j = \overline{1,\ N}$ is the digital grayscale image containing M lines and N

columns, presented in the matrix form with the quantized brightness levels in the appropriate image pixel; ν is the brightness quantization level; $\{\mathbf{I}_{\varphi_n}\}$ is the array (set) of the source image **I** turned copies, where $n = \overline{1, n_{\max}}$, $n_{\max} = \pi/\Delta\varphi$ is the number of turns (the image analysis directions); $\mathbf{U}_{\varepsilon,\varphi_n}$ and $\mathbf{B}_{\varepsilon,\varphi_n}$ is the «upper» and «lower» surfaces set created as a result of morphological processing $\{\mathbf{I}_{\varphi_n}\}$, where $\mathrm{U}_\varepsilon = [U_\varepsilon(i, j)]$ and $\mathrm{B}_\varepsilon = [B_\varepsilon(i, j)]$; $\mathbf{Y}_\varepsilon = [Y_1, Y_2, ..., Y_{\varepsilon_{\max}}]$ is the «plane» structural elements set in the horizontally oriented lines form of pixels which length w corresponds to the analyzable scale $w = 2\varepsilon + 1$; $Z(q,\varepsilon,\varphi_n)$ is the generalized statistical amount with number of the scaling moments orders $-\infty < q < \infty$, $q \neq 0$; $L(\varepsilon,\varphi_n)$ is the LMME created for n analysis directions; $\mathbf{S}_{q,n}$ is the directional morphological multifractal signature; $\mathbf{S}_{q,n}^{\mathrm{opt}}$ is the directional morphological multifractal signature with the reduced dimensionality which formation method explicitly was considered in the work [26], considering only the prevailing directions of a texture elements orientation at the different spatial scales.

Computation according to the source image **I** with size by $M \times N$ pixels of the upper and lower coverings is made for the nmax copies $\mathbf{I}_{\varphi_n}$ of the image turned on the angle $\Delta\varphi = \overline{\frac{\pi}{180}, \frac{\pi}{2}}$ of the size $M' \times M'$ using of the dilatation and the erosion operations by the modified set of the structural elements Y_ε, where $M' = \left\lceil \sqrt{(M+1)^2 + (N+1)^2} \right\rceil + 1$; $\lceil g \rceil$ is the rounding operator to the nearest whole to the big side. At the same time the upper $\mathbf{U}_{\varepsilon,\varphi_n}$ and lower $\mathbf{B}_{\varepsilon,\varphi_n}$ coverings values for scale $\varepsilon = 0$ are equal to the source images on an output of the turn operator $\mathcal{L}_R$

$$\mathbf{U}_{0,\varphi_n} = \mathbf{B}_{0,\varphi_n} = \mathbf{I}_{\varphi_n}, \tag{1}$$

and for the scales $\varepsilon > 1$ are defined by the expressions

$$U_{\varepsilon,\varphi_n}(i, j) = \max \mathbf{I}'_{\varphi_n};\ B_{\varepsilon,\varphi_n}(i, j) = \min \mathbf{I}'_{\varphi_n}, \mathbf{I}'_{\varphi_n} \in \mathbf{X}_\varepsilon, \tag{2}$$

where $\mathbf{X}_\varepsilon = \{I_{\varphi_n}(i, j')\}, |j' - j| \leq \lceil (w-1)/2 \rceil$ is the definition range of a structural element at the scale ε.

The surface area of the turned image $\mathbf{I}_{\varphi_n}$ sequentially is calculated for each scale on the basis of calculated the upper $\mathbf{U}_{\varepsilon,\varphi_n}$ and lower $\mathbf{B}_{\varepsilon,\varphi_n}$ coverings $S(\varepsilon, \varphi_n) = V(\varepsilon, \varphi_n)/2\varepsilon$, where $V(\varepsilon, \varphi_n) = \sum_{i=1}^{W} \sum_{j=1}^{W} (U_{\varepsilon,\varphi_n}(i, j) - B_{\varepsilon,\varphi_n}(i, j))$ is the image surface «volume» in the $W \times W$ window concluded between the lower and upper coverings.

The generalized statistical amount $Z(q, \varepsilon, \varphi_n)$ as a distribution function of a multifractal q-order set measure of at each analyzable scale ε for each turned image $\mathbf{I}_{\varphi_n}$ is determined by the equation

$$Z(q, \varepsilon, \varphi_n) = S(\varepsilon, \varphi_n) \sum_{i=1}^{W} \sum_{j=1}^{W} \left| U_{\varepsilon,\varphi_n}(i, j) - B_{\varepsilon,\varphi_n}(i, j) \right|^q V^{-1}(\varepsilon, \varphi_n), \tag{3}$$

where the scaling moment order q lies in value range $q \in \mathbb{Z}, q \neq 0$, $\mathbb{Z}$ is the integral numbers set. Here the $\mathbf{S}(\varepsilon,\varphi_n)$ provides computation of the morphological FD, entered in [27] (the similarity dimensionality analog $D_q\big|_{q=0}$ determined by the multifractal cellular method by means of the Renyi dimension spectrum finding), in the case of $Z_q(\varepsilon, \varphi_n)\big|_{q=1}$.

The MFS formation is carried out by the generalized statistical amount $Z(q, \varepsilon, \varphi_n)$ behavior determination between the adjacent analysis scales. The LMME calculation $Lq(\varepsilon, \varphi_n)$ according to expression is for this purpose made:

$$L(q, \varepsilon, \varphi_n) = \left(\log \frac{\varepsilon}{\varepsilon + 1}\right)^{-1} \log \frac{Z(q, \varepsilon + 1, \varphi_n)}{Z(q, \varepsilon)}. \tag{4}$$

The DMMFS array $\mathbf{S}_{q,n}$ values registers in the look

$$\mathbf{S}_{q,n} = \begin{bmatrix} \mathbf{L}_{-\infty}(\varepsilon_1) & \mathbf{L}_{-\infty}(\varepsilon_2) & \ldots & \mathbf{L}_{-\infty}(\varepsilon_{\max} - 1) \\ \vdots & \vdots & \ldots & \vdots \\ \mathbf{L}_{-1}(\varepsilon_1) & \mathbf{L}_{-1}(\varepsilon_2) & \cdots & \mathbf{L}_{-1}(\varepsilon_{\max} - 1) \\ \mathbf{L}_{1}(\varepsilon_1) & \mathbf{L}_{1}(\varepsilon_2) & \cdots & \mathbf{L}_{1}(\varepsilon_{\max} - 1) \\ \vdots & \vdots & \ddots & \vdots \\ \mathbf{L}_{\infty}(\varepsilon_1) & \mathbf{L}_{\infty}(\varepsilon_2) & \cdots & \mathbf{L}_{\infty}(\varepsilon_{\max} - 1) \end{bmatrix}, \tag{5}$$

where $\mathbf{L}_q(\varepsilon) = \left[L_q(\varepsilon, \varphi_1)\ L_q(\varepsilon, \varphi_2) \cdots L_q\left(\varepsilon, \varphi_{n_{\max}}\right)\right]^{\mathrm{T}}$ is the column vector LMME of dimensionality nmax of an order q of the given analysis scale ε, $\left[\mathrm{g}\right]^{\mathrm{T}}$ is the transposing operator.

In the work [26] for lowering of the DMMFS dimensionality and accounting only of the significant anisotropic properties of the analyzable image the procedure of the texture elements prevailing orientation directions determination based on approximations by the LMME values set ellipses $\left[L_q(\varepsilon, \varphi_n)\right]$ in the case of the given indices q and ε in a polar coordinate system and determination of the ellipticity coefficient $kec(q,\varepsilon)$ and the ellipse slope angle $\psi(q,\varepsilon)$ with the subsequent formation of the DMMFS with the reduced dimensionality $\mathbf{S}_{q,n}^{\mathrm{opt}}$ is offered. The accounting of the prevailing orientation direction of a texture elements in the DMMFS $\mathbf{S}_{q,n}$ is carried out by a choice from the LMME array $\left[L_q(\varepsilon, \varphi_n)\right]$ for each q and ε the measure value $L_q\left(\varepsilon\left(\varphi_{n_{opt}}\right)\right)$ in the turn number case $n_{\mathrm{opt}} = \left\lfloor \Delta\varphi^{-1}\left(\pi/2 + \psi(q, \varepsilon)\right)\right\rfloor$ provided that there $kec(q,\varepsilon)$ is the less threshold value ktr, where $\lfloor\cdot\rfloor$ is the rounding operator to the nearest whole to the smaller side.

As a result of the LMME $L_q(\varepsilon,\varphi_n)$ choice by the criterion of the texture elements prevailing (optimum) orientation direction, the DMMFS takes the form

$$\mathbf{S}_{q,n}^{\mathrm{opt}} = \begin{bmatrix} L_{q\,\min}\left(\varepsilon_1\left(\varphi_{n_{opt}}\right)\right) & \cdots & L_{q\,\min}\left(\varepsilon_{\max-1}\left(\varphi_{n_{opt}}\right)\right) \\ \vdots & \cdots & \vdots \\ L_{-1}\left(\varepsilon_1\left(\varphi_{n_{opt}}\right)\right) & \cdots & L_{-1}\left(\varepsilon_{\max-1}\left(\varphi_{n_{opt}}\right)\right) \\ L_{1}\left(\varepsilon_1\left(\varphi_{n_{opt}}\right)\right) & \cdots & L_{1}\left(\varepsilon_{\max-1}\left(\varphi_{n_{opt}}\right)\right) \\ \vdots & \ddots & \vdots \\ L_{q\,\max}\left(\varepsilon_1\left(\varphi_{n_{opt}}\right)\right) & \cdots & L_{q\,\max}\left(\varepsilon_{\max-1}\left(\varphi_{n_{opt}}\right)\right) \end{bmatrix}. \qquad (6)$$

Thus, the DMMFS and the DMMFS with reduced dimensionality receiving numerical evaluations peculiar properties by using the directional morphological multifractal blanket method ($\mathrm{DMBM_M}$) by the fractal signatures computation means for the given scaling moments orders q taking into account the prevailing orientation directions of the image texture elements are considered.

4 Conclusions

In this work the specificities of the new $\mathrm{DMBM_M}$ method for fractal features measurement of an image textures synthesized on the basis of two best the ABRG and the $\mathrm{MBM_M}$ methods in their groups, are considered. Simultaneous accounting of the multifractal, singular and anisotropic properties of the image texture with limited scaling character allowed to increase measuring accuracy both the FD, and he LFD at each analysis scale.

References

1. A.A. Potapov, *Fractals in radiophysics and radiolocation: sample topology*, Universitetskaya kniga, Moscow (2005), (in Russian)
2. A.A. Potapov, Postulate "the topology maximum at the energy minimum" for textural and fractal-and-scaling processing of multidimensional super weak signals against a back-ground of noises, in L.A. Uvarova, A.B. Nadykto, A.V. Latyshev, ed. by *Nonlinearity: Problems, Solutions and Applications*, 2nd edn. (Nova Science Publ., New York, 2017)
3. R.M. Haralick, K. Shanmugan, I. Dinstein, Textural features for image classification. IEEE Trans. SMC **3**(6), 610–621 (1973)
4. B.B. Mandelbrot, *The Fractal Geometry of Nature* (Freeman, New York, 1982)
5. A.P. Pentland, Fractal-based description of natural scenes. IEEE Trans. Pattern Annal. Mach. Intell. **6**(6), 661–674 (1984)
6. H.-O. Peitgen, B. Saupe, *The Science of Fractal Images* (Springer, New York, 1988)
7. K.J. Falconer, *The Geometry of Fractal sets, Cambridge Tracts in Mathematics*, Vol. 85, (University Press, Cambridge, 1985)
8. L.M. Novak, *Thinking in Patterns: Fractals and Related Phenomena in Nature* (World Scientific, Singapore, 2004)
9. R.F. Voss, Random fractals: characterization and measurement. Physica Scripta **13**, 27–32 (1986)

10. A.A. Potapov, Yu.V. Guliaev, S.A. Nikitov, A.A. Pakhomov, V.A. German, *Latest Techniques of Image Processing* (Fizmatlit, Moscow, 2008). (in Russian)
11. G.A. Andreev, A.A. Potapov, A.V. Gorbunov, Sravnitel'nyy analiz statisticheskikh priznakov opticheskikh i radioizobrazheniy pochvenno-rastitel'nykh ob'yektov, Issledovanie Zemli iz kosmosa **1**, 112–121 (1990) (in Russian)
12. A.A. Potapov, T.V. Galkina, T.I. Orlova, Ya.L. Khlyavich, Dispersion detection method of determined object at texture optical and radar images of earth surface. Radiotekhnika i Electronica **35**(11), 2295–2301 (1990) (in Russian)
13. A.A. Potapov, T.V. Galkina, A.I. Kolesnikov, O klassifikatsii izobrazheniy po ikh teksturnym priznakam, Issledovanie Zemli iz kosmosa, **2**, 91–96 (1990). (in Russian)
14. A.A. Potapov, V.A. German, Methods of measuring the fractal dimension and fractal signatures of a multidimensional stochastic signal. J. Commun. Technol. Electron. **49**(12), 1370–1391 (2004). (in Russian)
15. V.A. Kuznetsov, A.N. Pototsky, Method of measuring directional morphological multifractal signatures of the texture images. Achievements Modern Radio Electron. **3**, 39–52 (2017). (in Russian)
16. S. Peleg, J. Naor, R. Hartley, Multiple resolution texture analysis and classification, IEEE Trans. Pattern. Annal. Mach. Intell. **6**(4), 518–523 (1984)
17. Y. Xia, D. Feng, R. Zhao, Y. Zhang, Multifractal signature estimation for textured image segmentation. Pattern. Recognit Lett. **31**, 163–169 (2010)
18. A.A. Potapov, Topology of sample. J. Nonlinear World **2**(1), 4–13 (2004). (in Russian)
19. G. Jacquet, W. Ohley, C. Fortin, Bone texture characterization by oriented fractal analysis, in *Proceedings of the 18th IEEE on Bioengineering*, vol. 18, pp. 22–23 (1988)
20. W.J. Yi, M.S. Heo, S. Lee, Direct measurement of trabecular bone anisotropy using directional fractal dimension and principal axes of inertia. Oral Surg. Oral Med. Oral Pathol Oral Radiol Endod **104**, 110–116 (2007)
21. J. A. Lynch, D. J. Hawkess, J. C. Bukland-Wright, Analysis of texture in macro radiographs of osteoarthritic knees, using the fractal signature. Phys. Med. Biol. **36**(6), 709–722 (1991)
22. P. Podsiadlo, G.W. Stachowiak, The development of the modified hurst orientation transform for the characterization of surface topography of wear particles. Tribology Lett, **3**, 215–229 (1998)
23. M. Wolski, P. Podsiadlo, G.W. Stachowiak, Directional fractal signature analysis of trabecular bone: evaluation of different methods to detect early osteoarthritis in knee radiographs. Proc. Inst. Mech. Eng. **223**(2), 211–236 (2009)
24. M. Wolski, P. Podsiadlo, G.W. Stachowiak, Directional fractal signature analysis of self-structured surface textures. Tribology Lett. **47**, 323–340 (2012)
25. A.A. Potapov, V.V. Bulavkin, V.A. German, O.F. Vyacheslavova, Fractal signature methods for profiling of processed surfaces. Tech. Phys. **75**(5), 560–575 (2005). (in Russian)
26. A.N. Pototskiy, V.A. Kuznetsov, S.F. Galiev, Sposob izmereniya morfologicheskoy mul'tifraktal'noy signatury, Patent RU2647675, (2018). (in Russian)
27. J. Samarabandu, R. Acharya, E. Hausmann, Analysis of bone x-rays using morphological fractals. IEEE Trans. Med. Imaging **12**(3), 466–470 (1993)

The Role of the Angular Momentum in Shaping Collective Effects

E. Prozorova

Abstract The main goal of the work is to clarify the consequences arising from the disregard of the law of the angular momentum as an independent law. As a result, some of the collective effects in mechanics are not taken into account but they are essential. The main laws in physics and mechanics are the laws of conservation of mass, momentum, energy, angular momentum, charge, and some others. In the article it is shown that the sum of the forces is insufficient for a complete description of the interacting particles. Any redistribution of particles is accompanied by the emergence of collective effects, which is associated with the action of the angular momentum and, consequently, with the action of an additional force. The effect always manifests itself, regardless of the branch of science: the formation of fluctuations, structures, quantum mechanics and some others. When constructing a theory, it is impossible to restrict oneself to potential forces that depend only on the distance between particles, since when the particles move, the center of inertia shifts, forming a moment. In continuum mechanics, for example, the stress tensor loses its symmetry for this reason.

Keywords Conference · CHAOS · Chaotic Modeling · CMSIM style

1 Introduction

Classical mechanics deals with material points and, as a rule, with closed systems. The definition of material points in mathematics and physics is different. The main equation in theoretical mechanics is the Liouville equation, which describes the motion of a system of material points of a closed system. Collective interactions occur through an external force, but the main interaction is the binary interaction of particles. The initial and boundary conditions are not considered, although the impossibility of considering them is stipulated due to the huge number of particles.

E. Prozorova (✉)
Mathematical-Mechanical Department St. Petersburg State, University Av. 28, Peterhof 198504, Russia
e-mail: e.prozorova@spbu.ru

C. H. Skiadas and Y. Dimotikalis (eds.), *14th Chaotic Modeling and Simulation International Conference*, Springer Proceedings in Complexity,
https://doi.org/10.1007/978-3-030-96964-6_24

However, Hamilton's formalism is legitimate to use in the case of a no dissipative system, when there is no dependence on the velocity, which is not observed in the presence of disturbing surfaces or under conditions of large gradients of velocities, temperatures, densities, or other characteristics. Using the formalism of Bogolyubov [1], for certain conditions the Boltzmann equation is derived. When deriving the Boltzmann equation and other kinetic equations, the assumption is made that the process is "Markov", that is, there is no dependence on the "past". In reality, however, it is partially manifested through flows at the border. The effect of the boundary is essentially visible in the calculations by the molecular dynamics method and in the numerical solution of the Boltzmann equation [2].

The solutions coincide if a large number of particles are taken and there are no flows across the border. Thus, the Boltzmann equation takes into account the change in state only within the elementary volume. Therefore, it is only suitable for small gradients. In addition, the Boltzmann equation does not fulfill the law of conservation of angular momentum. In the same work, the validity of Hilbert's hypothesis is proved about the dependence of the distribution function on time only through the dependence on macroparameters. The Navier-Stokes (Barnett, etc.) equations are derived from the Boltzmann equation by the Chapman-Enskiy method or by some other method, and the continuity equation is determined, which coincides in form with the Liouville equation. Thus, the consistency of the whole theory is proved. The concepts of "closed" and "open" systems are introduced on the example of systems of "particles", the motion of which is described by the reversible Hamilton equations. These include, for example, the "Boltzmann" gas—a system of "structureless atoms" [3]. However, the "mathematical" and "physical" points are very different. While we are considering a "mathematical point" we are not very interested in whether it rotates or not. For a physical "point", both its rotation and the structure of the "point" under consideration are important. It is known that the moment of force (angular momentum) is responsible for rotation. The role of the angular momentum is manifested in all processes associated with the uneven distribution of particles or their physical parameters. The magnitude of the additional force is determined by the value of the gradient of physical quantities (density, speed, momentum, temperature). The action of the angular momentum, i.e. moment of forces essentially depends on the position of the axis of inertia (center of inertia). The angular momentum is a vector quantity. Additive schemes for calculating intermolecular interactions, in which non-additivity is included in the parameters of atom-atomic potentials, does not take into account the entire variety of conditions.

Analysis of the parameters included in the description of the rarefied gas flow showed that for the equilibrium distribution function the ratio of the gas mean free path l to the characteristic macro length d [4]:

for 37% of trajectories $l/d > 1.0$,
for 90% of trajectories $l/d > 0.1$,
for 99% of trajectories $l/d > 0.01$, etc.

The commonly used criterion $l/d > 1.0$, indicated above, takes the form $Kn > 1$; $Kn = 1.0$ means that $l/d > 1.0$ for only 37% of the trajectories, which does not satisfy the condition l/d > 1.0, while Kn = 10 satisfies the condition l/d > 1.0 for 90% of the trajectories, and Kn = 100—for 99%.

The theory originally proposed for the solution of relaxation problems is extrapolated to the solution of problems associated with gas dynamics, including for solving problems of gas flow near the surface. Limitations of the scheme N. N. Bogolyubov stipulated by the author himself and is associated with the fulfillment of the conditions for the weakening of correlations, the existence of four characteristic time scales (respectively, spatial scales), a particular class of solutions for the s-particle distribution function as a function that depends on time through a single-particle distribution function $f_s(t) = f_s(f)$, binary central interactions providing the law preserving the angular momentum in collision integral, the potential U rapidly falling with distance, the dominance of volume effects and neglect borders.

Potential of interaction of molecules $\Phi = \Phi|\boldsymbol{r} - \boldsymbol{r}_0|$. An additional implicit assumption is the weak deviation of the distribution function from the equilibrium state. For relaxation problems and elastic collisions, all assumptions are satisfied. The case is excluded when the characteristic relaxation times of the single-particle F_1 and the two-particle F_2 are commensurable. It should be noted that for molecules with a more complex interaction potential depending on the angle, averaging over the angle is performed before calculating the collision cross section (potential averaging). There are no studies concerning the influence of the permutation of the operations of averaging the collision cross section and the potential. Here f is the distribution function in the phase γ-space. When deriving the modified equation, the designations will remain generally accepted, that is, r is the radius vector; x—point coordinate; ξ is the velocity of the point, m is the molecular weight, and, according to the definition of the distribution function f_N, the probability of finding the system at the points (x_i, ξ_i) in the intervals $dx_i d\xi_i$ is $f_N(t, x_1, \ldots, x_N, \xi_1, \ldots, \xi_N)dx_1, \ldots, d\xi_1, \ldots, d\xi_N)$.

When calculating macroparameters through the distribution function and projecting values on the coordinate axis, the symmetry of some quantities may be violated. This can happen when calculating the pressure and the pressure tensor:

$P_{ij} = m \int \mathbf{c}_i c_j f(\mathrm{t}, \mathbf{x}, \boldsymbol{\xi}) d\boldsymbol{\xi}$. The symmetry of the stress tensor is postulated on the basis of this form.

In aeromechanics, the projections of the calculated values are used, and not the indices of the velocities included in the formula. Therefore, there is no way to speak unambiguously about symmetry. Symmetry will be observed provided that the rotation of the elementary volume is canceled. The Navier-Stokes equations are obtained under the indicated condition.

An important difference between the interaction of gas and plasma molecules is the long-range nature of the interaction of plasma molecules. A distinctive feature of plasma is a combination of properties characteristic of both a continuous medium (long-range nature of the Coulomb interaction) and systems of individual particles. Therefore, the kinetic theory of plasma differs from the kinetic theory for gas. As we have already noted, there are significant differences in the definitions of mathematical and physical points. Hence, it became necessary to develop a generalized kinetic theory. The need for general definitions of physically infinitesimal scales has matured and is currently given, for example, in [3]. Fluctuations of particles in a liquid play a separate and important role. Their behavior is also determined by collective interactions. The nature of the interaction differs from the interaction of molecules in a gas and from the interaction in a plasma. It should be noted that the generally accepted kinetic equations, by virtue of considering only the translational motion of the medium, without taking into account rotation and fluxes through the boundary, do not take into account the action of the moment of force and diffusion fluxes through the boundaries. The need to take into account certain effects depends on the specific task. For example, when considering waves in a "cold" isotropic plasma, it is not necessary to take into account the angular momentum and diffusion. In any case, the absence of motion of heavy particles also does not require taking into account the moment and diffusion. The movement of electrons alone does not create a change in the position of the center of inertia (due to the difference in masses) if there is no movement of the ions. When considering the Landau collision integral (the kinetic equation for a weakly interacting gas, including a Coulomb plasma), it is necessary to take into account the influence of the moment. The question of Landau damping, which consists in the damping of a perturbation in a plasma as it propagates from the point of origin, despite the collisionless (without binary collisions) nature of the interaction of molecules, requires additional research. This work is devoted to the study of the influence of the angular momentum in collective interactions.

2 Kinetic Equations

The classical derivation of the Boltzmann equation is to write the particle balance in terms of the relation for the one-particle distribution function

$$f(t+dt, \boldsymbol{r}+\boldsymbol{\xi}_i dt, \boldsymbol{\xi}_i+\boldsymbol{F}_i dt) d\boldsymbol{r} d\boldsymbol{\xi}_i = f(t, \boldsymbol{r}, \boldsymbol{\xi}_i) d\boldsymbol{r} d\boldsymbol{\xi}_i + \left(\frac{\partial f}{\partial t}\right)_{coll} dt$$

The latter is often written in the form

$$\boldsymbol{f}(\boldsymbol{t}+\boldsymbol{dt}, \boldsymbol{r}+\boldsymbol{\xi}_i \boldsymbol{dt}, \boldsymbol{\xi}_i+\boldsymbol{F}_i \boldsymbol{dt}) = \boldsymbol{f}(\boldsymbol{t}, \boldsymbol{r}, \boldsymbol{\xi}_i) + \widetilde{\left(\frac{\partial \boldsymbol{f}}{\partial \boldsymbol{t}}\right)}_{coll} \boldsymbol{dt}.$$

where $(\frac{\partial f}{\partial t})_{coll}$, $(\frac{\partial \tilde{f}}{\partial t})_{coll}$—are the collision integrals written in different phase spaces. Outwardly, these equalities are identical, however, the second relation is fulfilled at the times of interaction of molecules and all interactions are correlated. For gas-dynamic problems, the characteristic length of the elementary volume, for which equality (2) is written, equal to cm is small and the requirement for a large number of particles in the elementary volume is not fulfilled for altitudes of 120–300 km in the earth's atmosphere. Indeed, the required minimum size is 10^{-3} cm. Since, $N = \pi R^2 \cdot \boldsymbol{\xi} \cdot \boldsymbol{\tau} \cdot \boldsymbol{n}$, here R is the radius of the cylinder of the elementary volume; τ is the mean time of free path, then for statistical independence the number of particles

N must be at least 100. Then, i.e. see. In addition, the possibility of reducing the characteristic size is limited not only by the limited computer memory, but also by the limits of applicability of the model [5, 6], as well as by the growth of computational errors. In this equation, it is assumed that the elementary volume does not rotate and there are no incoming particles through the side surfaces.

When working with a physical elementary volume, it is necessary to take into account the action of the angular momentum responsible for rotation, and due to the finite value of the radius, it is necessary to take into account the arrival of molecules with a selected speed due to diffusion. We consider the hydrodynamic approximation, assuming the definition of a point in terms of the mean free path.

The usual transition to the Boltzmann equation involves expanding the function in a series and obtaining the following equation

$$\left(\frac{\partial}{\partial t} + \boldsymbol{\xi}_i \nabla_r + \boldsymbol{F}_i \nabla_i\right) f(t, \boldsymbol{r}, \boldsymbol{\xi}_i) = \left(\frac{\partial f}{\partial t}\right)_{coll} dt = \mathrm{I}$$

Taking into account rotation and diffusion, the equation () has the form

$$\begin{aligned} & f(t + dt, \boldsymbol{r} + \boldsymbol{\xi}_i dt + r \times \omega, \boldsymbol{\xi}_i + \boldsymbol{F}_i dt + \frac{\partial \mathrm{M}}{\partial \mathrm{r}} dt) d\boldsymbol{r} d\boldsymbol{\xi}_i \\ & + G_2\left(t + dt, \boldsymbol{r} + \boldsymbol{\xi}_i dt + r \times \omega, \boldsymbol{\xi}_i + \boldsymbol{F}_i dt + \frac{\partial \mathrm{M}}{\partial \mathrm{r}} dt\right) \\ & = f(t, \boldsymbol{r}, \boldsymbol{\xi}_i) d\boldsymbol{r} d\boldsymbol{\xi}_i + G_1(t, \boldsymbol{r}, \boldsymbol{\xi}_i) + \left(\frac{\partial f}{\partial t}\right)_{coll} dt. \end{aligned}$$

M is the moment associated with the collective action of all particles on each other as a result of the displacement of the center of inertia, which is the result of the movement of particles with different speeds. G_1 and G_2—flows through the boundaries of the considered elementary volume. Let's calculate these values.

$G_1 = m\boldsymbol{\xi}_i \frac{\partial f}{\partial \mathrm{r}}$.

Accounting for flows across the border (G_1, G_2) leads to the equations of Vallander [7, 8]

Here E is the internal energy, $E = c_v T$, where c_v is the heat capacity coefficient

$$Q_x = D_1 \frac{\partial \rho}{\partial x} + D_2 \frac{dT}{dx}, \; Q_y = D_1 \frac{d\rho}{dy} + D_2 \frac{dT}{dy}, \; Q_z = D_1 \frac{d\rho}{dz} + D_2 \frac{dT}{dz},$$

$$t_x = k_1 \frac{d\rho}{dx} + k_2 \frac{dT}{dx}, \;\; t_y = k_1 \frac{d\rho}{dy} + k_2 \frac{dT}{dy}, \; t_z = k_1 \frac{d\rho}{dz} + k_2 \frac{dT}{dz}$$

$$D_1 = \frac{\mu}{\rho} \alpha_1, \; D_2 = \frac{\mu}{T} \alpha_2, \; k_1 = \frac{\mu c_v T}{\rho}, \; k_2 = \mu c_v \beta_2, \; \lambda = \alpha \mu,$$

where $\alpha_1, \alpha_2, k_1, k_2$ are numerical constants depending on the type of gas.

Q_x, Q_y, Q_z are the mass fluxes across the face perpendicular to the coordinate axes of the moving gas with the velocity V, ρ-density, D_1, D_2 are the coefficients of self-diffusion and thermal diffusion,k_1, k_2 are the thermal conductivity coefficients, and R is the gas constant.

Here I consider it necessary to add to these equations a term related to the velocity gradient (bulk viscosity D_3), so that $Q_x = D_1 \frac{\partial \rho}{\partial x} + D_2 \frac{dT}{dx} + D_3 \frac{\partial u}{\partial x}$. The rest of the values change in the same way. Let us recall the difference between the values obtained through the distribution function and by the molecular dynamics method [9–12].

The general formula for the distribution function (dependence on r).

$$f = f(t, \boldsymbol{r}(t), \boldsymbol{\xi}(t))$$

$$\frac{\partial f}{\partial t}|_{r=const} == \frac{\partial}{\partial t} \frac{\sum_{i=1}^{n} \delta(\boldsymbol{r}_i - \boldsymbol{r})}{\sum_{i=1}^{N} \delta(\boldsymbol{r}_i - \boldsymbol{r})}.$$

By construction $\delta(\boldsymbol{r}_i - \boldsymbol{r})$– depends on t only through the $\boldsymbol{r}_i(t) - \boldsymbol{r}(t)$. Here n is number molecules in elementary volume, N– in full volume.

A more complex option when there are time-dependent flows across the border

1. Without flow across the border

$$\frac{F_1}{F_3} - \frac{F_2}{F_4} = \frac{\sum_{i=1}^{n} \delta(\boldsymbol{r}_i - \boldsymbol{r}) + \sum_i^n \Delta t \frac{\partial \delta(\boldsymbol{r}_i - \boldsymbol{r})}{\partial t} + \dots}{\sum_{i=1}^{N} \delta(\boldsymbol{r}_i - \boldsymbol{r}) + \sum_i^N \Delta t \frac{\partial \delta(\boldsymbol{r}_i - \boldsymbol{r})}{\partial t} + \dots} - \frac{\sum_{i=1}^{n} \delta(\boldsymbol{r}_i - \boldsymbol{r})}{\sum_{i=1}^{N} \delta(\boldsymbol{r}_i - \boldsymbol{r})}$$

$$\approx \frac{\sum_{i=1}^{n} \delta(\boldsymbol{r}_i - \boldsymbol{r}) + \sum_i^n \Delta t \frac{\partial \delta(\boldsymbol{r}_i - \boldsymbol{r})}{\partial t} + \dots}{\sum_{i=1}^{N} \delta(r_i - \boldsymbol{r})} \left(1 - \frac{\sum_i^N \Delta t \frac{\partial \delta(\boldsymbol{r}_i - \boldsymbol{r})}{\partial t} + \dots}{\sum_{i=1}^{N} \delta(\boldsymbol{r}_i - \boldsymbol{r})}\right)$$

$$- \frac{\sum_{i=1}^{n} \delta(\boldsymbol{r}_i - \boldsymbol{r})}{\sum_{i=1}^{N} \delta(r_i - \boldsymbol{r})}$$

$$\approx \frac{\sum_i^n \Delta t \frac{\partial \delta(\boldsymbol{r}_i - \boldsymbol{r})}{\partial t} + O((\Delta t)^2}{\sum_{i=1}^{N} \delta(\boldsymbol{r}_i - \boldsymbol{r})}$$

$\frac{\partial \delta(\boldsymbol{r}_i - \boldsymbol{r})}{\partial t}$ – thus, when solving the Boltzmann equation, the time derivative of distribution function will indeed be determined by the dependence through the macro parameters. This approximation, which is made in the theory of rarefied gas in the construction of the Enskog-Chapman solution

2. If there is a flow across the border, depending only on time. The force is not.

$$\frac{F_1}{F_3} - \frac{F_2}{F_4} = \frac{\sum_{i=1}^{n} \delta(\boldsymbol{r}_i - \boldsymbol{r}) + \sum_i^n \Delta t \frac{\partial \delta(\boldsymbol{r}_i - \boldsymbol{r})}{\partial t} + \sum_j \frac{p_j \Delta t}{m} \delta(r_j - r) + \sum_j \frac{p_j}{m} \Delta t^2 \frac{\partial \delta(r_j - r)}{\partial t} + \dots}{\sum_{i=1}^{N} \delta(r_i - \boldsymbol{r}) + \sum_i^N \Delta t \frac{\partial \delta(\boldsymbol{r}_i - \boldsymbol{r})}{\partial t} + \sum_j \frac{p_j}{m} \delta(r_j - r) + \sum_j \frac{p_j}{m} \Delta t \frac{\partial \delta(r_j - r)}{\partial t} + \dots}$$

$$- \frac{\sum_{i=1}^{n} \delta(\boldsymbol{r}_i - \boldsymbol{r})}{\sum_{i=1}^{N} \delta(r_i - \boldsymbol{r})}.$$

$\sum_j \frac{p_j}{m}\delta(r_j - r) = J_2 - J_1$—is a flow of fast molecules from neighboring cells. The first two terms correspond to the number of molecules in the volume and their motion. Thus, for large gradients the role of flows across the border is increasing. The distribution function can no give a correct contribution to the distribution of molecules. We need in large number particles in elementary volume. There remains the method of molecular dynamics with a very small time step.

Most often, the kinetic Boltzmann equation is taken as the initial one, and one of the variants of the perturbation theory in a small parameter is used to pass to the aeromechanical equations. As we can see, the Boltzmann equation, depending on the problem, requires modifications, since it does not fulfill one of the laws of theoretical mechanics, the conservation law of angular momentum. For the obtained equations, for example, for Navier-Stokes, additional assumptions are made: discarding the rotational velocity component and using Pascal's law obtained for the equilibrium case to nonequilibrium flows. As a result, the pressure becomes a scalar. Using the Boltzmann equation, we obtain an equation for the internal stress tensor. Here, the gas-dynamic functions ρ, u, T are the moments of the velocity v or the deviation of the velocity from its mean value: $\delta v = v - u$.

$$P_{ij}(r,t) = mn \int \delta v_i \delta v_j f(r,p,t)dp,$$

$$\left(\frac{\partial p}{\partial t} + u_k \frac{\partial p}{\partial r_k} + \frac{5}{3}\frac{\partial u_k}{\partial r_k}\right)\delta_{ij} + p\left(\frac{\partial u_i}{\partial r_j} + \frac{\partial u_j}{\partial r_i} - \frac{2}{3}\delta_{ij}\frac{\partial u_k}{\partial r_k}\right)$$
$$= mn \int \delta v_i \delta v_j I_B(r,p,t)dp,$$

$$P_{ij}(r,t) = \delta_{ij} p(r,t) + \pi_{ij}$$

Pascal's formula does not follow from the formula and pressure is not defined as 1/3 of the sum of the pressures on the coordinate pads. An interesting feature of the all research is the emphasis on the openness of the considered elementary volumes and, despite the "openness", the use of conservation laws for closed volumes. For example, the law of conservation of energy. We have already shown that the distribution function gives an idea of a probabilistic state in an elementary volume without the influence of boundaries and, therefore, information about flows across the boundary is lost. In addition, information about the "rearrangement" of the arrangement of molecules due to the influence of the motion of the center of inertia is lost. These collective effects should be taken into account when writing kinetic equations and for equations of a continuous medium. "A unified description of kinetic and hydrodynamic processes" [3] requires the same correction. In this case, there is no contradiction between the kinetic equations, the equations for fluctuations, the Fokker-Planck equation, and the Landau damping in plasma. It is essential that these terms are not included in the collision integral. Formally, the equation is without dissipation and is reversible, but in fact the diffusion flows have dissipative properties. It should be recalled that to satisfy Hilbert's hypothesis, one should take the macroparameters of the modified Navier-Stokes equation in the solution for the locally equilibrium function, but not Euler to match the orders of approximation in the Chapman-Enskiy solution. In addition, the definition of pressure must be changed and a torque gradient must be entered. Then the nonequilibrium.

Chapman-Enskiy solution implies the existence of a vector distribution function, which is observed in numerical calculations when solving the Boltzmann equation [10], the proof of this is the different temperature values along the coordinate axes.

Recall that the stress tensor is not symmetric and the symmetry condition for the stress tensor is one of the conditions for closing the problem; to fulfill the condition, it is required to discard the rotation of the elementary volume. For numerical calculation, the latter simplifies programming only slightly. The classic Chapman-Ensky solution is given below.

$$nf(r,p,t) = \frac{\rho/m}{(2\pi mk_bT)^{3/2}} exp\left[-\frac{(p-mu)^2}{2\pi mk_bT}\right] \times \left[1 + \frac{\pi_{ij}}{2p}\frac{m\delta v_i\delta v_j}{3k_bT} + \frac{m(\delta v_i q)}{pk_bT}\left(\frac{m(\delta v)^2}{3k_bT} - 1\right)\right].$$

Changes in the values will be in the macroparameters of the local equilibrium distribution function, the collision integral will not change. In kinetic theory, when considering the role of delay for rarefied gas, one must understand the question of what is measured in the experiment: instantaneous values or averages. If the experiment deals with average values, then it is important to choose the time and scale of averaging. At the agreed times, in this case, it is not necessary to take into account the delay, except for the cases of commensurability of the relaxation and retardation times.

3 Damping of Longitudinal Oscillations of an Electron Plasma (Landau Damping), Kinetic Equations of Langevin and Fokker-Planck

Let us consider oscillations in a plasma without collisions, that is, let us proceed to the study of waves propagating in a plasma, the frequency of which is high in comparison with the frequency of pair collisions of electrons and ions. In this case, there are several options to consider. Landau collisional damping for large Knudsen numbers; for small Knudsen numbers in unbounded plasma; for small Knudsen numbers in a confined plasma. They differ from each other. When studying oscillations, we will consider small deviations from equilibrium [3, 13–18].

Since we are interested in wave attenuation, we need to consider the plasma dielectric constant ε, which is determined by the attenuation coefficient γ. First, let's trace the waves in the "cold" isotropic plasma. The variant corresponds to the "collisionless" wave approximation. In this case, the Maxwell distribution functions

$$f_e^{(0)} = \frac{1}{(2\pi m_e k_b T)^{3/2}} exp\left(-\frac{p^2}{2m_e k_b T}\right),\ f_i(p) = \delta(p).\ \frac{r_D}{l} \ll \frac{\lambda}{l} \ll 1,$$

the damping is determined by diffusion, but not by the Landau damping. The influence of the thermal motion of plasma particles on such oscillations is always small [3].

Here $f_e^{(0)}$ is the equilibrium distribution function, l is the mean free path r_D is the Debye radius, λ is the wavelength, the rest of the notation is generally accepted. Consider an unbounded plasma for small Knudsen numbers $l \ll \lambda$. Diffusion works here as well. Let us consider the dispersion and damping of longitudinal oscillations of an electron plasma under the influence of the thermal motion of plasma particles. Let us investigate a variant of a limited plasma, a free-molecular flow with a region of wavelengths (values of wave numbers) for which the contribution corresponding to Landau damping is the main one.

$$l \to L(l \gg L), r_D \ll \gamma \ll \sqrt{r_D L}\ (r_D \ll L \ll l)$$

We must use the Vlasov kinetic equations [18] with a self-consistent field. Since we are interested in high-frequency oscillations, for which $\omega\tau \gg 1$, where τ is the average time between pair collisions of particles, we can ignore the integrals of particle collisions in the kinetic equations. Longitudinal oscillations of an electron plasma in the classical case are described by the following two equations (collisionless case, Vlasov equation)

$$\frac{\partial \delta f}{\partial t} + v\frac{\partial \delta f}{\partial r} + e\delta E\frac{\partial f_0}{\partial p} = 0,$$

$$div\delta E = 4\pi \int dp\delta f.$$

$$\varepsilon_l(\omega, k)e \int dp\delta f(p, k, \omega) = i \int dre^{-ikr} e \int dp\frac{\delta f(p, r, t_0)}{\omega - kv}.$$

Suggested variant is

$$\frac{\partial \delta f}{\partial t} + v\frac{\partial \delta f}{\partial r} + e\delta E\frac{\partial f_0}{\partial p} + \frac{\partial \delta M}{\partial r}\frac{\partial f_0}{\partial p} + \frac{\partial}{\partial r}D\frac{\partial \delta f}{\partial r} = 0,$$

$$div\delta E = 4\pi \int dp\delta f.$$

$$\varepsilon_l(\omega, k)e \int dp\delta f(p, k, \omega) = i \int dre^{-ikr} e \int dp\frac{\delta f(p, r, t_0)}{\omega - kv} + i \int dre^{-ikr} \int dp\frac{\delta M(p, r, t)}{\omega - kv}\frac{\partial f_0}{\partial p}.$$

Qualitatively, we can say that for this case, diffusion plays a small role and, since part of the energy is converted into rotational motions (the action of the moment), the reversible operator will act as a dissipative one. Note that at the initial moment, the distributed moment of force also exists and concentrates a certain amount of energy.

For monochromatic waves of large amplitude, the action can lead to the formation of a vertical velocity component, forming complex plane flows. Despite the collisionless nature of the movement binary collisions exist, as follows from the table of mean free paths presented in the introduction. They create additional dissipation. It should be noted that the generalized equation for a unified description of kinetic and gas-dynamic processes is suitable for "weak" interactions. As before, the contribution of the angular momentums in the motion of molecules is not taken into account. Most likely, the difference between the most probable and average values is due precisely to the lack of taking into account the rotational movements for which the moment is responsible. Similar effects will be essential for Brownian motion. The theory of Brownian motion is one of the main branches of the statistical theory of open systems. Fluctuation (from Latin fluctuation—fluctuation)—any random deviation of any value. In mechanics, a deviation from the mean value of a random variable characterizing a system of a large number of chaotically interacting particles. In the theory of Brownian motion elementary objects are small particles, while in kinetic theory, the main objects are molecules. Both models are macromodels, but the level of description of the structure of the environment is different. Fluctuations exist both in nonequilibrium states and in unsteady processes; in their absence, relaxation would be a "smooth" process and they could be described by single-valued functions of time. The presence of thermal fluctuations causes random deviations of real processes from such a "smooth" flow. The kinetic equation corresponds to a more detailed description. We believe that the environment is in equilibrium. We will consider two approaches to solving problems: the equation for a single particle and for an ensemble of particles (the Fokker-Planck equation) To take into account the atomic structure of a liquid, Langevin introduced an additional force into the equations of motion

$$F_L = My(t), F = -M\gamma v, \gamma = \frac{6\pi a}{M}\eta, \eta = \rho\nu.$$

Equations
$\frac{dr}{dt} = v, \frac{dp}{dt} + \gamma p = F_0 + My(t), F_0 = -gradU.$ F_0- external force.
$\langle y_i(t)\rangle, \langle y_i(t), y_i(t')\rangle = 2D\delta_{ij}(t - t')$, the coefficient D was determined by Einstein.

First, about a single particle. Let us repeat the reasoning performed in [3, 19], but replace y (t) with the moment of force M_i calculated for a given period of time. It can be calculated using the operation algorithm. As before, we assume that the characteristic correlation time of the values of the Langevin force is $\tau_{cor}^L \ll \tau_{rel} = \frac{1}{\gamma}$. As a result, we arrive at an expression for two time moments:

$$\langle M_i(t)\rangle = 0, \langle M_i(t)M_j(t')\rangle = 2D\delta_{ij}(t - t').$$

$D = \gamma\frac{k_bT}{m}$—Einstein's coefficient, parenthesis means a function from a function (functional).

When using the kinetic description of Brownian motion, it is necessary to introduce an ensemble of noninteracting Brownian particles—he corresponding Gibbs ensemble. In this case, we represent the ensemble of Brownian particles as a continuous medium. However, the difference lies in the use of the "Hamiltonian" formalism for moving particles; for a continuous medium, in this case, the Langevin equation is used. The kinetic classical Fokker-Planck equation has the form [18, 19]

$$\frac{\partial f}{\partial t}+v\frac{\partial f}{\partial r}-\frac{1}{M}\frac{\partial U}{\partial r}\frac{\partial f}{\partial v}=D\frac{\partial^2 y}{\partial v^2}+\frac{\partial}{\partial v}(\gamma v f).$$

The equation of A. Vlasova

$$\left\{\frac{\partial f}{\partial t}+v\frac{\partial f}{\partial r}+e\left(E+\frac{1}{c}[vB]\right)\frac{\partial}{\partial p}\right\}F(r,p,t)=0.$$

Here E, B are the total electric and magnetic fields, which are composed of external and self-consistent fields generated by plasma particles. They satisfy Maxwell's equations.

In the classical case, equilibrium is possible between Brownian particles and the medium; the particles can be distributed evenly [20]. However, such an assumption can be considered unlikely due to the distribution of particles over velocities and the formation of new moments for individual particles due to the motion of the center of inertia. The fact is that in this case the action of the moment creates a force that distributes the particles not only in terms of velocities, but also in coordinates. The proposed modified Fokker-Planck equation has the form:

$$\frac{\partial f}{\partial t}+v\frac{\partial f}{\partial r}-\frac{1}{M}\frac{\partial U}{\partial r}\frac{\partial f}{\partial v}+\frac{1}{M}\frac{\partial M}{\partial r}\frac{\partial f}{\partial v}=D\frac{\partial^2 y}{\partial v^2}+\frac{\partial}{\partial v}(\gamma v f).$$

Thus, in the kinetic theory for a gas, for the Landau damping and the motion of Brownian particles, the nonuniform distribution of particles in velocities and coordinates is supported by the angular momentum and creates fluctuations in physical quantities that must be taken into account. Consider the consequences associated with taking into account the moment in the mechanics of a continuous medium.

4 The Influence of the Angular Momentum in the Equations of Continuum Mechanics

Conservation laws were obtained experimentally and therefore were originally written in integral form. Differential laws are obtained in two ways: using the finite volume method for an elementary volume and using the Ostrogradsky Gauss theorem by replacing the surface integral to the volume integral, that is, taking the integral by parts with further use of the theorems on the conditions Integral turning in zero.

Usually the derivation of conservation laws is analyzed using the Ostrogradsky-Gauss theorem for a fixed volume without moving. The theorem is a consequence of the application of the integration in parts at the spatial case. In reality, in mechanics and physics gas and liquid move and not only progressively, but also rotate. Let us consider the consequences that arise from the generally accepted conservation laws in the mechanics of a continuous medium and which do not correspond to classical theoretical mechanics and mathematics. The speeds of various processes at the time of writing the equations were relatively small compared to modern ones. In further studies, the scope of the theory developed for potential flows to flows with significant gradients of physical parameters was expanded. It was based on the laws of balance of forces, the law of conservation of moment was considered as a consequence of the fulfillment of the law of balance of forces. Allocating the rotational velocity component and ignoring it leads to a symmetric stress tensor. The symmetric tensor is obtained only if the rotational velocity component is neglected. However, this variant of closing the problem is one of the possible variants of solving the system of three equations in the plane case for four unknowns [9–11]. A similar conclusion can be made for the three-dimensional case. For modern computer technology, it is possible to solve the complete equations of fluid mechanics, rather than truncated ones (like Navier-Stokes). From the definition of pressure, both from the classical Boltzmann equation and the modified one, it does not follow that the hydrostatic pressure is one third of the sum of the pressures on the coordinate areas. Using Pascal's law for equilibrium, the pressure is chosen equal to one third of the pressure on the coordinate pads. However, the theory remains the same when determining the different pressure on each of the sites, i.e.p_x, p_y, p_z. The use of one pressure is possible under equilibrium conditions (Pascal's law), but for nonequilibrium conditions the fact is not obvious. Neglecting outside the integral term when taking integrals by parts (the Ostrogradsky-Gauss theorem) is possible only for slow laminar flows. Writing out separately the law of equilibrium for forces and separately for moments of forces without taking into account the mutual influence, although the moment creates an additional force, we come to the conclusion about the symmetry of the stress tensor. If we consider different pressures in different directions, we lose a moment of force, but the pressure gradient is a force. The proposed modified equations of continuum mechanics include the action of the moment and are given in [9–11] and new equations:

$$\rho(\frac{\partial u}{\partial t} + u\frac{\partial u}{\partial x} + v\frac{\partial u}{\partial y} + w\frac{\partial u}{\partial z}) = \rho f_1 + \frac{\partial \sigma_{xx}}{\partial x} + \frac{\partial \sigma_{yx}}{\partial y} + \frac{\partial \sigma_{zx}}{\partial z} + \rho f_{M_x}$$

$$\rho(\frac{\partial v}{\partial t} + u\frac{\partial v}{\partial x} + v\frac{\partial v}{\partial y} + w\frac{\partial v}{\partial z}) = \rho f_2 + \frac{\partial \sigma_{xy}}{\partial x} + \frac{\partial \sigma_{yy}}{\partial y} + \frac{\partial \sigma_{zy}}{\partial z} + \rho f_{M_y},$$

$$\rho(\frac{\partial w}{\partial t} + u\frac{\partial w}{\partial x} + v\frac{\partial w}{\partial y} + w\frac{\partial w}{\partial z}) = \rho f_3 + \frac{\partial \sigma}{\partial x} + \frac{\partial P\sigma_{yz}}{\partial y} + \frac{\partial \sigma_{zz}}{\partial z} + \rho f_{M_z},$$

$$y(\frac{\partial \sigma_{xz}}{\partial x} + \frac{\partial \sigma_{yz}}{\partial y} + \frac{\partial \sigma_{zz}}{\partial z} + \rho f_3) - z(\frac{\partial \sigma_{xy}}{\partial x} + \frac{\partial \sigma_{yy}}{\partial y} + \frac{\partial \sigma_{zy}}{\partial z} + \rho f_2)$$

$$+\sigma_{zy}-\sigma_{zy}++M_x=0,$$

$$x(\frac{\partial\sigma_{xy}}{\partial x}+\frac{\partial\sigma_{yy}}{\partial y}+\frac{\partial\sigma_{zy}}{\partial z}+\rho f_2)-y\left(\frac{\partial\sigma_{xx}}{\partial x}+\frac{\partial\sigma_{yx}}{\partial y}+\frac{\partial\sigma_{zx}}{\partial z}+\rho f_1\right)$$
$$+\sigma_{yx}-\sigma_{xy}++M_y=0,$$

$$x(\frac{\partial\sigma_{xz}}{\partial x}+\frac{\partial\sigma_{yz}}{\partial y}+\frac{\partial\sigma_{zz}}{\partial z}+\rho f_1)-z(\frac{\partial\sigma_{xx}}{\partial x}+\frac{\partial\sigma_{yx}}{\partial y}+\frac{\partial\sigma_{zx}}{\partial z}+\rho f_2)$$
$$+\sigma_{zx}-\sigma_{xz}++M_z=0.$$

Here all designations are standard, $f_{M_x}, f_{M_y}, f_{M_z}$ forces created by the moment, M_x, M_y, M_z are external moments.

5 Conclusion

The paper proposes to take into account the influence of the angular momentum (force) in kinetic equations and in stochastic processes. The definitions of a material point in mathematics and physics are different. As a result, some of the collective effects in mechanics are not taken into account. The main laws in physics and mechanics are the laws of conservation of mass, momentum, energy, angular momentum, charge, and some others. In the article it is shown that not all of the forces are enter for a complete description of the interacting particles. Any redistribution of particles is accompanied by the emergence of collective effects, which is associated with the action of the angular momentum and, consequently, with the action of an additional force. The effect always manifests itself, regardless of the branch of science: the formation of fluctuations, structures, quantum mechanics and some others. When constructing a theory, it is impossible to restrict oneself to potential forces that depend only on the distance between particles, since when the particles move, the center of inertia shifts, forming an angular momentum. In continuum mechanics, for example, the stress tensor loses its symmetry for this reason. Some modification of the theory is suggested.

References

1. N.N. Bogolyubov, Problems of dynamic theory in statistical physics. M.: Gostekhizdat (1946), 146 p
2. K.P. Gurov, Foundations of the kinetic theory. M.: Nauka (1966), 350 p
3. Y.L. Klimontovich, Statistical theory of open systems. T. 1.2. Moscow: Lenand, (2019)
4. F. Goodman, G. Vachmann, Dynamics of gas scattering by a surface. M.: Mir, (1980), 423p
5. E.V. Prozorova, Mathematical modeling of the processes of mechanics with large gradients (St. Petersburg. St. Petersburg University, 339c, 2005)

6. E.V. Prozorova, The influence of dispersion in models of continuum mechanics (St. Petersburg. St. Petersburg University, 2013), 94 p
7. S.V. Vallander, Equations of motion of a viscous gas. Rep. Acad. of Sci. USSR **78**(1), 25–27 (1951)
8. S.V. Vallander, M.P. Elovskih, The theoretical dependence of the heat conductivity of gases on temperature. Rep. ANAS **79**(1), 37–40 (1951)
9. E.V. Prozorova, Features of the rarefied gas description in terms of a distribution function. APhM2018. IOP Conf. Ser. J. Phys. Conf. Ser. **1250**, 012023 (IOP Publishing, 2019). https://doi.org/10.1088/1742-6596/1250/1/0120232
10. E. Prozorova, Influence the form of writing conservation laws in computation. JP J. Heat Mass Transf. 28 Oct 2019. Accepted: 23 Nov 2019
11. E. Prozorova, The effect of angular momentum and Ostrogradsky-Gauss theorem in the equations of mechanics WSEAS TRANSACTIONS on FLUID MECHANICS. https://doi.org/10.37394/232013.2020.15.2
12. E. Prozorova, The influence of the no symmetric stress tensor on the flow separation WSEAS TRANSACTIONS on APPLIED and THEORETICAL MECHANICS. https://doi.org/10.37394/232011.2020.15.9
13. E. Prozorova, Consequences of the Ostrogradsky-Gauss theorem for numerical simulation in aeromechanics. Elsever Granthaalayah, **8**(6) (2020). https://doi.org/10.29121/granthaalayah.v8.i6.2020.549
14. R. Balescu, Equilibrium and nonequilibrium statistic mechanics. A Wiley-Intersciences Publication John Willey and Sons. New-yourk-London (1975)
15. N.G. Van Kampen, Stochastic processes in physics and chemistry. North-Holland (1984)
16. A.I. Akhiezer, I.A. Akhiezer, R.V. Polovin, A.G. Sitenko, K.N. Stepanov, Plasma electrodynamics. M.: Science (1974)
17. P. Silin, Introduction to the kinetic theory of gases. M.: Science (1971)
18. S. Ishimaru, Basic principles of plasma physics. M.: Atomizdat (1975)
19. A.A. Vlasov, Nonlocal statistical mechanics. M.: Science (1978)
20. P. M¨orters, Y.P. Brownian, Notes on Brownian motion and related phenomena Deb Shankar Ray*, Department of Physical Chemistry, Indian association for the cultivation of science, Jadavpur, Calcutta 700032 (2008)

Non-autonomous Two Channel Chaotic Generator: Computer Modelling, Analysis and Practical Realization

Volodymyr Rusyn, Christos H. Skiadas, and Aceng Sambas

Abstract Circuit of the non-autonomous two channel chaotic generator that contains two operational amplifiers, four capacitors, two resistors and two sinusoidal voltage sources is presented. Regimes of chaotic behavior was modeled by using NI's software MultiSim. Analysis of chaotic attractor, time series and spectra are shown. The layout and 3D model of realization of the non-autonomous two channel chaotic generator was designed by using software Proteus 8.

Keywords Non-autonomous · Chaotic generator · Two-channel · MultiSim · Proteus

1 Introduction

Chaos has great potential and useful in many different engineering areas, such as computer and information sciences, biomedical systems, optics, power systems, robotics, memristors, telecommunications, and cyber security [1–15].

Nonlinear theory is the most interdisciplinary areas; it includes nonlinear phenomena and complex analysis that have been intensively studied and regard in many different areas ranging from mathematics and engineering to natural sciences (biology, ecology, economy) [16–21].

Some nonlinear systems were realized using Arduino and FPGA boards [22–25].

V. Rusyn (✉)
Department of Radio Engineering and Information Security, Yuriy Fedkovych Chernivtsi National University, Chernivtsi, Ukraine
e-mail: rusyn_v@ukr.net

C. H. Skiadas
Technical University of Crete, University Campus, 73100 Chania, Crete, Greece
e-mail: skiadas@cmsim.net

A. Sambas
Department of Mechanical Engineering, Universitas Muhammadiyah Tasikmalaya, Tasikmalaya, Indonesia

C. H. Skiadas and Y. Dimotikalis (eds.), *14th Chaotic Modeling and Simulation International Conference*, Springer Proceedings in Complexity,
https://doi.org/10.1007/978-3-030-96964-6_25

There are many different scheme-technical realizations of chaotic generators [26–31]. Great interest are non-autonomous generators that demonstrate chaotic behavior.

In this paper, we present a new non-autonomous chaotic generator that was built as two channel generator. Chaotic behavior was detected due to the frequency ratio of the two sinusoidal generators.

The paper is organized as follows. In Sect. 2, computer modelling of the circuit, main information properties such as chaotic attractor, time series and spectrum using software MultiSim are presented. In the following section, the practical realization, i.e. layout and 3D model using Proteus 8 are presented. The conclusionsare summarized in the last section.

2 Computer Modelling of the Non-autonomous Chaotic Generator

Figure 1 shows proposed electrical scheme that realize non-autonomous two channel chaotic generator. This circuit was realized around two operational amplifiers, namely

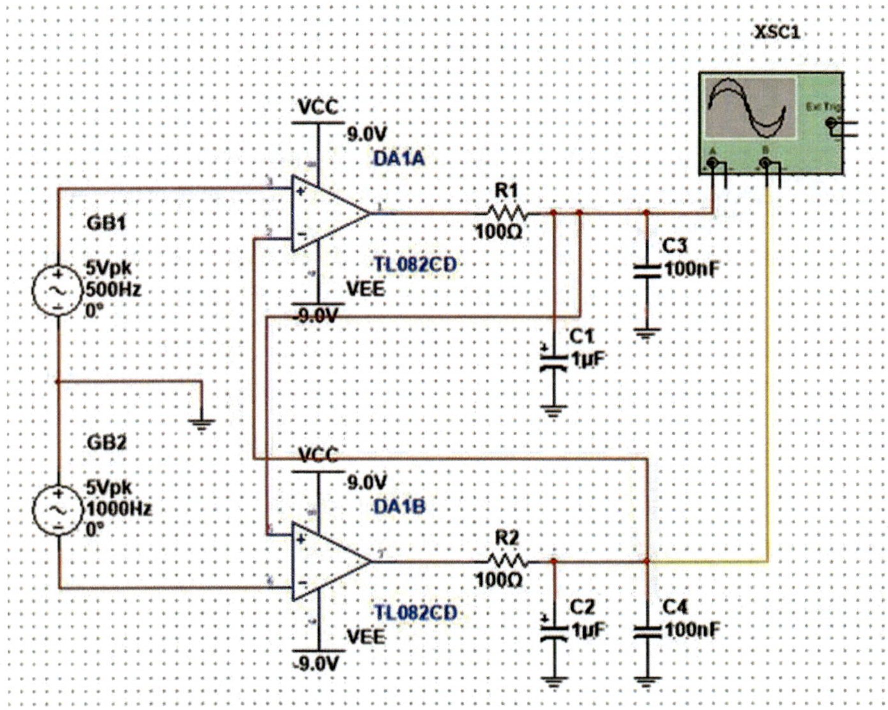

Fig. 1 Non-autonomous two channel chaotic generator

TL082. The elements used and their values were: capacitors C1 = C2 = 1 uF, C3 = C4 = 100 nF, resistors R1 = R2 = 100 Ω. The circuit was powered by a symmetrical power source of ± 9 V. Also, for realize chaotic behavior was used two sinusoidal generators with next parameters: GB1 (amplitude $U_1 = 5$ V, frequency f = 500 Hz) and GB2 (amplitude $U_2 = 5$ V, frequency f = 1000 Hz).

Simulations of the circuit behavior were carried out by using NI's MultiSim platform.

In Fig. 2 the generated phase portrait namely "heart scroll" based on the circuit's chaotic signals is presented on the platform's virtual oscilloscope. The *x*-axis corresponds to the voltage of capacitor C3 (U_{C3}), which will be called the *x-signal*; while the *y*-axis corresponds to the voltage of capacitor C4 (U_{C4}), which will be called the *y-signal.* It is noted that the channels' settings were for channel A, 5 V/div and channel B, $U_2 = 5$ V/div. The chaotic nature of the produced attractor, as this comes out of the its complex structure, is evident.

In Fig. 3 the timeseries of both *x*- and *y-signals* appear. Their non-periodic nature is evident. Figure 3 shows time dependences of the coordinates X (top) and Y (bottom) respectively (the channels' settings were for channel A, 10 V/div and for channel B, 10 V/div. Timescale 2 ms/div.

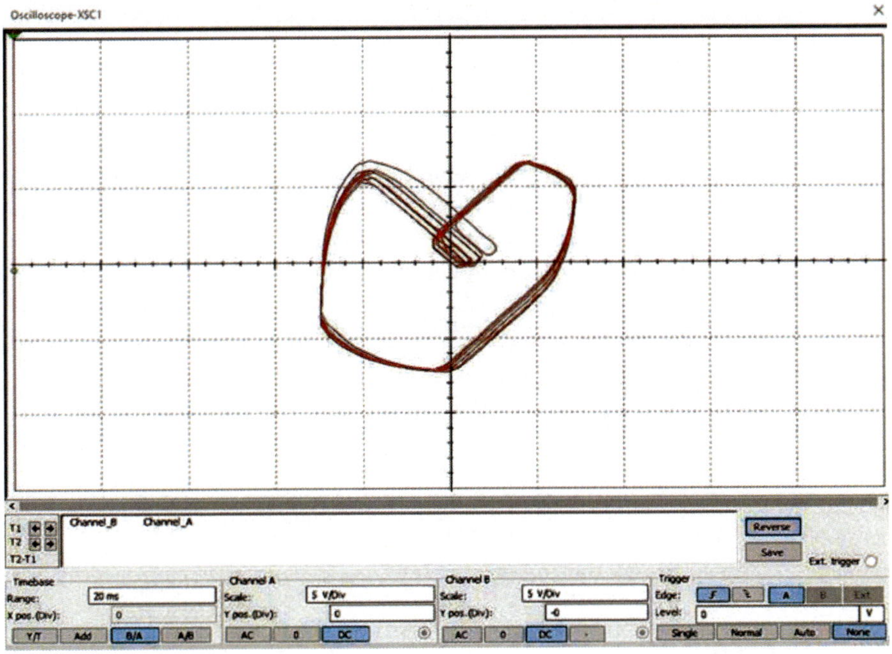

Fig. 2 The simulated chaotic attractor of the new non-autonomous chaotic oscillator

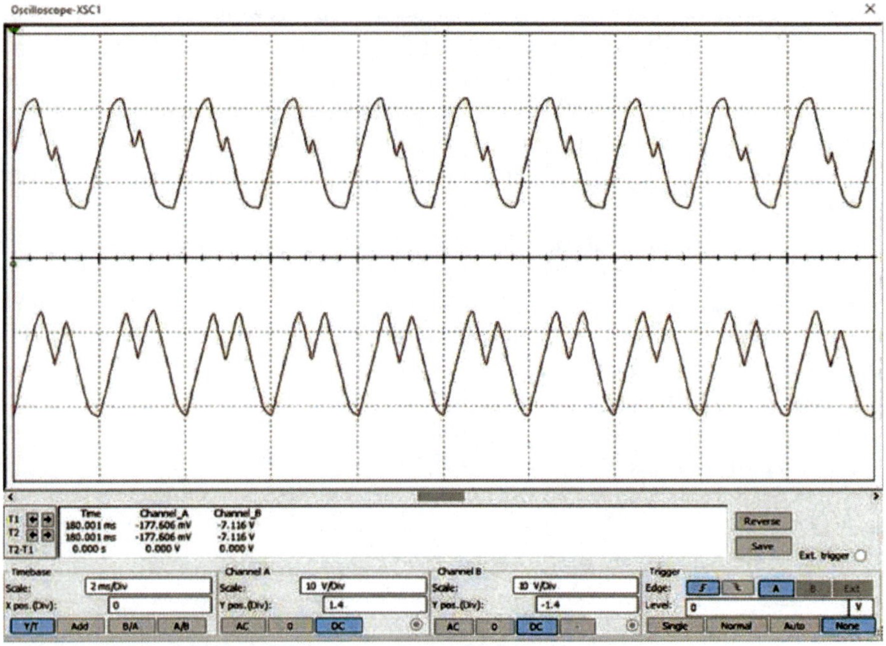

Fig. 3 The *x-signal* (upper) and the *y-signal* (lower) timeseries

Finally, in Figs. 4 and 5 the power spectrum for each of the two signals appears. Apparently, the power spectra of the produced signals are broadband, typical of chaotic signals. They span to a frequency range that goes beyond 5 kHz. The peak of the frequency spectrum was measured to be at 0.6 kHz, and it corresponds to a prevailing frequency of the implementing oscillating loop.

3 Practical Realization of the New Non-autonomous Chaotic Generator

For engineers is important a practical realization. In this Section we present designed layout (Fig. 6) and 3D model (Fig. 7) using Proteus 8. Layout sizes are 45 * 30 mm.

4 Conclusions

Designed new two channel non-autonomous chaotic generator is presented. Computer modelling results of the circuit realization and main information properties are shown. For demonstrate of these properties was using MultiSim software.

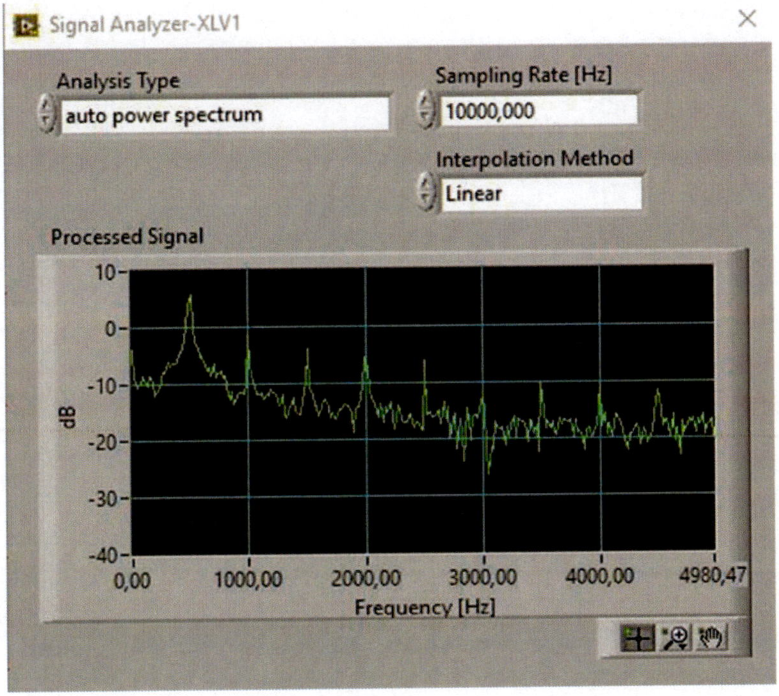

Fig. 4 The spectral distribution of the *x-signal*, typical of chaotic signals

Also, layout and 3D model of the new two channel non-autonomous chaotic generator using software Proteus are presented. This non-autonomous generator can be used as a one of the main portable part of the modern communication system for masking and decrypt of the information.

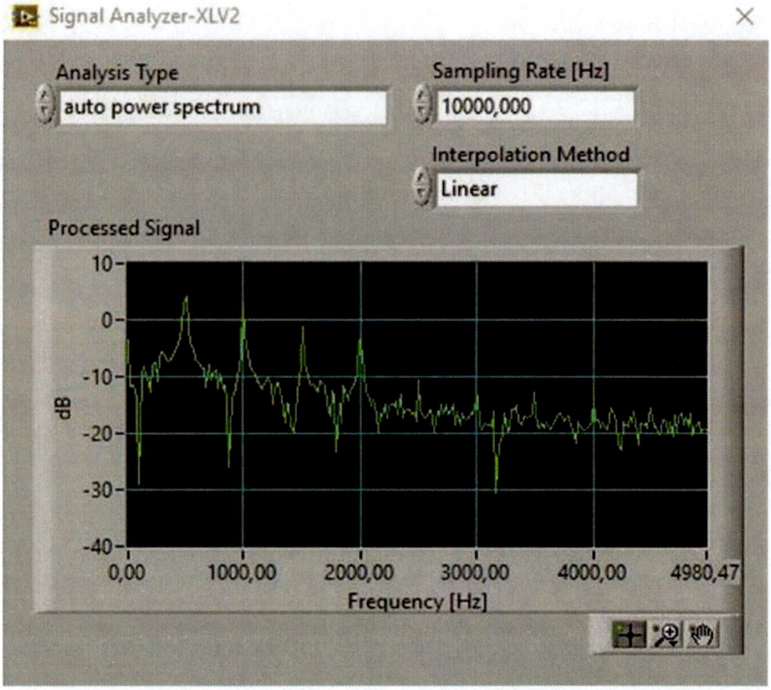

Fig. 5 The spectral distribution of the *y-signal*, typical of chaotic signals

Fig. 6 The designed layout of the new non-autonomous chaotic generator

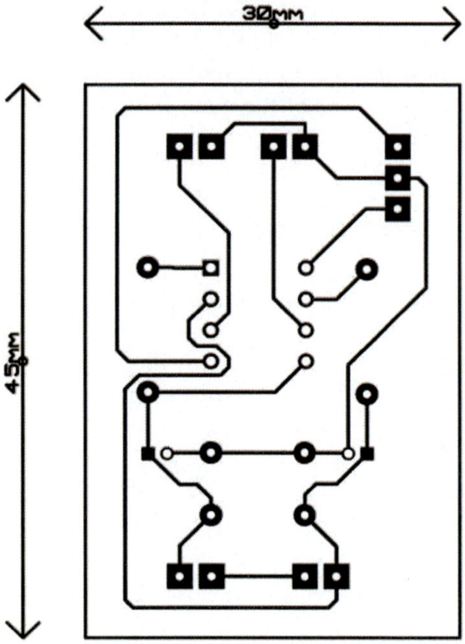

Fig. 7 3D model of the new non-autonomous chaotic generator

References

1. V. Rusyn, Modeling and Research *Information Properties of Rucklidge Chaotic System using LabView*. CHAOS 2017—Proceedings: 10th Chaotic Modeling and Simulation International Conference, (2017), pp. 739–744
2. V. Rusyn, A. Samila, Ch. Skiadas, Computer modeling and practical realization of chaotic circuit with a light-emitting diode. In: *Fourteenth International Conference on Correlation Optics, Chernivtsi*, 16–19 Sept 2019, pp. 113690D (2020)
3. A. Sambas, W.S.M. Sanjaya, M. Mamat, P.R. Putra, A.T. Azar, Mathematical modelling of chaotic jerk circuit and its application in secure communication system. Stud. Fuzziness Soft Comput. **337**, 133–153 (2016)
4. V. Rusyn, M. Mamat, F. Azharul, W.S. Mada Sanjaya, A. Sambas, E. Dwipriyoko, A. Sutoni, Computer modelling of the information properties of hyper chaotic lorenz system and its application in secure communication system. J. Phys. Conf. Ser. **1764**, 012205
5. L. Chua, Memristor—the missing circuit element. IEEE Trans. Circuit Theory **18**(5), 507–519 (1971)
6. V. Rusyn, S. Khrapko, Memristor: Modeling and research of information properties. *Springer Proceedings in Complexity*, 229–238 (2019)
7. I. Petras, Y. Chen, C. Coopmans, *Fractional-Order Memristive Systems*. ETFA 2009—2009 IEEE Conference on Emerging Technologies and Factory Automation, art. no. 5347142 (2009)
8. C. Coopmans, I. Petras, Y. Chen, *Analogue Fractional-Order Generalized Memristive Devices*. Proceedings of the ASME International Design Engineering Technical Conferences and Computers and Information in Engineering Conference 2009, DETC2009, vol. 4, PART B, (2010), pp. 1127–1136

9. M. Itoh, L. Chua, Dynamics of Hamiltonian systems and memristor circuits. Int. J. Bifurcat. Chaos **27**(2), 1730005 (2017)
10. D. Yu, C. Zheng, H.H.C. Iu, T. Fernando, L.O. Chua, A new circuit for emulating memristors using inductive coupling. IEEE Access **5**, 1284–1295 (2017)
11. I. Petras, Y. Chen, *Fractional-order Circuit Elements with Memory*. Proceedings of the 2012 13th International Carpathian Control Conference, ICCC 2012, art. no. 6228706 (2012), pp. 552–558
12. I. Petras, An effective numerical method and its utilization to solution of fractional models used in bioengineering applications. Advances in Difference Equations, art. no. 652789 (2011)
13. P.Milicka, P. Cížek, J. Faigl, On chaotic oscillator-based central pattern generator for motion control of hexapod walking robot. CEUR Workshop Proceedings, 1649, 131–137 (2016)
14. A. Sambas, M.W.S. Sanjaya, M. Mamat, H. Diyah, Design and analysis bidirectional chaotic synchronization of rossler circuit and its application for secure communication. Appl. Math. Sci. **7**(1), 11–21 (2013)
15. V. Rusyn, Ch.H. Skiadas, A. Sambas, M. Mamat, S. Vaidyanathan, Process of pulse transformation of the analog nonlinear signals. Telecommun. Radio Eng. **79**(13), 1141–1147 (2020)
16. A. Rhif, S. Vaidyanathan, A. Sambas, Mujiarto, Subiyanto, A fish biology chaotic system and its circuit design. IOP Conf. Ser. J Phys. **1179**, art. no. 012011 (2019)
17. S. Vaidyanathan, M. Feki, A. Sambas, C.H. Lien, A new biological snap oscillator: Its modelling, analysis, simulations and circuit design. Int. J. Simul. Process Model. **13**(5), 419–432 (2018)
18. V. Rusyn, O. Savko, *Modeling of Chaotic Behavior in the Economic Model*. CHAOS 2015—8th Chaotic Modeling and Simulation International Conference, Proceedings 2015, 705–712 (2015)
19. S. Vaidyanathan, A. Sambas, S. Kacar, U. Cavusoglu, A new finance chaotic system, its electronic circuit realization, passivity based synchronization and an application to voice encryption. Nonlinear Eng. **8**(1), 193–205 (2019)
20. C.H. Skiadas, C. Skiadas, *Chaotic Modelling and Simulation: Analysis of Chaotic Models, Attractors and Forms* (Taylor & Francis Group, LLC, 2008), pp. 1–345
21. C.H. Skiadas, Exact solutions of stochastic differential equations: Gompertz, generalized logistic and revised exponential. Methodol. Comput. Appl. Probab. **12**(2), 261–270 (2010)
22. V. Rusyn, S. Subbotin, A. Sambas, Analysis and experimental realization of the logistic map using Arduino Pro Mini. CEUR Workshop Proceedings, 2608, 300–310 (2020)
23. A. Sambas, S. Vaidyanathan, T. Bonny, S. Zhang, H.Y. Sukono, G. Gundara, M. Mamat, Mathematical model and FPGA realization of a multi-stable chaotic dynamical system with a closed butterfly-like curve of equilibrium points. Appl. Sci. **11**(2), 788 (2021)
24. E.-Z. Dong, R.-H. Li, S.-Z. Du, A multi-directional controllable multi-scroll conservative chaos generator: modelling, analysis and FPGA implementation. Chin. Phys. B **30**(2), art. no. 020505 (2020)
25. X. Peng, Y. Zeng, M. Wang, Z. Li, Generating multi-layer nested chaotic attractor and its FPGA implementation. Chin. Phys. B **30**(6), art. no. 060509 (2021)
26. C.-H. Lien, S. Vaidyanathan, A. Sambas, M. Mamat, W.S. Sanjaya, A new two-scroll chaotic attractor with three quadratic nonlinearities, its adaptive control and circuit design. IOP Conf. Ser. Mater. Sci. Eng. **332**(1), art. no. 012010 (2018)
27. A. Sambas, S. Vaidyanathan, M. Mamat, W.S. Mada Sanjaya, A six-term novel chaotic system with hidden attractor and its circuit design. Stud. Syst. Decis. Control **133**, 365–373 (2018)
28. V. Rusyn, Ch. Skiadas, Threshold method for control of chaotic oscillations. *Springer Proceedings in Complexity* (Springer, 2020), pp. 217–229
29. V. Rusyn, M.A. Mohamad, D. Purwandari, M. Mamat, J. Titaley, B. Pinontoan, Chaotic and controlling regimes of a new modified Chua's generator. J. Adv. Res. Dyn. Control Syst. **12**(02), 556–561 (2020)

30. V. Rusyn, M. Mohamad, J. Titaley, N. Nainggolan, M. Mamat, Design, computer modelling, analysis and control of the new chaotic generator. J. Adv. Res. Dyn. Control Syst. **12**(02), 2306–2311 (2020)
31. V. Rusyn, M. Sadli, M. Mamat, W.M. Sanjaya, Computer modelling of a new simple chaotic generator. J. Phys. Conf. Ser. **1477**, 022010 (2020)

External Synchronization of Solitary States and Chimeras in Unidirectionally Coupled Neural Networks

E. Rybalova, A. Zakharova, and G. Strelkova

Abstract We perform numerical simulation of the dynamics of a multiplex network consisting of two unidirectionally coupled rings of FitzHugh-Nagumo neurons with nonlocal interaction. When uncoupled, one ring demonstrates solitary state regimes and the other one exhibits chimera states. We explore in detail how the synchronization degree between the layers depends on the type of unidirectional interlayer coupling (via fast or slow variables) and on the structures in the driver layer. It is shown that the structure in the response layer can be suppressed and is replaced by the driver layer structure. However, the degree of external synchronization is higher in the case when the driver layer demonstrates solitary states and when the unidirectional coupling is executed via the fast variables. In the case of coupling via the slow variables, external synchronization of neither solitary states nor chimeras cannot be achieved in the considered network.

Keywords Synchronization · FitzHugh-Nagumo neuron · Multiplex network · Chimera state · Solitary state

1 Introduction

Exploring various properties of cooperative dynamics of multicomponent systems, as well as the effects observing in such systems and synchronization between their elements is one of the main part of nonlinear dynamics [1–6]. This is inextricably linked to the fact that most systems in the world are complex networks with various individual elements and types of coupling between them. There is a plenty of works devoted to synchronization phenomena in systems of completely different nature, such as physics [7–10], chemistry [11, 12], neuroscience [13–18], sociologybreak

E. Rybalova (✉) · G. Strelkova
Institute of Physics, Saratov State University, 83 Astrakhanskaya Street, Saratov 410012, Russia
e-mail: rybalovaev@gmail.com

A. Zakharova
Institut für Theoretische Physik, Technische Universität Berlin, Hardenbergstr. 36, 10623 Berlin, Germany

C. H. Skiadas and Y. Dimotikalis (eds.), *14th Chaotic Modeling and Simulation International Conference*, Springer Proceedings in Complexity,
https://doi.org/10.1007/978-3-030-96964-6_26

[19–21], etc., as well as in real-world systems, for instance, communication systems [22], power grids [23, 24], transportation networks [25].

Dynamics of ensembles of nonlocally coupled elements, when each node is coupled with a finite number of its nearest neighbors, has recently attracted much interest due to the discovery of a new spatiotemporal structure, later called "chimera state" [26, 27]. This structure is a striking example of cluster synchronization when a network dynamics spontaneously splits into coherent (synchronous behavior) and incoherent (desynchronized dynamics) clusters with well-defined boundaries in the network space. Although these structures have been found in networks with different individual elements [26–32], with different types of coupling between them [28, 33–37], as well as in real experiments [7–11, 38, 39], greater interest in these structures was caused by their connection with natural and man-made dynamics [14, 17, 22–25].

Solitary states are another example of partial synchronization [40]. With this type of synchronization, solitary nodes appear on the coherent profile of the system and are evenly distributed over the entire ensemble. Oscillators in the solitary state regime fundamentally differ in their dynamics from the other oscillators of the network. This kind of pattern has been observed in networks of the Kuramoto models [40–42], the discrete-time systems [43], the FitzHugh-Nagumo systems [44–46], and others. They have also been detected in experiments with mechanical pendulums [39]. Later, solitary state chimeras were revealed when an incoherent cluster includes several solitary states and coexists with coherent clusters [46, 47].

Studying interaction between different spatiotemporal structures is an important task in the numerical simulation of collective dynamics of complex systems. It was shown in [30] that chimera states can be observed in a ring of nonlocally coupled FitzHugh-Nagumo oscillators. Later, these studies were expanded in [46] and it was found out that this network can also demonstrate solitary states. The interaction between chimeras and solitary states was explored for the first time in [48] where two rings of nonlocally coupled FitzHugh-Nagumo oscillators were bidirectionally coupled either via fast or slow variables. The objective of the present paper is to study the peculiarities of external synchronization of chimeras and solitary states in a two-layer network of unidirectionally coupled rings of FitzHugh-Nagumo oscillators depending on the type of interlayer coupling (via activators or inhibitors) and of the spatiotemporal structures in a driver and a response layer. The identity of synchronous structures in the considered network is quantified using a global interlayer synchronization error.

2 Model Under Study

The model under study represents a multiplex network consisting of two unidirectionally coupled layers. Each layer is given by a ring of nonlocally coupled FitzHugh-Nagumo oscillators [49, 50]. The network is governed by the following system of equations:

$$\varepsilon \frac{du_{1i}}{dt} = u_{1i} - \frac{u_{1i}^3}{3} - v_{1i} + \frac{\sigma}{2P} \sum_{j=i-P}^{i+P} [b_{uu}(u_{1j} - u_{1i}) + b_{uv}(v_{1j} - v_{1i})] + c^u(u_{2i} - u_{1i}),$$

$$\frac{dv_{1i}}{dt} = u_{1i} + a + \frac{\sigma}{2P} \sum_{j=i-P}^{i+P} [b_{vu}(u_{1j} - u_{1i}) + b_{vv}(v_{1j} - v_{1i})] + c^v(v_{2i} - v_{1i}),$$

$$\varepsilon \frac{du_{2i}}{dt} = u_{2i} - \frac{u_{2i}^3}{3} - v_{2i} + \frac{\sigma}{2P} \sum_{j=i-P}^{i+P} [b_{uu}(u_{2j} - u_{2i}) + b_{uv}(v_{2j} - v_{2i})] + s^u(u_{1i} - u_{2i}),$$

$$\frac{dv_{2i}}{dt} = u_{2i} + a + \frac{\sigma}{2P} \sum_{j=i-P}^{i+P} [b_{vu}(u_{2j} - u_{2i}) + b_{vv}(v_{2j} - v_{2i})] + s^v(v_{1i} - v_{2i}). \quad (1)$$

Dynamical variables u_{li} correspond to the activators or the fast variables, and v_{li} are the inhibitors or the slow variables in each ring, where $l = 1, 2$ is the layer number, and $i = 1, 2, \ldots, N = 300$ is the node number in each ring (all indices are modulo N). Individual FitzHugh-Nagumo oscillators can demonstrate either excitable ($|a| > 1$) or oscillatory ($|a| < 1$) regimes, which depend on the excitability threshold parameter a. In the present study, all the FitzHugh-Nagumo oscillators in the network (1) operate in the oscillatory regime at $a = 0.5$ and the time-scale separation parameter is also fixed $\varepsilon = 0.05$ for all the network nodes.

The nonlocal intralayer coupling in each layer is given by the coupling strength σ and the coupling range P which denotes the number of nearest neighbors of the ith node from both sides in each layer. In our numerical simulation we choose $\sigma = 0.3$ and $P = 105$ in both rings. The intralayer interaction of the FitzHugh-Nagumo neurons in the system (1) has not only direct couplings between activator (u) and inhibitor (v) variables but also cross ones which are executed according to a rotational coupling matrix:

$$B = \begin{pmatrix} b_{uu} & b_{uv} \\ b_{vu} & b_{vv} \end{pmatrix} = \begin{pmatrix} \cos\phi & \sin\phi \\ -\sin\phi & \cos\phi \end{pmatrix}, \quad (2)$$

where $\phi \in [-\pi; \pi)$. In the work [30] this type of coupling was used for the first time and it has been shown that chimera states can be observed in the ring of nonlocally coupled FitzHugh-Nagumo neurons at $\phi = \pi/2 - 0.1$. This research was expanded in the paper [46] where the effect of parameter ϕ on the regimes observed in the FitzHugh-Nagumo ring was explored in detail. It was particularly shown that this ensemble can demonstrate not only chimera states but also solitary states. In the present research the parameter ϕ_l ($l = 1, 2$) values are set in such a way to observe a solitary state regime in the first ring and chimera states in the second one.

The interlayer coupling in the network (1) is organized to be unidirectional with coefficients c^u, c^v, s^u, and s^v. Therefore, when the first layer affects the second one (solitary states affect chimeras) we have $c^u = 0$, $c^v = 0$, $s^u \neq 0$, $s^v \neq 0$, where the superscripts correspond to the coupling via the fast (u) or the slow (v) variables. Vice versa, when the first layer is subjected to the second one (chimeras affect solitary states), the interlayer coupling is defined by $c^u \neq 0, c^v \neq 0, s^u = 0, s^v = 0$. In our simulations, initial conditions are chosen to be randomly distributed on circle $u^2 + v^2 \leqslant 2^2$. The layers are coupled from the initial time $t = 0$, and the equations (1) are integrated using the Runge-Kutta-Fehlberg method with step $h = 0.02$.

Figure 1 shows typical spatiotemporal structures which can be observed in uncoupled FitzHugh-Nagumo rings for the chosen intralayer coupling parameter values. The first layer demonstrates the solitary states (Fig. 1a–c), and the second layer exhibits the chimera state (Fig. 1d–f). As can be seen from Fig. 1a, the solitary nodes are evenly distributed along the coherent profile, while the mean phase velocity profile is rather flat (Fig. 1b) (this parameter is calculated with the formula $\omega_{li} = \frac{2\pi M_i}{\Delta T}$, where M_i is the number of complete rotations around the origin performed by the ith oscillator during the time interval ΔT [30], $l = 1, 2$ is the layer number). Differences in the dynamics of the solitary nodes and the oscillators from the coherent part can be observed in the (u_1, v_1) phase plane (Fig. 1c), where the attracting set with a large radius (black dots) corresponds to the oscillators from the coherent region, and the small cycle (red dots) to the solitary nodes. In the case of chimera states, the snapshot of the second ring dynamics splits into two clusters (Fig. 1d): one includes elements $150 \leqslant i \leqslant 250$ with coherent dynamics, and the other one consists of nodes $1 \leqslant i \leqslant 149$ and $251 \leqslant i \leqslant 300$ which behave incoherently. In the mean phase velocity profile (Fig. 1e), the coherent domain is characterized by a smooth distribution, while an arc-like dependence is characteristic for the nodes from the incoherent cluster. There are also two intersecting sets in the (u_2, v_2) phase plane (Fig. 1f): the green attractor reflects the dynamics of the elements from the incoherent cluster of the chimera state and the black set refers to the nodes from the coherent domain. As can be seen from the phase portraits, the green attracting set is essentially thick if compared with a limit cycle for a single FitzHugh-Nagumo system [49, 50], and unlike the solitary states, these sets are overlapping.

To analyze the degree of synchronous behavior (or identity of synchronous structures) of the coupled layers we apply a global interlayer synchronization error:

$$\delta = \frac{1}{N} \sum_{i=1}^{N} \left(\frac{1}{t_2 - t_1} \int_{t_1}^{t_2} (u_{1i} - u_{2i})^2 dt \right), \tag{3}$$

where $N = 300$. Since the coupled FitzHugh-Nagumo rings (1) are not identical, the external interlayer synchronization can be considered in its effective sense. In our numerical studies, imposing certain quantitative conditions for the global interlayer synchronization error we can distinguish effective external synchronization if $0.001 < \delta < 0.01$ and full (complete) external synchronization when $\delta < 0.001$.

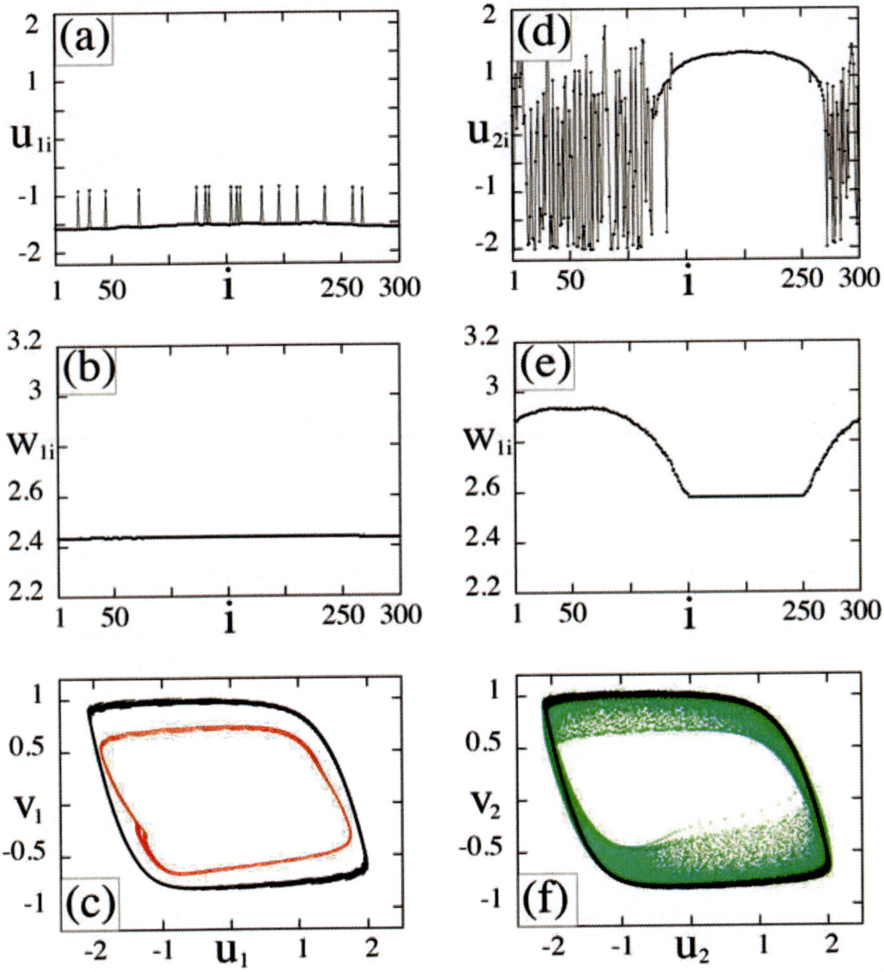

Fig. 1 Dynamics of uncoupled rings ($c^u = 0, c^v = 0, s^u = 0, s^v = 0$): the solitary states in the first ring (**a–c**) and the chimera state in the second ring (**d–f**). Snapshots of variables u_{1i} and u_{2i} (upper row), mean phase velocity profiles (middle row, ω_{1i}, ω_{2i}) and phase portraits for all elements of the rings (lower row, (u_1, v_1) and (u_2, v_2)). Black lines on the phase portraits correspond to elements in the coherent mode, red curves to the solitary nodes, and green ones to the incoherent cluster of the chimera state. Parameters: $\sigma = 0.3$, $P = 105$, $\phi_1 = \pi/2 - 0.2$, $\phi_2 = \pi/2 - 0.04$, $\varepsilon = 0.05$, $a = 0.5$, and $N = 300$

3 Unidirectional Interlayer Coupling via Fast Variables

We study numerically the case when the FitzHugh-Nagumo rings (1) are unidirectionally coupled via the fast variables, i.e., $c^u \neq 0$, $s^u \neq 0$ and $c^v = 0$, $s^v = 0$. It was shown in [48] that in the presence of this type of the interlayer coupling in a system of two symmetrically coupled rings, first chimera states are formed in both rings, then with an increase in the coupling strength, the rings are completely synchronized and their dynamics correspond to coherent spatial profiles. However, at certain values of the interlayer coupling strength, the regime of solitary states can also be observed in both rings.

3.1 Impact of Solitary States on Chimera

Let us first consider the possibility of suppressing the chimera structure in the second ring and the establishment of solitary states under the unidirectional influence of the first ring which demonstrates the solitary states. In this case the first FitzHugh-Nagumo ring is a driver ($c^u = 0$, $c^v = 0$), while the second one is a response ($s^u \neq 0$, $s^v = 0$). Figure 2 illustrates the dependence of the global interlayer synchronization error and the evolution of the second ring dynamics when the interlayer coupling strength s^u grows. As can see from Fig. 2b, already for a sufficiently weak coupling, the chimera state in the second ring completely disappears and is replaced by the regime of solitary states. However, the observed structure is not synchronous with that one in the driver (see Figs. 1a and 2b): the frequency of the solitary nodes is not equal to that of the elements from the coherent part of the ring (Fig. 2c), and the corresponding attracting set (red points) in the phase plane (inset in Fig. 2c) is wider than that shown in Fig. 1c. As follows from Fig. 2a, when the interlayer coupling is sufficiently weak ($s^u < 0.15$), the synchronization error δ does not satisfy the effective synchronization condition.

Even when s^u slightly increases, the observed solitary state regime in the second ring is still not synchronous to the mode in the first ring (Fig. 2d and e). However, as can be seen from the phase portrait in Fig. 2e, the set corresponding to the solitary nodes (red points) is separated from the oscillators from the coherent profile (black line). Starting from the region where the dependence $\delta(s^u)$ becomes smooth ($s^u > 0.33$ in Fig. 2a), the solitary nodes in the second ring begin to correspond to the solitary nodes in the first ring (Fig. 1c). In this case we have a smooth frequency profile and two phase portraits clearly separated in the phase space (Fig. 2f and g). On the other hand, already starting from $s^u > [0.15; 0.28]$ (the exact value depends on the initial conditions) the global interlayer synchronization error becomes less than 0.01 and we can talk about effective synchronization. Only when $s^u > 0.67$ (Fig. 2a), complete external synchronization ($\delta < 0.001$) occurs in the network (1).

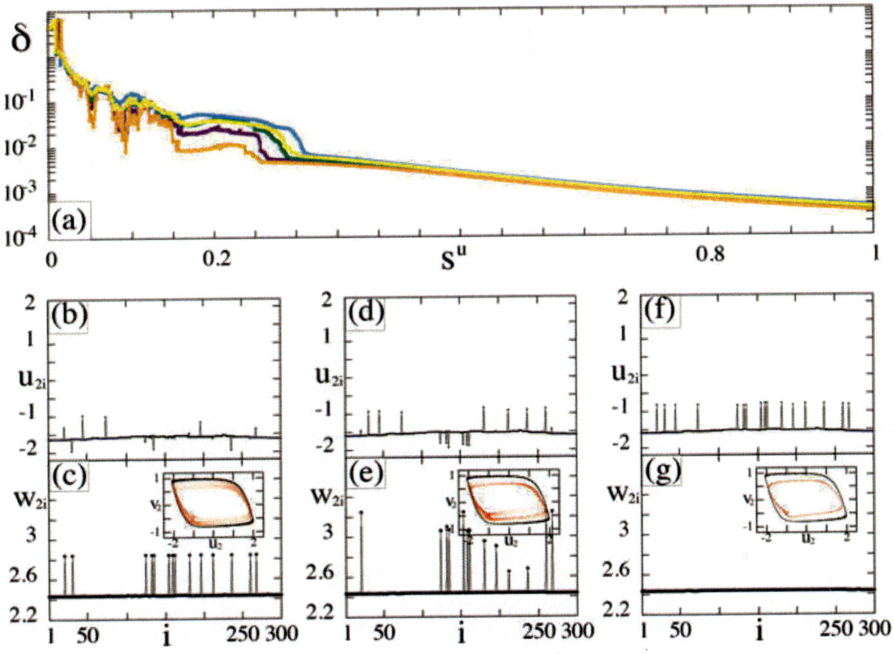

Fig. 2 Unidirectional impact of the first ring (solitary states) on the second one (chimera) via the fast variables: $s^u \neq 0$, $s^v = 0$, $c^u = 0$, $c^v = 0$ in the network (1). **a** Dependence of δ (3) on the interlayer coupling strength s^u plotted for 5 different sets of random initial conditions in each ring (marked by different colors). **b–g** Dynamics of the second ring for increasing s^u: 0.035 (**b**, **c**), 0.13 (**d**, **e**), 0.35 (**f**, **g**). **b**, **d**, **f** Snapshots of variables u_{2i}, **c**, **e**, **g** mean phase velocity profiles w_{2i} and phase portraits for all ring elements (insets (u_2, v_2)): black lines indicate the coherent dynamics, red curves correspond to the solitary nodes. Other parameters: $\sigma = 0.3$, $P = 105$, $\phi_1 = \pi/2 - 0.2$, $\phi_2 = \pi/2 - 0.04$, $\varepsilon = 0.05$, $a = 0.5$, and $N = 300$

3.2 *Impact of Chimera on Solitary States*

When the second ring exhibiting the chimera state is the driver, the global synchronization error δ demonstrates a smooth dependence on the interlayer coupling strength c^u over the entire interval of its variation (Fig. 3a). This can be explained by the fact that there is no need to synchronize individual elements (solitary nodes) which introduce deviations into dependence $\delta(c^u)$. In this case, the solitary states in the first ring also quickly disappear, and the snapshot splits into coherent and incoherent clusters (Fig. 3c). However, the arc-like dependence does not immediately appear on the frequency profile (Fig. 3c). Increasing c^u leads to the appearance of the arc in the frequency profile, which at first looks a bit noisy (Fig. 3e). Only when c^u grows ($c^u > 0.1$), the frequency profile becomes smooth (Fig. 3g). As follows from the snapshots (Fig. 3d and f) and the phase portraits (insets in Fig. 3e and g), already at a very weak interlayer coupling c^u, the first ring (response) starts behaving similarly to the second ring (driver) (see Figs. 1d, f and 3d, e).

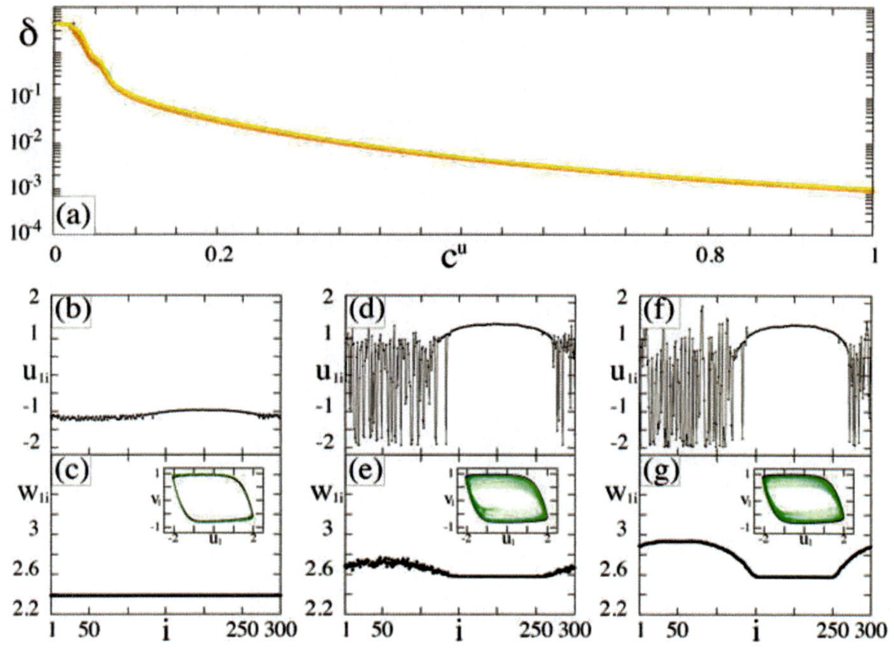

Fig. 3 Unidirectional impact of the second ring (chimera) on the first ring (solitary states) via the fast variables: $c^u \neq 0$, $s^u = 0$, $s^v = 0$, $c^v = 0$ in the network (1). **a** Dependence of δ (3) on the interlayer coupling strength c^u plotted for 5 different sets of random initial conditions in each ring (marked by different colors). **b–g** Dynamics of the first ring in (1) for increasing c^u: 0.014 (**b**, **c**), 0.06 (**d**, **c**), 0.235 (**f**, **g**). **b**, **d**, **f** Snapshots of variables u_{1i}, **c**, **e**, **g** mean phase velocity profiles w_{1i} and phase portraits for all ring elements (insets (u_1, v_1)): black lines indicate the coherent dynamics, red curves correspond to the solitary nodes. Other parameters are as in Fig. 2

In contrast to the previously considered case, in this situation the global interlayer synchronization error does not fall below the 0.001 level even for a rather strong unidirectional interlayer coupling (Fig. 3a). This means that only effective external synchronization of the chimera state takes place.

4 Unidirectional Interlayer Coupling via Slow Variables

We now turn to the case when the two rings (1) are unidirectionally coupled via the slow variables, i.e., $c^v \neq 0$, $s^v \neq 0$, while there is no coupling via the fast variables, $c^u = 0$, $s^u = 0$. Our previous studies [48] showed that with this type of coupling in a system of two bidirectionally coupled rings, firstly the chimera states in the second ring disappear and are replaced by uniformly distributed solitary nodes, but they are not synchronous with the solitary nodes in the first ring. At the same time, the solitary nodes in the first ring gradually disappear. A further increase of the coupling strength between the layers leads to a coherent regime in the first ring and the solitary state chimera in the second ring. By increasing the coupling strength, we can observe the classical chimera states in both rings, which behave quite synchronously. With a further increase in the coupling strength the dynamics of the two-layer network is similar to the dynamics of the rings which are coupled through the fast variables. The rings are completely synchronized and their behavior corresponds to coherent spatial profiles. Moreover, at certain values of the interlayer coupling strength (more than 1.0), the solitary state mode can be observed in both rings.

4.1 Impact of Solitary States on Chimera

Consider the case when the second ring in the chimera state is driven via the slow variables by the first ring in the solitary state regime. In this case the chimera state also quickly disappears and is replaced by the solitary nodes (Fig. 4b–e). However, the solitary nodes are distributed throughout the whole ring and their location does not coincide with that in the driver layer (see Figs. 1a and 4d). It is also evident that the frequencies of several solitary nodes are not equal to the frequency of the other oscillators, as it should be (Fig. 4e). Moreover, in the phase portrait, the trajectories of some solitary nodes do not lie separately from the phase portrait of the oscillators from the coherent domain, i.e., there is an intersection of the red and black sets (inset in Fig. 4e). When the interlayer coupling increases, the most part of the solitary nodes disappears but the remaining nodes are not synchronized with those in the first ring (Fig. 4f and g). A further increase in s^v does not lead to the observation of a more synchronous mode of oscillations of the second ring with the first one, but, on the contrary, leads to the fact that the phase portraits of the elements change greatly and the rings are never synchronized (for example, Fig. 5).

Let us pay attention to the change in the global interlayer synchronization error δ as s^v increases (Fig. 4a). Within the interval $s^v \in [0; 0.2]$, the dependence has several minima and maxima and does not smoothly decrease when s^v grows. This is due to the fact that initially, under the influence of the first ring, the chimera state in the second ring is gradually destroyed (the first minimum is at $s^v \approx 0.04$). Then a lot of solitary nodes appear in the second ring, which do not correspond to the solitary nodes in the first ring and are not synchronized with them (maximum is at $s^v \approx 0.055$) (see Figs. 1a and 4d). Afterwards, the solitary nodes gradually disappear with increasing coupling strength (minimum is at $s^v \approx 0.11$). Finally, the solitary nodes in the second ring correspond to the same oscillators as in the first ring (maximum is at $s^v \approx 0.2$) (see Figs. 1a and 4f), and the rings are partially synchronized with a further increase of the coupling strength s^v. However, even for a rather strong coupling strength s^v, even effective external synchronization is not observed in the network (1) since $\delta > 0.01$.

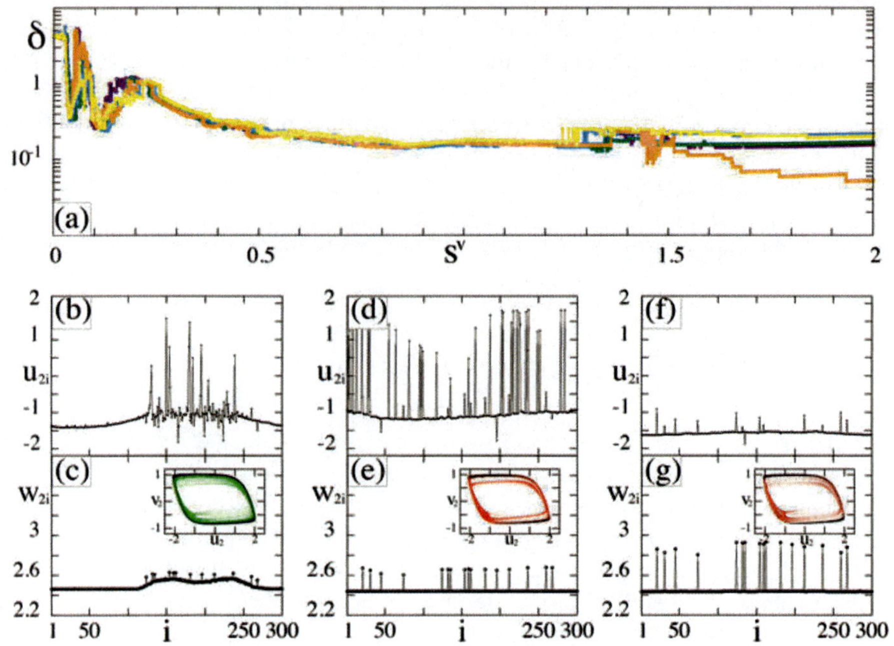

Fig. 4 Numerical results for the case when the first ring (solitary states) is unidirectionally coupled with the second one (chimera state) via the slow variables: $s^v \neq 0$, $s^u = 0$, $c^u = 0$, $c^v = 0$. **a** Dependence of δ (3) on the interlayer coupling strength s^v plotted for 5 different sets of random initial conditions in each ring (marked by different colors). **b–g** Dynamics of the second ring for increasing s^v: 0.025 (**b**, **c**), 0.062 (**d**, **c**), 0.5 (**f**, **g**). **b**, **d**, **f** Snapshots of variables u_{2i}, **c**, **e**, **g** mean phase velocity profiles w_{2i} and phase portraits for all ring elements (insets (u_2, v_2)): black lines indicate the coherent dynamics, red curves correspond to the solitary nodes. Other parameters: $\sigma = 0.3$, $P = 105$, $\phi_1 = \pi/2 - 0.2$, $\phi_2 = \pi/2 - 0.04$, $\varepsilon = 0.05$, $a = 0.5$, and $N = 300$

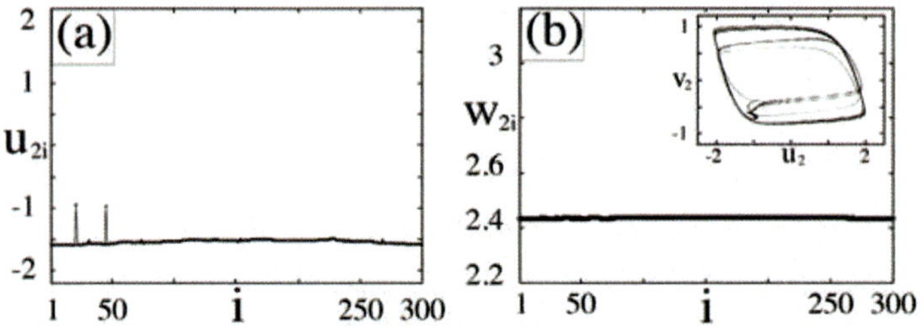

Fig. 5 Dynamics of the second ring under the unidirectional impact from the first ring at $s^v = 1.446$ ($s^u = 0, c^u = 0, c^v = 0$). **a** Snapshots of variables u_{2i}, **b** mean phase velocity profiles w_{2i} and phase portraits for all ring elements (insets (u_2, v_2)): black lines indicate the coherent dynamics, red curves correspond to the solitary nodes. Other parameters are as in Fig. 4

4.2 Impact of Chimera on Solitary States

Finally, we explore the network (1) dynamics when the driver layer (the second ring) exhibits the chimera state. In this case, the network dynamics is similar to that which is observed when the unidirectional coupling is executed via the fast variables. The solitary nodes in the first ring gradually disappear as the coupling strength c^v increases (Fig. 6b and c), and the snapshots of the first ring dynamics (Fig. 6d and e) consist of incoherent and coherent parts, that is related to the chimera state. However, even with a strong coupling, in the presence of a well-developed chimera state in the first layer, the frequency profile demonstrates only a barely noticeable arc-like structure (Fig. 6f and g). Synchronization between the rings is not achieved even for a very strong interlayer coupling: the global interlayer synchronization error never goes below 0.1 (see Fig. 6a). However, for certain sets of random initial conditions in each ring, the values of δ can be lower than for the other sets (see the blue line in Fig. 6a).

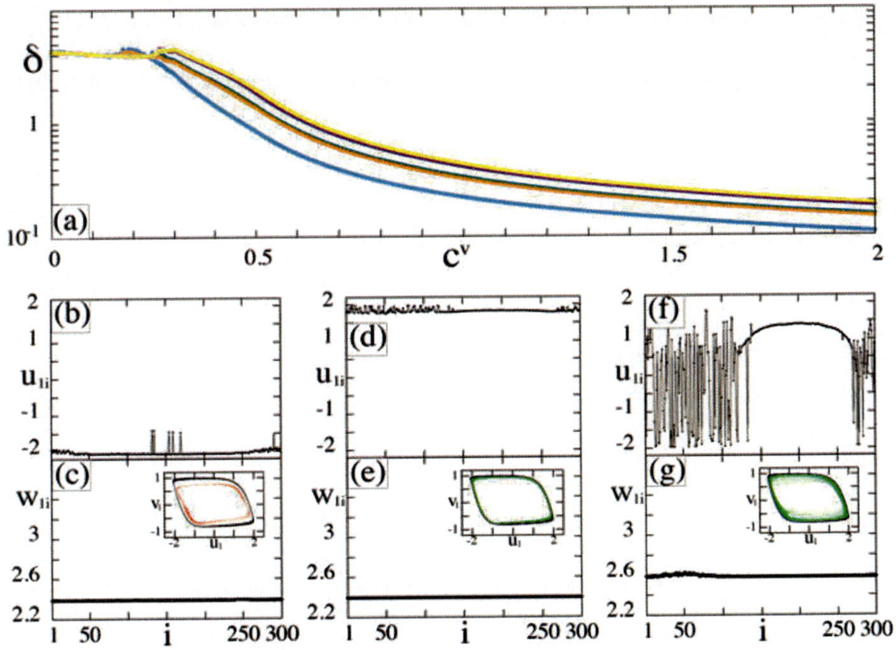

Fig. 6 Numerical results for the case when the first ring (solitary states) in the network (1) is driven by the second ring (chimera) via the slow variables: $c^v \neq 0, s^v = 0, s^u = 0, c^u = 0$. **a** Dependence of δ (3) on the interlayer coupling strength c^u plotted for 5 different sets of random initial conditions in each ring (marked by different colors). **b–g** Dynamics of the first ring in (1) for increasing c^v: 0.025 (**b, c**), 0.223 (**d, c**), 1.6 (**f, g**). **b, d, f** Snapshots of variables u_{1i}, **c, e, g** mean phase velocity profiles w_{1i} and phase portraits for all ring elements (insets (u_1, v_1)): black lines indicate the coherent dynamics, red curves correspond to the solitary nodes. Other parameters are as in Fig. 4

5 Conclusion

In this paper we have presented results of numerical simulation of a two-layer multiplex network of unidirectionally coupled rings of FitzHugh-Nagumo oscillators. Our studies have shown that in the case of unidirectional coupling via the fast variables (activators), it is possible to suppress both chimera states and solitary states and establish a different spatiotemporal regime. However, with external synchronization of solitary states, the global interlayer synchronization error shows a stronger similarity between the rings than in the case of synchronization of chimera states. This fact is easily explained by the structure of these states. Since the oscillators in the coherent cluster are synchronized more easily, it is natural to assume that the solitary states will demonstrate a higher degree of synchronization.

In the case of unidirectional coupling between the FitzHugh-Nagumo rings via the slow variables (inhibitors), although the initial structure of the ring is rapidly destroyed under external influence, the structure of the driver layer can be only partially reproduced in the response layer. This fact is confirmed by the global interlayer synchronization error which does not fall below 0.01.

Thus, our studies have shown that both solitary states and chimera states can be suppressed when the two layers are unidirectionally coupled via both the fast and the slow variables. The response layer reproduces the structure of the driver layer instead of its own. However, the effect of external synchronization (both effective and complete) is observed only when the layers are coupled via the activators. These studies can be useful in practical applications when it is needed to suppress one of the structures and establish another one. Thus, one can control the dynamics of multilayer networks.

Acknowledgements This work was supported by the Deutsche Forschungsgemeinschaft (DFG, German Research Foundation)-Projektnummer 163436311-SFB 910 and by the RFBR and the DFG according to the Research Project No. 20-52-12004.

References

1. V.S. Afraimovich, V.I. Nekorkin, G.V. Osipov, V.D. Shalfeev. *Stability, Structures and Chaos in Nonlinear Synchronization Networks* (World Scientific, 1995)
2. V.N. Belykh, I.V. Belykh, E. Mosekilde, Cluster synchronization modes in an ensemble of coupled chaotic oscillators. Phys. Rev. E **63**(3), 036216 (2001)
3. V. Nekorkin, M.G. Velarde, *Synergetic Phenomena in Active Lattices: Patterns, Waves, Solitons, Chaos* (Springer, 2002)
4. G.V. Osipov, J. Kurths, Ch. Zhou, *Synchronization in Oscillatory Networks* (Springer, 2007)
5. V.I. Nekorkin, A.S. Dmitrichev, D.V. Kasatkin, V.S. Afraimovich, Relating the sequential dynamics of excitatory neural networks to synaptic cellular automata. Chaos **21**(4), 043124 (2011)
6. S. Boccaletti, A.N. Pisarchik, C.I. Del Genio, A. Amann, *Synchronization: From Coupled Systems to Complex Networks* (Cambridge University Press, 2018)
7. A.M. Hagerstrom, T.E. Murphy, R. Roy, P. Hövel, I. Omelchenko, E. Schöll, Experimental observation of chimeras in coupled-map lattices. Nat. Phys. **8**(9), 658–661 (2012)

8. F. Rogister, R. Roy, Localized excitations in arrays of synchronized laser oscillators. Phys. Rev. Lett. **98**(10), 104101 (2007)
9. M. Wickramasinghe, I.Z. Kiss, Spatially organized partial synchronization through the chimera mechanism in a network of electrochemical reactions. Phys. Chem. Chem. Phys. **16**(34), 18360–18369 (2014)
10. D.P. Rosin, D. Rontani, N.D. Haynes, E. Schöll, D.J. Gauthier, Transient scaling and resurgence of chimera states in networks of Boolean phase oscillators. Phys. Rev. E **90**(3), 030902 (2014)
11. M.R. Tinsley, S. Nkomo, K. Showalter, Chimera and phase-cluster states in populations of coupled chemical oscillators. Nat. Phys. **8**(9), 662–665 (2012)
12. J.F. Totz, J. Rode, M.R. Tinsley, K. Showalter, H. Engel, Spiral wave chimera states in large populations of coupled chemical oscillators. Nat. Phys. **14**(3), 282–285 (2018)
13. A.E. Pereda, Electrical synapses and their functional interactions with chemical synapses. Nat. Rev. Neurosci. **15**(4), 250–263 (2014)
14. T. Chouzouris, I. Omelchenko, A. Zakharova, J. Hlinka, P. Jiruska, E. Schöll, Chimera states in brain networks: empirical neural vs. modular fractal connectivity. Chaos **28**(4), 045112 (2018)
15. P.J. Uhlhaas, W. Singer, Neural synchrony in brain disorders: relevance for cognitive dysfunctions and pathophysiology. Neuron **52**(1), 155–168 (2006)
16. P. Jiruska, M. De Curtis, J.G. Jefferys, C.A. Schevon, S.J. Schiff, K. Schindler, Synchronization and desynchronization in epilepsy: controversies and hypotheses. J. Physiol. **591**(4), 787–797 (2013)
17. R.G. Andrzejak, C. Rummel, F. Mormann, K. Schindler, All together now: analogies between chimera state collapses and epileptic seizures. Sci. Rep. **6**(1), 1–10 (2016)
18. C. Hammond, H. Bergman, P. Brown, Pathological synchronization in Parkinson's disease: networks, models and treatments. Trends Neurosci. **30**(7), 357–364 (2007)
19. M. Girvan, M.E. Newman, Community structure in social and biological networks. Proc. Natl. Acad. Sci. **99**(12), 7821–7826 (2002)
20. R. Amato, A. Díaz-Guilera, K.K. Kleineberg, Interplay between social influence and competitive strategical games in multiplex networks. Sci. Rep. **7**(1), 1–8 (2017)
21. R. Amato, N.E. Kouvaris, M. San Miguel, A. Díaz-Guilera, Opinion competition dynamics on multiplex networks. N. J. Phys. **19**(12), 123019 (2017)
22. S. Hong, Y. Chun, Efficiency and stability in a model of wireless communication networks. Social Choice Welfare **34**(3), 441–454 (2010)
23. P.J. Menck, J. Heitzig, J. Kurths, H.J. Schellnhuber, How dead ends undermine power grid stability. Nat. Commun. **5**(1), 1–8 (2014)
24. B. Wang, H. Suzuki, K. Aihara, Enhancing synchronization stability in a multi-area power grid. Sci. Rep. **6**(1), 1–11 (2016)
25. A. Cardillo, M. Zanin, J. Gómez-Gardenes, M. Romance, A.J. del Amo, S. Boccaletti, Modeling the multi-layer nature of the European Air Transport Network: resilience and passengers rescheduling under random failures. Eur. Phys. J. Spec. Topics **215**(1), 23–33 (2013)
26. Y. Kuramoto, D. Battogtokh, Coexistence of coherence and incoherence in nonlocally coupled phase oscillators. Nonlinear Phenom. Complex Syst. **5**(4), 380–385 (2002)
27. D.M. Abrams, S.H. Strogatz, Chimera states for coupled oscillators. Phys. Rev. Lett. **93**(17), 174102 (2004)
28. I. Omelchenk, Y. Maistrenko, P. Hövel, E. Schöll, Loss of coherence in dynamical networks: spatial chaos and chimera states. Phys. Rev. Lett. **106**(23), 234102 (2011)
29. M.J. Panaggio, D.M. Abrams, Chimera states: coexistence of coherence and incoherence in networks of coupled oscillators. Nonlinearity **28**(3), R67 (2015)
30. I. Omelchenko, E. Omelchenko, P. Hövel, E. Schöll, When nonlocal coupling between oscillators becomes stronger: patched synchrony or multichimera states. Phys. Rev. Lett. **110**(22), 224101 (2013)
31. A. Zakharova, M. Kapeller, E. Schöll, Chimera death: symmetry breaking in dynamical networks. Phys. Rev. Lett. **112**(15), 154101 (2014)
32. S.A. Bogomolov, A.V. Slepnev, G.I. Strelkova, E. Schöll, V.S. Anishchenko, Mechanisms of appearance of amplitude and phase chimera states in ensembles of nonlocally coupled chaotic systems. Commun. Nonlinear Sci. Numer. Simul. **43**, 25–36 (2017)

33. S. Ulonska, I. Omelchenko, A. Zakharova, E. Schöll, Chimera states in networks of Van der Pol oscillators with hierarchical connectivities. Chaos **26**(9), 094825 (2016)
34. E. Schöll, Synchronization patterns and chimera states in complex networks: interplay of topology and dynamics. Eur. Phys. J. Spec. Topics **225**(6), 891–919 (2016)
35. J. Sawicki, I. Omelchenko, A. Zakharova, E. Schöll, Chimera states in complex networks: interplay of fractal topology and delay. Eur. Phys. J. Spec. Topics **226**(9), 1883–1892 (2017)
36. S.I. Shima, Y. Kuramoto, Rotating spiral waves with phase-randomized core in nonlocally coupled oscillators. Phys. Rev. E **69**(3), 036213 (2004)
37. M.J. Panaggio, D.M. Abrams, Chimera states on a flat torus. Phys. Rev. Lett. **110**(9), 094102 (2013)
38. E.A. Martens, S. Thutupalli, A. Fourriere, O. Hallatschek, Chimera states in mechanical oscillator networks. Proc. Natl. Acad. Sci. **110**(26), 10563–10567 (2013)
39. T. Kapitaniak, P. Kuzma, J. Wojewoda, K. Czolczynski, Y. Maistrenko, Imperfect chimera states for coupled pendula. Sci. Rep. **4**, 6379 (2014)
40. Y. Maistrenko, B. Penkovsky, M. Rosenblum, Solitary state at the edge of synchrony in ensembles with attractive and repulsive interactions. Phys. Rev. E **89**(6), 060901 (2014)
41. H. Wu, M. Dhamala, Dynamics of Kuramoto oscillators with time-delayed positive and negative couplings. Phys. Rev. E **98**(3), 032221 (2018)
42. P. Jaros, S. Brezetsky, R. Levchenko, D. Dudkowski, T. Kapitaniak, Y. Maistrenko, Solitary states for coupled oscillators with inertia. Chaos **28**(1), 011103 (2018)
43. E. Rybalova, N. Semenova, G. Strelkova, V. Anishchenko, Transition from complete synchronization to spatio-temporal chaos in coupled chaotic systems with nonhyperbolic and hyperbolic attractors. Eur. Phys. J. Spec. Topics **226**(9), 1857–1866 (2017)
44. L. Schülen, D.A. Janzen, E.S. Medeiros, A. Zakharova, Solitary states in multiplex neural networks: onset and vulnerability. Chaos Solitons Fractals **145**, 110670 (2021)
45. L. Schülen, S. Ghosh, A.D. Kachhvah, A. Zakharova, S. Jalan, Delay engineered solitary states in complex networks. Chaos Solitons Fractals **128**, 290–296 (2019)
46. E. Rybalova, V.S. Anishchenko, G.I. Strelkova, A. Zakharova, Solitary states and solitary state chimera in neural networks. Chaos **29**(7), 071106 (2019)
47. A. Bukh, E. Rybalova, N. Semenova, G. Strelkova, V. Anishchenko, New type of chimera and mutual synchronization of spatiotemporal structures in two coupled ensembles of nonlocally interacting chaotic maps. Chaos **27**(11), 111102 (2017)
48. E.V. Rybalova, A. Zakharova, G.I. Strelkova, Interplay between solitary states and chimeras in multiplex neural networks. Chaos Solitons Fractals **148**, 111011 (2021)
49. R. FitzHugh, Impulses and physiological states in theoretical models of nerve membrane. Biophys. J. **1**(6), 2061–2070 (1961)
50. J. Nagumo, S. Arimoto, S. Yoshizawa, An active pulse transmission line simulating nerve axon. Proc. IRE **50**(10), 2061–2070 (1962)

On a Cournot Duopoly Game with Relative Profit Maximization

Georges Sarafopoulos and Kosmas Papadopoulos

Abstract In this article, the authors investigate the dynamics of an oligopoly game in which, they consider a nonlinear Cournot-type duopoly game with homogeneous goods and heterogeneous expectations. The authors investigate the case, where managers have a variety of attitudes toward relative performance that are indexed by their type. In this game they suppose a linear demand and quadratic cost functions. The game is modeled with a system of two difference equations. Existence and stability of equilibrium of the system are studied. The authors show that the models gives more complex, chaotic and unpredictable trajectories, as a consequence of change in the parameter k of speed of the player's adjustment, the parameter d of the horizontal product differentiation and the relative profit parameter μ. The chaotic features are justified numerically via computing Lyapunov numbers and sensitive dependence on initial conditions.

Keywords Cournot duopoly game · Relative profit maximization · Discrete dynamical system · Nash equilibrium · Stability · Bifurcation diagrams · Lyapunov numbers · Strange attractors · Chaotic behavior

1 Introduction

Oligopoly is a market structure between monopoly and perfect competition in which there are only a few number of firms producing homogeneous products. The dynamic of an oligopoly game is more complex because the players (sellers) must consider not only the consumers' behavior but also, the competitors' reactions, i.e., their expectations concerned in how their rivals will act. In 1838 Antoine Augustin Cournot was the first that introduced a formal theory of oligopoly. Joseph Louis Francois

G. Sarafopoulos (✉) · K. Papadopoulos
Department of Economics, Democritus University of Thrace, Komotini, Greece
e-mail: gsarafop@econ.duth.gr

K. Papadopoulos
e-mail: kpapa@econ.duth.gr

C. H. Skiadas and Y. Dimotikalis (eds.), *14th Chaotic Modeling and Simulation International Conference*, Springer Proceedings in Complexity,
https://doi.org/10.1007/978-3-030-96964-6_27

Bertrand, the French mathematician in 1883 modified Cournot's game suggesting that the players (sellers) actually choose prices rather the quantities. Cournot and Bertrand models originally were based on the premise that all players take decisions by naïve way, so that in every step, each player assumes the last values were taken by the competitors without an estimation of their future reactions. However, under the conditions of real market, such an assumption is very unlikely since not all players share naïve beliefs. Different approaches to firm behavior were proposed. Some of the authors considered duopolies under homogeneous expectations and found a variety of complex dynamics in their games, such as appearance of strange attractors [1, 4–6, 10, 18, 23, 25–27, 36]. Also, models with heterogeneous agents were studied [2–4, 11, 14, 15, 17, 28–30, 32, 34, 35].

In real market, producers do not know the entire demand function, though it is possible that they have a perfect knowledge of technology, represented by the cost function. Here it is more likely that firms employ some local estimate of the demand. This issue has been previously analyzed [7–9, 20, 22, 24, 31, 33]. Bounded rational players (sellers) update their strategies based on discrete time periods and by using a local estimate of the marginal profit. With such local adjustment mechanism, the players are not requested to have a complete knowledge of the demand and the cost functions [2, 8, 12, 21, 35].

In this paper the concept of generalized relative profit in a Cournot—type duopoly game with homogeneous goods, linear demand and quadratic cost functions is introduced. The paper is organized as follows: In Sect. 2, the dynamics of the Cournot duopoly game with homogeneous goods and generalized relative profit maximization for two players are analyzed. Heterogeneous expectations, linear demand and quadratic cost functions are supposed. The existence and local stability of the equilibrium points are also analyzed. In Sect. 3 numerical simulations are used to show complex dynamics via computing Lyapunov numbers, bifurcations diagrams, strange attractors and sensitive dependence on initial conditions. Finally, the paper is concluded in Sect. 4.

2 The Game

2.1 The Construction of the Game

Two firms offer their products at discrete-time periods ($t = 0, 1, 2, \ldots$) on a common market. A simple Cournot-type duopoly market where firms (players) produce homogeneous goods and their production decisions are taken at discrete-time periods is considered. This study contains heterogeneous players and more specifically, the first firm chooses its production quantity in a rational way following an adjustment mechanism (bounded rational player) and the second firm decides by naïve way choosing this production quantity that maximizes its utility function (naïve player). At each period t, every firm must form an expectation of the rival's output in the next time period in order to determine the corresponding profit-maximizing quantities for period $t + 1$. The variables q_1, q_2 are the production quantities of each firm, and the inverse demand function (as a function of quantities) for each i firm is given by the following equation:

$$p_i = \alpha - Q, \text{ with } \; i = 1, 2 \tag{1}$$

where p_i is the product price of firm i, $Q = q_1 + q_2$ and α is a positive parameter which expresses the market size.

Specifically, for each firm the inverse demand functions are given by the following equations:

$$p_i = \alpha - q_1 - q_2 \text{ and } p_2 = \alpha - q_2 - q_1 \tag{2}$$

In this duopoly game quadratic cost functions are supposed:

$$C(q_i) = c \cdot q_i^2, \text{ with } \; i = 1, 2 \tag{3}$$

where $c > 0$, is the same cost parameter for two firms.

With these assumptions the profits of the firms are given by:

$$\Pi_1(q_1, q_2) = p_1 q_1 - C(q_1) = p_1 q_1 - cq_1^2 = [\alpha - (q_1 + q_2)]q_1 - cq_1^2 \tag{4}$$

and

$$\Pi_2(q_1, q_2) = p_2 q_2 - C(q_2) = p_2 q_2 - cq_2^2 = [\alpha - (q_1 + q_2)]q_2 - cq_1^2 \tag{5}$$

Potential managers take on a continuum of attitudes toward relative performance which is captured by their type μi. The utility function of a manager of type μi puts weight of $(1 - \mu i)$ on own profits and a weight μi on the difference between own profits and the profits of the firm's rival. This is equivalent to putting unit weight on own profits and weight—μi on the rival's profit. Hence we can write the objective function of a type μi manager working for firm i as:

$$\begin{aligned} U_i &= (1 - \mu_i) \cdot \Pi_i + \mu_i \cdot (\Pi_i - \Pi_j) \\ &= \Pi_i - \mu_i \cdot \Pi_j, \quad i,j = 1, 2, \quad i \neq j \end{aligned} \tag{6}$$

The parameter $\mu_i \in [0, 1]$ is formed by the profile of each player i and the higher (lower) value of μi, the more (less) the player i takes into account the profit of player j. To make our calculations easier it supposed that $\mu = \mu i$, which means that the two player have the same profile in their utility function.

Then the marginal utilities are given by:

$$\frac{\partial U_1}{\partial q_1} = \alpha - 2(1 + c)q_1 - (1 - \mu)q_2 \tag{7}$$

and

$$\frac{\partial U_2}{\partial q_2} = \alpha - 2(1 + c)q_2 - (1 - \mu)q_1 \tag{8}$$

It is supposed that the first firm decides to increase its level of adaptation if it has a positive marginal utility, or decreases its level if the marginal utility is negative (bounded rational player). If $k > 0$ the dynamical equation of player i is:

$$\frac{q_1(t+1) - q_1(t)}{q_1(t)} = k \cdot \frac{\partial U_1}{\partial q_1} \tag{9}$$

$k > 0$ the speed of adjustment of player 1, it is a positive parameter which gives the extent of production variation of the firm following a given profit signal. Moreover it captures the fact that relative effort variations are proportional to the marginal utility.

On the other hand the second firm decides by naïve way choosing a production that maximizes its profits (naïve player):

$$q_2(t+1) = \arg\max_{y} U_2\big(q_1(t), q_2(t)\big) \tag{10}$$

The dynamical system of the players is described by:

$$\begin{cases} q_1(t+1) = q_1(t) + k \cdot q_1(t) \dfrac{\partial U_1}{\partial q_1} \\ q_2(t+1) = \dfrac{\alpha + (\mu - 1) \cdot q_1(t)}{2(1+c)} \end{cases} \tag{11}$$

The dynamics of this system focus on the parameter k (first player's speed of adjustment) and the parameter μ (relative profit parameter).

2.2 *Dynamical Analysis*

2.2.1 The Equilibriums of the Game

The equilibriums of the dynamical system (11) are obtained as nonnegative solutions of the algebraic system:

$$\begin{cases} q_1^* \cdot \dfrac{\partial U_1}{\partial q_1} = 0 \\ q_2^* = \dfrac{\alpha + (\mu - 1) \cdot q_1^*}{2(1 + c)} \end{cases} \tag{12}$$

which obtained by the following settings: $q_1\ (t + 1) = q_1\ (t) = q^*_1$ and $q_2\ (t + 1) = q_2\ (t) = q^*_2$.

- If $q_1^* = 0$, and $\frac{\partial U_2}{\partial q_2} = 0$, then it gives $q_2^* = \frac{\alpha}{2(1+c)}$ and the equilibrium position is:

$$E_0 = \left(0, \frac{\alpha}{2(1 + c)}\right) \tag{13}$$

- If $\frac{\partial U_1}{\partial q_1} = \frac{\partial U_2}{\partial q_2} = 0$, then it gives $q_1^* = q_2^* = \frac{\alpha}{2(1+c)+(1-\mu)}$ and the Nash equilibrium is:

$$E_* = \left(q_1^*, q_2^*\right) = \left(\frac{\alpha}{2(1 + c) + (1 - \mu)}, \frac{\alpha}{2(1 + c) + (1 - \mu)}\right) \tag{14}$$

The effect of the parameter k (speed of adjustment) and the parameter μ (relative profit parameter) on the dynamics of this system is investigated.

2.2.2 Stability of Equilibriums

To study the stability of game's equilibriums, the Jacobian matrix is used. The Jacobian matrix J (q_1, q_2) along the variable strategy (q_1, q_2) is:

$$J(q_1, q_2) = \begin{bmatrix} f_{q_1} & f_{q_2} \\ g_{q_1} & g_{q_2} \end{bmatrix} \tag{15}$$

where:

$$f(q_1, q_2) = q_1 + k \cdot q_1 \cdot \frac{\partial U_1}{\partial q_1} \tag{16}$$

and

$$g(q_1, q_2) = \frac{\alpha + (\mu - 1) \cdot q_1}{2(1 + c)} \tag{17}$$

The Jacobian matrix becomes as:

$$J(q_1^*, q_2^*) = \begin{bmatrix} 1 + k \cdot \left(\frac{\partial U_1}{\partial q_1} + q_1^* \cdot \frac{\partial^2 U_1}{\partial q_1^2}\right) & k \cdot (\mu - 1) \cdot q_1^* \\ \frac{\mu-1}{2(1+c)} & 0 \end{bmatrix} \tag{18}$$

For the E_0 the Jacobian matrix becomes as:

$$J(E_0) = \begin{bmatrix} 1 + k \cdot \frac{\alpha(2c+1+\mu)}{2(1+c)} & 0 \\ \frac{\mu-1}{2(1+c)} & 0 \end{bmatrix} \tag{19}$$

with $Tr = 1 + k \cdot \frac{\alpha(2c+1+\mu)}{2(1+c)}$ and $Det = 0$.

The characteristic equation of J (E_0), gives the nonnegative eigenvalue:

$$r_1 = Tr = 1 + k \cdot \frac{\alpha(2c + 1 + \mu)}{2(1 + c)}$$

it's clearly seems that $|r_1| > 1$ and the E_0 equilibrium is unstable.

For the E_* the Jacobian matrix becomes as:

$$J(E_*) = \begin{bmatrix} 1 - 2k \cdot (1 + c) \cdot q_1^* & -k \cdot (1 - \mu) \cdot q_1^* \\ \frac{\mu-1}{2(1+c)} & 0 \end{bmatrix} \tag{20}$$

with

$$Tr = 1 - 2k \cdot (1 + c) \cdot q_1^* \text{ and } Det = -k \cdot \frac{(1 - \mu)^2}{2(1 + c)} \cdot q_1^* \tag{21}$$

To study the stability of Nash equilibrium the method of three conditions is used and the equilibrium position is locally asymptotically stable when they are satisfied simultaneously:

$$\begin{array}{ll} (i) & 1 - Det > 0 \\ (ii) & 1 - Tr + Det > 0 \\ (iii) & 1 + Tr + Det > 0 \end{array} \tag{22}$$

The first condition (i) gives:

$$1 - Det > 0 \Leftrightarrow 1 + k \cdot \frac{(1 - \mu)^2}{2(1 + c)} q_1^* > 0 \tag{23}$$

which is always satisfied.

The second condition (ii) gives:

$$1 - Tr + Det > 0 \Leftrightarrow k \cdot \left[4(1 + c)^2 - (1 - \mu)^2\right] \cdot q_1^* > 0 \tag{24}$$

and it's always satisfied because $4\,(1 + c)^2 - (\mu - 1)^2 > 0$, for $c > 0$ and $\mu_i \in [0, 1]$.

Finally, the condition (iii) becomes as:

$$1 + \mathrm{Tr} + \mathrm{Det} > 0 \Leftrightarrow 2 - k\frac{\alpha \cdot \left[4(1+c)^2 + (1-\mu)^2\right]}{2(1+c) \cdot [2(1+c) + (1-\mu)]} > 0 \quad (25)$$

and it gives the stability condition of Nash Equilibrium E_*.

Proposition *The Nash equilibrium of the discrete dynamical system* Eq. (11) *is locally asymptotically stable if:*

$$2 - k\frac{\alpha \cdot \left[4(1+c)^2 + (1-\mu)^2\right]}{2(1+c) \cdot [2(1+c) + (1-\mu)]} > 0$$

3 Numerical Simulations

3.1 Stability Space

At first the stability space (Fig. 1) is made including the main two parameters that the dynamical analysis focuses on, the parameters k (speed of adjustment) and μ (relative profit parameter). This two-dimensional space is obtained by the stability condition that is described above in Proposition, setting specific values for the other parameters $\alpha = 5$ and $c = 0.20$.

3.2 Focusing on the Parameter k (Speed of Adjustment)

In this section various numerical results focusing on the parameter k, including bifurcation diagrams, strange attractors, Lyapunov numbers and sensitive dependence on initial conditions (Kulenovic, M. and Merino, O.) are presented. Focusing on the parameter k the stability condition becomes as:

$$0 < k < \frac{4(1+c) \cdot [2(1+c) + (1-\mu)]}{\alpha \cdot \left[4(1+c)^2 + (1-\mu)^2\right]} \quad (26)$$

Choosing the specific values of the parameters: $\alpha = 5$, $c = 0.20$ and $\mu = 0.50$ the coordinates of Equilibrium position can be calculated as: $q^*_1 = q^*_2 \simeq 1.72$ and the stability space for the parameter is described as:

$$0 < k < 0.48$$

It is verified by the bifurcation diagrams of the parameter k against the variables q^*_1 (left) and q^*_2 (right) that are shown in Figs. 2 and 3. These two figures show

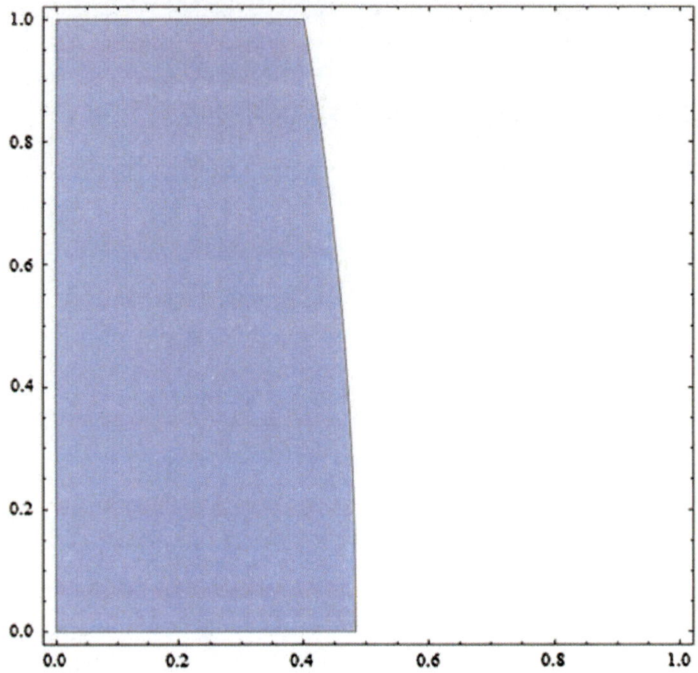

Fig. 1 Region of stability between k (horizontal axis) and μ (vertical axis) for $\alpha = 5$ and $c = 0.20$

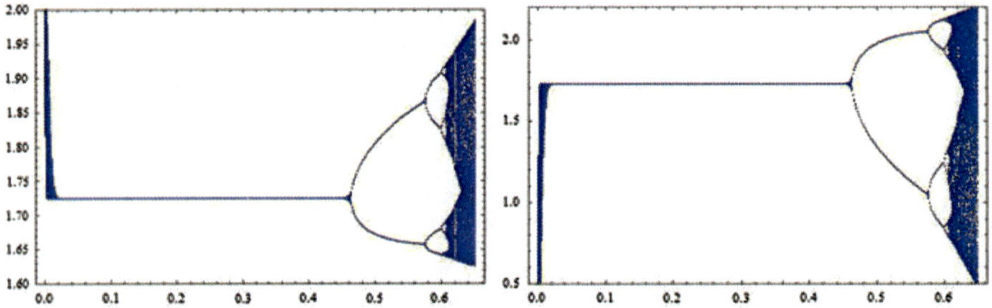

Fig. 2 Bifurcation diagrams with respect to the parameter k against the variables q_1 (left) and q_2 (right) with 400 iterations of the map Eq. (11) for $\alpha = 5$, $c = 0.20$ and $\mu = 0.50$

that the equilibrium undergoes a flip bifurcation at $k = 0.48$. Then a further increase in speed of adjustment implies that a stable two-period cycle emerges for $0.48 < k < 0.58$. As long as the parameter k reduces a four-period cycle, cycles of highly periodicity and a cascade of flip bifurcations that ultimately lead to unpredictable (chaotic) motions are observed when k is larger than 0.62.

This unpredictable (chaotic) behavior of the system Eq. (11) is visualized in Fig. 4 (left) with the strange attractor for $k = 0.65$. This is the graph of the orbit of (0.1,0.1)

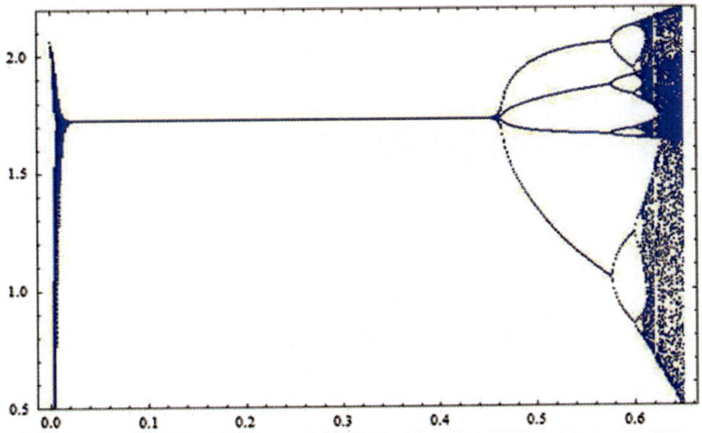

Fig. 3 Two bifurcation diagrams of Fig. 2 are plotted in one

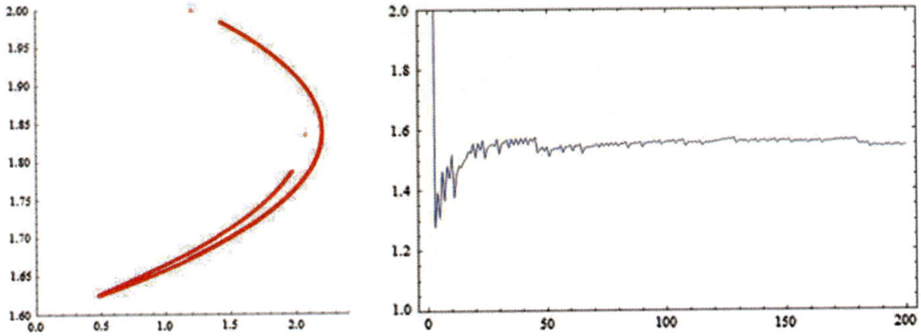

Fig. 4 Phase portrait (strange attractor) (left) and Lyapunov numbers (right) of the orbit of (0.1,0.1) with 8000 iterations of the map Eq. (11) for $\alpha = 5$, $c = 0.20$, $\mu = 0.50$ and $k = 0.65$

with 8000 iterations of the map Eq. (11) for $\alpha = 5$, $c = 0.20$, $\mu = 0.50$ and $k = 0.65$. Also, we use the useful tool of Lyapunov numbers [Fig. 4 (right)] (i.e. the natural logarithm of Lyapunov exponents) as a function of the parameter of interest. Figure 4 (right) shows the Lyapunov numbers of the same orbit. It is known that if the Lyapunov number is greater than 1, one has evidence for chaos.

Another characteristic of deterministic chaos is the sensitivity dependence on initial conditions. In order to show the sensitivity dependence on initial conditions of the system Eq. (11), we have computed two orbits with initial points (0.1,0.1) and (0.101,0.1) respectively. Figure 5 shows the sensitivity dependence on initial conditions for q_1—coordinate of the two orbits, for the system Eq. (11), plotted against the time with the parameter values $\alpha = 5$, $c = 0.20$, $\mu = 0.50$ and $k = 0.65$. At the beginning the time series are indistinguishable; but after a number of iterations, the difference between them builds up rapidly. From these numerical

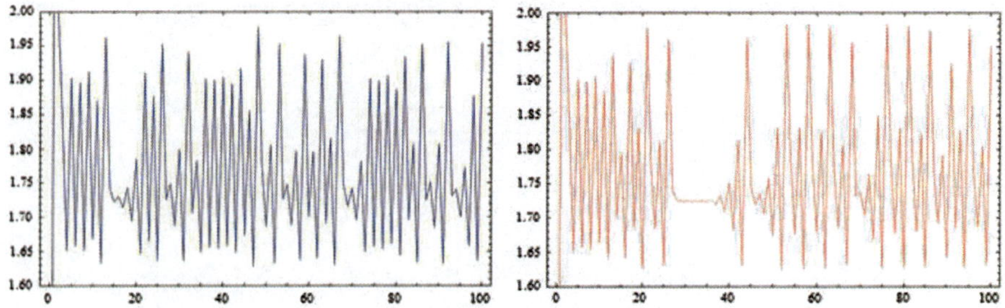

Fig. 5 Sensitive dependence on initial conditions for q_1-coordinate plotted against the time: the orbit of (0.1,0.1) (left) and the orbit of (0.101,0.1) (right) of the system Eq. (11) for $\alpha = 5$, $c = 0.20$, $\mu = 0.50$ and $k = 0.65$

results when all parameters are fixed and only k is varied the structure of the game becomes complicated through period doubling bifurcations, more complex bounded attractors are created which are aperiodic cycles of higher order or chaotic attractors.

3.3 *Focusing on the Parameter μ (Relative Profit Maximization)*

As it seems the Nash Equilibrium Eq. (14) is independent of the parameter k (the speed of adjustment) but it depends on the values of the other parameters. As a result when the values of the parameters α, c and k remain constant and only the parameter μ varies, this makes the Nash Equilibrium not to be constant, but it changes for each different value of the parameter μ. Focusing on the parameter μ the stability condition becomes as:

$$k\alpha\mu^2 + [4(1 + c) - 2k\alpha]\mu + k\alpha - 4(1 + c) + 4(1 + c)^2(k\alpha - 2) > 0 \quad (27)$$

with

$$\Delta_1 = 16(1 + c)^2 \cdot \left(1 + 2k\alpha - k^2\alpha^2\right) \quad (28)$$

The inequality (27) can be satisfied if:

$$\Delta_1 > 0 \Rightarrow -k^2\alpha^2 + 2k\alpha + 1 > 0 \quad (29)$$

and

$$\mu \in (\mu_1, \mu_2), \text{ where } \mu_{1,2} = \frac{-2(1 + c) \pm 2(1 + c)\sqrt{1 + 2k\alpha - k^2\alpha^2}}{k\alpha} + 1 \quad (30)$$

The Eq. (29) becomes true if:

$$\left.\begin{array}{l} k \in (-\infty, k_1) \cup (k_2 + \infty) \\ k > 0 \\ k_{1,2} = \dfrac{1 \pm \sqrt{2}}{\alpha} \end{array}\right\} \Rightarrow 0 < k < \frac{1+\sqrt{2}}{\alpha}$$

Finally, the stability condition Eq. (27) is satisfied if:

$$\left.\begin{array}{l} \mu \in (\mu_1, \mu_2) \\ \mu > 0 \\ 0 < k < \dfrac{1+\sqrt{2}}{\alpha} \end{array}\right\} \Rightarrow \left.\begin{array}{l} \mu \in (0, \mu_2) \\ 0 < k < \dfrac{1+\sqrt{2}}{\alpha} \end{array}\right\} \Rightarrow \left.\begin{array}{l} 0 < \mu < \dfrac{2(1+c)\sqrt{1+2k\alpha - k^2\alpha^2} - 1}{k\alpha} + 1 \\ 0 < k < \dfrac{1+\sqrt{2}}{\alpha} \end{array}\right\}$$

Setting the specific values of the parameters: $\alpha = 5$, $c = 0.20$ and $k = 0.475$ it gives that the stability space focusing on the parameter μ becomes as:

$$0 < \mu < 0.32$$

Using the stability space (Fig. 1) when $\alpha = 5$, $c = 0.20$ and $k = 0.475$, it can be verified that there is a stable equilibrium for $\mu \in (0, 0.32)$ and it is also verified by the bifurcation diagrams of μ against q_1 (left) and q_2 (right) (Fig. 6). Also, the chaotic behavior for the system Eq. (11) appears only for values of the parameter μ (relative profit parameter) larger than 1 so if $k = 0.475$ in this market the parameter μ cannot make the system unpredictable. Finally, the stability space between the main parameters k and μ (Fig. 1) gives the useful result that for small values of the parameter k, the Nash Equilibrium remain stable for every value of the parameter μ and in this area the parameter μ cannot destabilize the economy (Fig. 7).

Now, if the other parameter take the values: $\alpha = 5$, $c = 0.20$ and $k = 0.60$, it seems that first a fixed period 2 trajectory is created and then it enters a period doubling and as the parameter μ takes higher values the system enters a chaotic orbit and becomes unpredictable. (Figs. 8 and 9). The larger the values of the parameter μ more Strange

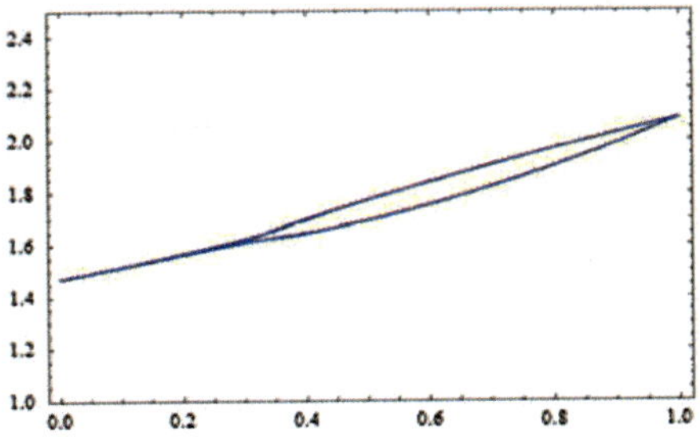

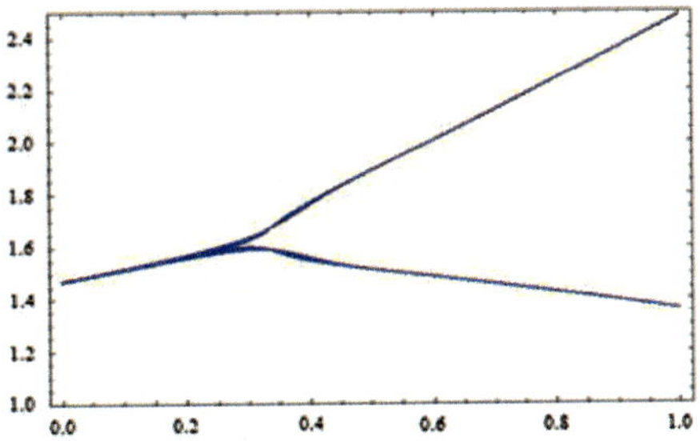

Fig. 6 Bifurcation diagrams with respect to the parameter μ against the variables q_1 (left) and q_2 (right) with 400 iterations of the map Eq. (11) for $\alpha = 5$, $c = 0.20$ and $k = 0.475$

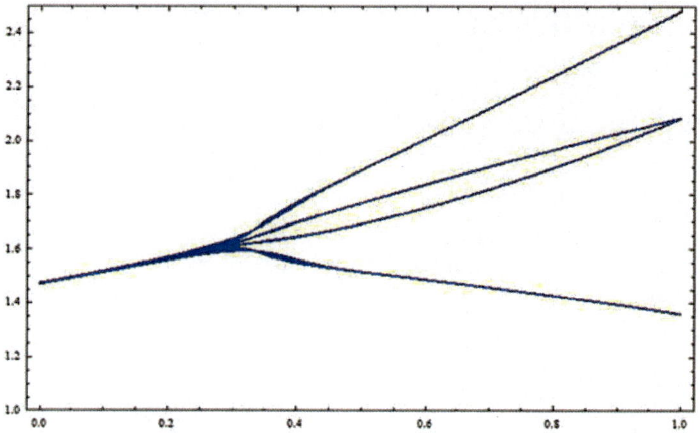

Fig. 7 Two bifurcation diagrams of Fig. 9 are plotted in one

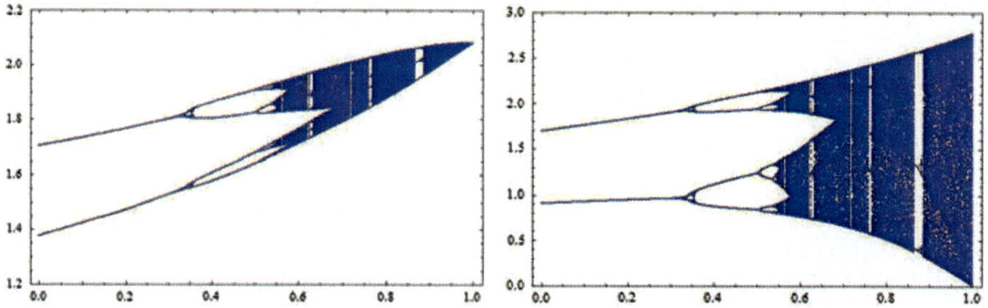

Fig. 8 Bifurcation diagrams with respect to the parameter μ against the variables q_1 (left) and q_2 (right) with 400 iterations of the map Eq. (11) for $\alpha = 5$, $c = 0.20$ and $k = 0.60$

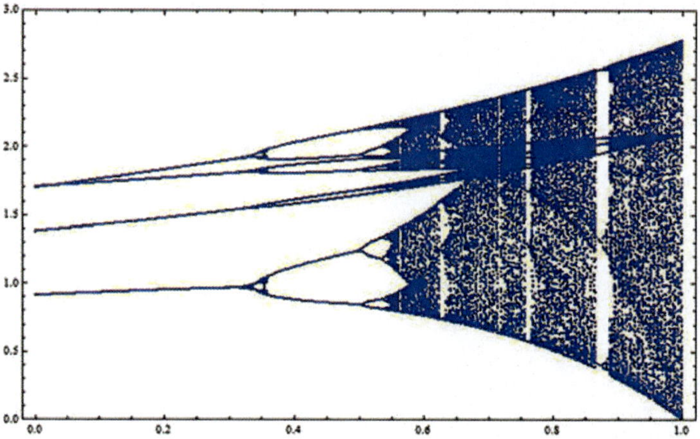

Fig. 9 Two bifurcation diagrams of Fig. 11 are plotted in one

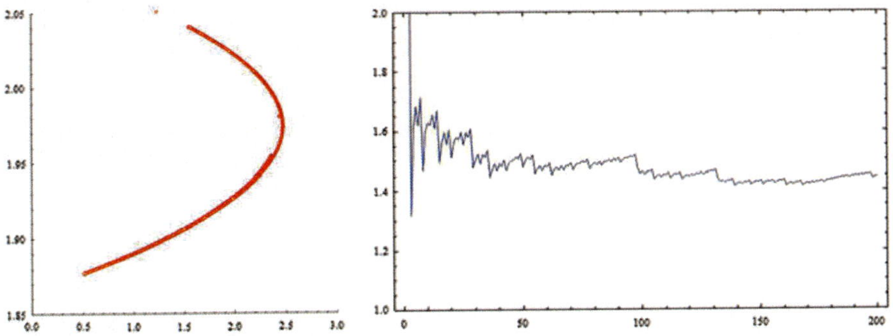

Fig. 10 Phase portrait (strange attractor) (left) and Lyapunov numbers (right) of the orbit of (0.1,0.1) with 8000 iterations of the map Eq. (11) for $\alpha = 5$, $c = 0.20$, $k = 0.60$ and $\mu = 0.80$

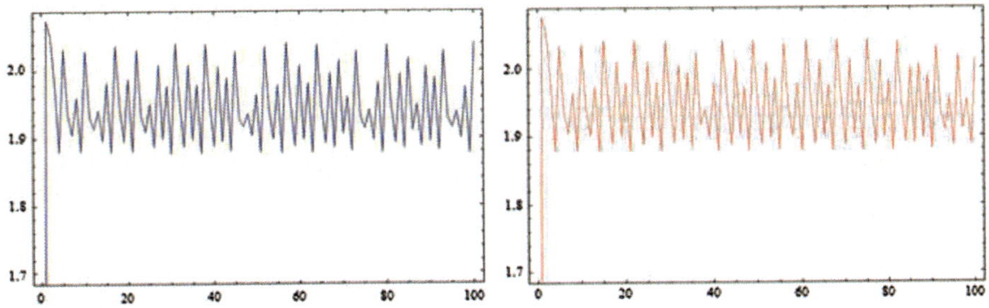

Fig. 11 Sensitive dependence on initial conditions for q_1-coordinate plotted against the time: the orbit of (0.1,0.1) (left) and the orbit of (0.101,0.1) (right) of the system Eq. (11) for $\alpha = 5$, $c = 0.20$, $k = 0.60$ and $\mu = 0.80$

attractors and Lyapunov numbers larger than 1 (Fig. 10) are appeared for the same values of the paramaters α, c and k. Also, the system becomes sensitive on initial condition (Fig. 11) for these large values of the parameter μ (outside the stability space of μ).

4 Conclusions

In this paper the dynamics of a differentiated Cournot duopoly with heterogeneous expectations, linear demand and quadratic cost functions are analyzed. By assuming that at each time period each firm maximizes its expected relative profit function U under bounded rationality expectation, a discrete dynamic system was obtained. Existence and stability of equilibrium of this system are studied. It is numerically shown that the model gives chaotic and unpredictable trajectories. The main result is that higher values of the speed of adjustment and relative profit parameter may destabilize the Cournot–Nash equilibrium. Finally, it is proved that for lower values of the speed of adjustment the equilibrium is stable for every value of the relative profit parameter.

References

1. H.N. Agiza, On the stability, bifurcations, chaos and chaos control of Kopel map. Chaos Solitons Fractals **11**, 1909–1916 (2004)
2. H.N. Agiza, A.A. Elsadany, Chaotic dynamics in nonlinear duopoly game with heterogeneous players. Appl. Math. Comput. **149**, 843–860 (2004)
3. H.N. Agiza, A.A. Elsadany, Nonlinear dynamics in the Cournot duopoly game with heterogeneous players. Physica A **320**, 512–524 (2003)
4. H.N. Agiza, A.S. Hegazi, A.A. Elsadany, Complex dynamics and synchronization of duopoly game with bounded rationality. Math. Comput. Simulat. **58**, 133–46 (2002)
5. A. Agliari, L. Gardini, T. Puu, Some global bifurcations related to the appearance of closed invariant curves. Math. Comput. Simul. **68**, 201–219 (2005)
6. A. Agliari, L. Gardini, T. Puu, Global bifurcations in duopoly when the Cournot point is destabilized via a subcritical Neimark bifurcation. Int. Game Theor. Rev. **8**, 1–20 (2006)
7. S.S. Askar, On complex dynamics of monopoly market. Econ. Model. **31**, 586–589 (2013)
8. S.S. Askar, Complex dynamic properties of Cournot duopoly games with convex and log-concave demand function. Oper. Res. Lett. **42**, 85–90 (2014)
9. W.J. Baumol, R.E. Quandt, Rules of thumb and optimally imperfect decisions. Am. Econ. Rev. **54**(2), 23–46 (1964)
10. G.I. Bischi, M. Kopel, Equilibrium selection in a nonlinear duopoly game with adaptive expectations. J. Econ. Behav. Org. **46**, 73–100 (2001)
11. W.J. Den Haan, The importance of the number of different agents in a heterogeneous asset-pricing model. J. Econom. Dynam. Control **25**, 721–746 (2001)
12. A.A. Elsadany, Dynamics of a Cournot duopoly game with bounded rationality based on relative profit maximization. Appl. Math. Comput. **294**, 253–263 (2017)
13. S. Elaydi, *An Introduction to Difference Equations*, 3rd edn. (Springer-Verlag, New York, 2005)
14. L. Fanti, L. Gori, The dynamics of a differentiated duopoly with quantity competition. Econ. Model. **29**, 421–427 (2012)
15. C.H. Hommes, Heterogeneous agent models in economics and finance, in *Handbook of Computational Economics, Agent-Based Computational Economics*, ed. by L. Tesfatsion, K.L. Judd, vol. 2. Elsevier Science B.V (2006), pp. 1109–1186
16. G. Gandolfo, *Economic Dynamics* (Springer, Berlin, 1997)
17. Y. Gao, Complex dynamics in a two dimensional noninvertible map. Chaos Solitons Fractals **39**, 1798–1810 (2009)
18. M. Kopel, Simple and complex adjustment dynamics in Cournot duopoly models. Chaos Solitons Fractals **12**, 2031–2048 (1996)
19. M. Kulenonic, O. Merino, *Discrete Dynamical Systems and Difference Equations with Mathematica.* Chapman & Hall/Crc. (2002)
20. A.K. Naimzada, G. Ricchiuti, Complex dynamics in a monopoly with a rule of thumb. Appl. Math. Comput. **203**, 921–925 (2008)
21. A. Naimzada, L. Sbragia, Oligopoly games with nonlinear demand and cost functions: two boundedly rational adjustment processes. Chaos Solitons Fractals **29**(3), 707–722 (2006)
22. T. Puu, The chaotic monopolist. Chaos, Solitons Fractals **5**(1), 35–44 (1995)
23. T. Puu, The chaotic duopolists revisited. J. Econom. Behav. Org. **37**, 385–394 (1998)
24. T. Puu, Chaos in duopoly pricing. Chaos Solitons Fractals **1**, 573–581 (1991)
25. T. Puu, Complex oligopoly dynamics, in *Nonlinear Dynamical Systems in Economics* ed. by M. Lines (Springer Wien NewYork, CISM) (2005), pp. 165–186

26. G. Sarafopoulos, On the dynamics of a duopoly game with differentiated goods. Procedia Econ. Finan. **19**, 146–153 (2015)
27. G. Sarafopoulos, Complexity in a duopoly game with homogeneous players, convex, log linear demand and quadratic cost functions. Procedia Econ. Finan. **33**, 358–366 (2015)
28. G. Sarafopoulos, K. Papadopoulos, On a Cournot duopoly game with differentiated goods, heterogeneous expectations and a cost function including emission costs. Sci. Bull. Econ. Sci. **16**(1), 11–22 (2017)
29. G. Sarafopoulos, K. Papadopoulos, Complexity in a Bertrand duopoly game with heterogeneous players and differentiated goods, in *Springer Proceedings in Business and Economic*, article 2 (Springer Nature Switzerland AG, ISBN: 978-3-030-12169-3, 2019), pp. 15–26
30. G. Sarafopoulos, K. Papadopoulos, On a Bertrand dynamic game with differentiated goods, heterogeneous expectations and asymmetric cost functions, in *Springer Proceedings in Business and Economic*, article 14 (Springer Nature Switzerland AG, (ISBN: 978-3-030-39927-6, 2020), pp. 223–241
31. N. Singh, X. Vives, Price and quantity competition in a differentiated duopoly. RAND J. Econ. **15**, 546–554 (1984)
32. F. Tramontana, Heterogeneous duopoly with isoelastic demand function. Econ. Model. **27**, 350–357 (2010)
33. F. Westerhoff, Nonlinear expectation formation, endogenous business cycles and stylized facts. Stud. Nonlinear Dyn. Econ. **10** (4) (Article 4) (2006)
34. W. Wu, Z. Chen, W.H. Ip, Complex nonlinear dynamics and controlling chaos in a Cournot duopoly economic model. Nonlinear Anal. Real World Appl. **11**, 4363–4377 (2010)
35. J. Zhang, Q. Da, Y. Wang, Analysis of nonlinear duopoly game with heterogeneous players. Econ. Model. **24**, 138–148 (2007)
36. J. Zhang, Q. Da, Y. Wang, The dynamics of Bertrand model with bounded rationality. Chaos, Solitons Fractals **39**, 2048–2055 (2009)

Multifractal Analysis of Bioenergy Transport in a Protein Nanomotor

Narmin Sefidkar, Samira Fathizadeh, and Fatemeh Nemati

Abstract The potential application of biological molecules as functional devices has been heralded as the dawn of a new field in biotechnology and medicine. In this regard, molecular motors have attracted the most attention for decades. In the current study, we have studied the bioenergy transfer in a protein chain as a self-powered nanomotor. The effect of different factors on the energy transfer in protein is studied to obtain the best functional condition for the protein machine. The temperature plays the critical role in the control of energy transfer in the system. On the other hand, the external mechanical tension as a vibrator can increase the energy flowing in our system. The chaos theory tools can verify and estimate the results. Generally, one can engineer a self-powered nanomotor based on protein chains and control bioenergy transfer.

Keywords Nanomotor · Protein · Biological energy · Energy transfer · Multifractal analysis

1 Introduction

Biological molecules, an exciting field with continuous and robust growth for about half a century of existence, encompass the medical, biological, chemical, and material sciences [1]. In the current study, we have studied a protein chain as a biological molecule and its energy transfer. Protein, as the largest macromolecule in the body, is one of the most important components of the life systems that plays vital role in the body. Proteins are composed of amino acid chain sequences that the number of the amino acids can reach several thousand depending on the type of protein [2].

In the current study, we investigate the bioenergy transfer in protein molecules as an automated nanomotor. Automated nanomotor uses the bioenergy in the protein as self-burning energy and converts it into autonomous motion [3]. These nanomotors, which derive their kinetic energy from the biological energy contained in proteins,

N. Sefidkar · S. Fathizadeh (✉) · F. Nemati
Department of Physics, Urmia University of Technology, Urmia, Iran
e-mail: s.fathizadeh@sci.uut.ac.ir

C. H. Skiadas and Y. Dimotikalis (eds.), *14th Chaotic Modeling and Simulation International Conference*, Springer Proceedings in Complexity,
https://doi.org/10.1007/978-3-030-96964-6_28

are called protein nanomotors. Protein motors have potential medical applications and can collect, transport, and release drug carriers of various sizes. Internal order in eukaryotic cells is created by protein motors that transport organs and molecules along the cytoskeletal pathways of self-assembled proteins such as tubulin and shuttle actin. Three known families of cytoskeletal protein motors include kinesin, dynein, and myosin [4]. Kinesins and dyneins, which are the first and second types of protein motors, respectively, move in tubules or microtubules [5]. But Myosins, the third family of protein motors, move on actin filaments and are responsible for muscle contraction in the heart and skeletal muscles [6].

2 Model and Methods

In the current work, the Hamiltonian model used to study the energy transfer in protein molecules is based on the Peng model as follows:

$$H = H_1 + H_2 + H_3 \tag{1}$$

where H_1 indicates that a boson-type Frankit exciton is excited in protein molecules using the energy released in ATP hydrolysis written as follows:

$$H_1 = \sum_n \left[\varepsilon_0 a_n^\dagger a_n - J\left(a_n^\dagger a_{n+1} + a_n a_{n+1}^\dagger\right)\right] \tag{2}$$

H_2 defines the harmonic properties of the remaining amino acids:

$$H_2 = \sum_n \left[\frac{P_n^2}{2m} + \frac{1}{2}W(u_n - u_{n-1})^2\right] \tag{3}$$

H_3 introduces the interaction between the two modes of motion [7]:

$$H_3 = \sum_n \left[\chi_1(u_{n+1} - u_{n-1})a_n^\dagger a_n + \chi_2(u_{n+1} - u_n)\left(a_{n+1}^\dagger a_n + a_n^\dagger a_{n+1}\right)\right] \tag{4}$$

The parameters used in the calculations are shown in Table 1 [8]:

Here, $a_n and a_n^\dagger$ are the creation and annihilation operators for exciton. $u_n and p_n$ are the displacement and momentum operators for the amino acid residue at site n, respectively. ε_0 is the energy of the exciton. χ_1 and χ_2 are nonlinear coupling constants. m is the mass of amino acid residue. W is the elasticity constant of the proteins, and J is the dipole–dipole interaction energy between neighboring amino acids [9].

In this work, we have used the classical chaos theory. In this case, the evolution equations for the classical part are derived using the Hamiltonian equation

Table 1 Parameters used

Parameter	Unit	Value
ε_0	eV	0.2035
χ_1	N	6.2×10^{-11}
χ_2	N	$(10–18) \times 10^{-12}$
m	Amu	115
W	$\frac{N}{m}$	13
J	eV	9.68×10^{-4}

$\left(\dot{p}_n = -\frac{\partial H}{\partial q_n}\right)$. Also, the evolution equations for the quantum part are analyzed using the Heisenberg equation $\left(\dot{a}_n = -\frac{i}{\hbar}[a_n, H]\right)$ as follows:

$$\ddot{u}_n = \frac{1}{m}\sum_n\Big[W(u_{n+1} - 2u_n + u_{n-1}) + \chi_1\left(a^\dagger_{n+1}a_{n+1} - a^\dagger_{n-1}a_{n-1}\right) + \chi_2\left(a^\dagger_{n+1}a_n + a^\dagger_n a_{n+1} - a^\dagger_n a_{n-1} - a^\dagger_{n-1}a_n\right)\Big] \quad (5)$$

$$\dot{a}_n = -\frac{i}{\hbar}\sum_n\{-J(a_{n-1} + a_{n+1}) + \varepsilon_0 a_n + \chi_1(u_{n+1} - u_{n-1})a_n + \chi_2[(u_n - u_{n-1})a_{n-1} + (u_{n+1} - u_n)a_{n+1}]\} \quad (6)$$

To study the energy transfer in protein molecules, we have extracted the energy flux. The energy flux obtains using the continuity equation $\left(\mathrm{I} = -\frac{i}{\hbar}\left[a^\dagger_n a_n, H\right]\right)$ as follows:

$$I = -\frac{i}{\hbar}\sum_n\Big[-J\left(a^\dagger_n - a_{n-1} - a^\dagger_{n-1}a_n + a^\dagger_n a_{n+1} - a^\dagger_{n+1}a_n\right) - \chi_2(u_n - u_{n-1})a^\dagger_{n-1}a_n + \chi_2(u_{n+1} - u_n)\left(a^\dagger_n a_{n+1} - a^\dagger_{n+1}a_n\right)\Big] \quad (7)$$

3 Results and Discussion

Different factors effect on energy transfer in biological systems. One of the vital parameters on energy transfer in the protein system is the effect of temperature. We have used the Nosé Hoover thermostat to apply the temperature to the system. The evolution equation of thermostat is written as follows [10]:

$$\dot{\xi} = \frac{1}{M}\left(\sum_n m\dot{u}_n^2 - NK_B T_0\right) \quad (8)$$

where ξ describes a thermodynamic coefficient of friction. T_0 is the temperature of the system. K_B is the constant of Boltzmann, and M is the thermostat constant.

A biological system can be affected by mechanical shocks. Therefore, we have investigated the effect of mechanical stress on energy transfer in protein molecules. The Hamiltonian of mechanical stress is written as follows [11]:

$$H_{str} = -\delta_{1,n}(u_{n+1} - u_n)F_0 \sin \omega t \tag{9}$$

where F_0 and ω are the amplitude and frequency of the external mechanical stress.

We can also examine the best range of parameters affecting energy transfer in protein molecules using multifractal system analysis. Multifractal analysis of the system can confirm the obtained results. In addition, it can classify system parameters to determine the desired results.

In this regard, we use the Renyi dimension spectrum, analogous of the free energy and analogous of the thermodynamic specific heat.

3.1 Simultaneous Effect of Ambient Temperature and the Range of Force Exerting Mechanical Stress

Temperature and external stress as the very influential factors in the performance of biological systems can be considered as the control parameters in the energy transfer in the protein system. We have considered the simultaneous effect of ambient temperature and the range of force exerting mechanical stress on the energy flux in the protein system (Fig. 1). The variation of the ambient temperature from 300 to

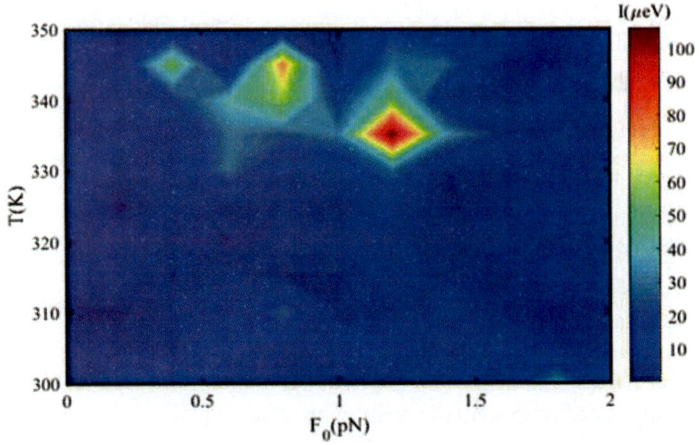

Fig. 1 The energy flux through the system concerning the simultaneous variation of the amplitude of applied stress and the ambient temperature

330 K and at the same time changing the amplitude of the force from 0 to 2 pN has negligible effect on the amount of energy flux transmitted in the system. In this region, the energy flux fluctuates at a minimum of about 10 to 30 μeV. But, when the temperature rises above 330 K, in the force range of about 1.2 pN, we can see a considerable peak in the energy flux, so that the energy flux reaches about 100 μeV. Similar to this phenomenon is also observed at a temperature of about 345 K and a force range of about 0.8 pN. Therefore, it can be said that temperatures above 330 and the applied stress are two critical parameters in the transfer of bioenergy in our protein system.

3.2 Simultaneous Effect of Temperature and Applied Stress Frequency

The frequency of applied stress simultaneously with the ambient temperature can be another compelling factor in regulating the energy flux through the system. As shown in Fig. 2, at 300–350 K, the energy flux is more than 50 μV. But there are characteristic frequencies that can significantly increase the energy flux through the system. A frequency of about 0.09 THz at 310 K causes the energy flux of the system to reach about over 200 μeV, and this shows that energy transport in the protein system is highly sensitive to several frequencies that can be used to optimize the performance of the biological system.

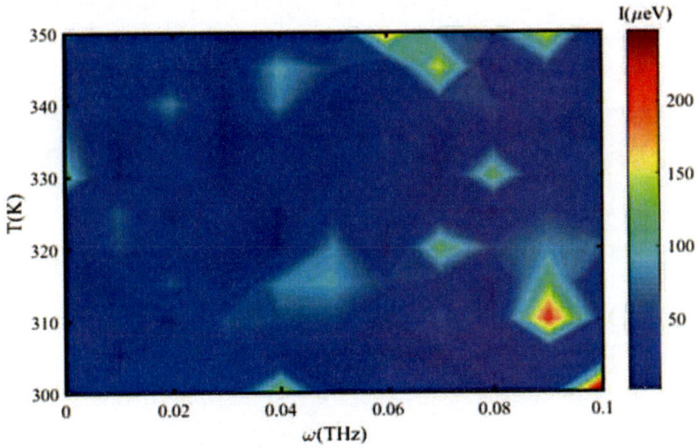

Fig. 2 Energy flux relative to a simultaneous variation of applied stress frequency and temperature (N = 150, F_0 = 3 pN)

3.3 Multifractal Analysis

We have studied the multifractal spectrum of the energy flux through the protein system by using Renyi dimension. In this regard, we consider the d dimensional phase space of system and divide it into cubes of size land calculate the Renyi dimension (Dq) as follows [12]:

$$D_q = \frac{1}{q-1} \lim_{l \to 0} \frac{\sum_i^M P_i^q}{\ln l} \tag{10}$$

Figure 3 shows the Renyi dimension spectrum for four different frequencies of applied stress. D_q shows the descending behavior by increasing q which indicates that the system is multifractal. If we go back to the previous results, we observe that the temperature of 300 K for a force amplitude of 3 pN, the frequencies ω = 0.01 THz, ω = 0.03 THz, ω = 0.05 THz are in the blue region where the energy flux crossing the system is minimal (Fig. 2). Also, the frequency ω = 0.09 THz is in the red region where the energy flux passing through the system is maximum. As shown in Fig. 3, for negative q, the Renyi dimensions at 300 K and the force amplitude of 3 pN corresponds to the frequencies ω = 0.01THz, ω = 0.03THz, ω = 0.05THz are coincident with almost the same amount. But the Renyi dimension of the frequency ω = 0.09THz shows a greater value for negative values of q. The Renyi dimension distinguishes the regions with the highest energy flux from the regions with the lowest energy flux.

To have a similarity between a multifractal system and a thermodynamic state, relations are usually equated with their thermodynamic equivalents. We consider τ(q) as analogous the thermodynamic free energy of the system, and check the

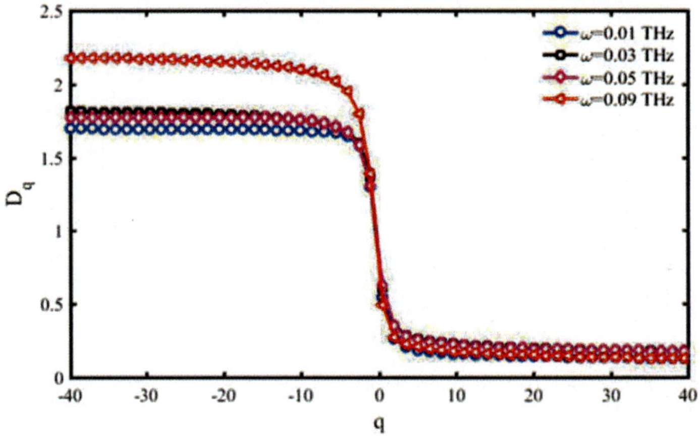

Fig. 3 Renyi dimension spectrum in different values of the stress frequency (T = 310 K, F_0 = 3 pN, N = 150)

multifractality of the system. The analogous of the free energy is written as follows [13]:

$$\tau(q) \equiv (q-1)D_q \tag{11}$$

The system is a homogenous fractal when $\tau(q)$ have a linear dependence to q. On the other hand, $\tau(q)$ shows a deviation from the linear state when the system is multifractal and the higher deviation from the linear state shows the more multifractality of the system. According the Fig. 4, $\tau(q)$ is deviated from the linear state in terms of q for all frequencies at the point $q = 0$, and this confirms the multifractality of the system. On the other hand, for negative q, $\tau(q)$ is the same for the frequencies in the blue region. But the frequency ω = 0.09 THz, which is in the red area, has the highest deviation from linear behavior. Thus, the analogous of the thermodynamic free energy shows the distinct regions for negative q at 300 K and for the force range of 3 pN.

On the other hand, the analogous of the specific heat *C(q)* which is obtained from a second-order derivative of the analogous of free energy concerning q can be analyze the multifractal behavior of system through the following equation:

$$\frac{\partial^2\tau}{\partial q^2} \approx \tau(q+1) - 2\tau(q) + \tau(q-1) \tag{12}$$

Analogous of specific heat is one of the measures to check the multifractality of a system. As shown in Fig. 5, the frequencies of the blue region overlap in the peak area. The frequency ω = 0.09 THz, which is in the red region, shows a higher peak. Therefore, a analogous of thermodynamic specific heat also shows the distinct regions and confirms the previous results.

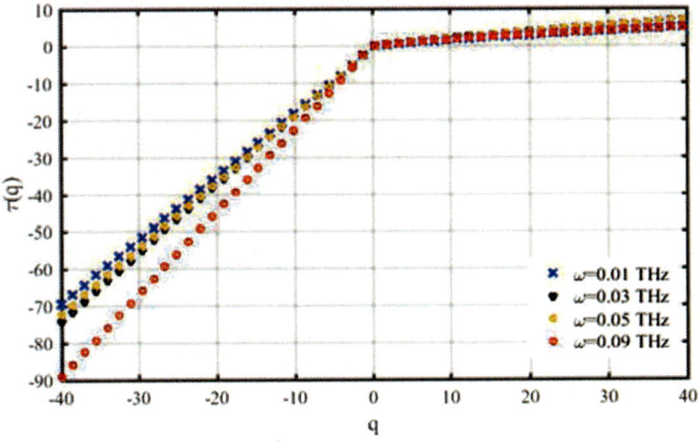

Fig. 4 Analogous of the thermodynamic free energy in multifractal systems for different frequencies of the external stress (N = 150, F_0 = 3 pN, T = 310 K)

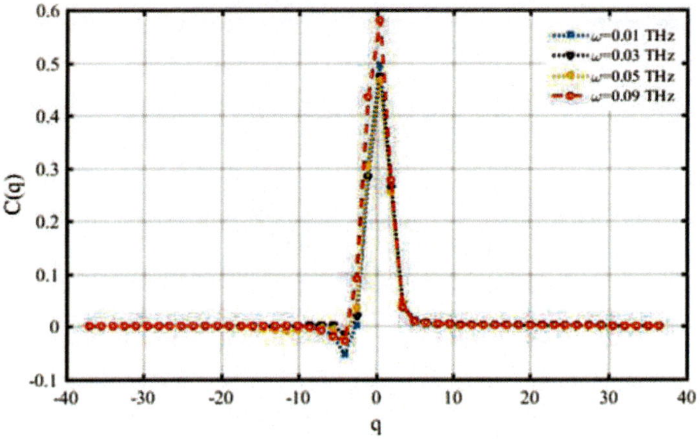

Fig. 5 Analogous of thermodynamic specific heat in multifractal systems for different frequencies (N = 150, F_0 = 3 pN, T = 310 K)

4 Conclusions

We have studied the bioenergy transfer in a protein system to design a nanomotor. In this regard, we have investigated the effect of various parameters such as temperature and mechanical stress on the energy transfer in protein nanomotors. We have analyzed the multifractal nature of the system using the multifractal analysis methods. Using the multifractal analysis, we are able to confirm the previous results and show the distinct areas.

References

1. I. Kulinets, Biomaterials and their applications in medicine, in *Regulatory Affairs for Biomaterials and Medical Devices (2015)* (Woodhead Publishing), pp. 1–10
2. D.S. Miller, P.R. Payne, A theory of protein metabolism. J. Theor. Biol. **5**(3), 398–411 (1963)
3. F. Zha, T. Wang, M. Luo, J. Guan, Tubular micro/nanomotors: Propulsion mechanisms, fabrication techniques and applications. Micromachines **9**(2), 78 (2018)
4. P.D. Vogel, Nature's design of nanomotors. Eur. J. Pharm. Biopharm. **60**(2), 267–277 (2005)
5. R.D. Vale, T.S. Reese, M.P. Sheetz, Identification of a novel force-generating protein, kinesin, involved in microtubule-based motility. Cell **42**(1), 39–50 (1985)
6. M.E. Brown, P.C. Bridgman, Myosin function in nervous and sensory systems. J. Neurobiol. **58**(1), 118–130
7. X.F. Pang, M.J. Liu, The influences of temperature and chain-chain interaction on features of solitons excited in A-helix protein molecules with three channels. Int. J. Mod. Phys. B **23**(10), 2303–2322 (2009)
8. X.F. Pang, M.J. Liu, The effects of damping and temperature of medium on the soliton excited in α-helix protein molecules with three channels. Mod. Phys. Lett. B **23**(01), 71–88 (2009)

9. X.F. Pang, H.W. Zhang, Y.H. Luo, The influence of the heat bath and structural disorder in protein molecules on soliton transported bio-energy in an improved model. J. Phys. Condens. Matter **18**(2), 613 (2005)
10. M. Peyrard, A.R. Bishop, Statistical mechanics of a nonlinear model for DNA denaturation. Phys. Rev. Lett. **62**(23), 2755 (1989)
11. S. Fathizadeh, S. Behnia, Control of a DNA based piezoelectric biosensor. J. Phys. Soc. Jpn. **89**(2), 024004 (2020)
12. S. Behnia, S. Fathizadeh, E. Javanshour, F. Nemati, Light-driven modulation of electrical current through DNA sequences: Engineering of a molecular optical switch. J. Phys. Chem. B **124**(16), 3261–3270 (2020)
13. S. Fathizadeh, Behnia, Control of a DNA based piezoelectric biosensor. J. Phys. Soc. Jpn. **89**(2), 024004 (2020)

Forced van der Pol Oscillator—Synchronization from the Bifurcation Theory Point of View

Jan Ševčík and Lenka Přibylová

Abstract The contribution presents a bifurcation theory point of view to synchronization of a forced van der Pol oscillator, which is coupled to a master oscillator as a system with a stable limit cycle corresponding to harmonic oscillation. We present bifurcation manifolds, 3D sections of the phase space and its Poincaré sections for parameters close to these manifolds providing a clear visualization of the dynamics of the 4D system. Among other things, we present the coexistence of a stable torus and a stable cycle arising from q-fold bifurcation on an Arnold tongue.

Keywords Synchronization · Van der Pol oscillator · Bifurcations of limit cycles · Neimark–Sacker bifurcation · q-fold bifurcation · Arnold tongues

1 Introduction

Synchronization of coupled systems of oscillators is an important phenomenon that touches a large class of nonlinear dynamical systems. Synchronization is ubiquitous and methods of applied nonlinear dynamics can thus help to solve problems and create new technologies in neuroscience [1, 5, 14], chemistry [6], biology [12], superconducting electronics [3, 10], spintronics [9], computing [8], or even particle physics [2]. Since these nonlinear systems exhibit complex and sometimes even counterintuitive dynamics, the most commonly used methods to study synchronizations are simulations.

Although the theory of bifurcations offers a suitable apparatus for the analysis of the systems mentioned above, it is usually not used. The highly abstract thinking and mathematically generalized view of dynamics needed for such an analysis are not

J. Ševčík (✉) · L. Přibylová
Department of Mathematics and Statistics, Faculty of Science, Masaryk University, Kotlářská 2, 61137 Brno, Czech Republic
e-mail: 460534@mail.muni.cz

L. Přibylová
e-mail: pribylova@math.muni.cz

C. H. Skiadas and Y. Dimotikalis (eds.), *14th Chaotic Modeling and Simulation International Conference*, Springer Proceedings in Complexity,
https://doi.org/10.1007/978-3-030-96964-6_29

the only obstacles to using bifurcation analysis methods. Another problem occurs because the phase variables present in such models usually enter as harmonic terms. Due to that, the systems are typically stiff, and standard numerical continuation techniques fail.

Our contribution brings a suitable method for analyzing dynamics of forced oscillators concerning synchronization. We present this method on the forced van der Pol oscillator example. In addition, it also allows excellent visualization of the state space in the neighborhood of bifurcation manifolds that belong to the onset of synchronization. All nonlinear phenomena that are closely related to it, as torus birth, resonances, or complex dynamics near double Hopf bifurcation, can be visualized in 3D space which greatly simplifies their explanation. This approach can be used for much more complex systems of coupled oscillators as you can see in Záthurecký and Přibylová [13].

2 Forced van der Pol Oscillator Representation

Consider the widely known van der Pol oscillator driven by an external harmonic force represented by the equation

$$\ddot{x} - \mu\left(1 - x^2\right)\dot{x} + \omega_0^2 x + A\sin\omega t = 0, \tag{1}$$

where $x \in \mathbb{R}$ is a time-dependent position coordinate, $\mu > 0$ denotes a parameter indicating the nonlinearity (the strength of the damping), and $\omega_0 \in \mathbb{R}$ is the natural frequency. The last term represents the external driving force with amplitude $A > 0$ and frequency $\omega \in \mathbb{R}$.

This second-order differential equation can be expressed in the following form of two-dimensional non-autonomous system

$$\dot{x} = y + \varepsilon\cos\omega t, \tag{2a}$$

$$\dot{y} = \mu\left(1 - x^2\right)y - \omega_0^2 x, \tag{2b}$$

where $\varepsilon = \frac{A}{\omega}$. To obtain an autonomous system, it is usually convenient to rewrite the time-dependent term $\varepsilon\cos\omega t$ in (2a) using a pair of new variables, specifically

$$\dot{x} = y + \varepsilon u, \tag{3a}$$

$$\dot{y} = \mu\left(1 - x^2\right)y - \omega_0^2 x, \tag{3b}$$

$$\dot{u} = -\omega v, \tag{3c}$$

$$\dot{v} = \omega u. \tag{3d}$$

Unfortunately, this system is stiff, and the continuation is impossible since the periodic solution of (3c), (3d) is not asymptotically stable. Therefore, we replace this subsystem with a normal form of supercritical Hopf bifurcation

$$\dot{u} = ru - \omega v - u\left(u^2 + v^2\right),$$
$$\dot{v} = \omega u + rv - v\left(u^2 + v^2\right)$$

with an exponentially stable driving cycle allowing a stable continuation of limit cycles and their bifurcations. Note that the added parameter r provides a possibility to investigate bifurcations connected to the birth of an invariant torus.

Hence, one can examine the forced van der Pol oscillator (1) as two interacting master-slave oscillators in the form

$$\dot{x} = y + \varepsilon u, \tag{4a}$$
$$\dot{y} = \mu\left(1 - x^2\right) y - \omega_0^2 x, \tag{4b}$$
$$\dot{u} = ru - \omega v - u\left(u^2 + v^2\right), \tag{4c}$$
$$\dot{v} = \omega u + rv - v\left(u^2 + v^2\right). \tag{4d}$$

This step also provides an opportunity to clearly visualize synchronization phenomena of the famous van der Pol oscillator since variables u and v are complementary, and one of them can be omitted in the state space description.

3 Basic Bifurcation Analysis

The studied system (4) is evidently uncoupled for zero coupling, i.e., $\varepsilon = 0$. In this case, one can investigate both subsystems separately. Assuming $\mu = 0$, there is no damping in the van der Pol system, and thus the system exhibits simple conservative harmonic oscillations with frequency ω_0. It is known that the unforced van der Pol oscillator undergoes a supercritical Hopf bifurcation that gives rise to a stable limit cycle while crossing $\mu = 0$ as well as the forcing, master system while crossing $r = 0$. It follows that a double Hopf bifurcation manifold (i.e., parameter subspace $\mu = 0, r = 0, \varepsilon = 0$) can be detected as a transversal intersection of these two Hopf hyperplanes.

4 Torus Birth and Synchronization

Double Hopf bifurcation leads to complex dynamics that is related to other bifurcations for nearby parameters. Generically, two branches of Neimark–Sacker bifurcation of a cycle, resulting in a torus birth, emanate from the double Hopf point.

The system (4) gives birth to the stable invariant torus for positive μ and r near zero obviously since supercritical bifurcations appear at $\mu = 0$ and $r = 0$, respectively. An

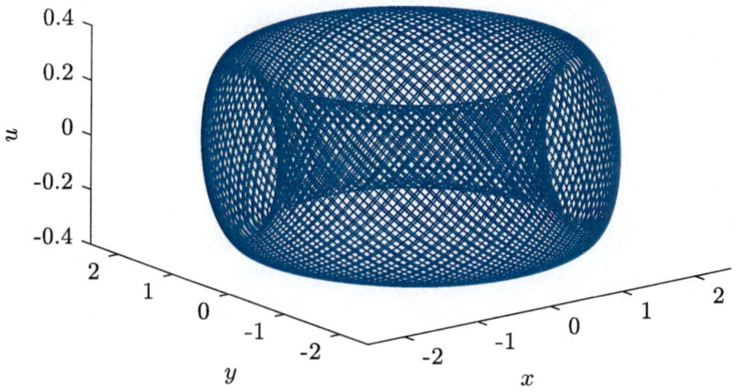

Fig. 1 A segment of a quasiperiodic orbit on a stable invariant torus of system (4) for $\mu = 0.1$, $r = 0.1$, $\varepsilon = 2.5$, $\omega_0 = 1$ and $\omega = \sqrt{5}$

example of a quasiperiodic trajectory densely covering the torus is presented in Fig. 1. The trajectories of the system (4) on the invariant torus can become periodic since the torus is described by a pair of frequencies that can be in a mutually rational proportion. In that moment, the synchronization appears in terms of phase- or frequency-locking. For given external harmonic forcing with nonzero amplitude r and natural frequency ω_0, zero damping μ and zero coupling ε, it is exactly for ratios $\frac{\omega_0}{\omega}$ that are rational. These points are resonances (two-parametric cusp bifurcations of cycles or q-fold bifurcation points in the notation of the bifurcation theory) that correspond to cusp Arnold tongues emanating from Neimark–Sacker bifurcation manifold $\varepsilon = 0$. The Arnold tongues' borders are fold bifurcation manifolds of a stable cycle and a saddle cycle that coincide with each other. The stable cycle persists inside the Arnold tongue and corresponds to the synchronization. Notice that Neimark–Sacker bifurcation, the torus, and fold bifurcation of a cycle manifold continue to positive ε. Since the cusp bifurcation has a typical V-shape, more coupling strength makes the synchronization easier.

From now on, we will consider fixed values $\mu = 0.1$, $r = 0.1$ and $\omega_0 = 1$. The following results are independent of the choice in the sense that we can choose any small μ and r to start with quasiperiodic orbit on a torus. The natural frequency ω_0 is taken as normalized, but it can be easily reparametrized. Let us study the effect of parameters ω and ε on the synchronization in system (4). Using numerical continuation methods in MATCONT toolbox by Dhooge et al. [4], one can compute bifurcation curves of (4) in the parameter space (ω, ε).

Since the natural frequency of the van der Pol oscillator is chosen as $\omega_0 = 1$, the Arnold tongues emanate from all rational numbers on the ω-axis, i.e., points $(\omega, \varepsilon) = \left(\frac{p}{q}, 0\right)$ for coprime $p, q \in \mathbb{N}$. Figure 2 shows several Arnold tongues $\mathcal{A}_{p:q}$ in space (ω, ε) representing parameter values, for which the synchronization $p : q$ takes place in the studied system (4) ($p : q$ is a ratio between the two frequencies on the torus, $q : p$ is the period ratio). As usual, most of the Arnold tongues are

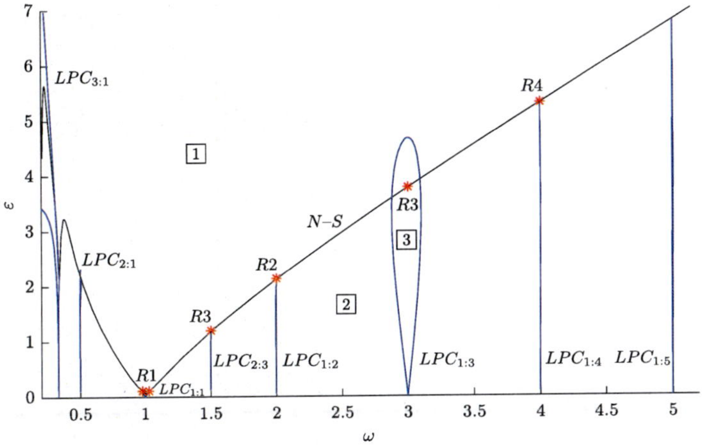

Fig. 2 Bifurcation diagram of system (4) in the parameter space (ω, ε) for $\mu = 0.1$, $r = 1$ and $\omega_0 = 1$

relatively narrow and hence difficult to be manually detected. Notice that we found a non-trivial branch of Neimark–Sacker bifurcation that is different from $\mu = 0, r > 0$ or $r = 0$, $\mu > 0$, respectively, in this parameter space. Dynamics near this branch for $1 : 3$ resonance is shortly mentioned in Sect. 6.

5 Visualizations of the Torus Birth

In addition to the analysis itself, we focused on visualization of dynamics near bifurcation manifolds. One dimension of the 4D state space of the system (4) can be omitted easily as a complement due to harmonic forcing. The 3D invariant torus that appears in the state space for positive r and μ is projected to a two-dimensional torus. Its natural section in a given phase is a Poincaré 2D plane section of a trajectory on the torus. This situation makes it possible to explicitly show qualitative changes in the neighborhood of bifurcation manifolds in the plane and 3D space.

At first, let's look at the transition between regions $\boxed{1}$ and $\boxed{2}$ (see Fig. 3). The system possesses a stable limit cycle in the region $\boxed{1}$ (see Fig. 4). When crossing the non-trivial Neimark–Sacker curve into region $\boxed{2}$, the corresponding Neimark–Sacker bifurcation of a cycle causes a loss of the cycle's stability. It gives rise to a stable invariant torus in its neighborhood (see Figs. 5 and 6). As these figures show, using Poincaré section determined by zero u-coordinate, for example, one can visualize bifurcations of limit cycles via specific orbit topological change of the discrete dynamical system (see Neimark–Sacker bifurcation of maps in Kuznetsov [7]) on the corresponding Poincaré section.

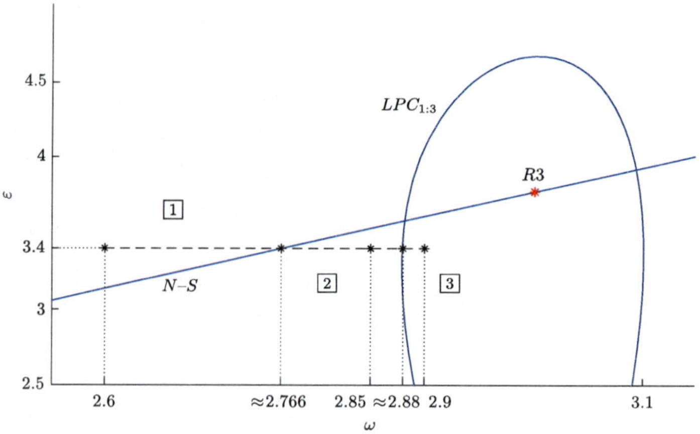

Fig. 3 Considered transitions between regions $\boxed{1}$, $\boxed{2}$ and $\boxed{3}$ in the parameter space (ω, ε) for $\mu = 0.1$, $r = 1$ and $\omega_0 = 1$

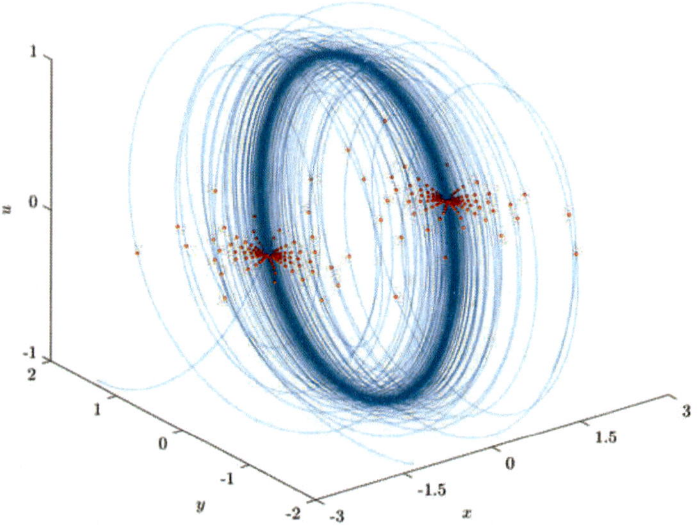

Fig. 4 Poincaré section $\{u = 0\}$ of system (4) for $(\omega, \varepsilon) = (2.6, 3.4)$, region $\boxed{1}$

6 Bistability of the Forced van der Pol Oscillator

Finally, let's look closely to qualitative changes of dynamics near 1 : 3 resonance point R3 on the non-trivial Neimark–Sacker branch depicted in Figs. 2 or 3 (for the positive r, μ, ε and ω). Figure 7 shows a typical symmetric dynamic structure near 1 : 3 resonance (see Kuznetsov [7]). It visualizes the transition between regions $\boxed{2}$ and $\boxed{3}$. As we have just seen, in the region $\boxed{2}$ (outside the Arnold tongue), the

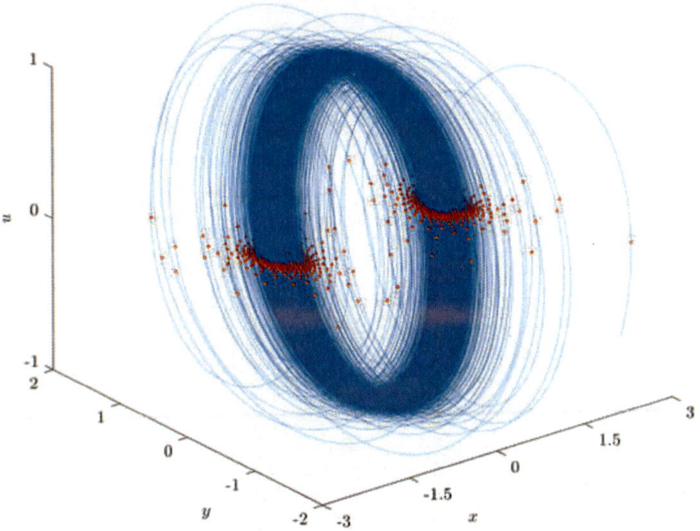

Fig. 5 Poincaré section $\{u = 0\}$ of system (4) $(\omega, \varepsilon) = (2.766, 3.4)$, near N–S manifold

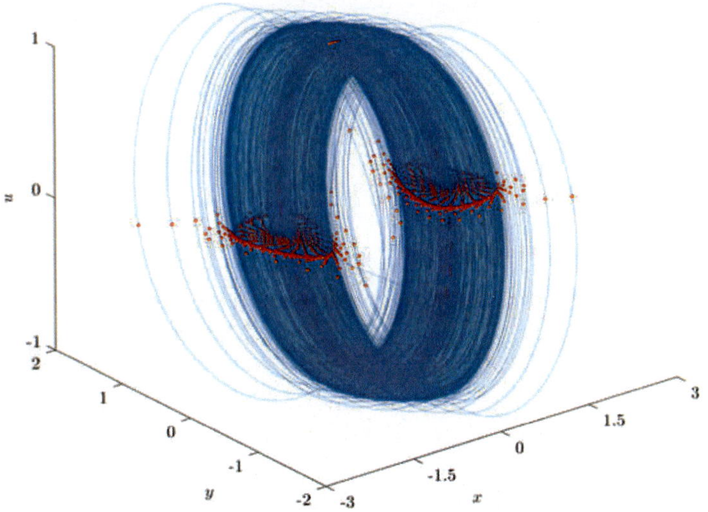

Fig. 6 Poincaré section $\{u = 0\}$ of system (4) for $(\omega, \varepsilon) = (2.85, 3.4)$, region $\boxed{2}$

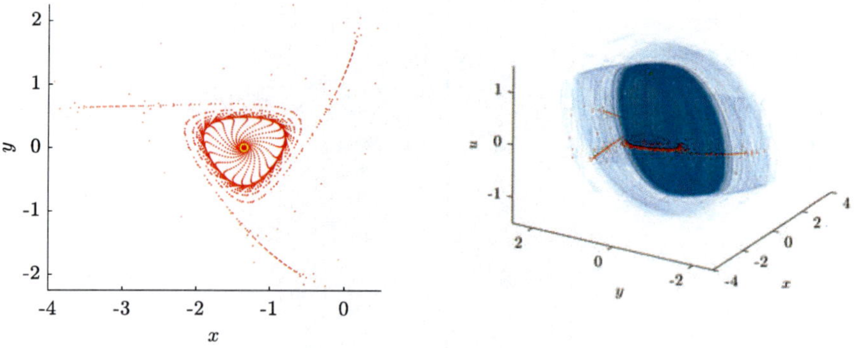

(a) Region $\boxed{2}$: $(\omega, \varepsilon) = (2.85, 3.4)$, before crossing $LPC_{1:3}$

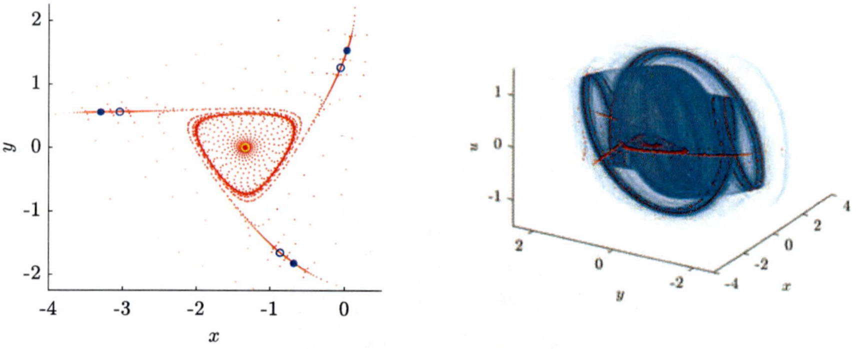

(b) Region $\boxed{3}$: $(\omega, \varepsilon) = (2.88, 3.4)$, right after crossing $LPC_{1:3}$

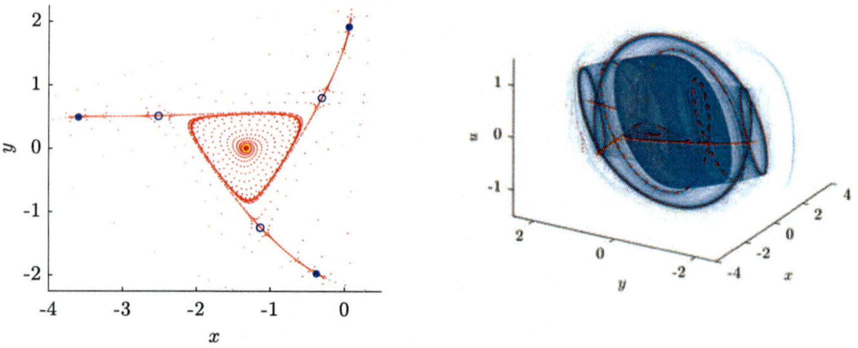

(c) Region $\boxed{3}$: $(\omega, \varepsilon) = (2.90, 3.4)$, after crossing $LPC_{1:3}$

Fig. 7 The onset of synchronization 1 : 3 in system (4) visualized using Poincaré section $\{u = 0, v = -1\}$ in the state space for parameters (ω, ε) from regions $\boxed{2}$ and $\boxed{3}$, i.e., for the crossing of the $LPC_{1:3}$ curve corresponding to fold bifurcation of limit cycles (see Fig. 3) system possesses a stable quasiperiodic invariant torus. When crossing the LPC curve (entering the Arnold tongue, $\boxed{3}$), the fold bifurcation gives rise to a pair of limit cycles—stable and saddle, respectively. The forced van der Pol oscillator evince bistable behavior for these parameters inside and close to the Arnold tongue border since there are two stable attractors—an outside stable limit cycle and a stable invariant torus. The torus may be destructed via a heteroclinic bifurcation. In the Poincaré section depicted red in Fig. 7b, c, you can see a symmetric triplet of saddles approaching the invariant loop that belongs to the inside torus. The coincidence of the saddle cycle with the loop destroys the stable torus.

7 Discussion and Conclusions

To summarize and outline the possible research connected to synchronizations of forced oscillators, we mention the topics we would like to focus on.

There is much more to study since Arnold tongues interfere with each other, and symmetries near resonances give birth to various types of synchronizations. Also, there is usually a period-doubling cascade inside the Arnold tongues, and this route to a chaotic attractor is possible and likely. Study of all these phenomena is allowed only using a suitable representation (4) of forced van der Pol oscillator (1). The proper transformation of the original system and using Poincaré sections give possibility to use continuation methods of bifurcation theory, and also visualize in 3D the hidden phenomena behind synchronizations of limit cycles.

We plan to continue with an analysis of bifurcation manifolds near the mentioned double Hopf bifurcation, as well as near resonances. Very interesting dynamics could be found near the torus break on the heteroclinic orbit for parameters inside the Arnold tongues. We would like to focus also on bistability in the case of coupled oscillators. We are convinced that this phenomenon is closely related to chimera-like dynamics, as well as to routes to complexity and chaos.

Acknowledgements The work has received financial support from Mathematical and Statistical modelling project MUNI/A/1615/2020.

References

1. H. Alinejad, D. Yang, P.A. Robinson, Mode-locking dynamics of corticothalamic system response to periodic external stimuli. Physica D **402** (2020)
2. C. Beck, Possible resonance effect of axionic dark matter in Josephson junctions. Phys. Rev. Lett. **111**, 23 (2013)
3. A.I. Braginski, Superconductor electronics: status and outlook. J. Supercond. Novel Magn. **32**, 23–44 (2019)
4. A. Dhooge, W. Govaerts, Y.A. Kuznetsov, MATCONT: a MATLAB package for numerical bifurcation analysis of ODEs. ACM Trans. Math. Softw. (TOMS) **29**(2), 141–164 (2003)

5. H. Ju, A.B. Neiman, A.L. Shilnikov, Bottom-up approach to torus bifurcation in neuron models. Chaos **28**, 10 (2018)
6. Y. Kuramoto, *Chemical Oscillations, Waves, and Turbulence*. Chemistry Series (Dover, New York, 1984)
7. Y.A. Kuznetsov, *Elements of Applied Bifurcation Theory*, vol. 112 (Springer Science & Business Media, 2013)
8. A. Mallick, M. Bashar, D. Truesdell, B. Calhoun, S. Joshi, N. Shukla, Using synchronized oscillators to compute the maximum independent set. Nat. Commun. **11**, 1–7 (2020)
9. B. Sturgis-Jensen, P.-L. Buono, A. Palacios, J. Turtle, V. In, P. Longhini, On the synchronization phenomenon of a parallel array of spin torquenano-oscillators. Physica D **396**, 71–81 (2019)
10. U. Welp, K. Kadowaki, R. Kleiner, Superconducting emitters of THz radiation. Nat. Photonics **7**, 702–710 (2013)
11. S. Wiggins, *Introduction to Applied Nonlinear Dynamical Systems and Chaos*, vol. 2 (Springer, New York, 1990)
12. A.T. Winfree, *The Geometry of Biological Time* (Springer, New York, 2001)
13. J. Záthurecký, L. Přibylová, Coupled FitzHugh-Nagumo type neurons driven by external voltage stimulation. In: *14th Chaotic Modeling and Simulation International Conference* (2021)
14. W. Zhen, S.A. Campbell, Phase models and clustering in networks of oscillators with delayed, all-to-all coupling. IFAC-PapersOnLine **48**(12), 105–110 (2015)

Fractal Nanoparticles of Phase-Separating Solid Solutions: Nanoscale Effects on Phase Equilibria, Thermal Conductivity, Thermoelectric Performance

Alexander V. Shishulin, Alexander A. Potapov, and Anna V. Shishulina

Abstract In recent years, thermoelectric materials and energy converters have attracted considerable attention, especially as a part of advanced "green" energy and space technologies. One of the most promising ways of obtaining high values of the thermoelectric figure of merit is the formation of nanostructured 3D materials with nanoparticles of phase-separating alloys. In this chapter, using the example of low-temperature thermoelectric Bi_{1-x}-Sb_x alloys for the application in space engineering, we have shown how nanoscale effects on phase equilibria in nanoparticles influence on their thermoelectric properties. Such effects consist in nonlinear changes in mutual solubilities of components at a given temperature, phase transition temperatures and even the total suppression of the phase separation depending on the morphology of a nanoparticle as well as on some other factors. The combination of thermodynamic and *ab initio* approaches has been used while the nanoparticle shape has been determined using the methods of fractal geometry. A method has been suggested in order to calculate the optimal morphology of nanoparticles, at which their equilibrium phase composition leads to a dramatic reduction of the phonon thermal conductivity, favoring the growth of the thermoelectric figure of merit. A decrease in the phonon thermal conductivity in nanoparticles of a pure substance depending on their

A. V. Shishulin (✉)
G.A. Razuvaev Institute of Organometallic Chemistry, Russian Academy of Sciences, Tropinin str., 49, Nizhny Novgorod 603137, Russia
e-mail: chichouline_alex@live.ru

A. A. Potapov
V. A. Kotelnikov Institute of Radio Engineering and Electronics, Russian Academy of Sciences, Mokhovaya str., 11-7, Moscow 125009, Russia
e-mail: potapov@cplire.ru

JNU-IREE RAS Joint Laboratory of Information Technology and Fractal Processing of Signals, JiNan University, Guangzhou 510632, China

A. V. Shishulina
R.E. Alekseev Nizhny, Novgorod State Technical University, Minin str., 24, Nizhny Novgorod 603950, Russia

N.I. Lobachevsky Nizhny Novgorod State University, Gagarin ave., 23, Nizhny Novgorod, Russia

C. H. Skiadas and Y. Dimotikalis (eds.), *14th Chaotic Modeling and Simulation International Conference*, Springer Proceedings in Complexity,
https://doi.org/10.1007/978-3-030-96964-6_30

morphology as well as an approach of calculating the equilibrium size and shape distribution within a nanoparticle ensemble have also been discussed.

1 Introduction

Thermoelectric materials and energy converters on their basis have been an object of considerable interest among researchers in recent years [1]. The expanding field of their application includes energy generators which operate in extreme conditions (e.g. radioisotope thermoelectric generators for Voyager-2 and other space modules [1]), thermoelectric converters for utilizing the waste heat dispersed into environment [2], cooling and temperature-control facilities based on the Peltier effect etc. [3]. Despite the intensive development of multiple approaches to obtain thermoelectric materials with promising properties based on low-dimension structures (nanofilms [4, 5], quantum wires [6] etc.), highly-effective and low-cost thermoelectrics can be produced on the basis of 3D nanocrystalline structures [7, 8].

The key characteristic parameter determining the materials thermoelectric efficiency is the dimensionless figure of merit, *ZT*, which can be expressed as a function of thermal conductivity κ, electrical conductivity σ and Seebeck coefficient α: $ZT = \alpha^2\sigma\overline{T}/\kappa$ where $\overline{T}$ is the average temperature between the "hot" and "cold" sides of a thermoelectric converter and $\kappa \approx \kappa_{car} + \kappa_{ph}$[1, 2, 7]. Here, κ_{car} and κ_{ph} are the contributions of charge carriers (electrons, holes) and phonons to the thermal conductivity, respectively (several other contributions, e.g. the ones of excitons or spin–orbit coupling, are neglected, being much lower). In every single material, as a rule, electrical conductivity σ and the contribution of carriers to the total thermal conductivity, κ_{car}, cannot be varied separately, and optimal thermoelectric properties can be obtained primarily through the reduction of the phonon contribution κ_{ph}. In nanostructured polycrystalline materials, κ_{ph} is generally decreased through formation of multiple interfaces (grain boundaries) which scatter thermal phonons [9], add to this, the energy filtration of charge carriers at grain boundaries, which reduces the contribution of low-energy carriers to transport properties, results in an additional increase in the Seebeck coefficient [2]. Nanostructured polycrystalline thermoelectric materials can be produced using various technologies of up-to-date powder metallurgy, e.g. spark plasma sintering, selective laser sintering, selective laser melting etc. [7, 8].

It is necessary to take into account that several specific properties of nanoscale particles provide an additional "knob" which can be used to tune the functional properties of a nanostructured material. On the one hand, promising thermoelectric properties are associated with nanoparticles of bi- and polycomponent solutions (e.g., Si-Ge, Bi-Sb, Bi-Sb–Te, Bi-Te-Se etc.) while *ab-initio* calculations demonstrate in many cases a dramatic decrease in phonon thermal conductivity κ_{ph} with an increase in the concentration of a dopant [10]. In several systems including the Bi_{1-x}-Sb_x alloy (which is considered to be one of the most efficient low-temperature thermoelectric materials [10] with multiple possible applications, for example, in space engineering), however, the solid solutions with high dopant concentrations are thermodynamically unstable at the operation temperatures, according to the available reference data for macroscopic phases, and undergo the phase separation. On the other hand, all characteristics of phase equilibria including the thermodynamical stability of solid solutions with high dopant concentrations as well as phase transition temperatures become morphology-dependent at the nanoscale [11–18]. In the case of stratifying solid solutions, such effects manifest themselves in dramatic changes solubility limits, temperature ranges of the thermodynamical stability of various heterogeneous and homogeneous states including the value of the upper critical dissolution temperature (UCDT) depending of the geometric characteristics of a nanoparticle under consideration, its external environment [19] and some other factors [20–22]. These effects are interpreted as the realization of several mechanisms reducing the free energy of a nanoscale system, which can also be competing in several cases leading to specific non-monotonous dependences of phase equilibria characteristics [15, 23]. Being observed experimentally [24], at the particle sizes appropriate for the additive technologies, the mentioned effects can be simulated using the thermodynamical approach (the size limits for the thermodynamics applicability have been discussed in [25]). For much smaller particles, the molecular dynamics approach can be applied which demonstrate a perfect accordance with the thermodynamically obtained results [26].

In this chapter, we present and summarize the brand-new findings on the realization of nanoscale phase equilibria effects in nanostructured thermoelectric materials and their possible influence on the thermoelectric characteristics. The materials morphology has been determined using the methods of fractal geometry which allows expressing the materials geometric characteristics in the most general form while the fractal characteristics of nanoparticles can be measured experimentally [27].

2 How to Simulate Phase Equilibria at the Nanoscale: Mathematical Formulations

As the object of modeling below, we consider differently shaped nanoparticles of the Bi_{1-x}-Sb_x system with various atomic fractions of Sb (x). The volume of such a particle is set by its effective diameter d_{eff}-diameter of the sphere whose volume is equal to the volume of the particle. We assume that, at the phase equilibrium in the temperature range below the upper critical dissolution temperature (UCDT), such a particle contains a spherical inclusion of a solid solution (*core*-phase) surrounded by a layer of a solid solution with a different composition (*shell*-phase). In a closed binary system with a *core–shell* configuration, the volume of a particle, total amount n of the matter in the system, numbers n_i of moles of each component (subscripts i = 1, 2 refer to Sb and Bi, respectively), and concentrations x_{ij} of components i in phases j (subscripts $j = c, s$ correspond to *core*- and *shell*-phases, respectively) are interrelated through the conservation conditions of matter:

$$\pi d_{\text{eff}}^3/6=\sum_j V_j,\ n_1=xn,\ n_2+n_1=n,\ n_i=\sum_j n_{ij},$$

$$V_j=\sum_i n_{ij}V_i,\ x_{ij}=n_{ij}\Big/\sum_i n_{ij}.$$

Here, V_j is the volume of phase j while V_i stands for the molar volume of component i. The molar volumes are $V_1 = 18.4$ cm^3/mol and $V_2 = 21.3$ cm^3/mol [28, 29].

In order to describe the irregular morphology of real nanoparticles, it is convenient to use the methods of fractal geometry. In the terms of the given approach, the phase of a nanoparticle is characterized by fractal dimension D which correlates its volume V and surface area A_s: $A_s = CV^{2/D}$, where C is the numerical coefficient which also matches units. For the sake of convenience, we assume $C = 4\pi$ below (this assumption is not accompanied by any losses of generality). For simple geometric structures, $D \equiv 3.00$ while for structures with a complicated and irregular morphology, $D < 3.00$ and can be non-integer. The examples of nanoparticles with various fractal dimensions can be found in [30, 31]. Fractal dimension D can also be related to so-called *shape coefficient k* which has been used as a calculation parameter in some previous papers of us [23, 28–30] and other authors [32]: $k(V, D) = V^{2/D}/(3V/4\pi)^{2/3}$. The shape coefficient is equal to the ratio between the surface areas of the particle under consideration and the sphere of the same volume, for example, for nanoparticles with the effective diameter of 40 nm and fractal dimensions of 2.60, 2.75 and 2.90, the shape coefficients are equal to 2.51, 1.72 and 1.23, respectively.

Thus, the geometrical parameters of the *core–shell* structure can be described as follows: $A_c = 4\pi(3V_c/4\pi)^{2/3}$, $A_s = C(\pi d_{\text{eff}}^3/6)^{2/D}$.

The criterion for an equilibrium state of the system is the minimum of Gibbs function g including the energy contribution of all interfaces:

$$\begin{aligned} g &= \sum_j (n_{1j}+n_{2j})G_j(x_{1j},T)+\sigma_s A_s+\sigma_{cs}A_s, \\ G_j(x_{1j},T) &= A_{\text{I}}(T)x_{1j}(1-x_{1j})+A_{\text{II}}(T)x_{1j}+A_{\text{III}}(T)(1-x_{1j}) \\ &\quad + RT\left(x_{1j}\ln x_{1j}+(1-x_{1j})\ln(1-x_{1j})\right) \end{aligned} \tag{1}$$

where R is the universal gas constant, σ_s and σ_{cs} are the surface energies on the outer (*shell*-) and inner (*core–shell*) interfaces. $A_{\text{I}}(T) = 6500 - 2.6T$, $A_{\text{II}}(T) = -2.645R(903.7 - T)$, $A_{\text{III}}(T) = -2.458R(544.6 - T)$. Without any losses of generality, the following approximation has been used to calculate the values of σ_s and σ_{cs} (see also [33]): $\sigma_{cs} = 0.5\sum_j \sigma(x_{1j})$, $\sigma(x_{1j}) = \sigma_1 x_{1j} + \sigma_2(1 - x_{1j})$, $\sigma_s = \sigma_1 x_{1s} + \sigma_2(1 - x_{1s})$ where $\sigma_1 = 0.300$ J/m^2, $\sigma_2 = 0.521$ J/m^2 [19, 28, 29].

3 Nanoscale Phase Equilibria and Lattice Thermal Conductivity: Specific Phenomena

For a *core–shell* structure, Gibbs function (1) has two minima which correspond to heterogeneous states with different mutual arrangements of co-existing solid solutions. The state where the *shell*-phase is formed with a Bi-based solid solution is hereinafter called *state 1*. In *state 2*, in turn, Sb prevails in the *shell*-phase. In the case of a two-component system, simulation results and the main regularities can be illustrated with the θ-diagrams for the Gibbs function. The example of such diagrams is given in Fig. 1 where Gibbs function (1) is plotted using dimensionless variables $\theta_i = n_{ic}/n_1$, corresponding to the atomic fraction of component i in the *core*-phase relative to the total amount of component i in the system ($\theta_i \in [0, 1]$). The minima are indicated with a darker color.

In the heterogeneous state of a macroscopic system (Fig. 1a), the minima of the Gibbs function are symmetric, have equal energies and correspond to the equal compositions of co-existing phases which match the reference data (the mutual arrangement of phases, the morphology of the system and its total composition (x) have no influence on the phase composition at each temperature as well as on the upper critical dissolution temperature i.e. the temperature value at which the phase separation terminates and all the compositions of a particle become thermodynamically stable). At the nanoscale, at the same time, a notable energy contribution of all the interfaces leads to shifts of the minima in comparison with a macroscopic system while the energy of state 1 becomes somewhat higher than the energy of another one and a high-energy metastable homogeneous state can also emerge in the system (see Fig. 1b). As a result, the equilibrium compositions and relative volumes of co-existing phases are different in states 1 and 2, differ from the "macroscopic" values and depend on the system morphology.

In both states 1 and 2, a decrease in the particle size and/or a decrease in its fractal dimension lead to a considerable reduction of the UCDT. However, the UCDT values differ in different states, the upper critical dissolution temperature in state 1 being always lower than the one in state 2. At the temperatures above the UCDT for state 1, the heterogeneous state with a Bi-based *shell*-phase is replaced with a homogeneous state which co-exists with the heterogeneous one with a Sb-based shell (see Fig. 1c). Above the UCDT for state 2, the homogeneous state becomes the only one in the system. An example of the dependence of the UCDT on the particle morphology is plotted in Fig. 2b for an equiatomic particle of $d_{\text{eff}} = 40$ nm in state 1. At $D < 2.74$ the UCTD of such a particle is reduced below 100 K while for a particle of the same volume in state 2 this takes place at $D < 2.57$.

The dependences of the phonon contribution to the thermal conductivity coefficient exhibit its dramatic jumpwise reduction at high concentrations of dopants (an example of those dependences is plotted in Fig. 2b for two crystallographic directions in the Bi-Sb crystal lattice at $T = 100$ K. At other temperatures, the characteristic view of such dependences is expected to remain the same). A decrease in the UCDT down to the low operating temperatures leads to the thermodynamical stabilization

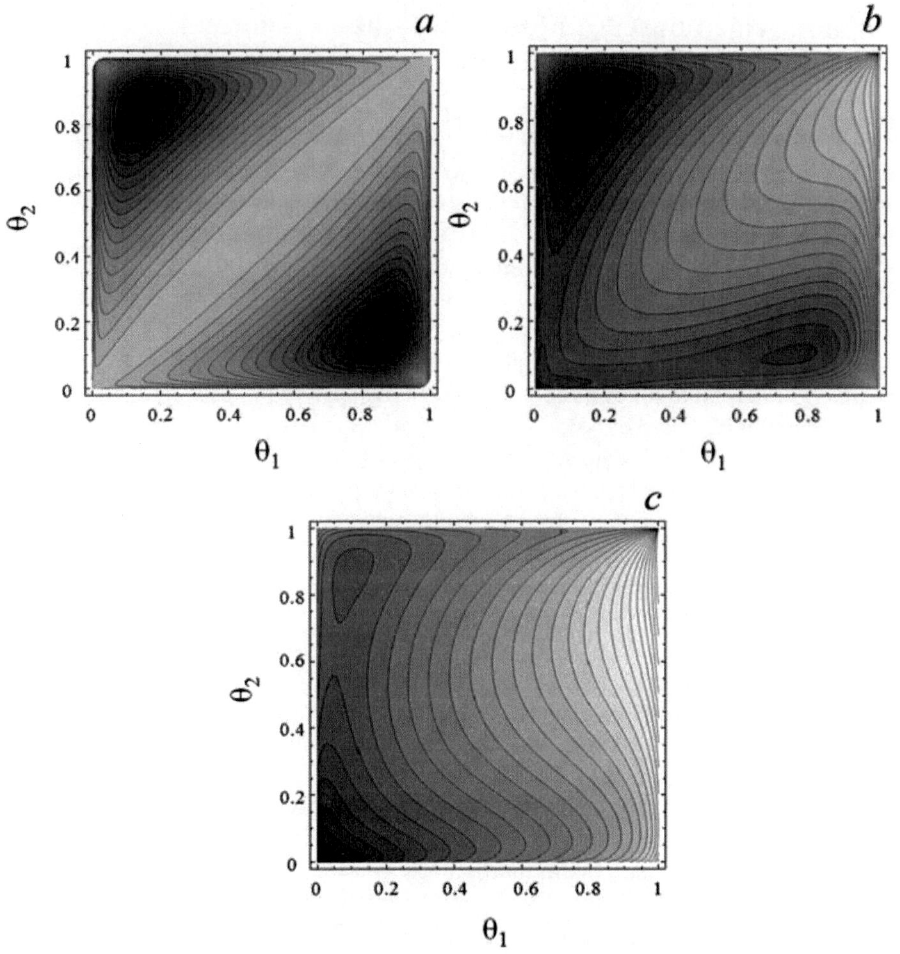

Fig. 1 Examples of θ-diagrams: for a macroscopic structure (**a**), for a nanoscale particle at "low" temperatures with both heterogeneous states and a highly metastable homogeneous one (**b**). for a nanoscale particle upper the UCDT for state 1 (**c**). The minima in the lower right corner and in the upper left one correspond to states 1 and 2, respectively. The homogeneous state corresponds to the minimum located in the lower left corner at $\theta_1 = \theta_2 = 0$

of phases with equiatomic or near-equiatomic compositions with significantly (by more than 50%) reduced phonon thermal conductivities (for example, in equiatomic nanoparticles of $d_{\mathrm{eff}} = 40$ nm in state 1, decreasing the fractal dimension below 2.74 at $T = 100$ K leads to a decrease in the phonon thermal conductivity by 71% as compared to the value of the *shell*-phase of a spherical 40-nm-diameter particle in the same thermodynamical conditions or even by 88% as compared to the one of its *core*-phase).

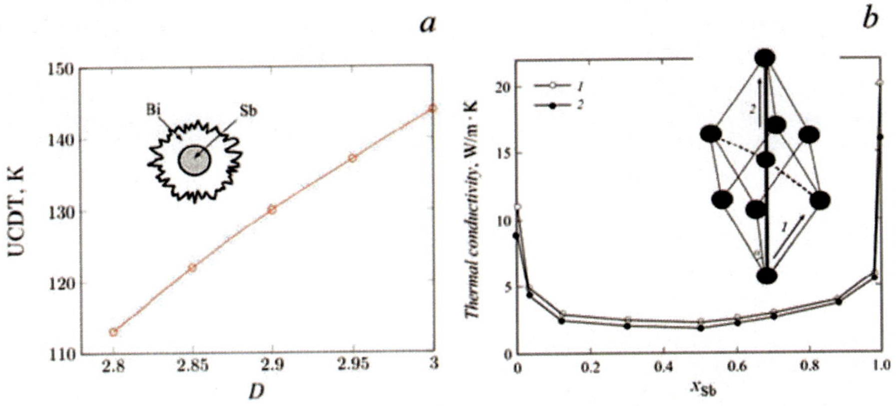

Fig. 2 The dependence of the upper critical dissolution temperature (UCDT) on the fractal dimension for a nanoparticle ($d_{\text{eff}} = 40$ nm) in state 1 (**a**) and the influence of the Sb content on the phonon thermal conductivity of Bi-Sb solid solutions for binary (*1*) and trigonal (*2*) crystallographic directions (**b**). In the box, the crystalline structure of Bi and Sb. Angle α is 57° 30′ for Bi and 75° 84′ for Sb

Morphology-dependent changes in the mutual solubilities below the UCTD can lead to smooth variations of phonon thermal conductivity. The temperature dependences of the solubilities in nanoparticles of the Bi-Sb alloy of different volumes and shapes have been obtained in [19, 28, 29]. Such dependences result from the implementation of two mechanisms of lowering the free energy of a nanoscale system. In the most general case, these mechanisms have been described in [15, 23] along with a specific case when the mechanisms (so-called "molar-volume-controlled segregation" and "surface-energy-controlled segregation") are competitive. These results should be accompanied by the consideration of one of the most interesting and non-trivial effects in small-volume structures—the effect of the chemical composition. In macroscopic structures, a variation of the chemical composition of the system leads to changes in the volume fraction of co-existing phases (according to the lever rule) but does not affect the mutual solubility of components, in small-volume systems the chemical composition of the system determines not only the volume fractions of the phases, but also their equilibrium composition (previously such effect has been considered in the case of systems with the liquid–liquid phase separation [20] and nanoparticles above the solidus temperature [21, 22] and explained according to the abovementioned mechanisms of lowering the free energy). In a spherical 70-nm-diameter nanoparticle containing 50 at. % Sb ($x = 0.5$), for example, the solubility limits are ~ 1.01 at. % for Bi in Sb, ~ 8.00 at. % for Sb in Bi (state 1), ~ 1.49 at. % for Bi in Sb, ~ 1.56 at. % for Sb in Bi (state 2). But for $x = 0.3$, we have: ~ 0.96 at. % for Bi in Sb, ~ 6.98 at. % for Sb in Bi (state 1), ~ 1.13 at. % for Bi in Sb, ~ 1.59 at. % for Sb in Bi (state 2). For $x = 0.6$, it turn, the solubilities are: ~ 0.99 at. % for Bi in Sb, ~ 10.47 at. % for Sb in Bi (state 1), ~ 1.82 at. % for Bi in Sb, ~ 1.53 at. % for Sb in Bi (state 2). Note that despite the fact the term "nanoscale effects" is widely used, there is a broad class of systems, especially polymeric ones with great molecular volumes and masses, in which such effects manifest themselves at characteristic sizes even several thousand times higher [16, 20] (and such effects should probably be called "small-amounts-of-matter effects").

All the demonstrated regularities for individual nanoparticles are expected to be realized in nanoparticle-fabricated alloys where the role of interfaces is played by grain boundaries. In such cases, however, some other geometrical configurations of co-existing phases may be preferable (e.g. the nucleation and growth of a new phase in a grain boundary) while the values of grain boundary energies can be estimated using different approaches (see [34], for example) which may require minor changes in the presented mathematical formulations. In polycrystalline structures, the further optimization of the materials thermoelectric performance can be expected due to additional scattering of phonons at grain boundaries, tunneling of carriers between nanograins and energy filtering of carriers (see [29] and Refs. within). One of the models which describe the phenomenon of grain boundary scattering can be found in [9].

4 Conclusions and Additional Remarks

The obtained results can be accompanied by the estimates which allow predicting an additional decrease in the phonon thermal conductivity coefficients in nanoparticles of a pure substance corresponding to the changes in phonon dynamics in nanoscale particles. In order to establish the morphology-based model to calculate the corresponding changes in the lattice thermal conductivity parameters, one of the possible approaches consists in using one of the models for the morphology-dependent melting temperature as well as the Debye theory and the Lindemann criterion of melting which correlates the melting and Debye temperatures of the crystal, estimating the changes in the average phonon velocity and phonon mean-free path on the basis of the size-dependent Debye temperature, and in applying one of the models which describes phonon scattering effects. An attempt of such calculations is presented in [35] where the expression of G. Guisbiers for the melting temperature of nanoscale structures has been used, and the obtained results demonstrate a good agreement with the experimental data for pure Si and GaAs nanostructures (especially, nanorods and nanofilms). For nanoparticles, however, we suggest using the more convenient model of W. H. Qi and M. P. Wang (experimentally justified, for example, for pure Bi nanoscale structures [32]) for the nanoparticle melting behavior combined with the fractal geometry approach. The Qi-Wang model is based on the correlation between the melting temperature and the cohesive energy of the crystal (note that the cohesive energy is also related to the temperatures of several other types of first-order and second-order phase transitions including the magnetic ones [36]) and leads to the following equation: $T_{\mathrm{m}}^{\mathrm{nano}} = T_{\mathrm{m}}^{\mathrm{bulk}}\left(1 - 6kr_{\mathrm{at}}/d_{\mathrm{eff}}\right)$ where $T_{\mathrm{m}}^{\mathrm{nano}}$ and $T_{\mathrm{m}}^{\mathrm{bulk}}$ are the melting temperatures of a nanoparticle and the bulk material, respectively, r_{at} is the atomic radius and k and d_{eff} are the shape coefficient and the effective diameter of a nanoparticle as they have been introduced above. All these considerations allow expressing the morphology-dependent phonon contribution to the thermal conductivity as follows:

$$\kappa_{\text{ph}}^{\text{nano}} = \eta \exp(1 - l_0/d_{\text{eff}}) \left(1 - C(\pi/6)^{2/D-1} d_{\text{eff}}^{3(2/D - 2/3)} r_{\text{at}} \Big/ d_{\text{eff}} \right) \kappa_{\text{ph}}^{\text{bulk}}$$

where pre-term $\eta \exp(1 - l_0/d_{\text{eff}})$ is included as a quite simplified approach to take into account the phonon scattering which increases with a decrease in the volume of a particle, an increase in the surface-to-volume ratio and surface roughness [35]. $\eta \in (0, 1]$ represents the surface roughness parameter while l_0/d_{eff} is the Knudsen number, l_0 is the bulk phonon mean-free path value [35]. Coefficient C which matches the units has been introduced above. The larger the value of η is, the smoother the surface of the particle is. Note that the correlation between the surface roughness and fractal dimension of the nanoparticle is not unique (the fractal dimension value is associated with the surface-to-volume ratio while the surface roughness corresponds primarily to the number of edges). Several notes on the dependence of fractal dimension on the surface roughness are given in [37] (see also their graphical representation in Fig. 2 of [37]).

In the case of nanoparticle ensembles, the average phase composition and functional properties depend on the size and shape distributions in an ensemble. In [31], we have suggested a method for calculating such distributions based on the combined usage of number theory, fractal geometry and statistical thermodynamics. For example, the equilibrium size distributions for nanoparticles with fractal dimension D in a free-dispersed system can be expressed as follows:

$$f_D(\phi_p, D, N) \sim \exp\left(-\frac{\sigma A_{sp}(D) + RT \ln f_p}{RT} \right),$$

$$f_p = \frac{N}{N - \phi_p} \exp\left\{ \pi \left(\sqrt{\frac{2}{3}}(N - \phi_p) - \sqrt{\frac{2}{3} N} \right) \right\}.$$

Here, $\phi_p = \omega\left(d_{eff}/d_{at}\right)^3$ is the number of atoms in a nanoparticle, ω is the lattice packing density, N is the total number of atoms in the system, $A_{sp}(D)$ is the specific surface area of the ensemble. The presented estimates are in perfect accordance with the experimental data (see [31] and Refs. within) and make it possible to model the thermodynamical conditions for the realization of optimal average characteristics of nanoparticles (equilibrium compositions, phase transition temperatures, thermoelectric properties etc.) as well to predict the degree at which such characteristic are "blurred" in an ensemble.

References

1. D.M. Rowe (ed.), *Thermoelectric Handbook: Macro to Nano* (CRC Press, Boca Raton, 2006), 1008 p
2. I.A. Tambasov, A.S. Voronin, N.P. Evsevskaya, M.N. Volochaev, Y.V. Fadeev, M.M. Simunin, A.S. Aleksandrovsky, T.E. Smolyarova, S.R. Abelian, E.V. Tambasova, M.O. Gornakov, V.A. Eremina, Yu.M. Kuznetsov, M.V. Dorokhin, E.D. Obraztsova, Thermoelectric properties of low-cost transparent single wall carbon nanotube thin films obtained by vacuum filtration. Phys. E. **114**, 113619 (2019). https://doi.org/10.1016/j.physe.2019.113619
3. J.-Z. Hu, B. Liu, J. Zhou, B. Li, Y. Wang, Enhanced thermoelectric cooling performance with graded thermoelectric materials. Jpn. J. Appl. Phys. **57**, 71801–71806 (2018). https://doi.org/10.7567/JJAP.57.071801
4. Z. Li, N. Miao, J. Zhou, Z. Sun, Z. Liu, H. Xu, High thermoelectric performance of few-quintuple Sb_2Te_3 nanofilms. Nano Energy **43**, 285–290 (2018). https://doi.org/10.1016/j.nanoen.2017.11.043
5. I.V. Erofeeva, M.V. Dorokhin, V.P. Lesnikov, Y.M. Kuznetsov, A.V. Zdoroveyshchev, E.A. Pitirimova, Thermoelectric effects in nanoscale layers of manganese silicide. Semiconductors **51** (11), 1403–1408 (2017). https://doi.org/10.1134/S1063782617110112
6. O. Caballero-Calero, M. Martín-González, Thermoelectric nanowires: a brief prospective. Scripta Mater. **111**, 54–57 (2016). https://doi.org/10.1016/j.scriptamat.2015.04.020
7. M.V. Dorokhin, I.V. Erofeeva, Y.M. Kuznetsov, M.S. Boldin, A.V. Boryakov, A.A. Popov, E.A. Lantsev, P.B. Demina, A.V. Zdoroveyshchev, Investigation of the initial stages of spark-plasma siltering of Si–Ge-based thermoelectric materials. Nanosyst. Phys. Chem. Math. **9**(5), 622–630 (2018). https://doi.org/10.17586/2220-8054-2018-9-5-622-630
8. Y.M. Kuznetsov, M. Bastrakova, M.V. Dorokhin, I.V. Erofeeva, P.B. Demina, E. Uskova, A.A. Popov, A.V. Boryakov, Molecular dynamics studies on spark-plasma sintering of Si–Ge-based thermoelectric materials. AIP Adv. **10**(6), 065219 (2020). https://doi.org/10.1063/5.0011740
9. L.P. Bulat, I.A. Drabkin, V.V. Karatayev, V.B. Osvenskii, D.A. Pshenay-Severin, Effect of boundary scattering on the thermal conductivity of a nanosctructured semiconductor material based on the $Bi_xSb_{2-x}Te_3$ solid solution. Phys. Solid State **52**(9), 1836–1841 (2010). https://doi.org/10.1134/S1063783410090088
10. S. Lee, K. Esfarjani, J. Mendoza, M.S. Dresselhaus, G. Chen, Lattice thermal conductivity of Bi, Sb, and Bi–Sb alloy from first principles. Phys. Rev. B. **89**, 85206–85215 (2014). https://doi.org/10.1103/PhysRevB.89.085206
11. S. Bajaj, M.G. Haverty, R. Arróyave, W.A. Goddard, S. Shankar, Phase stability in nanoscale material systems: extension from bulk phase diagrams. Nanoscale **7**(9868) (2015). https://doi.org/10.1039/C5NR01535A
12. G. Guisbiers, S. Khanal, F. Ruiz-Zapeda, J. Roque de la Puente, M.J. Yakamán, Cu-Ni nano-alloy: mixed, *core-shell* or *janus* nano-particle? Nanoscale **6**, 14630–14635 (2014). https://doi.org/10.1039/C4NR05739B
13. R. Mndoza-Pérez, S. Muh, Phase diagrams of refractory bimetallic nanoalloys. J. Nanopart. Res. **22**(36) (2020). https://doi.org/10.1007/s11051-020-05035-x
14. L.-D. Geoffrion, G. Guisbiers, Chemical ordering in $Bi_{1-x}Sb_x$ nanostructures: alloy, *janus* or *core-shell*. J. Phys. Chem. C. **124**(25), 14061 (2020). https://doi.org/10.1021/acs.jpcc.0c04356
15. A.V. Shishulin, V.B. Fedoseev, On some peculiarities of stratification of liquid solutions within pores of fractal shape. J. Mol. Liq. **278**, 363–367 (2019). https://doi.org/10.1016/j.molliq.2019.01.050
16. A.V. Shishulin, V.B. Fedoseev, Thermal stability and phase composition of stratifying polymer solutions in small-volume droplets. J. Eng. Phys. Thermophys. **93**(4), 802–809 (2020). https://doi.org/10.1007/s10891-020-02182-9
17. V.B. Fedoseev, A.V. Shishulin, E.K. Titaeva, E.N. Fedoseeva, On the possibility of the formation of a NaCl–KCl solid-solution crystal from an aqueous solution at room temperature in small-volume systems. Phys. Solid State **58**(10), 2095–2100 (2016). https://doi.org/10.1134/S1063783416100152

18. A.V. Shishulin, V.B. Fedoseev, Size effect in the phase separation of Cr-W solid solutions. Inorg. Mater. **54**(6), 546–549 (2018). https://doi.org/10.1134/S0020168518050114
19. A.V.Shishulin, V.B. Fedoseev, A.V. Shishulina, Environment-dependent phase equilibria in a small-volume system in the case of decomposition of Bi-Sb solid solutions. Butlerov. Commun. **51**(7), 31–37 (2017) [in Russian]
20. A.V. Shishulin, V.B. Fedoseev, Features of the influence of the initial composition of organic stratifying mixtures in microsized pores on the mutual solubility of components. Tech. Phys. Lett. **46**(9), 938–941 (2020). https://doi.org/10.1134/S1063785020090291
21. A.V. Shishulin, V.B. Fedoseev, Effect of initial composition on the liquid-solid phase transition in Cr–W alloy nanoparticles. Inorg. Mater. **55**(1), 14–18 (2019). https://doi.org/10.1134/S0020168519010138
22. A.V. Shishulin, A.V. Shishulina, Several peculiarities of high-temperature phase equilibria in nanoparticles of the Si_x–Ge_{1-x} system, in *Physical and Chemical Aspects of the Study of Clusters, Nanostructures and Nanomaterials*, vol. 11 (2019), pp. 380–388. https://doi.org/10.26456/pcascnn/2019.11.268 [in Russian]
23. A.V.Shishulin, V.B. Fedoseev, Stratifying polymer solutions in microsized pores: phase transitions induced by deformation of a porous material. Tech. Phys. **65**(3), 340–346 (2020). https://doi.org/10.1134/S1063784220030238
24. B. Straumal, B. Baretzky, A. Mazilkin, S. Protasova, A. Myatiev, P. Straumal, Increase of Mn solubility with decreasing grain size in ZnO. J. Eur. Ceram. Soc. **29**, 1963–1970 (2009). https://doi.org/10.1016/j.jeurceramsoc.2009.01.005
25. V.M. Samsonov, D.E. Demenkov, V.I. Karacharov, A.G. Bembel, Fluctuation approach to the problem of thermodynamics' applicability to nanoparticles. Bull. Russ. Acad. Sci. Phys. **75**(8), 1073 (2011). https://doi.org/10.3103/S106287381108034X
26. Y. Magnin, A. Zappelli, H. Amara, F. Ducastelle, C. Bichara, Size-dependent phase diagrams of nickel-carbon nanoparticles. Phys. Rev. Lett. **115**(205502) (2015). https://doi.org/10.1103/PhysRevLett.115.205502
27. J. Li, Q. Du, C. Sun, An improved box-counting method for image fractal dimension estimation. Pattern Recognit. **42**, 2460–2469 (2009). https://doi.org/10.1016/j.patcog.2009.03.001
28. V.B. Fedoseev, A.V. Shishulin, Shape effect in layering of solid solutions in small volume: bismuth-antimony alloy. Phys. Solid State. **60**(7), 1398–1404 (2018). https://doi.org/10.1134/S1063783418070120
29. A.V. Shishulin, V.B. Fedoseev, A.V. Shishulina, Phonon thermal conductivity and phase equilibria of fractal Bi–Sb nanoparticles. Tech. Phys. **64**(4), 512–517 (2019). https://doi.org/10.1134/S1063784219040200
30. A.V. Shishulin, A.A. Potapov, V.B. Fedoseev, Phase equilibria in fractal *core-shell* nanoparticles of $Pb_5(VO_4)_3Cl$–$Pb_5(PO_4)_3Cl$ system: the influence of size and shape, in *Advances in artificial systems for medicine and education II* (Springer, Cham, 2019), pp. 405–413. https://doi.org/10.1007/978-3-030-12082-5_37
31. V.B. Fedoseev, A.V. Shishulin, On the size distribution of dispersed fractal particles. Tech. Phys. **66**(1), 34–40 (2021). https://doi.org/10.1134/S1063784221010072
32. W.H. Qi, M.P. Wang, Size and shape-dependent melting temperature of metallic nanoparticles. Mater. Chem. Phys. **88**, 280–284 (2004). https://doi.org/10.1016/j.matche,phys.2004.04.026
33. D. Hourlier, P. Perrot, Au–Si and Au–Ge phases diagrams for nanosystems. Mater. Sci. Forum. **653**, 77–85 (2010). https://doi.org/10.4028/www.scientific.net/MSF.653.77
34. W. Tyson, W. Miller, Surface free energies of solid metals. Estimation from liquid surface tension measurements. Surf. Sci. **62**(267), 267–276 (1977). https://doi.org/10.1016/0039-6028(77)90442-3
35. M. Goyal, Shape, size and phonon scattering effect on the thermal conductivity of nanostructures. Pramana: J. Phys. 281 **91**(87). https://doi.org/10.1007/s12043-018-1660-8

36. A.V. Shishulin, V.B. Fedoseev, A.V. Shishulina, Variation of the Curie temperature in porous materials. Tech. Phys. Lett. **46**(7), 680–682 (2020). https://doi.org/10.1134/S106378502007024X
37. V.B. Fedoseev, The use of fractal geometry for the thermodynamic description of the free-dimensional crystal structure elements. Lett. Mater. **2**, 78–83 (2012)

Maximal Attractors in Nonideal Hydrodynamic Systems

Aleksandr Shvets and Serhii Donetskyi

Abstract Some nonideal hydrodynamic systems of the type "tank with fluid - source of excitation of oscillation" are considered. New types of limit sets of such systems, so called maximal attractors, have been discovered and described. It was found that the maximal attractors can be both regular and chaotic. Main characteristics of the described maximal attractors are analyzed in details. Transitions to deterministic chaos in such systems are considered. Despite the fact that maximal attractors are not attractors in the traditional sense of this term, it is shown that the transition from regular maximal attractors to chaotic maximal attractors can occur by known before scenarios transition to chaos for "usual" attractors.

Keywords Nonideal hydrodynamic systems · Maximal attractors · Scenarios of transition to chaos

1 Introduction

Many modern machines, mechanisms and technical devices as structural elements contain cylindrical tanks partially filled with fluid. Therefore, the study of oscillations free surface of fluid in cylindrical tanks over the past decades has been attracting close attention [1–4].

Since the end of 70s years of the last century, there have been the so-called "low-dimensional" mathematical models describing such oscillations [5–8]. These models allow us to describe oscillations of the free surface of the liquid in the tank using nonlinear systems of ordinary differential equations instead of partial differential equations that arise when describing the problem in the general setting. "Low-

A. Shvets (✉) · S. Donetskyi
National Technical University of Ukraine "Igor Sikorsky Kyiv Polytechnic Institue", Kyiv, Ukraine
e-mail: aleksandrshvetskpi@gmail.com

S. Donetskyi
e-mail: dsvshka@gmail.com

C. H. Skiadas and Y. Dimotikalis (eds.), *14th Chaotic Modeling and Simulation International Conference*, Springer Proceedings in Complexity,
https://doi.org/10.1007/978-3-030-96964-6_31

dimensional" models allow you to get a fairly adequate description of the problem in cases where the power produced by of the source of excitation of oscillations significantly exceeds the power consumed by the oscillating load (cylindrical tank with fluid). These cases are called ideal by Sommerfeld–Kononenko. However, in practice, most often there are cases in which the power source of excitation of oscillations is comparable to the power consumed by the oscillatory load. Such cases are called nonideal. In these cases, it is imperative to take into account the interaction between the source of excitation of oscillations and the oscillatory load, which leads to essential refinement of the mathematical models used in ideal cases. The neglect of the interaction between the excitation source and the oscillatory load leads to gross errors in the description of the dynamics of the studied systems [9–17].

2 Evolution Equation

Consider dynamic system, the layout of which is shown in Fig. 1. The electric motor shaft is connected to the platform through the crank mechanism, on which a rigid cylindrical tank of radius R is fixed, partially filled with liquid.

When the crank a rotates through an angle Ψ, the platform makes a vertical movement of the form $v(t) = a\cos\Psi(t)$. To describe the vibrations of the free surface of a liquid, we introduce a cylindrical coordinate system $Oxr\theta$ with origin in the tank axis, on the undisturbed fluid surface. The relief equation of free surface of the fluid we write down in the form $x = \eta(r, \theta, t)$. Suppose liquid inviscid and incompressible with density ρ and fills a cylindrical tank of cross-section S section to the depth $x = -d$.

We will find function of the relief of surface of liquid in the form of an eigenmode expansion:

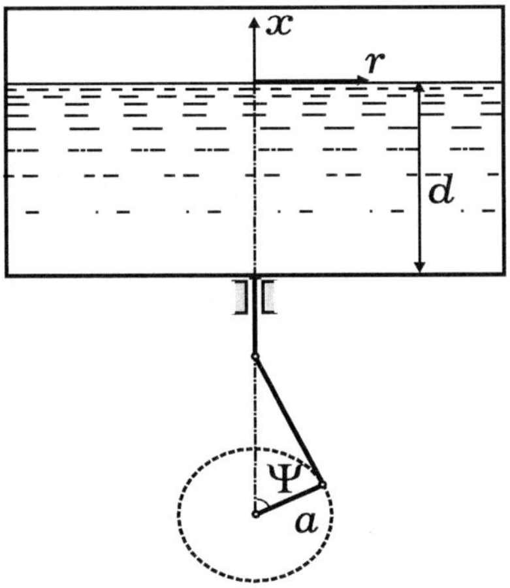

Fig. 1 Scheme of system

$$\eta(r,\theta,t) = \sum_{i,j} [q_{ij}^c(t)k_{ij}(r)\cos i\theta + q_{ij}^s(t)k_{ij}(r)\sin i\theta]. \tag{1}$$

Then we write the kinetic energy of the total system in the form [15, 16]:

$$T = \frac{1}{2}I\dot{\Psi}^2 + \frac{1}{2}m_0\dot{v}^2 + \frac{1}{2}\rho S \sum_{i,j,m,n} a_{ijmn}\dot{q}_{ij}^{c,s}\dot{q}_{mn}^{c,s}. \tag{2}$$

Here I is the moment of inertia of the motor shaft; m_0—mass of the liquid tank; a_{ijmn}—nonlinear functions of $q_{ij}^{c,s}(t)$, $q_{mn}^{c,s}(t)$.

In turn, the potential energy of movements of the free surface of the liquid is Shvets [15, 16]

$$V = \rho \int\int_S dS \int_0^\eta (g + \ddot{v})x dx = \frac{1}{2}\rho S(g + \ddot{v}) \sum_{i,j} q_{ij}^{c,s} q_{ij}^{c,s}, \tag{3}$$

where g is the acceleration of gravity.

Therefore, the Lagrangian of the system takes the form

$$\begin{aligned} L = {} & \frac{1}{2}I\dot{\Psi}^2 + \frac{1}{2}m_0 a^2\dot{\Psi}^2 \sin^2\Psi + \frac{1}{2}\rho S \sum_{i,j,m,n} a_{ijmn}\dot{q}_{ij}^{c,s}\dot{q}_{mn}^{c,s} \\ & + \frac{1}{2}\rho S a(\dot{\Psi}^2\cos\Psi + \ddot{\Psi}\sin\Psi)\sum_{i,j} q_{ij}^{c,s}q_{ij}^{c,s} - \frac{1}{2}\rho S g \sum_{i,j} q_{ij}^{c,s}q_{ij}^{c,s}. \end{aligned} \tag{4}$$

As a result, for $\Psi(t)$ we obtain the following evolution equation

$$\begin{aligned} I\ddot{\Psi} = {} & -2m_0 a^2\dot{\Psi}^2 \sin\Psi\cos\Psi - m_0 a^2\ddot{\Psi}\sin^2\Psi + a\rho S(\dot{\Psi}^2\sin\Psi \\ & - \ddot{\Psi}\cos\Psi)\sum_{i,j} q_{ij}^{c,s}q_{ij}^{c,s} - 2a\rho S\dot{\Psi}\cos\Psi \sum_{i,j} q_{ij}^{c,s}\dot{q}_{ij}^{c,s} + \Phi(\Psi) - H(\Psi). \end{aligned} \tag{5}$$

The last two terms on the right side of (5) are the driving moment and the moment internal forces of resistance of the electric motor.

Suppose that the speed of rotation of the shaft $\dot{\Psi}(t)$ in steady state conditions of the engine is close to $2\omega_1$, where ω_1 is natural frequency of main tone of oscillations of the free surface, which corresponds to the modes $q_{11}^c(t)k_{11}(r)\cos\theta$ and $q_{11}^s(t)k_{11}(r)\sin\theta$.

Let us introduce into consideration a small positive parameter

$$\varepsilon = \omega_1\sqrt{\frac{a}{g}}. \tag{6}$$

Also assume that

$$\dot{\Psi} - 2\omega_1 = \varepsilon^2 \omega_1 \beta. \tag{7}$$

The oscillations of the free surface of the liquid are approximated by oscillations in the main and secondary modes, whose amplitudes are defined as [15, 16]

$$\begin{aligned}
q_{11}^c(t) &= \varepsilon\upsilon\left[p_1(\tau)\cos\frac{\Psi}{2} + q_1(\tau)\sin\frac{\Psi}{2}\right];\\
q_{11}^s(t) &= \varepsilon\upsilon\left[p_2(\tau)\cos\frac{\Psi}{2} + q_2(\tau)\sin\frac{\Psi}{2}\right];\\
q_{01}(t) &= \varepsilon^2\upsilon\left[A_{01}(\tau)\cos\Psi + B_{01}(\tau)\sin\Psi + C_{01}(\tau)\right];\\
q_{21}^{c,s}(t) &= \varepsilon^2\upsilon\left[A_{21}^{c,s}(\tau)\cos\Psi + B_{21}^{c,s}(\tau)\sin\Psi + C_{21}^{c,s}(\tau)\right].
\end{aligned} \tag{8}$$

Here τ is slow time, $\tau = \frac{1}{4}\varepsilon^2\Psi$, $\upsilon = \frac{R}{1.8412}\tanh\left(\frac{1.8412}{R}d\right)$. Having determined the dimensionless amplitudes $A_{ij}^{c,s}(\tau)$, $B_{ij}^{c,s}(\tau)$, $C_{ij}^{c,s}(\tau)$ secondary modes by Miles method [5, 6, 8] through the amplitudes $p_1(\tau)$, $q_1(\tau)$, $p_2(\tau)$, $q_2(\tau)$ and applying the procedure of averaging the Lagrangian over the explicitly entering fast time $\Psi(t)$, for the amplitudes of dominant modes, we obtain the following system of equations [15, 16]:

$$\begin{aligned}
\frac{dp_1}{d\tau} &= \alpha p_1 - \left[\beta + \frac{A}{2}(p_1^2 + q_1^2 + p_2^2 + q_2^2)\right]q_1 + B(p_1q_2 - p_2q_1)p_2 + 2q_1;\\
\frac{dq_1}{d\tau} &= \alpha q_1 + \left[\beta + \frac{A}{2}(p_1^2 + q_1^2 + p_2^2 + q_2^2)\right]p_1 + B(p_1q_2 - p_2q_1)q_2 + 2p_1;\\
\frac{d\beta}{d\tau} &= N_3 + N_1\beta + \mu_1(p_1q_1 + p_2q_2);\\
\frac{dp_2}{d\tau} &= \alpha p_2 - \left[\beta + \frac{A}{2}(p_1^2 + q_1^2 + p_2^2 + q_2^2)\right]q_2 - B(p_1q_2 - p_2q_1)p_1 + 2q_2;\\
\frac{dq_2}{d\tau} &= \alpha q_2 + \left[\beta + \frac{A}{2}(p_1^2 + q_1^2 + p_2^2 + q_2^2)\right]p_2 - B(p_1q_2 - p_2q_1)q_1 + 2p_2.
\end{aligned} \tag{9}$$

In the system of (9), we have the following designations: $\alpha = -\frac{\delta}{\omega_1}$ – reduced factor of damping, N_0, N_1 – constants of linear static characteristic of the electric motor, $N_3 = \frac{1}{\omega_1}\left(N_0 + 2N_1\omega_1\right)$; $\mu_1 = \frac{\rho S\upsilon R^2}{(1.8412)^2(2I + m_0a^2)\omega_1^2}$; A and B – constants ranging from the diameter of the tank and filling it with liquid. The system of evolutionary equations (9) is used as the main mathematical model for study the dynamics of oscillations of a tank with a liquid, excited by an electric motor of limited power.

The main aim of the research is to study the possible types of limit sets of the system (9). Since this system is a rather complex nonlinear system of equations of the fifth order, then for constructing its limit sets, a whole complex of numerical methods and algorithms were used. The technique for carrying out such numerical calculations for system with limited excitation is described in detail in Shvets [18] and Krasnopolskaya and Shvets [16].

3 Numerical Studies of Steady-State Regimes of Oscillations

Initially, we define the conditions under which the system is dissipative. Let us denote by $\boldsymbol{F}$ the vector field generated by the system of equations (9). Accordingly, by F_1, F_2, F_3, F_4, F_5 we denote the components of this vector field, that is, the right-hand sides of the system of equations (9). Then the divergence of this vector field can be found by the formula

$$\begin{aligned}\operatorname{div}\boldsymbol{F} &= \frac{\partial F_1}{\partial p_1}+\frac{\partial F_2}{\partial q_1}+\frac{\partial F_3}{\partial \beta}+\frac{\partial F_4}{\partial p_2}+\frac{\partial F_5}{\partial q_2} = \alpha - Ap_1q_1 + Bp_2q_2 + \alpha \\ &+Ap_1q_1 - Bp_2q_2 + N_1 + \alpha - Ap_2q_2 + Bp_1q_1 + \alpha + Ap_2q_2 - Bp_1q_1 \\ &= 4\alpha + N_1.\end{aligned} \tag{10}$$

So the divergence of the vector field $\mathbf{F}$ is constant. The dissipativity condition for the system of equations has the form,

$$4\alpha + N_1 < 0. \tag{11}$$

The quantities included formula (11), α (coefficient of damping) and N_1 (angle of inclination of the static characteristic of electricmotor) are always negative. Therefore, the divergence of the vector field generated by the system of equations (9) will always be negative. Thus, this system will always be dissipative.

We will begin the study of the dynamics of the system (9) by finding its equilibrium positions. Obviously, that

$$p_1 = 0, q_1 = 0, \beta = -\frac{N_3}{N_1}, p_2 = 0, q_2 = 0 \tag{12}$$

is one of those equilibrium position. This equilibrium position is isolated one. The conditions of the asymptotic stability of this equilibrium position may be obtained by using Liénard-Chipart theorem [19]).

In addition to the isolated equilibrium position (12), there is an infinite number of non-isolated equilibrium positions. These equilibrium positions form a family of non-isolated equilibrium positions, which exists in a form of some closed line. These equilibrium positions can be found only using numerical methods, for example,

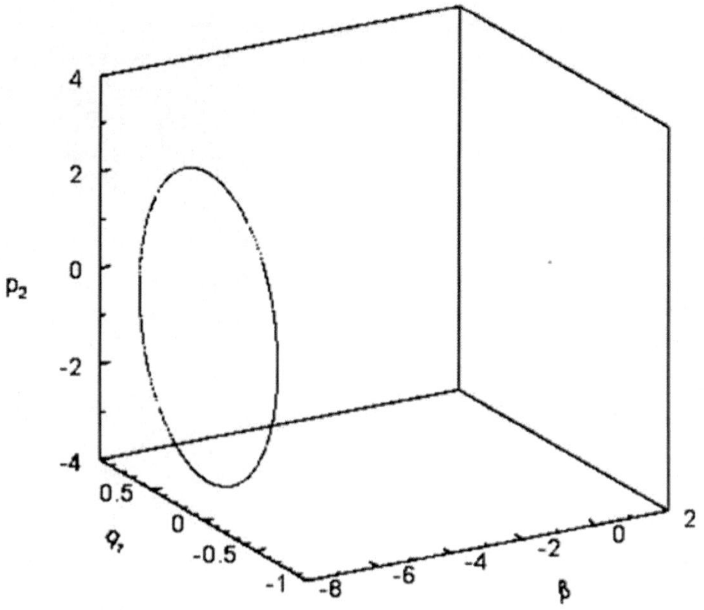

Fig. 2 Family of equilibrium positions

Newton's method. In Fig. 2, an example of such family of equilibrium positions for one concrete values of parameters of the system (9) is shown. Conditions for the stability of such family can be obtained using the Liénard-Chipart theorem. True, these conditions are extremely cumbersome. Their analysis can be carried out in reality only by using computers. Note that all equilibrium positions shown in Fig. 2 may be stable, but can not be asymptotically stable. In the case of stability of these non-isolated equilibrium positions, each of them belongs to the limit set of system (9), but is not an attractor in the traditional sense of this term. We will give a description of the attractive properties of this family below.

There are sufficiently large regions in space of parameters of the system (9), in which all equilibrium positions are unstable. In these areas, extremely interesting limit sets of this system arise, which can be as regular, and as chaotic.

Limit sets of the first type may be periodic. In this case they form family of an infinite number of closed trajectories (cycles), all of which exists simultaneously. Any cycle neighbourhood contains other cycles of the family, that is, they are not isolated. However, such cycles do not have tangency or intersection points. Each such closed trajectory is itself a limit set. This is due to fact that almost any trajectory that starts in some large area of phase space tends to one of the cycles of the family. But none of cycles is an attractor in the traditional sense of this term. So, each of these cycles is not limit cycle. Moreover, every single cycle has same period, same Lyapunov's characteristic exponents and similar Poincare sections. It is worth noting that the cardinality of this family is equal to continuum.

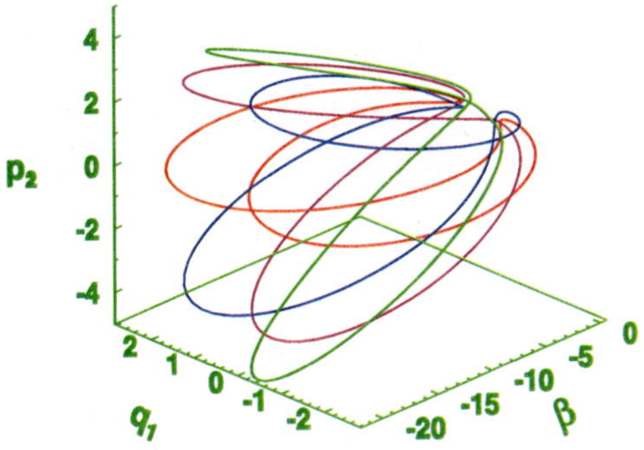

Fig. 3 Four representatives of maximal regular attractor.

In Fig. 3 regular periodic limit sets of system (9) are constructed at $\alpha = -0.8$, $A = 1.12$, $B = -1.531$, $N_1 = -1.25$, $N_3 = 2$, $\mu_1 = -5.15$. Each cycle is plotted in different color. There are four cycles in total, each of which is a representative of the infinite family of cycles. We emphasize once again that each of the cycles forming the family is not an attractor in the traditional sense of this concept. In our opinion, the most suitable term for describing such family is the concept of maximal attractor.

A clear definition of the concept of maximal attractor is given by Milnor [20], by Anischenko and Vadivasova [21], as well as by Sharkovsky [22]. Thus, two different families that are shown in Figs. 2 and 3, are essentially two different types of regular maximal attractors.

With an increase in the value of the parameter μ_1, family of chaotic trajectories arises in the system. The arising family includes an infinite number of chaotic trajectories. It is known that the "traditional" chaotic attractor consists of an infinite number of unstable trajectories. The resulting family, at first glance, is a union of an infinite number of chaotic attractors. However, each member of this family is not an attractor in the "traditional" sense. Here, as before, to define such union, the concept of maximal attractor can be proposed. All trajectories of the chaotic maximal attractor have same spectrum of Lyapunov's characteristic exponents, including positive one. The Poincare sections of each of the trajectories of the family are structurally similar chaotic sets consisted of an infinite number of points.

In Fig. 4 for the values $\alpha = -0.8$, $A = 1.12$, $B = -1.531$, $N_1 = -1.25$, $N_3 = 2$, $\mu_1 = -4.6463$ limit sets of second type of the system (9) is constructed. Each represenative of the chaotic maximal attractor is plotted in its own color. In total, there are three chaotic trajectories of the family are presented in Fig. 4.

In Fig. 5 three representatives of another kind of chaotic maximal attractor, constructed at $\mu_1 = -4.6462$, are shown. On the whole, the chaotic maximal attractor of the second kind are characterized by a much denser filling of the localization region with trajectories. This two kinds of chaotic maximal attractor are typical for system (9).

We note one more feature of the constructed maximal attractors, both regular and chaotic. Some trajectories of those families are localized in the three-dimensional subspace of the five-dimensional phase space of system (9). So the trajectories shown

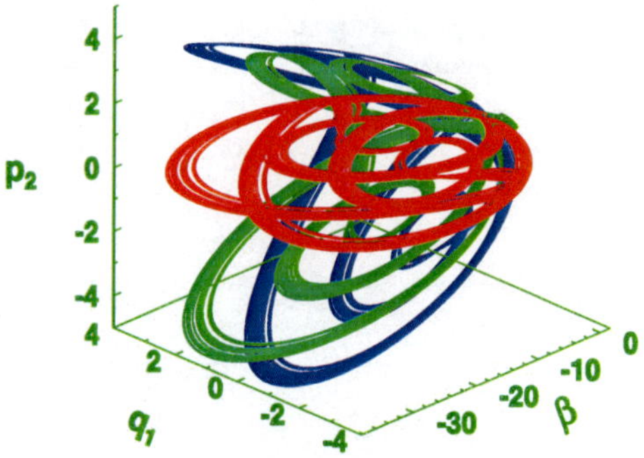

Fig. 4 Three representatives of maximal chaotic attractor of first kind

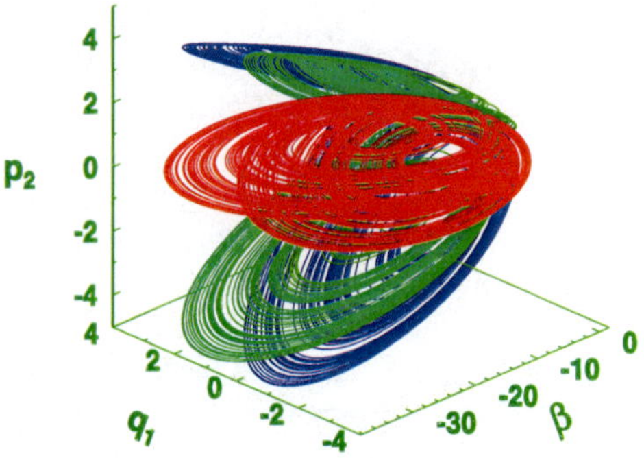

Fig. 5 Three representatives of maximal chaotic attractor of second kind

in red in Figs. 3, 4 and 5 are localized in three-dimensional space. This means that the coordinates p_2 and q_2 of the "red" trajectories are equal to zero. That is, there are no oscillations in the second dominant mode.

Shortly we underscore one more interesting feature of the maximal attractors of the system (9). Although these attractors are not attractors in the traditional sense, the transition from regular to chaotic regimes and the "chaos-chaos" transitions follow the scenarios inherent in such transitions for traditional attractors. Thus transitions according to the Feigenbaum scenario [23, 24], Manneville–Pomeau scenario were found [25, 26] along with various scenarios of generalized intermittency [16, 17, 27, 28].

Let us briefly consider the features of the transition to chaos according to the Feigenbaum and Manneville-Pomeau scenarios for maximal attractors. One of the possible scenarios is the transition from regular regime to a chaotic one is a cascade of bifurcations of period doubling of the cycles. At the same value of the bifurcation parameter the period of all cycles, that form the maximal attractor, is doubled. Then, at the next bifurcation point, the period of all cycles of the maximal attractor is again doubled, and so on. This endless process of period doubling bifurcations ends with the emergence of a chaotic maximal attractor. That is, the transition from a periodic limit set to a chaotic limit set is realized according to the classical Feigenbaum's scenario.

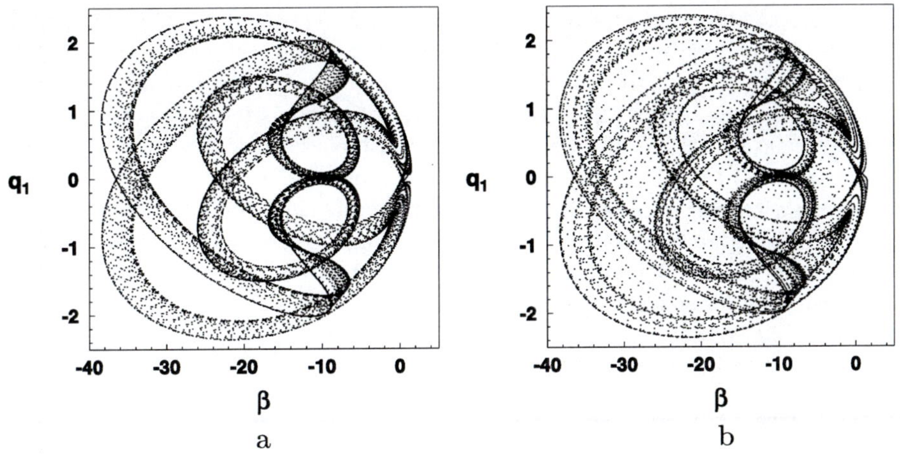

Fig. 6 Distribution of invariant measure over projections of maximal attractors

The transition to chaos through intermittency (the Manneville-Pomeau scenario) for the maximal attractors occurs as follows. The system has a maximal attractor consisting of an infinite set of simultaneously existing cycles. Moreover, all the trajectories of the family have the same period. When passing through the bifurcation point, all cycles of the family disappear and a chaotic maximal attractor arises in the system. The movement along the trajectories of all representatives of this maximum attractor consists of two phases - laminar and turbulent. That is, for all representatives of the family of cycles, there is a simultaneous transition to chaos, through one rigid bifurcation.

In conclusion, let us illustrate the transition from the maximal chaotic attractor of one type to the maximal chaotic attractor of another type through generalized intermittency. In Fig. 6a the distribution of the invariant measure over the phase portrait projection of the representative of the chaotic maximal attractors of the system (9) constructed at $\alpha = -0.8$, $A = 1.12$, $B = -1.531$, $N_1 = -1.25$, $N_3 = 2$, $\mu_1 = -4.6463$ is shown. At $\mu_1 = -4.6462$ (other parameters unchanged) maximal attractor disappears and chaotic maximal attractor of new type is born in the system. The distribution of the invariant measure over the projection of the phase portrait of the representative of the new chaotic maximal attractor is shown in Fig. 6b. The transition from one type of chaotic maximal attractor to the chaotic maximal attractor of another type occurs according to the scenario of generalized intermittency, which was described for attractors in the traditional sense of this term. At such transition, the scenario of generalized intermittency is simultaneously fulfilled for all representatives of chaotic maximal attractor that presented in Fig. 4. For each representative of the new chaotic maximal attractor, the motion along the trajectory consists of two alternating phases, namely rough-laminar phase and turbulent phase. In the rough-laminar phase, the trajectory makes chaotic movements in the neighborhood of the trajectories of the representative of the disappeared chaotic maximal attractor. Then, at an unpredictable moment of time, the trajectory leaves the localization region of the representative of the disappeared maximal attractor and moves to distant regions of the phase space. Rough-laminar phase corresponds to the much blacker areas in fig in Fig. 6. These areas in Fig. 6a are nearly the same as the distribution of the invariant measure from in Fig. 6b. In turn, turbulent phase corresponds to much less darkened areas in Fig. 6b. After some time, the movement of the trajectory returns to the rough-laminar phase again. Then, trajectories switch to turbulent phase again. Such transitions are repeated an infinite number of times. Note that the duration of both rough-laminar and turbulent phases is unpredictable as are the moments of times of transition from one phase to another.

References

1. G.S. Narimanov, L.V. Dokuchaev, I.A. Lukovsky, *Nonlinear Dynamics of Flying Apparatus with Liquid* (Mashinostroenie, Moscow, 1977)
2. R.A. Ibrahim, *Liquid Sloshing Dynamics: Theory and Applications* (Cambridge University Press, Cambridge, 2005)
3. I.A. Lukovsky, *Nonlinear Dynamics* (De Gruyter, Mathematical models for rigid bodies with a liquid, 2015)
4. I. Raynovskyy, A. Timokha *Sloshing in Upright Circular Containers: Theory, Analytical Solutions and Applications* (CRC Press/Taylor & Fransis Group, 2021)
5. J.W. Miles, Nonlinear surface waves in closed basins. J. Fluid Mech. **75**, 419–448 (1979)
6. J.W. Miles, Resonantly forced surface waves in circular cylinder. J. Fluid Mech. **149**, 15–31 (1984)
7. E. Meron, J. Procaccia, Low-dimensional chaos in surface waves: theoretical analysis of an experiment. Phys. Rev. A **34**, 3221–3237 (1986)
8. J.W. Miles, D. Henderson, Parametrically forced surface waves. Ann. Rev. Fluid Mech. **22**, 143–165 (1990)
9. A. Sommerfeld, Beitrage zum dynamischen Ausbau der Festigkeitslehre. Physikalische Zeitschrift **3**, 266–271 (1902)
10. V.O. Kononenko, *Vibrating System with a Limited Power-Supply* (Iliffe, London, 1969)
11. K.V. Frolov, T.S. Krasnopol'skaya, Sommerfeld effect in systems without internal damping. Sov. Appl. Mech. **23**, 1122–1126 (1987)
12. T.S. Krasnopolskaya, Acoustic chaos caused by the Sommerfeld effect. J. Fluids Struct. **8**(7), 803–815 (1994)
13. T.S. Krasnopolskaya, Chaos in acoustic subspace raised by the Sommerfeld-Kononenko effect. Meccanica **41**(3), 299–310 (2006)
14. T.S. Krasnopol'skaya, A.Yu. Shvets, Prorerties of chaotic oscillations of the liquid in cylindrical tanks. Prikladnaya Mekhanika **28**(6), 52–61 (1992)
15. T.S. Krasnopol'skaya, A.Y. Shvets, Parametric resonance in the system: Liquid in tanks + electric motor. Int. Appl. Mech **29**(9), 722–730 (1993)
16. T.S. Krasnopolskaya, A.Yu. Shvets, Chaotic surface waves in limited power-supply cylindrical tank vibrations. J. Fluids Struct. **8**(1), 1–18 (1994)
17. T.S. Krasnopolskaya, A.Yu. Shvets, Dynamical chaos for a limited power supply for fluid oscillations in cylindrical tanks. J. Sound Vibr. **322**(3), 532–553 (2009)
18. A.Yu. Shvets, Deterministic chaos of a spherical pendulum under limited excitation. Ukr. Math. J. **59**, 602–614 (2007)
19. A. Liénard, M.H. Chipart, Sur le signe de la partie r'eelle des racines d'une quation alg'ebrique. J. Math. Pures Appl. **10**(4), 291–346 (1914)

20. J. Milnor, On the concept of attractor, it Commun. Math. Phys. **99**, 177–195 (1985)
21. V.S. Anischenko, T.E. Vadivasova, *Lectures on Nonlinear Dynamics* (R&C Dynamics, Moskow, 2011)
22. A.N. Sharkovsky, *Attractors of Trajectories and Their Basins* (Naukova Dumka, Kiev, 2013)
23. M.J. Feigenbaum, Quantative universality for a class of nonlinear transformations. J. Stat. Phys. **19**(1), 25–52 (1978)
24. M.J. Feigenbaum, The universal metric properties of nonlinear transformations. J. Stat. Phys. **21**(6), 669–706 (1979)
25. P. Manneville, Y. Pomeau, Different ways to turbulence in dissipative dynamical systems. Physica D. Nonlinear Phenom. **1**(2), 219–226 (1980)
26. Y. Pomeau, P. Manneville, Intermittent transition to turbulence in dissipative dynamical systems. Comm. Math. Phys. **74**(2), 189–197 (1980)
27. A. Shvets, V. Sirenko, Hyperchaos in Oscillating Systems with Limited Excitation, in *11th Chaotic Modeling and Simulation International Conference* (CHAOS 2018). Springer Proceedings in Complexity, ed. by C. Skiadas, I. Lubashevsky (Springer, Cham, 2019), pp. 265–273
28. A.Yu. Shvets, V.A. Sirenko, Scenarios of transitions to hyperchaos in nonideal oscillating systems. J. Math. Sci. **243**(2), 338–346 (2019)

Universality of Boltzmann Statistical Mechanics, Thermodynamics, Quantum Mechanics, and Shannon Information Theory

Siavash H. Sohrab

Abstract Universal nature of Boltzmann statistical mechanics, generalized thermodynamics, quantum mechanics, spacetime, black hole mechanics, Shannon information theory, Faraday lines of force, and Banach-Tarski paradox (BTP) are studied. The nature of matter and Dirac anti-matter are described in terms of states of compression and rarefaction of physical space, Aristotle fifth element, or Casimir vacuum identified as a compressible tachyonic fluid. The model is in harmony with perceptions of Plato who believed that the world was formed from a formless primordial medium that was initially in a state of *total chaos* or "*Tohu Vavohu*" (Sohrab in Int J Mech 8:73–84, [1]). Hierarchies of statistical fields from photonic to cosmic scales lead to universal scale-invariant Schrödinger equation thus allowing for new perspectives regarding connections between classical mechanics, quantum mechanics, and chaos theory. The nature of *external* physical time and its connections to *internal* thermodynamics time and Rovelli *thermal time* are described. Finally, some implications of renormalized Planck distribution function to economic systems are examined.

Keywords Thermodynamics · Quantum mechanics · Anti-matter · Spacetime · Thermal time · Information theory · Faraday lines of force · Banach-Tarski paradox · T.O.E.

1 Introduction

The universal nature of turbulence and mathematical similarities amongst transport laws shared by stochastic quantum fields [2–18] and classical hydrodynamic fields [19–31] resulted in introduction of a scale-invariant model of Boltzmann statistical mechanics and its applications to the fields of thermodynamics [32, 33], fluid mechanics [34, 35], and quantum mechanics [36–38] at intermediate, large, and small scales.

S. H. Sohrab (✉)
Department of Mechanical Engineering, Robert R. McCormick School of Engineering and Applied Science, Northwestern University, 2145 Sheridan Road, Evanston, IL 60208-3111, USA
e-mail: s-sohrab@northwestern.edu

C. H. Skiadas and Y. Dimotikalis (eds.), *14th Chaotic Modeling and Simulation International Conference*, Springer Proceedings in Complexity,
https://doi.org/10.1007/978-3-030-96964-6_32

The present study begins with a brief description of invariant model of Boltzmann statistical mechanics and the invariant forms of conservation equations. Next, generalized thermodynamics and Helmholtz decomposition of energy and momentum, and definitions of dark-energy, dark-matter, and dark-momentum are discussed. The concept of internal spacetime versus external independent space and time and their connection to Rovelli *thermal time* are presented. Invariant Schrödinger equation recently derived from invariant Bernoulli equation [38] and some of its implications to a new paradigm for physical foundation of quantum mechanics are described next. In particular, the nature of wave function, wave-particle duality, and hierarchies of quantum mechanics wave functions and particles as de Broglie wave-packets [4] are studied. Also, the implication of the model to entropy and the problem of information loss in black hole is addressed. A universal hydrodynamic model of Faraday line of force applicable from very small scale of stochastic chromodynamics to the exceedingly large cosmic scale is presented. Finally, some implications of the model to Banach-Tarski paradox are examined.

2 Scale–Invariant Model of Boltzmann Statistical Mechanics

The scale-invariant model of statistical mechanics for equilibrium galactic-, planetary-, hydro-system-, fluid-element-, eddy-, cluster-, molecular-, atomic-, subatomic-, kromo-, and tachyon-dynamics corresponding to the scale β = g, p, h, f, e, c, m, a, s, k, and t is schematically shown on the left hand side of Fig. 1.

For each statistical field, one defines particles that form the background fluid and are viewed as point-mass or "*atom*" of the field. Next, the *elements* of the field are defined as finite-sized composite entities composed of an ensemble of "*atoms*". Finally, ensemble of a large number of "*elements*" is defined as the statistical "*system*" at that particular scale. The most-probable element of scale β defines the "atom" (system) of the next higher $\beta + 1$ (lower $\beta - 1$) scale.

Following the classical methods [20, 39–43], the invariant definitions of the density ρ_β, and the velocity of *atom* $\mathbf{u}_\beta$, *element* $\mathbf{v}_\beta$, and *system* $\mathbf{w}_\beta$ at the scale β are given as [37]

$$\rho_\beta = n_\beta m_\beta = m_\beta \int f_\beta du_\beta, \quad \mathbf{u}_\beta = \mathbf{v}_{w\beta-1} \tag{1}$$

$$\mathbf{v}_\beta = \rho_\beta^{-1} m_\beta \int \mathbf{u}_\beta f_\beta du_\beta, \quad \mathbf{w}_\beta = \mathbf{v}_{w\beta+1} \tag{2}$$

Similarly, the invariant definitions of the peculiar and diffusion velocities are introduced as

$$\mathbf{V}'_\beta = \mathbf{u}_\beta - \mathbf{v}_\beta, \quad \mathbf{V}_\beta = \mathbf{v}_\beta - \mathbf{w}_\beta, \quad \mathbf{V}_\beta = \mathbf{V}'_{\beta+1} \tag{3}$$

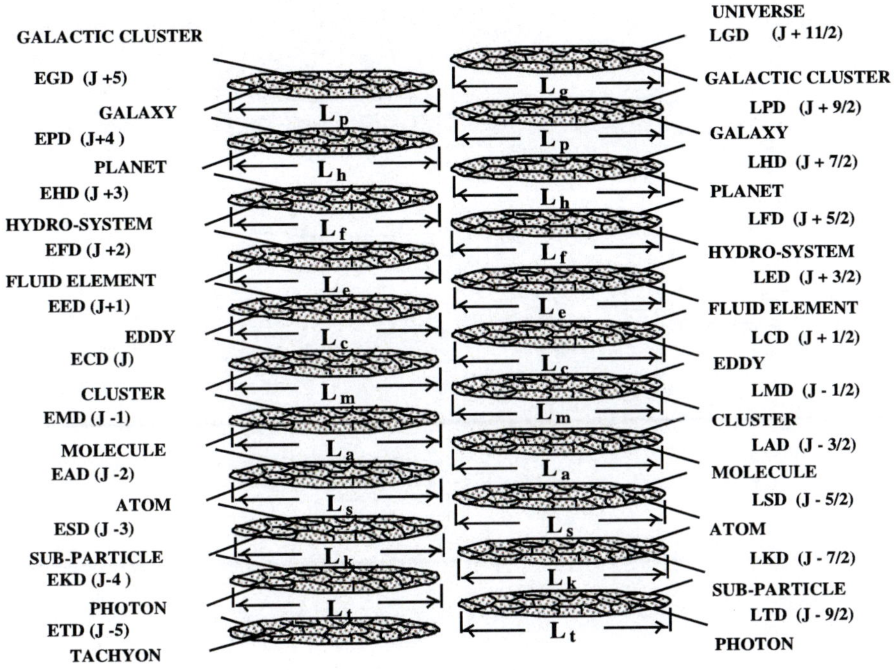

Fig. 1 A scale-invariant model of statistical mechanics. Equilibrium-β-dynamics on the left-hand-side and non-equilibrium laminar-β-dynamics on the right-hand-side for scales β = g, p, h, f, e, c, m, a, s, k, and t as defined in [37]. Characteristic lengths of (system, element, "atom") are (L_β, λ_β, ℓ_β) and λ_β is the mean-free-path

A magnified view of part of hierarchy of statistical fields in Fig. 1 is shown in Fig. 2 where atomic, element, and system velocities of stochastic fields are better revealed.

Following the classical methods [20, 39–41], the scale-invariant forms of mass, thermal energy, linear and angular momentum conservation equations at scale β are given as [34, 35]

$$\frac{\partial \rho_{i\beta}}{\partial t_\beta} + \nabla.(\rho_{i\beta}\mathbf{v}_\beta) = \Re_{i\beta} \tag{4}$$

$$\frac{\partial \varepsilon_{i\beta}}{\partial t_\beta} + \nabla.(\varepsilon_{i\beta}\mathbf{v}_\beta) = 0 \tag{5}$$

$$\frac{\partial \mathbf{p}_{i\beta}}{\partial t_\beta} + \nabla.(\mathbf{p}_{i\beta}\mathbf{v}_\beta) = -\nabla.\mathbf{P}_{ij\beta} \tag{6}$$

$$\frac{\partial \boldsymbol{\pi}_{i\beta}}{\partial t_\beta} + \nabla.(\boldsymbol{\pi}_{i\beta}\mathbf{v}_\beta) = \rho_{i\beta}\boldsymbol{\omega}_\beta.\nabla\mathbf{v}_\beta \tag{7}$$

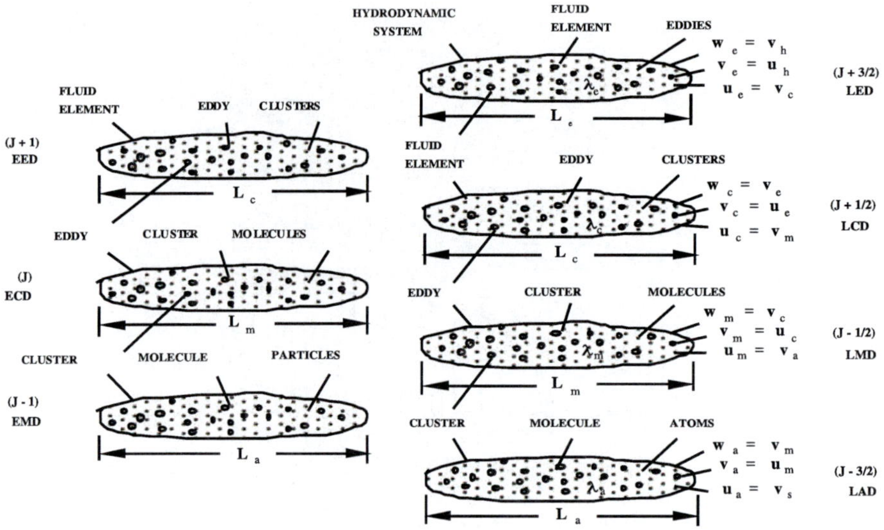

Fig. 2 Hierarchy of embedded statistical fields from LAD to LED scales with atom $\mathbf{u}_\beta$, element $\mathbf{v}_\beta$, and system $\mathbf{w}_\beta$ velocity

involving the *volumetric density* of thermal energy $\varepsilon_{i\beta} = \rho_{i\beta}\tilde{h}_{i\beta}$, linear momentum $\mathbf{p}_{i\beta} = \rho_{i\beta}\mathbf{v}_{i\beta}$, and angular momentum $\boldsymbol{\pi}_{i\beta} = \rho_{i\beta}\boldsymbol{\omega}_{i\beta}$ (since $r_{a\beta-1} = 1$). Also, $\Re_{i\beta}$ is the chemical reaction rate and $\tilde{h}_{i\beta} = \hat{h}_{i\beta}/m_\beta$ is the absolute enthalpy [37]. It is noted that the time coordinates in (4–7) also have a scale subscript β.

3 Generalized Thermodynamics, Helmholtz Decomposition of Thermal Energy and Momentum

To arrive at scale-invariant model of Boltzmann statistical mechanics with (atom, element, system) velocities ($\mathbf{u}_{i\beta}$, $\mathbf{v}_{i\beta}$, $\mathbf{w}_{i\beta}$), one requires a stable "atom" stabilized by an external pressure acting as *Poincaré stress* [36, 37]. Next, atoms with different energy $\varepsilon_j = h\nu_j$ are grouped in atomic-clusters or elements (energy levels) of various sizes. Atoms of various energy are under constant motion and allowed to make transition between elements by emitting/absorbing sub-particles. The question is, given the total number of atoms N of an ideal gas and the total energy H, with mean atomic enthalpy $\hat{h}_w = 4kT$ hence $H = 4NkT$ [33], what distribution of element sizes corresponding to various atomic energy, speed, and momenta leads to *stochastically stationary field.* Such state of thermodynamic equilibrium corresponds to energy, speed, and velocity of particles being governed by invariant Planck, Maxwell–Boltzmann, and Gauss (Maxwell) distribution functions [37].

$$\frac{\varepsilon_\beta dN_\beta}{V} = \frac{8\pi h}{u_\beta^3} \frac{\nu_\beta^3}{e^{h\nu_\beta/kT} - 1} d\nu_\beta \tag{8}$$

$$\frac{dN_{u\beta}}{N} = 4\pi \left(\frac{m_\beta}{2\pi k T_\beta} \right)^{3/2} u_\beta^2 e^{-m_\beta u_\beta^2/2kT_\beta} du_\beta \tag{9}$$

$$f_\beta(v_\beta) = \left(\frac{m_\beta}{2\pi k T_\beta} \right)^{3/2} e^{-m_\beta u_\beta^2/2kT_\beta} \tag{10}$$

Also, due to the closure of the gap between photon gas in Planck equilibrium radiation theory and ideal gas in Boltzmann kinetic theory, just like photon gas, the potential and internal energy of ideal gas are related as [38]

$$p_\beta V_\beta = \frac{U_\beta}{3} = \frac{3Nk T_\beta}{3} = Nk T_\beta \tag{11}$$

Hence, Sommerfeld [44] "total thermal energy" or *enthalpy* H_β for an ideal gas is the sum of internal energy and potential energy [33]

$$H_\beta = U_\beta + p_\beta V_\beta = U_\beta + U_\beta/3 = \frac{4}{3} U_\beta = 4N_\beta k T_\beta \tag{12}$$

By (11)–(12), enthalpy could also be expressed as

$$\begin{aligned} H_\beta &= U_\beta + \frac{U_\beta}{3} = U_\beta + \frac{H_\beta}{4} = U_\beta + \frac{1}{4}\left(U_\beta + \frac{U_\beta}{3} \right) = \cdots \\ &= U_\beta \left(1 + \frac{1}{4} + \frac{1}{4^2} + \frac{1}{4^3} + \cdots \right) = \frac{4}{3} U_\beta \end{aligned} \tag{13}$$

involving Archimedes [45] theorem on the area under a parabolic segment

$$1 + \frac{1}{4} + \frac{1}{4^2} + \frac{1}{4^3} + \cdots = \sum_{n=0}^{\infty} \frac{1}{4^n} = \frac{4}{3} \tag{14}$$

With frequency made dimensionless through division by the most probable or Wien frequency ν_w, re-normalized Planck [46] distributions at two adjacent scales appear as shown in Fig. 3.

It is known that precisely 3/4 and 1/4 of the total thermal energy under Planck distribution curve in Fig. 3 fall on $\nu > \nu_w$ and $\nu < \nu_w$ sides of Wien frequency ν_w [33]. In his pioneering study, Helmholtz [47] decomposed the total thermal energy or enthalpy H into what he called *free-heat* and *latent-heat* that were recently identified as *internal energy U* and *potential energy pV* [33]

$$H_\beta = U_\beta + p_\beta V_\beta \tag{15}$$

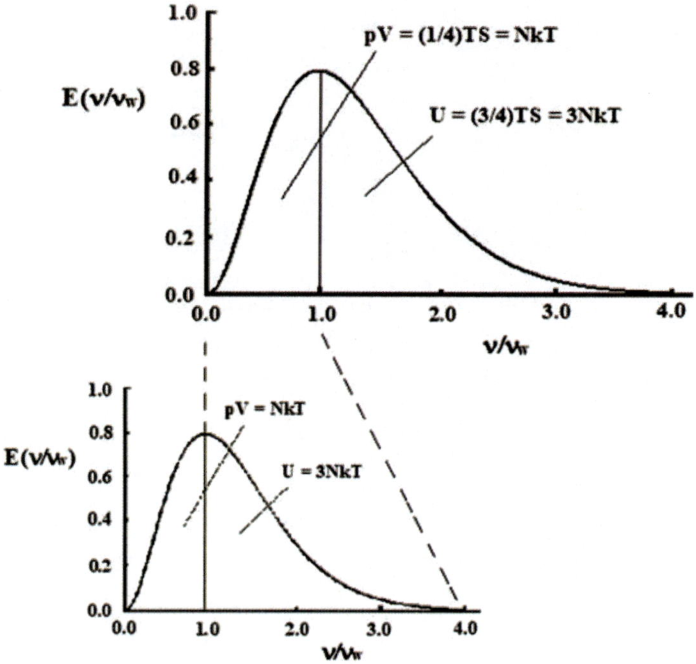

Fig. 3 Re-normalized Planck energy distributions as a function of ν/ν_w

Hence, the result in (14) is in exact agreement with the pioneering prediction by Hasenöhrl [48, 49] of the relation between total energy and *electromagnetic energy* or *dark-energy* expressed as

$$H_\beta = \mathrm{M_t c^2} = \frac{4}{3}\mathrm{M_{EM}c^2} \tag{16}$$

such that total mass relates to electro-magnetic mass by [38]

$$\mathrm{M_t} = \frac{4}{3}\mathrm{M_{EM}} \tag{17}$$

Since in equilibrium radiation within enclosures photons are at "stationary state", their speed is Wien speed

$$\mathrm{v_{kw}^2 = v_{kw^+}^2 + v_{kw^-}^2 = 2v_{kw+}^2} \tag{18}$$

that is related to the root-mean square speed by [33]

$$\mathrm{c^2 = c_{rk}^2 = 3v_{kw+}^2} \tag{19}$$

Therefore, Lorentz [50] relativistic mass also leads to prediction of Hasenöhrl [48, 49]

$$M_t = \frac{M_{EM}}{(1 - v_{kw}^2/c^2)^{1/2}} \simeq M_{EM}\left(1 + \frac{v_{kw}^2}{2c^2}\right) = M_{EM}\left(1 + \frac{1}{3}\right) = \frac{4}{3}M_{EM} \quad (20)$$

As discussed in [32], if *potential energy* is identified as *dark-matter* or *gravitational mass*

$$p_\beta V_\beta = M_{DM\beta}c^2 = \frac{1}{3}M_{EM\beta}c^2 = \frac{1}{4}M_{t_\beta}c^2 = N_\beta k T_\beta \quad (21)$$

total mass becomes the sum of *dark energy* or *electromagnetic mass* M_{EM}, and *dark matter* or *gravitational mass* M_{GM} [32, 51–54]

$$M_{t\beta} = M_{EM\beta} + M_{GM\beta} = \frac{3}{4}M_{t\beta} + \frac{1}{4}M_{t\beta} \quad (22)$$

It is most interesting to note that, as discussed in [32], Helmholtz [47] decomposition of system thermal energy at thermodynamic equilibrium in (19) also extends to cosmological scale (Fig. 1) and is in accordance with predictions of general theory of relativity [55, 56] as described by Pauli [56]

> The energy of a spatially finite universe is three-quarters electromagnetic and one-quarter gravitational in origin

In addition, predicted fractions 3/4 and 1/4 for *dark energy* and *dark matter* in (19) are in good agreement with recent cosmological observations [57–60].

On the other hand, according to Planck [46] energy distribution (Fig. 3) and (18), dark matter of scale β is the total energy or enthalpy of the next lower scale $\beta - 1$.

$$(Dark\ Matter)_\beta = (Total\ Energy)_{\beta-1}$$

Also, it is known that when particles form "cooper pairs" and behave as composite bosons [61, 62], one can derive Schrödinger equation from invariant Bernoulli equation for potential incompressible flow [38]. Hence, following classical methods [63, 64], quantum mechanics wave function may be expressed as products of translational, rotational, vibrational, and potential (internal) wave functions as [62]

$$\boldsymbol{\Psi}_\beta = \boldsymbol{\Psi}_{t\beta}\boldsymbol{\Psi}_{r\beta}\boldsymbol{\Psi}_{v\beta}\boldsymbol{\Psi}_{p\beta} = \boldsymbol{\Psi}_{t\beta}\boldsymbol{\Psi}_{r\beta}\boldsymbol{\Psi}_{v\beta}\boldsymbol{\Psi}_{\beta-1} = \cdots \quad (23)$$

At cosmological scale Ψ_g, the wave-particle duality of galaxies is evidenced by their observed quantized red-shifts [65]. Therefore, the scale-invariance of the model (Fig. 1) and (23) lead to hierarchy of embedded statistical fields with translational TKE, rotational RKE, vibrational VKE kinetic energy (dark energy) and potential energy PE (dark matter) resulting in energy cascade from cosmic to photonic scales shown in Fig. 4.

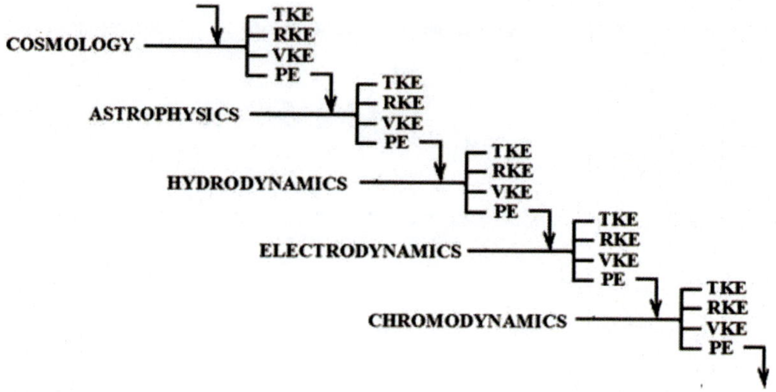

Fig. 4 Cascades of kinetic energy TKE, RKE, VKE (dark energy), and potential energy PE (dark matter) from cosmic to photonic scales

Following Helmholtz [47], one can consider *decomposition of momentum* in Normalized Maxwell–Boltzmann (NMB) distribution as a function of dimensionless speed with respect to the most-probable or Wien speed $v/v_w = \lambda_w/\lambda$ shown in Fig. 5.

As compared with (3/4, 1/4) division of energy in Planck curve in Fig. 3, the division of momentum on either side of Wien speed in Fig. 5 is (2/3, 1/3). In view of the equality of translation kinetic and potential energy due to Boltzmann equipartition principle, $m_\beta v_{wx+\beta-1}^2 = m_\beta V_{x\beta}'^2$ [38], three components of momentum are equal due to what is called *principle of equipartition of translational momenta*

$$\overline{p}_{x+} = mv_{x+} = \overline{p}_{x-} = mv_{x-} = \overline{p}_x' = mV_x' \tag{24a}$$

Therefore, for an ideal gas, of the total dimensionless translational momentum $\overline{P}_x = (\overline{p}_{x+} + \overline{p}_{x-} + \overline{p}'_x)/\overline{p}_{xw} = 3$ under NMB curve in Fig. 5, 2/3 is on $v > v_w$ side of the Wien speed and is associated with root-mean-square momenta due to atomic translational velocity in (x_+, x_-) directions, and 1/3 is on $v < v_w$ side of the Wien speed and is associated with peculiar translational momentum hence,

$$\overline{p}_x = \overline{p}_{x+} + \overline{p}_{x-} + \overline{p}_x' = \frac{2}{3}\overline{P}_x + \frac{1}{3}\overline{P}_x \tag{24b}$$

Parallel to the concept of *dark-matter* in Helmholtz energy decomposition, for decomposition of momentum the second part of (24b) may be referred to as *dark-momentum.*

As discussed in a recent study [66], once physical variables are made dimensionless, *particular* problems of physics become *universal* problems of mathematics and the nature of the specific physical entities being studied no longer matters. As an example, the distribution of annual personal income in economic systems is considered. In a recent study by Roper [67], it is suggested that the *log-Verhulst* distribution function fits the data better than does the *lognormal* distribution function. In view

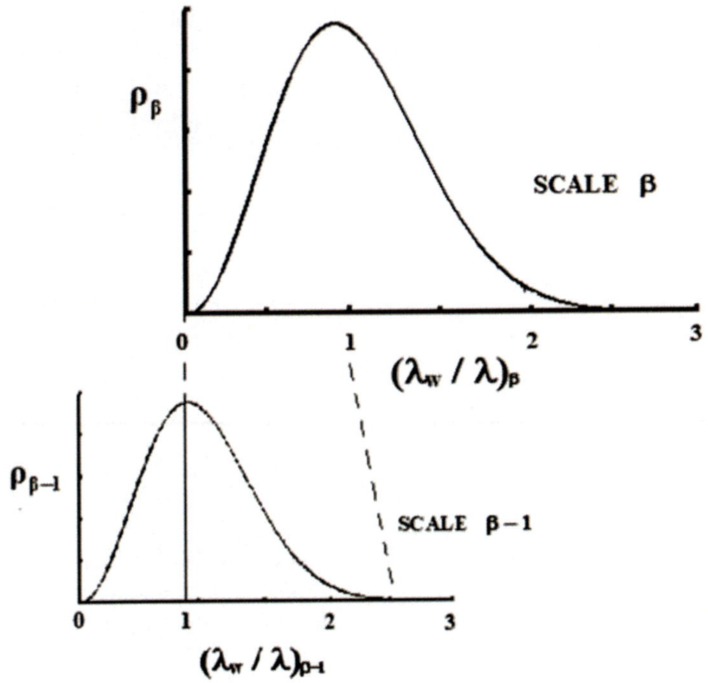

Fig. 5 Normalized Maxwell–Boltzmann distribution as a function of dimensionless speed $v/v_w = \lambda_w/\lambda$ [37]

of random nature of economic systems, in some economic literature Gauss normal distribution is considered as "ideal" or optimal income distribution. However, typical data of annual personal income distribution [68] shown in Fig. 6 clearly indicate the non-Gaussian character of actual income distribution.

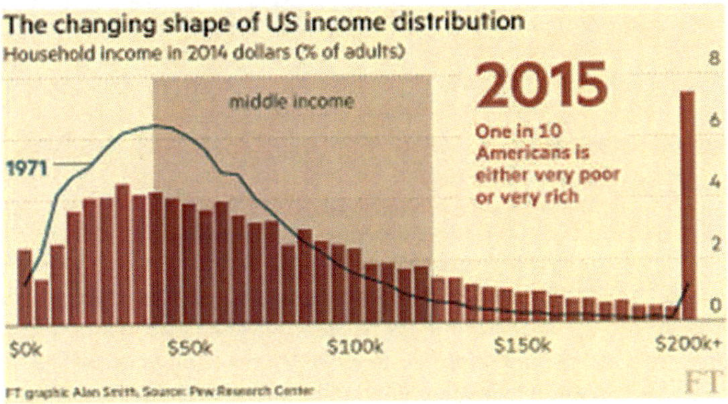

Fig. 6 Comparison between annual income distributions in 1971 and 2015 [68]

If the income in Fig. 6 is made dimensionless by division with the most probable income I_{mp}, the distribution will become similar to Planck energy distribution since I/I_{mp} is in the range (1–4) in Fig. 3 rather than the range (1–3) in Fig. 5. Hence, with dimensionless personal annual income (I/I_{mp}) viewed as dimensionless frequency (ν/ν_w), renormalized Planck distribution could be considered as the optimum or "*ideal*" income distribution because it corresponds to an equilibrium i.e., *maximum entropy* state as discussed below. It is reasonable to anticipate that Gauss normal distribution will govern the vector field corresponding to "velocity" of money flow between various income levels (energy levels), in analogy to kinetic theory of ideal gas [37]. Rather than individuals, at larger scales of companies (corporations), one expects similar normalized Planck distribution of income (like Fig. 3) with thousands (millions) of dollars instead of dollars as "atomic" units exchanging between various income-levels of companies (corporations).

As discussed in [32, 33, 37, 38], in accordance with Boltzmann kinetic theory of ideal gas and Planck theory of photon gas [46], one asks the question: given a total amount of money M and total number of income earners N, what is the distribution of number of income earners N_j with income I_j that results in *stochastically stationary* i.e., equilibrium economic system. On the other hand, entropy of an ideal gas was recently identified [32, 33] as the maximum number of Heisenberg-Kramers [69] *virtual oscillators* $S = 4N\text{k}$, given the total system energy or Hamiltonian i.e., enthalpy $TS = H = 4N\text{k}T$. Therefore, maximization of entropy in Planck [46] distribution ensures that the total energy (total monitory wealth) is distributed amongst *maximum number of oscillators* (income levels). In such *quantum mechanical economic model*, the transfer of energy (money) between different energy levels (income levels) will be governed by Schrödinger equation such that at equilibrium all income levels will be in "*stationary states*".

In Fig. 6, one notes the shifting of income from middle-class to upper-class from 1971 to 2015 that constitutes a departure from equilibrium thus having a destabilizing effect on the socio-economic system. The unfortunate *delta function* at the maximum income level in Fig. 6 is even more embarrassing departure from Planck optimal distribution thus further enhancing economic instabilities that may lead to future revolutionary (*quantum*) change in the socio-economic system.

4 Thermodynamic Definition of Spacetime and the Nature of Rovelli Thermal Time

Since Aristotle [70] and St Augustine [71], the nature of physical time has remained a mysterious problem of physics. The central insight of Aristotle namely "*the concept of time without change is meaningless*", although correct remained puzzling and circular since the concept of *change* itself involves the notion of time. The hierarchy of time durations encountered from cosmic to photonic scales (Fig. 1) is described in an excellent recent book by 't Hooft and Vandoren [72]. Although the pioneering insights of Poincaré [73–76], Lorentz [50], Einstein [77], and Minkowski [78] resulted in introduction of the concept of spacetime as a 4-dimentional manifold, the exact physical nature of such mathematical concept remained obscure. Also, even though Einstein [79] general theory of relativity (GTR) attributed a dynamic character to spacetime, the very notion of existence of time was questioned in what is known as the "*time problem*" of GTR [80–90].

In a recent study [91], the nature of physical space and time was investigated and the concepts of *internal spacetime* versus *external space and time* were introduced. Assuming that a statistical field at scale β is in thermodynamic equilibrium with the physical space at scale $(\beta - 1)$ within which it resides, both fields will have a homogenous constant temperature $T_\beta = T_{\beta-1}$ defined in terms of Wien wavelength of particle thermal oscillations as [33]

$$m_\beta u_\beta^2 = m_{\beta-1} v_{w\beta-1}^2 = kT_{\beta-1} = k\lambda_{w\beta-1} \tag{25}$$

Hence, by definition of most-probable or Wien speed $v_{ws} = \lambda_{ws}\nu_{ws} = \lambda_{ws}/\tau_{ws}$, one can associate constant internal measures of (extension, duration)

$$\begin{cases} \lambda_{ws} \text{ Internal measure of extension} \\ \tau_{ws} \text{ Internal measure of duration} \end{cases} \tag{26}$$

with every "point" or "atom" of space in a universe at constant temperature $T_\beta = T_{\beta-1}$. Next, external space and time that are independent of each other are defined in terms of the internal spacetime in (26). For example, at cosmic scale $\beta = g$, one employs internal (ruler, clock) of the lower scale of astrophysics $\beta = s$ to define *external space and time* coordinates defined as [91]

$$(x_\beta, y_\beta, z_\beta) = (N_{x\beta}, N_{y\beta}, N_{z\beta})\lambda_{wx\beta-1}, \quad t_\beta = N_{t\beta}\tau_{w\beta-1} \tag{27}$$

with the four numbers $(N_{x\beta}, N_{y\beta}, N_{z\beta}, N_{t\beta})$ being independent numbers. Whereas internal spacetime in (26) provides *local* structure of spacetime, the external space and time in (27) describe *global* dynamics of the system and are irreversible. Also, according to (26–27) both internal and external space and time are quantized. The four dimensions $(x_\beta, y_\beta, z_\beta, t_\beta)$ with three real space and one imaginary time coordinates represent Poincaré [75] and Minkowski [78] four-dimensional spacetime manifold.

Recently, the author became aware of a number of wonderful books and articles by Rovelli [92–96] and consequently learned about his much earlier pioneering contributions to the understanding of the nature of time in general and what he called *thermal time* in particular. Clearly, the definition of spacetime in (26) is in accordance with the perceptions of Rovelli [95]

> The theory seems to indicate that there is no explanation for the peculiar properties of the time variable at the mechanical level. We suggest that such an explanation should be searched at

the thermodynamical level. We propose the idea that the very concept of time is meaningful only in the thermodynamic context.

It is emphasized that the definition of internal spacetime in (26) is based on *thermodynamic equilibrium* corresponding to *stochastically stationary* thus *time-reversible* state. The objective is to define what the variable called physical time represents as noted by Rovelli [95]

It may seem strange that time is determined by an equilibrium state, since in an equilibrium state the system loses track of the direction (the versus, the arrow) of time. However, we are not concern here with versus of the time flow: we are concerned with definition of a variable that represents time, which is a very different problem.

Therefore, the external or *physical time* quantitatively defined in (27) is called *Rovelli thermal time*. Of course, because entropy generation due to various dissipations in all real systems lead to change in temperature, the internal measures of spacetime in (26) will also change. For example, in cosmology, the internal measure of spacetime change extremely slowly (eons) due to dissipations and the expansion of universe.

In another recent investigation by Rovelli [96] concerning general relativistic statistical mechanics, thermodynamic temperature was related to the ratio between the thermal time τ and physical time t as

$$T = \frac{\tau}{t} \tag{28}$$

Since dimension of absolute temperature is meters $T = \lambda_w$, (28) *appears* to be dimensionally non-homogeneous. To reveal the nature of dimensional homogeneity of (28) we consider the velocity ratio

$$\frac{v}{v_w} = \frac{x/t}{\lambda_w/\tau_w} = \frac{\tau_w}{t}\frac{x}{\lambda_w} = N \tag{29}$$

When the external spatial extension or length is defined as $x = N\ell_x m$, (29) simplifies as

$$\frac{\tau_w}{t} = \frac{N\lambda_w}{x} = \frac{N\lambda_w}{N\ell_x} = \frac{\lambda_w}{\ell_x} = \lambda_w = T \tag{30}$$

Equation (30) that is *dimensionally homogeneous* becomes identical to (28) because of the choice of the metric or *unit of length* $\ell_x = 1$ m. Therefore, (30) in effect requires that the *unit of length* (say meters) for external spatial coordinate x be *identical* to the internal unit employed to express absolute temperature $T = \lambda_w m$.

According to (27), since external (ruler, clock) = (x_β, t_β) at scale β within the hierarchy (Fig. 1) are always defined in terms of internal (ruler, clock) = $(\lambda_{w\beta-1}, \tau_{w\beta-1})$ at the next lower scale $\beta - 1$, the definition of (extension, duration) = (space, time) could be relegated to lower scales ad infinitum. This is because infinite divisibility of both extension and duration must follow the philosophy of Anaxagoras [97]

> Neither is there a smallest part of what is small, but there is always a smaller, for it is impossible that what is should ever cease to be

Therefore, both absolute zero and absolute infinite (extension, duration) are singularities as ideal Aristotle potential limits never actualized as discussed in [66]. The fundamental and quantum nature of both space and time and their role in constitution of matter in quantum field theory and GTR will be further discussed in the following section.

5 Universality of Quantum Mechanics and the Nature of Wave-Particle Duality

The fact that Boltzmann anticipated quantum mechanics by about three decades is evidenced by the following quotation taken from his pioneering and often neglected 1872 paper [98]

> We wish to replace the continuous variable x by a series of discrete values ε, 2ε, 3ε ... pε. Hence, we must assume that our molecules are not able to take up a continuous series of kinetic energy values, but rather only values that are multiples of a certain quantity ε. Otherwise, we shall treat exactly the same problem as before. We have many gas molecules in a space R. They are able to have only the following kinetic energies:
>
> $$\varepsilon, 2\varepsilon, 3\varepsilon, 4\varepsilon, \ldots, p\varepsilon$$
>
> No molecule may have an intermediate or greater energy. When two molecules collide, they can change their kinetic energies in many different ways. However, after the collision the kinetic energy of each molecule must always be a multiple of ε. I certainly do not need to remark that for the moment we are not concerned with a real physical problem. It would be difficult to imagine an apparatus that could regulate the collisions of two bodies in such a way that their kinetic energies after a collision are always multiples of ε. That is not a question here.

Although more recent theoretical understanding of quantum mechanics based on fundamental contributions of its founders Planck [46, 99], Einstein [100], Bohr [101], de Broglie [2–4], Heisenberg [102], Dirac [103], Schrödinger [104], Pauli [101], and Born [105] is fully established, its physical and intuitive understanding encounter difficulties due to abstract nature of its mathematical foundation. As a result, the theory confronts many problems associated with its *physical interpretation* such as

1. The nature of wave function and its probabilistic interpretation.
2. Wave-particle-duality.
3. Particle–particle entanglement.
4. Double-slit experiment.
5. EPR and problem of action-at-a-distance.
6. Quantum-jumps and trajectory problems.
7. Schrödinger cat.

among others.

The problem of lack of intuitive understanding of quantum mechanics mentioned above extends to Newton [106] law of gravitation, Einstein [79] general theory of relativity, and Dirac [107] equation of quantum field theory. This is because, similar to quantum mechanics, such mathematical theories were based on certain desired mathematical properties, such as the inverse square law, the equivalence principle, or relativistic wave equation with positive probability, rather than derivation from the first principles. As a result, in spite of excellent predictive power of the theories, the exact connection between abstract mathematical models and the physical phenomena they aim to reveal remain obscure.

Recent investigations [36, 37] were focused on connections between energy spectrum of photon gas given by Planck [99] distribution and both energy and dissipation spectrum of isotropic stationary turbulence. Thus, the gap between the problems of quantum mechanics and hydrodynamics was closed through connections between Cauchy, Euler, Bernoulli equations of hydrodynamics, Hamilton–Jacobi equation of classical mechanics, and finally Schrödinger equation of quantum mechanics. This resulted in recent derivation of invariant time-independent and time-dependent Schrödinger equations from invariant Bernoulli equation for potential incompressible flow [38]

$$\frac{\hbar^2}{2\mathrm{m}_\beta}\nabla_\xi^2\psi_\beta + (\tilde{H}_\beta - \tilde{V}_\beta)\psi_\beta = 0 \tag{31}$$

$$i\hbar\frac{\partial\Psi_\beta}{\partial t_\beta} + \frac{\hbar^2}{2\mathrm{m}_\beta}\nabla_\xi^2\Psi_\beta - \tilde{V}_\beta\Psi_\beta = 0 \tag{32}$$

The quantum mechanics wave function is defined as [38]

$$\Psi_\beta(\xi, t) = \rho^{1/2}\Phi'_\beta(\xi, t) = \rho^{1/2}\mathrm{e}^{ik_r\xi}\mathrm{e}^{-i\tilde{H}t_\beta/\hbar} \tag{33}$$

such that $\Psi_\beta\Psi_\beta^* = \rho_\beta$ accounts for *particle localization* in accordance with classical results [108]. The velocity potential Φ'_β of peculiar velocity that is *complex* accounts for normalization and hence the success of Born [105] *probabilistic interpretation* of Ψ_β. In the following, some implications of the model to the resolution of problems in the list (1–7) above concerning interpretation of quantum mechanics will be briefly examined.

According to the invariant model of Boltzmann statistical mechanics, each equilibrium statistical field (Fig. 1) is composed of a spectrum of cluster or wave-packet sizes containing “atoms” with velocity, speed, and energy respectively following Gauss, Maxwell–Boltzmann, and Planck distribution functions. As discussed in [38], the conventional field of fluid dynamics does not correspond to equilibrium molecular dynamics EMD $\beta = \mathrm{m}$ but rather to the next higher scale of equilibrium cluster-dynamics ECD $\beta = \mathrm{c}$. Hence, in stationary fluid at ECD scale, Maxwell–Boltzmann distribution function governs the spectrum of cluster sizes that are *stochastically stationary*. Random motion of clusters accounts for the Brownian motion of small

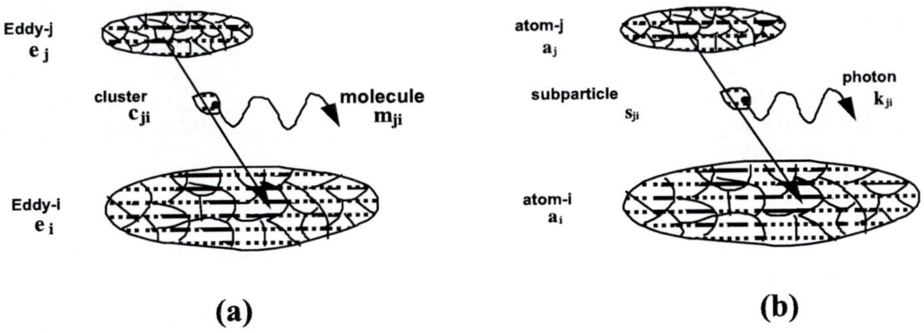

Fig. 7 **a** Transition of cluster c_{ij} from eddy-j to eddy-i leading to emission of molecule m_{ij}. **b** Electron transitions with emission of photon γ_{ji} [38]

suspensions that is known to be a non-dissipative stationary phenomenon [36]. Transition of a cluster from a small fast-oscillating "eddy" (high energy-level-j) to a large slowly-oscillating "eddy" (low energy-level-i) results in emission of a "sub-particle" that is a molecule to carry away the excess energy in harmony with Bohr [101] frequency formula $\Delta\varepsilon_{ji\beta} = h(\nu_{j\beta} - \nu_{i\beta})$ as schematically shown in Fig. 7a.

Similarly, but at the much smaller scale of ESD $\beta = s$ or stochastic electrodynamics (SED), transition of an electron from high to low energy atoms lead to emission of a sub-particle namely photon γ_{ji} as shown in Fig. 7b.

As described in [38], derivation of invariant Schrödinger equation from invariant Bernoulli equation results in a new paradigm of physical foundation of quantum mechanics. Considering the case of stationary fluid or equilibrium cluster-dynamics ECD $\beta = c$, the quantum mechanics wave function Ψ_c relates to the velocity potential of particle peculiar velocity. However, particle or "atom" of ECD field namely cluster is the most-probable molecular cluster by definition $\mathbf{u}_c = \mathbf{v}_{wm}$ in (1). Therefore, particle of scale β is identified as the most probable *wave-packet* of the lower scale $\beta - 1$. Hence, each statistical field will have a quantum mechanics wave function Ψ_β and particle P_β that is stationary wave-packet of the lower scale

$$\begin{cases} \Psi_\beta = (\text{Wave Function}) = \text{WF}_\beta \\ P_\beta = (\text{Particle})_\beta = (\text{Wave Packet})_{\beta-1} = \text{WP}_{\beta-1} \end{cases} \tag{34}$$

In harmony with de Broglie [2–4] picture of quantum mechanics, motion of "particle" P_β as *local* singularity identified as *wave-packet* $= \text{WP}_{\beta-1}$ of lower scale is *guided* by a *global* external quantum mechanics wave function Ψ_β as schematically shown in Fig. 8.

Hence, at any scale within the hierarchy of statistical fields in Fig. 1, the solution of Schrödinger equation gives the energy spectrum of "atomic" clusters that represent Bohr [101] *stationary states* or energy levels of the field.

When one moves to the next lower scale of equilibrium molecular dynamics EMD $\beta = m$, derivation of Schrödinger equation [38] involves a stationary coordinate

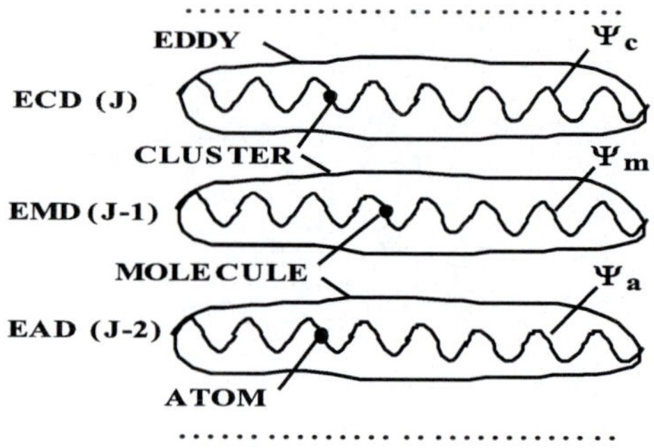

Fig. 8 Macroscopic wave functions Ψ_β or de Broglie guidance waves at (ECD), (EMD), and (EAD) scales that guide particles identified as wave-packets $(\mathrm{WP})_{\beta-1}$ or de Broglie matter-waves [38]

moving with velocity $\mathbf{v}_{wa}$ since $\mathbf{v}_m = \mathbf{u}_m - \varepsilon \mathbf{V}'_m = \mathbf{v}_{wa} - \varepsilon \mathbf{V}'_m$ by (1–2). Because by (33) Ψ_m relates to the velocity potential of molecular peculiar velocity $\mathbf{V}'_m$, under thermodynamic equilibrium $\mathbf{v}_{wm} = \mathbf{u}_c$ will also be related to $\mathbf{V}'_m$ thus connecting Ψ_m with P_c. As a result, particle P_β of the upper scale is identified as quantum mechanics wave function of the lower scale $\Psi_{\beta-1}$ and one arrives at a hierarchy of embedded wave functions expressed as

$$\begin{cases} P_\beta = \Psi_{\beta-1} = (\text{Wave Function})_{\beta-1} = \mathrm{WF}_{\beta-1} \\ P_{\beta-1} = (\text{Wave Packet})_{\beta-2} = \mathrm{WF}_{\beta-2} \end{cases} \tag{35}$$

According to (35) and Figs. 1 and 6, the universe is composed of hierarchies of embedded waves suggesting that the entire universe was formed when the Almighty decided to make some waves!

The results in (34–35) and Fig. 8 may help in the understanding of the first and second problems in the list 1–7 above. The wave-particle duality problem is understood in terms of wave function Ψ_β that globally guides motion of particles identified as wave-packet of lower scale in accordance with the perceptions of de Broglie [4]. New perspectives provided by the results in (34–35) and Fig. 8 concerning problems 1–2 are also expected to facilitate the resolution of the remaining problems 3–7. For example, problem number 6 namely absence of particle trajectories in quantum mechanics becomes understandable because as shown in Fig. 7, any particle from cluster j can make a transition to cluster i through any arbitrary trajectory with exactly the same final results, thus accounting for success of Feynman method of sum-over-paths. Regarding number 7 problem concerning Schrödinger cat, in view of probabilistic aspect of Φ'_β hence Ψ_β by (33), it is clear that any interference with the field due to a measuring instrument will disturb the velocity potential thus leading to *collapse of the wave function* Ψ_β.

Schrödinger cat problem is more challenging since it involves the phenomenon of *life* and *death* that are not understood. Since as discussed in [1] all living systems are composed of living elements, and living elements are in turn composed of living cells, one may speculate if such infinite regression leads to an "atom of life" or Leibniz "*living Monad*"! Although at present such questions are *metaphysical* and hence outside of jurisdiction of science, some aspects of the problem may be considered within the framework of quantum mechanics.

To introduce the required concepts, we need to consider an example from cosmology. It is well known that sometimes around 380,000 years after the explosion of Lemaître [109] "atom" of our universe, the Big Bang, there was decoupling of radiation field from the baryonic matter field and the present Penzias-Wilson [110] cosmic background microwave radiation temperature of 2.73 [m] is remnant of the cooling of Casimir [111] vacuum. It is also reasonable to anticipate that a living system will involve complex dynamics at EMD, EAD, ESD, EKD, END, ETD, ... scales (Fig. 1) with END denoting *equilibrium-neutrino-dynamics* at scale $\beta = n$ (not shown in Fig. 1). By invariant Schrödinger (31) and (35), hierarchies of wave functions and particles will be associated with such fields. Therefore, due to the well-known decoupling of radiation from matter field in cosmology, one cannot rule out possible *decoupling* of some fields say neutrino-dynamics (END) or lower scale of tachyon-dynamics (ETD) from the baryonic field of molecular-dynamics of living systems at the moment of death $t = t_f$. In view of the hierarchies of wave functions in (23), there will be a critical sub-photonic decoupling scale $\beta = \beta_d$ at which the cascade of wave functions in (23) will be broken

$$\ldots \Psi_{\beta=d+2}\Psi_{\beta=d+1} \Longleftrightarrow \Psi_{\beta=d}\Psi_{d-1}\Psi_{d-2}\ldots \tag{36}$$

Such an event may correspond to what Hegel referred to as the moment in which *the spirit transcends temporality* [94]. It is ironic that according to such a model, death or decoupling of Schrödinger cat corresponds to the *collapse of wave function of cat's life*! Of course, strictly speaking, according to the present model (Fig. 1), complete decoupling hence *total isolation* of any part of the universe from the rest should be impossible as noted by Boltzmann [37].

Interestingly, Feynman [112] suggested that Schrödinger equation might in fact explain life

> Often people in some unjustified fear of physics say you cannot write an equation for life. Well, perhaps we can. As a matter of fact, we very possibly already have the equation to a sufficient approximation when we write the equation of quantum mechanics:
>
> $$H\Psi = -\frac{\hbar}{i}\frac{\partial\Psi}{\partial t}$$

Although decoupling of sub-photonic statistical fields from living system at molecular-dynamic scale is plausible, regarding its connection to the mind–body problem of Descartes or Hegel's transcendence of spirit from corporal temporality, I respond by borrowing a quotation from Rovelli [93] about Plato's account of a statement by Socrates: "*I am not sure*".

At the important scale of LKD (Fig. 1) physical space or Casimir vacuum [111] is identified as a compressible fluid, Planck compressible ether [113], as discussed in [114]. A schematic diagram of physical space as states of a compressible fluid from infinite rarefaction (*white hole* WH) to infinite compression (*black hole* BH) is shown in Fig. 9.

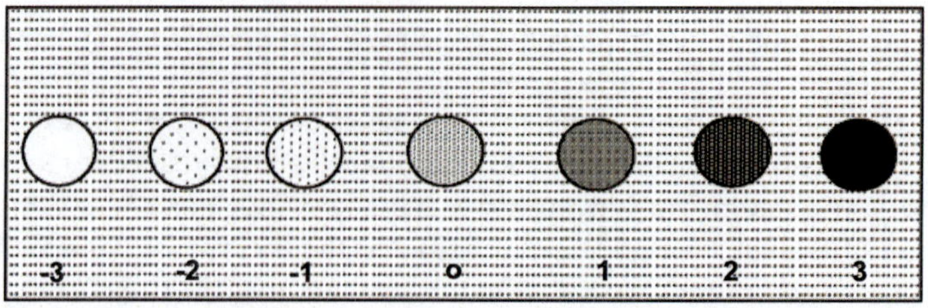

Fig. 9 Different degrees of rarefaction and compression of Casimir vacuum identified as a compressible fluid. (**−3**) White hole $\rho_{WH} = 0$(**−2, −1**) Anti-matter $\rho_{AM} < \rho_v$ (**0**) Casimir vacuum $\rho = \rho_v$ (**1, 2**) Matter $\rho_{MA} > \rho_v$ (**3**) Black hole $\rho_{BH} = \infty$ [38]

Compressibility of physical space was recently shown [115] to account for relativistic effects when Michelson number $Mi = v/c$ approaches unity thus revealing the *causal* [56] nature of Lorentz-FitzGerald contraction in accordance with Poincaré-Lorentz *dynamic theory of relativity* as opposed to Einstein *kinematic theory of relativity* in harmony with ideas of Darrigol [116] and Galison [117].

When physical space or Casimir [111] vacuum is identified as superfluid photon or Bose–Einstein condensate, it is reasonable to anticipate that when heated to a critical vaporization or *boiling temperature* T_b, the vacuum or space will nucleate what could be called photon gas bubbles that following Dirac could be also referred to as holes or anti-matter particles. Similarly, if space or Casimir vacuum cools below certain critical fusion or *melting temperature* T_m liquid photon undergoes phase transition and becomes solidified thus forming *solid-light* that was defined as black hole [118]. Therefore, in such a model, Hawking *evaporation* of BH will instead correspond to Hawking *melting* or *sublimation* of BH. Loss of mass due to melting of black hole could be caused by heating due to absorption of photon gas bubbles, anti-matter particles, that give their excess energy to melt part of BH and convert it to Casimir vacuum, i.e., space. This is in accordance with absorption of negative *curvature energy* in classical model of quantum gravity described by 't Hooft [119]

> When a black hole loses energy, it is primarily because it absorbs negative amounts of 'curvature energy'. Clearly, our primordial model must allow for the presence of negative amounts of energy. Actually, this is obviously true for quantum mechanical energy, because, after diagonalization, the total Hamiltonian has zero eigenvalue. Prior to diagonalization of the total H, the Hamiltonian density $\mathcal{H}(x)$ must have negative eigenstates. We now see that, since the black hole must lose weight, the primordial model must also have local fluctuations with negative 'curvature energy'. Black holes absorb negative amounts of energy, allowing positive energy to scape to infinity.
>
> It is due to the postulated thermodynamical stability that the fluctuations surviving at spatial infinity may only have positive energy. Since the total energy balances out, the black hole will therefore receive net amounts of negative energy falling in. Hence it loses weight and decays.

It is reasonable to expect two surfaces of event-horizon (BHH, WHH) to surround (BH, WH) with the corresponding surface temperatures (T_m, T_b). Under such a paradigm of physical space, Casimir vacuum with constant density $\rho = \rho_v$ will correspond to constant measure (*zero curvature*) Euclidean space, colder and denser $\rho_m > \rho_v$ regions correspond to matter (*positive curvature*) called Riemannian space, and finally hotter and lighter $\rho_{am} < \rho_v$ regions correspond to Dirac anti-matter (*negative curvature*) and called Lobachevskian space [38].

The new perspectives concerning the nature of physical space described in Fig. 9 as well and the identification of dimension of absolute thermodynamic temperature as length [m] associated with Wien wavelength of thermal oscillation will have a major impact on cosmology in general and physical interpretation of Einstein [79] GTR in particular. Compressible nature of physical space (Fig. 9) with "atomic" or quantum volume $\hat{v} = T^3 = \lambda_w^3$ may facilitate bridging the gap between QM and GTR since it is harmonious with modern paradigms of quantum gravity [119–121]. For example, it is reasonable to anticipate that gravitational forces will be associated with gradient of Casimir [111] vacuum density (scalar curvature) hence pressure of physical space.

In a recent study [33] it was shown that entropy of black hole is $S = 4N\mathrm{k}$ in exact agreement with prediction of Major and Setter [122]. The entropy of black hole according to Rovelli and Vidotto [123] is

$$S = \mathrm{k}\frac{4}{3}\sqrt[4]{\frac{\pi^2 \mathrm{L}^3 \mathrm{E}^3}{15(\mathrm{c}\hbar)^3}} \tag{37}$$

However, one notes that the power of four in the expression

$$\tilde{N}^4 = \frac{\pi^2 \mathrm{L}^3 \mathrm{E}^3}{15(\mathrm{c}\hbar)^3} \tag{38}$$

is due to the four degrees of freedom per oscillator associated with its translational, rotational, vibrational, and potential energy such that the actual number of oscillators is

$$\tilde{N} = \frac{\pi^2 \mathrm{L}^3 \mathrm{E}^3}{15(\mathrm{c}\hbar)^3} \tag{39}$$

From a recent study [37] on closure of the gap between photon gas in Planck equilibrium radiation and Boltzmann kinetic theory of ideal monatomic gas, the number of photons in volume V of Casimir [111] vacuum is

$$N = \frac{8\pi^5 V}{45}\left(\frac{\mathrm{kT}}{\mathrm{hc}}\right)^3 = \frac{\pi^2 \mathrm{L}^3}{45}\left(\frac{\mathrm{E}}{\mathrm{c}\hbar}\right)^3 \tag{40}$$

The results in (37), (39), and (40) give

$$S = 4N\mathrm{k} \tag{41}$$

in exact agreement with [33, 122].

The result in (41) is also in agreement with general relativistic statistical mechanics of Rovelli [96]

$$S = -\int_{\Gamma} \rho_{\beta}(\beta \ln \rho - \ln Z) = \beta E + \ln Z \tag{42}$$

when mean value of energy

$$h = -\int_{\Gamma} \ln \rho \tag{43}$$

is identified as the internal energy U that is related to Hamiltonian (enthalpy) H as

$$E = h = \frac{3}{4} H = U = 3N\mathrm{k}T \tag{44}$$

Substituting for $\beta = 1/\mathrm{k}T = 1/T$, with the assumption $\mathrm{k} = 1$ [96], the translational partition function $Z_{\mathrm{t}} = \mathrm{e}^{N}$, and the mean energy E from (44) into (42) gives the black hole entropy in (41).

Alternatively, the partition function Z in [96]

$$Z(\beta) = \int \mathrm{e}^{-\beta H} = \mathrm{e}^{-\beta F} \tag{45}$$

is the translational partition function $Z_{\mathrm{t}} = \mathrm{e}^{-\beta F}$ and $F = -N\mathrm{k}T$ is Helmholtz free energy of ideal gas [32]. Inclusion of translational, rotational, and vibrational degrees of freedom gives $Z = Z_{\mathrm{t}} Z_{\mathrm{r}} Z_{\mathrm{v}} = \mathrm{e}^{-3\beta F} = \mathrm{e}^{\beta U}$ such that the mean energy E [96]

$$E = -\frac{1}{\beta} \frac{\mathrm{d} \ln Z}{\mathrm{d}\beta} \tag{46}$$

becomes the internal energy $U = -3F = 3N\mathrm{k}T$ as in (40). Therefore, the result in (41) is also in exact agreement with entropy given by Rovelli [96] formula

$$S = \beta E - \beta F \tag{47}$$

after substitution for $\beta = 1/T$, $F = -N\mathrm{k}T$, and $E = U = 3N\mathrm{k}T$ from (46).

An outstanding problem regarding connection between thermodynamics and black hole mechanics [124–130] concerns Shannon information theory [131]

$$I = -\mathrm{K} \sum_{\mathrm{j}} \mathrm{p_j} \ln \mathrm{p_j} \tag{48}$$

and what happens to the information as matter crosses the event horizon into a black hole. For ideal monatomic gas with four degrees of freedom namely translational, rotational, vibrational, and potential, $Z = Z_r Z_r Z_v Z_p = e^{4N}$, the atomic, element, and system entropy relate to the number of Heisenberg-Kramers [69] virtual oscillators as [33]

$$\hat{s} = -k \ln \hat{p}, \quad \hat{p} = e^{-4} \tag{49a}$$

$$S_j = -k \sum_j \ln \hat{p} = -k \ln p_j, \quad p_j = e^{-4N_j} \tag{49b}$$

$$S = \sum_j S_j = -k \sum_j \ln p_j = -k \ln p, \quad p = e^{-4N} \tag{49c}$$

For a multicomponent mixture, the atomic mixture entropy is [1]

$$\hat{s}_{mix} = \frac{S}{N} = -k \sum p_j \ln p_j \tag{50}$$

Therefore, Shannon formula in (48) becomes identical to (50) of thermodynamics if one defines Shannon *measure* K in terms of Boltzmann constant as $K = N k$ and (48) gives “atomic” information as

$$\hat{i} = \frac{I}{N} = -k \sum_j p_j \ln p_j \tag{51}$$

With entropy identified as the number of Heisenberg-Kramers [69] virtual oscillators, the problem of information loss in black hole is resolved since loss of number of oscillators could be attributed to coarse-graining as photons freeze from liquid to solid phase when they cross black hole event-horizon BHH. In other words, as temperature decreases, atoms of space i.e., photons collect in larger and larger clusters, thus decreasing the number of oscillators leading to loss of entropy by (41) [1]. On the other hand, when anti-matter bubbles enter the black hole, their excess thermal energy leads to melting of part of black hole from solid into liquid photon at BHH increasing entropy and producing more Casimir [111] vacuum that accounts for observed accelerative expansion of the universe [57–60].

In view of the model of physical space in Fig. 9 and entropy of black hole in (41), it is reasonable to assume that Lemaître [109] primordial “atom” of our universe was in a state of *solid light* extremely close to absolute zero temperature hence having nearly zero entropy as Planck perfect crystal [46, 99]. This is in harmony with perceptions of Plato who believed that the world was formed from a formless state of *total chaos* or “*Tohu Vavohu*” [1]. Since according to Fig. 9 matter and anti-matter annihilate each other leaving Casimir [111] vacuum of a flat universe, in harmony with perceptions of Aristotle there is no “void” except the singularity called *white hole* (Fig. 9).

6 Universal Hydrodynamic Model of Faraday Line of Force from Cosmic to Photonic Scales

Because of the scale-invariant nature of the model (Fig. 1) the physical insights available at ordinary scales can help in the understanding of nature at much larger or much smaller scales that are less accessible to ordinary human intuition. For example, it is known that a rotating sphere in an otherwise quiescent fluid causes polar inflow jets (IJ) and equatorial outflowing disk (OD) [132] as shown in Fig. 10.

However, if the rigid sphere is stationary but instead the surrounding fluid is rotating, Huygens centrifugal forces will reverse direction resulting in accretion by inflowing disk (ID) and polar outflowing jets (OJ). Occurrence of outflow jets (OJ) from black holes is well established in cosmology.

The flow fields in otherwise stationary background fluid induced by rotating spherical particles are shown in Fig. 11.

Because of finite available energy and momentum, such flows cannot extend to infinity and instead form a finite spherical volume by outflowing equatorial disk turning around and joining the inflowing polar jets as shown in Fig. 11b resembling magnetic field lines in electrodynamics. When multiple spinning spheres are present, the hydrodynamic forces cause spinning particles to form a chain of alternating particle and "anti-particle" also called "hole" that is spherical regions of rarefaction, called *hydrodynamic model of Faraday line of force* [114] schematically shown in Fig. 12.

In a pioneering study, Dirac [133] introduced the mathematical concept of Faraday line of force as a directional line with an electron at one end and a positron at the other,

> This leads us to a picture of discrete Faraday lines of force, each associated with a charge, −e or +e. There is a direction attached to each line, so that the ends of a line that has two ends are not the same, and there is a charge −e at one end and a charge +e at the other. We may have lines of force extending to infinity, of course, and then there is no charge.

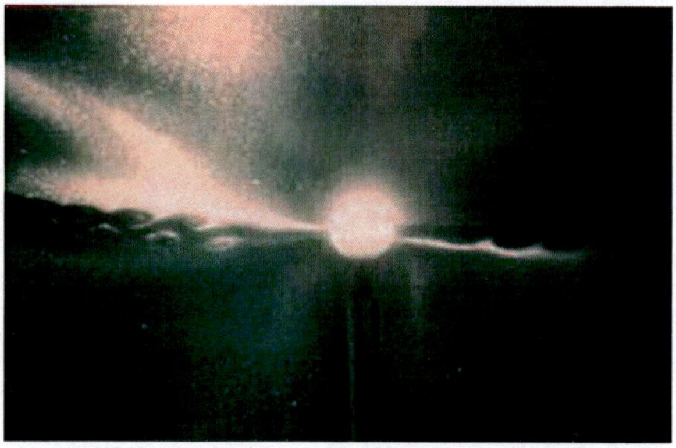

Fig. 10 Direct photograph of swirling equatorial disc outflow (DO) created by a rotating rigid sphere in otherwise stationary silicon oil with aluminum powder illuminated by laser sheet light [132]

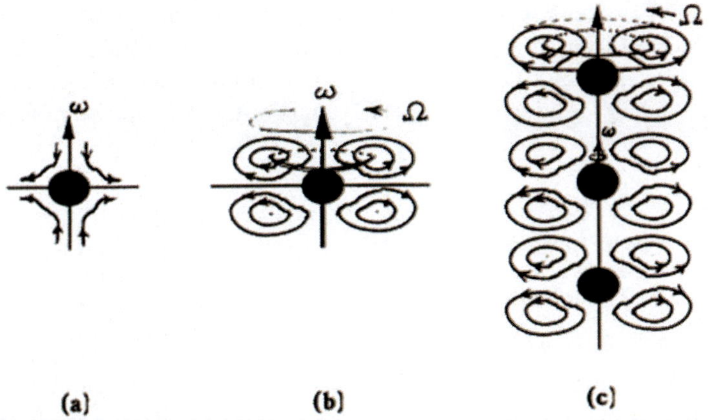

Fig. 11 Schematic model (**a**) flow near a spinning particle (**b**) locally conserved flow streamlines (**c**) formation of Faraday line of force from a row of co-spinning particles and the associated vortex field within the subquantum background fluid [132]

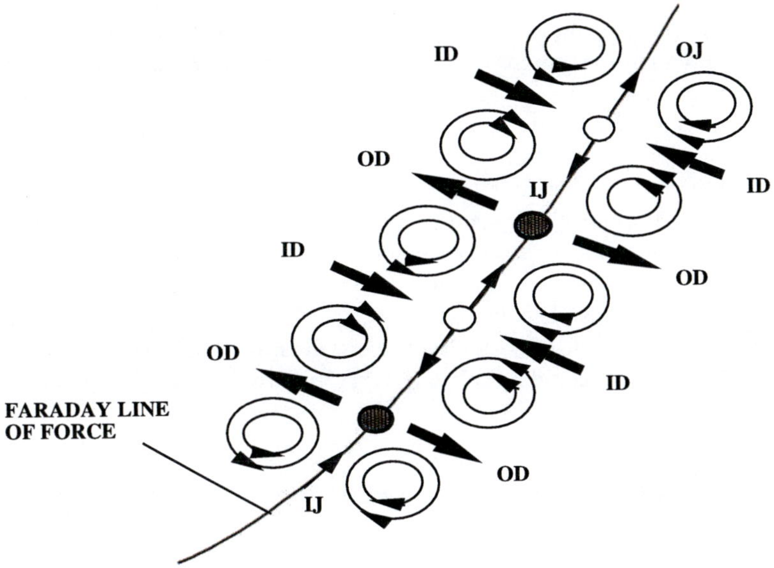

Fig. 12 Faraday line of force as electron (black) and positron (white) string with inflow jet (IJ) of one matching the outflow jet (OJ) of its neighbor. Also shown are alternating outflow (OD) and inflow discs (ID) [114]

The fluid or Casimir vacuum between two spinning spherical particles is expected to cavitate, because of strong equatorial outflowing disks from spinning particles (Figs. 11 and 12), thus forming a spherical region of vapor called "hole" or Dirac "anti-matter" particle. For example, the Faraday line of force in stochastic electrodynamics at ESD scale (Fig. 1) will be composed of an alternating chain of electron–positron as shown in Fig. 12. The breakage of such stable vortex lines is analogous to the following description of Dirac [133] concerning the breaking of Faraday line of force:

> This process - the breaking of the line of force - would be the picture for creation of an electron (e−) and a positron (e+). It would be quite a reasonable picture, and if one could develop it, it would provide a theory in which e appears as a basic quantity. I have not yet found any reasonable system of equations of motion for these lines of force, and so I just put forward the idea as a possible physical picture we might have in the future.

Similarly, but at much smaller chromodynamics (SU3) or EKD scale, the chromodynamic Faraday line of force will be identified as strings of quark-antiquark as described by 't Hooft [134]

> It took several years before it became clear that these are exactly the expressions obtained if each of these mesons is viewed as being a kind of rope with quark at one end and an antiquark at the other. The ropes can be stretched ad infinitum, because stretching them adds energy to them, which will be turned into more matter: that is, more rope.

At large hydrodynamic scales turbulent eddies are known to form vortex tubes. By Kelvin circulation theorem, it is known that such vortex lines cannot abruptly end within potential flows and instead must be either pinned to a solid boundary or else close on themselves thus forming a closed vortex "loop" in harmony with LQG [120, 121] models of quantum gravity. Such behavior is well known in superfluid helium-3 experiments revealing "quantized" vortex lines discussed in [132].

An example at molecular-dynamics scale is combustion of acetylene that results in large amounts of soot or carbon particles that form many centimeters long chains. At the much larger scale of astrophysics, it has recently been observed that our galaxy the Milky Way contains large numbers of very long star streams [135, 136]. Finally, at exceedingly large scale of cosmology, it is well known that rotating galaxies form very long streams that could be called *cosmic Faraday lines of force*.

7 Implication to Banach-Tarski Paradox

In this section, application of invariant model of Helmholtz vorticity equation [34, 35] to the interesting mathematical problem of Banach-Tarski [137] paradox (BTP) is examined. To begin, let us consider a spherical flow within a fluid droplet located at the stagnation point of axisymmetric gaseous counterflow jets as shown in Fig. 13. As seen in this figure, induced by the outer flow, two cylindrically closed recirculation flows, or two tori, form in the upper and lower hemispheres of the spherical droplet.

The streamlines for such a flow (Fig. 14a) obtained from solution of modified Helmholtz vorticity equation [34].

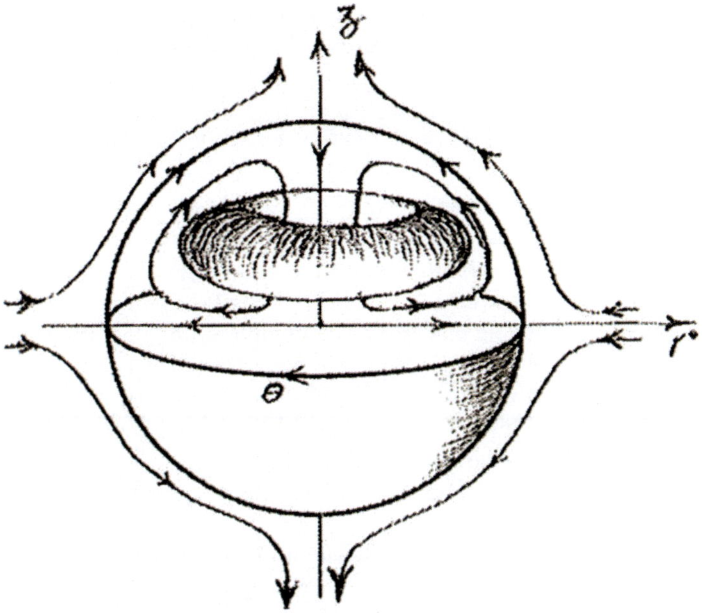

Fig. 13 Flow in liquid droplet composed of two semi-spherical Hill vortices at stagnation point of gaseous axisymmetric counterflow [34]

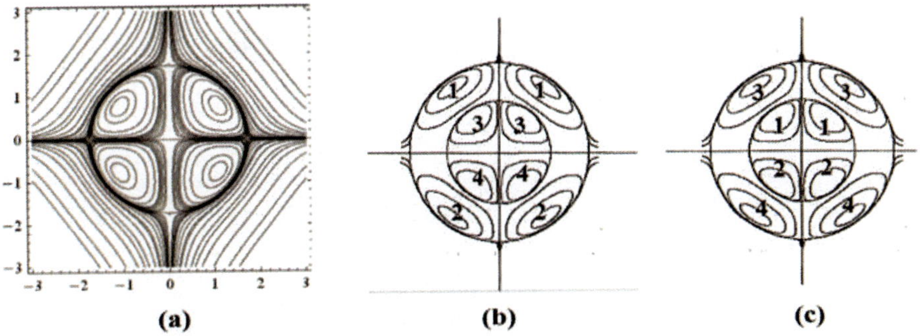

Fig. 14 **a** Streamlines from (53). Turning a sphere from (54) inside out: **b** 1-1 and 2-2 as outer tori, **c** 1-1 and 2-2 as inner tori [34]

$$w_r \frac{\partial \omega_\theta}{\partial \xi} + w_z \frac{\partial \omega_\theta}{\partial \zeta} = \frac{\omega_\theta v_r}{\xi} + \frac{\partial^2 \omega_\theta}{\partial \xi^2} + \frac{1}{\xi}\frac{\partial \omega_\theta}{\partial \xi} - \frac{\omega_\theta}{\xi^2} + \frac{\partial^2 \omega_\theta}{\partial \zeta^2} \tag{52}$$

is given by the stream function

$$\Psi = \zeta \xi^2 (1 - \xi^2/R^2 - \zeta^2/R^2) \tag{53}$$

Because of linearity of Helmholtz vorticity (52), one can apply the *superposition principle* and introduce *product solutions* for flow within two immiscible liquid droplets given by the stream function

$$\Psi = \zeta^2 \xi^4 (1 - \xi^2/R^2 - \zeta^2/R^2)(3 - \xi^2/R^2 - \zeta^2/R^2) \tag{54}$$

leading to the flow configurations shown in Fig. 14b.

First, the problem of turning a sphere inside-out is considered by looking at flow fields within two concentric spherical flows shown in Fig. 14b. Each hemisphere is composed of two semi-spherical tori, the outer toros 1-1 and the inner toros 3-3 in Fig. 14b. It is easy to imagine that one could compress the inner toros 3-3 towards vertical axis and move it upwards, while the outer toros 1-1 is stretched outwards and pushed down. When one imagines each toros as a cylindrical balloon, the above procedure changes the position of outer 1-1 and inner 3-3 tori thus turning the sphere 14b inside out as shown in Fig. 14c.

A flow configuration that results in fusion of two spheres into one sphere of identical volume is shown in Fig. 15.

As two spherical Hill vortices (S_1, S_2) in uniform flow (Fig. 15a) in opposing cylindrically-symmetric jets approach each other (Fig. 15b), they form two *semi-spherical Hill vortices* [35] and merge into a single spherical flow S_3 at the stagnation point as shown in Fig. 15c. It is possible to adjust the flow conditions such that spheres (S_1, S_2) containing $N_1 = N_2$ molecules of ideal gas $p_1 V_1 = N_1 kT_1$ and $p_2 V_2 = N_2 kT_2$ form sphere S_3 at temperature T_3

$$T_3 = T_1/2 = T_2/2, \quad N_3 = N_1 + N_2 = 2N_1 \tag{55}$$

such that,

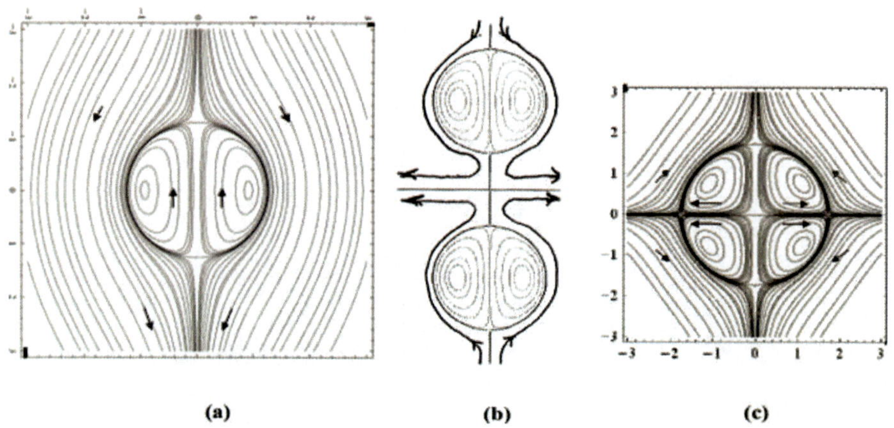

Fig. 15 **a** Hill spherical vortex. **b** Fusion of two Hill spherical vortices into **c** a spherical flow composed of two semi-spherical Hill vortices [35]

$$p_1 V_1 = p_2 V_2 = p_3 V_3 = N_3 kT_3 = E_1 \tag{56}$$

In the above fusion process, density is doubled, temperature is halved, and energy of S_3 is half that of the original two spheres

$$E_3 = p_3 V_3 = E_1, \quad p_1 V_1 + p_2 V_2 = E_1 + E_2 = 2E_3 \tag{57}$$

Hence, conservation of energy requires that energy E_1 be removed from the system (*exothermic*). The inverse process of "fission", namely S_3 splitting into two spheres (S_1, S_2) will be *endothermic* and require absorption of energy E_1.

The above transformations of spherical geometry, though related, do not correspond to the mathematical problem of *Banach-Tarski paradox* (BTP) [137] since in BTP problem a sphere is shown to transform to two spheres with identical volume and density as the original. Hence, BTP constitutes a clear violation of rational foundation of mathematics in the spirit emphasized by Nelson [138]. As a result, some mathematicians have raised objection, and justifiably so, against the *assumptions* underlying the set theoretical foundation of BTP. It is important to emphasize that the two spheres generated in BTP are known to be "measureless". In the following, some implications of invariant model of Boltzmann statistical mechanics (Fig. 1) and the associated laws of generalized thermodynamics to BTP are examined.

In the above fusion of spherical objects, whereas the role of number of particles N and volume V as mathematically concepts are clear, that of temperature is not. According to (25) and (26), absolute temperature is the most-probable or *Wien wavelength* of particle thermal oscillation thus constituting a "*measure*" of spatial extension. It is most interesting that in his description of *hyperbolic geometry*, Poincaré [139] anticipated absolute thermodynamic temperature as a "measure" or metric of physical space:

> Suppose, for example, a world enclosed in a large sphere and subject to the following laws: The temperature is not uniform; it is greatest at the center, and gradually decreases as we move towards the circumference of the sphere, where it is absolute zero. The law of this temperature is as follows: If R be the radius of the sphere, and r the distance of the point considered from the center, the absolute temperature will be proportional to $R^2 - r^2$. Further, I shall suppose that in this world all bodies have the same coefficient of dilatation, so that the linear dilatation of any body is proportional to its absolute temperature. Finally, I shall assume that a body transported form one point to another of different temperature is instantaneously in thermal equilibrium with its new environment. There is nothing in these hypotheses either contradictory or unimaginable. A moving object will become smaller and smaller at it approaches the circumference of the sphere. Let us observe, in the first place, that although from the point of view of our ordinary geometry this world is finite, to the inhabitants it will appear infinite. As they approach the surface of the sphere, they become colder and at the same time smaller and smaller. The steps they take are therefore smaller and smaller, so that they can never reach the boundary of the sphere. If to us geometry is only the study of the laws according to which invariable solids move, to these imaginary beings it will be the study of the laws of motion of solids deformed by the differences of temperature alluded to.

Therefore, $T_\beta = \lambda_{w\beta}$ is called Poincaré *thermal measure* of extension.

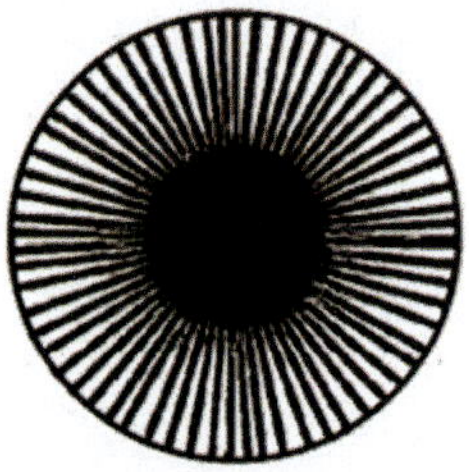

Fig. 16 Formation of *black-core* due to coalescence of radial lines

To address BTP problem, we begin with the following thought experiment concerning geometry. Let us consider a circle with 360 m circumference and at spacings of 1-m around the circumference, let straight lines of equal and uniform thickness t = 1 mm be drawn to the origin of the circle as schematically shown in Fig. 16.

Clearly, due to their finite thickness, adjacent lines will coalesce on a circle with approximate circumference of 360 mm beyond which the merged lines form a "*black core*" around the origin as shown in Fig. 16. This thought experiment suggests that there exists a circle with critical minimum radius corresponding to maximum number density even in the limit of infinitesimal thickness of radial lines. Clearly, the diameter of such a "black core" could be used as a measure of maximum resolution of printers.

Therefore, the question is what happens if the above procedure (Fig. 16) is considered in the limits of a circle of infinite radius when infinite number of radial lines of zero thickness are drawn to the origin. For example, at cosmological scale (Fig. 1) the problem corresponds to infinite radial lines from circumference of a spherical universe converging on a galaxy as "atom" at the center like a hologram. As discussed in a recent study on the gap between physics and mathematics [66], the invariant model of statistical mechanics leads to coordinates shown in Fig. 17.

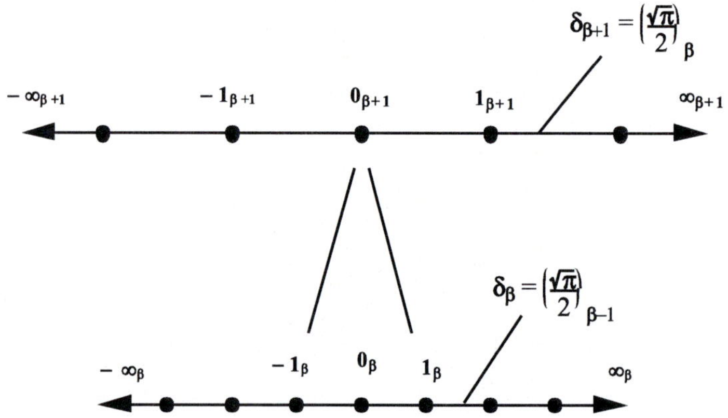

Fig. 17 Hierarchy of normalized coordinates associated with embedded statistical fields [37]

According to this figure, interval $x_\beta = (0_\beta, 1_\beta)$ of upper (outer) scale corresponds to interval $x_{\beta-1} = (1_{\beta-1}, \infty_{\beta-1})$ of next lower (inner) scale (Fig. 17). The location of the new origin $x_{\beta-1} = 0_{\beta-1}$ is defined by the new *unit of length* or *measure* at the lower scale.

According to Fig. 5, invariant Maxwell–Boltzmann distribution when *re-normalized* with respect to the *most-probable state* leads to transformation ($x_{mp} \to 1, x_\infty \to 4/\sqrt{\pi} = 2.567$). For example, three consecutive scales within the hierarchy of coordinates with ($0_\beta = 10^{-120}, 1_\beta = (4/\sqrt{\pi})10^{-100}, \infty_\beta = (4/\sqrt{\pi})^2 10^{-80}$) and the size of zero and infinity relative to unity taken as ($10^{-20}, 1, 10^{20}$) is shown in Fig. 18 from the previous study [66].

Applying the hierarchy of normalized coordinates $x_\beta = (0_\beta, 1_\beta, \infty_\beta)$, results in circular geometry corresponding to three generation as shown in Fig. 19.

In Fig. 19, three generations covering the range of radii $(0_\beta, 1_\beta)$, $(0_{\beta+1}, 1_{\beta+1})$, $(0_{\beta+2}, 1_{\beta+2})$ are revealed. Following Aristotle, it is assumed that *potentially infinite* statistical fields are compactified within the *black hole* singularity at the origin. According to (25) and Fig. 18, since temperature is Wien wavelength or Poincaré *thermal measure* $T_\beta = \lambda_{w\beta}$, absolute zero of adjacent scales relate as $0_\beta = 1_{\beta-1} = \infty_{\beta-2}$.

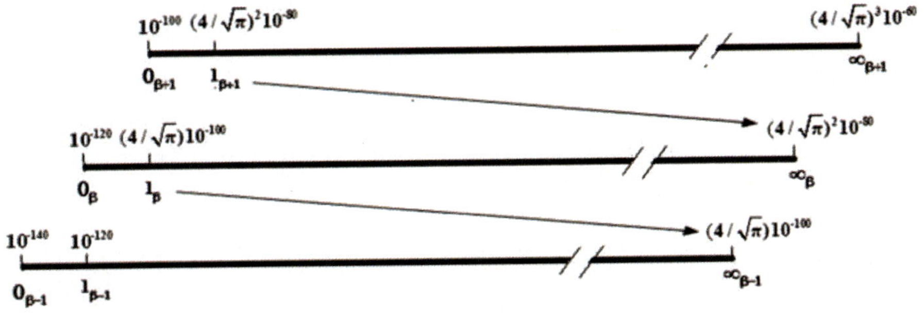

Fig. 18 Hierarchy of coordinates for hierarchy of embedded statistical fields [66]

Fig. 19 Formation of hierarchy of embedded black holes due to infinite number of compactified statistical fields at the origin [118]

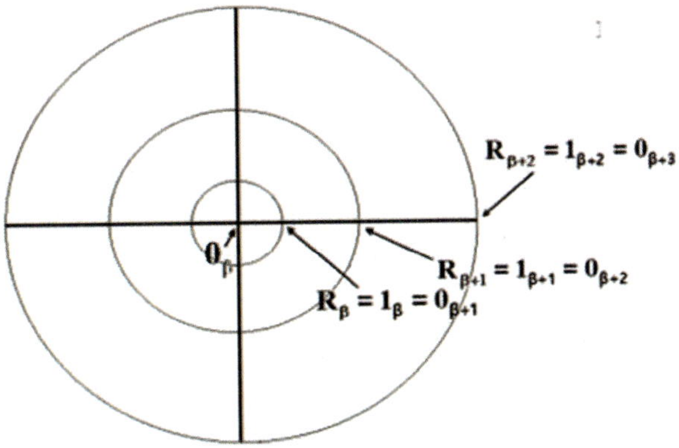

As a second thought experiment, let us consider a spherical volume S_1 at scale β composed of two semi-spherical tori as shown in Fig. 14a. The central singularity or *black hole* is at "absolute zero" $r_\beta = 0_\beta$ and located at the origin. Let the sphere S_1 contain total of N_1 atoms of ideal gas ($N_1/2$ in each toros) at the average temperature $T_{ave} = \overline{T}$, thus the total energy or enthalpy $E_1 = H_1 = 4N_1 k\overline{T}$ [33]. Let thermal energy E_1 be added to S_1 resulting in N_1 *new atoms* being decompactified from "black hole" singularity, due to phase transition such as melting, forming a smaller internal spherical volume (tori 3-3 and 4-4 in Fig. 14b) at temperature $T_3 = T_4 = T_L$. It is assumed that half of the added energy E_1 is absorbed by gases in original sphere S_1 (tori 1-1 and 2-2 in Fig. 14b) raising their temperature from $T_{ave} = \overline{T}$ to $T_1 = T_2 = T_H$. Finally, let the two embedded spheres in Fig. 14b undergo fission and split into two spheres S_2 = Tori (1-1)-(3-3) and S_3 = Tori (2-2)-(4-4). It is assumed that during the fission process the heat exchange between tori (1-1) and (3-3) results in their temperatures being respectively lowered and raised to the average temperature $(T_H + T_L)/2 = T_{ave} = \overline{T}$. Exactly similar heat exchange is assumed between (2-2) and (4-4) tori also leading to the same final average temperature $\overline{T}$.

Therefore, the original sphere S_1 has undergone an *endothermic* fission process (absorbed energy E_1) creating two identical spheres S_2 and S_3 with exactly the same number of atoms N_1, volume V_1, pressure p_1, and temperature $\overline{T}$. This process could be repeated ad infinitum, as long as energy E_1 is added to the system each time, due to Aristotle potentially infinite statistical fields that are compactified in the central black hole singularity [118]. Satisfaction of energy conservation principle as well as clarification of "*measureless*" nature of generated spheres in BTP problem, through introduction of what is called Poincaré *thermal measure* $T_\beta = \lambda_{w\beta}$, may help in resolution of the *paradoxical* aspect of Banach-Tarski problem [137].

In connection to BTP problem, it is further noted that due to Poincaré *thermal measure* $T_\beta = \lambda_{w\beta}$, macroscopic (extensive) *system* volume V_β and microscope (intensive) most probable *atomic* volume $\hat{v}_{w\beta}$ are related as

$$V_\beta = N_\beta \hat{v}_{w\beta} = N_\beta \lambda_{w\beta}^3 = N_\beta T_\beta^3 \tag{58}$$

Also, at thermodynamic equilibrium, by NMB distribution in Fig. 5 and in view of speed versus wavelength ratio relation [37]

$$v_j / v_w = \lambda_w / \lambda_j \tag{59}$$

the statistical field is composed of a *spectrum* of cluster or "element" volumes

$$V_{j\beta} = \sum_i \hat{v}_{ij\beta} = \sum_i \lambda_{ij\beta}^3 \tag{60}$$

such that

$$V_\beta = \sum_j V_{j\beta} \tag{61}$$

The quantum nature of physical space with spectrum of "atomic" volumes $\hat{v}_{ij\beta}$ forming a spectrum of cluster volumes $V_{j\beta}$ is in harmony with modern concepts of quantum gravity [119–123].

8 Concluding Remarks

Some implications of a scale-invariant model of Boltzmann statistical mechanics to generalized thermodynamics, Helmholtz decomposition of energy and momentum, and definitions of dark-energy, dark- matter, and dark-momentum were investigated. The model resulted in internal spacetime and external space and time having quantum nature in accordance with thermodynamic *thermal time* of Rovelli [95]. Invariant Schrödinger equation resulted in introduction of hierarchies of quantum mechanics wave functions and particles as de Broglie wave-packets from cosmic to photonic scales.

Physical space, Casimir [111] vacuum, was identified as a compressible fluid with density ρ_v varying from infinite rarefaction $\rho_{WH} = 0$ (white hole) to infinite compression $\rho_{BH} = \infty$ (black hole) as its two singularities. With space curvature viewed as deviation of density from density of Casimir vacuum $\kappa = \rho_\beta - \rho_v$, the states ($\kappa > 0, \kappa = 0, \kappa < 0$) of (matter, vacuum, anti-matter) fields were associated with (*Riemannian, Euclidean, Lobachevskian*) geometry. An invariant hydrodynamic model of Faraday line of force was introduced and shown to be in harmony with observations from cosmic to photonic scales. Finally, some implications of the model to black hole entropy and information loss as well as Banach-Tarski paradox were examined. The results were found to be in harmony with quantum gravity considered as dissipative deterministic dynamic system [119].

Acknowledgements This research was in part supported by NASA grant No. NAG3-1863.

References

1. S.H. Sohrab, Boltzmann entropy of thermodynamics versus Shannon entropy of information theory. Int. J. Mech. **8**, 73–84 (2014)
2. L. de Broglie, Interference and corpuscular light. Nature **118**, 441–442 (1926); Sur la Possibilité de Relier les Phénomènes d'Interférence et de Diffraction à la Théorie des Quanta de Lumière. C. R. Acad. Sci. Paris **183**, 447–448 (1927); La Structure Atomique de la Matière et du Rayonnement et la Mécanique Ondulatoire **184**, 273–274 (1927); Sur le Rôle des Ondes Continues en Mécanique Ondulatoire **185**, 380–382 (1927)
3. L. de Broglie, *Non-Linear Wave Mechanics: A Causal Interpretation* (Elsevier, New York, 1960)
4. L. de Broglie, The reinterpretation of wave mechanics. Found. Phys. **1**(5), 5–15 (1970)
5. E. Madelung, Quantentheorie in Hydrodynamischer Form. Z. Physik. **40**, 332–326 (1926)
6. E. Schrödinger, Über die Umkehrung der Naturgesetze. Sitzber Preuss Akad Wiss Phys-Math Kl **193**, 144–153 (1931)

7. R. Fürth, Über Einige Beziehungen zwischen klassischer Staristik und Quantenmechanik. Z. Phys. **81**, 143–162 (1933)
8. D. Bohm, A suggested interpretation of the quantum theory in terms of "hidden variables" I. Phys. Rev. **85**(2), 166–179 (1952)
9. T. Takabayasi, On the foundation of quantum mechanics associated with classical pictures. Prog. Theor. Phys. **8**(2), 143–182 (1952)
10. D. Bohm, J.P. Vigier, Model of the causal interpretation of quantum theory in terms of a fluid with irregular fluctuations. Phys. Rev. **96**(1), 208–217 (1954)
11. E. Nelson, Derivation of the Schrödinger equation from Newtonian mechanics. Phys. Rev. **150**(4), 1079–1085 (1966)
12. E. Nelson, *Quantum Fluctuations* (Princeton University Press, Princeton, NJ, 1985)
13. L. de la Peña, New foundation of stochastic theory of quantum mechanics. J. Math. Phys. **10**(9), 1620–1630 (1969)
14. L. de la Peña, A.M. Cetto, Does quantum mechanics accept a stochastic support? Found. Phys. **12**(10), 1017–1037 (1982)
15. A.O. Barut, Schrödinger's interpretation of ψ as a continuous charge distribution. Ann. der Phys. **7**(4–5), 31–36 (1988)
16. A.O. Barut, A.J. Bracken, Zitterbewegung and the internal geometry of the electron. Phys. Rev. D **23**(10), 2454–2463 (1981)
17. J.P. Vigier, De Broglie waves on dirac aether: a testable experimental assumption. Lett. Nuvo Cim. **29**(14), 467–475 (1980); C. Cufaro Petroni, J.P. Vigier, Dirac's aether in relativistic quantum mechanics. Found. Phys. **13**(2), 253–286 (1983); J.P. Vigier, Derivation of inertia forces from the Einstein-de Broglie-Bohm (E.d.B.B) causal stochastic interpretation of quantum mechanics. Found. Phys. **25**(10), 1461–1494 (1995)
18. F.T. Arecchi, R.G. Harrison, *Instabilities and chaos in quantum optics* (Springer, Berlin, 1987)
19. O. Reynolds, On the dynamical theory of incompressible viscous fluid and the determination of the criterion. Phil. Trans. Roy. Soc. A **186**(1), 123–164 (1895)
20. D. Enskog, Kinetische Theorie der Vorgange in Massig Verdunnten Gasen (Almqvist and Wiksells Boktryckeri-A.B., Uppsala, 1917). English translation: G.S. Brush, *Kinetic Theory* (Pergamon Press, New York, 1965), pp. 125–225
21. G.I. Taylor, Statistical theory of turbulence-parts I-IV. Proc. Roy. Soc. A **151**(873), 421–478 (1935)
22. T. Kármán, L. Howarth, On the statistical theory of isotropic turbulence. Proc. Roy. Soc. A **164**(917), 192–215 (1938)
23. H.P. Robertson, The invariant theory of isotropic turbulence. Proc. Camb. Phil. Soc. **36**, 209–223 (1940)
24. A.N. Kolmogoroff, Local structure on turbulence in incompressible fluid. C. R. Acad. Sci. U. R. S. S. **30**, 301–305 (1941); A refinement of previous hypothesis concerning the local structure of turbulence in a viscous incompressible fluid at high Reynolds number. J. Fluid Mech. **13**, 82–85 (1962)
25. A.M. Obukhov, On the distribution of energy in the spectrum of turbulent flow. C. R. Acad. Sci. U. R. S. S. **32**, 19–22 (1941); Some specific features of atmospheric turbulence. J. Fluid Mech. **13**, 77–81 (1962)
26. S. Chandrasekhar, Stochastic problems in physics and astronomy. Rev. Mod. Phys. **15**(1), 1–89 (1943)
27. S. Chandrasekhar, *Stochastic, Statistical, and Hydrodynamic Problems in Physics and Astronomy*, Selected Papers, vol. 3 (University of Chicago Press, Chicago, 1989), pp. 199–206
28. W. Heisenberg, On the theory of statistical and isotropic turbulence. Proc. Roy. Soc. A **195**, 402–406 (1948); Zur Statistischen Theorie der Turbulenz. Z. Phys. **124**(7–12), 628–657 (1948)
29. G.K. Batchelor, *The Theory of Homogeneous Turbulence* (Cambridge University Press, Cambridge, 1953)
30. L.D. Landau, E.M. Lifshitz, *Fluid Dynamics* (Pergamon Press, New York, 1959)
31. H. Tennekes, J.L. Lumley, *A First Course in Turbulence* (MIT Press, Cambridge, 1972)

32. S.H. Sohrab, Some implications of a scale invariant model of statistical mechanics to classical and relativistic thermodynamics. Int. J. Thermodyn. **17**(4), 233–248 (2014)
33. S.H. Sohrab, On a scale-invariant model of statistical mechanics and the laws of thermodynamics. J. Energy Resour. Technol. **138**(3), 1–12 (2016)
34. S.H. Sohrab, Invariant forms of conservation equations and some examples of their exact solutions. J. Energy Resour. Technol. **136**, 1–9 (2014)
35. S.H. Sohrab, Solutions of modified equation of motion for laminar flow across (within) rigid (liquid) and sphere and cylinder and resolution of Stokes paradox, in *AIP Conference Proceedings*, vol. 1896 (2017), p. 130004
36. S.H. Sohrab, Quantum theory of fields from Planck to cosmic scales. WSEAS Trans. Math. **9**, 734–756 (2010)
37. S.H. Sohrab, On a scale invariant model of statistical mechanics, kinetic theory of ideal gas, and Riemann hypothesis. Int. J. Mod. Commun. Tech. Res. **3**(6), 7–37 (2015)
38. S.H. Sohrab, Connecting Bernoulli and Schrödinger equations and its impact on quantum-mechanic wave function and entanglement problems, in *13th Chaotic Modeling and Simulation International Conference*, ed. by C.H. Skiadas, Y. Dimotikalis. Springer Proceedings in Complexity (2021)
39. R.S. de Groot, P. Mazur, *Nonequilibrium Thermodynamics* (North-Holland, 1962)
40. H. Schlichting, *Boundary-Layer Theory* (McGraw Hill, New York, 1968)
41. F.A. Williams, *Combustion Theory*, 2nd edn. (Addison Wesley, New York, 1985)
42. J.O. Hirschfelder, C.F. Curtiss, R.B. Bird, *Molecular Theory of Gases and Liquids* (Wiley, New York, 1954)
43. S. Chapman, T.G. Cowling, *The Mathematical Theory of Non-uniform Gases* (Cambridge University Press, Cambridge, 1953)
44. A. Sommerfeld, *Thermodynamics and Statistical Mechanics* (Academic Press, New York, 1956)
45. T. Dantzig, *Mathematics in Ancient Greece* (Dover, New York, 1954), p. 153
46. M. Planck, On the law of the energy distribution in the normal spectrum. Ann. der Phys. **4**, 553–558 (1901)
47. H. Helmholtz, Uber der Ehaltung der Kraft, Eine Physikalische Abhandlung (G. Reiner, Berlin). English Translation: G.S. Brush, *Kinetic Theory*, vol. 1–3 (Pergamon Press, New York, 1965)
48. F. Hasenöhrl, Zur Theorie der Strahlung in bewegten Körpern. Ann. der Phys. **16**, 589–592 (1905)
49. S.H. Sohrab, Invariant laws of thermodynamics and validity of Hasenöhrl mass-energy equivalence formula $m = (4/3)\ E/c^2$ at photonic, electrodynamic, and cosmic scales. Bull. Am. Phys. Soc. **62**(1), 124 (2017)
50. A. Lorentz, Electromagnetic phenomena in a system moving with any velocity less than that of light. Proc. Acad. Sci. Amst. **6**, 809–831 (1904)
51. M.S. Turner, Dark matter in the universe. Phys. Scr. **T36**, 167 (1991)
52. L.M. Krauss, M.S. Turner, The cosmological constant is back. arXiv:astro-ph/9504003v1 (1995)
53. M.S. Turner, The case for $\Omega M = 0.33 \pm 0.035$. Astrophys. J. **576**, L101 (2002)
54. Z.G. Dai, E.W. Liang, D. Xu, Constraining ΩM and dark energy with gamma-ray bursts. Astrophys. J. **612**, L102 (2004)
55. A. Einstein, Do gravitational fields play an essential part in the structure of the elementary particles of matter, in *The Principles of Relativity* (Dover, New York, 1952)
56. W. Pauli, *Theory of Relativity* (Dover, New York, 1958), p. 204
57. A.G. Riess et al., Observational evidence from supernovae for an accelerating universe and a cosmological constant. Astron. J. **116**, 1009 (1998)
58. A.G. Riess et al., Type Ia supernova discoveries at $z > 1$ from the *Hubble Space Telescope*: evidence for past deceleration and constraints on dark energy evolution. Astrophys. J. **607**, 665 (2004)

59. B.G. Schmidt et al., The high-Z supernovae search: measuring cosmic deceleration and global curvature using type Ia supernovae. Astrophys. J. **507**, 46 (1998)
60. S. Perlmutter et al., Measurements of Ω and Λ from 42 high-redshift supernovae. Astrophys. J. **517**, 565 (1999)
61. F. Laloe, J.H. Freed, The effects of spin in gases. Sci. Am. **258**, 94–101 (1988)
62. S.H. Sohrab, The physical foundation of a grand unified statistical theory of fields and the invariant Schrödinger equation. WSEAS Trans. Circ. Syst. **3**(4), 1017–1025 (2004)
63. R.E. Sonntag, G.J. van Wylen, *Fundamentals of Statistical Thermodynamics* (Wiley, New York, 1966)
64. M. Kardar, *Statistical Physics of Particles* (Cambridge University Press, New York, 2007)
65. M. Dersarkissian, Does wave-particle duality apply to galaxies? Lett. Nuovo Cim. **40**, 390 (1984)
66. S.H. Sohrab, Some implications of invariant model of Boltzmann statistical mechanics to the gap between physics and mathematics, in *12th Chaotic Modeling and Simulation International Conference*, ed. by C.H. Skiadas, Y. Dimotikalis. Springer Proceedings in Complexity (2020), pp. 231–243
67. L.D. Roper, Income distribution in the United States; A quantitative study. http://arts.bev.net/RoperLDavid/. 29 November 2007
68. S. Fleming, S. Donnan, America's middle-class meltdown: core shrinks to half of US homes. Financial Times, December 9, 2015. https://www.ft.com/content/98ce14ee-99a6-11e5-95c7-d47aa298f769
69. B.L. van der Waerden, Towards quantum mechanics, in *Sources of Quantum Mechanics*. ed. by B.L. van der Waerden (Dover, New York, 1967), pp. 1–59
70. Aristotle, *Time*, ed. by J. Westphal, C. Levenson (Hackett Publishing Company, Indianapolis, Indiana, 1993)
71. St. Augustine, *Time*, ed. by J. Westphal, C. Levenson (Hackett Publishing Company, Indianapolis, Indiana, 1993)
72. G. 't Hooft, S. Vandoren, *Time in Powers of Ten* (World Scientific, Singapore, 2014)
73. H. Poincaré, La mesure du temps. Rev. Metaphys. Morale **6**, 1 (1898); English translation: H. Poincaré, The measure of time, in *The Value of Science*
74. H. Poincaré, Sur la dynamique de l'électron. C. R. Acad. Sci. Paris **140** (1905)
75. H. Poincaré, Sur la dynamique de l'electron. Rend. Circ. Mat. Palermo **21**(12), 9–175 (1906)
76. A.A. Logunov, *On the Articles by Henri Poincaré, "On the Dynamics of the Electron"* (JINR, Dubna, 2001)
77. A. Einstein, Zur Elecrodynamik bewegter Körper. Ann. der Phys. (Leipzig) **17**, 891–921 (1905)
78. H. Minkowski, Space and time, in *Theory of Relativity* (Dover, New York, 1952), p. 75
79. A. Einstein, The foundation of general theory of relativity, in *The Principles of Relativity* (Dover, New York, 1952), pp. 111–164
80. G. Jaroszkiewicz, *Images of Time* (Oxford, 2016)
81. S. Hawking, *A Brief History of Time* (Bantam Book, New York, 1988)
82. H.D. Zeh, *The Physical Basis of the Direction of Time* (Springer, Berlin, 1999)
83. P. Yourgrau, *A World Without Time* (Basic Books, Perseus Books Group, NY, 2005)
84. C. Kiefer, Does time exist in quantum gravity? www.fqxi.org/communitty/forum/topic/265 (2008)
85. C. Rovelli, Forget time, www.fqxi.org/communitty/forum/topic/237 (2008)
86. J. Barbour, The nature of time, www.fqxi.org/communitty/forum/topic/360 (2008)
87. L. Smolin, *The Life of the Cosmos* (Oxford University Press, Oxford, 1997); in *Conceptual Problems of Quantum Gravity*, ed. by A. Ashtekar, J. Stachel (Birkhäuser, Boston, 1991)
88. S.B. DeWitt, Quantum mechanics and reality.Phys. Today **23**, 9 (1970)
89. S.B. DeWitt, Quantum theory of gravity I. The canonical theory. Phys. Rev. **160**, 1113 (1967)
90. J.A. Wheeler, Superspace and the nature of quantum geometro-dynamics, in *Battelle Rencontres*, ed. by C.M. DeWitt, J.A. Wheeler (Benjamin, New York, 1968)

91. S.H. Sohrab, An invariant model of Boltzmann statistical mechanics and some of its implications to thermodynamics and quantum nature of space and time. WSEAS Trans. Appl. Theor. Mech. **13**, 199–212 (2018)
92. C. Rovelli, *The Order of Time* (Riverhead Books, New York, 2018)
93. C. Rovelli, *Reality Is Not What It Seems* (Riverhead Books, New York, 2017)
94. C. Rovelli, Carlo Rovelli on the meaning of time. *Financial Times*, April 20, 2018
95. C. Rovelli, Statistical mechanics of gravity and the thermodynamical origin of time. Class. Quantum Gravity **10** (1993)
96. C. Rovelli, General relativistic statistical mechanics. *General Relativity and Quantum Cosmology*. http://doi.org/10.1103/PhysRevD.87.084055, arXiv:1209.0065v2 (2012)
97. J.L. Bell, *The Continuous and the Infinitesimal in Mathematics and Philosophy* (Polimetrica, Milano, Italy, 2006)
98. L. Boltzmann, Weitere Studien uber das Warmegleichgewicht unter Gasmoleculen, in *Sitzungsberichte*, Part II, vol. 66 (Akad.Wiss., Vienna, 1872), pp. 275–370. English translation: G.S. Brush, *Kinetic Theory* (Pergamon Press, New York, 1965), pp. 88–175
99. M. Planck, *The Theory of Heat Radiation* (Dover, New York, 1991)
100. A. Einstein, Ist die Trägheit eines Körpers vondessen Energieinhalt abhängig? Ann. der Phys. **18**, 639–643 (1905)
101. B.L. van der Waerden, *Sources of Quantum Mechanics*, ed. by B.L. van der Waerden (Dover, New York, 1967)
102. W. Heisenberg, *The Physical Principles of Quantum Theory* (Dover, New York, 1949)
103. P. Dirac, On the theory of quantum mechanics. Roy. Soc. Proc. A **110**, 561 (1926)
104. E. Schrödinger, Quantization as a problem of proper values, Part I. Ann. der Phys. **79**(4), 361–376 (1926). Quantization as a problem of proper values, Part II. Ann. der Phys. **79**(4), 489–527 (1926); Part III. Ann. der Phys. **81**, 109–139 (1926)
105. M. Born, Zur Quantenmechanik der Stoßvorgänge. Z. Physik **37**, 863 (1926)
106. I. Newton, *Philosophiae Naturalis Principia Mathematica* (London, 1687)
107. A.M.P. Dirac, *Directions in Physics* (Wiley, New York, 1978)
108. W. Schommers, Evolution of quantum mechanics, in *Quantum Theory and Pictures of Reality*, ed. by W. Schommers (Springer, Berlin, 1989), pp. 1–48
109. G.H.J.É. Lemaître, A homogeneous universe of constant mass and growing radius accounting for the radial velocity of extragalactic nebulae. Mon. Not. Roy. Astron. Soc. **xci**, 483 (1931); PNAS **20**, 12 (1934)
110. A.A. Penzias, R.W. Wilson, A measurement of excess antenna temperature at 4080 Mc/s. Astrophys. J. **142**, 419–421 (1965)
111. H.B.G. Casimir, On the attraction between two perfectly conducting plates. Proc. Kon. Ned. Akad. Wet. **51**, 793–795 (1948)
112. R. Feynman, R.B. Leighton, M. Sands, *Lectures on Physics*, vol. II (Addison-Wesley Publishing, New York, 1964), pp. 41–11
113. H.A. Lorentz, *Selected Works of L.H. Lorentz*, vol. 5, ed. by N.J. Nersessian, H.F. Cohen (Palm Publications, Nieuwerkerk, 1987), p. 7
114. S.H. Sohrab, Invariant model of statistical mechanics, quantum mechanics, and physical nature of space and time, in *Proceedings of the International Conference on 8th CHAOS* (Henri Poincare Institute, Paris, France, 2015), pp. 769–801, 26–29
115. S.H. Sohrab, Scale invariant model of statistical mechanics and quantum nature of space, time, and dimension. Chaotic Model. Simul. (CMSIM) **3**, 231–245 (2016)
116. O. Darrigol, The mystery of the Einstein Poincare' connection. Isis **95**, 614–626 (2004)
117. P. Galison, *Einstein's Clocks, Poincaré's Maps* (W. W. Norton & Company, New York, 2003)
118. S.H. Sohrab, Some implications of a scale-invariant model of statistical mechanics to classical and black hole thermodynamics. Bull. Am. Phys. Soc. **62**(1), 124 (2016)
119. G. 't Hooft, Quantum gravity as a dissipative deterministic system. Class. Quantum Gravity **16**, 3263–3279 (1999)
120. A. Ashtekar, The winding road to quantum gravity. Curr. Sci. **89**(12), 2064 (2005)

121. C. Rovelli, L. Smolin, Loop space representation of quantum general relativity. Nucl. Phys. B **331**, 80 (1990)
122. S.A. Major, K.L. Setter, Gravitational statistical mechanics: a model. Class. Quantum Gravity **18**(23), 5125–5142 (2001)
123. C. Rovelli, F. Vidotto, *Covariant Loop Quantum Gravity* (Cambridge University Press, Cambridge, 2015), p. 198
124. J.M. Bardeen, B. Carter, S.W. Hawking, The four laws of black hole mechanics. Commun. Math. Phys. **31**(2), 161–170 (1973)
125. S.W. Hawking, Black hole explosions. Nature **248**(5443), 30–31 (1974)
126. S.W. Hawking, Black holes and thermodynamics. Phys. Rev. D **13**(2), 191–197 (1976)
127. J.D. Bekenstein, Black holes and entropy. Phys. Rev. D **7**(8), 2333–2346 (1973)
128. R.M. Wald, Black holes and thermodynamics, in *Black Holes and Relativistic Stars*, ed. by R.M. Wald (University of Chicago Press, Chicago, IL, 1998), pp. 155–176
129. G. 't Hooft, On the quantum structure of a black hole. Nucl. Phys. B **256**, 727–745 (1985)
130. D. Grumiller, R. McNees, J. Salzer, Black hole thermodynamics: the first half century, in *Quantum Aspects of Black Holes*, ed. by X. Calmet. Fundamentals of Theoretical Physics 178 (Springer, Switzerland, 2015)
131. C.E. Shannon, The mathematical theory of communication. Bell Syst. Tech. J. **27**, 379–423 and 623–656 (1948)
132. S.H. Sohrab, Some implications of a scale invariant model of statistical mechanics to transport phenomena, in *ICS'09: Proceedings of the 13th WSEAS International Conference on Systems*, July 2009, pp. 557–568
133. A.M.P. Dirac, The evolution of the physicist's picture of nature. Sci. Am. **208**, 45 (1963)
134. G. 't Hooft, *Search of the Ultimate Building Blocks* (Cambridge University Press, Cambridge, 1998), p. 161
135. A. Bonaca, D.W. Hogg, The information content in cold stellar streams arXiv:1804.06854v1 [astro-ph.GA] 18 April 2018.
136. A.H.W. Küpper et al., Globular cluster streams as galactic high-precision scales—the poster child Palomar 5. Astrophys. J. **803**, 80 (26p) (2015)
137. S. Banach, A. Tarski, Sur la d'ecomposition des ensembles de points en parties respectivement congruentes. Fund. Math. **6**, 244–277 (1924)
138. E. Nelson, Warning signs of a possible collapse of contemporary mathematics, in *Infinity, New Research Frontiers*, ed. by M. Heller, W. Hugh Woodin (Cambridge University Press, Cambridge, 2011), pp. 76–85
139. H. Poincaré, *Science and Hypothesis* (Dover, New York, 1952), p. 65

D-Entropy in Classical Mechanics

V. M. Somsikov

Abstract The work is devoted to a new concept in physics—D entropy, defined as the relative increment of the internal energy of a body due to its energy of motion. D-entropy follows from the body's motion equation, which is derived based on the principle of dualism of symmetry (PDS). According to the PDS, the evolution of bodies is determined by both the symmetry of space and the symmetry of the body and the motion equation is derived from the expression of energy, which is the sum of the body's internal energy and the energy of its motion. Such a representation of energy is carried out in micro- and macro-variables that determine the motions of the elements of the body and the body itself, as a whole, respectively. This made it possible to take into account dissipative forces in the body's motion equation, which depend on micro- and macro-variables, arising when the body moves in an inhomogeneous field of forces, and determining the transformation of its energy of motion into internal energy. The D-entropy for large equilibrium systems, like the Clausius entropy, only increases. For small systems, the D-entropy can decrease. The main advantage of D-entropy is that it is determined through the dynamic parameters of the body. This makes it possible to use it to study the processes of evolution of objects, for example, the Universe, since it takes into account the role of relative motions in changing their internal states within the framework of the fundamental laws of physics.

Keywords Entropy · Symmetry · Nonequilibrium · Evolution · Mechanics

1 Introduction

The modern physics can be divided into two sections. In the first section, the motion of bodies in space is studied. This research is carried out in the frame of classical mechanics. In this case, the role of the structure of bodies in the dynamics and the processes of changes in the internal state of the body due to its motion are not taken

V. M. Somsikov (✉)
Al-Farabi Kazakh National University, Almaty 050040, Kazakhstan
e-mail: vmsoms@rambler.ru

C. H. Skiadas and Y. Dimotikalis (eds.), *14th Chaotic Modeling and Simulation International Conference*, Springer Proceedings in Complexity,
https://doi.org/10.1007/978-3-030-96964-6_33

into account. In the second section, the internal state of the body and its change due to change in external restrictions is studied. This study is carried out on the basis thermodynamics, statistical physics, and kinetics. In this case, the processes motion of bodies in the space are not taken into account. However, in the nature, there is no such separation of processes. It is clearly visible on the example of the Universe, which consists of moving galaxies, stellar systems, stars, etc. In the Universe, it is practically impossible to separately describe the dynamics of systems and the change in their internal states. The development of the Universe occurs in the interdependence of changes in internal states and the movement of objects in their inhomogeneous fields of interaction forces.

Recently, new approach for complex studying of matter was offered. In this approach, the motion of a body is studied together with changes of his internal state. The essence of this approach is that, unlike Newton's mechanics for an unstructured body, it is based on the motion equation for the structured bodies (SB). In this new approach, the role of structure of bodies in their dynamics is taken into account. Such approach led to a possibility of creation of "physics of evolution" [1, 2].

Let us define the general task of "physics of evolution" as a description of the processes of origin, development and destruction of natural systems within the framework of the fundamental laws of physics based on a new approach. New approach is constructed within laws of classical mechanics for the description of evolution of the moving and interacting bodies by association of two key concepts of natural sciences—of the motion energy and entropy.

Let us briefly define the role of the concepts of motion energy and entropy in physics. The concept of the motion energy was arisen in classical mechanics. It is characterizes dynamic properties of a body in space. The concept of entropy arose phenomenologically in thermodynamics. After that, the entropy was proved in the frame of statistical physics for models of the systems, close to equilibrium. This model was presented in the form of statistical ensembles of not strongly interacting subsystems.

The statistical concept of entropy faces with the problems of its application for non-equilibrium systems. Because of the researches, it became clear that in order to solve these problems, it is necessary to find a type of entropy that takes into account the role of motion and interaction of systems in changes in their internal states [3]. To search for this type of entropy, it was necessary to solve the following problems: what is entropy from positions of laws of classical mechanics; how entropy can follow from fundamental laws of classical mechanics; how entropy of bodies can change due to the motion energy; what essence of interrelation of "Order" and "Chaos".

In this paper we will show how these and other problems were solved by creation of mechanics of system of potentially interacting material points (MP) and how within this mechanics was appeared a new concept, which we called D-entropy. We will also consider the physical essence of the concept of D-entropy.

2 Dissipative Equation of the System's Motion

The definition of D-entropy follows from the system's motion equation of potentially interacting MP, which takes into account the role of the internal dynamics of the system's elements in its motion [2]. Therefore, in order to explain the essence of D-entropy, let us briefly explain why and how this motion equation was obtained.

According to classical mechanics, built based on the Newton's motion equation, the dynamics of matter must be reversible. However, all processes in nature are dissipative and therefore irreversible. This contradiction leads to an important problem for physics: how to explain the irreversibility of the dynamics of bodies, if Newton's equation of motion is reversible [4, 5]. At the beginning of solving this problem, a probabilistic mechanism of irreversibility was discovered [4]. A necessary condition for the occurrence of this irreversibility is the presence of arbitrarily small random external influences on the system. Therefore, let us call this mechanism probabilistic. However, this mechanism did not allow answers many questions. For example, how the second law of thermodynamics relates to the fundamental laws of physics; how "order" arises from "disorder", etc. Therefore, further searches for a solution to the problem of irreversibility, but strictly within the framework of the basic laws of physics, were continued.

Our search for a solution to this problem began with the derivation of the motion equation of a structural particle in the framework of the laws of classical mechanics without any restrictions, which were used to obtain the Lagrange and Hamilton equations [2]. The main idea was to try to account for the role of body structure in its dynamics. Obtaining this equation was carried out based on fundamental laws and principles that apply both to the system and to its elements. These laws and principles include the laws of conservation of energy and momentum, Galileo's principle. But since we considered the motion of a structured body, instead of a MP, in addition to these principles, ***the principle dualism of symmetry*** (PDS) was used. This principle claims that *the motion of the structured bodies is defined not only by the space symmetry, as in case of a MP, but also by the body's internal symmetry.*

Before proceeding to explain the derivation of the deterministic motion equation of the MP system, taking into account dissipative processes, we consider such an equation obtained in the framework of the theory of fluctuations. In accordance with statistical laws, the motion of a body, taking into account friction, is described by an equation in which the friction forces are proportional to the velocity [6]:

$$M\dot{V}_0 = -F_0 - \mu V_0, \tag{1}$$

where M—is the body's mass, V_0—is the velocity of the center of mass, $-F_0$—is the force acting on the center of mass, μ—is the effective coefficient of friction.

In accordance with this equation, the motion energy of the body is converted into the internal energy of the relative motion of its elements. This result is achieved within the framework of the laws of statistics that govern the molecular kinetic theory [6].

Let us show how one can explain the mechanism of transformation of the motion energy into the internal energy of the body, taking into account the role of the structure of the body in its motion [7, 8].

Consider the motion of a body along an inclined rough surface under the action of gravity. For this purpose, we take a body model as an equilibrium system consisting from a sufficiently large number of potentially interacting MP. At the initial moment of time, the equilibrium system has potential energies. During the sliding of the body, part of potential energy is converted into its kinetic energy. Another part goes to increase the internal energy because of the work of the friction force. This means that each MP of the body participates in two types of motion: in motion together with the center of mass of the system and in motion relative to its center of mass. Therefore, *the invariant of motion is the sum of the motion energies and the internal energy of MPs. Dissipation is associated with a part of the system's motion energy, which is converted into its internal energy*. Thus, if we want the system's motion equation to describe dissipative processes, it is necessary that it take into account the transformation of the motion energy into internal energy. Below we will show how to do this.

The total energy of the system can be represented as the sum of the energy of motion and internal energy using two groups of variables [2]. The group of variables that determine the internal energy are called micro-variables. The group of variables that determine the system's motion energy are called macro variables. The key point for the possibility of representing the system's energy in the form of this sum is that the following equality holds for the scalar sum of quadratic functions of vectors [2]:

$$N\sum_{i=1}^{N} v_i^2 = NM_N V_N^2 + \sum_{i=1}^{N-1}\sum_{j=i+1}^{N} v_{ij}^2 \tag{2}$$

The vector v_i determines the velocity of the MP in the laboratory coordinate system; $i, j = 1, 2, 3, \ldots, N$—numbers of MPs where the values i, j run from 1 to N and $i \neq j$; $v_{ij} = v_i - v_j$; vector $V_N = (\sum_{i=1}^{N} v_i)/N$ is the velocity of the system's center of mass; $M_N = Nm$; $m = 1$, and therefore $M_N = N$.

It is very important to emphasize that the first term on the right hand side of (2) corresponds to the measure of "Order", and the second term corresponds to the measure of "Chaos". Indeed, the first term is maximum when all MPs' vectors of velocity are equal in magnitude and coincide in direction. And vice versa, when the sum of all vectors is equal to zero, then the second term is maximal.

The existence of this equality proves the independence of micro- and macro variables [2]. The total energy in these variables is decomposed into the system's internal energy and the motion energy. That is, micro- and macro-variables belong to two symmetry groups. In the dual coordinate system in which the total energy is presented according to (2), the system's energy has the form:

$$E_N = E_N^{tr} + E_N^{ins} = const \tag{3}$$

Here $E_N^{ins} = T_N^{ins} + U_N^{ins}$—is the system's internal energy, where $T_N^{ins} = \sum_{i=1}^{N} m\tilde{v}_i^2/2$ is the kinetic component of internal energy; $E_N^{tr} = T_N^{tr} + U_N^{tr}$—is the motion energy, T_N^{tr}—is the system's kinetic energy, depending on the macro-variables, U_N^{tr}—is the system's potential energy in the field of external forces.

In a dual coordinate system, the internal energy is determined by micro-variables. This is because the sum of the energies of the relative motions of MPs and the energy, determined by the sum of the kinetic energies of their motion relative to the center of mass, coincides. The body's motion energy is associated with a group of macro-variables. The system's motion energy is characterized by the fact that the sum of the impulses of its elements is equal to the total impulse of the system, but the internal energy of the system is characterized by the fact that the sum of the impulses of all elements is equal to zero. *The law of conservation of a system's energy is that the sum of the energy of motion of the system and the internal energy is invariant along its trajectory, but each of these types of energy is not an invariant of motion.*

The system's motion equation follows from (3) by differentiating in relation to time, by summing scalar values of energy changes for each MP. It has the form [7, 8]:

$$M_N \dot{V}_N = -F_N^0 - \mu^d V_N, \tag{4}$$

where $F_N^0 = \sum_{i=1}^{N} F_i^0$; F_i^0—is external force acting on the *i*th MP; $\mu^d = \dot{E}_N^{\text{int}}/(V_N^{max})^2$; F_{ij}—is the strength of interaction *i* and *j* MPs; $F_{ij}^0 = F_i^0 - F_j^0$; $\dot{E}_N^{\text{int}} = \sum_{i=1}^{N-1} \sum_{j=i+1}^{N} v_{ij}(m\dot{v}_{ij} + F_{ij}^0 + NF_{ij})$; $V_N^{max} = -\dot{E}_N^{\text{int}}/F_N^0$.

Equation (4) already takes into account the relationship between the body's motion energy and it internal energy during the motion of the system.

In the right hand side of (4), the first term determines the external forces, which applied to the center of mass. These potential forces change the system's velocity.

The second term is nonlinear and bisymmetric, because it depends simultaneously from micro- and macro-variables. This term defines the role of the structure of a system in its dynamics. The coefficient "μ^d" determines the part of the system's motion energy, which is transformed into internal energy. The work of external forces spent on increasing internal energy is nonzero only when we have: $F_{ij}^0 = F_i^0 - F_j^0 \neq 0$, or when the field of external forces is non-homogeneous. This term was call as the *evolutionary nonlinearity*, because it links two symmetry groups, which depended from of micro- and macro-variables and leads to a decrease of the motion energy and evolution [9].

The nature of the evolutionary nonlinearity can be explained by that that due to the non-homogeneity of the external field of forces, the linking of the vectors from the different symmetry groups is appeared. For the structured bodies these groups of body's symmetry and space symmetry. It is determine the conversion of the body's motion energy into its internal energy and lead to a violation of the conservation of motion energy *when a body moves in an external non-uniform force field.*

An example of inhomogeneous external forces are gravitational fields in the Universe. Its leads to the transformation of the motion energy of Universe objects into their internal energy as a result of the work of forces proportional to the gradients of gravitational fields these objects. This effect can be called "gravitational friction". "Electromagnetic friction" can be defined in a similar way.

In the cases of homogeneity field of the external forces or in approximation of a solid body, the second term in the right hand side is equal to zero and (4) becomes the Newton's motion equation.

The dynamics of an equilibrium system in a weak inhomogeneous field of external forces is irreversible [2, 7]. Indeed, in accordance with (4), the magnitude of the change in the internal energy is of the second order of smallness. Therefore, the disturbance of equilibrium of the system can be neglected. However, according to Galileo's principle, the motion energy of an equilibrium system cannot increase due to internal energy. Consequently, we have a decrease in the motion energy of an equilibrium system along its trajectory in an inhomogeneous space.

The existence of dissipation is a necessary condition for formation of attractors [10]. However, the dissipation is possibly only for the structured bodies. This means that according to the laws of classical mechanics, the matter should be an infinite hierarchy of systems. That is, any arbitrarily small selected part of the body is a system of elements.

According to (4), the efficiency of increasing of the system's internal energy is determined by the ratio of its increment to the value of the internal energy itself. Thus we have [3]:

$$\Delta S_N^d = \Delta E_N^{\text{int}} / E_N^{\text{int}} \tag{5}$$

Here ΔS_N^d—is a D-entropy for system from N elements. This expression was called D-entropy [3]. The symbol "D" was introduced because this entropy determines the measure of the transformation of the ordered energy of motion of the system into the chaotic motion energy of its elements relative to the center of mass.

Below it will be briefly explained how the D-entropy is determined for an open nonequilibrium dynamical system (ONDS), what are the properties of D-entropy and what is the relationship between D-entropy and existing definitions of entropy.

D-entropy for nonequilibrium systems

A description of the evolutionary processes of matter is impossible without taking into account the fact that in general case all bodies in nature are ONDS. The nature of such processes is determined by D-entropy, which determines the relationship of external influences on the system with its internal structure in according with (4).

It turned out that when a sufficiently small system moves in inhomogeneous fields of external forces; its internal energy could either increase or decrease. For example, calculations showed that for an oscillator with $N = 2$ which in motion in an inhomogeneous field of external forces, internal energy can transformed into the motion energy depending on the initial phase of its oscillation and D-entropy can be negative [11]. However, with an increase of the number of particles in the system, the part of the internal energy that can be transformed into the system's motion energy is decreasing. When $N_1 > 100$, the internal energy could only increase and we have: $\Delta S_N^d > 0$. For $N_2 > 10^3$ the increment of internal energy growth do not increase [12]. Thus, $N_2 \sim 10^3$ determines the range of applicability of the thermodynamic description for the system. In the general case, these critical numbers depend on the parameters of the task. This means [1], that the irreversibility is qualitative: the more particles in the system, the more irreversibly it behaves (that is, the more unlikely reversibility). Thus, D-entropy allows us to determine the area of applicability of thermodynamics based on the laws of classical mechanics.

In the approximation of local thermodynamic equilibrium, with a sufficient degree of generality, ONDS can be submitted by a set of equilibrium subsystems moving relative to each other [6]. The motion of each subsystem is equivalent to its motion in an inhomogeneous field of forces, created by all other subsystems. When a system is in equilibrium, the relative velocities of its subsystems and the resulting forces acting from other subsystems are zero [6]. If to take a nonequilibrium closed system, then its total energy is invariant and the change of the D-entropy is determined by the sum of the increments of the entropies of each subsystem. Therefore, we have [13]:

$$\Delta S_N^d = \Delta E_N^{\text{int}}/E_N^{\text{int}} = \sum_{L=1}^{a} \left\{ N_L \sum_{k=1}^{N_L} [\int \sum_s F_{ks}^L v_k dt]/E_L \right\} \tag{6}$$

E_L—is internal energy of L-subsystem; F_{ks}^L—is a force, acting on the kth particles of the subsystem from the side of the particles of the other subsystems; s—is external particles with respect to L-subsystem, interacting with its kth particles; v_k—is a speed of the ith particles; N_L—is a number of particles in L-subsystems; $L = 1, 2, 3, \ldots$; a—is a number of subsystems in nonequilibrium system.

The calculations of (2) showed that the magnitude of the fluctuations of the system's internal energy due to changes of number of particles, obeys the law [12]:

$$\delta E^{tr}\ 1/\sqrt{N}. \tag{7}$$

Since this statistical law follow from (2), it can be argued that it follow from the deterministic laws of physics. A similar conclusion was made in [14]. This is also confirmed by the fact that the principle of maximum entropy corresponds to the principle of least action [13]. It follows that the fundamental laws of physics determine the field of application of statistical laws, and these laws can be considered as possible simplifications of the analysis of the dynamics of systems [15]. The proof of equilibrating closed non-equilibrium dynamical systems can be reduced to the proof that the energy of the relative motions of subsystems is irreversibly transformed into their internal energy. Let us show that in accordance with (4) such a transformation takes place. This can be done by assessing the energy flows between subsystems [2].

Obviously, for a non-equilibrium system consisting from equilibrium subsystems, the mechanism for the formation of direct and reverse energy flows for subsystems is associated with the mutual transformation of the energies of the relative motions of the subsystems and their internal energies. Consequently, the proof of the irreversibility of a nonequilibrium system is reduced to the proof that the inflow of internal energy of the subsystems is greater than the outflow.

Let us ΔE^{tr} is the energy of the relative motion of the subsystems, which is transformed into its internal energy. According to (4), ΔE^{tr} is determined by a bilinear term whose value is equal to the second order of smallness. Let us notice that the value ΔE^{tr} is also a second order of smallness in according with the statistical estimations of an increment of entropy [6]. Therefore we can write: $\Delta E^{tr}\ \chi^2$, where χ is a small parameter, for example, the disturbance of the internal energy. If it so, then $\Delta E^{tr}/E^{\text{int}} \ll 1$ and the violation of the equilibrium of the subsystems can be neglected. In this case, irreversibility takes place, since the transformation of the internal energy of the equilibrium subsystem into of its motion energy is impossible due to the law of conservation of momentum.

Let us consider the second case. If the equilibrium subsystems' interaction forces or their gradients are the great enough, the equilibrium of the subsystems can be disturbed. Then the subsystems can be represented as a set of equilibrium systems moving relative to each other. In this case, to increase the internal energy of the subsystems, one can write: $\Delta E^{tr} = \Delta E^{tr}_{ins} + \Delta E^h$, where ΔE^{tr}_{ins} is the increment of the energy of the relative motions and ΔE^h is the increment of the internal energies of the subsystems.

That is, $\Delta E^{tr}_{ins} < \Delta E^h$. The energy of the equilibrium subsystems cannot be transformed into their motion energy. Therefore, we will proceed from the fact that only the energy of the relative motions of sub-subsystems can be transformed back into the motion energy of the subsystems. Let us denote such a reverse flow of the subsystems internal energy, as: ΔE^{tr}_{ret}.

According to (4), the value ΔE^{tr}_{ret} is determined by the bilinear function of the sub-subsystems variables, which determined its motion energies and the internal energies. These are terms of the second order of smallness of their micro- and macro variables. But because: $\Delta E^{tr}\ \chi^2$, we will have that $\Delta E^{tr}_{ret}\ \chi^4$. Thus, the return flow of the internal energy of subsystems into the energy of its motion cannot be more than the fourth order of smallness. The decrease in the motion energy of the subsystems can be determined by the following equation:

$$\Delta E^{tr}_{dec} = \alpha\chi^2 - \beta\chi^4. \tag{8}$$

and we have:

$$\Delta S^d\ \Delta E^{tr}_{dec}/E_{\text{int}} \tag{9}$$

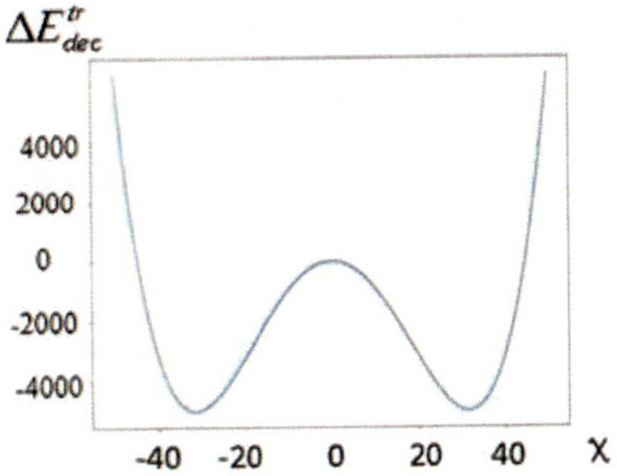

Fig. 1 The graph of the formula 8

Here the α, β constants can be determined using of (4).

Figure 1 shows a graph of ΔE^{tr}_{dec}. For values: $|\chi| \prec \chi_0$, where $\pm\chi_0$—are roots of (8), the irreversibility takes place. In general, for $N \gg 1$, we have $\Delta E^{tr}_{dec} > 0$. This corresponds to the second law of thermodynamics. For the stationarity of the system it is necessary to fulfill the equality: $\Delta E^{tr}_{dec} = 0$. However, this state is unstable and determined by micro variables [13].

Thus, the concept of D-entropy arose in the mechanics of structured particles, which operates with the total energy of the system, including the system's motion energy. In this mechanics, internal energy is defined, as the energy of motion of the system's elements relative to the center of mass.

The work of external forces, acting on the system, is divided into mechanical work to move it and work to change its internal energy. The energy of the relative motion of interacting subsystems due to the presence of gradients of external forces can pass into their internal energy, but the reverse process is forbidden by the Galilean principle, since the momentum of the system cannot change due to its internal energy.

D-entropy is valid for any equilibrium and nonequilibrium systems from any number of elements. At its determination, the absent of the interaction between subsystems was not required, as it is required in the case of statistical physics [6]. However, the interaction energy of subsystems, which is not taken into account in statistical physics, determines the processes of evolution of systems.

In thermodynamics, in contrast to the mechanics of structured particles, the concept of internal energy is defined as the total energy of a system minus its motion energy.

D-entropy in quantum mechanics is defined in the same way as in classical mechanics. D-entropy follows from the expanded Schrödinger equation obtained from the principle of dualism of energy [16].

The entropy for open non-equilibrium systems can be obtained also with a help of distribution function. This function, $f_p = f_p(r, p, t)$, is found using the extended Liouville equation, which has the form [17]:

$$df_p/dt = \partial f_p/\partial t + \sum_{i=1}^{N} \{v_i(\partial f_p/\partial r_i) + \dot{p}_i(\partial f_p/\partial p_i)\} = -f_p \sum_{i=1}^{N} \partial F_i^P/\partial p_i \quad (10)$$

Here $i = 1, 2, 3, \ldots, N$—is a number of subsystem, F_i^p—is a forces acted on the ith subsystems, p_i—is a momentum of the subsystem.

Extended Liouville (10), which can be used to describe ONDS, was obtained using (4). This equation, given on the phase plane of coordinates and momenta of a system of structured particles, differs from the canonical prototype in that the phase volume of the system is not conserved due to the dissipative processes.

From (10), it follows that only non-potential forces contribute to the change in the distribution function of particles of the system. In accordance with (4), the magnitude of the change in the distribution function is proportional to the gradients of forces. For a closed non-equilibrium system, the value of $\sum_{i=1}^{N} \partial F_i^p/\partial p_i$ decreases with a decrease in the energy of the relative motions of the subsystems due to its transformation into the internal energy of the subsystems [8].

Formal solution of (10) can be written like this:

$$f_p = f_p^o \exp \int \left(\sum_{i=1}^{N} \partial F_i^p/\partial p_i \right) dt. \tag{11}$$

The generality of $f_p = f_p(r, p, t)$ lies in the fact that it was obtained taking into account the work of dissipative forces. That is, this distribution function directly follows from the system's motion equation. Therefore, it can be used to analyze of evolution.

It is known that for entropy can be wright [6]: $S^B = -\int (f \ln f) dp dq$. From here and (10), we can obtain:

$$dS^B/dt + \sigma S^B = f\sigma \tag{12}$$

where $\sigma = \sum_{i=1}^{N} \partial F_i^p/\partial p_i$.

Thus, if $\sigma = 0$, then we have: $dS^B/dt = 0$. The S^B has a maximum when the subsystems do not have relative speeds. This is correspond to the equilibrium.

S^d is more general than S^B. This is due to the fact that S^d is acceptable for describing the evolution of ONDS moving in inhomogeneous fields of external forces, without using averaging any statistical hypothesize.

D-entropy for open nonequilibrium dynamical systems

The emergence and existence of all objects in nature is possible only due to dissipative processes as a result of the interaction of bodies, the exchange of energy, momentum and matter [10, 18]. Therefore, to describe evolutionary processes, it is necessary to take into account the openness of bodies. In addition, it must be borne in mind that dissipative processes arise only if the bodies, as well as their elements, have a structure. Hence it follows that matter must be infinitely divisible [8]. The infinite divisibility of matter or the impossibility of the existence of bodies with zero internal energy, follows from the mechanics of structural particles. That is, if all bodies possess all these properties and have arisen as a result of evolution, then they must be ONDS. The idea that the main element of matter is ONDS was also expressed in [13, 18]. Then *matter is a hierarchy of ONDS*. One way or another, the model of the body, as an ONDS, should be used to study the processes of self-organization of systems, the emergence of "Order" from "Chaos" and evolution [8]. To cover general qualitative properties of the structure and dynamics of matter, the chain of the structure of matter can be written as [8]: $MP \Rightarrow SB \Rightarrow ONDS$. Therefore, according to the PDS, to describe the dynamics at all hierarchical levels of the structure of matter its energy should be represented as the sum of the motion energy and internal energy.

The change in D-entropy at an each hierarchical level consists of the increments of the motion energies and internal energies for the constituent parts of this level. These increments are carried out due to the energy of the external hierarchical level. Let us assume that the system is near a stationary state. Let outside forces begin to work on the system. According to the principle of dualism of energy, this will lead to a change in the motion energy and internal energies of the elements of the first hierarchical level. Their change, in turn, will lead to a change in the energy of the second hierarchical level, and so on. So we have [8]:

$$\Delta E_0 = \Delta E_1^m + \Delta E_1^{\text{in}};\quad \Delta E_1 = \Delta E_2^m + \Delta E_2^{\text{in}}; \ldots; \Delta E_{N-1} = \Delta E_N^m + \Delta E_N^{\text{in}};$$
$$\Delta E_N^{\text{in}} = \Delta E_N^m \tag{13}$$

$$\Delta S_i^d = \Delta E_i^{\text{in}} / E_i^{\text{in}}. \tag{14}$$

Here ΔE_0—the work of the external energy. The energies of the corresponding hierarchical levels of matter consist of the sum of the motion energies of elements and their internal energies, denoted by the symbols "m" and "in", respectively.

Equations (13) and (14) are chains of energy and D-entropy increments for all hierarchical levels of matter due to the work of external forces. These equations can be called *the principle of relativity of energy and D-entropy for the steps of the hierarchical ladder of matter*. The motion equation for ONDS can be obtained from its energy [8].

The internal dissipative processes lead to a decrease in the motion energy of elements of a levels due to its transformation into their internal energy. As a result, the change in the motion energy at each hierarchical levels can be determined by the condition:

$$\delta E_i^m = \Delta E_i^m - \Delta E_i^{dis} \tag{15}$$

The quantity $\Delta W_i^{ND} = \delta E_i^m / \Delta E_i^{in}$ we will call *D-negentropy which is necessary for the stationary state of ONDS*. Thanks to ΔW_i^{ND}, an ONDS system can be in a stationary non-equilibrium state. Thus, the stationary state takes place when for each hierarchical level of ONDS the the next equality have a place:

$$\delta E_i^m = 0 \tag{16}$$

Of course, not all the factors that determine the stationary state of the system are taken into account here. For example, an external influence on a system can directly affect several hierarchical levels of matter. For example, in the case of a flow of solar radiation to the Earth. The spectrum of this flow is wide enough change to directly the state of terrestrial matter at many of its hierarchical levels [19]. However, the nature of such an impact always obeys to the PDS. Moreover, the positive flux of entropy can be compensated by the Planck radiation [20]. However, a complete description of the energy balance goes beyond the scope of classical mechanics.

3 Conclusions

Taking into account the structure of a body in its dynamics leads to a joint description of the dynamics and changes in the state of bodies and to the emergence of the concept of D-entropy. D-entropy follows from the body's motion equation and expressed through dynamic variables (micro- and macro variables).

D-entropy is proportional to the gradient of the external inhomogeneous force field in which it moves.

The physical meaning of D-entropy is that it determines the efficiency of dissipative transformation of the motion energy into internal energy.

D-entropy is applicable for bodies with a large and small number of elements. For large equilibrium systems, it transforms into the Clausius entropy.

D-entropy can be used to substantiate thermodynamics, statistical physics, and kinetics within the framework of the fundamental laws of physics.

Mathematically, D-entropy is due to the nonlinear interaction of micro- and macro-variables. The description of the dynamics of the body based on micro- and macro-variables we will call "*complete description*". It is called as complete, because it takes into account the role of influence of the internal dynamics on its dynamics as a whole. Complete description connects the violation of time-symmetry with a change in the internal states of systems due to their motion.

D-entropy can be used to analyze the nature of evolution of systems and to substantiate empirical branches of physics based on a "complete description" of the dynamics of systems. The advantage of the D-entropy is it applicable to the analyses of ONDS. Consequently, it can be used to study the processes of evolution of matter (Universe), as a hierarchy of the ONDS within the framework of the basic laws of physics.

Acknowledgements The work was carried out with the financial support of the Committee of Science of the Ministry of Education and Science of the Republic of Kazakhstan grant project AP09259554.

References

1. R. Penrose, *The Path to Reality or the Laws Governing the Universe. Full Guide* (Izhevsk, Moscow, 2007)
2. V.M. Somsikov, Deterministic mechanism of irreversibility. JAP **14**, 5708–5733 (2018)
3. V.M. Somsikov, The dynamical entropy. Int. J. Sci. **4**, 30–36 (2015)
4. G.M. Zaslavsky, *Stochasticity of Dynamical Systems* (Nauka, Moscow, 1984)
5. I. Prigogine, *From Being to Becoming* (Nauka, Moscow, 1980)
6. L.D. Landau, E.M. Lifshits, *Statistical Physics* (Nauka, Moscow, 1976)
7. V.M. Somsikov, Transition from the mechanics of material points to the mechanics of structured particles. Mod. Phys. Lett. B **4**, 1–11 (2016)
8. V.M. Somsikov, The irreversibly mechanics of the structured particles systems, in *2nd Chaotic Modeling and Simulation International Conference*, Chania, Greece, 1–6, 2009
9. V.M. Somsikov, Non-linearity of dynamics of the non-equilibrium systems. World J. Mech. **2**(7), 11–23 (2017)
10. A. Yu Loskutov, A.S. Mikhailov, *Introduction to Synergetics* (Nauka, Moscow, 1990)
11. V.M. Somsikov, A. Mokhnatkin, Non-linear forces and irreversibility problem in classical mechanics. J. Mod. Phys. **5**(1), 17–22 (2014)
12. V.M. Somsikov, A.B. Andreev, On the criteria for the transition to a thermodynamic description of the dynamics of systems. Izv. Vuzov. Phys. Ser. **7**, 30–39 (2015)
13. V.M. Somsikov, *To the Basics of Physics Evolution* (Nauka, Almaty, 2016)
14. F. Baldovin, L.G. Moyano, C. Tsallis, Boltzmann-Gibbs thermal equilibrium distribution descends from Newton laws: a computational evidence. arXiv:cond-mat/0402635v1 (2004)
15. V. Famourzadeh, M. Sefidkhosh, Straddling between determinism and randomness: chaos theory vis-à-vis Leibniz. arXiv:1909.13635v1 [physics.hist-ph] (2019)
16. V.M. Somsikov, Limitation of classical mechanics and ways it's expansion. PoS (Baldin ISHEPP XXII-047), September JINR, Dubna, 1–12, 2014
17. V.M. Somsikov, The equilibration of a hard-disks system. IJBC **14**, 4027–4033 (2004)
18. Yu.L. Klimontovich, *Introduction to the Physics of Open Systems* (Janus, Moscow, 2002)
19. A.B. Andreev, V.M. Somsikov, S.N. Mukasheva et al., Nonequilibrium effects in atmospheric perturbations caused by solar radiation flux. Geomagn. Aeron. **58**, 106–112 (2018)
20. Y.B. Rumer, M.S. Rivkin, *Thermodynamics. Stat. Phys. and Kinetics* (Nauka, Moscow, 1977)

Some Aspects of Rainbows and Black Hole Linked to Mandelbrot Set and Farey Diagram

Alberto Tufaile, Lori-Anne Gardi, and Adriana Pedrosa Biscaia Tufaile

Abstract We present in this work some experiments involving the optics of systems where the paraxial optics approximation cannot be applied, as in the case of rainbow formation by raindrops and the case of light deflection by massive objects, such as in the vicinity of black holes. The first experiment is the injection of laser light into a glass cylinder, while the other is a circular billiard formed by a circular mirror and a laser beam. We use the theory of dynamical systems and the Mandelbrot set as an analogy to represent the paths of the light beam, as well as the properties involving the Farey sequence.

Keywords Farey mediant · Optics · Dynamical systems · Mandelbrot set

1 Introduction

This work was developed due to the authors' interest in different aspects of dynamical systems that involve optics [1, 2], gravitation and the aesthetic appeal of fractal forms [3, 4]. In this way we will present some connections between concepts of optics and Mandelbrot sets resulting from recursion formulas, feedback processes or systems in which we have the repeated application of some mapping rule.

How to compare trajectories of light rays with a recursive mapping normally used in dynamic systems? In Fig. 1 we have a luminous ray inside a glass plate that undergoes multiple reflections. A change in the path of this ray can be associated with the series of events shown in Fig. 1b. This series of events is reproduced topologically with the "circle map" of the graphic diagram in Fig. 1c, with the Poincaré section

A. Tufaile (✉) · A. P. B. Tufaile
Soft Matter Lab, EACH, University of São Paulo, São Paulo 03828-000, Brazil
e-mail: tufaile@usp.br

A. P. B. Tufaile
e-mail: atufaile@usp.br

L.-A. Gardi
Western University, 1151 Richmond St, London, ON N63K7, Canada

C. H. Skiadas and Y. Dimotikalis (eds.), *14th Chaotic Modeling and Simulation International Conference*, Springer Proceedings in Complexity,
https://doi.org/10.1007/978-3-030-96964-6_34

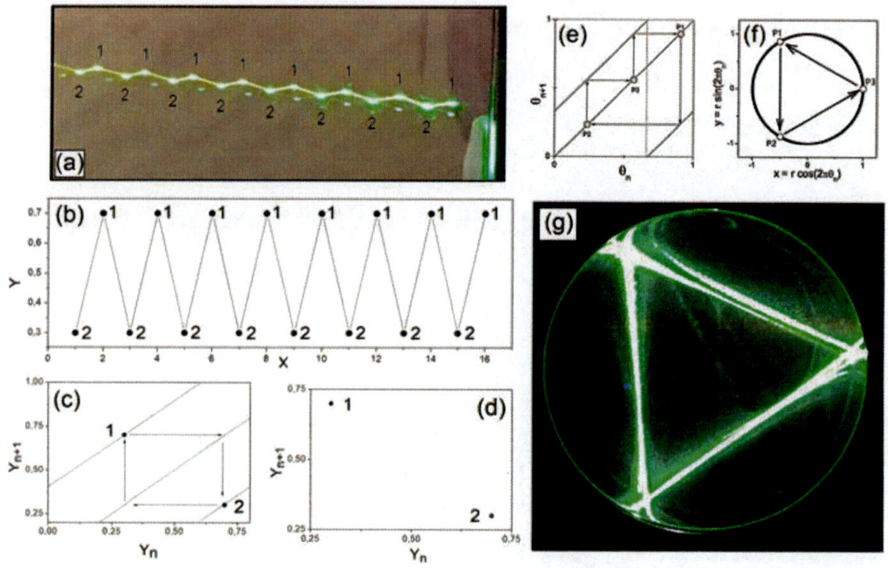

Fig. 1 Multiple reflections of a laser beam in a glass plate in (a). b Sequence of events. c Iterations in the circle map for a period-2. d Poincaré section of this period-2. Circle map in a period-3 in a circle map and its respective orbits in a circle in (f). Experiment showing multiple reflections in a period-3 in a reflective cylinder in (g)

shown in Fig. 1d [1]. We can see the same thing applied to period-3 in Fig. 1e for the case of the circle map, and the same dynamical system in a circular billiard in (f) and (g). For the cases found in nature, we know from optics that light rays in a raindrop of Fig. 2, which can behave like particles in a circular billiard, forming multiple rainbows, through recurring internal reflections. In this case, in addition to the rays of light that hit the droplet wall several times, obeying the law of reflection, we have rays that escape the droplet by refraction forming multiple rainbows, which give us some information about the shape of the droplet, as in the case of twin rainbows, in which we see an ellipsoidal rainbow between two circular rainbows, which appears to be a bifurcation, as it also occurs in other atmospheric optics phenomena, as in the case of the circumscribed halo surrounding the 22° solar halo of Fig. 3.

In the context of relativity, massive objects can "bend" light, forming Einstein's rings [1], which behave like lenses, as in the case of the lens in the form of a pseudo-sphere. In the case of a black hole acting as a lens in Fig. 3b, the properties of Minkowsky's space–time are altered by the presence of the mass, causing it to curve around itself, as if it were an optical lens. We also observed the reverse shape of this lens causing light to be distorted in Fig. 3c.

In our previous work on this subject, we studied rainbows and massive objects in the formation of luminous halos, starting from an optical system composed of a cylinder with the injection of a laser beam of Fig. 4, where we use the circular section

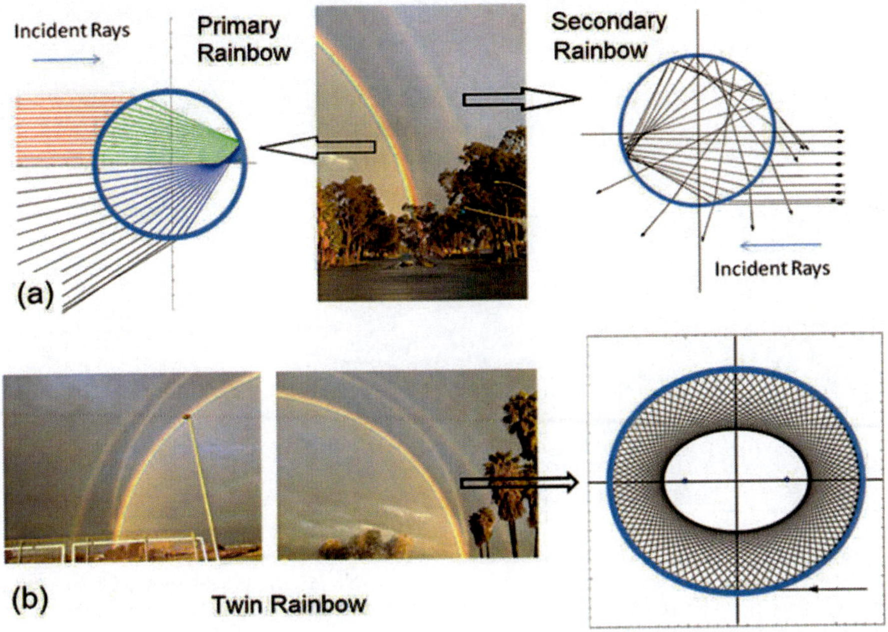

Fig. 2 In (**a**) reflections in a rain drop and primary rainbow, and reflections in a drop and secondary rainbow. In (**b**) the curious case of twin rainbows that could be explained considering elliptical drops

of the cylinder and compare it with the circular section of a drop. The modification of the cylinder topology led to the case of the pseudo-sphere.

This optical system formed by the laser hitting obliquely in the glass cylinder of Fig. 4c, d, behaves like an open billiard, with which we can see the formation of curious patterns associated with the star polygons, which due to the fact of restrictions to the total internal reflection, follow some rules related to a Farey mediant of Fig. 4f. This system is related to the phenomenon known as period locking (mode-locking), which can be observed in patterns projected on a screen of Fig. 5. This mode-locking is also seen in Mandelbrot set systems.

Increasing the inclination of the laser beam in the cylinder, we perceive a process of "optical deformation" of the circular section of the cylinder, for an elliptical billiard. With that, we recognize some interesting effects like the formation of caustics in the rainbow angle, which unfolded in separated branches [1]. This suggested to us the possibility that elliptical rainbows may be related to caustics unfolded from the cylinder, as discussed in the literature for ellipsoidal drops of Fig. 2b.

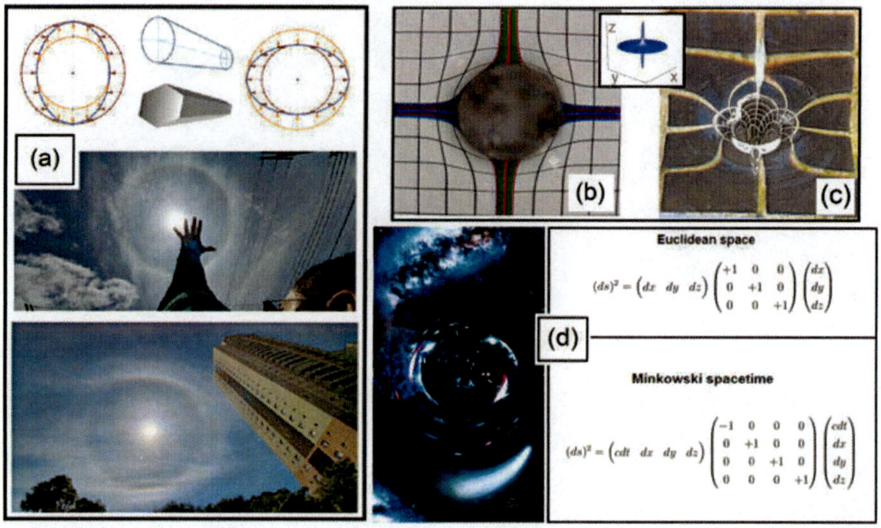

Fig. 3 In **a** cylinders and hexagonal bars can be used as lens with a laser to generate patterns. For the case of cylinder, we can associate elliptical profile with two circles. In 22° halo formed by hexagonal column of ice crystals can be measured with a hand flat. The 22° halo and the circumscribed halo. A pseudosphere lens in (**b**) and inverse pseudosphere lens from a pool drain in (**c**). Comparing the three-dimensional Euclidean space and Minkowski spacetime in (**d**) with a representation of the optical effects of a massive object such as black hole. In the four-dimensional Minkowski spacetime used in general relativity equations, time and space expand in opposite direction, with one solution of these equations being the black hole depicted in (**d**)

2 Mandelbrot Set and Star Polygons

Motivated by the observation of geometric patterns in the laser/cylinder system, we realized that the formation of caustics in cylinders may be related to the recursive rotations of maps in the complex plane, as in the case of the Mandelbrot Set, with the formation of star polygons for different initial conditions. The evolution of the iterations in Mandelbrot set presents a rich dynamics, with some complete routes to chaos, like Feigenbaum route and mode-locking route to chaos, besides intermittency. The Mandelbrot can be used as a toy model to explore the dynamics of the light in the cylinder, in order to help to explore this system and the pseudo-sphere used to simulate black holes or massive objects bending light.

For example, in Fig. 6, we present the existence of star polygons in the Mandelbrot set, with period-2, period-3, period-5 and period-7. For this last case we can compare the dynamics of ray in a billiard forming a period-7 with the Mandelbrot set.

The orbits of the Mandelbrot system are directly related to the Farey mediant, as shown in the diagram in Fig. 7a. The stability of the orbits can be observed with the cobweb diagram for values of the Mandelbrot set with real numbers. We can observe the stability of the system converging to a fixed point in Fig. 7b. When the control parameter is changed in Fig. 7c, we have different behaviors such as intermittency

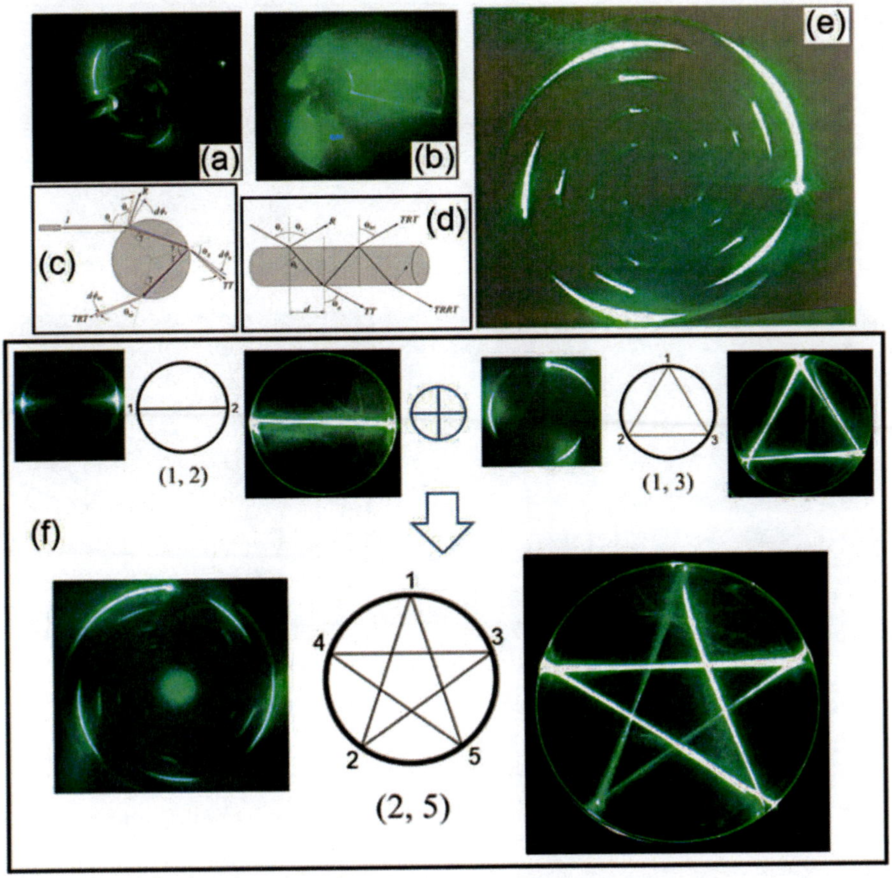

Fig. 4 Images of light scattering in cylinders in (**a**) and (**b**). Diagrams of light rays in the glass cylinder in (**c**) and (**d**). Light spiral in (**e**) obtained from this experiment. Comparison between stellar polygons in a billiard and the case of the glass cylinder, in which we can observe the case of Farey's mediant in (**f**)

in Fig. 7b, or this fixed point becoming a saddle point in which a doubling of period occurs in Fig. 7e. Changing the control parameter further, we have a period-4 in Fig. 7f, followed by chaotic behavior in Fig. 7g.

Another interesting relationship between the Mandelbrot set and the formation of caustics in a circular billiards is shown in Fig. 8. This relationship can be better understood by watching the video cited in [5] along with the papers describing the process of caustic formation [6–8]. The central idea is that these caustics are related to the fractal pattern profile associated with different polynomials of the Mandelbrot set.

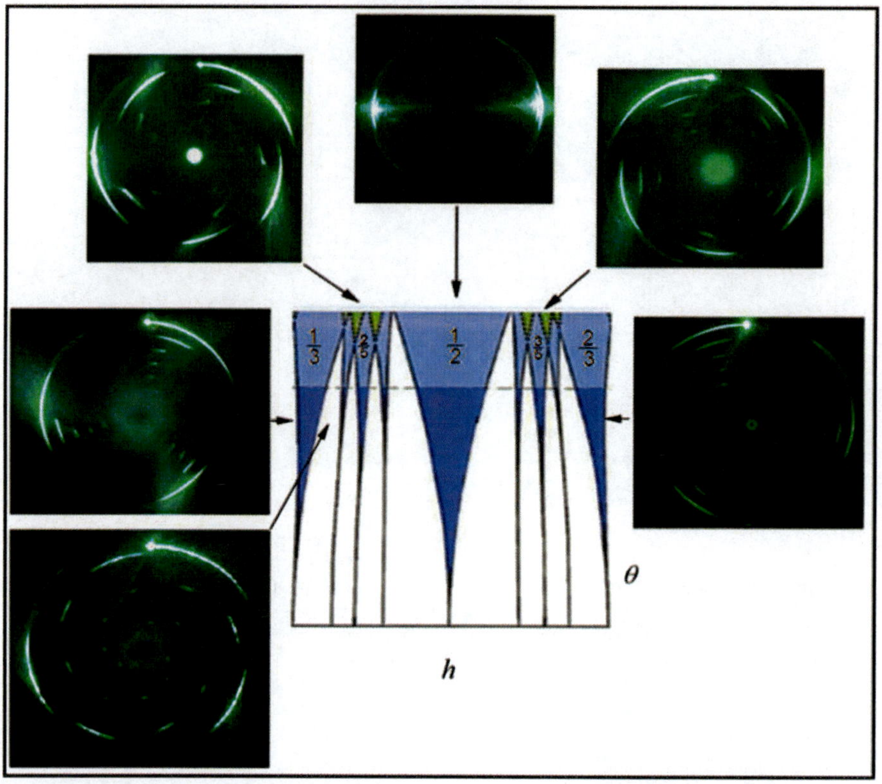

Fig. 5 Mode-locking and Farey sequence in the Arnold tongues with some light patterns

Finally, we show in Fig. 9 two examples of initial conditions in the Mandelbrot set for the case of orbits forming spiral patterns and compare them with the case of the experiment with the laser beam in the glass cylinder. We can see that the rotation orientation of the spiral depends on the position of choice of the initial condition, represented by the blue dot, with respect to the horizontal axis, which makes the spiral clockwise or counterclockwise. The same happens with the experimentally observed dynamical system, with the choice of the starting point that the laser beam touches the glass cylinder with respect to the horizontal axis, with the red spiral rotating clockwise and the green spiral rotating counterclockwise. For more information about these experiments, we recommend this video in [9].

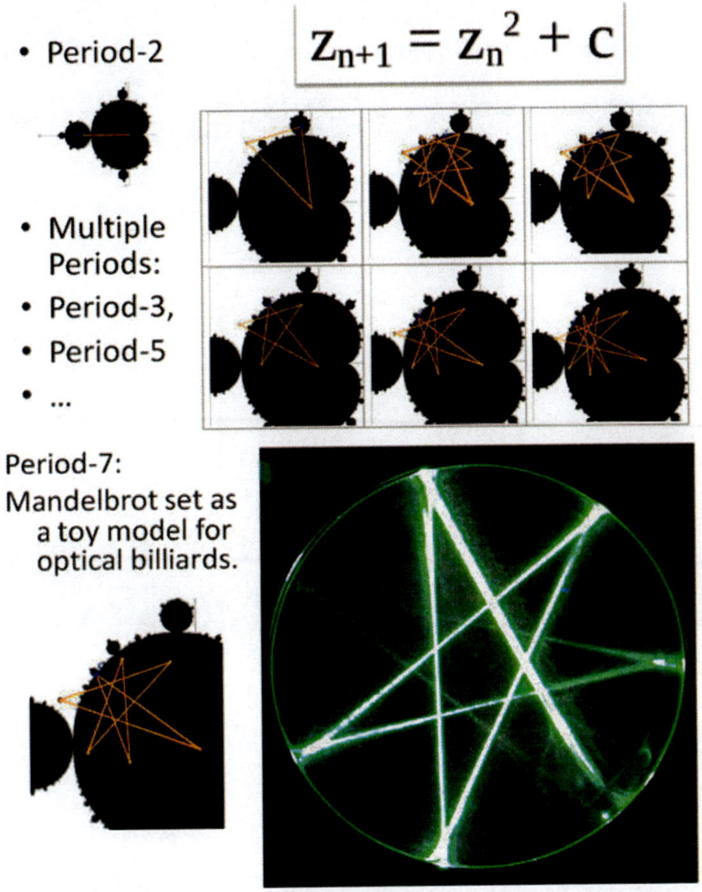

Fig. 6 Some star polygons in the Mandelbrot set. Period-7 in the Mandelbrot set and in the circular billiard

3 Conclusions

These connections between billiard systems, Optics, Mandelbrot set and caustics were explored due to their interesting aesthetic appeal, which facilitates the investigation of the optical system with the well-known system formed by Mandelbrot set. Multiple reflections of a laser inside a cylinder, which escapes by refraction and can projected in a screen, can be compared to the trajectories of the iterations of Mandelbrot map for certain regions in the Argand plane. In systems like the ones discussed in this paper, we can observe mode-locking and the formation of star polygons. These comparisons allow us to extract the essence of the dynamics existing in experimental optical systems that we cannot use the well-known paraxial optics, as happens in real physical systems such as light in raindrops or in a black hole. In

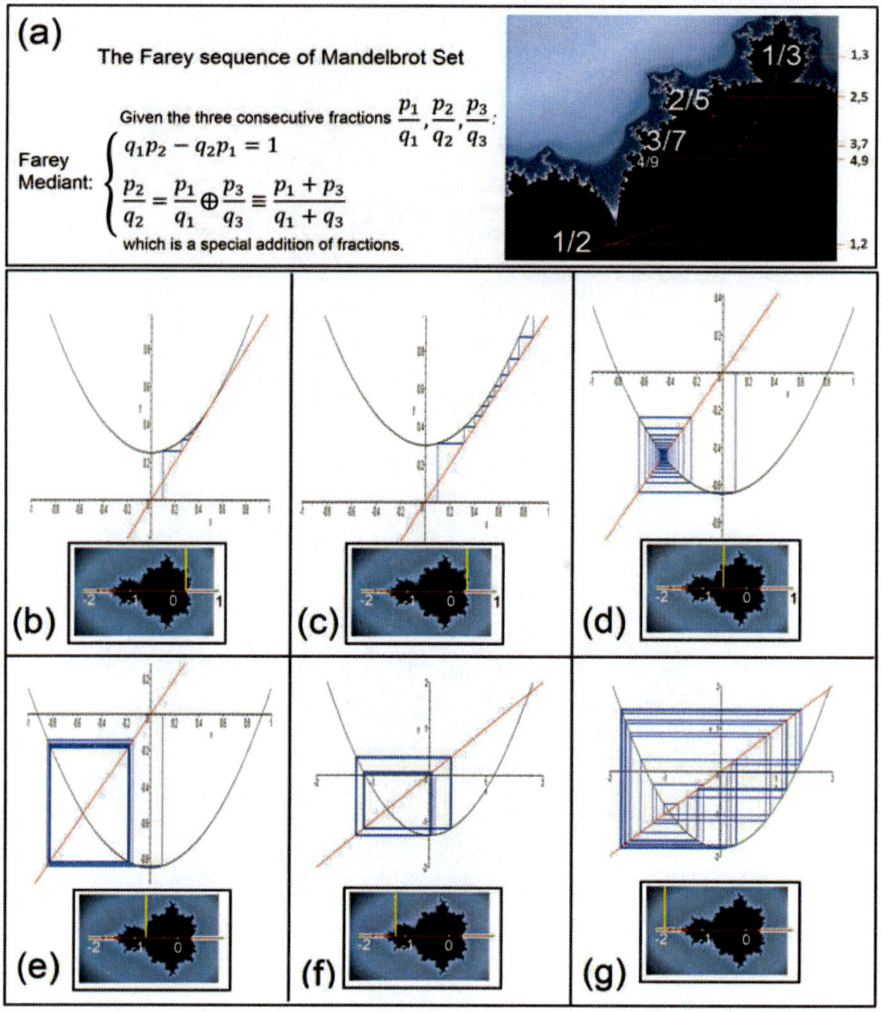

Fig. 7 Example of Farey sequence in the Mandelbrot set and the stability for some initial conditions in the Mandelbrot set

the case of massive object optics, such as the bending of light around black holes, matrix operations involving Minkowski space can generate rotations that cause the doubling of light rays from distant stars, as well as the formation of halos, such as Einstein rings, which were discussed in the case of light patterns in the laser/glass cylinder system.

Nonlinear dynamical systems that involve some kind of rotation as the case of the map of the circle presented in our previous works, or the Mandelbrot set, or matrix operations for the case of general relativity, can lead to certain results that can be observed experimentally with a non-paraxial optic that involves multiple

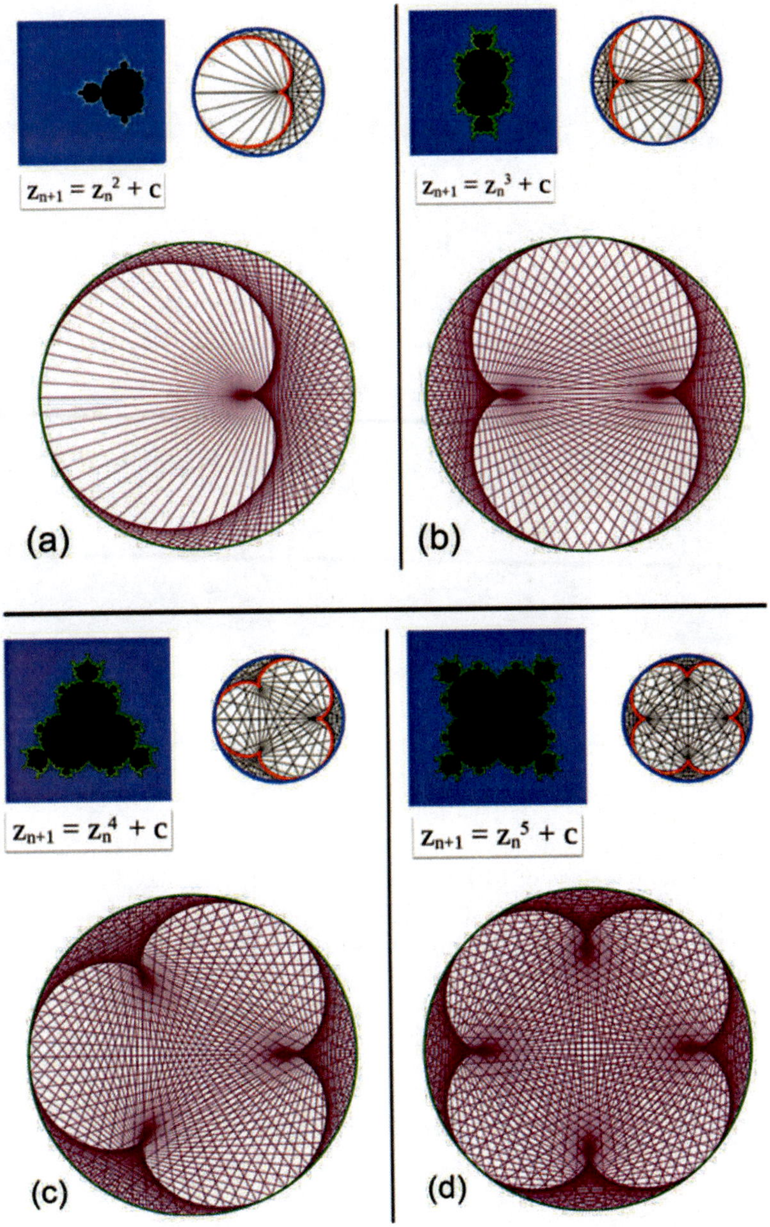

Fig. 8 Mandelbrot set and some caustics obtained with multiplication table applied to ray dynamics in a circle: in **a** the cardioid, in **b** the nephroid, with three lobes in (**c**) and with four lobes in (**d**)

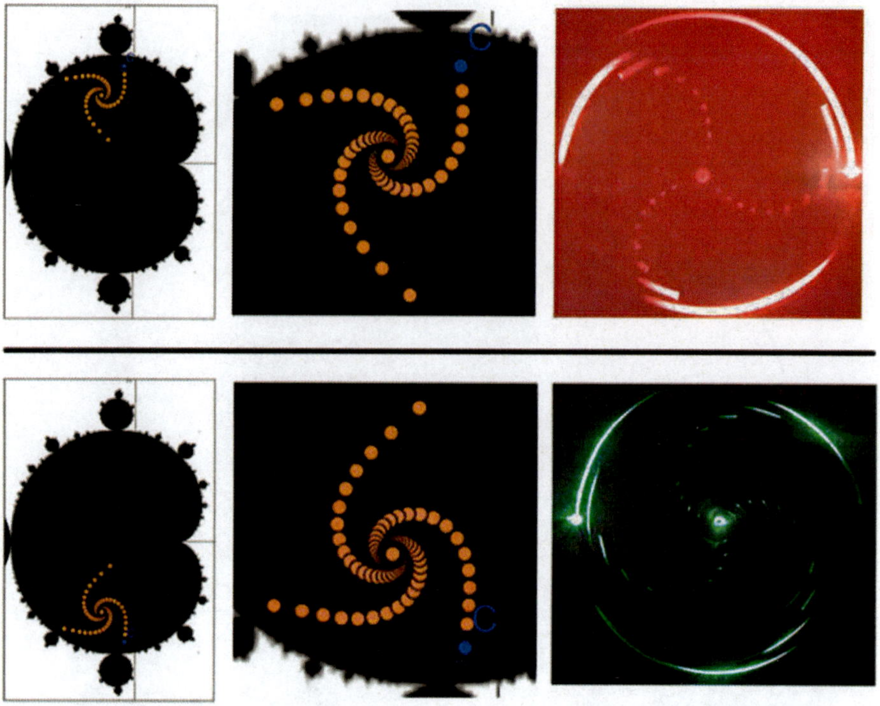

Fig. 9 Dynamics of spiral formation in the Mandelbrot set compared to the laser in the glass cylinder

reflections as in the case of the cylinder/laser and circular billiards. In this work, we emphasize how the use of Mandelbrot systems helps us to explore the different behaviors in a practical way, observing the different types of orbits based on the choice of initial conditions, which allows us to analyze the stability behavior of these orbits. Furthermore, we reproduce some interesting results of the formation of caustics in circular billiards through the multiplication rules, which are directly connected with Mandelbrot polynomials and their remarkable fractal representation. Overall, we believe that a lot can still be explored in the context of these analogies.

Acknowledgements The authors would like to thank to Lan Ma for picture of the rainbow (Twitter user @LanMisty).

References

1. A. Tufaile, A.P.B. Tufaile, Halo dynamics: from rainbows to black holes, in *Proceedings of 13th Chaotic Modeling and Simulation International Conference*. https://www.researchgate.net/publication/344929431_Halo_Dynamics_From_Rainbows_to_Black_Holes
2. L. Gardi, The mandelbrot set as a quasi-black hole, in *Proceedings of CNPS*. http://www.theomparticle.com/OMparticle/Paper.pdf
3. A. Tufaile, A.P.B. Tufaile, Parhelic-like circle from light scattering in plateau borders. Phys. Lett. A **379**, 529–534 (2015). https://doi.org/10.1016/j.physleta.2014.12.006
4. A. Tufaile, A.P.B. Tufaile, The dynamics of diffracted rays in foams. Phys. Lett. A **379**, 3059–3098 (2015). https://doi.org/10.1016/j.physleta.2015.10.011
5. Mathologer, *Multiplication Table, Mandelbrot and the Heart of Mathematics* (Tabuada de Multiplicação, Mandelbrot e o Coração da Matemática) (YouTube). https://youtu.be/qhbuKbxJsk8
6. E. Pegg Jr., *Modular Multiplication on a Circle* (Wolfram Demonstration Project). https://demonstrations.wolfram.com/ModularMultiplicationOnACircle/
7. J. Kowszun, *Mandelbrot Set*. https://www.geogebra.org/m/VbRZXqDV
8. S. Plouffe, The reflection of light rays in a cup of coffee or the curves obtained with b^n mod p. http://xahlee.info/SpecialPlaneCurves_dir/Cardioid_dir/_p/LightsRaysReflections.pdf
9. A. Tufaile, *Chaos, Rainbows and Black Holes* (YouTube). https://youtu.be/zljw0GxNvc4

Sequences of PRN's from Algebraic Curves over the Ring $\mathbb{Z}_{p^m}$

Sergey Varbanets and Yakov Vorobyov

Abstract In this work there is considered the method of producing the sequences of pseudorandom numbers basing on solutions of congruences of two variables modulo the power of prime number. The estimates of discrepant function of constructed sequences of pseudorandom numbers have been obtained.

Keywords Pseudorandom numbers · Elliptic curve and Exponential sum · Discrepancy

1 Introduction

Following the revelation of public-key cryptography that arose at the last quarter of twentieth century, in 1985 Nil Koblitz and Victor Miller have found that the elements over the group of points from elliptic curve over finite field are able to store the secrete information due to of complexity on addition operation. And it would serve as motive to study the cryptography on elliptic curves. The sequences of pseudorandom number at every time was being intrinsic part of cryptography, and therefore for the last 20 years the theory of elliptic curves has application in problem of generating of sequences of pseudorandom numbers. The useful survey in this direction belongs to Shparlinskii [4].

In our paper we consider the algorithm of producing the sequences of pseudorandom numbers from algebraic curves over the ring $\mathbb{Z}_{p^m}$ of residue classes of prime power modulus. The according elements of such sequences accept the polynomial representation over $\mathbb{Z}_{p^m}$. We demonstrate this concept to construct the sequences of

S. Varbanets (✉)
Department of Computer Algebra and Discrete Mathematics, Odessa I.I. Mechnikov National University, Dvoryanskaya street 2, Odessa, Ukraine
e-mail: varb@sana.od.ua

Y. Vorobyov (✉)
Department of Mathematics, Informatics and Information Activities, Izmail State University of Humanities, Repina Street 12, Odessa, Ukraine
e-mail: yashavoro@gmail.com

C. H. Skiadas and Y. Dimotikalis (eds.), *14th Chaotic Modeling and Simulation International Conference*, Springer Proceedings in Complexity,
https://doi.org/10.1007/978-3-030-96964-6_35

pseudorandom numbers of algebraic curves

$$y^2 \equiv x^3 + ax + b \pmod{p^m}$$

and

$$ax^3 + y^3 \equiv 1 \pmod{p^m}.$$

The constructed sequences have the fixed period $\tau = p^{m-1}$ that can be grown as for the growth of prime number p or factor m.

Notations. The letter p denotes a prime number, $p \geq 3$. For $n \in \mathbb{N}$ the notations $\mathbb{Z}_{p^m}$(accordingly, $\mathbb{Z}^*_{p^m}$) denote the complete (accordingly, reduced) system of residues modulo p^m. We write (a, b) for notation a great common divisor of a and b. For $z \in \mathbb{Z}$, $(z, p) = 1$ let z' or z^{-1} be the multiplicative inverse of a modulo p^m. We write $\nu_p(A)$ if $p^{\nu_p(A)}|A$, $p^{\nu_p(A)+1} \nmid A$. Landau symbol "O" is equivalent to Vinogradov symbol "$\ll$". The notation $f(x) \ll g(x)$ means that for $x \to \infty$ the inequality $|f(x)| \leq C \cdot g(x)$ holds with arbitrary constant C. Through $[x]$ we will denote the integral part of real number x.

2 Auxiliary Results

Let $E(\mathbb{F}_p)$ be an elliptic curve defined over $\mathbb{F}_p$ given by an affine Weierstraβ equation of the form

$$Y^2 + (a_1X + a_3)Y = X^3 + a_2X^2 + a_4X + a_6,$$

where $a_1, a_2, a_3, a_4, a_6 \in \mathbb{F}_p$ such that the partial derivations $\frac{\partial F}{\partial X}$ and $\frac{\partial F}{\partial Y}$ for the function

$$F(X, Y) = Y^2 + (a_1X + a_3)Y - X^3 - a_2X^2 - a_4X - a_6$$

do not become zero simultaneously at the points of the curve $(x, y) \in E(\overline{\mathbb{F}}_p)$ over the algebraic closure $\overline{\mathbb{F}}_p$ of $\mathbb{F}_p$.

For the case $p > 3$ the previous equation can be deduce to form

$$Y^2 = X^3 + ax + b \tag{1}$$

for some $a, b \in \mathbb{F}_p$ with $4a^3 + 27b^2 \neq 0$.

We recall that the set of points of curve $E(\mathbb{F}_p)$ together with point at infinity $\mathcal{O}$, relatively to a special operation $\oplus$, forms the abelian group E_p of order $\mathcal{N}(E_p)$ which satisfies inequality

$$\left|\mathcal{N}(E_p) - p - 1\right| \leq 2p^{\frac{1}{2}}.$$

For a point $Q \in E(\mathbb{F}_p)$ we use $x(Q)$, $y(Q)$ to denote its coordinates, that is, $(x(Q), y(Q))$.

For $m > 1$ we denote $E_p(m)$ as the set of solutions (x, y) satisfying to the congruence

$$y^2 \equiv x^3 + ax + b \pmod{p^m} \tag{2}$$

The set $E_p(m)$ we will call the elliptic curve over the ring $\mathbb{Z}_{p^m}$ and $\mathcal{N}(E_p(m))$ be a number of solutions of (2) with condition $(y, p) = 1$.

Lemma 1 *Let (x_0, y_0) be a solution of (2) with $(y_0, p) = 1$ and $m = 1$. Then for any integer t the congruence*

$$y^2(t) \equiv (x_0 + pt)^3 + a(x_0 + pt) + b \pmod{p^m} \tag{3}$$

has just two incongruent solutions modulo p^m for every positive m.

The assertion of this lemma follows from the fact that any solution (x_0, y_0) of congruence (3) with $m = 1$ we can grow to the solutions $y_1(t) = y(t), y_2(t) = -y(t)$.

Denote by $y_i(t), i = 1, 2$ the solution of congruence (3).

Lemma 2 *Let $p > 2$ be a prime, $m \geq 3$ be an integer, $s = \left[\frac{p-1}{p-2}m\right]$. There exist the polynomial $\varphi(t) \in \mathbb{Z}_{p^m}[t]$ of degree s*

$$\varphi(t) = \phi_0(x_0) + p^{\lambda_1}\phi_1(x_0)t + \cdots + p^{\lambda_s}\phi_s(x_0) \cdot t^s,$$

where $(\phi_i(x_0), p) = 1$, $i = 0, 1, \ldots, s$, and $\lambda_1, \lambda_2, \ldots, \lambda_s \in \mathbb{N}$, moreover

$$\lambda_j \geq j\frac{p-2}{p-1}, \quad j = 1, \ldots, s.$$

such that

$$y_i(t) = y_i(0)\varphi(t) \pmod{p^m}, \quad i = 1, 2,$$

and the points $(x_0 + pt, y_i(t))$, $i = 1, 2$, belong to the elliptic curves (2).

Proof Let (x_0, y_0) is the solution of (2) for $m = 1$, $(y_0, p) = 1$. For every t, $0 \leq t \leq p^{m-1} - 1$, we denote $y_1(t), y_2(t)$ as two different solutions of the congruence

$$y^2(t) \equiv (x_0 + pt)^3 + a(x_0 + pt) + b \pmod{p^m}.$$

Denote by x_0' the multiplicative inverse of $x_0^3 + ax_0 + b$, i.e.

$$x_0'(x_0^3 + ax_0 + b) \equiv 1 \pmod{p^m}.$$

Such solution exists since $(y_0, p) = 1$.

Hence, we find that (3) is equivalent to

$$y^2(t) \equiv (x_0^3 + ax_0 + b)(1 + (3ptx_0^2 + 3p^2t^2x_0 + p^3t^3)x_0').$$

Let $U^2(\omega) = (1 + (3\omega x_0^2 + 3\omega^2 x_0 + \omega^3)x_0')$.
Expanding the function $U(\omega)$ to series in powers of ω

$$U(\omega) = \sum_{i=0}^{\infty} X_i(x_0, x_0')\omega^i$$

and its logarithmic derivation

$$\frac{d \log U(\omega)}{d\omega} = \frac{U'(\omega)}{U(\omega)} = \frac{\sum_{i=1}^{\infty} iX_i(x_0, x_0')\omega^{i-1}}{\sum_{i=0}^{\infty} X_i(x_0, x_0')\omega^i}$$

gives the following recursion formulas for $j = 2, 3, \ldots$:

$$\begin{aligned} X_{j+1} = & -\frac{2j+1}{2(j+1)}(3x_0^2x_0' + ax_0)X_j \\ & -\frac{3(j-2)}{j+1}x_0x_0'X_{j-1} \\ & -\frac{2j-7}{2(j+1)}x_0'X_{j-2}, \\ X_0 = & \ 1, \\ X_1 = & \ \frac{1}{2}(3x_0^2x_0' + ax_0), \\ X_2 = & \ \frac{1}{2}3x_0x_0' - \frac{1}{8}(3x_0^2x_0' + ax_0')^2. \end{aligned} \tag{4}$$

Let show that the formal p-adic series for $U(pt)$ converges in p-adic metric and modulo p^m the congruence

$$U(pt) \equiv \varphi(t) \pmod{p^m},$$

where

$$\varphi(t) = \phi_0(x_0) + p^{\lambda_1}\phi_1(x_0)t + \cdots + p^{\lambda_s}\phi_s(x_0) \cdot t^s, \tag{5}$$

and $\varphi_j(x_0) \in \mathbb{Z}$, $\lambda_j \in \mathbb{N}$ and $\lambda_j \geq m$ for $j > s$. holds.

In our reasoning we will use p-adic analysis by schema of Postnikova [3].

Let us introduce the variables Y_j, Z_j, $j = 1, 2, \ldots, s$ defined by the conditions

$$Y_1 = 0,\ Y_2 = 1,\ Y_3 = \frac{1}{2}(3x_0^2x_0' + ax_0'),$$
$$Z_1 = 0,\ Z_2 = 0,\ Z_3 = 1$$

and for $j \geq 4$ Y_j, Z_j be determined by recursion formulas of type (4).

Let us consider determinants

$$\Delta_j = \begin{vmatrix} X_{j-2} & X_{j-1} & X_j \\ Y_{j-2} & Y_{j-1} & Y_j \\ Z_{j-2} & Z_{j-1} & Z_j \end{vmatrix}, \quad j = 3, 4, \ldots, s.$$

In particular, we have modulo p^m

$$\Delta_3 = \frac{1}{2}(3x_0x_0' + ax_0').$$

From this moment on, we suppose that $-3a$ is the non-quadratic residue modulo p. Therefore, we have

$$(x_0', p) = 1,\ (3x_0^2 + a, p) = 1.$$

(since otherwise the congruence $x^2 \equiv -3a \pmod p$ has the solution).

But then $\nu_p(\Delta_3) = 0$.

Also for $j \geq 4$ we easily obtain

$$\begin{aligned} \Delta_j &= -\frac{2j-9}{2j}x_0'\Delta_{j-1} \\ \cdots &= (-x_0')^{j-s}\frac{(2j-9)(2j-11)\cdots 3\cdot 1\cdot(-1)}{2^{j-s}j(j-1)\cdots 4}\Delta_3 \\ &= (-x_0')^{j-3}\frac{(2j-9)!\cdot 6}{2^{2j-7}j!(j-4)!}\Delta_3. \end{aligned}$$

Let $\nu_p(X_jp^j) = \lambda_j$, $\nu_p(Y_jp^j) = \mu_j$, $\nu_p(Z_jp^j) = \tau_j$.

Now let take out a common factor $p^{\min(\lambda_{j-1},\lambda_j,\lambda_{j-2})}$ from the first row of determinant Δ_j. From the second and third rows let do the same with $p^{\min(\mu_{j-1},\mu_j,\mu_{j-2})}$ and $p^{\min(\tau_{j-1},\tau_j,\tau_{j-2})}$, respectively.

It easy prove that

$$\lambda_j \geq j\frac{p-2}{p-1},\quad \mu_j \geq j\frac{p-2}{p-1},\quad \tau_j \geq j\frac{p-2}{p-1}.$$

Now, taking into account the relation between Δ_j and Δ_3 we easily find

$$\min(\lambda_j, \lambda_{j-1}, \lambda_{j-2}) \leq 3j - 3 - 2(j-2)\frac{p-2}{p-1}$$
$$+ \sum_{k=1}^{\infty}\left[\frac{2j-9}{p^k}\right] - \sum_{k=1}^{\infty}\left[\frac{j}{p^k}\right] - \sum_{k=1}^{\infty}\left[\frac{j-4}{p^k}\right].$$

Also take into account that $[2x] \leq 2[x] + 1$ for $x \geq 0$, and the quantity of nonzero summand in sum $\sum_{k=1}^{\infty}\left[\frac{2j-9}{p^k}\right]$ be at most $\frac{2j-9}{p} < \frac{2j}{p-1}$.

Then we have

$$\min(\lambda_j, \lambda_{j-1}, \lambda_{j-2}) \leq j + 1 + \frac{4(j-1)}{p-1}.$$

Bringing up the definition for $\varphi(t)$ (5) we at once obtain the proof of Lemma 2. □

Corollary 1 *In the conditions of Lemma 2 we obtain p-adic description of the solutions of the congruence*

$$y^2 \equiv x^3 + ax + b \pmod{p^m}$$

in the form

$$x = x_0 + pt, \ \ y_i(t) = y_i(0)(1 + A_1pt + A_2p^2t^2 + A_3p^{\lambda_3}t^3 + \cdots) \ (\pmod{p^m}),$$

where

$$\lambda_1 = 1, \ \lambda_2 = 2, \ \lambda_3 \geq 3, \ j = 3, 4, \ldots;$$
$$A_0 = 1, \ A_1 = 2^{-1}(3x_0^2x_0' + ax_0'), \ A_2 = 3\cdot 2^{-1}x_0x_0' - 2^{-3}(3x_0^2x_0' + ax_0')^2;$$
$$(A_i, p) = 1, \ i = 1, 2, 3, \ldots.$$

(here 2^{-1} be the multiplicative inverse for 2 modulo p^m).

Corollary 2 *For the fixed $x_0, y_0 \in E_p$ and $y_i(0)$, $i = 1, 2$ we have*

$$y_i(t_1) \equiv y_i(t_2) \pmod{p^m}$$

if and only if $t_1 \equiv t_2 \pmod{p^{m-1}}$. And hence, the sequences $y_i(t)$, $t = 0, 1, \ldots,$ $p^{m-1} - 1$ have the least period $\tau = p^{m-1}$ (here $i = 1$ or 2, $y_2(t) = -y_1(t)$). Thus we obtain the family of different sequences $\{y(t)\}$, which define by selection of initial point (x_0, y_0) on the curve E_p and by selection of index $i \in \{1, 2\}$.

Bellow we will show that the sequence of real numbers $\{\frac{y(t)}{p^m}\}$, $t = 0, 1, \ldots,$ $p^{m-1} - 1$ be the sequence of real numbers from [0, 1) that may be considered as the sequence of pseudorandom numbers passes the serial test on pseudorandomness.

Note that the same point (x_0, y_0) of elliptic curve E_p generate two sequences $y_i(t)$ defined by Lemma 2, the selection of which defines by the values of $y_i(0)$ as the solution of congruence

$$y^2 \equiv x^3 + ax + b \pmod{p^m}.$$

If $0 < y(0) < \frac{p}{2}$ then $y_i(t)$ denotes by $y_1(t)$, otherwise we have $y_2(t)$.

Over constructed set of sequences $\{y(t)\}$ we can define operation "*" by the following way:

$$y'(t) * y''(t) = y'''(t),$$

where $y'''(t)$ defines by sum of two points (x_0', y_0') and (x_0'', y_0'') of elliptic curve E_p

$$(x_0', y_0') \oplus (x_0'', y_0'')$$

and by Lemma 2, where $0 < y'''(0) < \frac{p}{2}$ if $y'(0)$ and $y''(0)$ simultaneously belong to $\left[0, \frac{p}{2}\right]$ or $\left[\frac{p}{2}, p\right]$. Otherwise, $y'''(0)$ is selected from interval $\left[\frac{p}{2}, p\right]$.

Similarly, we can construct the sequence $\{y(t)\}$ same to the sequence from Lemma 2 produced by the congruence

$$y^\ell \equiv f(x) \pmod{p^m},$$

where $f(x)$ be the polynomial with integer coefficients of degree ≥ 3.

In particular, let see the congruence

$$ax^3 + y^3 \equiv 1 \pmod{p^m}. \tag{6}$$

We will assume that p be the prime number of form $6k - 1$.

Define by $y(t)$ the solution of congruence

$$y^3 \equiv 1 - a(x_0 + pt)^3 \pmod{p^m}. \tag{7}$$

where (x_0, y_0) be the anyone solution of congruence

$$y^3 \equiv 1 - ax^3 \pmod{p}.$$

with $1 - ax_0^3 \not\equiv 0 \pmod{p}$. Every of such x_0 uniquely define the respective y_0. So, the solution $y(t)$ of congruence (7) defines uniquely.

Lemma 3 *Let* $s = \left[\frac{p-1}{p-2}m\right]$. *There exists the polynomial of degree* s

$$\varphi(t) = \Phi_0(x_0) + p^{\lambda_1}\Phi_1(x_0)t + \cdots + p^{\lambda_s}\Phi_s(x_0)t^s,$$

where $(\Phi_i(x_0), p) = 1$, $i = 0, 1, \ldots, s$; $\lambda_1, \ldots, \lambda_s$ *are the natural numbers satisfy the inequalities* $\lambda_j \geq j\frac{p-2}{p-1}$, *such that*

$$y(t) \equiv y(0)\varphi(t) \pmod{p^m}.$$

The proof of this lemma passes simultaneously to proof of Lemma 2 and the respective coefficients $\Phi_j(x_0)$ define by recurrent relation

$$\Phi_{j+1} = \frac{3j-1}{j+1} a x_0^2 x_0' \Phi_j + \frac{3j-5}{j+1} a x_0 x_0' \Phi_{j-1} + \frac{j-3}{j+1} a x_0' \Phi_{j-2},$$

moreover,

$$\Phi_0 = 1, \ \Phi_1 = -a x_0^2 x_0', \ \Phi_2 = -a x_0 x_0' - a^2 x_0^4 x_0'^2.$$

Here, x_0' is the multiplicative inverse modulo p^m for $1 - a x_0^3$.

3 Discrepancy

Let $\{x_n\}$ be the sequence of points from $[0, 1)$. As characteristic property of equidistribution of such sequences the following discrepant function D_N is used

$$D_N(x_0, x_2, \ldots, x_{N-1}) = D_N := \sup_{\Delta \subset [0,1)} \left| \frac{A_N(\Delta)}{N} - |\Delta| \right|,$$

where $A_N(\Delta)$ is the number of points among $x_0, x_2, \ldots, x_{N-1}$ falling into Δ, and $|\Delta|$ denotes the length of Δ.

In the same way there is defined the discrepancy for the sequence of s-dimensional points $X_n \subset [0, 1)^s$.

From definition of equidistribution of sequences of s-dimensional points we can conclude that for $D_N^{(s)} \to 0$ with $N \to \infty$ we can obtain better uniformly distributed sequences $\{X_n^{(s)}\}$.

Every sequence $\{x_n\}$, $x_n \in [0, 1)$ defines the sequence of s-dimensional points $X_n^{(s)}$, where $X_n^{(s)} = (x_n, x_{n+1}, \ldots, x_{n+s-1})$.

It is clear that for every equidistributed sequence $\{x_n\}$, which elements are statistically independent (unpredictable) for every integer $s \in \mathbb{N}$, the according sequence $\{X_n^{(s)}\} = \{x_n, x_{n+1}, \ldots, x_{n+s-1}\}$ be the equidistributed sequence.

We say that the sequence $\{x_n\}$, $x_n \in [0, 1)$ passes s-dimensional test on pseudorandomness if every sequence $\{X_n^{(s)}\}$, $s = 1, 2, \ldots, s$ be the equidistributed on s-dimensional unit interval $[0, 1)^s$.

To estimate the s-dimensional discrepant function of sequence $\{X_n^{(s)}\}$ the following lemmas is used.

For integers $s \geq 1$ and $q \geq 2$, let $C_s(q)$ be the set of all nonzero lattice points $\mathbf{h} = (h_1, \ldots, h_s) \in \mathbb{Z}^s$ with $-\frac{q}{2} < h_j \leq \frac{q}{2}$ for $1 \leq j \leq s$. Define for $\mathbf{h} \in C_s(q)$

$$r(h,q) = \begin{cases} 1 & if\ h = 0, \\ q\sin(\pi\frac{|h|}{q}) & if\ h \neq 0, \end{cases}$$
$$r(\mathbf{h},q) = \prod_{j=1}^{s} r(h_j,q) \tag{8}$$

Lemma 4 *Let $N \geq 1$ and $q \geq 2$ be integers. Suppose that $\mathbf{y_0}, \mathbf{y_1}, \ldots, \mathbf{y_{N-1}} \in \mathbb{Z}_q^s$. Then the discrepancy of the points $\mathbf{t_k} = \frac{\mathbf{y_k}}{q} \in [0,1)^s$, $k = 0, 1, \ldots, N-1$, satisfies*

$$D_N(\mathbf{t_0}, \mathbf{t_1}, \ldots, \mathbf{t_{N-1}}) \leq \frac{s}{q} + \frac{1}{N}\sum_{\mathbf{h}\in C_s(q)} \frac{1}{r(\mathbf{h},q)}\left|\sum_{k=0}^{N-1} e(\mathbf{h}\cdot\mathbf{t_k})\right| \tag{9}$$

(Proof of this lemma see in [1],[2]).

From the last statement it follows the classical statement of Turan-Erdös-Koksma inequality.

Lemma 5 *Let $T \geq N \geq 1$ and $q \geq 2$ be integers, $\mathbf{y_k} \in \{0, 1, \ldots, q-1\}^s$ for $k = 0, 1, \ldots, N-1$; $\mathbf{t_k} = \frac{\mathbf{y_k}}{q} \in [0,1)^s$. Then*

$$D_N(\mathbf{t_0}, \mathbf{t_1}, \ldots, \mathbf{t_{N-1}}) \leq \frac{s}{q} + \frac{1}{N}\sum_{\mathbf{h}\in C_s(q)}\sum_{h_0\in(-\frac{T}{2},\frac{T}{2}]} \frac{1}{r(\mathbf{h},q)r(h_0,T)} \times \left|\sum_{k=0}^{T} e(\mathbf{h}\cdot\mathbf{t_k} + \frac{kh_0}{T})\right| \tag{10}$$

This assertion follows from Lemma 4 and from an estimate of incomplete exponential sum through complete exponential sum.

Lemma 6 (Niederreiter, [1]). *Let $q \geq 2$, $T > 1$ be integers. Then*

$$\sum_{\substack{\mathbf{h}\in C_s(q) \\ \mathbf{h}\equiv 0 \pmod v}} r(\mathbf{h},q) < \frac{1}{v}\left(\frac{2}{\pi}\log q + \frac{7}{5}\right)^s$$

for any divisor v of q with $1 \leq v < q$, and

$$\sum_{h_0\in(-\frac{T}{2},\frac{T}{2}]} \frac{1}{r(h_0,T)} \leq \frac{2}{\pi}\log T + \frac{7}{5} \tag{11}$$

Lemma 7 *The discrepancy of N arbitrary points $\mathbf{t_0}, \mathbf{t_1}, \ldots, \mathbf{t_{N-1}} \in [0,1)^2$ satisfies*

$$D_N(\mathbf{t_0}, \mathbf{t_1}, \ldots, \mathbf{t_{N-1}}) \geq \frac{1}{2(\pi+2)|h_1 h_2| N} \left| \sum_{k=0}^{N-1} e(\mathbf{h} \cdot \mathbf{t_k}) \right| \tag{12}$$

for any lattice point $\boldsymbol{h} = (h_1, h_2) \in \mathbb{Z}^2$ *with* $h_1 h_2 \neq 0$.

(It is the special version of Niederreiter result in [1]).

From these lemmas we can to see that the character of equidistribution of sequence $\{x_n\}$, $x_n \in [0, 1)$ completely defines by estimate of exponential sum

$$S_N := \sum_{n=1}^{N} e^{2\pi i h x_n}, \quad h \in \mathbb{N}.$$

In Sect. 2 we constructed two sequences $\{x_t\}$, $x_t = \frac{y(t)}{p^m}$ that were being produced by the algebraic curves over the ring $\mathbb{Z}_{p^m}$ defined by the congruences (2) and (6). From Lemmas 2 and 3 it is clear to see that $y(t)$ are defining by special polynomials from the ring $\mathbb{Z}_{p^m}[t]$. These polynomials have the form

$$y(t) = A_0 + A_1 pt + A_2 p^2 t^2 + A_3 p^{\lambda_3} t^3 + \cdots,$$

moreover, $\lambda_j \geq 3$, $(A_j, p) = 1$ for $j \geq 3$.

The according sums S_N can be estimated by use of the generalized Gauss sums and the last can be estimated using the following lemma.

Lemma 8 (see, [5], Lemma 3). *Let* $p > 2$ *be a prime number,* $m \geq 2$ *be a positive integer,* $m_0 = \left[\frac{m}{2}\right]$, $f(x)$, $g(x)$, $h(x)$ *be polynomials over* $\mathbb{Z}$

$$f(x) = A_1 x + A_2 x^2 + \cdots,$$
$$g(x) = B_1 x + B_2 x^2 + \cdots,$$
$$h(x) = C_\ell x + C_{\ell+1} x^{\ell+1} + \cdots, \quad \ell \geq 1,$$

$$\nu_p(A_j) = \lambda_j, \quad \nu_p(B_j) = \mu_j, \quad \nu_p(C_j) = \nu_j,$$

and, moreover,

$$k = \lambda_2 < \lambda_3 \leq \cdots, \quad 0 = \mu_1 < \mu_2 < \mu_3 \leq \cdots,$$
$$\nu_p(C_\ell) = 0, \quad \nu_p(C_j) > 0, \quad j \geq \ell + 1.$$

Then the following bounds occur

$$\left|\sum_{x\in\mathbb{Z}_{p^m}} e_m(f(x))\right| \le \begin{cases} 2p^{\frac{m+k}{2}} & if\ \nu_p(A_1)\ge k,\\ 0 & if\ \nu_p(A_1)<k;\end{cases}$$

$$\left|\sum_{x\in\mathbb{Z}^*_{p^m}} e_m(f(x)+g(x^{-1}))\right| \le I(p^{m-m_0})p^{\frac{m}{2}}$$

$$\left|\sum_{x\in\mathbb{Z}^*_{p^m}} e_m(h(x))\right| \le \begin{cases} 1\ if\ \ell=1,\\ 0\ if\ \ell>1,\end{cases}$$

where $I(p^{m-m_0})$ is a number of solutions of the congruence

$$y\cdot f'(y)\equiv g'(y^{-1})\cdot y^{-1} \pmod{p^{m-m_0}},\ \ y\in\mathbb{Z}^*_{p^{m-m_0}}.$$

This lemma is the estimation of complete generalized Gauss sum. The incomplete generalized Gauss sum

$$\sum_{t=1}^{N} e^{2\pi i\frac{f(t)}{p^m}},\ \ 1\le N\le p^m$$

we can estimate by using the inequality

$$\left|\sum_{t=1}^{N} e^{2\pi i\frac{f(t)}{p^m}}\right| \le \sum_{k=1}^{p^m}\frac{1}{\max(k,p^m-k)}\left|\sum_{t=1}^{p^m} e^{2\pi i\frac{f(t)+kt}{p^m}}\right| =$$

$$= \max_{1\le k\le p^m}\left|\sum_{t=1}^{p^m} e^{2\pi i\frac{f(t)+kt}{p^m}}\right| \log p^m \ll p^{\frac{m}{2}}\log p^m.$$

Now we can obtain the estimate of discrepancy for sequences generated in Lemmas 2 and 3.

Indeed, the function $y(t)$ for the sequence generated by elliptic curve (2) as the function $y(t)$ for the sequence generated by (6) both satisfy for all conditions of Lemma 8 and so the sum $\sum_{t=1}^{p^m} e^{2\pi i\frac{y(t)}{p^m}}$ can be estimated as $O(p^{\frac{m+1}{2}}\log p^m)$. And now using Lemmas 4 and 5 we obtain the estimate of discrepancy for the sequence $\{x_t\}$, where $x_t=\frac{y(t)}{p^m}$, $t=1,2,\ldots,N$, $N\le p^{m-1}$

$$D_N^{(1)} \le \frac{3p^{\frac{m+1}{2}}}{N}\log N$$

This proves the equidistribution of the sequence $\{x_t\}$. Moreover, $h_1y(t) + h_2y(t + 1) + \cdots + h_sy(t + s - 1)$ be the polynomial which for the nontrivial set of coefficients $h_1, \ldots, h_s$ generates the polynomial $Y(t)$ that satisfies to condition of Lemma 8 and so the discrepancy of s-dimensional sequence $\{X_n^{(s)}\}$ has an estimate

$$\frac{s}{N} + \frac{p^{\frac{m+1}{2}}}{N}(3\log N)^s.$$

Therefore, the sequences produced by congruences (2) and (6) pass serial test for $s \leq p - 2$.

To obtain the lower bounds for discrepancy of sequences generated from elliptic curve we apply Lemma 7.

From Corollary 1 we can write

$$y(t) = y(0)(1 + A_1pt + A_2p^2t^2 + A_3p^{\lambda_3}t^3 + \cdots) \ (\text{mod} p^m)$$

Therefore, we have

$$\begin{aligned} y(t + k) = y(0)(1 &+ A_1p + 2A_2p^2 + 3A_3p^{\lambda_3} + \cdots)t \\ &+ (A_2t^2 + 3A_3p^{\lambda_3} + \cdots)t^2 \\ &+ (A_3p^{\lambda_3} + 2p^{\lambda_4}A_4 + \cdots)t^3 + \cdots \end{aligned}$$

And hence,

$$\begin{aligned} h_1y(t) + h_2y(t + 1) = &\textit{free term} + (A_1h_1 + A_1h_2 + 2A_2h_2p)pt \\ &+ (A_2h_1 + A_2h_2 + 3A_3h_2p)p^2t^2 \\ &+ p^{\lambda_3}t^3\psi(t) \end{aligned}$$

where $\psi(t)$ is a polynomial with coefficients from $\mathbb{Z}_{p^m}$.

By form of coefficients for A_1 and A_2 it is clear that we can find x_0 such that the coefficient at t in the last equality is divided at least by p^2 but the coefficient at t^2 exactly divided by p^2. Let define this conditions as (*).

Now Lemma 8 gives

$$\left| \sum_{t=0}^{p^{m-1}-1} e^{2\pi i \frac{h_1y(t)+h_2y(t+1)}{p^{m-1}}} \right| = \begin{cases} p^{\frac{m+\nu}{2}} & \textit{if} \text{ conditions (*) hold}, \\ 0 & \textit{otherwise}. \end{cases}$$

Theorem 1 *Let* $\{x_t\}$ *be the sequence of PRN's produced by elliptic curve* $y^2 \equiv x^3 + ax + b \pmod{p^m}$. *There exists the point* (x_0, y_0), $y_0 \neq 0, \infty$ *on the curve* $y^2 \equiv x^3 + ax + b \pmod{p}$ *such that the sequence of two-dimensional points* $\{X_t\}$, $X_t = (x_t, x_{t+1})$ *has discrepancy* $D_\tau^{(2)}$, $\tau = p^{m-1}$ *for which the following inequalities*

$$\frac{1}{4(\pi+2)h^*}p^{-\frac{m-1}{2}} \le D_\tau^{(2)} \le 3p^{-\frac{m-1}{2}}\log^2 p^m,$$

hold, where $h^* = \min(h_1, h_2)$, (h_1, h_2) *is a point from* $(\mathbb{Z}^*_{p^{m-1}})^2$ *with conditions (*).*

This theorem together with Lemma 8 shows that the obtained upper bound is, in general, the best possible up to the logarithmic factor for any inversive congruential sequence $\{(x_t, x_{t+1})\}$, $t \ge 0$ (defined by the congruence (2)).

Hence, on the average, the discrepancy $D_\tau^{(2)}$ has an order of magnitude between $p^{-(\frac{m-1}{2}-\nu)}$ and $p^{-(\frac{m-1}{2}-\nu)}\log^2 p^m$. In the certain sense, inversive congruential pseudorandom numbers model the random numbers very closely.

4 Conclusion

In conclusion let introduce the step by step algorithm of constructing the sequences of PRN's with a period $\tau = p^{m-1}$, associated with elliptic curve over finite ring $\mathbb{Z}_{p^m}$, $p > 3$ be a prime, $m \ge 3 \in \mathbb{N}$, that can be described by the following way.

First of all for $(x_0, y_0) \in E_p$, $(y_0, p) = 1$, i.e. for the point of elliptic curve $y^2 \equiv x^3 + ax + b \pmod{p}$ over $\mathbb{Z}_p$ with non-quadratic residue $-3a$ we construct the points $(x(t), y(t))$, $0 \le t \le p^{m-1} - 1$ which belongs to elliptic curve over $\mathbb{Z}_{p^m}$. Then

(1) we select (x_0, y_0), where $y_0 \ne 0$ and $y_0 \ne \infty$;
(2) calculate $x(t) \equiv x_0 + pt \pmod{p^m}$;
(3) calculate $y_i(0)$, $i = 1, 2$ as the solutions of congruence

$$y^2 \equiv x_0^3 + ax_0 + b \pmod{p^m};$$

(4) we will use the Taylor series for the function of ω at the point $\omega = 0$ in form

$$\sqrt{1 + (3\omega x_0^2 + 3\omega^2 x_0 + \omega^3)x_0'} = X_0 + X_1\omega + X_2\omega^2 + \cdots . \quad (13)$$

(here x_0' is the multiplicative inverse modulo p^m for $x_0^3 + ax_0^2 + b$).
(5) In (13) we put $\omega = pt$ and then modulo p^m we construct the following polynomial:

$$\begin{aligned}\varphi(t) &\equiv 1 + X_1 pt + X_2 p^2 t^2 + \cdots + X_s p^s t^s \\ &\equiv \Phi_0(x_0) + p^{\lambda_1}\Phi_1(x_0)t + \cdots + p^{\lambda_s}\Phi_s(x_0)t^s \pmod{p^m},\end{aligned}$$

where $\Phi_j(x_0) \in \mathbb{Z}$, $(\Phi_j(x_0), p) = 1$, $\lambda_j \in \mathbb{N}$, $\lambda_j \ge j\frac{p-2}{p-1}$, $j = 1, 2, \ldots, s$.
(6) This polynomials and the solutions $y_i(0)$, $i = 0, 1$ we use to construct the following representations modulo p^m:

$$y_i(t) \equiv y_i(0)(\Phi_0(x_0) + \Phi_1(x_0)p^{\lambda_1 t} + \cdots + \Phi_s(x_0)p^{\lambda_s t^s})$$

$$\equiv y_i(0)(1 + A_1 pt + A_2 p^2 t^2 + A_3 p^{\lambda_3} t^3 + \cdots + A_s p^{\lambda_s} t^s)$$

for each $i = 1, 2$, which produce two sequences of PRN's

$$\left\{ \frac{y_i(t)}{p^m} \right\}, \quad t = 0, 1, \ldots$$

with the period $\tau = p^{m-1}$.

Using the results obtained in previous sections we can say that the constructed sequence of PRN's, associated with elliptic curve $y^2 \equiv x^3 + ax + b \pmod{p^m}$, passes the serial test on pseudorandomness, and therefore may be used in cryptographic applications.

References

1. H. Niederreiter, *Random Number Generation and Quasi-Monte Carlo Methods* (SIAM, Philadelphia, 1992)
2. H. Niederreiter, Quasi-Monte Carlo methods and pseudorandom numbers. Bull. Amer. Math. Soc. **84**, 957–1041 (1978)
3. L.P. Postnikova, Distribution of solutions of the congruence $x^2 + y^2 \equiv 1\ (mod\ p^n)$. Matem. sb. **65**(2), 228–238 (1964) (in Russian)
4. I. Shparlinski, Pseudorandom number generators from elliptic curves. Contemp. Math. **477**, 121–141 (2009)
5. S. Varbanets, Exponential sums over the sequences of PRN's produced by inversive generators. Annales Univ. Sci. Budapest. Sect. Comp. **48**, 225–232 (2018)

Fractional Chaotic System Solutions and Their Impact on Chaotic Behaviour

Chunxiao Yang, Ina Taralova, and Jean Jacques Loiseau

Abstract This paper is devoted to the analysis of calculation methods for solving fractional chaotic systems and the impact of these different approaches on the behavior of the fractional chaotic system. Two widely used time domain fractional differential equations solving approaches are discussed, the fractional ABM corrector-predictor method based on Caputo fractional derivative definition, and the long memory calculation approach based on Grunwald fractional derivative. These numerical solutions calculation methods are employed to depict the phase portrait of a class of commensurate fractional chaotic systems. The Lyapunov exponent and bifurcation diagrams of the systems over various fractional orders and parameters are illustrated to detect the impact on the dynamics of the chaotic system applying different calculation approaches.

Keywords Fractional calculus · Numerical solution · Fractional Chaotic system · Non linear dynamics

1 Introduction

Chaos is a random-like behavior exhibited by many nonlinear dynamic systems. The very first proponent of this topic can be dated back to 1880 while the three body problem was studied [1]. Eighty years later, when Edward Lorenz worked on weather prediction, the so-called 'Lorenz attractor' was found [2]. By giving it a description and a poetic name of 'butterfly effect', the gate of the mathematical and

C. Yang (✉) · I. Taralova · J. J. Loiseau
Laboratoire des Sciences du Numérique de Nantes LS2N, UMR CNRS 6004 Ecole Centrale de Nantes, Nantes, France
e-mail: chunxiao.yang@ls2n.fr

I. Taralova
e-mail: ina.taralova@ls2n.fr

J. J. Loiseau
e-mail: jean-jacques.loiseau@ls2n.fr

C. H. Skiadas and Y. Dimotikalis (eds.), *14th Chaotic Modeling and Simulation International Conference*, Springer Proceedings in Complexity,
https://doi.org/10.1007/978-3-030-96964-6_36

scientific world in Chaos was opened. Since then, many researchers have tried to uncover the deterministic laws behind the apparently random states of disorder of different chaotic systems.

One of the characteristics of the chaotic system is that it is very sensitive to the initial conditions as described by the butterfly effect. This sensitivity can be measured by Lyapunov Exponent(LE) which calculates the rate of exponential divergence of trajectories starting from two close initial conditions. This characteristic also contributes to the application of chaotic systems in many domains of science and engineering, such as biology [3], economics [4], finance [5], cryptography [6, 7] and etc.

In the meantime, fractional calculus is considered as the generalization of classical integer-order integration and differentiation operators to real, or complex orders [8]. Many mathematicians have discussed the fractional calculus since 1695 by introducing different mathematical characterisations (definitions) for fractional derivative and integration. In many cases, these characterisations are equivalent if the initial conditions are ad hoc [9], and the most well known three are Rienmann-Liouville(RL), Grunwald-Letnikov(GL), and Caputo characterisations.

The analysis and discussion of fractional calculus remained purely in the domain of mathematics for centuries. It was not until the 1980s that the application of fractional calculus in the domain of science and engineering has started to be studied and explored. Due to the memory effect possessed by fractional calculus, it is considered to be suitable to model many real-life systems. After years of research, the fractional differential equations have now been used in diverse disciplines like physics, biology, economics, etc. [10, 11].

The fractional chaotic system also attracts a lot of attention. The difficulties for this research owes to the intricate geometric interpretation of fractional derivatives [12] and the fact that there exist, as mentioned above, different definitions for fractional derivatives. One basically considers continuous systems, and uses numerical methods to approximate the solution. In the case of a fractional system, the discrete approximating system may inherit the chaotic behaviour of the initial continuous system, but this relationship is somehow complex. What adds to the intricacy is that the chaotic behavior of the approximating can be different for different numerical methods employed to solve the fractional differential equations [13]. Therefore, the understanding of the impact on the chaoticity of the system applying one or another numerical calculation approaches is of great importance, in order to choose the most appropriate one for a given application.

In the following, two numerical calculation methods under GL and Caputo characterisation for fractional differential equations are recalled. Then, we employ both methods to obtain the states of two fractional chaotic systems extended from classical integer order chaotic system. The impact on the chaoticity of the systems applying the two approaches has been analyzed in terms of LE and from the aspect of bifurcation diagram and time responses.

2 Preliminaries on Fractional Calculus and Fractional Systems

In this section, some preliminaries on fractional calculus and fractional systems are introduced to give a rough idea on the topic. The widely-accepted stability criteria for a commensurate fractional system is also illustrated.

2.1 Fractional Calculus

As mentioned before, the fractional calculus studies the fractional derivative and integral which can be considered as the extension of classical integer order differentiation and integration to real or complex orders. In the long history of the study of fractional calculus, many mathematicians have contributed and introduced different characterisations(referred as 'definitions' in many papers) towards the topic. Here after, we give two well-known definitions Grünwald-Letnikov (GL) and Caputo definitions [14, 15].

The fractional derivatives under GL characterisation can be writen as

$$ {}_aD_t^\alpha f(t) = \lim_{h\to 0} \frac{1}{h^\alpha} \sum_{j=0}^{[\frac{t-a}{h}]} (-1)^j \binom{\alpha}{j} f(t-jh) \tag{1} $$

The term $[\frac{t-a}{h}]$ in (1) stands for the integer part of $\frac{t-a}{h}$; a and t are the bounds of the derivative operation ${}_aD_t^\alpha$ for $f(t)$; α represents the fractional derivative order. The term $\binom{\alpha}{j}$ in (1) is defined in (2), where $\Gamma(.)$ is the Euler Gamma function in the form of (3).

$$ \binom{\alpha}{j} = \frac{\Gamma(\alpha+1)}{\Gamma(j+1)\Gamma(\alpha-j+1)} \tag{2} $$

$$ \Gamma(\alpha) = \int_0^\infty \frac{t^{\alpha-1}}{e^t} dt \tag{3} $$

The Caputo type fractional derivative holds the form as following,

$$ {}_a^cD_t^\alpha f(t) = \frac{1}{\Gamma(n-\alpha)} \int_a^t \frac{f^{(n)}(\tau)}{(t-\tau)^{\alpha-n+1}} d\tau, \text{ for } n-1 < \alpha < n \tag{4} $$

where α denotes the fractional derivative order; a and t are the bounds for the operation; n is the smallest integer greater than α; $\Gamma(.)$ is the Euler Gamma function in (3); and $f^{(n)}(t)$ is the n-th derivative of $f(t)$.

The Caputo type fractional derivative is often used for engineering application since the fractional differential equations with this type of derivative can provide the applied problem with an interpretable initial condition.

2.2 Fractional System

A fractional system is a dynamic system which can be modeled by fractional differential equations [16]. A general form of fractional system is as follows,

$$\begin{aligned} {}_0D_t^{\alpha_i} x_i(t) &= f_i(x_1(t), x_2(t), ..., x_n(t), t) \\ x_i(0) &= c_i, i = 1, 2, ..., n. \end{aligned} \tag{5}$$

In (5), $x_i(0)(i = 1, 2, ...n)$ denotes the initial conditions for each component constituting the state vectors; $\alpha_i(i = 1, 2, ...n)$ is the fractional derivative order for i-th differential equations consisting the system, and f_i is a linear or non-linear function.

The equilibrium points of system (5) can be obtained by solving the equation $f_i(x) = 0(i = 1, 2, ..., n)$. If a commensurate system with $\alpha_i = \alpha, i = 1, 2, ..., n$ is considered, then, according to the stability theorem defined in [17], the equilibrium points are locally asymptotically stable if the eigenvalue of the Jacobian matrix of system (5) satisfies the following equation evaluated at equilibria.

$$|\arg(\text{eig}(\mathrm{J}))| = |\arg(\lambda_i)| > \alpha\frac{\pi}{2}, i = 1, 2, ..., n \tag{6}$$

where J denotes the Jacobian matrix of (5), $\lambda_i(i = 1, 2, ...n)$ are its eigenvalues.

3 Numerical Calculation Methods for Fractional Differential Equations

In this section, two numerical solutions calculation methods for fractional differential equations are introduced. The methods are based on Grünwald-Letnikov and Caputo fractional derivative characterisations.

3.1 Grünwald-Letnikov Calculation Method

The explicit numerical approximation of q-th derivative under GL characterisation at the points kh, $(h = 1, 2, ...)$ is expressed as follows [14]

$$_{(k-L_m)/h}D^{\alpha}_{t_k} f(t) \approx h^{-\alpha} \sum_{j=0}^{k} (-1)^j \binom{\alpha}{j} f(t_{k-j}). \tag{7}$$

In expression (7), L_m is the memory length; $t_k = kh$, where h is the calculation time step; the binomial coefficient $(-1)^j \binom{\alpha}{j}$ can be denoted as $c_j^{(\alpha)} (j = 0, 1, ...)$ which is expressed using the following expression [18],

$$c_0^{(\alpha)} = 1, c_j^{(\alpha)} = \left(1 - \frac{1+\alpha}{j}\right) c_{j-1}^{(\alpha)}. \tag{8}$$

Thus, the general numerical solution of the fractional differential equation described by (9) can be expressed as given in (10).

$$_aD_t^{\alpha} y(t) = f(y(t), t) \tag{9}$$

$$y(t_k) = f(y(t_k), t_k)h^{\alpha} - \sum_{j=\nu}^{k} c_j^{(\alpha)} y(t_{k-j})) \tag{10}$$

The sum in (10) stands for the memory term. If a 'long memory effect' is considered, then the lower index $\nu = 1$ for all k, otherwise $\nu = 1$ for $k < (L_m/h)$ and $\nu = k - L_m$ for $k > (L_m/h)$.

3.2 *Fractional ABM Corrector-Predictor Method*

The fractional ABM corrector-predictor method is another widely used time domain numerical calculation method in the domain of engineering. It is a generalization of the classical Adams-Bashforth-Moulton integrator which is used for the numerical calculation of classical first order problem.

From the analytical point of view, the fractional differential equations under Caputo characterization with initial conditions $y^k(0) = y_0^k, k = 0, 1, 2...m-1$ where $m := \lceil \alpha \rceil$, is equivalent to Volterra integral equation expressed as follows,

$$y(x) = \sum_{k=0}^{\lceil \alpha \rceil - 1} y_0^{(k)} \frac{x^k}{k!} + \frac{1}{\Gamma(\alpha)} \int_0^x (x-t)^{\alpha-1} f(t, y(t)) dt \tag{11}$$

The algorithm is developed on a uniform grid $\{t_n = nh : n = 0, 1, ...N\}$. The basic idea of the algorithm is to obtain the approximation of the latter point on the grid from the former point. Detailed formula derivation for the algorithm can be found in [19]. Here, we only give out the derived equations for the next states values in (12)–(15).

$$y_h(t_{n+1}) = \sum_{k=0}^{\lceil\alpha\rceil-1} \frac{t_{n+1}^k}{k!} y_0^{(k)} + \frac{h^\alpha}{\Gamma(\alpha+2)} f(t_{n+1}, y_h^P(t_{n+1})) \\ + \frac{h^\alpha}{\Gamma(\alpha+2)} \sum_{j=0}^{n} a_{j,n+1} f(t_j, y_h(t_j)), \tag{12}$$

$$y_h^P(t_{n+1}) = \sum_{k=0}^{\lceil\alpha\rceil-1} \frac{t_{n+1}^k}{k!} y_0^{(k)} + \frac{1}{\Gamma(\alpha)} \sum_{j=0}^{n} b_{j,n+1} f(t_j, y_h(t_j)). \tag{13}$$

$$a_{j,n+1} = \begin{cases} n^{\alpha+1} - (n-\alpha)(n+1)^\alpha, & \text{if } j = 0, \\ (n-j+2)^{\alpha+1} + (n-j)^{\alpha+1} - 2(n-j+1)^{\alpha+1}, & \text{if } 1 \le j \le n, \\ 1, & \text{if } j = n+1. \end{cases} \tag{14}$$

$$b_{j,n+1} = \frac{h^\alpha}{\alpha}((n+1-j)^\alpha - (n-j)^\alpha) \tag{15}$$

In the above equations, $y_h(t_{n+1})$ stands for the next state, $y_h^P(t_{n+1})$ denotes the predictor value for the next state, a and b are coefficients.

4 Fractional Chaotic Chen and Lu Systems

4.1 Fractional Chaotic Chen Systems

The system equation for fractional Chen system can be expressed as following[20],

$$f_c(x) = \begin{cases} D^{\alpha_c} x_1(t) = a_c(x_2(t) - x_1(t)) \\ D^{\alpha_c} x_2(t) = (c_c - a_c) x_1(t) - x_1(t) x_3(t) + c_c x_2(t) \\ D^{\alpha_c} x_3(t) = x_1(t) x_2(t) - b_c x_3(t) \end{cases} \tag{16}$$

In the equation, D^{α_c} denotes the fractional derivative with order α_c, (a_c, b_c, c_c) are the parameters of the system. The system is an extension from integer order chaotic Chen system studied in [21].

The equilibria of the system can be obtained through the same way as its original integer order system, by setting the right-hand side system equation equal to zero $f_c(x^*) = 0$ as given below,

$$\begin{cases} a_c(x_2(t) - x_1(t)) = 0 \\ (c_c - a_c) x_1(t) - x_1(t) x_3(t) + c_c x_2(t) = 0 \\ x_1(t) x_2(t) - b_c x_3(t) = 0 \end{cases} \tag{17}$$

The singularity of the equilibrium points can also be acquired through the classical method as given below, by evaluating the eigenvalue of the jacobian matrix of the system at equilibrium points.

$$\det(\lambda_c \mathrm{I} - \mathrm{J}_c) = \begin{bmatrix} \lambda_c + a_c & -a_c & 0 \\ -c_l + a_l + x_3^* & \lambda_c - c_c & x_1^* \\ -x_2^* & -x_1^* & \lambda_c + b_c \end{bmatrix} = 0 \tag{18}$$

$\mathrm{J_c}$ in (18) represents the Jacobian matrix of the system equation, I is the identity matrix, λ_c denotes the eigenvalue, and (x_1^*, x_2^*, x_3^*) stands for the equilibrium point. The singularity of the three equilibrium points of fractional Chen system for system parameters $(a_c, b_c, c_c) = (35, 3, 28)$ can be obtained through above analytical expressions and are given in Table 1.

4.2 Fractional Chaotic Lu System

The system equation for fractional chaotic Lu system extended from integer order Lu system can be described as follows [22],

$$f_l(x) = \begin{cases} D^{\alpha_l} x_1(t) = a_l(x_2(t) - x_1(t)) \\ D^{\alpha_l} x_2(t) = -x_1(t)x_3(t) + c_l x_2(t) \\ D^{\alpha_l} x_3(t) = x_1(t)x_2(t) - b_l x_3(t) \end{cases} \tag{19}$$

where D^{α_l} denotes the fractional derivative with order α_l, a_l, b_l, and c_l are the parameters of the system. The equilibrium points of the system can be acquired calculating the solutions of the following system of equations,

$$\begin{cases} a_l(x_2(t) - x_1(t)) = 0 \\ -x_1(t)x_3(t)x_1(t) + c_l x_2(t) = 0 \\ x_1(t)x_2(t) - b_l x_3(t) = 0 \end{cases} \tag{20}$$

The singularity of the equilibria can be obtained the same way as discussed previously for the fractional Chen system through the following identities,

$$\det(\lambda_l \mathrm{I} - \mathrm{J}_l) = \begin{bmatrix} \lambda_l + a_l & -a_l & 0 \\ x_3^* & \lambda_l - c_c & x_1^* \\ -x_2^* & -x_1^* & \lambda_l + b_l \end{bmatrix} = 0 \tag{21}$$

where $\mathrm{J_l}$ in (21) represents the Jacobian matrix of the fractional Lu system, λ_l denotes the eigenvalue, and (x_1^*, x_2^*, x_3^*) stands for the equilibrium point. When the parameters of the system is set to $(a_c, b_c, c_c) = (36, 3, 20)$, three equilibrium points $E_1^* = (0, 0, 0)$, $E_2^* = (7.460, 7.460, 20)$ and $E_3^* = (-7.460, -7.460, 20)$ can be obtained applying (20). The singularity of the equilibria is also given in Table 1.

Table 1 Fractional Chen and Lu systems' equilibria and their singularity

System	Equilibrium	Eigenvalue		Singularity
		λ_1	λ_2, λ_3	
Fractional Chen system	(0, 0, 0)	– 30.8359	23.8359, – 3	Saddle
	(−7.9379, −7.9379, 21)	– 18.4280	4.2140±14.8846i	Saddle focus
	(7.9379, 7.9379, 21)	– 18.4280	4.2140±14.8846i	Saddle focus
Fractional Lü system	(0, 0, 0)	−36	20, −3	Saddle
	(−7.460, −7.460, 20)	−22.6516	$1.8258 \pm 13.6887i$	Saddle focus
	(7.460, 7.460, 20)	−18.4280	$1.8258 \pm 13.6887i$	Saddle focus

5 Solutions for the Chaotic Systems Applying Different Approaches

In this section, the solutions for fractional Chen and Lu solutions are obtained applying both GL method and fractional ABM corrector-predictor method discussed in Sect. 3. The impact of the two approaches on the chaotic behavior of the systems are also discussed.

5.1 *Chaotic System Applying GL Method*

With the numerical solution of fractional differential equation calculated under GL method derived as in (10), the calculation for the states of fractional Chen system and fractional Lu system (expression (16) and (19)) can be expressed by the following identities (22) and (23), respectively.

$$\begin{cases} x_1(n) = (a_c(x_2(n) - x_1(n-1)))h^{\alpha_c} - \sum_{j=\nu}^{n} c_j^{(\alpha_c)} x_1(n-j) \\ x_2(n) = ((c_c - a_c)x_1(n) - x_1(n)x_3(n-1) + c_c x_3(n-1))h^{\alpha_c} - \sum_{j=\nu}^{n} c_j^{(\alpha_c)} x_2(n-j) \\ x_3(n) = (x_1(n)x_2(n) - b_c x_3(n-1))h^{\alpha_c} - \sum_{j=\nu}^{n} c_j^{(\alpha_c)} x_3(n-j) \end{cases} \tag{22}$$

$$
\begin{cases}
x_1(n) = (a_l(x_2(n-1) - x_1(n-1)))h^{\alpha_l} - \sum\limits_{j=\nu}^{n} c_j^{(\alpha_l)} x_1(n-j) \\
x_2(n) = (-x_1(n)x_3(n-1) + c_l x_2(n-1))h^{\alpha_l} - \sum\limits_{j=\nu}^{n} c_j^{(\alpha_l)} x_2(n-j) \\
x_3(n) = (x_1(n)x_2(n) - b_l x_3(n-1))h^{\alpha_l} - \sum\limits_{j=\nu}^{n} c_j^{(\alpha_l)} x_3(n-j)
\end{cases}
\quad (23)
$$

To be mentioned is that in our work, the 'long memory effect' is adopted applying GL method which means that the number ν in (22) and (23) is equal to 1. The time step h in the above equations is set to a fixed value 0.001.

We plotted the phase portraits of the two systems with fractional orders $\alpha_c = 0.9$ and $\alpha_l = 0.95$ in Fig. 1a and b, respectively. The parameters and initial conditions for Chen system are (35, 3, 28) and (−9, −5, 14). Those of Lu system are chosen to be (36, 3, 20) and (0.2, 0.5, 0.3).

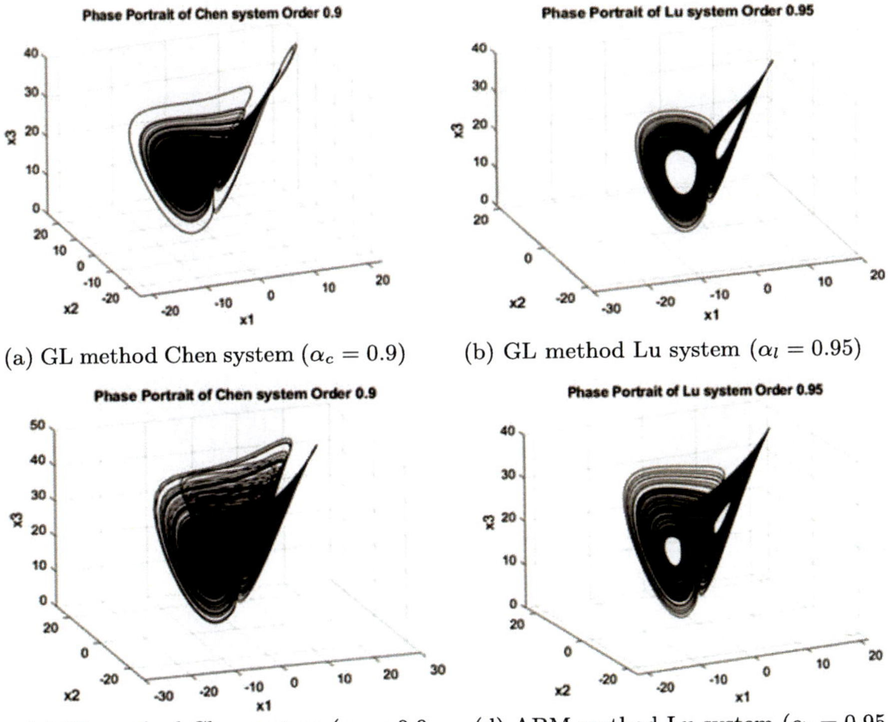

(a) GL method Chen system ($\alpha_c = 0.9$)
(b) GL method Lu system ($\alpha_l = 0.95$)
(c) GL method Chen system ($\alpha_c = 0.9$
(d) ABM method Lu system ($\alpha_l = 0.95$

Fig. 1 Phase portrait of fractional Chen and Lu systems characterized by GL and ABM method

5.2 Chaotic Systems Applying ABM Corrector-Predictor Approach

Based on the fractional ABM corrector-predictor numerical calculation approach for the solution of fractional differential equations given in (12)–(15), the states of fractional Chen system applying ABM predictor corrector approach can be expressed as follows,

$$\begin{aligned} X_c(n+1) &= X_c(0) + \frac{h^{\alpha_c}}{\Gamma(\alpha_c+2)} f_c(X_c^P(n+1)) \\ &+ \frac{h^{\alpha_c}}{\Gamma(\alpha_c+2)} \sum_{j=0}^{n} a_{j,n+1} f_c(X_c(j)) \end{aligned} \tag{24}$$

$$X_c^P(n+1) = X_c(0) + \frac{1}{\Gamma(\alpha_c)} \sum_{j=0}^{n} b_{j,n+1}^1 f_c(X_c(j))$$

$$\begin{aligned} a_{j,n+1} &= \begin{cases} n^{\alpha_c+1} - (n-\alpha_c)(n+1)^{\alpha_c}, & \text{if} j = 0, \\ (n-j+2)^{\alpha_c+1} + (n-j)^{\alpha_c+1} - 2(n-j+1)^{\alpha_c+1}, & \text{if} 1 \le j \le n, \\ 1, & \text{if} j = n+1. \end{cases} \\ b_{j,l+1} &= \frac{h^{\alpha_c}}{\alpha_c}((n+1-j)^{\alpha_c} - (n-j)^{\alpha_c}) \end{aligned} \tag{25}$$

In the above expressions, $X_c(n+1)$, $X_c(n)$ and $X_c^P(n+1)$ are state vectors composed of all the state components x_1, x_2, and x_3; α_c is the fractional order between $(0, 1)$; f_c stands for the Chen system equations.

The formula for the calculation of the states of fractional Lu system can be obtained by substituting the state vectors, fractional order and system equations in (24)–(25) with X_l, α_l and f_l where $0 < \alpha_l < 1$. The phase portraits of the two systems acquired employing the corrector-predictor approach are given in Fig. 1c and d, respectively. The fractional orders, parameters and initial conditions are the same as those for the GL method.

5.3 Impact on System Chaoticity with Chosen Methods

For the work in this section, we used the same parameters and initial conditions for the two systems as adopted in the previous section, which are $(a_c, b_c, c_c) = (35, 3, 28)$, $X_c(0) = (-9, -5, 14)$; $(a_l, b_l, c_l) = (36, 3, 20)$, $X_l(0) = (0.2, 0.5, 0.3)$, respectively. The time step h is set to 0.005. The MATLAB code [23] for ABM corrector-predictor method and [24] is employed for the following simulation and the calculation of LE.

According to the stability criteria introduced by the inequality (6) , the reference [17] states that for a fractional system $D^\alpha x = f(x)$ to remain chaotic, a necessary condition is keeping the eigenvalues λ in the unstable region, which gives the following equation for the fractional derivative order α.

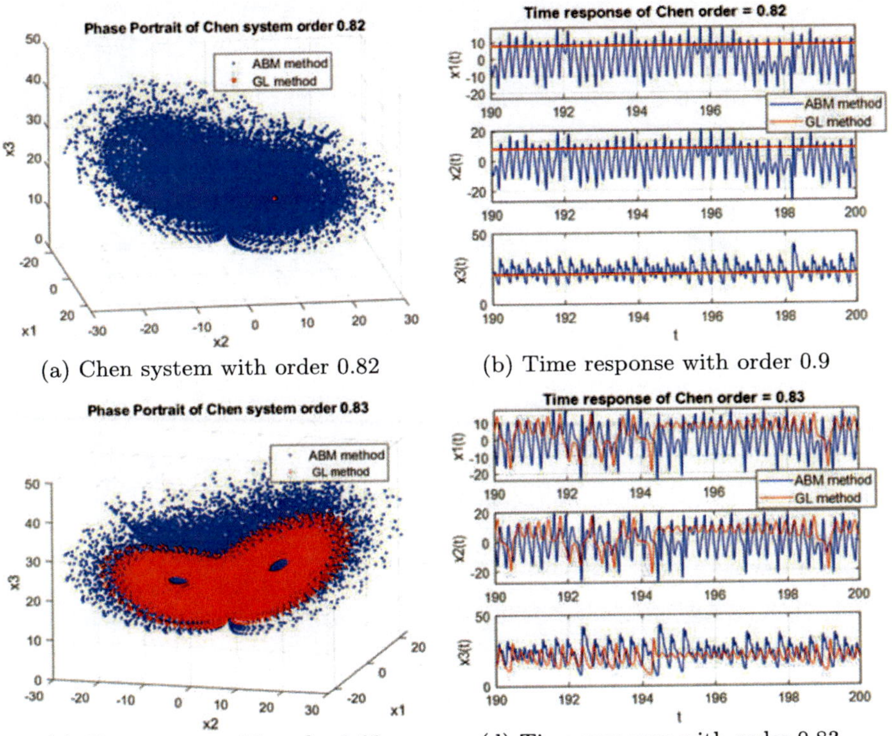

(a) Chen system with order 0.82 (b) Time response with order 0.9

(c) Chen system with order 0.83 (d) Time response with order 0.83

Fig. 2 Phase portrait and time response of Chen system at boundary fractional order values

$$\alpha > \frac{2}{\pi} \tan^{-1}\left(\frac{|Im(\lambda)|}{Re(\lambda)}\right) \tag{26}$$

where λ denotes the eigenvalues of the Jacobian matrix of the system, α is the commensurate fractional order. Therefore, for the given parameter values, the fractional chaotic Chen system should have a fractional order α_c greater than or equal to 0.8244.

In Fig. 2, we plot the phase portrait of fractional Chen system at boundary fractional values 0.82 and 0.83 applying both GL and ABM corrector-predictor methods. The time response of the last 2000 states obtained through both methods are also given. The states calculated by GL method is in red and ABM corrector-predictor in blue. It is not difficult to observe from Fig. 2a and c that with order 0.82 there are only one red point in the figure, which indicates that the states stay at the same fixed point applying GL method. Whereas for the applied ABM method(blue dots), they appear to have the shape of the attractors. When the system order is equal to 0.83, both methods display the shapes with attractors. This indicates that when applying GL calculation method with long memory effect, the system's dynamic behavior is in accordance with the stability criteria given by equation (26). While the ABM cal-

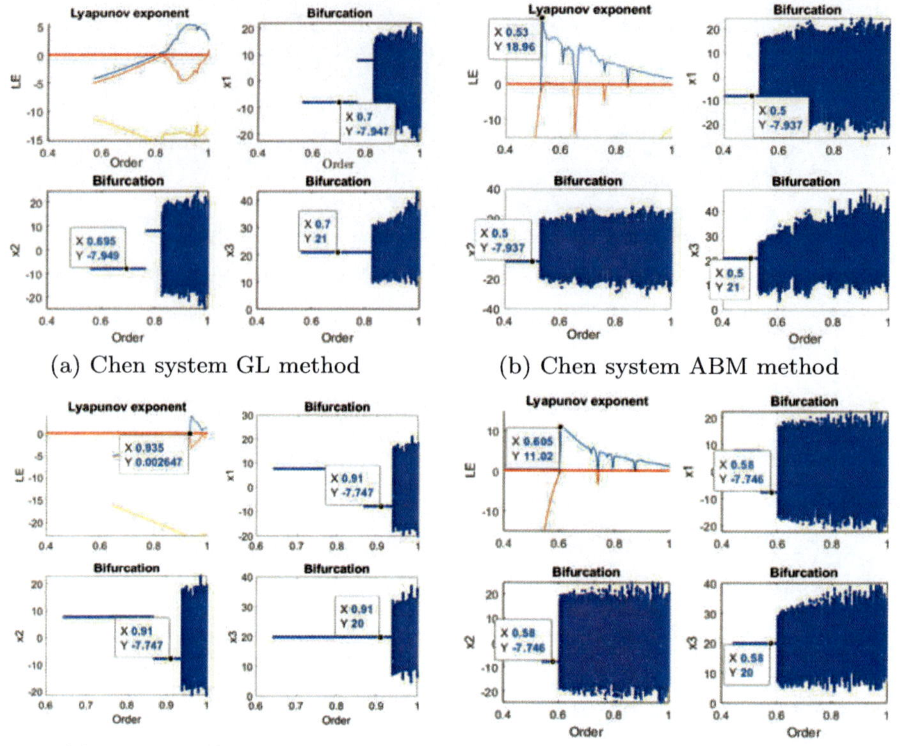

(a) Chen system GL method

(b) Chen system ABM method

(c) Lu system GL method

(d) Lu system ABM method

Fig. 3 LE and bifurcation results for Chen and Lu systems over different fractional derivatives employing different methods $(a_c, b_c, c_c) = (35, 3, 28)$, $(a_l, b_l, c_l) = (36, 3, 20)$

culation method applied in this paper provides the system with a smaller derivative order for the system to be

The time response figures given by Fig. 2b and d confirm the founding. The blue curve stands for the states obtain through ABM method and red for GL. It is clear that for derivative order 0.82, the red attractors stays at the same value for the three state vector components x_1, x_2 and x_3, while the blue curves appear to be oscillating.

We also give the Lyapunov exponent and bifurcation diagrams over different fractional orders of fractional Chen and Lu systems in Fig. 3. For each fractional derivative orders, 10^4 states were generated and the LEs were calculated throughout the iterations. The LE spectrum curves in 3a and 3b are obtained by combining LE values of the last iteration for every evaluated order. The plots show that only x_1 component possesses LE value greater than 0 applying both methods. It can be observed that applying ABM corrector-predictor approach, for the fractional Chen system, the LE of x_1 greater than 0 appears before order 0.53, whereas for GL method, the LE exceeds 0 after fractional order of 0.8. The LEs for fractional Lu system calculated using both methods show the similar results, with ABM method

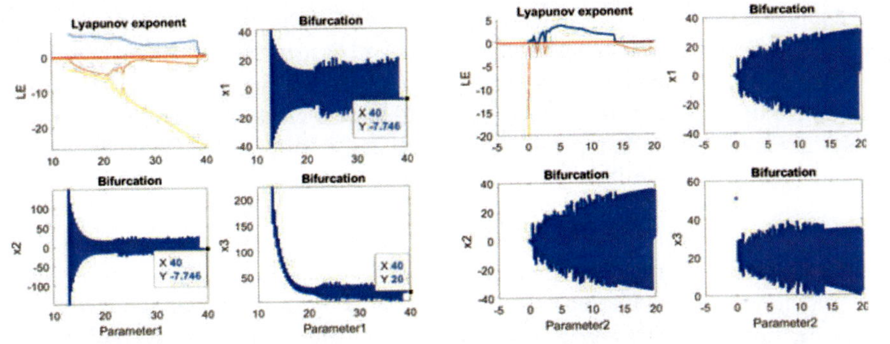

(a) Lu system GL method a_l LE results (b) Lu system ABM method a_l LE results

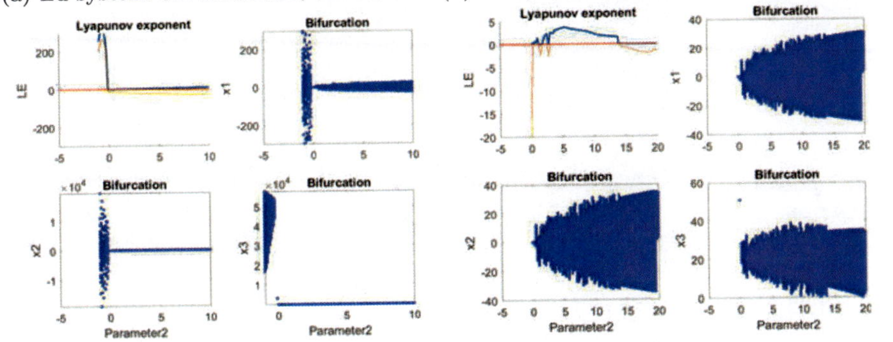

(c) Lu system GL method b_l LE results (d) Lu system ABM method b_l LE results

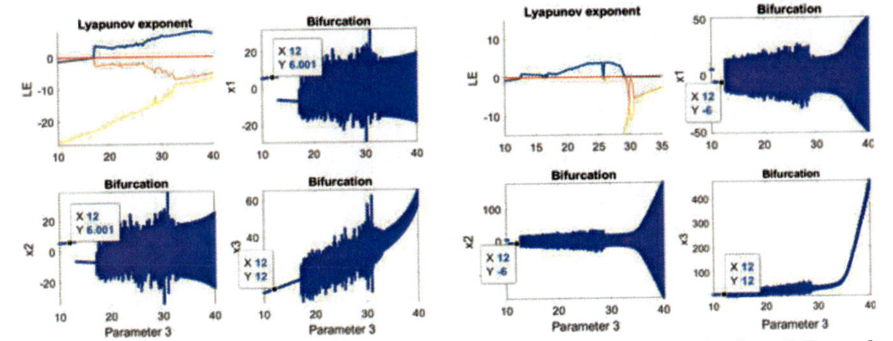

(e) Lu system GL method c_l LE results (f) Lu system ABM method c_l LE results

Fig. 4 LE and bifurcation results for Chen and Lu systems over different fractional parameters employing different methods having a smaller chaotic fractional derivative value. This is in accordance with our previous findings concerning the phase portrait and time response which draws to the conclusion that GL method give a more accurate approximation of original fractional system. Apart from this, from the y-coordinates of the bifurcation diagram where the system is non-chaotic, it can be observed that the solution obtained using ABM method stays at the equilibrium point as obtained through analytical analysis.

The LEs results and bifurcation diagram over different parameters of the fractional Lu system are also given in Fig. 4 to illustrate the dynamics possessed by the system. We set the system fractional order fixed to 0.9. It can be observed that applying different numerical calculation methods, the system dynamics is quite different. It is worth mentioning that the results for different parameters are conducted by changing one parameter at a time and fixing the other two unchanged.

6 Conclusion

In this paper, we recalled two numerical solutions calculation methods for fractional differential equations adopting Grünward-Leinikov and Caputo characterization of fractional derivative, respectively. Two fractional chaotic systems, fractional Chen system and fractional Lu system are discussed and their discretized states were calculated employing both methods. The results show that compared to the adopted ABM corrector-predictor method, the GL approach with long memory effect provide the original fractional system with a better approximation in coherence with the analytical studies. On the contrary, employing ABM method, the approximation accuracy appears to be deteriorated. However, in terms of chaoticity, it has a greater chaotic range for fractional derivatives.

References

1. H. Poincaré. Sur le problème des trois corps et les équations de la dynamique. Divergence des séries de M. Lindstedt. Acta Mathematica **13**(1–2): 1–270 (1890)
2. E.N. Lorenz, The predictability of hydrodynamic flow. Trans. New York Acad. Sci. **25**(4), 409–432 (1963)
3. E. Liz, A. Ruiz-Herrera, Chaos in discrete structured population models. SIAM J. Appl. Dyn. Syst. **11**(4), 1200–1214 (2012)
4. C. Kyrtsou, W. Labys, Evidence for chaotic dependence between US inflation and commodity prices. J. Macroecon. **28**(1), 256–266 (2006)
5. J. Fernando, Applying the theory of chaos and a complex model of health to establish relations among financial indicators. Procedia Computer Sci. **3**, 982–986 (2011)
6. Z. Qiao, I. Taralova, S. El Assad, Efficient pseudo-chaotic number generator for cryptographic applications. Int. J. Intell. Computing Res. **11**, 1041–1048 (2020)
7. M. Babaei, A novel text and image encryption method based on chaos theory and DNA computing. Nat. Comput. **12**(1), 101–107 (2013)
8. I. Petráš, *Fractional-Order Nonlinear Systems* (Springer, Berlin, Heidelberg, 2011)

9. K. Diethelm, *The Analysis of Fractional Differential Equations* (Springer, Berlin, Heidelberg, 2010)
10. F. Mainardi, *Fractional Calculus and Waves Linear Viscoelasticity: An Introduction to Mathematical Models* (Imperial College Press, London, UK, 2010)
11. V.E. Tarasov, V.V. Tarasova, Macroeconomic models with long dynamic memory: Fractional calculus approach. Appl. Math. Comput. **338**, 466–486 (2018)
12. T. Li and M. Yang et al., A novel image encryption algorithm based on a fractional-order hyperchaotic system and DNA computing. Complexity **2017** (Special issue, 2017)
13. C. Yang, I. Taralova et al., Design of a fractional pseudo-chaotic random number generator. Int. J. Chaotic Comput. **7**(1), 166–178 (2021)
14. I. Podlubny, *Fractional Differential Equations* (Academic Press, San Diego, 1999)
15. M. Caputo, Linear models of dissipation whose Q is almost frequency independent-II. Geophys. J. Int. **13**(5), 529–539 (1967)
16. B.J. West, M. Bologna, P. Grigolini. *Physics of Fractal Operators* (Springer, New York, 2003), pp. 235–270
17. M.S. Tavazoei, M. Haeri, A necessary condition for double scroll attractor existence in fractional-order systems. Phys. Lett. A. **367**, 102–113 (2007)
18. L. Dorcak. Numerical models for the simulation of the fractional-order control systems, in *UEF-04-94, The Academy of Sciences, Inst. of Experimental Phsic, Kosice, Slovakia* (1994)
19. K. Diethelm, N.J. Ford, A. Freed, A predictor-corrector approach for the numerical solution of fractional differential equations. Nonlinear Dyn. 2**9**, 3–22 (2002)
20. J. Lu, G. Chen, A note on the fractional-order Chen system. Chaos, Solitons and Fractals. **27**, 685–688 (2006)
21. J. Lu, G. Chen, A new chaotic attractor coined. Int. J. Bifurcat. Chaos **12**, 659–661 (2002)
22. W.H. Deng, C.P. Li, Chaos synchronization of the fractional L*ü* system. Physica A. **353**, 61–72 (2005)
23. R. Garrappa, *Predictor-Corrector PECE Method for Fractional Differential Equations*. MATLAB Central File Exchange. Retrieved 29 June 2021
24. M.F. Danca, N. Kuznetsov, Matlab code for Lyapunov exponents of fractional-order systems. Int. J. Bifurcat. Chaos **28**(5), 1850067 (2018)

Coupled FitzHugh-Nagumo Type Neurons Driven by External Voltage Stimulation

Jakub Záthurecký and Lenka Přibylová

Abstract We have extended some results of previous works on coupled FitzHugh-Nagumo type neurons stimulated by an external alternate voltage source. At first a few electronic circuits modeling the influence of brain waves on particular groups of coupled neurons are constructed. Bifurcation analysis of limit cycle dynamics is carried out, and route to chaotic dynamics is described with respect to coupling strength and forcing amplitude. This analysis enables constructing systems of coupled oscillators with specified dynamic behavior. We show that for appropriate values of parameters, a chimera-like state can be observed in a system of four coupled forced oscillators. A similar approach can generally be used, and the chimera state in coupled systems may be explained by means of bifurcation theory.

Keywords FitzHugh-Nagumo · Neuronal cells · Brain waves · Chaos · Chimera state

1 Introduction

This short text is inspired by the works Kyprianidis et al. [7] and Kyprianidis and Makri [8] about coupled FitzHugh-Nagumo oscillators. Nonlinear dynamics, especially chaos theory, is becoming a very important tool in understanding brain activity, as can be seen, for example, in Faure and Korn [2] and Korn and Faure [5]. Here we attempt to demonstrate this fact by using bifurcation theory to explain some kinds of behavior in systems of coupled FHN type neurons. At first, we recall basic results about dynamics of a single oscillator driven by external periodic voltage source. Such a system can model a neuron influenced by brain waves (an externally driven

J. Záthurecký (✉) · L. Přibylová
Department of Mathematics and Statistics, Faculty of Science, Masaryk University, Kotlářská 2, 61137 Brno, Czechia
e-mail: 437099@mail.muni.cz

L. Přibylová
e-mail: pribylova@math.muni.cz

C. H. Skiadas and Y. Dimotikalis (eds.), *14th Chaotic Modeling and Simulation International Conference*, Springer Proceedings in Complexity,
https://doi.org/10.1007/978-3-030-96964-6_37

FHN type neuron). We describe an electric circuit that is characterised by the studied equations. Next, we investigate two coupled oscillators driven by the same voltage source as a model of two externally driven FHN type coupled neurons influenced by brain waves. In every case, the coupling is realized by a resistor, hence we model gap junctions between cells.

Finally, we use these results to construct a system of four coupled oscillators exhibiting a stable state similar to a chimera state. As a model of four neurons coupled in a ring topology, we construct a minimum system that exhibits such dynamics with two groups of two mutually synchronized neurons with two different dynamic behaviors. Bifurcation analysis (using continuation program MATCONT, Dhooge et al. [1]) explains the choice of parameters for presented synchronizations and chimera-like states.

2 Dynamics of Externally Driven FHN Type Neuron

The starting point of the work Kyprianidis et al. [7] is an electric circuit depicted in Fig. 1. The component denoted by N_R is a resistor with a cubic $i - v$ characteristic. The dynamics of this system is described by the following equations, that can be derived using Kirchhoff's laws and dimensional analysis:

$$\begin{aligned} x' &= x - \frac{x^3}{3} - y + \varepsilon(u - x), \\ y' &= k(a - by + x), \end{aligned}$$

where x is proportional to the voltage v, y is proportional to the current i_L, ε is proportional to the reciprocal of the resistance R_S and finally u is proportional to the voltage v_S. The alternate voltage u is given by the formula $u(t) = U_0 \cos 2\pi\omega t$. To make the system autonomous, non-stiff, and suitable for bifurcation continuations, we attach two more equations:

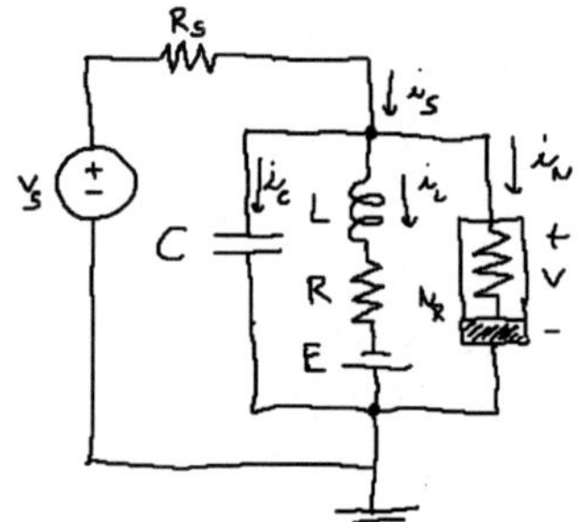

Fig. 1 Electric circuit modelling FitzHugh-Nagumo equations

$$u' = 2\pi\omega(\mu u - w - u(u^2 + w^2)),$$
$$w' = 2\pi\omega(u + \mu w - w(u^2 + w^2)).$$

Since these equations exhibit a stable solution given by the formula

$$(u, w)(t) = (\sqrt{\mu}\cos(2\pi\omega t), \sqrt{\mu}\sin(2\pi\omega t)),$$

the parameter $\mu, \mu \geq 0$, corresponds to the earlier introduced parameter $U_0, \mu = {U_0}^2$. From now we assume the following values of parameters: $a = 0.7, b = 0.8, k = 0.1$ and $\omega = 0.16$. This is a typical choice that leads to oscillatory dynamics (from regular neuron spiking to more complex firing patterns). The nonlinear phenomenon is generic and another parameter choice in domain of the limit cycle stability can be taken with analogous results. We briefly reconstruct a bifurcation analysis from the original paper. A stable (globally attractive) limit cycle undergoes period-doubling cascade of bifurcations with respect to parameter μ that corresponds to variation in amplitude of the external forcing. The results are illustrated in Figs. 2, 3, 4, and 5.

Two-parameter bifurcation curves with respect to parameters μ and ε are depicted at Fig. 6 (numerically computed with MATCONT, Dhooge et al. [1]). Evidently, the occurrence of oscillations with different frequencies depends both on the amplitude of the external waves and on the strength of the coupling. The selection of only one of the parameters is not sufficient to create a complete cascade of period-doubling. Various levels of coupling give birth to bifurcation diagrams with bubbles that correspond to bounded parts of the Sharkovski sequence of periods with respect to parameter μ. For a deeper insight into the theory of two-parameter bifurcations of limit cycles, see Kuznetsov [6] or Wiggins [11].

According to the mentioned figures, a chaotic bubble appears for $\varepsilon = 0.15$. Especially, an increase of the parameter μ from initial values around 10 or a decrease of this parameter from initial values around 110 leads to an infinite cascade of period-doubling bifurcations. Positive values of maximal Lyapunov exponent indicate the

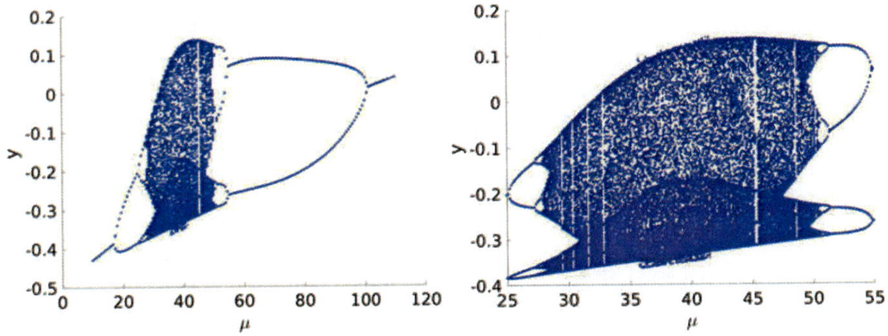

Fig. 2 Poincaré section by the hypersurface $x = -0.75$ for different values of the parameter μ, values of y are plotted. The value of the parameter ε is 0.15

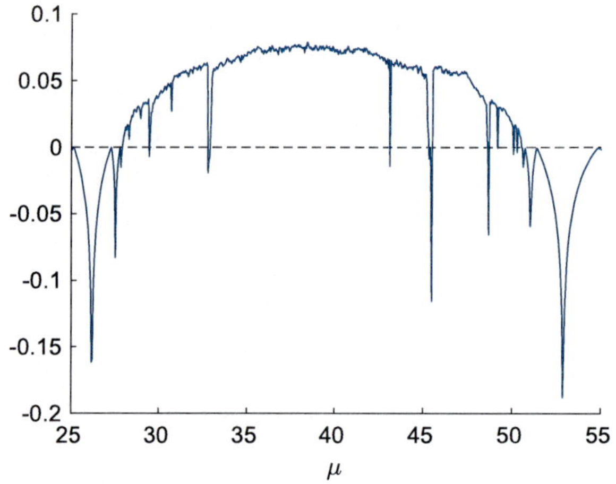

Fig. 3 Maximal Lyapunov exponent for different values of the parameter μ, $\varepsilon = 0.15$

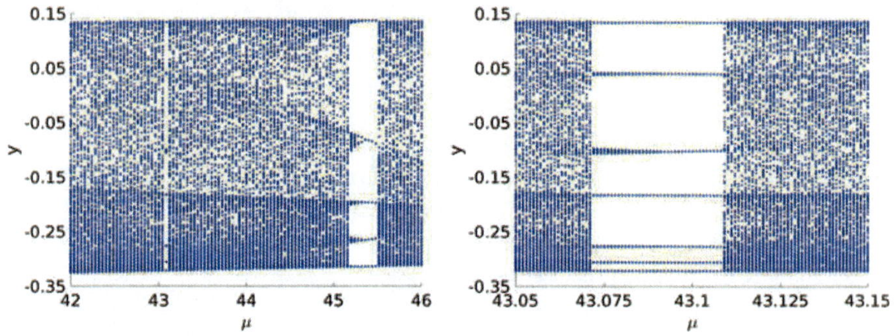

Fig. 4 Zoomed windows of stable cycles of odd period. The value of the parameter ε is 0.15

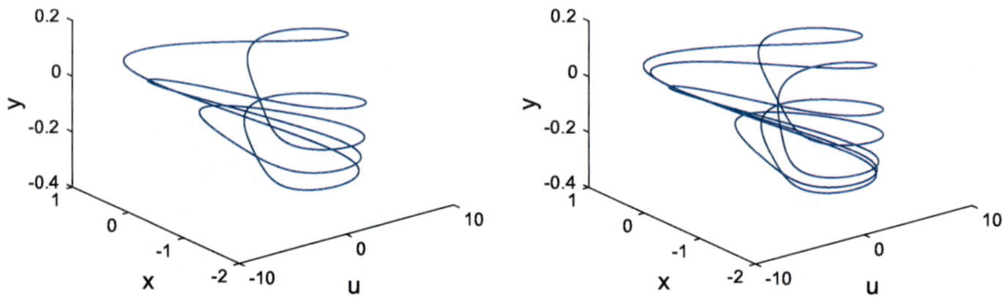

Fig. 5 Left: 5-cycle for $\mu = 45.4$, right: 7-cycle for $\mu = 43.1$. The value of the parameter ε is 0.15

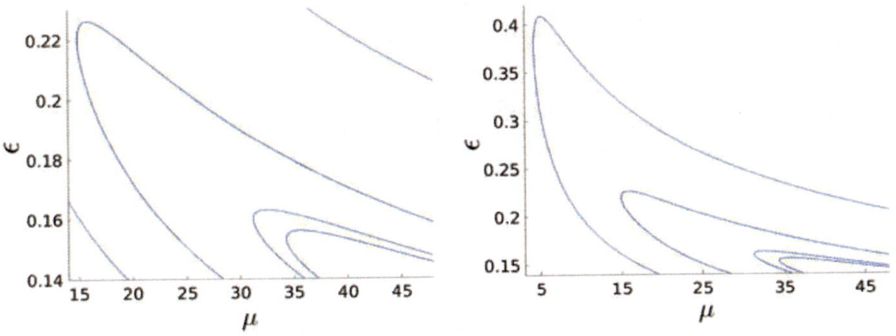

Fig. 6 Right: Bifurcation curves of a stable limit cycle (from the left upper corner to the right lower corner) of period-doubling (2-cycle birth), period-doubling (4-cycle birth), fold bifurcation of 5-cycle, and fold bifurcation of 7-cycle, continued in MATCONT. Left: Magnification of the right picture

presence of a chaotic attractor for appropriate values of the parameter μ. Visible windows corresponding to stable periodic orbits with odd period can be observed after magnification (recall Sharkovski ordering). Particularly, a 5-cycle and a 7-cycle windows are large enough to be detected easily. Curves of fold bifurcations of these cycles can be used for specifying borders of the stable oscillatory dynamics and domains of the chaotic attractor in $\varepsilon - \mu$ plane. Evidently, the cell exhibits chaotic behaviour for example for the values $(\varepsilon, \mu) = (0.15, 38)$ (see also Fig. 4), whereas the choice $(\varepsilon, \mu) = (0.3, 38)$ lead to periodic oscillations. We can possibly use a different pair of parameter values. Bifurcation diagrams in Fig. 6 agree with the results of the original paper Kyprianidis et al. [7], which demonstrate an extinction of the chaotic bubble with the increase of the parameter ε.

3 Dynamics of Two Coupled Externally Driven FHN Type Neurons

Further, we study a system of two coupled oscillators, both of which are driven by the same alternate voltage source. This situation can be interpreted as an influence of brain waves on two coupled neurons and is represented by the electric circuit depicted in Fig. 7. Again, Kirchhoff's laws and dimensional analysis give us the following set of equations which describes dynamics of the circuit:

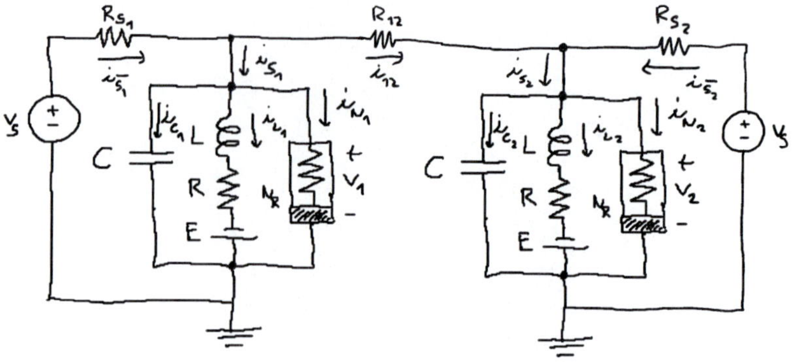

Fig. 7 Electric circuit modelling two coupled FitzHugh-Nagumo oscillators

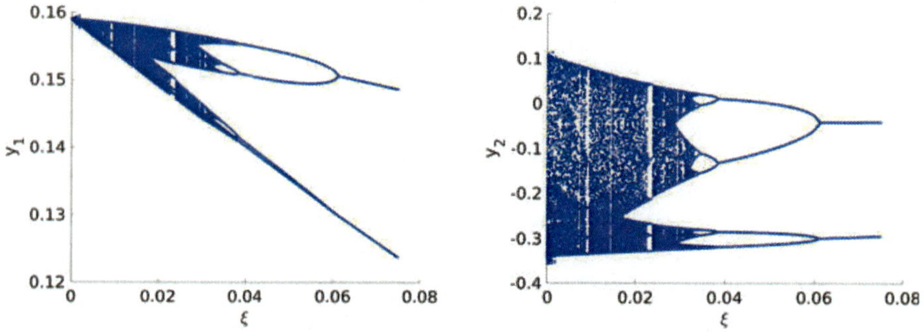

Fig. 8 Left: Poincaré section by the hypersurface $x_1 = -0.75$ for different values of the parameter ξ, the values of y_1 are plotted. Right: Poincaré section by the hypersurface $x_2 = -0.75$ for different values of the parameter ξ, the values of y_2 are plotted. The remaining parameters are $(\varepsilon_1, \varepsilon_2, \mu) = (0.3, 0.15, 38)$

$$
\begin{aligned}
x_1' &= x_1 - \frac{x_1^3}{3} - y_1 + \varepsilon_1(u - x_1) + \xi(x_2 - x_1),\\
y_1' &= k(a - by_1 + x_1),\\
x_2' &= x_2 - \frac{x_2^3}{3} - y_2 + \varepsilon_2(u - x_2) + \xi(x_1 - x_2),\\
y_2' &= k(a - by_2 + x_2).
\end{aligned}
$$

The new parameter ξ is proportional to the reciprocal of the resistance R_{12}. Now, we investigate an influence of the coupling strength ξ on attenuation of chaos in the first cell, hence we choose $(\varepsilon_1, \varepsilon_2, \mu) = (0.3, 0.15, 38)$. The results can be seen in Figs. 8, 9 and 10.

For sufficiently large ξ, the first cell makes the second one oscillate periodically, which is a natural state of the first one. The second cell would behave chaotically without the coupling. On the other hand, we can say that the strength of coupling is not sufficient for changing the natural behavior, if ξ is small. Poincaré sections

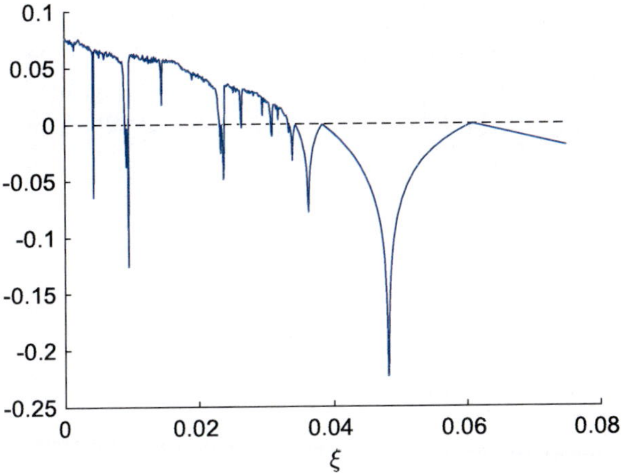

Fig. 9 Maximal Lyapunov exponent for different values of the parameter ξ

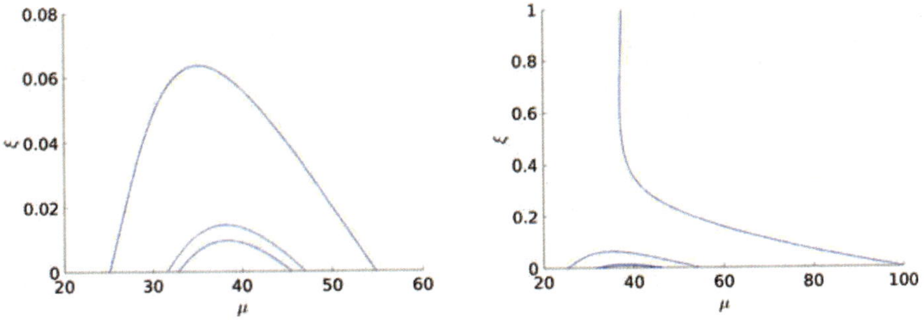

Fig. 10 Right: Bifurcation curves of a stable limit cycle (from the right upper corner to the left lower corner) of period-doubling (2-cycle birth), period-doubling (4-cycle birth), fold bifurcation of 7-cycle and fold bifurcation of 5-cycle. Left: Magnification of the right picture

indicate a presence of small deviations from periodic oscillation in the case of the first cell. However, they do not change behaviour qualitatively and can be considered as fluctuations caused by the environment. Bifurcation curves demonstrate an extinction of the chaotic bubble with the increase of the parameter ξ, see Fig. 11.

Before we proceed to the case of four oscillators, we focus on the influence of the parameter ξ on emerging of synchronization. We deal with two situations in which we couple a chaotic cell together with a periodically oscillating cell and two chaotic cells together, respectively. The dependence of synchronization on the parameter ξ is depicted in Fig. 12. In the first case, we couple two cells driven by different forces, which leads to coupling between different oscillators. It is not surprising that there will always be some difference between phases of these oscillators, no matter how strong the coupling is. On the other hand, we know that two identical chaotic

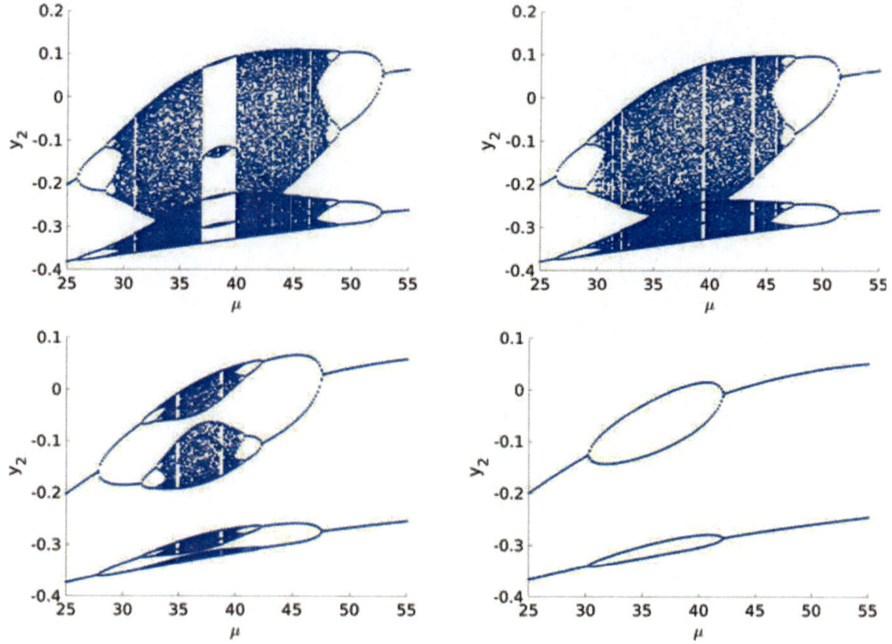

Fig. 11 Poincaré section by the hypersurface $x_2 = -0.75$ for different values of the parameter μ, the values of y_2 are plotted. The values of the parameter ξ are as follows $\xi_1 = 0.009$, $\xi_2 = 0.014$, $\xi_3 = 0.03$ and $\xi_4 = 0.05$. The remaining parameters are chosen to be $(\varepsilon_1, \varepsilon_2) = (0.3, 0.15)$

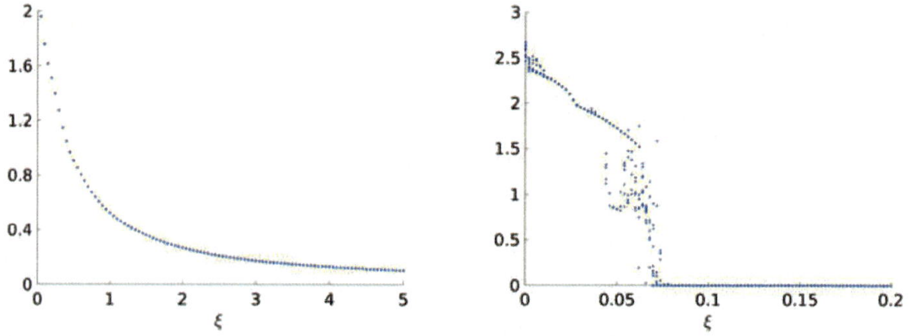

Fig. 12 Dependence of synchronization on the parameter ξ. For the values of the parameter ξ from appropriate intervals ten sets of initial conditions were chosen randomly. Uniform distribution on the interval $[-2, 2]$ has been used. Maxima of deviations $|x_1(t) - x_2(t)|$ for t in the interval $[1000, 2000]$ are plotted. The values of parameters are $(\varepsilon_1, \varepsilon_2, \mu) = (0.3, 0.15, 38)$ for the left diagram and $(\varepsilon_1, \varepsilon_2, \mu) = (0.15, 0.15, 38)$ for the right diagram

oscillators are able to achieve complete synchronization if the coupling is strong enough. A comprehensive treatment of a concept of synchronization can be found in Pikovsky et al. [9].

4 Chimera-Like States

The main result of this contribution is the explanation of the chimera-like behavior of a system of coupled oscillators by means of bifurcation theory. We can easily use previous bifurcation analysis results to construct various systems of four coupled cells exhibiting synchronizations/non-synchronizations, chaotic behavior and partial synchronizations, and especially stable chimera state, which means that two groups of oscillators have qualitatively different dynamics. One group of two oscillators behave chaotically (in a synchronized or non-synchronized state), the other group exhibits synchronized oscillations. The same approach can be analogously used for other systems of Hodgkin-Huxley type, Morris-Lecar type, Hindmarsh-Rose type, or others.

We are interested in an electric circuit depicted in Fig. 13 with the following equations governing its dynamics:

$$
\begin{aligned}
x_1' &= x_1 - \frac{x_1^3}{3} - y_1 + \varepsilon_1(u - x_1) + \xi_{12}(x_2 - x_1) + \xi_{13}(x_3 - x_1),\\
y_1' &= k(a - by_1 + x_1),\\
x_2' &= x_2 - \frac{x_2^3}{3} - y_2 + \varepsilon_2(u - x_2) + \xi_{12}(x_1 - x_2) + \xi_{24}(x_4 - x_2),\\
y_2' &= k(a - by_2 + x_2),\\
x_3' &= x_3 - \frac{x_3^3}{3} - y_3 + \varepsilon_3(u - x_3) + \xi_{13}(x_1 - x_3) + \xi_{34}(x_4 - x_3),\\
y_3' &= k(a - by_3 + x_3),\\
x_4' &= x_4 - \frac{x_4^3}{3} - y_4 + \varepsilon_4(u - x_4) + \xi_{24}(x_2 - x_4) + \xi_{34}(x_3 - x_4),\\
y_4' &= k(a - by_4 + x_4).
\end{aligned}
$$

Three different situations are going to be considered all of which are illustrated in Figs. 14, 15 and 16. The choice of parameters in Fig. 14 represents a situation in which the natural state of the first cell is periodic oscillations, while the remaining cells behave chaotically. Moreover, the coupling between the first two cells is strong enough to suppress chaos in the second cell. Finally, chaotic cells are not able to synchronize, and respective groups do not influence each other too much. Since the first two cells are different oscillators, they are not able to achieve complete synchronization. However, frequency locking appears obviously. If the natural state of the

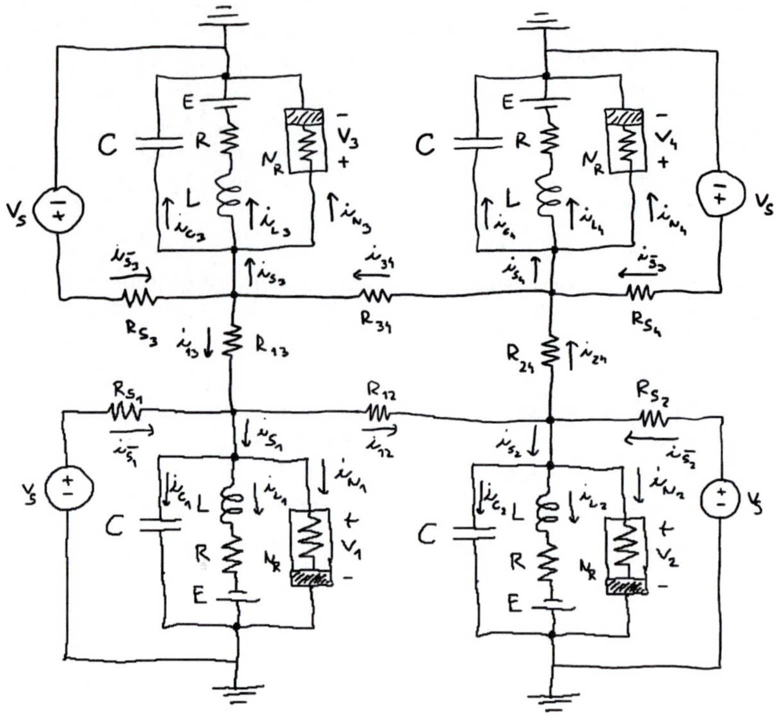

Fig. 13 Electric circuit modelling four coupled FitzHugh-Nagumo oscillators

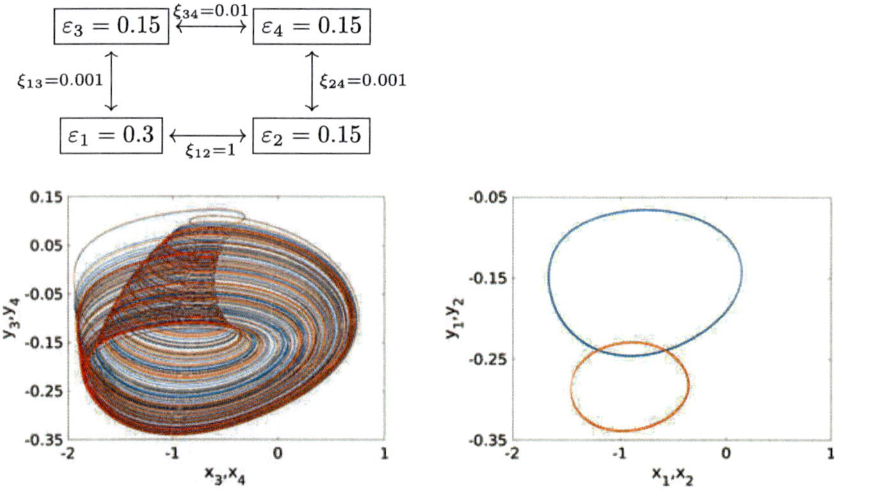

Fig. 14 Bottom: Projection of the trajectory on the phase space of the third and of the fourth cell on the left and on the phase space of the first and of the second cell on the right, respectively

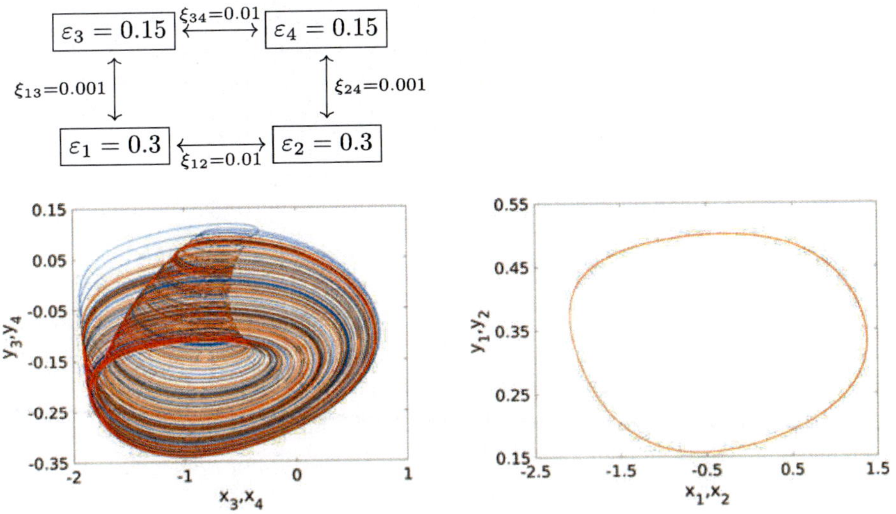

Fig. 15 Bottom: Projection of the trajectory on the phase space of the third and of the fourth cell on the left and on the phase space of the first and of the second cell on the right, respectively

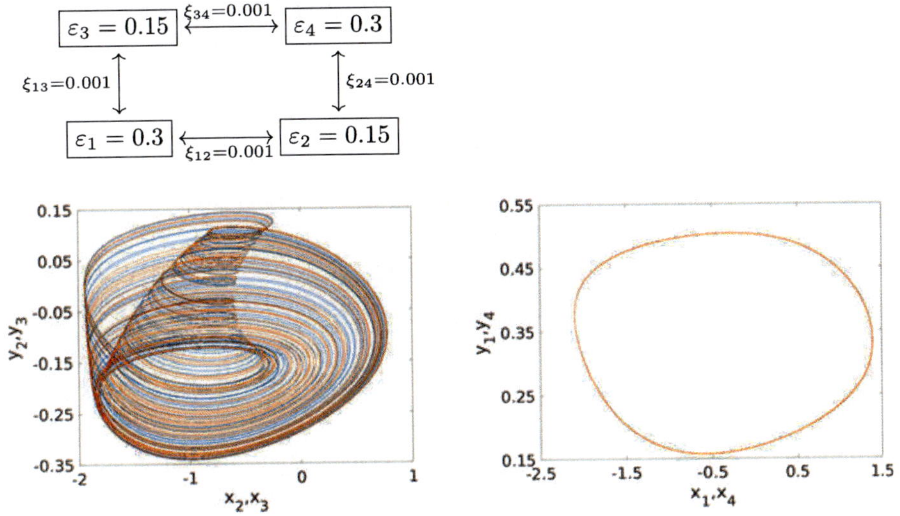

Fig. 16 Bottom: Projection of the trajectory on the phase space of the second and of the third cell on the left and on the phase space of the first and of the fourth cell on the right, respectively

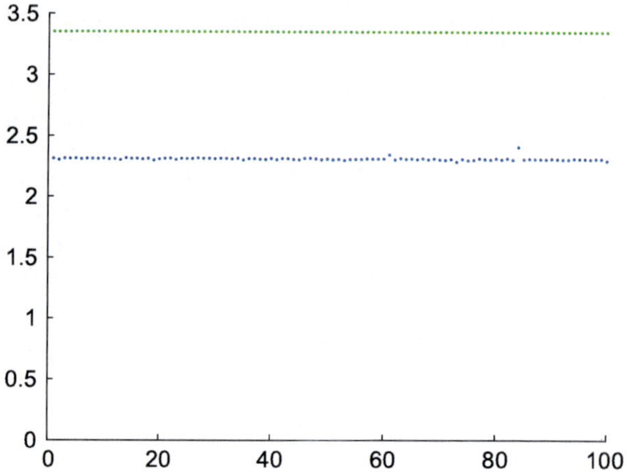

Fig. 17 Deviations between pairs of oscillators for one hundred initial conditions chosen randomly from the interval $[-2, 2]$ with uniform distribution. Maxima of deviations $|x_1(t) - x_2(t)|$ for $t \in [1000, 2000]$ are displayed in green colour and maxima of deviations $|x_3(t) - x_4(t)|$ for $t \in [1000, 2000]$ are displayed in blue colour. Parameters are the same as in Fig. 15

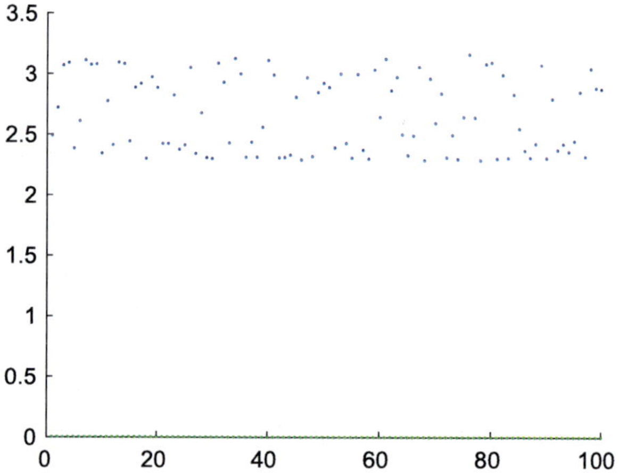

Fig. 18 Deviations between pairs of oscillators for one hundred initial conditions chosen randomly from the interval $[-2, 2]$ with uniform distribution. Maxima of deviations $|x_1(t) - x_4(t)|$ for $t \in [1000, 2000]$ are displayed in green colour and maxima of deviations $|x_2(t) - x_3(t)|$ for $t \in [1000, 2000]$ are displayed in blue colour. Parameters are the same as in Fig. 16

second cell is periodic oscillations, then the choice of parameters in Fig. 15 leads to complete synchronization in the first group. On the other hand, the fluctuations caused by the group of chaotic cells are still present, but they are much weaker. Finally, we can consider a situation in which cells are connected to cells of the other group only, which is illustrated in Fig. 16. Since periodically oscillating cells are driven by the same force, they do not need to be coupled in order to achieve synchronization. The stability of partial synchronization in the second and the third situation is demonstrated in Figs. 17 and 18.

5 Conclusion

The majority of this work can be considered as a straightforward extension of ideas and analyses published in Kyprianidis et al. [7] and Kyprianidis and Makri [8]. On the other hand, the system of coupled oscillators can be seen as a prototype for the description of chimera-like dynamics, which can be explained by bifurcation analysis. It is possible to explain changes in dynamics with respect to the coupling force between the oscillators or the external excitation (brain waves) by crossing bifurcation manifolds. We have observed two main types of cell behavior: a cell can exhibit periodic oscillations that lead to participation in the generation of brain waves, or it can behave chaotically. There could also be various ratios of phase-locking observed as the period-doubling route to chaos takes place. It thus explains the phenomenon when a neuron can contribute both to a synchronized brain wave activity but also have different and even chaotic dynamics [5]. Analysis of phase-locking is connected with fold bifurcations on a torus, more precisely birth of torus via Neimark-Sacker bifurcation of a cycle and near Arnold tongues existence. This analysis is not included in this contribution, but a possible analysis approach is explained in Ševčík and Přibylová [10].

All studied systems of differential equations describe dynamics of some electric circuit which is presented in detail. We have seen that the resulting type of behavior strongly depends on the topology of the network, the strength of coupling between cells, and the amplitude of the forcing signal. Our further research will be focused on applications of functional analysis, singularity theory, and group theory to the investigation of the origin of partial synchronization based on ideas from Golubitsky and Schaeffer [3] and Golubitsky et al. [4] since in the case of similar coupled neurons, symmetry can be used to study and describe idealized and perfectly same coupled neurons, and breaking symmetry can lead to chimera states that could be explained by means of equivariant bifurcation theory.

Acknowledgements The work has received financial support from Mathematical and Statistical modelling project MUNI/A/1615/2020.
Conflict of interest The authors declare that they have no conflict of interest.

References

1. A. Dhooge, W. Govaerts, Y.A. Kuznetsov, MATCONT: a MATLAB package for numerical bifurcation analysis of ODEs. ACM Trans. Math. Softw. (TOMS) **29**(2), 141–164 (2003)
2. P. Faure, H. Korn, Is there chaos in the brain? I. Concepts of nonlinear dynamics and methods of investigation. Comptes Rendus de l'Académie des Sciences-Series III-Sciences de la Vie **324**(9), 773–793 (2001)
3. M. Golubitsky, D.G. Schaeffer, Singularities and groups in bifurcation theory, vol. I, vol. 51 of *Applied Mathematical Sciences* (1985)
4. M. Golubitsky, I. Stewart, D.G. Schaeffer, Singularities and groups in bifurcation theory, vol. II, vol. 69 of *Applied Mathematical Sciences* (1988)

5. H. Korn, P. Faure, Is there chaos in the brain? II. Experimental evidence and related models. Comptes Rendus Biologies **326**(9), 787–840 (2003)
6. Y.A. Kuznetsov, *Elements of Applied Bifurcation Theory*, vol. 112 (Springer Science & Business Media, 2013)
7. I.M. Kyprianidis et al., Dynamics of coupled chaotic Bonhoeffer-van der Pol oscillators. WSEAS Trans. Syst. **11**(9), 516–526 (2012)
8. I.M. Kyprianidis, A.T. Makri, Complex dynamics of FitzHugh-Nagumo type neurons coupled with gap junction under external voltage stimulation. J. Eng. Sci. Technol. Rev. **6**(4), 104–114 (2013)
9. A. Pikovsky, M. Rosenblum, J. Kurths, *Synchronization: A Universal Concept in Nonlinear Sciences*, no. 12 (Cambridge University Press, Cambridge, 2003)
10. J. Ševčík, L. Přibylová, Forced van der Pol oscillator—synchronization from the bifurcation theory point of view. In: *14th Chaotic Modeling and Simulation International Conference* (2021)
11. S. Wiggins, *Introduction to Applied Nonlinear Dynamical Systems and Chaos*, vol. 2 (Springer, New York, 1990)

Demographic Dynamics of Inhomogeneous Economic Communities as an Institutional Trap

V. G. Zhulego and A. A. Balyakin

Abstract In this article we discuss the demographic dynamics modelling in communities of countries with different levels of economic development. Our approach is based on the stratum model of population growth, proposed by the authors earlier. The observed processes of depopulation of the periphery of such communities were studied within the framework of the model. The phenomenon of institutional trap is considered as an explanatory principle of the functioning of complex socio-economic structures. Its main traits are discussed. Based on the proposed model, the forecasts of population growth in several countries were calculated. Within the proposed model of institutional trap a set of measures to overcome the negative demographic trends were formulated.

Keywords Simulation · Demographics · Institutional trap · Stratum model · Forecast

1 Introduction

Significant progress has been made in the field of creating mathematical models for complex social systems in recent decades [1]. The results of scientific forecast (foresight) of socio-economic processes of countries and the world as a whole by methods of social, humanitarian and natural sciences are used both in the field of public administration and strategic planning, and in large business when developing a growth strategy [2]. Mathematical modeling of social processes, and, in particular, the population growth forecasts should be recognized as an integral element of foresight that enlightens the trends of economic development of the society [3]. The topic of forecasting the population growth of countries and the whole world continues

V. G. Zhulego
NPO ANEK, 16 Maksimova, Moscow, Russia

V. G. Zhulego · A. A. Balyakin (✉)
NRC Kurchatov Institute, 1, ac. Kurchatov sq., Moscow, Russia
e-mail: Balyakin_AA@nrcki.ru

C. H. Skiadas and Y. Dimotikalis (eds.), *14th Chaotic Modeling and Simulation International Conference*, Springer Proceedings in Complexity,
https://doi.org/10.1007/978-3-030-96964-6_38

to be relevant not only because of the limited life resources and the prospect of overpopulation of the planet, but also because countries tend to build management decisions based on reliable long-term forecasts. The study of this problem leads to the conclusion that such forecasts can be made on the basis of adequate mathematical models.

One of the first and most famous experiences of successful modeling in the field of social and economic sciences was the work of Malthus [4], which caused sharp criticism at the time. The main idea of T. R. Malthus was the point that the difference in the growth of the population and the productive forces (the wealth of society) leads to a complication of the social situation, producing wars, crises and diseases. The discussion of the "overpopulation crisis" predicted by Malthus, which was expected by 2004, led to the correction of the growth model and the creation of several variants of such model. In general, the proposed approaches can be divided into 3 groups:

- models-concepts based on the identification and analysis of general historical patterns and their representation in the form of cognitive schemes describing logical connections between various factors affecting historical processes (ideas of J. Goldstein, I. Wallerstein, L. N. Gumilev, N. S. Rosov, etc.). These models have a high degree of generalization, but they are not mathematical, but purely logical, conceptual in nature;
- pure mathematical models of the simulation type devoted to the description of specific historical events and phenomena (Yu. N. Pavlovsky, L. I. Borodkin, D. Meadows, J. Forrester, etc.). The applicability of such models is usually limited to a fairly narrow space–time interval since they are related to a specific geopolitical situation;
- "intermediate" mathematical models between the two specified types. Their task is to identify the basic patterns that characterize the flow of processes of the type under consideration.

Our work aimed at studying the relationship between socio-economic life and demographic processes, thus relating to the third type of models. This approach involves both conducting mathematical modeling, and taking into account and describing the factors and processes that affect the phenomena under consideration.

One of the first steps in this direction that should be named is the Verhulst logistics model [5] and the concept of "world-systems" [6]. Within the framework of these approaches, the unified system (the world) was divided into subsystems (economic subsystem, and/or social and demographic ones). Later, dynamic models were proposed that go beyond the neoclassical model of economic growth by R. Solow [7], based on equilibrium, when in a stationary state the rate of labor productivity growth is equal to the rate of technological progress, and the rate of economic growth is the sum of the rate of technological progress and the rate of population growth.

At the same time, the Solow model could not explain many problems related to economic growth, which was caused by the fact that many parameters of the model were set exogenously. The next step was the Cobb–Douglas model [8], the Ramsey–Cass–Koopmans model [9] and the Mankiw Romer and Vail model [10].

All the models considered assumed a different format of "combining" social, economic and demographic parameters [6, 11]. In practice, we should mention the Gushchin–Malkov model of macroeconomic dynamics (which describes the economic cycles of US GDP growth, see [12]) and the Korotaev model of great divergence/convergence [13], as well as the Kapitsa population growth model [11]. The last work proposed an exponential model of world population growth, showing the "limits of growth" beyond which a global catastrophe can await the planet. Based on this model, the trend change point or "transition point" (2005) was also predicted, when exponential growth is replaced by a slowdown in growth. Although this model made it possible to predict global trends quite well, it was completely unsuitable for calculating population growth forecasts for individual countries, in particular because migration processes play an important role in these processes, which were not taken into account in the model in any way.

Methodologically, our study continues the approach of dividing a single system into a number of subsystems, and in this sense, the approach can be called hierarchical. The subject of our study will be the population growth in the subsystems of a single "economic community", taking into account the socio-economic development of this system. It is assumed that in the system under consideration, it is possible to explicitly allocate the Center and the Periphery (i.e., to allocate subsystems). A similar problem was solved earlier in the course of mathematical modeling of the population size based on the stratum model [14].

Our special attention was attracted by the Korotaev's model [13] of great divergence/convergence, in which an attempt is explicitly made to take into account population growth in the economic model of the development of countries. The authors drew the conclusion about the "inevitable convergence of heterogeneous economic systems of the Periphery-Center type". Meanwhile, this conclusion contradicts the trends observed in some similar situations, in particular in the EU-Baltic states, where not only convergence is not observed, but rather divergence occurs, accompanied by the process of depopulation of the Periphery.

The obvious disadvantage of mentioned model is the lack of migration in it—and, as it seems to us, this process leads to the opposite effect: to the growth of the economic gap between these parts of the system and, ultimately, leads to the depopulation of the Periphery. The most important resource of economic growth is the labor force leaving depressed regions, which leads to a significant decrease in the economic growth potential of these regions.

The way out of this situation, in our opinion, may be the stratum model of population growth proposed by us in 2014 [14], which can be generalized to the case of a heterogeneous economic system/commonwealth of countries with different levels of development. An additional argument in favor of the attempt to combine the convergence model with the stratum model of population growth was our analysis of population growth forecasts for 2014 for several countries in comparison with statistical data for these countries over the past 4 years, which showed a good agreement of these forecasts with statistics.

The essence of the stratum model is that the population of a country is considered not as homogeneous, but as consisting of several strata. We used following denotations: x(t)—the number of urban population, y(t)—the number of rural population. The parameters that determine the dynamics of changes in each stratum are different, in particular, both the birth rate and mortality in the strata can differ significantly, in addition, there is a significant migration of the population (almost always it is the move of the rural population to the city). Taking into account these circumstances, the model of population growth in a particular country can be presented in this form:

$$\frac{dx}{dt} = a_x x(t) - d_x x^2(t) + c_x \frac{x(t)y(t)}{x^2 + \alpha^2}$$
$$\frac{dy}{dt} = a_y y(t) - d_y y^2(t) - c_y \frac{x(t)y(t)}{x^2 + \alpha^2} \quad (1)$$

The meaning of the parameters a_x and a_y is that they are determined by the balance of instantaneous fertility and mortality in each stratum. Since the economic conditions for the existence of strata and the way of life within strata are different, the characteristic coefficients a_x and a_y can vary greatly. The parameters d_x and d_y conditionally determine the "capacity of the corresponding niche", i.e. they reflect the limited life resources, and the ratio of the parameters (a_x, a_y) and (d_x, d_y) determines the linear "transition point".

The system (1) is written in a symmetric form, the coefficients (a_x, a_y) and (d_x, d_y) determine the internal dynamics of the stratum, and c_x and c_y determine the migration between the strata. The migration flow may depend on many factors, but for each specific country it is a fairly stable parameter, and the last term in the equations should be proportional to the frequency of meetings of residents of the city and village, it is this part of the equation that can significantly accelerate the dynamics of changes in the system.

This is also connected with the possibility of economic growth of the country exceeding the population growth rate, since the migration of the rural population to the city provides additional needs for industrial labor resources. For a particular country, the migration rate is determined by the coefficients $c_{x,y}$ (in the simplest case, these coefficients are equal and opposite in sign, which means that all those who left the "village" ended up in the "city"). If we start from the stratum model of population growth of one country (1), then it is easy to build a model of world population growth based on the principle of hierarchy: for this it is necessary to determine the coefficients (a_x, a_y), (d_x, d_y), (c_x, c_y) for each individual country, moreover, an additional term describing the emigration of the population from one country to other countries should be introduced, this term will be similar to the third in the system (1). The forecast calculated in this order for each country allows us to find the total population of the world.

It should be said that in this approach, the amount of calculations increases significantly, but the accuracy of the forecast also increases. It should be emphasized that it was the forecasts for these countries that led us to the idea of the existence of so-called "institutional traps" that individual countries fall into, i.e. such situations that do not disappear by themselves, but require a purposeful restructuring of the institutional environment. It is for this reason that below we present the results of calculations of forecasts for a group of countries, the choice of which is due to the fact that, using the example of these countries, we will try to construct and test a mathematical model of an “institutional trap”, i.e. a situation caused by the institutional characteristics of countries and a way out of which is possible only as a result of serious institutional changes.

In the next section we present the results of numerical simulation for population growth of countries that can be considered as good axample of center–periphery system.

2 The Dynamics of Population Growth as an Institutional Trap. Former USSR Republics Case

Let us discuss the demographics dynamics of former soviet republics, precisely—Russia, Belarus, Ukraine and Baltic states. Here we do not take into account the pandemic influence, hoping its negative consequences will be overcome soon, and the population trend will restore.

A characteristic feature of the demographic dynamics of Russia was a change in the trend from neutral (fluctuation in the population near 147 million) to moderately optimistic growth since 2010 year. The trend change was mainly due to the growth of urban population, while the rural population continued to decline. The graph on the right shows statistics for 4 years from 2014 to 2018. Thus, it can be argued that the managerial decision made in Russia on cash payments at the birth of the second and third child has already influenced the dynamics of population growth (Fig. 1).

Since the collapse of the USSR, Belarus is the only post-Soviet republic where the population began to grow. Urban growth continued after the collapse of the USSR, the rural population continued to decline at a constant rate, while the total population began to grow only in 2014. These changes however are not stable and can be reversed due to political tensions.

Since 1991, Ukraine has shown a steady population decline dynamics: population was decreasing at a constant rate of 1% per year. Moreover, the decrease in the population is taking place both in the countryside and in the city at almost the same rate. This dynamic indicates the ongoing economic crisis in the country. Data for 4 years (from 2014 to 2018) do not provide any indication of a change in trend (Fig. 2).

After the collapse of the USSR in the period up to 2000, the population decline in Estonia was the fastest among the post-Soviet countries, and then it diminished. The shift of the dynamics in recent years (positive growth) should be considered separately, since there is no obvious explanation for such a change.

After 1991, the most dramatic situation developed in Lithuania—depopulation was stable both in the city and in the countryside, and there is no need to talk about

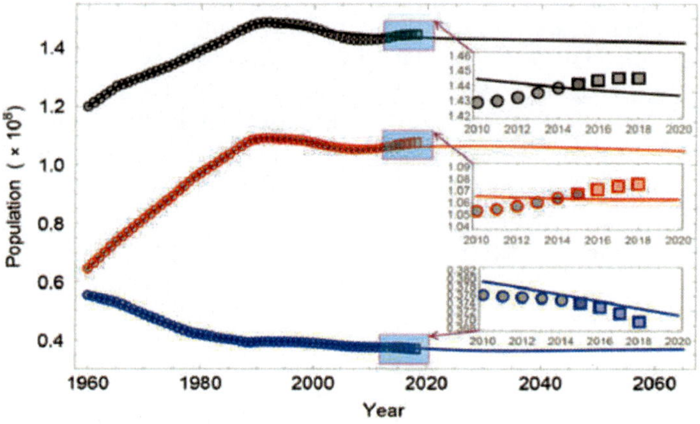

Fig. 1 Russia population forecast and statistical data

a change in trend. In Latvia the situation is almost completely similar to Lithuania: depopulation is going on at a constant pace, and no change in the trend is expected.

The graphs indicate the presence of two groups of countries: these are the post-Soviet countries that remained outside the economic blocs and the post-Soviet countries that entered the EU (or are associated with the EU). The demographic situation in these groups is radically different.

All this suggests that a mathematical model should be created that would take into account migration between countries and at the same time reveal the reasons for such large-scale population migrations. In our opinion stratum model is the best possible starting point on the way to such a socio-demographic model. As an explanatory principle we rely on institutional reasons. Since the ideas to link demographic variables with economic and social variables have appeared for a long time, we decided to make an attempt to create an economic-socio-demographic model that would allow us to consider the problems of demography in connection with economic and social ones. Moreover, one of these attempts raised the problem of convergence of countries with different levels of economic development.

Our explanation of the differences in dynamics boils down to the following causal chain: joining the EU (or association) opens borders for migration of the population, the existing gap in the standard of living and education leads to the flow of migrants from conditionally "poor countries" towards rich ones. Such migration deprives poor countries of human development resources, and then their economic development slows down.

The problem of heterogeneity or uneven development of countries has already been considered in the above mentioned work [13]. Obviously, without taking into account migration, such a model leads to "great convergence"—a completely fair goal, which was set by the countries that joined the EU. However, in fact, the entry of the Baltic countries into the EU led to depopulation and an increase in the income gap. A similar process is underway in the case of Ukraine (although with some delay).

We can expect the same situation with Belarus—Russia case, or Ukraine—Russia (if political tensions declined).

All this allows us to conclude that accounting for migration in the Center–Periphery model is necessary. Otherwise, the model will not adequately describe the dynamics of the system (Figs. 2, 3 and 4).

Of course, we will model a system that is simplified compared to the real one, but we will keep its most important features: the free movement of people (migration) and the heterogeneity of the system. Therefore, we will assume that the system consists of a developed Center and a backward Periphery. From the point of view of the stratum model, the new EU countries represent the same "city" and "rural" in the country, and there is unlimited and practically unregulated migration between

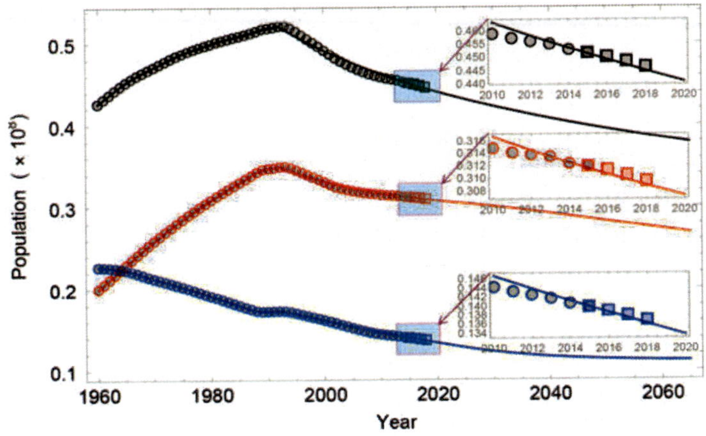

Fig. 2 Ukraine population forecast and statistical data

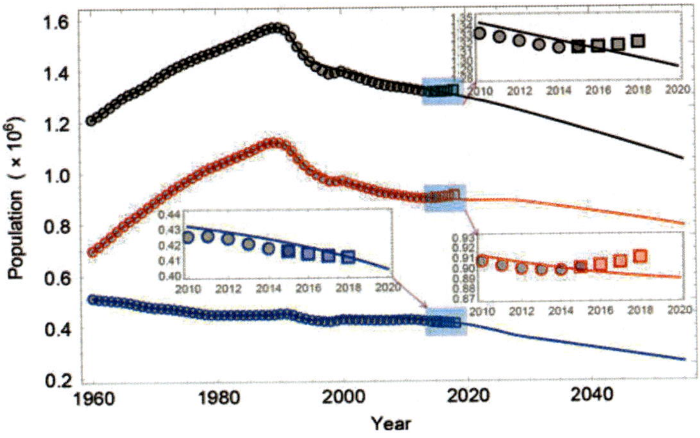

Fig. 3 Estonia population forecast and statistical data

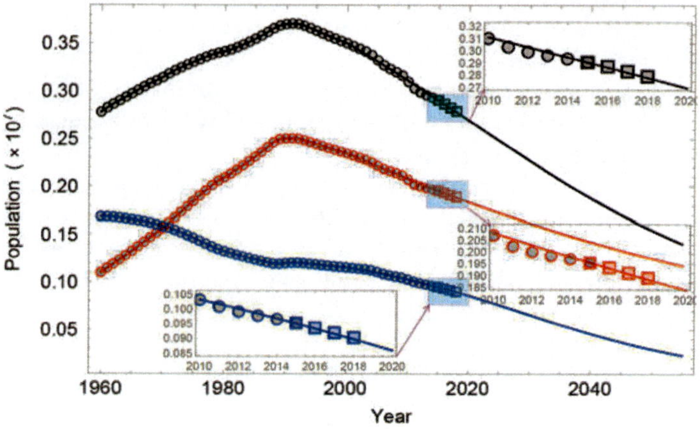

Fig. 4 Lithuania population forecast and statistical data

these strata. We note that the proposed model should transform into previous model of [14] in the case of low migration process.

3 Model Improvement: Inclusion of Social and Economic Factors

For the sake of generality, we will keep the previous designations: x(t)—the population of the Center, y(t)—the population of the Periphery. Let's supplement system (1) by adding equations to take into account the socio-economic development of the regions. By analogy with the work [13], we will introduce the level of "wealth"—$S_{x,y}$ and "education" $E_{x,y}$.

In practice, these factors reflect material wealth and other intangible benefits. The meaning of the parameters a_x and a_y is that they are determined by the balance of instantaneous fertility and mortality in each of the subsystems: in the Center and Periphery. Since the economic conditions of existence in the Center and on the Periphery, as well as the way of life within each subsystem are different, the characteristic coefficients ax and ay can be very different. The parameters d_x and d_y conditionally determine the "capacity of the corresponding niche", i.e. they reflect the limited life resources in the subsystems.

$$\frac{dx}{dt} = a_x x(t) - d_x x^2(t) + \left(A\frac{S_x - S_y}{S_x + S_y} + B\frac{E_x - E_y}{E_x + E_y}\right)\frac{x(t)y(t)}{x^2 + \alpha^2}$$
$$\frac{dy}{dt} = a_y y(t) - d_y y^2(t) - \left(A\frac{S_x - S_y}{S_x + S_y} + B\frac{E_x - E_y}{E_x + E_y}\right)\frac{x(t)y(t)}{x^2 + \alpha^2}$$

$$\frac{dS_x}{dt} = b_x S_x E_x \left(1 - \frac{G}{G_{\lim}}\right)$$
$$\frac{dS_y}{dt} = b_y S_y E_y \left(1 - \frac{G}{G_{\lim}}\right) + \varphi$$
$$\frac{dE_x}{dt} = f_x E_x (1 - E_x)$$
$$\frac{dE_y}{dt} = f_y E_y (1 - E_y) + \psi \quad (2)$$

The first two equations of the system (2) is written in a symmetric form, the coefficients (a_x, a_y) and (d_x, d_y) determine the internal dynamics of the subsystem Center and the Periphery, here, instead of the coefficients c_x and c_y, which determine the migration of the population from one part of the subsystem to another, some functions of the variables $S_{x,y}$ and $E_{x,y}$ are selected, which characterize the per capita income and the level of education in each part of the system: S_x is a relatively excess product per capita of the Center population, and S_y is a relatively excess product per capita of the Periphery population; E_x is the level of education of the population in the Center, and E_y is the level of education of the population in the Periphery. Differences in income level and in the level of education in the Center and on the Periphery will induce migration of the population to the Center. Additionally, the following notation is introduced in system (2): $G(t) = x(m + S_x) + y(m + S_y)$ is the GDP of the Center–Periphery system, m is the minimum necessary product (estimated as $ 440), $G_{\lim}$ is a certain fundamental limitation and a normalization term that defines a fundamental constraint in the system. In the model [14] describing the world-system, $G_{\lim}$ = $ 400 trillion, in the model we propose, $G_{\lim}$ should coincide in order with the EU GDP, i.e. about $ 100 trillion.

Let us pay attention to the choice of signs in these terms in the first two equations of the system (2) that fixes the direction of migration from the Periphery to the Center. Thus, to describe the dynamics of interaction of the heterogeneous Center–Periphery system, a socio-economic demographic mathematical model (2) of the system is proposed, which takes into account both the dynamics of the population of individual parts of the system, and the migration of the population from one part of the system to another, due to the difference in income in different subsystems and the difference in the level of education. Equation (2) contain φ, ψ which we will call "convergence functions", which show the relationship between the Center and the periphery. Unlike [13], we do not postulate the form of these functions, moreover, in our opinion, their form needs serious refinement.

At the same time, the choice of convergence functions should reflect the main trends in the modern world. Thus, according to the authors [12, 13], the gap between highly and medium-developed countries has been decreasing at a particularly rapid pace in recent years, and the gap between highly and underdeveloped countries is decreasing at a noticeably slower pace, at the same time they show an increase in the gap between medium and underdeveloped countries. In practice, it turns out that advanced economies are "going into isolation", medium-developed countries receive the greatest benefits from globalization, catching up with developed countries, but underdeveloped countries are moving to increasingly worse positions. According to the author [15], we are talking about the reconfiguration of the world-system and the trend towards the concentration of income. There is a discrepancy between the richest and the poorest people in the world, despite the general convergence of average incomes.

These conclusions, in our opinion, are very controversial. The existing practice shows that a country's participation in a successful economic community does not guarantee its automatic convergence, does not automatically raise it to the level of the Center. In our opinion, correct accounting of migration can lead simultaneously to depopulation and economic degradation of the Periphery, and "on average" (or "per capita") there will be an increase in welfare.

In the model we propose, in the simplest case:

$$\varphi = 0$$
$$\psi = -\gamma E_x E_y$$

Note that the function ψ must be non-zero, otherwise the resulting solutions for E_y will tend to 1, i.e. the village becomes fully educated, which will lead to a drop in the birth rate. Practically, in the system (2), the presence of the term ψ ensures, with the simplest choice of the convergence function, the presence of a stable stationary solution of the form

$$(E_x, E_y) = \left(1, 1 - \frac{\gamma}{c_y}\right)$$

It corresponds to full education in the Center and some non-zero (but not 100%) education in the Periphery. The proposed (2) are essentially nonlinear, and may contain complex dynamics, such as periodic oscillations, periodic oscillations with attenuation or increase in amplitude, or, conversely, an asymptotic output to constant values.

Stability analysis for the system (2) reveals that there are always trivial solutions for $E_{x,y} = 0$ or 1 (totally educated or fully uncivilized strata). Assuming two extreme cases $S_x \gg S_y$ (the welfare of the Center significantly exceeds the welfare of the Periphery) and $S_x = S_y$ (the convergence occurred) we obtain following results. For $S_x \gg S_y$ we get $A_> A_{\text{(critical value)}} = x^{(0)} a_y = a_y a_y/d_x$.

For the case $E_x = 1$, $E_y = 0$ (developed center and backward periphery) we get the restriction $(A + B)_{\text{(critical value)}} = x(0)\, a_y$. The system loses stability if the sum of the parameters $(A + B)$ exceeds the critical value. Thus interesting realistic solutions system (2) will be played out around the adiabatic values of variables x(t) and y(t).

A similar approach (adiabatic change of parameters) can be applied in the case of convergence functions, assuming that they are a small perturbation that has the greatest impact on the population of the periphery.

It seems that the next step would be reasonable to choose the convergence functions by analogy with the terms in the first two equations in the form

$$\varphi = \delta\left(A\frac{S_x - S_y}{S_x + S_y} + B\frac{E_x - E_y}{E_x + E_y}\right)S_x S_y$$
$$\psi = -\gamma\left(A\frac{S_x - S_y}{S_x + S_y} + B\frac{E_x - E_y}{E_x + E_y}\right)E_x E_y \tag{3}$$

The obtained (2) are characterized by the following properties. Firstly, there are no exogenous variables in them, taking into account external factors is contained only in the parameters (coefficients) of the equations. Secondly, the meaning of the parameters contained in the model follows from the equations themselves, and the values of these parameters can be determined from the analysis of statistical data for a certain period. Third, the type of convergence functions is not defined a priori.

It seems that this system of equations will allow us to study various modes of behavior of the Center–Periphery system depending on the values of the parameters, as well as to predict the behavior of the Center–Periphery system, in the case when the parameters are determined. This model is a development of the stratum model, taking into account the ideas about the functioning of the "world-system" [6]. The authors suggest that the developed approach will allow us to consider complex systems where simplified approaches do not work.

4 Conclusion

The proposed model of the institutional trap is described by a system of (2), which is characterized by the following properties:

Firstly, there are no exogenous variables in them, external factors are taken into account only in the parameters (coefficients) of the equations.
Secondly, the meaning of the parameters contained in the model follows from the equations themselves, and the values of these parameters can be determined from the analysis of statistical data for a certain period.
Third, the type of convergence functions is not defined a priori and may vary depending on the task.

It seems that this system of equations will allow us to study various modes of behavior of the Center–Periphery system depending on the values of the parameters, as well as to predict the behavior of the Center–Periphery system, in the case when the parameters are determined.

We emphasize once again that this model is a development of the stratum model [14], taking into account the ideas about the functioning of the "world-system" [6, 13]. The authors suggest that the developed approach will allow us to consider complex systems where simplified approaches do not work. The results obtained, however, should be used with caution: their applicability to specific situations is limited by both the initial conditions and the current operating conditions of the system under consideration. In terms of the institutional trap, this is equivalent to the destruction of the trap in the course of institutional restructuring (reform). Note that a similar effect can be achieved by a sharp change in the initial parameters of the system (the population in the Periphery, which has changed dramatically, for example, in the results of uncontrolled migration), which is equivalent to the "transfer" of the system to the pool of attraction of another attractor.

Given the almost unlimited labor migration within the EU, it is assumed that the proposed model will adequately describe the case of an institutional trap that occurs in the Center–Periphery system, in which economic integration does not lead to an equalization of the level of per capita income in the subsystems, but leads to the depopulation of the Periphery. The results of computer modeling should make it possible to estimate the characteristic times of the development of unfavorable dynamics (the "half-life" of the Periphery countries). The authors also hope that the study of this model will allow us to formulate recipes for getting out of emerging institutional traps (by controlling the parameters of the system).

Acknowledgements The work was supported by the RFBR grant No. 20-010-00576.

References

1. *Complex Systems and Society—Modeling and Simulation* (Springer, 2013)
2. M. Eberlin, *Foresight: How the Chemistry of Life Reveals Planning and Purpose* (Discovery Institute. 2019)
3. Unido Technology Foresight Manual, *United Nations Industrial Development Organization*, vol. 1 (Vienna, 2005), p. 8
4. G. Clark, Malthusian economy, in *The New Palgrave Dictionary of Economics* (Palgrave Macmillan UK, London, 2018), pp. 8148–8155
5. P.F. Verhulst, Notice sur la loi que la population poursuit dans son accroisment. Corres. Math. Phys. **1838**(10), 113–121
6. J.W. Forrester, *World Dynamics* (Wright-Allen Press, 1971)
7. H. Uzawa, Models of growth, in *The New Palgrave Dictionary of Economics* (Palgrave Macmillan UK, London, 2018), pp. 8885–8893. ISBN 978-1-349-95188-8
8. G. Renshaw, *Maths for Economics* (Oxford University Press, New York, 2005), pp. 516–526. ISBN 0-19-926746-4
9. S.E. Spear, W. Young, Optimum savings and optimal growth: Ramsey–Malinvaud–Koopmans nexus. Macroecon. Dynam. **18**(1), 215–243 (2014). https://doi.org/10.1017/S1365100513000291
10. P.M. Romer, Human capital and growth: theory and evidence. NBER Working Paper No. 3173 (1989). https://doi.org/10.3386/w3173
11. S.P. Kapitsa, *Global Population Blow-Up and After* (The Demographic Revolution and Information Society, Moscow, 2006)
12. A. Korotayev, A. Malkov, D. Khaltourina, *Introduction to Social Macrodynamics: Secular Cycles and Millennial Trends* (KomKniga/URSS, Moscow, 2006)
13. A. Korotayev, L. Grinin, J. Goldstone, *Great Divergence and Great Convergence. A Global Perspective* (Springer. 2015)

14. V.G. Zhulego, A.A. Balyakin, 2-Phase Model for Population Growth. Chaos Model. Simul. (CMSIM) **3**, 193–204 (2015)
15. P. Turchin, *Complex Population Dynamics: A Theoretical/Empirical Synthesis* (Princeton University Press, Princeton, NJ, 2003)

Author Index

C. H. Skiadas and Y. Dimotikalis (eds.), *14th Chaotic Modeling and Simulation International Conference*, Springer Proceedings in Complexity,
https://doi.org/10.1007/978-3-030-96964-6